RNA Nanotechnology and Therapeutics

RNA Nanotechnology and Therapeutics

Second Edition

Edited by
Peixuan Guo
and
Kirill A. Afonin

CRC Press
Taylor & Francis Group
Boca Raton London New York

CRC Press is an imprint of the
Taylor & Francis Group, an **informa** business

Second edition published 2022
by CRC Press
6000 Broken Sound Parkway NW, Suite 300, Boca Raton, FL 33487-2742

and by CRC Press
2 Park Square, Milton Park, Abingdon, Oxon, OX14 4RN

CRC Press is an imprint of Taylor & Francis Group, LLC

© 2022 selection and editorial matter, Peixuan Guo and Kirill A. Afonin; individual chapters, the contributors

First edition published by CRC Press 2013

Library of Congress Cataloging-in-Publication Data
Names: Guo, Peixuan, editor. | Afonin, Kirill A., editor.
Title: RNA nanotechnology and therapeutics / edited by Peixuan Guo, KirillA. Afonin.
Other titles: RNA nanotechnology and therapeutics (CRC Press)
Description: Second edition. | Boca Raton : CRC Press, 2022. | Includes bibliographical references and index. |
Summary: "Edited by the world's foremost experts, this thoroughly updated, comprehensive, state-of-the-art reference, details the latest research developments and challenges in the biophysical and single molecule approaches to RNA nanotechnology"-- Provided by publisher.
Identifiers: LCCN 2021050100 (print) | LCCN 2021050101 (ebook) | ISBN 9781138312869 (hardback) |
ISBN 9781032224930 (paperback) | ISBN 9781003001560 (ebook) Subjects: MESH: Nanotechnology--methods |
RNA--therapeutic use | Nanomedicine--methods | Nanoparticles Classification: LCC QP623 (print) | LCC QP623 (ebook)
| NLM QT 36.5 | DDC 572.8/8--dc23/eng/20211101 LC record available at https://lccn.loc.gov/2021050100 LC ebook
record available at https://lccn.loc.gov/2021050101

ISBN: 978-1-138-31286-9 (hbk)
ISBN: 978-1-032-22493-0 (pbk)
ISBN: 978-1-003-00156-0 (ebk)

DOI: 10.1201/9781003001560

Typeset in Times
by SPi Technologies India Pvt Ltd (Straive)

Contents

PART I   Concepts and Definitions in RNA Nanotechnology

PART II   Design, Synthesis, and Characterization Methods in RNA Nanotechnology

PART III   Immunorecognition of RNA Nanoparticles

PART IV   Delivery of Functional RNA Nanoparticles

PART V   Application and Exploitation in RNA Nanotechnology

Preface

RNA has previously been categorized into transfer RNA (tRNA), ribosomal RNA (rRNA), and messenger RNA (mRNA). However, the concept was subsequently rewritten with the discovery of self-replicating viral RNA, shortly followed by ribozyme and nuclear RNA. It was then predicted that there was a large assortment of small RNA species, known as sRNAs, in cells with obscure and novel functions but at the time was not been identified (Guo et al. Science, 1987, 236:690-4. A small viral RNA is required for in vitro packaging of bacteriophage phi 29 DNA). The implication of potential application of small interfering RNA (siRNA), microRNA (miRNA), ribozyme, aptamer, riboswitch, antisense RNA has enhanced the visibility of and stimulate interest in the RNA. Likewise, the field of RNA nanotechnology was proven in 1998 with the controlled self-assembly of RNA nanorings from multiple copies of Phi29 bacteriophage packaging RNAs. Witnessing the growth in the field of RNA nanotechnology, it was predicted in 2014 that RNA as either a drug or drug target will become the third milestone in the pharmaceutical industry. The approval of several RNA drugs by FDA, the widely used mRNA vaccine for the prevention of COVID-19 pandemic, the popular diagnosis of COVID-19 using viral RNA has evidenced the coming of the third milestone in drug development using RNA.

Since the first edition of this book *RNA Nanotechnology and Therapeutics* that reports the emergence of RNA nanotechnology, the field of RNA nanotechnology has boomed and advanced rapidly with many exciting new findings. The large volume of literature in RNA nanotechnology and therapeutics necessities an update of the book *RNA Nanotechnology and Therapeutics* as the second edition.

The intended style of this book is one that can engages researchers from undergraduates all the way to post-doctoral researchers and professors, in engineering and other sciences to further enhance this field. It is imperative that we work together to propel it to all that it can be, and more.

We are delighted to present this book and hope that readers will find it very resourceful and as exciting as we do.

Peixuan Guo
The Ohio State University, Columbus, OH

Acknowledgments

We would not have been able to accomplish our goal without the tremendous efforts of leading experts in the field, who have taken the time to contribute chapters. We express our gratitude to Lora McBride and Daniel Binzel for their tireless work in assisting the editing of this book. Without their valuable contributions, this book would have never come to fruition. We would like to dedicate this book to the memory of Neocles Leontis, a great mentor and scientist, who left us too soon. We also thank Cheyenne Wagner for all her administrative help and support in editing process of this book.

Introduction

RNA nanotechnology is the bottom-up self-assembly of RNA structures at the nanometer scale. In RNA nanoparticles, their major frame is mainly composed of RNA and often includes the associated scaffolding, ligands, and regulators and is composed of small building blocks akin to LEGOs. Ideal building material should have the following properties: (1) versatility and controllability in shape and stoichiometry; (2) spontaneous self-assembly when mixed together; and (3) thermodynamic, chemic, and enzymatic stability with a long shelf-life. RNA building blocks exhibit each of the above properties. In addition, RNA nanoparticles hold many unique and favorable properties that further enhance their applicability. RNA can be designed and manipulated with a level of simplicity of DNA while displaying versatile structure and enzymatic function attributed to proteins. RNA folds into a large variety of single-stranded loops or bulges suitable for intermolecular or domain interactions. These loops or bulges can serve as mounting dovetails alleviating the need for external linking dowels in nanoparticle construction. Building blocks that compose RNA nanoparticles are less than 80 nucleotides, allowing for chemical synthesis. As a result, their structure, shape, and physical/chemical properties are well characterized; homogeneity can be achieved and quality control is simpler.

The recent discovery that RNA nanoparticles display rubber- and amoeba-like properties adds a flying color to the special color of RNA in therapeutics. As demonstrated by optical tweezers, RNA nano-architectures are stretchable and shrinkable through multiple repeats like rubber. The rubbery- and amoeba-like property of RNA nanoparticles leads to compelling vessel extravasation to enhance passive tumor targeting and fast renal excretion to reduce toxicities. Since their rubber- or amoeba-like deformation property enables them to squeeze through leaky vasculature. The efficient tumor passive targeting, in addition to the active receptor-mediated endocytosis, makes RNA more valuable in cancer therapeutics since both the fast tumor targeting and body clearance lead to little accumulation in vital organs, thus little toxicity has been detected.

RNA nanotechnology has become an exciting field and overcomes several challenges. Recently, simple chemical modifications such as 2'-modification or 5'-3' linkage have made the resulting RNAs resistant to degradation without changing their folding properties, and in some cases even their biological functions. The powerful products formed by stable RNA have developed the vision of RNA nanotechnology into a reality. Secondly, the increased thermostability of nucleic acid nanoparticles is needed to ensure they remain assembled at low concentrations commonly seen during in vivo circulation in therapeutic applications. Compared with DNA, RNA is unique in its higher thermodynamic stability, typical and atypical base-pairing ability, base stacking, and a variety of loops and bulges are suitable for intramolecular and intermolecular interactions. Finally, nanoparticles needed to reduce or eliminate toxicity and immunogenicity. RNA nanoparticles have demonstrated no detectable toxicity in mice and have tunable and controllable immune responses. As a result, RNA nanoparticles now can be applicable for clinical applications.

While classical RNA biology studies focus on intramolecular interactions and folding, RNA nanotechnology instead focuses on inter-RNA interactions and quaternary (4D) structure. However, the simple inclusion of functional RNA modules onto gold particles, liposomes, polymers, or nanoparticles composed of polymers is not constituted as part of RNA nanotechnology. On the contrary, RNA nanotechnology is typically based on a bottom-up approach with RNA as the primary component of the nanoparticle itself. Therefore, RNA nanotechnologists have evolved from RNA biologists. In the past several decades, research on the folding and structure of RNA motifs and the development of three-dimensional calculations from traditional intramolecular interactions to intermolecular interactions have laid a solid foundation for the future development of RNA nanotechnology. RNA cell and molecular biologists are vital for the growth and development of RNA nanotechnology.

They also can apply RNA nanotechnology to their research or application. Therefore, RNA nanotechnology is an extension of RNA studies utilizing disciplines from several different fields.

This text strives to assemble information, engage its readers, and inspire scientists all over the world. It covers a wide range of topics, including principles and fundamentals of RNA nanotechnology; RNA folding, structure, and motifs in RNA nanoparticle assembly; RNA computation and structure prediction for RNA nanoparticle construction; nucleotide chemistry for nanoparticle synthesis, conjugation, and labeling; single-molecule and biophysical techniques in RNA nanostructure analysis; methods for the assembly of RNA nanoparticles; and RNA nanoparticles for therapy and diagnostic applications.

Peixuan Guo

Contributors

Sherine Abdelmawla
Kylin Therapeutics
West Lafayette, Indiana

Kirill A. Afonin
Nanoscale Science Program
Department of Chemistry
and
The Center for Biomedical Engineering and
    Science
The University of North Carolina at Charlotte
Charlotte, North Carolina

Yelixza I. Avila
Nanoscale Science Program
Department of Chemistry
The University of North Carolina at Charlotte
Charlotte, North Carolina

Damian Beasock
Nanoscale Science Program
Department of Chemistry
The University of North Carolina at Charlotte
Charlotte, North Carolina

Eckart Bindewald
Basic Science Program
Frederick National Laboratory for Cancer
    Research
Frederick, Maryland

Daniel W. Binzel
Center for RNA Nanobiotechnology and
    Nanomedicine
Division of Pharmaceutics and
    Pharmacology
College of Pharmacy
The Ohio State University
Columbus, Ohio

Alice Braga
Department of Clinical Neurosciences
University of Cambridge
Cambridge, United Kingdom

Nicolas D. Burns
Center for RNA Nanobiotechnology and
    Nanomedicine
and
College of Pharmacy
The Ohio State University
Columbus, Ohio

Morgan Chandler
Nanoscale Science Program
Department of Chemistry
University of North Carolina at Charlotte
Charlotte, North Carolina

Shi-Jie Chen
Department of Physics
Department of Biochemistry, and Informatics
    Institute
University of Missouri
Columbia, Missouri

Yi Cheng
Department of Physics
Department of Biochemistry, and Informatics
    Institute
University of Missouri
Columbia, Missouri

Ciara E. Conway
Nanoscale Science Program
Department of Chemistry
The University of North Carolina at
    Charlotte
Charlotte, North Carolina

Marina A. Dobrovolskaia
Nanotechnology Characterization
    Laboratory
Cancer Research Technology Program
Leidos Biomedical Research Inc.
Frederick National Laboratory for Cancer
    Research
Frederick, Maryland

Nikolay V. Dokholyan
Department of Pharmacology
and
Department of Biochemistry & Molecular
  Biology
Penn State College of Medicine
Hershey, Pennsylvania

Lili Du
Center for RNA Nanobiotechnology and
  Nanomedicine
and
College of Pharmacy
The Ohio State University
Columbus, Ohio

Sam Eisen
Nanoscale Science Program
Department of Chemistry
University of North Carolina at
  Charlotte
Charlotte, North Carolina

Dana Elasmar
Nanoscale Science Program
Department of Chemistry
University of North Carolina at Charlotte
Charlotte, North Carolina

Satheesh Ellipilli
Center for RNA Nanobiotechnology and
  Nanomedicine
and
College of Pharmacy
The Ohio State University
Columbus, Ohio

Yangwu Fang
ExonanoRNA Biomedicine LTD
Foshan, China

Paloma H Giangrande
Internal Medicine
University of Iowa
Iowa City, Iowa
and
Platform Discovery Sciences, Biology
Wave Life Sciences Inc.
Cambridge, Massachusetts

Bin Guo
Department of Pharmacological and
  Pharmaceutical Sciences
College of Pharmacy
University of Houston
Houston, Texas

Peixuan Guo
Center for RNA Nanobiotechnology and
  Nanomedicine
College of Pharmacy
James Comprehensive Cancer Center
Dorothy M. Davis Heart and Lung Research
  Institute
Department of Physiology and Cell
  Biology
College of Medicine
The Ohio State University
Columbus, Ohio

Sijin Guo
Center for RNA Nanobiotechnology and
  Nanomedicine
and
Division of Pharmaceutics and
  Pharmaceutical Chemistry, College
  of Pharmacy
and
Dorothy M. Davis Heart and Lung Research
  Institute
and
James Comprehensive Cancer Center
College of Medicine
The Ohio State University
Columbus, Ohio

Brent Hallahan
Nanobiotechnology Center, Markey Cancer
  Center and Department of Pharmaceutical
  Sciences
University of Kentucky
Lexington, Kentucky

Justin Halman
Nanoscale Science Program
Department of Chemistry
The University of North Carolina at
  Charlotte
Charlotte, North Carolina

Farzin Haque
Nanobiotechnology Center
Markey Cancer Center and Department of
    Pharmaceutical Sciences
University of Kentucky
Lexington, Kentucky
and
Nanobio Delivery Pharmaceutical Company
    Limited
Columbus, Ohio

Jordan Hill
ExonanoRNA LLC
Columbus, Ohio

Frank Hsiung
System Biosciences
Palo Alto, California

Erik D. Holmstrom
Department of Molecular Biosciences
and
Department of Chemistry
University of Kansas
Lawrence, Kansas

Haibo Hu
Nanobio Delivery Pharmaceutical Company
    Limited
Columbus, Ohio

Yuanyu Huang
School of Life Science
Advanced Research Institute of
    Multidisciplinary Science
Institute of Engineering Medicine
Key Laboratory of Molecular Medicine and
    Biotherapy
Beijing Institute of Technology
Beijing, P. R. China

Travis Hurst
Department of Physics
Department of Biochemistry and Informatics
    Institute
University of Missouri
Columbia, Missouri

Kaori Ishiguro
Department of Transplantation
Mayo Clinic
Jacksonville, Florida

M. Brittany Johnson
Department of Biological Sciences
University of North Carolina at
    Charlotte
Charlotte, North Carolina

W.K. Kasprzak
Basic Science Program
Frederick National Laboratory for Cancer
    Research
Frederick, Maryland

Balveen Kaur
Department of Neurosurgery
McGovern Medical School
University of Texas Health Science Center at
    Houston
Houston, Texas

Weina Ke
Nanoscale Science Program, Department of
    Chemistry
The University of North Carolina at
    Charlotte
Charlotte, North Carolina

Emil Khisamutdinov
Department of Chemistry
Ball State University
Muncie, Indiana

Joanna K. Krueger
Nanoscale Science Program
Department of Chemistry
The University of North Carolina at
    Charlotte
Charlotte, North Carolina

Seraphim S. Kozlov
Nanoscale Science Program, Department of
    Chemistry
The University of North Carolina at
    Charlotte
Charlotte, North Carolina

Tae Jin Lee
Department of Neurosurgery
McGovern Medical School
University of Texas Health Science Center at
    Houston
Houston, Texas

Marissa Leonard
Department of Cancer and Cell Biology
Vontz Center for Molecular Studies
University of Cincinnati College of Medicine
Cincinnati, Ohio

Dongsheng Li
ExonanoRNA Biomedicine LTD
Foshan, China

Hui Li
Nanobiotechnology Center, Markey Cancer
    Center and Department of Pharmaceutical
    Sciences
University of Kentucky
Lexington, Kentucky

Xin Li
Center for RNA Nanobiotechnology and
    Nanomedicine
and
College of Pharmacy
The Ohio State University
Columbus, Ohio

Zhefeng Li
Center for RNA Nanobiotechnology and
    Nanomedicine
and
College of Pharmacy
The Ohio State University
Columbus, Ohio

Chenxi Liang
Center for RNA Nanobiotechnology and
    Nanomedicine
and
College of Pharmacy
The Ohio State University
Columbus, Ohio

Jingwen Liu
Department of Pharmacological and
    Pharmaceutical Sciences
College of Pharmacy
University of Houston
Houston, Texas

Mei Lu
School of Life Science
Advanced Research Institute of
    Multidisciplinary Science
Institute of Engineering Medicine
Key Laboratory of Molecular Medicine and
    Biotherapy
Beijing Institute of Technology
Beijing, P. R. China

Giulia Manferrari
Department of Clinical Neurosciences
University of Cambridge
Cambridge, United Kingdom

Jessica McMillan
Department of Biological Sciences
University of North Carolina at
    Charlotte
Charlotte, North Carolina

Daniel Miller
Department of Chemistry
Ball State University
Muncie, Indiana

Senny Nordmeier
System Biosciences
Palo Alto, California

Christina Oh
Biochemistry and Cell Biology
Department of Biosciences
Rice University
Houston, Texas

Martin Panigaj
Faculty of Science
Institute of Biology and Ecology
Pavol Jozef Safarik University in
    Kosice
Kosice, Slovak Republic

Lorena Parlea
RNA Structure and Design Section
RNA Biology Laboratory
National Cancer Institute
Frederick, Maryland

Tushar Patel
Department of Transplantation
Mayo Clinic
Jacksonville, Florida

Fengmei Pi
ExonanoRNA LLC
Columbus, Ohio

Stefano Pluchino
Department of Clinical Neurosciences
University of Cambridge
Cambridge, United Kingdom

Victoria Portnoy
System Biosciences
Palo Alto, California

Anu Puri
RNA Structure and Design Section
RNA Biology Laboratory
National Cancer Institute
Frederick, Maryland

Randall Reif
Nanobiotechnology Center, Markey Cancer
    Center and Department of Pharmaceutical
    Sciences
University of Kentucky
Lexington, Kentucky

Lewis A. Rolband
Nanoscale Science Program
Department of Chemistry
The University of North Carolina at Charlotte
Charlotte, North Carolina

John J. Rossi
Irell & Manella Graduate School of Biological
    Sciences
Center for RNA Biology and Therapeutics
City of Hope Comprehensive Cancer Center
Duarte, California

Marwa Ben Haj Salah
Department of Molecular and Cellular
    Biology
City of Hope
Duarte, California

Luke Schmedake
Nanoscale Science Program
Department of Chemistry
University of North Carolina at Charlotte
Charlotte, North Carolina

Saptaswa Sen
Department of Molecular Biosciences
University of Kansas
Lawrence, Kansas

Bahar Seremi
Department of Physics
Texas Tech University
Lubbock, Texas

Bruce A. Shapiro
RNA Structure and Design Section
RNA Biology Laboratory
National Cancer Institute
Frederick, Maryland

Oleg A. Shevchenko
Nanoscale Science Program
Department of Chemistry
The University of North Carolina at Charlotte
Charlotte, North Carolina

Hua Shi
Department of Biological Sciences and the
    RNA Institute
University at Albany
Albany, New York

Dan Shu
Center for RNA Nanobiotechnology and
    Nanomedicine
and
College of Pharmacy, Division of
    Pharmaceutics and Pharmaceutical
    Chemistry
and
College of Medicine, Dorothy M. Davis Heart
    and Lung Research Institute and James
    Comprehensive Cancer Center
The Ohio State University
and
Nanobio Delivery Pharmaceutical Co. Ltd.
Columbus, Ohio

Yi Shu
Center for RNA Nanobiotechnology and
    Nanomedicine
The Ohio State University
Columbus, Ohio

Jayden A. Smith
Cambridge Innovation Technologies Consulting
    (CITC) Ltd.
Cambridge, United Kingdom

Caryn D. Striplin
Nanoscale Science Program
Department of Chemistry
The University of North Carolina at Charlotte
Charlotte, North Carolina

Mubin Tarannum
Department of Chemistry
The University of North Carolina at Charlotte
Charlotte, North Carolina

Mario Vieweger
ExonanoRNA LLC
Columbus, Ohio

Juan L. Vivero-Escoto
Department of Chemistry
The University of North Carolina at Charlotte
Charlotte, North Carolina

Hongzhi Wang
Center for RNA Nanobiotechnology and
    Nanomedicine
The Ohio State University
Columbus, Ohio

Jian Wang
Department of Pharmacology
Penn State College of Medicine
Hershey, Pennsylvania

Caroline M. West
Nanoscale Science Program
Department of Chemistry
The University of North Carolina at Charlotte
Charlotte, North Carolina

Haonan Xing
Beijing Institute of Technology
Beijing, P. R. China

Congcong Xu
Center for RNA Nanobiotechnology and
    Nanomedicine
and
Division of Pharmaceutics and Pharmaceutical
    Chemistry, College of Pharmacy
and
Dorothy M. Davis Heart and Lung Research
    Institute
and
James Comprehensive Cancer Center
College of Medicine
The Ohio State University
Columbus, Ohio

Margaret Yeh
Department of Neurosurgery
McGovern Medical School
University of Texas Health Science Center at
    Houston
Houston, Texas

Hongran Yin
Center for RNA Nanobiotechnology and
    Nanomedicine
and
Division of Pharmaceutics and Pharmaceutical
    Chemistry, College of Pharmacy
and
Dorothy M. Davis Heart and Lung Research
    Institute
and
James Comprehensive Cancer Center
College of Medicine
The Ohio State University
Columbus, Ohio

Ji Young Yoo
Department of Neurosurgery
McGovern Medical School
University of Texas Health Science Center at
    Houston
Houston, Texas

Paul Zakrevsky
RNA Structure and Design Section
RNA Biology Laboratory
National Cancer Institute
Frederick, Maryland

Long Zhang
Center for RNA Nanobiotechnology and
    Nanomedicine
and
College of Pharmacy
The Ohio State University
Columbus, Ohio

Dong Zhang
Department of Physics
Department of Biochemistry, and Informatics
    Institute
University of Missouri
Columbia, Missouri

Xiaoting Zhang
Department of Cancer and Cell Biology
Vontz Center for Molecular Studies
University of Cincinnati College of Medicine
Cincinnati, Ohio

Xiaoqin Zou
Department of Physics
Department of Biochemistry, and Informatics
    Institute
University of Missouri
Columbia, Missouri
and
Dalton Cardiovascular Research Center
University of Missouri
Columbia, Missouri

Part I

Part I

Concepts and Definitions in RNA Nanotechnology

1 RNA Structure and Folding

Xin Li and Peixuan Guo
The Ohio State University, Columbus, Ohio, USA

CONTENTS

RNA is a biopolymer composed of ribose sugar based on nucleic acids and made up of primarily four bases: adenine (A), cytosine (C), guanine (G), and uridine (U). RNA structure is a concept of the shape determined by interactions of the four bases. The sequence order of the bases in the polymer and the base paring, G pairs with C, and A pairs with U lead to the folding of the RNA.

RNA CHEMISTRY

RNA is composed of three parts: a phosphate group, a 5-carbon sugar, and a nitrogenous base. The 2′-OH group in the ribose makes RNA less enzymatically stable compared to DNA, but facilitates its versatile biological functions. RNA possesses not only Watson-Crick base pairing (A−U, G−C) but also noncanonical base pairing (G−U wobble, sheared G−A pair, G−A imino pair, A−U reverse Hoogsteen), which promotes folding into rigid structural motifs distinct from the structure of single-stranded DNA.

RNA STRUCTURE

RNA can form primary, secondary (2D), and tertiary (3D) structures. RNA primary structure refers to the linear RNA sequence of nucleotides that are linked by phosphodiester bonds. RNA secondary structure is formed by the Watson-Crick canonical base pairs, which contains four basic elements: helices, loops, bulges, and junctions. RNA tertiary structure is defined as the three-dimensional arrangement of RNA building blocks with both canonical and noncanonical base pair interactions. Such motifs are used as building blocks to construct nanoparticles with diverse size and shape by engineering parameters such as motif angle and sequence length.

RNA MODIFICATION

The characteristic of RNA that defines and differentiates it from DNA is the 2′-hydroxyl on each ribose sugar of the backbone. The 2′-OH group offers RNA a special property, which can be either an advantage or a disadvantage. From a structural point of view, the advantage of this additional hydroxyl group is that it locks the ribose sugar into a 3′-endo chair conformation. As a result, it is structurally favorable for the RNA double helix to adopt the A-form which is ~20% shorter and wider rather than the B-form that is typically present in the DNA double helix. Moreover, the 2′-OH group in RNA is chemically active and is able to initiate a nucleophilic attack on the adjacent 3′ phospho-diester bond in an SN2 reaction. This cleaves the RNA sugar-phosphate backbone, and this chemical mechanism underlies the basis of catalytic self-cleavage observed in ribozymes. The disadvantage is that the 2′-OH group makes the RNA susceptible to nuclease digestion since many RNases recognize

DOI: 10.1201/9781003001560-2

the structure of RNAs including the 2′-OH group as specific binding sites. However, such enzymatic instability has been overcome by applying chemical modification of the 2′-OH group. Modifications to RNA's natural structure can overcome the susceptibility of RNA therapeutics to serum exo- and endonucleases. Some of the most popular strategies include chemical modifications to the ribose sugar of the bases, base modification, and modification of the link between bases (backbone modification). Chemical modifications can impart higher chemical and thermal stability to RNA structure, thereby allowing *in vivo* use of RNA therapeutics and nanoparticles.

2 RNA Regulation and Function in Nature

Zhefeng Li, Daniel W. Binzel, and Peixuan Guo
The Ohio State University, Columbus, Ohio, USA

CONTENTS

RNA is the key genetic component in molecular biology central dogma that not only serves as the bridge between DNA (genome) and protein (function) but also has the property of both by itself (Guo, 2010, Li, Lee et al., 2015). Produced during transcription, RNA then serves as the template for protein synthesis during translation. RNA classically was believed to be a temporary product as part of the production of proteins that is then degraded at the end of translation. However, in today's world it is well known that RNA plays a much larger role in nature than simply protein translation. This is evidenced by the fact that only ~1.5% of the human genome is translated into proteins, while 98.5% of the genome is transcribed into other functioning RNAs, known as non-coding RNA (ncRNA).

RNA PRODUCTION AND FUNCTION FOR PROTEIN PRODUCTION

As the middle component of the central dogma, in nature, RNA is produced through transcription of DNA. Following the central dogma, the most well-known function of RNA is messenger RNA (mRNA) in which RNA serves as the template for protein production (Beelman & Parker, 1995, Weissman & Kariko, 2015). In eukaryotes, following the transcription of DNA into RNA, the RNA is then processed into mRNA through splicing, capping, and polyadenylation of the raw RNA production. This processes the RNA into mRNA by selecting exons and merging them together, while introns are excluded and processed into ncRNAs. Matured mRNA carrying the sequence information and guide ribosome machinery is then used to synthesize its coded protein. At the conclusion of the protein synthesis, the mRNA is then degraded in order to prevent overexpression of any given gene (Beelman & Parker, 1995). Additionally, in some so-called ribovirus (e.g., Ebola virus, SARS, hepatitis C, etc.), RNA by itself can serve as the genome that can self-replicate through several pathways according to the form of RNA genome. These genomic RNAs can be either double-stranded or single-stranded and can be replicated by RNA replicase or hijack the host for reproduction by reverse transcription mechanism.

NONCODING RNAs AND THEIR FUNCTION: MicroRNAs AND GENE REGULATION

Besides carrying genetic information, RNA can also directly exercise biological functions and gene regulation in cells without protein translation; these RNAs are known as ncRNA (Cech & Steitz, 2014, Liang & Wang, 2013, Mercer, Dinger et al., 2009, Palazzo & Lee, 2015, Zhang, 2009).

The majority of ncRNAs are derived from the intron regions of transcribed RNAs that cannot translate into protein. The function variety of RNA molecules depend on sequence, length, structure, modification, intra-molecule interaction, and spatial-temporal distribution. Based on the length, ncRNA can be divided into short ncRNA (<200 nt) (Mattick & Makunin, 2005, Zhang, 2009) and long ncRNA (>200nt) (Mercer et al., 2009, Morlando, Ballarino et al., 2014, Quinn & Chang, 2016). In the early stage, discovery and studies of ncRNAs were highly associated with mRNA. Transfer RNA (tRNA) was the first characterized ncRNA in the 1960s (Cole, Yang et al., 1972, Holley, 1965), followed by the discovery of ribosomal RNA (rRNA) in the 1980s (Kruger, Grabowski et al., 1982, Oostergetel, Wall et al., 1985). tRNA is a single-stranded RNA around 76–90 nt long and contains anticodon region-recognized mRNA and acceptor stem-attached amino acids. tRNA physically "transfers" particular amino acids that match the mRNA to the ribosome machinery and guide peptide synthesis (Stein & Crothers, 1976). tRNA can be considered as the earliest short ncRNA study while the rRNA would be the earliest for long ncRNA. rRNA contains thousands of nucleotides in general, and it comprises ~80% of RNA in cells. As the term of "ribosomal", rRNA is the key component in ribosome machinery and around 60% by mass of ribosome is rRNA. The majority of rRNA serves as building block or structural scaffold of the ribosome (Cech and Steitz, 2014, Dunkle and Cate, 2010).

Small ncRNA was also found to play important role in genome defense. The representative is the guide RNA that involved in the infection defense system in bacteria called the Clustered Regularly Interspaced Short Palindromic Repeats (CRISPR) (Fu, Sander, et al., 2014, Travis, 2015, Wiedenheft, Sternberg, and Doudna, 2012). These guide RNAs can lead the endonuclease complex to cleave the infected sequences out of genome. Most of RNAs exist in linear manner, until recently, circular RNA was discovered. These single-stranded RNAs covalently form closed loop so that they do not have 5' and 3' end as normal RNA and which also makes them more resistant to enzymatic digestion. Although the knowledge of circular RNA is still limited, studies had revealed circular RNA involves many disease- development and biological activities. Additional categories of ncRNAs exist but will not be discussed here, including, but not limited to, snoRNA (small nucleolar RNA) (Bachellerie, Cavaille et al., 2002), snRNA (small nuclear RNA) (Kiss, 2004, Matera, Terns et al., 2007), Y RNA, and piRNA (piwi-interacting RNA) (Zhang, 2009).

MicroRNAs AND GENE REGULATION

In human genome, 98.5% of DNA encode for ncRNA and for a very long period; these ncRNA are considered as "junk". With the advancement in physical biology, biochemistry, molecular biology, bioinformatics, and many interdisciplinary sciences, the mystery of RNA world is currently being revealed. With the research in last decades, numerous types of ncRNA have been discovered and classified to play essential role in regulating gene expression (Keene, 2010, Liang & Wang, 2013, Lieberman, Slack et al., 2013, Taft, Pang et al. 2010). The most typical is microRNA (miRNA); a type of small ncRNA, length in ~22 nt discovered as early as the 1990s, was found to be able to bind the 3'UTR region of mRNA by sequence complementary and suppress the translation level of the particular protein encoded by mRNA (Bartel, 2009, Chekulaeva & Filipowicz, 2009, Duchaine & Slack, 2009, Fire, Xu et al., 1998). This post-transcriptional regulation is known as RNA interference which was credited by the Nobel Prize in 2006. To date thousands of miRNAs have been discovered which are able to specifically bind to different genes and regulate their protein expression levels. The expressions of miRNAs and their function have proven to be very important in living organisms, as changes in their expression levels have been tied to diseases. Specifically, miRNAs have shown to play a large role in cancer development, progression, aggressiveness, and even chemo-resistance (Croce, 2008, Croce & Calin, 2005, Di Leva, Garofalo, and Croce, 2014, Garzon, Calin, and Croce, 2009, Lovat, Valeri, and Croce, 2011). Therefore, miRNAs have become the focus of extensive research in the areas of disease diagnosis and therapy.

CONCLUSIONS

In summary, RNA is the center of central dogma that not only serves as a bridge between DNA (genomic storage) and protein (function module) but also has overlap functionality with both. RNAs have a diverse function within cells that demonstrates the diverse structure and sequences of RNAs within living organisms.

REFERENCES

Bachellerie JP, Cavaille J, Huttenhofer A (2002) The expanding snoRNA world. *Biochimie* 84: 775–790.

Bartel DP (2009) MicroRNAs: target recognition and regulatory functions. *Cell* 136: 215–233.

Beelman CA, Parker R (1995) Degradation of mRNA in eukaryotes. *Cell* 81: 179–183.

Cech TR, Steitz JA (2014) The noncoding RNA revolution-trashing old rules to forge new ones. *Cell* 157: 77–94.

Chekulaeva M, Filipowicz W (2009) Mechanisms of miRNA-mediated post-transcriptional regulation in animal cells. *Current Opinion in Cell Biology* 21: 452–460.

Cole PE, Yang SK, Crothers DM (1972) Conformational changes of transfer ribonucleic acid. Equilibrium phase diagrams. *Biochemistry* 11: 4358–4368.

Croce CM (2008) Oncogenes and cancer. *New England Journal of Medicine* 358: 502–511.

Croce CM, Calin GA (2005) miRNAs, cancer, and stem cell division. *Cell* 122: 6–7.

Di Leva G, Garofalo M, Croce CM (2014) MicroRNAs in cancer. *Annual Review of Pathology* 9: 287–314.

Duchaine TF, Slack FJ (2009) RNA interference and micro-RNA-oriented therapy in cancer: rationales, promises, and challenges. *Current Oncology* 16: 265–270.

Dunkle JA, Cate JH (2010) Ribosome structure and dynamics during translocation and termination. *Annual Review of Biophysics* 39: 227–244.

Fire A, Xu S, Montgomery MK, Kostas SA, Driver SE, Mello CC (1998) Potent and Specific Genetic Interference by Double-stranded RNA in Caenorhabditis Elegans. *Nature* 391: 806–811.

Fu Y, Sander JD, Reyon D, Cascio VM, Joung JK (2014) Improving CRISPR-Cas nuclease specificity using truncated guide RNAs. *Nature Biotechnology* 32: 279–284.

Garzon R, Calin GA, Croce CM (2009) MicroRNAs in Cancer. *Annual Review of Medicine* 60: 167–179.

Guo P (2010) The emerging field of RNA nanotechnology. *Nature Nanotechnology* 5: 833–842.

Holley RW (1965) Structure of an alanine transfer ribonucleic acid. *JAMA* 194: 868–871.

Keene JD (2010) Minireview: global regulation and dynamics of ribonucleic Acid. *Endocrinology* 151: 1391–1397.

Kiss T (2004) Biogenesis of small nuclear RNPs. *Journal of Cell Science* 117: 5949–+.

Kruger K, Grabowski PJ, Zaug AJ, Sands J, Gottschling DE, Cech TR (1982) Self-splicing RNA: Autoexcision and Autocyclization of the Ribosomal RNA Intervening Sequence of Tetrahymena. *Cell* 31: 147–157.

Li H, Lee T, Dziubla T, Pi F, Guo S, Xu J, Li C, Haque F, Liang XJ, Guo P (2015) RNA as a Stable Polymer to Build Controllable and Defined Nanostructures for Material and Biomedical Applications. *Nano Today* 10: 631–655.

Liang Z, Wang XJ (2013) Rising from ashes: non-coding RNAs come of age. *J Genet Genomics* 40: 141–142

Lieberman J, Slack F, Pandolfi PP, Chinnaiyan A, Agami R, Mendell JT (2013) Noncoding RNAs and cancer. *Cell* 153: 9–10.

Lovat F, Valeri N, Croce CM (2011) MicroRNAs in the pathogenesis of cancer. *Seminars in Oncology* 38: 724–733.

Matera AG, Terns RM, Terns MP (2007) Non-coding RNAs: lessons from the small nuclear and small nucleolar RNAs. *Nature Reviews Molecular Cell Biology* 8: 209–220.

Mattick JS, Makunin IV (2005) Small regulatory RNAs in mammals. *Human Molecular Genetics* 14: R121–R132.

Mercer TR, Dinger ME, Mattick JS (2009) Long non-coding RNAs: insights into functions. *Nature Reviews Genetics* 10: 155–159.

Morlando M, Ballarino M, Fatica A, Bozzoni I (2014) The role of long noncoding RNAs in the epigenetic control of gene expression. *ChemMedChem* 9: 505–510.

Oostergetel GT, Wall JS, Hainfeld JF, Boublik M (1985) Quantitative structural analysis of eukaryotic ribosomal RNA by scanning transmission electron microscopy. *Proceedings of the National Academy of Sciences* 82: 5598–5602.

Palazzo AF, Lee ES (2015) Non-coding RNA: what is functional and what is junk? *Frontiers in Genetics* 6: 2.

Quinn JJ, Chang HY (2016) Unique features of long non-coding RNA biogenesis and function. *Nature Reviews Genetics* 17: 47–62.

Stein A, Crothers DM (1976) Conformational changes of transfer RNA. The role of magnesium (II). *Biochemistry* 15: 160–167.

Taft RJ, Pang KC, Mercer TR, Dinger M, Mattick JS (2010) Non-coding RNAs: regulators of disease. *Journal of Pathology* 220: 126–139.

Travis J (2015) Making the cut. *Science* 350: 1456–1457.

Weissman D, Kariko K (2015) mRNA: Fulfilling the Promise of Gene Therapy. *Molecular Therapy* 23: 1416–1417.

Wiedenheft B, Sternberg SH, Doudna JA (2012) RNA-guided genetic silencing systems in bacteria and archaea. *Nature* 482: 331–338.

Zhang C (2009) Novel functions for small RNA molecules. *Current Opinion in Molecular Therapeutics* 11: 641–651.

3 Principles and Fundamentals of RNA Nanotechnology

Long Zhang and Peixuan Guo
The Ohio State University, Columbus, Ohio, USA

CONTENTS

PHYLOGENETIC ANALYSIS OF RNAs

The packaging RNA (pRNA) has significantly shaped the current RNA nanotechnology landscape (Guo, 2010). The pRNA is an RNA molecule that derives from the phi29 bacteriophage or phi29-like family (Hill et al., 2016). In 1998, Guo et al. demonstrated the most-studied pRNA prototype: phi29 pRNA, sequence self-assembles *in vitro* (Guo et al., 1998). The full-length pRNA of phi29 is 174 nt, but for *in vitro* packaging, a 120 nt-long truncated version is enough. Different related species of bacteriophage share only 12% similarity in pRNA sequences (Bourassa & Major, 2002, Garver & Guo, 1997). All pRNAs from strains of the phylogenetically related phi29 family can form similar secondary structures, containing several stem-loops and a three-way junction (Figure 3.1) (Guo et al., 1987). More recent studies on phylogenetically related pRNAs have shown that *in vitro* self-assembly behavior varies by sequence and depends on the nucleotide composition at the 3WJ (Gu & Schroeder, 2011, Hao & Kieft, 2014).

FEATURES OF NUCLEIC ACIDS PROMOTING RNA NANOTECHNOLOGY

Nucleic acids-based nanotechnology generally encompasses the characterization and use of nucleic acids as a construction material to build homogeneous nanostructures on the nanoscale level by bottom-up self-assembly (Guo, 2005). Biological macromolecules, like DNA, RNA, and proteins, intrinsically have defined features at the nanometer scale and can serve as unique and powerful building blocks for the bottom-up fabrication of nanostructures and nanodevices (Jaeger et al., 1993, Ke et al., 2018, Mao et al., 2000, Moll et al., 2002, Zuker, 1989). DNA is structurally defined on the nanometer scale, which possesses remarkable binding specificity and thermodynamic stability. However, geometric flexibility and instability are mainly limitations of DNA branched junctions (Ke et al., 2018). Compared to DNA, RNA is a particularly interesting candidate for nanotechnology applications. RNA has versatile flexibility in structure and function, a property similar to that of proteins. For example, ribozymes are composed of RNAs but have the enzymatic property of proteins. Another is RNA aptamers, which are similar to antibodies in that they can bind small molecules as specific targeting moieties (Pleij & Bosch, 1989). Typically, RNA molecule contains large variety of single-stranded loops for inter- and/or intra-molecular interaction. These loops can serve

DOI: 10.1201/9781003001560-4

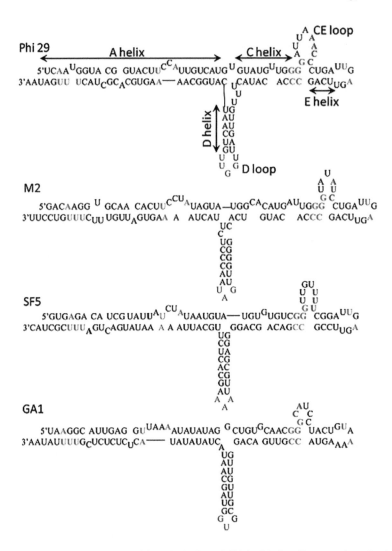

FIGURE 3.1   Secondary structures of φ29, M2, SF5, and GA1 pRNAs. Conserved nucleotides are green. Nucleotides that participate in potential Watson-Crick intermolecular tertiary and quaternary interactions between the CE bulge loop and the D loop hairpin are in red (Gu and Schroeder 2011).

as mounting dovetails; thus external linking dowels might not be needed in fabrication and assembly (Hansma et al., 2003). RNA molecules are polymers made up of four nucleotides: A, U, G, and C. Thus, a 30-nucleotide RNA polymer can generate as many as 1018 different RNA molecules. Three-dimensional RNA structures are of nanometer scale, and hence construction of RNA nanoparticles is feasible by a bottom-up approach. Meanwhile, RNA molecules can be designed and manipulated at a level of simplicity characteristic of DNA (Jaeger & Leontis, 2000, Shu et al., 2004a).

RNA MOTIFS USED IN RNA NANOTECHNOLOGY

RNA typically contains a series of RNA motifs, such as kissing loops, dovetails, pseudoknots, kink turns, and multiway junctions (Figure 3.2) (Jasinski et al., 2017). These RNA motifs allow for the construction of a more complicated secondary structure. Various RNA nanoparticles with diverse size and shape can be constructed by such motifs as building blocks. Furthermore, functional RNA

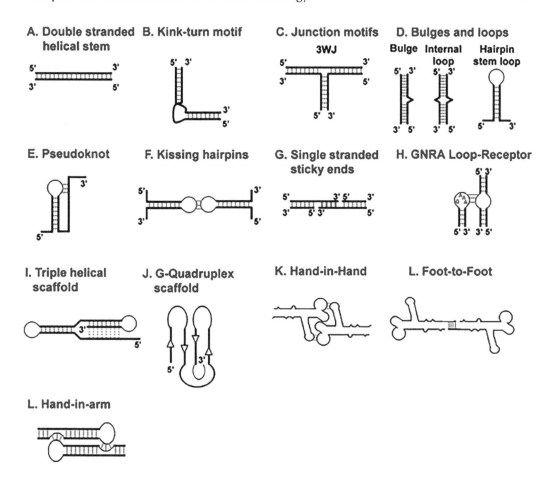

FIGURE 3.2   Motifs for constructing RNA nanoparticles (Jasinski et al. 2017).

molecules such as aptamers, ribozymes, short interfering RNA (siRNA), miRNA, riboswitches, and noncoding RNAs can be simply incorporated into the core sequences of RNA scaffold. Thus, the RNA nanoparticles can be designed with different functions (targeting, imaging, and therapeutic purpose) for diverse applications in nanobiotechnology.

STOICHIOMETRY DETERMINATION BY BINOMIAL DISTRIBUTION, BY LOG/LOG PLOT WITH A SERIAL DILUTION; BY FINDING THE COMMON MULTIPLE OF 2 AND 3

Taking advantage of a highly sensitive *in vitro* phage phi29 assembly and assay system, the stoichiometry study has been well developed in Guo lab since 1994 (Lee & Guo, 1994). The RNA stoichiometry has been determined using the methods of binomial distribution, which results in the discovery of a novel mechanism in viral DNA packaging (Guo et al., 1998). It has been found that pRNA interact hand-in-hand via two loops to form a hexagonal complex to gear the DNA translocation machine (Chen & Guo, 1997). A log/log plot method based on dilution factors and relative concentrations was developed for the stoichiometric quantification of molecules in bio-nanostructures. By using this method, Shu et al. exploited the correlation between the stoichiometry of the component and influence of the dilution factor on the reaction (Shu et al., 2004b). And also, Guo lab found the stoichiometry of pRNA monomer, dimer and trimer is a set of six interlocking pRNAs, two interlocking pRNAs and three interlocking pRNAs, respectively. In 1998, Dr. Guo found the number of pRNAs needed for the packaging of one viral genome is a common multiple of 2 and 3. The findings also supported hexamer formation by the stoichiometry determination using a competition inhibition method (Guo et al., 1998).

HISTORY AND EVOLUTION OF RNA NANOTECHNOLOGY

The emergence and advancement of the RNA nanotechnology field, demonstrated in 1998 by Prof. Peixuan Guo (Guo, et al. *Mol Cell*. 1998; featured in Cell) (Guo et al., 1998), has now developed into a mature field with high potential to serve as a novel therapeutic platform. The evolution of RNA nanotechnology is a collective effort of many insightful individuals. In 1998, Peixuan submitted a manuscript to *Cell*, reporting his finding of the assembly of pRNA dimers, trimers, and hexamers using re-engineered RNA fragments via bottom-up self-assembly; the *Cell* associate editor, Vivian Siegel, and the founding editor of *Cell*, Benjamin Lewin, immediately were intrigued. They strongly recommend publishing this significant discovery on newly initiated journal *Molecular Cell* and Roger Hendrix as the reviewer. This paper firstly showed the concept of RNA nanotechnology.

In the early 2000s, Eric Westhof et al. firstly predicted one of the RNA motifs, kissing loop, could promote the formation of special RNA structures (Leontis et al., 2002). In 2004, the editor and reporters of MSNBC reported the news of one paper about RNA structures in *Nano Letters* and published a groundbreaking news story entitled "Scientists build tiny structures out of RNA" to promote this concept of RNA nanotechnology. Subsequently, *Science* published Luc Jaeger's paper on tectoRNA with commentary by Hao Yan (Yan, 2004). More importantly, when three papers exemplifying the use of RNA nanotechnology to treat cancer were published, the NCI Alliance in Cancer Nanotechnology, led by Piotr Grodzinski, recognized the potential of RNA nanotechnology in cancer treatment (http://nano.cancer.gov/action/news/featurestories/monthly_feature_2006_august.pdf). As part of the NCI's effort to promote the RNA nanotechnology field, a workshop on RNA and disease was organized by the pioneer of computational RNA nanotechnology, Bruce Shapiro. A strong boost to the RNA nanotechnology field can be credited to an invited review by *Nature Nanotechnology* and a subsequent publication detailing the finding of a stable phi29 pRNA three-way junction implemented as an *in vivo* delivery system (Guo, 2010, Shu et al., 2011). Nowadays, the RNA nanotechnology is developed as a platform for (i) solubilizing chemical drugs delivery (Guo et al., 2020), (ii) siRNA delivery avoiding endosome trapping (Zheng et al., 2019), and (iii) construction of ligand displaying extracellular vesicles (Pi et al., 2018).

REFERENCES

Bourassa N, Major F (2002) Implication of the prohead RNA in phage phi29 DNA packaging. *Biochimie* 84: 945–951.

Chen C, Guo P (1997) Sequential action of six virus-encoded DNA-packaging RNAs during phage phi29 genomic DNA translocation. *J Virol* 71: 3864–3871.

Garver K, Guo P (1997) Boundary of pRNA functional domains and minimum pRNA sequence requirement for specific connector binding and DNA packaging of phage phi29. *RNA* 3: 1068–1079.

Gu X, Schroeder SJ (2011) Different sequences show similar quaternary interaction stabilities in prohead viral RNA self-assembly. *J Biol Chem* 286: 14419–14426.

Guo P (2005) RNA nanotechnology: engineering, assembly and applications in detection, gene delivery and therapy. *J Nanosci Nanotechnol* 5: 1964–1982.

Guo P (2010) The emerging field of RNA nanotechnology. *Nat Nanotechnol* 5: 833–842.

Guo P, Erickson S, Anderson D (1987) A small viral RNA is required for *in vitro* packaging of bacteriophage phi29 DNA. *Science* 236: 690–694.

Guo P, Zhang C, Chen C, Garver K, Trottier M (1998) Inter-RNA interaction of phage phi29 pRNA to form a hexameric complex for viral DNA transportation. *Mol Cell* 2: 149–155.

Guo S, Vieweger M, Zhang K, Yin H, Wang H, Li X, Li S, Hu S, Sparreboom A, Evers BM et al (2020) Ultra-thermostable RNA nanoparticles for solubilizing and high-yield loading of paclitaxel for breast cancer therapy. *Nat Commun* 11: 972.

Hansma HG, Oroudjev E, Baudrey S, Jaeger L (2003) TectoRNA and 'kissing-loop' RNA: atomic force microscopy of self-assembling RNA structures. *J Microsc* 212: 273–279.

Hao Y, Kieft JS (2014) Diverse self-association properties within a family of phage packaging RNAs. *RNA* 20: 1759–1774.

Hill AC, Bartley LE, Schroeder SJ (2016) Prohead RNA: a noncoding viral RNA of novel structure and function. *Wiley Interdiscip Rev RNA* 7: 428–437.

Jaeger JA, SantaLucia J, Jr., Tinoco I, Jr. (1993) Determination of RNA structure and thermodynamics. *Annu Rev Biochem* 62: 255–287.

Jaeger L, Leontis NB (2000) Tecto-RNA: One-Dimensional Self-Assembly through Tertiary Interactions This work was carried out in Strasbourg with the support of grants to N.B.L. from the NIH (1R15 GM55898) and the NIH Fogarty Institute (1-F06-TW02251-01) and the support of the CNRS to L.J. The authors wish to thank Eric Westhof for his support and encouragement of this work. *Angew Chem Int Ed Engl* 39: 2521–2524.

Jasinski D, Haque F, Binzel DW, Guo P (2017) Advancement of the Emerging Field of RNA Nanotechnology. *ACS Nano* 11: 1142–1164.

Ke Y, Castro C, Choi JH (2018) Structural DNA Nanotechnology: Artificial Nanostructures for Biomedical Research. *Annu Rev Biomed Eng* 20: 375–401.

Lee CS, Guo P (1994) A highly sensitive system for the in vitro assembly of bacteriophage phi 29 of Bacillus subtilis. *Virology* 202: 1039–1042.

Leontis NB, Stombaugh J, Westhof E (2002) The non-Watson-Crick base pairs and their associated isostericity matrices. *Nucleic Acids Res* 30: 3497–3531.

Mao C, LaBean TH, Relf JH, Seeman NC (2000) Logical computation using algorithmic self-assembly of DNA triple-crossover molecules. *Nature* 407: 493–496.

Moll D, Huber C, Schlegel B, Pum D, Sleytr UB, Sara M (2002) S-layer-streptavidin fusion proteins as template for nanopatterned molecular arrays. *Proc Natl Acad Sci U S A* 99: 14646–14651.

Pi F, Binzel DW, Lee TJ, Li Z, Sun M, Rychahou P, Li H, Haque F, Wang S, Croce CM et al (2018) Nanoparticle orientation to control RNA loading and ligand display on extracellular vesicles for cancer regression. *Nat Nanotechnol* 13: 82–89.

Pleij CW, Bosch L (1989) RNA pseudoknots: structure, detection, and prediction. *Methods Enzymol* 180: 289–303.

Shu D, Huang L, Guo P (2004a) A simple mathematical formula for stoichiometry quantification of viral and nanobiological assemblage using slopes of log/log plot curves. *J Virol Methods* 115: 19–30.

Shu D, Moll WD, Deng Z, Mao C, Guo P (2004b) Bottom-up Assembly of RNA Arrays and Superstructures as Potential Parts in Nanotechnology. *Nano Lett* 4: 1717–1723.

Shu D, Shu Y, Haque F, Abdelmawla S, Guo P (2011) Thermodynamically stable RNA three-way junctions for constructing multifuntional nanoparticles for delivery of therapeutics. *Nature Nanotechnology* 6: 658–667.

Yan H (2004) Materials science. Nucleic acid nanotechnology. *Science* 306: 2048–2049.

Zheng Z, Li Z, Xu C, Guo B, Guo P (2019) Folate-displaying exosome mediated cytosolic delivery of siRNA avoiding endosome trapping. *J Control Release* 311–312: 43–49.

Zuker M (1989) On finding all suboptimal foldings of an RNA molecule. *Science* 244: 48–52.

4 Computation and Folding Predictions

Lili Du and Peixuan Guo

The Ohio State University, Columbus, Ohio, USA

CONTENTS

RESOURCES FOR RNA STRUCTURE COMPUTATION

RNA structure prediction is very important to understand the physical mechanism of RNA functions and to design RNA-based therapies. The RNA structure prediction includes primary sequence, secondary structures, and three-dimensional (3D) structures. RNA 3D structure involves long-range tertiary interactions such as kissing interactions between the different loops; therefore, accurate evaluation of the energetic parameters for tertiary interactions is critical for 3D structure prediction. There are many programs predicting RNA 3D structures, like NAST (Jonikas et al., 2009), BARNACLE (Choi & Farach-Colton, 2003), FARFAR (Das & Baker, 2007), and others. NAST is a molecular dynamics simulation tool consisting of a knowledge-based statistical potential function applied to a coarse-grained model with a resolution of one bead nucleotide residue. Its greatest strength is to allow modeling of large RNA molecules. NAST requires secondary structure information and accepts tertiary contacts to direct the folding. When only the secondary structure information is considered, accurate prediction is limited, but when input information of tertiary contacts is taken into account, prediction accuracy can be dramatically improved. BARNACLE generates reasonable RNA-like structures using secondary structure information for small RNA molecules (<50 nt), but not for longer than 50 nt due to an increase in complexity of the probabilistic model. BARNACLE allows for efficient sampling of RNA conformations in continuous space and with related probabilities. FARFAR is only applicable to small RNA and has variable accuracy relying on the framework found successful for proteins using atomistic models and empirical potential functions.

There are also other programs that are used to predict RNA structures. RNA composer treats secondary structural motifs in a more detailed way by using slightly distorted helical conformations and tertiary contacts in secondary structure motifs (Purzycka et al., 2015). CovaRna and CovStat were developed to explore long-range co-varying RNA interaction networks using whole genome alignments (Bindewald & Shapiro, 2013). RNA Junction is a database containing structure and sequence information for known RNA helical junctions and kissing loop interactions (Bindewald et al., 2008). Nano Folder is another web-based software tool for RNA nanostructure by predicting the structure and sequence attributes of multi-stranded RNA constructs (Bindewald et al., 2011). RAG (RNAAs-Graphs) is an RNA topology resource developed by the Schlick group and is used

DOI: 10.1201/9781003001560-5

to classify/analyze topological characteristics of existing RNAs and design and predict novel RNA motifs (Izzo et al., 2011). GraPPLE applies secondary graph representations to identify and classify non-coding RNAs from sequences by using graph properties that capture structural features of functional RNAs (Childs et al., 2009).

2D/3D MODELING/PREDICTION OF DESIRED RNA NANOPARTICLES

MC-sym builds all-atom structures using the 3D version of the nucleotide cyclic motif (NCM) fragments which are not only extracted from known RNA 3D structures but also built on the fly if necessary. But this method is limited to short RNAs requiring 2D structures as input due to the limited NCM fragments for large, complex NCM motifs, such as six-way junctions and kissing loops. There are several programs that combine 2D and 3D analyses together, such as iFoldRNA, Vfold, and FR3D. iFoldRNA uses 3 three-bead models to predict RNA 3D structures by applying discrete molecular dynamics simulations (Sharma et al., 2008). The usage of coarse-grained representation and DMD approach significantly reduces the computational complexity and enhances the conformational sampling. It does not require secondary structure information and can rapidly predict structures for small RNAs. Vfold is a model used to predict RNA 2D and 3D structures and the folding stability from the sequence (Cao & Chen, 2006). The model distinguishes itself from other models by two unique features: physics-based modeling of conformational entropy for 2D structure prediction and template-based multiscale modeling for 3D structure prediction. Vfold3D is a template-based coarse-grained structure prediction model, which can divide 2D structure into different motifs and assemble the 3D motif templates from known RNA structures to predict 3D structures. VfoldLA is a new model for the assembly of RNA 3D structures using loop templates. The model divides a 2D structure into helices and single-stranded loops and classifies the loops into four different types. The model assembles RNA 3D structures from loop templates according to the loop-helix connections. FR3D is a server for finding and superimposing RNA 3D motifs. FR3D Align is a global pairwise alignment of RNA 3D structure using local superpositions (Sarver et al., 2008). JAR3D is predicting RNA 3D motifs in sequences. R3D-2-MSA Server is for accessing alignments from 3D structures. ASSEMBLE is another automated program using secondary or tertiary structure information from homologous RNAs to build a first-order approximation RNA 3D model.

FOLDING PREDICTIONS vs. STRUCTURAL BUILDING BLOCKS

Nowadays, scientists have discovered various functions of RNA, such as mechanisms of protein synthesis, enzymatic activities, gene regulation, and targeting. RNA functions highlight the biological significance of RNA folding since the function is determined by the structure. Hence, it is of great need to have a predictive model for RNA folding.

The recent development in statistical mechanical modeling of RNA folding has led to success in predicting RNA structures, including folding stabilities and folding kinetics for structures with increasing complexity. However, there are still several key issues unsolved in this field. Some examples are the computation of the entropy for RNA tertiary folds and the extraction of the energy/entropy parameters for non-canonical tertiary interactions from thermodynamic data and known structures. The early computational methods for RNA pseudoknot prediction were based on the genetic algorithm. These methods, such as the STAR and the MPGA fold models, predict the structures with the optimal kinetic accessibility instead of the ones with the lowest free energies. The structural building blocks approach involves several steps. First, the fragment length should be decided. Then, a set of training data should be given. All possible fragments of the specified length should be compiled to the whole set. Next, a clustering method is used to divide these fragments into clusters and pick up the center of each cluster to be a building block. If these building blocks are good enough, they can be used to represent all original fragments within a tolerable limit and reconstruct the 3D structure of a whole RNA within some tolerance.

SEQUENCE SEARCH AND ALIGNMENT OF RNA

A sequence alignment is a way of arranging the sequences of DNA, RNA, or protein to identify regions of similarity. Those similar regions may be a consequence of functional, structural, or evolutionary relationships between the sequences. Aligned sequences of nucleotide or amino acid residues are typically represented as rows within a matrix. In some cases, gaps are inserted between the residues so that identical or similar characters are aligned in successive columns. Here we are going to discuss two categories of tools: the Basic Local Alignment Search Tool and the ClustalW.

Basic Local Alignment Search Tool, as known as BLAST, is an algorithm for comparing primary biological sequence information, such as the amino-acid sequence of proteins or the nucleotides of DNA and/or RNA sequences. A BLAST search enables a researcher to compare a subject protein or nucleotide sequence with a library or database of sequences and identify library sequences that resemble the query sequence above a certain threshold. It finds regions of similarity between biological sequences. The program compares nucleotide or protein sequences to sequence databases and calculates the statistical significance of matches. BLAST can be used to infer functional and evolutionary relationships between sequences as well as help identify members of gene families. BLAST can be used for several purposes, including identifying species, locating domains, establishing phylogeny, DNA mapping, and comparison.

ClustalW, like the other Clustal tools, is used for aligning multiple nucleotide or protein sequences efficiently. ClustalW is a matrix-based algorithm, whereas tools like T-coffee and Dialign are consistency-based. ClustalW has a fairly efficient algorithm that competes well against other software. This program requires three or more sequences in order to calculate a global alignment, for pairwise sequence alignment. The algorithm of ClustalW provides a close-to-optimal result most of the time. However, it does exceptionally well when the data set contains sequences with varying degrees of divergence. This is because in a data set like this, the guide tree becomes less sensitive to noise. ClustalW was one of the first algorithms to combine pairwise alignment and global alignment in an attempt to be speed efficient at the price of accuracy compared to other software (Sauder et al., 2000).

REFERENCES

Bindewald, E. and B.A. Shapiro, *Computational detection of abundant long-range nucleotide covariation in Drosophila genomes*. Rna, 2013. **19**(9): p. 1171–1182.

Bindewald, E., et al., *RNAJunction: a database of RNA junctions and kissing loops for three-dimensional structural analysis and nanodesign*. Nucleic Acids Research, 2008. **36**: p. D392–D397.

Bindewald, E., et al., *Multistrand RNA secondary structure prediction and nanostructure design including pseudoknots*. ACS Nano, 2011. **5**(12): p. 9542–9551.

Cao, S. and S.J. Chen, *Predicting RNA pseudoknot folding thermodynamics*. Nucleic Acids Res, 2006. **34**(9): p. 2634–2652.

Childs, L., et al., *Identification and classification of ncRNA molecules using graph properties*. Nucleic Acids Res, 2009. **37**(9): p. e66.

Choi, V. and M. Farach-Colton, *Barnacle: an assembly algorithm for clone-based sequences of whole genomes*. Gene, 2003. **320**: p. 165–176.

Das, R. and D. Baker, *Automated de novo prediction of native-like RNA tertiary structures*. Proc Natl Acad Sci U S A, 2007. **104**(37): p. 14664–14669.

Izzo, J.A., et al., *RAG: an update to the RNA-As-Graphs resource*. BMC Bioinformatics, 2011. **12**: p. 219.

Jonikas, M.A., et al., *Coarse-grained modeling of large RNA molecules with knowledge-based potentials and structural filters*. Rna, 2009. **15**(2): p. 189–199.

Purzycka, K.J., et al., *Automated 3D RNA structure prediction using the RNAComposer method for ribo-switches*. Methods Enzymol, 2015. **553**: p. 3–34.

Sarver, M., et al., *FR3D: finding local and composite recurrent structural motifs in RNA 3D structures*. Journal of Mathematical Biology, 2008. **56**(1–2): p. 215–252.

Sauder, J.M., J.W. Arthur, and R.L. Dunbrack, Jr., *Large-scale comparison of protein sequence alignment algorithms with structure alignments*. Proteins, 2000. **40**(1): p. 6–22.

Sharma, S., F. Ding, and N.V. Dokholyan, *iFoldRNA: three-dimensional RNA structure prediction and folding*. Bioinformatics, 2008. **24**(17): p. 1951–1952.

5 Enzymatic Synthesis and Modification of RNA Nanoparticles

Daniel W. Binzel and Peixuan Guo
The Ohio State University, Columbus, Ohio, USA

CONTENTS

RNA nanotechnology utilizes the bottom-up construction of multiple RNA component strands into predefined shapes, structures, and stoichiometry based on computer-aided designs (Guo, 2010). The construction of these RNA component strands are accomplished in two manners: (1) Chemical synthesis (discussed in Chapter 6) and (2) Enzymatic synthesis (Jasinski et al., 2017, Xu et al., 2018). While chemical synthesis allows for large-scale production of RNA oligos, it is limited in the number of nucleotides it can incorporate into a strand due to the chemical reaction efficiencies (Xu et al., 2018). Therefore, RNA production by enzymatic synthesis is valuable in that it can theoretically produce RNA strands of any length to overcome the limitations of chemical synthesis.

Researchers have developed *in vitro* techniques of RNA synthesis to allow for the reliable production of RNA using the T7 phage RNA polymerase, which is discussed in detail below (Cheetham & Steitz, 1999, Beckert & Masquida, 2011, Chamberlin & Ring, 1973a, Davanloo et al., 1984, Kochetkov et al., 1998, Milligan, et al., 1987, Milligan & Uhlenbeck, 1989). RNA enzymatic production requires several bottom-up construction steps, involving the construction of a DNA template, *in vitro* transcription, and purification of RNA products.

CONSTRUCTION OF DNA TEMPLATES

In order to complete enzymatic RNA synthesis, a DNA template is required for the polymerase enzymes to read and transcribe into RNA. Therefore, the desired sequence to be transcribed or produced must be designed in order to hybridize into the RNA nanoparticle upon completion of transcription. First and foremost, the DNA sequence must include the polymerase promoter sequence. Each polymerase has its own requirements to start transcription of a DNA strand; and when using the T7 RNA polymerase, a 17 nucleotide promoter sequence is needed to bind the polymerase to the DNA strand (Chamberlin & Ring, 1973a, Cheetham & Steitz, 1999). The end of the promoter sequence includes the "Tata box", named due to its TATA sequence; immediately following the "Tata box", the DNA sequence requires a GC or GG transcription start sequence in which the T7 polymerase will begin transcribing the rest of the DNA template sequence into RNA (Cheetham &

Steitz, 1999). Therefore the enzymatic synthesis of the RNA is somewhat limited by sequence constraints as each strand must begin with either a GC or GG.

Once a desired sequence of the RNA oligo is designed, the DNA template needs to be constructed (Guo & Guo, n.d., Guo et al., 2012, Shu et al., 2013). The total DNA template may be short enough to chemically synthesized; chemical synthesis of two DNA oligos is limited to ~150 nucleotides in length. If this is the case, one DNA strand should incorporate the T7 promoter sequence, while the second is the complement sequence of the desired transcription product (Cheetham & Steitz, 1999). The T7 RNA polymerase will read along the strand while placing the complementary RNA nucleotide to the DNA strand. However, if the desired RNA strand is longer than the production limits of DNA synthesis, smaller shorter strands can be synthesized or purchased and one long DNA template strand can be produced using polymer chain reaction (PCR) (Saiki et al., 1985, Shu et al., 2013). PCR is normally used for the isolation and amplification of a shorter DNA sequence from a long DNA (Saiki et al., 1988, Saiki et al., 1985); however, by using PCR primers that overhang from a shorter DNA template, the template strand can be elongated during the PCR amplification (Shu et al., 2013). If needed, several rounds of PCR with further extending primers can be completed to elongate the DNA templates to the desired final DNA template. At the completion of the PCR, the DNA template should be purified from DNA polymerase and free nucleotides, prior to moving onto transcription.

IN VITRO TRANSCRIPTION OF RNA OLIGO STRANDS

The T7 phage RNA polymerase is most commonly used polymerase due to its high rate of transcription and a very low error rate when reading template DNA (Chamberlin & Ring, 1973a, 1973b). The T7 RNA polymerase (RNAP) protein has been isolated from the phage and expressed into plasmids that have been stably place in bacteria such as *E. coli*. This has allowed for easy purification of the polymerase to be used *in vitro* applications (Afonin et al., 2012, Beckert & Masquida, 2011, Davanloo et al., 1984, Gondai, et al., 2008, Milligan et al., 1987). Large quantities of the T7 RNA polymerase are easily created by scaling up bacteria production and purification of the polymerase through common molecular biology techniques.

Upon the purification of RNA polymerase and the DNA template, transcription of DNA into RNA is completed through a simple biological reaction. T7 polymerase requires the presence of magnesium ions; thus a reaction buffer is needed to initiate the transcription (Shu et al., 2013, Shu et al., 2009). In addition to mixing of the RNAP, buffer, and DNA template, ribonucleic acid nucleotides are needed for the polymerase to add one by one onto the RNA strand being synthesized. The RNA polymerase then identifies and binds transcription promoter sequences and begins transcription, synthesizing RNA strands reading along the DNA at an estimated rate of ~230 nuc/sec (Kochetkov et al., 1998). As the polymerase reaches the end of the DNA template, it falls off ending transcription until it is able to identify a new DNA template strand in the solution. The reaction will occur until the polymerase runs out of activity due to misfolding, RNA nucleotides are depleted, or buffer and temperature conditions change enough to prevent polymerase activity.

At the completion of enzymatic RNA synthesis, RNA products are needed to be isolated and purified from the transcription mixture. Due to a similar size, the DNA template can be degraded by the addition of DNase I into the reaction. Furthermore, the removal of salts, nucleotides, and enzyme proteins can be removed by RNA isolation kits, using size cutoffs or chemical extraction. However, such methods isolate a total RNA population that includes incomplete transcription products and other byproduct. Therefore, it is common to complete size exclusion chromatography, polyacrylamide gel purification, or ultracentrifugation techniques to retain the desired, pure RNA product (Guo & Guo, n.d., Guo et al., 2018, Jasinski et al. 2015, Shu et al., 2013, Shu et al., 2009). At this point individual RNA strands can be used to complete the construction of RNA nanoparticles either through homogenous interactions of repeat structures or mixing with other synthesized RNA strands.

INCLUSION OF CHEMICAL MODIFICATIONS INTO ENZYMATIC SYNTHESIS OF RNA OLIGOS

It is well known that RNA is susceptible to nucleases within the body and can be quickly degraded. The chemical modification of RNA backbones creates a slight conformational change in the RNA structuring and folding to disguise itself from RNases (Liu et al., 2011). The chemical modifications (i.e. $2'$-Fluoro and $2'$-OCH$_3$) have become very prevalent in the field of RNA nanotechnology and used in FDA-approved RNA-based therapies (Guo, 2010, Shu et al., 2014, Xu et al., 2018). They provide benefits in improving the thermodynamic stability of RNA helical structuring and folding and significantly improve the stability in serums and RNases to beyond 24 hours (Binzel et al., 2016, Haque et al., 2017, Liu et al., 2011). As such native T7 RNA polymerase is not able to incorporate these modified nucleotides during transcription. However, researchers have developed a single amino acid mutated (Y639F) T7 RNA polymerase variant that is able to incorporate $2'$-fluoro modified nucleotides (Chelliserrykattil & Ellington, 2004). Additional RNA polymerases have been modified in similar ways such as the Syn5 polymerase to allow for these important modifications on RNA nanoparticles (Meyer et al., 2015, Zhu et al., 2015). Therefore, by using a modified *in vitro* transcription, enzymatic production of RNA nanoparticles for clinically relevant approaches can be completed.

ROLLING CIRCLE TRANSCRIPTION FOR LARGE-SCALE PRODUCTION OF RNA

Using *in vitro* enzymatic transcription for the commercial-scale production of RNA and RNA nanoparticles has been limited by the overall yield of production that cannot meet the commercial-scale needs. There has been great interest in creating new technologies that are able to increase the yield of production of RNA transcriptions. Recently, researchers developed rolling circle transcription (RCT) that is able to do just that (Figure 5.1) (Mohsen & Kool, 2016). RCT is completed by developing a circularized DNA template, joining the two ends of the template with the transcription

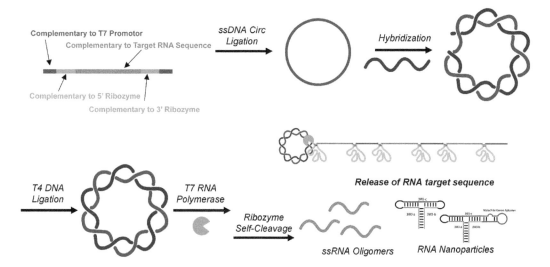

FIGURE 5.1 Overview of RNA Rolling Circle Transcription for RNA Nanoparticle Production. DNA templates are designed, including transcription promotor sequence (red) followed by desired RNA sequence. For rolling circle transcription, the DNA template is then circularized and made into dsDNA. Incubation with RNA polymerase with the DNA template allows for transcription of RNA into desired products. In rolling circle transcription RNA product (orange) is self-released by RNA processing by ribozymes (light blue).

promoter sequence creating a short dsDNA region. Therefore, T7 RNAP binds to the promoter and initiates transcription just as normal linear transcription is completed; however, due to the circular DNA template, the polymerase does not fall off the template completing transcription of the strand. Instead the T7 polymerase continues reading and transcribing the DNA template over and over creating long RNA concatemer productions (Jasinski et al., 2019). Using RCT has proven to increase RNA transcription yields and efficiencies due to the fact that once bound, the polymerase no longer needs to identify and bind to the promoter sequence (Jasinski et al., 2019). The high transcription rate of the T7 RNAP makes RCT highly efficient.

Researchers have implemented RCT for large-scale production of RNA molecules. RNA microsponges elongated to include siRNA sequences were produced by the Hammond research group using RCT (Roh et al., 2016, Roh et al., 2014, Shopsowitz et al. 2014, Roh et al., 2014). Additionally, micrometer-scale nanowires were produced by RCT (Zheng et al., 2014). Both of these products utilized rolling circle to produce long RNAs consisting of repeat sequences to self-assemble into the microsponges and nanowires. Additionally, shorter RNA sequences have been produced by RCT by incorporating self-cleaving RNA ribozymes in the DNA template sequence (Hsu, Hagerman et al., 2014, Roh et al., 2016, Shopsowitz et al., 2014). As the RNA is produced the ribozyme folds and cleaves the targeted sequence for release of repeats. Furthermore, RNase H has been included in transcriptions to cleave and release shorter RNA transcripts from the long RNA concatemers (Wang et al., 2015).

Additionally, the Guo lab recently developed a method of producing a large scale of RNA nanoparticles composed of multiple short RNA strands using RCT (Jasinski et al., 2019). For the first time, a completely double-stranded, circular DNA template was used for RCT. This prevented the folding of the DNA template, thus preventing the RNAP from reading the template. Additionally, ribozyme sequences were included in between target RNAs for the release of the short RNAs to build the RNA nanoparticles. By co-transcription, three short RNA strands were produced and self-assembled into the pRNA-3WJ nanoparticle. Additionally, Jasinski, et al. displayed that RCT reactions out produced linear, run off transcriptions of the same particles, both as total RNA product and RNA nanoparticle product (Jasinski et al., 2019). These results, combined with the large-scale purification using a polyacrylamide gel column demonstrated the movement of enzymatic production towards commercial applications.

CONCLUSIONS

Enzymatic production of RNA and RNA nanoparticles has been around for numerous years and is a well-studied method. While there are limitations to RNA transcription methods, advances in technologies allow for this technique to move closer to large-scale production levels required for commercial applications. The lack of limit of nucleotide length allows for enzymatic production of RNA nanoparticles to continue to stay relevant and a popular method within research applications.

REFERENCES

Afonin, K.A., et al., *Co-transcriptional assembly of chemically modified RNA nanoparticles functionalized with siRNAs*. Nano Letters, 2012 **12**(10): p. 5192–5195.

Beckert, B. and B. Masquida, *Synthesis of RNA by in vitro transcription*. Methods in Molecular Biology, 2011 **703**: p. 29–41.

Binzel, D.W., et al., *Specific Delivery of MiRNA for High Efficient Inhibition of Prostate Cancer by RNA Nanotechnology*. Molecular Therapy, 2016. **24**(7): p. 1267–1277.

Chamberlin, M. and J. Ring, *Characterization of T7–specific ribonucleic acid polymerase. 1. General properties of the enzymatic reaction and the template specificity of the enzyme*. Journal of Biological Chemistry, 1973a. **248**(6): p. 2235–2244.

Chamberlin, M. and J. Ring, *Characterization of T7-specific ribonucleic acid polymerase. II. Inhibitors of the enzyme and their application to the study of the enzymatic reaction.* Journal of Biological Chemistry, 1973b. **248**(6): p. 2245–2250.

Cheetham, G.M. and T.A. Steitz, *Structure of a transcribing T7 RNA polymerase initiation complex.* Science, 1999. **286**(5448): p. 2305–2309.

Chelliserrykattil, J. and A.D. Ellington, *Evolution of a T7 RNA polymerase variant that transcribes 2'-O-methyl RNA.* Nature Biotechnology, 2004. **22**(9): p. 1155–1160.

Davanloo, P., et al., *Cloning and expression of the gene for bacteriophage T7 RNA polymerase.* Proceedings of the National Academy of Sciences, 1984. **81**(7): p. 2035–2039.

Gondai, T., et al., *Short-hairpin RNAs synthesized by T7 phage polymerase do not induce interferon.* Nucleic Acids Research, 2008.

Guo, P., *The emerging field of RNA nanotechnology.* Nature Nanotechnology, 2010. **5**(12): p. 833–842.

Guo, P. and P. Guo, *Methods: RNA Nanotechnology.* Elsevier. 201–294.

Guo, P., et al., *Synthesis, Conjugation, and Labeling of Multifunctional pRNA Nanoparticles for Specific Delivery of siRNA, Drugs and Other Therapeutics to Target Cells.* Methods in Molecular Biology, 2012. **928**: p. 197–219.

Guo, S., et al., *Methods for construction and characterization of simple or special multifunctional RNA nanoparticles based on the 3WJ of phi29 DNA packaging motor.* Methods, 2018. **143**: p. 121–133.

Haque, F., et al., *Using Planar Phi29 pRNA Three-Way Junction to Control Size and Shape of RNA Nanoparticles for Biodistribution Profiling in Mice.* Methods in Molecular Biology, 2017. **1632**: p. 359–380.

Hsu, B.B., et al., *Multilayer films assembled from naturally-derived materials for controlled protein release.* Biomacromolecules, 2014. **15**(6): p. 2049–2057.

Jasinski, D.L., D.W. Binzel, and P. Guo, *One-Pot Production of RNA Nanoparticles via Automated Processing and Self-Assembly.* ACS Nano, 2019. **13**(4): p. 4603–4612.

Jasinski, D., et al., *Large Scale Purification of RNA Nanoparticles by Preparative Ultracentrifugation.* Methods in Molecular Biology. 2015. **1297**: p. 67–82.

Jasinski, D., et al., *Advancement of the Emerging Field of RNA Nanotechnology.* ACS Nano, 2017. **11**(2): p. 1142–1164.

Kochetkov, S.N., E.E. Rusakova, and V.L. Tunitskaya, *Recent studies of T7 RNA polymerase mechanism.* FEBS Letters, 1998. **440**(3): p. 264–267.

Liu, J., et al., *Fabrication of stable and RNase-resistant RNA nanoparticles active in gearing the nanomotors for viral DNA packaging.* ACS Nano, 2011. **5**(1): p. 237–246.

Meyer, A.J., et al., *Transcription yield of fully 2'-modified RNA can be increased by the addition of thermostabilizing mutations to T7 RNA polymerase mutants.* Nucleic Acids Research, 2015. **43**(15): p. 7480–7488.

Milligan, J.F. and O.C. Uhlenbeck, *Synthesis of small RNAs using T7 RNA polymerase.* Methods in Enzymology, 1989. **180**: p. 51–62.

Milligan, J.F., et al., *Oligoribonucleotide synthesis using T7 RNA polymerase and synthetic DNA templates.* Nucleic Acids Research, 1987. **15**(21): p. 8783–8798.

Mohsen, M.G. and E.T. Kool, *The Discovery of Rolling Circle Amplification and Rolling Circle Transcription.* Accounts of Chemical Research, 2016. **49**(11): p. 2540–2550.

Roh, Y.H., et al., *Layer-by-layer assembled antisense DNA microsponge particles for efficient delivery of cancer therapeutics.* ACS Nano, 2014. **8**(10): p. 9767–9780.

Roh, Y.H., et al., *A Multi-RNAi Microsponge Platform for Simultaneous Controlled Delivery of Multiple Small Interfering RNAs.* Angewandte Chemie (International Ed. in English), 2016. **55**(10): p. 3347–3351.

Saiki, R.K., et al., *Enzymatic amplification of beta-globin genomic sequences and restriction site analysis for diagnosis of sickle cell anemia.* Science, 1985. **230**(4732): p. 1350–1354.

Saiki, R.K., et al., *Primer-directed enzymatic amplification of DNA with a thermostable DNA polymerase.* Science, 1988. **239**(4839): p. 487–491.

Shopsowitz, K.E., et al., *RNAi-microsponges form through self-assembly of the organic and inorganic products of transcription.* Small, 2014. **10**(8): p. 1623–1633.

Shu, Y., et al., *Fabrication of Polyvalent Therapeutic RNA Nanoparticles for Specific Delivery of siRNA, Ribozyme and Drugs to Targeted Cells for Cancer Therapy.* IEEE/NIH Life Science Systems and Applications Workshop, 2009: p. 9–12.

Shu, Y., et al., *Assembly of multifunctional phi29 pRNA nanoparticles for specific delivery of siRNA and other therapeutics to targeted cells.* Methods. 2011. **54**: p. 204–214.

Shu, Y., et al., *Fabrication of pRNA nanoparticles to deliver therapeutic RNAs and bioactive compounds into tumor cells.* Nature Protocols. 2013. **8**(9): p. 1635–1659.

Shu, Y., et al., *Stable RNA nanoparticles as potential new generation drugs for cancer therapy.* Advanced Drug Delivery Reviews, 2014. **66**: p. 74–89.

Wang, X., et al., *Preparation of Small RNAs Using Rolling Circle Transcription and Site-Specific RNA Disconnection.* Molecular Therapy--Nucleic Acids, 2015. **4**: p. e215.

Xu, C., et al., *Favorable biodistribution, specific targeting and conditional endosomal escape of RNA nanoparticles in cancer therapy.* Cancer Letters, 2018. **414**: p. 57–70.

Zheng, H.N., Y.Z. Ma, and S.J. Xiao, *Periodical assembly of repetitive RNA sequences synthesized by rolling circle transcription with short DNA staple strands to RNA-DNA hybrid nanowires.* Chemical Communications, 2014. **50**(17): p. 2100–2103.

Zhu, B., et al., *Synthesis of 2'-Fluoro RNA by Syn5 RNA polymerase.* Nucleic Acids Research, 2015. **43**(14): p. e94.

6 Synthetic and Enzymatic Methods for RNA Labeling and Modifications

Satheesh Ellipilli and Peixuan Guo
University of Cincinnati, Cincinnati, Ohio, USA

CONTENTS

The ability to alter RNA composition has enabled the advances in RNA nanotechnology; however, the progress in the RNA nanotechnology hinges on the ability to determine and alter RNA's structure to engineer alternate RNA functions for nanotechnological applications. The functional tools/labels that are essential to develop the useful RNA nanotechnological applications can be introduced into RNA by incorporating the modified nucleotides or 5′/3′ terminal modifiers into the RNA strand. The modifications on the nucleotide can be done at nucleobase (U/G/C/T) or to the ribose, particularly at 2′-position.

RNA SYNTHESIS

RNA is usually synthesized by the following two methods: 1) chemical synthesis and 2) enzymatic synthesis using a double-stranded DNA as a template.

1. ***Chemical synthesis of RNA:*** Solid-phase synthesis is a regular employed method for the synthesis of RNA using the phosphoramidite chemistry [1, 2]. The process has been automated and the RNA oligonucleotide synthesis is carried out by a stepwise addition of nucleobase phosphoramidites to the 3′-terminus of the growing chain until the desired sequence is synthesized. Each addition is referred to as a synthetic cycle and consists of four chemical reactions: Step i: de-blocking: The 4,4′-Dimethoxytrityl (DMT) group present on solid support is removed with a solution of 2% trichloroacetic acid (TCA) in solvent dichloromethane leaves a free 5′-terminal hydroxyl group. Step ii: Coupling: The nucleoside phosphoramidite in acetonitrile is activated by an acidic azole catalyst; the activated phosphoramidite is then brought in contact with the starting solid support to react with the 5′-hydroxy group to form a phosphite triester linkage. Step iii: Capping: The capping step is performed by treating the solid support-bound material with a mixture of acetic anhydride and 1-methylimidazole to block unreacted 5′-hydroxy groups and is acetylated by the capping mixture. Step iv: Oxidation: The treatment

of the support-bound material with iodine and water in the presence of a weak base, usually pyridine, oxidizes the phosphite group into a stable phospho-diester bond. The steps i–iv will be repeated until the desired sequence synthesis is completed in the 3' to 5' direction.

2. ***Enzymatic Synthesis of RNA***: Enzymatic synthesis of RNA is catalyzed by RNA polymerase using DNA as a template, and the process is called transcription. Initiation of transcription begins with the binding of the enzyme to a promoter sequence in the DNA. The enzyme then progresses along the template strand in the 3' to 5' direction, synthesizing a complementary RNA strand with elongation occurring in the 5' to 3' direction [3–5].

SYNTHETIC METHODS FOR RNA LABELING

The site-specific incorporation of labels and reactive groups can be introduced through synthetic chemistry. Solid-phase synthesis is the widely used method to incorporate the labels/functional tools by incorporating modified nucleotides/terminal modifiers having various functional groups such as alkyne [6–11], thiol [12–18], aldehyde [19–22], amine [23–28], carboxylic acid [29–34], and biotin [13, 35–38] into the extending RNA site specifically. These functional groups can be modified with drug molecules or imaging markers having orthogonal functional groups using appropriate chemistries; they include copper-catalyzed click reaction, N-hydroxysuccinimide (NHS) chemistry, thiol chemistry, periodate chemistry, copper-free click chemistry, imine formation, and 5'-phosphate activation [39–50]. Among them, the click reaction, NHS, thiol, and copper-free click chemistries are widely used for both internal and terminal modifications, whereas 5'-phosphate activation, periodate chemistry are mainly employed for terminal modifications. The functional tools are further utilized for labeling and conjugation of various molecular probes and drug molecules in order to engineer RNA nanoparticles for applications in diagnosis and medicine.

Synthetic Methods for Ligation

The synthetic chemistry methodologies extend for ligation of RNA to another RNA/DNA strand by installing appropriate functional groups within two nucleic acid molecules. This process can be enhanced by bringing the two molecules together with a 'splint' [51]. Various functional group combinations are used in joining two nucleic acid molecules, and they include 5'-phosphate/3'-OH [52, 53], 3'disulfide/5'SH, [54–56] 2'NH2/5'aldehyde [57–59], and 3'-O-propargyl/5'azide [60–62]. The functional groups are activated using appropriate activators to obtain the ligated full-length nucleic acid products. The chemistries also extended to ligate ribozymes, morpholino linkages, and sulfur chemistries are often employed to ligate small ribozymes and two pieces of the Varkud satellite (VS) ribozymes, respectively [63–67]. Click chemistry is also utilized in ligating ribozymes, and they are functionally indistinguishable from a transcribed form of the ribozyme.

ENZYMATIC METHODS FOR RNA LABELING

The enzymatic incorporation of labels allows for the modification of RNA where synthetic methods are not feasible. Various molecular probes, markers, radiolabels have been conjugated to the 5'/3' terminus of the RNA using different enzymes. T4 polynucleotide kinase (PNK) is traditionally used for 5'-radiolabeling [68], whereas Terminal transferase and T4 RNA ligase are used to incorporate radioactive nucleotides at the 3'-end of the RNA [69, 70]. Also, the Terminal transferase and T4 RNA ligase together have been used to incorporate biotin and fluorophore functionalities to the RNA as molecular probes [71, 72]. T7 polymerase and its variants are widely used to incorporate various modified nucleotides into RNA transcripts that facilitate the modification of the RNA at internal and terminal sites. The T7 polymerase is also used to incorporate various modified molecular probes or fluorescence markers at the 5' end of the RNA transcripts using 5'-modified guanosine analogs in transcriptional initiation site [73–75]. With the use of a Φ 2.5 promoter, the T7 polymerase can even accommodate adenosine monophosphate (AMP) analogs used to generate 5′-Cy5, biotin, or folate-labeled transcripts [76–78]. Similarly, poly (A) polymerase has been exploited for its activity to include 3′-azido-, 8-azido-, and 2′-azido-2′,3′-deoxyadenosine into the 3′-end of RNA useful for subsequent 'click' labeling and conjugations [79–81]. The enzymes also used to ligate two RNA/DNA strands together. Most of the ligases isolated from different organisms are used to ligate DNA/RNA *in vitro*. T4 DNA and RNA ligases can ligate both DNA and RNA hybridized onto DNA and RNA splints. T4 RNA ligase 1 is utilized in intramolecular ligations that lead to the creation of circular ribozymes. Besides, the T4 RNA ligase 1 can ligate single-stranded RNA or DNA to double-stranded RNA; this is particularly useful to amplify double-stranded RNA.

ARTIFICIAL ENZYMES FOR LIGATION

Nucleic acid-based ligases employed to ligate various nucleic acid strands. The ligases are called either DNAzyme or RNAzyme based on their native nucleic acid type and are created through *in vitro* selection. The ligation reaction happens between two different functional groups such as 3′-OH/5′-phosphate, 2′, 3′-cyclic phosphate/5′-OH, or 2′, 3′-diol/5′-triphosphate [82, 83]. The *P. carinii* group I intron-derived ribozyme catalyzes the insertion of six nucleotides of an exogenous nonamer, which can be amenable to introduce phosphorothioate, aminopurine, and thiouridine modifications into the target RNA. On the other hand, DNAzymes catalyze the formation of an unnatural 2′-5′ linkage from a 2′-3′-cyclic phosphate and a 5′-OH. These DNAzymes further adapted to label RNA molecules where the DNAzyme ligates an RNA carrying a label to the 2′-carbon of a target RNA sequence [84, 85]. Thus, the method is useful to install various labels by tagging different molecular probes to the RNA such as biotin, fluorescein, or TAMRA [86].

CONCLUSIONS

The tools/labels needed to engineer RNA functions for nanotechnological applications can be introduced into RNA by incorporating the modified nucleotides into the RNA strands either using synthetic methodologies or enzymes. Solid-phase synthesis enables for the incorporation of various modified nucleosides with plenty of functional groups essential to engineer RNA nanoparticles for applications in diagnosis and medicine by labeling and/or conjugating them to molecular probes and drug molecules. On the other hand, the enzymes play a pivotal role to modify long RNA transcripts where synthetic methods are not amenable. The RNA transcripts are labeled with various molecular probes, markers, radiolabels to the 5′/3′ terminus as well as at internal sites. The artificial enzymes such as DNAzyme or RNAzyme are also utilized to label different molecular probes to the RNA. Thus, the developed synthetic and enzymatic methods are wisely utilized in developing various modified RNAs in an effort to engineer various RNA nanotechnological applications.

REFERENCES

1. Kierzek, R., et al., Polymer-supported RNA synthesis and its application to test the nearest-neighbor model for duplex stability. *Biochemistry*, 1986. **25**(24): p. 7840–6.
2. Sakatsume, O., T. Yamaguchi, and H. Takaku, Solid phase synthesis of mRNA by the phosphoramidite approach using 2′-O-1-(2-chloroethoxy) ethyl protection and its stability in E. coli system. *Nucleic Acids Symp Ser*, 1991(24): p. 33–6.
3. Korencic, D., D. Soll, and A. Ambrogelly, A one-step method for in vitro production of tRNA transcripts. *Nucleic Acids Res*, 2002. **30**(20): p. e105.
4. Sastry, S.S. and J.E. Hearst, Studies on the interaction of T7 RNA polymerase with a DNA template containing a site-specifically placed psoralen cross-link. I. Characterization of elongation complexes. *J Mol Biol*, 1991. **221**(4): p. 1091–110.

5. Sharmeen, L. and J. Taylor, Enzymatic synthesis of RNA oligonucleotides. *Nucleic Acids Res*, 1987. **15**(16): p. 6705–11.

6. Chittepu, P., V.R. Sirivolu, and F. Seela, Nucleosides and oligonucleotides containing 1,2,3-triazole residues with nucleobase tethers: synthesis via the azide-alkyne 'click' reaction. *Bioorg Med Chem*, 2008. **16**(18): p. 8427–39.

7. Fujino, T., et al., Duplex-forming oligonucleotide of Triazole-linked RNA. *Chem Asian J*, 2019. **14**(19): p. 3380–3385.

8. Husken, N., et al., "Four-potential" ferrocene labeling of PNA oligomers via click chemistry. *Bioconjug Chem*, 2009. **20**(8): p. 1578–86.

9. Schwechheimer, C., L. Doll, and H.A. Wagenknecht, Synthesis of dye-modified oligonucleotides via copper(I)-catalyzed alkyne azide cycloaddition using on- and off-bead approaches. *Curr Protoc Nucleic Acid Chem*, 2018. **72**(1): p. 4 80 1-4 80 13.

10. van Delft, P., et al., Synthesis of oligoribonucleic acid conjugates using a cyclooctyne phosphoramidite. *Org Lett*, 2010. **12**(23): p. 5486–9.

11. Zewge, D., et al., High-throughput chemical modification of oligonucleotides for systematic structure-activity relationship evaluation. *Bioconjug Chem*, 2014. **25**(12): p. 2222–32.

12. Boysen, R.I. and M.T. Hearn, The metal binding properties of the CCCH motif of the 50S ribosomal protein L36 from *Thermus thermophilus. J Pept Res*, 2001. **57**(1): p. 19–28.

13. Fan, F., et al., Protein profiling of active cysteine cathepsins in living cells using an activity-based probe containing a cell-penetrating peptide. *J Proteome Res*, 2012. **11**(12): p. 5763–72.

14. Geiermann, A.S. and R. Micura, Native chemical ligation of hydrolysis-resistant 3'-NH-cysteine-modified RNA. *Curr Protoc Nucleic Acid Chem*, 2015. **62**: p. 4 64 1-4 64 36.

15. Lee, D.J., et al., Tumoral gene silencing by receptor-targeted combinatorial siRNA polyplexes. *J Control Release*, 2016. **244**(Pt B): p. 280–291.

16. Nitsche, C., et al., Biocompatible macrocyclization between cysteine and 2-cyanopyridine generates stable peptide inhibitors. *Org Lett*, 2019. **21**(12): p. 4709–4712.

17. Paul, R. and M.M. Greenberg, Independent generation and reactivity of uridin-2'-yl radical. *J Org Chem*, 2014. **79**(21): p. 10303–10.

18. Roberts, W.J., et al., Synthesis of the p10 single-stranded nucleic acid binding protein from murine leukemia virus. *Pept Res*, 1988. **1**(2): p. 74–80.

19. Biyani, M., Y. Husimi, and N. Nemoto, Solid-phase translation and RNA-protein fusion: a novel approach for folding quality control and direct immobilization of proteins using anchored mRNA. *Nucleic Acids Res*, 2006. **34**(20): p. e140.

20. De Pasquale, I., et al., Microbial ecology dynamics reveal a succession in the core microbiota involved in the ripening of pasta filata caciocavallo pugliese cheese. *Appl Environ Microbiol*, 2014. **80**(19): p. 6243–55.

21. Ede, N.J., et al., Solid-phase synthesis and screening of a library of C-terminal arginine peptide aldehydes against Murray Valley encephalitis virus protease. *J Pept Sci*, 2012. **18**(11): p. 661–8.

22. Halasz, L., et al., RNA-DNA hybrid (R-loop) immunoprecipitation mapping: an analytical workflow to evaluate inherent biases. *Genome Res*, 2017. **27**(6): p. 1063–1073.

23. Amirkhanov, N.V., et al., Design of (Gd-DO3A)n-polydiamidopropanoyl-peptide nucleic acid-D (Cys-Ser-Lys-Cys) magnetic resonance contrast agents. *Biopolymers*, 2008. **89**(12): p. 1061–76.

24. Barka, T., et al., Induction of the synthesis of a specific protein in rat submandibular gland by isoproterenol. *Lab Invest*, 1986. **54**(2): p. 165–71.

25. Bortolin, S. and T.K. Christopoulos, Time-resolved immunofluorometric determination of specific mRNA sequences amplified by the polymerase chain reaction. *Anal Chem*, 1994. **66**(23): p. 4302–7.

26. Klein, P.M., et al., Efficient shielding of polyplexes using heterotelechelic polysarcosines. *Polymers* (Basel), 2018. **10**(6): p. 689.

27. Menegatti, S., et al., mRNA display selection and solid-phase synthesis of Fc-binding cyclic peptide affinity ligands. *Biotechnol Bioeng*, 2013. **110**(3): p. 857–70.

28. Miao, S., et al., Duplex stem replacement with bPNA+ triplex hybrid stems enables reporting on tertiary interactions of internal RNA domains. *J Am Chem Soc*, 2019. **141**(23): p. 9365–9372.

29. Artigas, G. and V. Marchan, Synthesis of Janus compounds for the recognition of G-U mismatched nucleobase pairs. *J Org Chem*, 2013. **78**(21): p. 10666–77.

30. Hnedzko, D., D.W. McGee, and E. Rozners, Synthesis and properties of peptide nucleic acid labeled at the N-terminus with HiLyte Fluor 488 fluorescent dye. *Bioorg Med Chem*, 2016. **24**(18): p. 4199–4205.

31. Shemesh, Y. and E. Yavin, Postsynthetic conjugation of RNA to carboxylate and dicarboxylate molecules. *Nucleosides Nucleotides Nucleic Acids*, 2015. **34**(11): p. 753–62.

32. Sriwarom, P., P. Padungros, and T. Vilaivan, Synthesis and DNA/RNA Binding Properties of Conformationally Constrained Pyrrolidinyl PNA with a Tetrahydrofuran Backbone Deriving from Deoxyribose. *J Org Chem*, 2015. **80**(14): p. 7058–65.

33. Verheijen, J.C., et al., Transition metal derivatives of peptide nucleic acid (PNA) oligomers-synthesis, characterization, and DNA binding. *Bioconjug Chem*, 2000. **11**(6): p. 741–3.

34. Zhang, Z., et al., Rotational dynamics of HIV-1 nucleocapsid protein NCp7 as probed by a spin label attached by peptide synthesis. *Biopolymers*, 2008. **89**(12): p. 1125–35.

35. Fujita, K. and J. Silver, Surprising lability of biotin-streptavidin bond during transcription of biotinylated DNA bound to paramagnetic streptavidin beads. *Biotechniques*, 1993. **14**(4): p. 608–17.

36. Gorin, D.J., A.S. Kamlet, and D.R. Liu, Reactivity-dependent PCR: direct, solution-phase in vitro selection for bond formation. *J Am Chem Soc*, 2009. **131**(26): p. 9189–91.

37. Kempe, T., et al., Chemical and enzymatic biotin-labeling of oligodeoxyribonucleotides. *Nucleic Acids Res*, 1985. **13**(1): p. 45–57.

38. Ma, Y., F. Teng, and M. Libera, Solid-phase nucleic acid sequence-based amplification and length-scale effects during RNA amplification. *Anal Chem*, 2018. **90**(11): p. 6532–6539.

39. Marfin, Y.S., et al., Recent advances of individual BODIPY and BODIPY-based functional materials in medical diagnostics and treatment. *Curr Med Chem*, 2017. **24**(25): p. 2745–2772.

40. Odeh, F., et al., Aptamers chemistry: chemical modifications and conjugation strategies. *Molecules*, 2019. **25**(1): p. 3.

41. Prakash, T.P., et al., Synergistic effect of phosphorothioate, 5′-vinylphosphonate and GalNAc modifications for enhancing activity of synthetic siRNA. *Bioorg Med Chem Lett*, 2016. **26**(12): p. 2817–2820.

42. Shiraishi, T., R. Hamzavi, and P.E. Nielsen, Subnanomolar antisense activity of phosphonate-peptide nucleic acid (PNA) conjugates delivered by cationic lipids to HeLa cells. *Nucleic Acids Res*, 2008. **36**(13): p. 4424–32.

43. Tailhades, J., et al., Solid-phase synthesis of difficult purine-rich PNAs through selective Hmb incorporation: application to the total synthesis of cell penetrating peptide-PNAs. *Front Chem*, 2017. **5**: p. 81.

44. Zhang, L., et al., Photomodulating gene expression by using caged siRNAs with single-aptamer modification. *Chembiochem*, 2018. **19**(12): p. 1259–1263.

45. Hansenova Manaskova, S., et al., *Staphylococcus aureus* sortase A-mediated incorporation of peptides: effect of peptide modification on incorporation. *PLoS One*, 2016. **11**(1): p. e0147401.

46. Kang, L., et al., Noninvasive visualization of RNA delivery with 99mTc-radiolabeled small-interference RNA in tumor xenografts. *J Nucl Med*, 2010. **51**(6): p. 978–86.

47. Ni, X., et al., Nucleic acid aptamers: clinical applications and promising new horizons. *Curr Med Chem*, 2011. **18**(27): p. 4206–14.

48. Sun, D., et al., Silver nanoparticles-quercetin conjugation to siRNA against drug-resistant Bacillus subtilis for effective gene silencing: in vitro and in vivo. *Mater Sci Eng C Mater Biol Appl*, 2016. **63**: p. 522–34.

49. Wen, A.M., et al., Interior engineering of a viral nanoparticle and its tumor homing properties. *Biomacromolecules*, 2012. **13**(12): p. 3990–4001.

50. Yang, Y. and C.Y. Zhang, Sensitive detection of intracellular sumoylation via SNAP tag-mediated translation and RNA polymerase-based amplification. *Anal Chem*, 2012. **84**(3): p. 1229–34.

51. Zhovmer, A. and X. Qu, Proximal disruptor aided ligation (ProDAL) of kilobase-long RNAs. *RNA Biol*, 2016. **13**(7): p. 613–21.

52. Graifer, D. and G. Karpova, General approach for introduction of various chemical labels in specific RNA locations based on insertion of amino linkers. *Molecules*, 2013. **18**(12): p. 14455–69.

53. Pradere, U., F. Halloy, and J. Hall, Chemical synthesis of long RNAs with terminal 5′-phosphate groups. *Chemistry*, 2017. **23**(22): p. 5210–5213.

54. Liu, Y., et al., SIRPbeta1 is expressed as a disulfide-linked homodimer in leukocytes and positively regulates neutrophil transepithelial migration. *J Biol Chem*, 2005. **280**(43): p. 36132–40.

55. Lopez-Sanchez, L.M., et al., Inhibition of nitric oxide synthesis during induced cholestasis ameliorates hepatocellular injury by facilitating S-nitrosothiol homeostasis. *Lab Invest*, 2010. **90**(1): p. 116–27.

56. Page, M.L., et al., A homolog of prokaryotic thiol disulfide transporter CcdA is required for the assembly of the cytochrome b6f complex in Arabidopsis chloroplasts. *J Biol Chem*, 2004. **279**(31): p. 32474–82.

57. Biswas, S., et al., Hydrophobic oxime ethers: a versatile class of pDNA and siRNA transfection lipids. *ChemMedChem*, 2011. **6**(11): p. 2063–9.
58. DuBois, J.L. and J.P. Klinman, The nature of O_2 reactivity leading to topa quinone in the copper amine oxidase from Hansenula polymorpha and its relationship to catalytic turnover. *Biochemistry*, 2005. **44**(34): p. 11381–8.
59. He, Z., et al., DDPH: improving cognitive deficits beyond its alpha 1-adrenoceptor antagonism in chronic cerebral hypoperfused rats. *Eur J Pharmacol*, 2008. **588**(2–3): p. 178–88.
60. He, X., X. Yang, and E. Jabbari, Combined effect of osteopontin and BMP-2 derived peptides grafted to an adhesive hydrogel on osteogenic and vasculogenic differentiation of marrow stromal cells. *Langmuir*, 2012. **28**(12): p. 5387–97.
61. Jana, S.K., et al., 2′-O-methyl- and 2′-O-propargyl-5-methylisocytidine: synthesis, properties and impact on the isoCd-dG and the isoCd-isoGd base pairing in nucleic acids with parallel and antiparallel strand orientation. *Org Biomol Chem*, 2016. **14**(21): p. 4927–42.
62. Pujari, S.S., P. Leonard, and F. Seela, Oligonucleotides with "clickable" sugar residues: synthesis, duplex stability, and terminal versus central interstrand cross-linking of 2′-O-propargylated 2-aminoadenosine with a bifunctional azide. *J Org Chem*, 2014. **79**(10): p. 4423–37.
63. Hobartner, C., et al., Syntheses of RNAs with up to 100 nucleotides containing site-specific 2′-methylseleno labels for use in X-ray crystallography. *J Am Chem Soc*, 2005. **127**(34): p. 12035–45.
64. Sasaki, S., K. Onizuka, and Y. Taniguchi, Oligodeoxynucleotide containing s-functionalized 2′-deoxy-6-thioguanosine: facile tools for base-selective and site-specific internal modification of RNA. *Curr Protoc Nucleic Acid Chem*, 2012. **Chapter 4**: p. Unit 4 49 1–16.
65. Lilley, D.M., The Varkud satellite ribozyme. *RNA*, 2004. **10**(2): p. 151–8.
66. Wilson, T.J. and D.M. Lilley, A mechanistic comparison of the Varkud satellite and hairpin ribozymes. *Prog Mol Biol Transl Sci*, 2013. **120**: p. 93–121.
67. Zhao, Z.Y., et al., Nucleobase participation in ribozyme catalysis. *J Am Chem Soc*, 2005. **127**(14): p. 5026–7.
68. Goldrick, M. and D. Kessler, RNA analysis by nuclease protection. *Curr Protoc Neurosci*, 2003. **Chapter 5**: p. Unit 5 1.
69. Porecha, R. and D. Herschlag, RNA radiolabeling. *Methods Enzymol*, 2013. **530**: p. 255–79.
70. Schaad, M.C. and R.S. Baric, Evidence for new transcriptional units encoded at the 3′ end of the mouse hepatitis virus genome. *Virology*, 1993. **196**(1): p. 190–8.
71. Gumport, R.I., et al., T4 RNA ligase as a nucleic acid synthesis and modification reagent. *Nucleic Acids Symp Ser*, 1980(7): p. 167–71.
72. Moritz, B. and E. Wahle, Simple methods for the 3′ biotinylation of RNA. *RNA*, 2014. **20**(3): p. 421–7.
73. Eberwine, J., et al., Analysis of gene expression in single live neurons. *Proc Natl Acad Sci U S A*, 1992. **89**(7): p. 3010–4.
74. Forghani, B., G.J. Yu, and J.W. Hurst, Comparison of biotinylated DNA and RNA probes for rapid detection of varicella-zoster virus genome by in situ hybridization. *J Clin Microbiol*, 1991. **29**(3): p. 583–91.
75. Khan, H., et al., CRISPR-Cas13a mediated nanosystem for attomolar detection of canine parvovirus type 2. *Chin Chem Lett*, 2019. **30**(12): p. 2201–2204.
76. Huang, F., et al., Synthesis of biotin-AMP conjugate for 5′ biotin labeling of RNA through one-step in vitro transcription. *Nat Protoc*, 2008. **3**(12): p. 1848–61.
77. Laing, B.M., P. Guo, and D.E. Bergstrom, Optimized method for the synthesis and purification of adenosine – folic acid conjugates for use as transcription initiators in the preparation of modified RNA. *Methods*, 2011. **54**(2): p. 260–6.
78. Folsom, V., et al., Detection of DNA targets with biotinylated and fluoresceinated RNA probes. Effects of the extent of derivitization on detection sensitivity. *Anal Biochem*, 1989. **182**(2): p. 309–14.
79. Custer, T.C. and N.G. Walter, In vitro labeling strategies for in cellulo fluorescence microscopy of single ribonucleoprotein machines. *Protein Sci*, 2017. **26**(7): p. 1363–1379.
80. Jaju, M., W.A. Beard, and S.H. Wilson, Human immunodeficiency virus type 1 reverse transcriptase. 3′-Azidodeoxythymidine 5′-triphosphate inhibition indicates two-step binding for template-primer. *J Biol Chem*, 1995. **270**(17): p. 9740–7.
81. Ma, Q.F., et al., The observed inhibitory potency of 3′-azido-3′-deoxythymidine 5′-triphosphate for HIV-1 reverse transcriptase depends on the length of the poly(rA) region of the template. *Biochemistry*, 1992. **31**(5): p. 1375–9.

82. Semlow, D.R. and S.K. Silverman, Parallel selections in vitro reveal a preference for 2′-5′ RNA ligation upon deoxyribozyme-mediated opening of a 2′,3′-cyclic phosphate. *J Mol Evol*, 2005. **61**(2): p. 207–15.

83. Hoadley, K.A., et al., Zn2+-dependent deoxyribozymes that form natural and unnatural RNA linkages. *Biochemistry*, 2005. **44**(25): p. 9217–31.

84. Camden, A.J., et al., DNA Oligonucleotide 3'-Phosphorylation by a DNA Enzyme. *Biochemistry*, 2016. **55**(18): p. 2671–6.

85. Flynn-Charlebois, A., et al., In vitro evolution of an RNA-cleaving DNA enzyme into an RNA ligase switches the selectivity from 3′-5′ to 2′-5′. *J Am Chem Soc*, 2003. **125**(18): p. 5346–50.

86. Samanta, B., D.P. Horning, and G.F. Joyce, 3′-End labeling of nucleic acids by a polymerase ribozyme. *Nucleic Acids Res*, 2018. **46**(17): p. e103.

7 Methods and Assembly of RNA Nanotechnology

Hongzhi Wang, Chenxi Liang, and Peixuan Guo
The Ohio State University, Columbus, USA

CONTENTS

RNA nanotechnology is conceptualized as the bottom-up self-assembly of nanometer-scale RNA architectures. Two major approaches have been used for the construction of RNA nanoparticles. The first approach is the sequence-dependent self-folding of RNA nanoarchitectures that is based on computational algorithms and prediction of secondary or tertiary structures (Afonin et al., 2010, Yingling & Shapiro, 2007). Another approach employs the natural RNA motifs as core building blocks to assemble RNA nanoparticles (Dibrov et al., 2011, Haque et al., 2012, Severcan et al., 2010, Shu et al., 2011, Westhof et al., 1996). The self-assembly process of RNA nanoparticles relies on base-pairing. The RNA nanoparticles are assembled spontaneously using highly programmable and predictable building blocks in a pre-defined manner that is very different from traditional routes, which is normally formed by covalent polymerization and hydrophobic interactions (Liechty et al., 2010). The construction of functional RNA nanoparticles integrates the knowledge of RNA chemistry, RNA biology, and computational approaches.

Bottom-up RNA nanoparticle synthesis results in consistent assembly with narrow size and shape distributions, contributing to a favorable and reproducible set of PK/PD factors. The homogeneous RNA nanoparticle assembly with defined size, shape, and stoichiometry necessitates the purification to acquire clinically used RNA nanoparticles. Two major purification techniques have been established. Polyacrylamide gels provide a fast and simple method which is widely used in laboratory. HPLC with high resolution and consistent result is used in a variety of industrial and scientific applications.

METHODS FOR THE CONSTRUCTION OF RNA NANOPARTICLES

The field of RNA nanotechnology can be traced back to the first construction of RNA nanoparticles from the pRNA (prohead RNA) (Guo et al., 1998). The pRNA molecules were originally derived from bacteriophage phi29 DNA packaging motor (Guo et al., 1987a, Guo et al., 1987b). They contain two interlocking loops that enable the formation of dimers, trimers, hexamers, and patterned superstructures via intermolecular interactions (Chen et al., 1999, Shu et al., 2013b). Particularly, the pRNA three-way junction (pRNA- 3WJ) derived from pRNA core has been extensively used as a central core for the construction of multifunctional RNA nanoparticles (Binzel et al., 2016, Cui et al., 2015, Shu et al., 2011). The branched feature of the 3WJ motif allows different functional modules to be conveniently incorporated to the three helical regions, making it an ideal scaffold for targeted drug delivery. This multivalent property suggests that pRNA-3WJ nanoparticles have enormous potential for therapeutic, imaging, and diagnostic applications. Generally, in order to construct

DOI: 10.1201/9781003001560-8

multifunctional pRNA-3WJ nanoparticles, the types of functional modules and functionalities need to be well defined, and the global structure needs to be established before the RNA synthesis and construction. The general strategies for the multifunctionalization of RNA nanoparticles include both covalent and non-covalent methods, such as end extensions, hybridization, and chemical conjugation or labeling. The rigidity of the pRNA-3WJ scaffold allows all incorporated modules to be spaced apart from each other, retaining their authentic folding and independent functionalities (Shu et al., 2013a). Recently, the integration of multiple functional modules into one nanoparticle has been used for the treatments of various cancer models (Guo et al., 2019, Piao et al., 2019).

Computer modeling can also assist with the design of novel nanostructures and with the optimization of the building blocks and final nanostructures according to requirements of the function and fabrication techniques. Moreover, it is a relatively inexpensive and a fast way to create different structural designs and assess their properties. Also, computer-aided RNA nanoarchitectures can use a shape-based approach similar to that used for proteins, where the desired shape guides the choice of the specific building blocks.

IN VITRO NANOPARTICLE ASSEMBLY

RNA nanoparticles can be constructed with defined sizes, shapes, and stoichiometry by bottom-up self-assembly based on the intra- and inter-RNA interactions. This is due to the RNA structure provides a high thermal stability and versatility which are well beyond the simplistic canonical Watson–Crick base pairing in DNA nanostructures. To assemble the RNA nanoparticles *in vitro*, a simple heating–cooling method is normally used. After the mixing of different components at equal molar ratios, the mixtures are first heated to be fully denatured, then slowly cooled. During this process, the thermodynamic stable RNA structure will be self-assembled. The ion concentrations need to be adjusted based on the structure stability. A well-characterized ultra-stable 3WJ motif derived from the central domain of the natural pRNA of bacteriophage phi29 usually served as core scaffold to construct RNA architectures. It can be constructed from three short ssRNA fragments with unusually high efficiency by a bottom-up self-assembly approach. Briefly, these three fragments are mixed together at equal molar concentration in TMS buffer, followed by heating to 85°C for 5 min and slowly cooled over 40 min to 4°C. Generally, this assembly process is highly efficient even in the absence of Mg^{2+}, as long as the three RNA fragments are mixed at equal molar ratio.

In vitro transcription and assembly are simple and fast; however, there are several limitations such as low yield, time efficiency, homogeneity, and purity (discussed in details above) (Chamberlin & Ring, 1973, Maslak & Martin, 1993). Rolling circle transcription (RCT) is increasing in popularity as a result of its high production capabilities and efficiency (Han et al., 2014, Lee et al., 2012, Mohsen & Kool, 2016, Guo, 2012). One-stranded or multi-stranded nanoparticle assembly can occur co-transcriptionally, reducing the total number of steps required for RNA nanoparticle preparation (Afonin et al., 2012, Shu et al., 2011). Recently, Guo lab has developed a novel method for the construction of circular dsDNA templates that code for self-cleaving ribozymes and self-assembled 3WJ RNA nanoparticles (Jasinski et al., 2019). Upon *in vitro* transcription, the ribozymes self-cleave with high efficiency, producing large amounts of product RNA and RNA nanoparticles. Single-stranded 3WJ RNA nanoparticles were synthesized via RCT by addition of loops to link adjacent 3WJ strands. RCT produced a 3.2 times higher yield of fully assembled RNA nanoparticles compared to that of traditional *in vitro* runoff transcription.

IN VIVO NANOPARTICLE ASSEMBLY

Compared with *in vitro* assembly, *in vivo* assembly approaches for targeting delivery have certain advantages, including rapid accumulation of small molecular agents in tumors, shorter circulation time requirements, and nanoparticles accumulated in tumors can be utilized to alter the pharmacokinetics of therapeutic agents (Perrault & Chan, 2010). The success of DNA nanotechnology has allowed designers to develop RNA nanotechnology as a growing discipline which facilitates the possible *in vivo* assembly strategy. RNA origami is a new concept derived from DNA origami (Amir et al., 2014) and has great potential for applications in nanomedicine and synthetic biology (Krissanaprasit et al., 2019). The method was developed to allow new creations of large RNA nanostructures that create defined scaffolds for combining RNA-based functionalities (Han et al., 2017, Liu et al., 2014, Sparvath et al., 2017). Because of the infancy of RNA origami, many of its potential applications are still in the process of discovery. Its structures are able to provide a stable basis to allow functionality for RNA components. These structures include riboswitches, ribozymes, interaction sites, and aptamers. Aptamer structures allow the binding of small molecules which gives possibilities for construction of future RNA-based nanodevices. RNA origami is further useful in areas such as cell recognition and binding for diagnosis. Chengde Mao has developed a programmable ssRNA folding strategy for the preparation of complex nanostructures. Each nanostructure is primarily composed of RNA duplexes and is folded from one long ssRNA. Thus, the formation of the nanostructures is independent of stoichiometry and RNA concentration. This strategy allows cloning and expression of RNA nanostructures inside cells in the same way as those of recombinant proteins, pointing to a cost-effective way for large-scale production of nucleic acid nanostructures. This strategy would facilitate a wide range of *in vivo* biomedical applications of RNA nanotechnology, such as integrating multiple functional RNA moieties to regulate cellular processes.

CONCLUSIONS

RNA nanoparticles with defined size, shape and stoichiometry, as well as multivalency, provide an ideal platform for constructing multifunctional drug delivery platform, especially in targeted cancer therapy. By incorporating different functional modules to the branched region and modified nucleic acids, RNA nanoparticles can be constructed either via a bottom-up self-assembly approach with controllable structures or computational algorithms to predict the secondary and more complicated structures. The recent advancement in RNA origami and RCT technique makes it possible to synthesize and assemble the functional nanoparticles *in vivo*. These RNA nanoparticles will play a significant role in biomedical sciences.

REFERENCES

Afonin, K. A.; Bindewald, E.; Yaghoubian, A. J.; Voss, N.; Jacovetty, E.; Shapiro, B. A.; Jaeger, L., In vitro assembly of cubic RNA-based scaffolds designed in silico. *Nat. Nanotechnol* 2010, *5* (9), 676–682.

Afonin, K. A.; Kireeva, M.; Grabow, W. W.; Kashlev, M.; Jaeger, L.; Shapiro, B. A., Co-transcriptional assembly of chemically modified RNA nanoparticles functionalized with siRNAs. *Nano. Lett* 2012, *12* (10), 5192–5195.

Amir, Y.; Ben-Ishay, E.; Levner, D.; Ittah, S.; bu-Horowitz, A.; Bachelet, I., Universal computing by DNA origami robots in a living animal. *Nature Nanotechnology* 2014, *9* (5), 353–357.

Binzel, D.; Shu, Y.; Li, H.; Sun, M.; Zhang, Q.; Shu, D.; Guo, B.; Guo, P., Specific Delivery of MiRNA for High Efficient Inhibition of Prostate Cancer by RNA Nanotechnology. *Molecular Therapy* 2016, *24*, 1267–1277.

Chamberlin, M.; Ring, J., Characterization of T7-specific ribonucleic acid polymerase. 1. General properties of the enzymatic reaction and the template specificity of the enzyme. *J Biol Chem* 1973, *248* (6), 2235–2244.

Chen, C.; Zhang, C.; Guo, P., Sequence requirement for hand-in–hand interaction in formation of pRNA dimers and hexamers to gear phi29 DNA translocation motor. *RNA* 1999, *5*, 805–818.

Cui, D.; Zhang, C.; Liu, B.; Shu, Y.; Du, T.; Shu, D.; Wang, K.; Dai, F.; Liu, Y.; Li, C.; Pan, F.; Yang, Y.; Ni, J.; Li, H.; Brand-Saberi, B.; Guo, P., Regression of gastric cancer by systemic injection of RNA nanoparticles carrying both ligand and siRNA. *Scientific reports* 2015, *5*, 10726.

Dibrov, S. M.; McLean, J.; Parsons, J.; Hermann, T., Self-assembling RNA square. *Proc. Natl. Acad. Sci. U. S. A* 2011, *108* (16), 6405–6408.

Guo, P. Rolling Circle Transcription of Tandem siRNA to Generate Spherulitic RNA Nanoparticles for Cell Entry. *Mol Ther-Nucleic Acids* 2012, *1*, e36.

Guo, P.; Bailey, S.; Bodley, J. W.; Anderson, D., Characterization of the small RNA of the bacteriophage phi29 DNA packaging machine. *Nucleic Acids Res* 1987b, *15*, 7081–7090.

Guo, P.; Erickson, S.; Anderson, D., A small viral RNA is required for *in vitro* packaging of bacteriophage phi29 DNA. *Science* 1987a, *236*, 690–694.

Guo, S.; Vieweger, M.; Zhang, K.; Yin, H.; Wang, H.; Li, X.; Li, S.; Dong, Y.; Chui, W.; Guo, P., Ultra-thermostable RNA nanoparticles for solubilizing and high yield loading of paclitaxel for breast cancer therapy with undetectable toxicity. *Nature Communications* 2019.

Guo, P.; Zhang, C.; Chen, C.; Trottier, M.; Garver, K., Inter-RNA interaction of phage phi29 pRNA to form a hexameric complex for viral DNA transportation. *Molecular Cell* 1998, *2*, 149–155.

Han, D.; Park, Y.; Kim, H.; Lee, J. B., Self-assembly of free-standing RNA membranes. *Nature Communications* 2014, *5*, 4367.

Han, D.; Qi, X.; Myhrvold, C.; Wang, B.; Dai, M.; Jiang, S.; Bates, M.; Liu, Y.; An, B.; Zhang, F.; Yan, H.; Yin, P., Single-stranded DNA and RNA origami. *Science* 2017, *358* (6369).

Haque, F.; Shu, D.; Shu, Y.; Shlyakhtenko, L.; Rychahou, P.; Evers, M.; Guo, P., Ultrastable synergistic tetravalent RNA nanoparticles for targeting to cancers. *Nano Today* 2012, *7*, 245–257.

Huang, Z.; Lin, C. Y.; Jaremko, W.; Niu, L., HPLC purification of RNA aptamers up to 59 nucleotides with single-nucleotide resolution. *Methods Mol Biol* 2015, *1297*, 83–93.

Jasinski, D. L.; Binzel, D. W.; Guo, P., One-Pot Production of RNA Nanoparticles via Automated Processing and Self-Assembly. *ACS Nano* 2019, *13* (4), 4603–4612.

Keefe, A. D.; Cload, S. T., SELEX with modified nucleotides. *Curr. Opin. Chem Biol* 2008, *12* (4), 448–456.

Krissanaprasit, A.; Key, C.; Fergione, M.; Froehlich, K.; Pontula, S.; Hart, M.; Carriel, P.; Kjems, J.; Andersen, E. S.; LaBean, T. H., Genetically Encoded, Functional Single-Strand RNA Origami: Anticoagulant. *Adv Mater* 2019, *31* (21), e1808262.

Largy, E.; Mergny, J. L., Shape matters: size-exclusion HPLC for the study of nucleic acid structural polymorphism. *Nucleic Acids Res* 2014, *42* (19), e149.

Lee, J. B.; Hong, J.; Bonner, D. K.; Poon, Z.; Hammond, P. T., Self-assembled RNA interference microsponges for efficient siRNA delivery. *Nat Mater* 2012, *11* (4), 316–322.

Liechty, W. B.; Kryscio, D. R.; Slaughter, B. V.; Peppas, N. A., Polymers for drug delivery systems. *Annu. Rev. Chem Biomol. Eng* 2010, *1*, 149–173.

Liu, Y.; Zhou, J.; Pan, J. A.; Mabiala, P.; Guo, D., A novel approach to block HIV-1 coreceptor CXCR4 in nontoxic manner. *Mol Biotechnol* 2014, *56* (10), 890–902.

Maslak, M.; Martin, C. T., Kinetic analysis of T7 RNA polymerase transcription initiation from promoters containing single-stranded regions. *Biochemistry* 1993, *32* (16), 4281–4285.

Mohsen, M. G.; Kool, E. T., The Discovery of Rolling Circle Amplification and Rolling Circle Transcription. *Acc. Chem Res* 2016, *49* (11), 2540–2550.

Perrault, S. D.; Chan, W. C., In vivo assembly of nanoparticle components to improve targeted cancer imaging. *Proc Natl Acad Sci U S A* 2010, *107* (25), 11194–11199.

Piao, X.; Yin, H.; Guo, S.; Wang, H.; Guo, P., RNA nanotechnology to solubilize hydrophobic antitumor drug for targeted delivery. *Advanced Science* 2019, *In Press*.

Severcan, I.; Geary, C.; Chworos, A.; Voss, N.; Jacovetty, E.; Jaeger, L., A polyhedron made of tRNAs. *Nat Chem* 2010, *2* (9), 772–779.

Shu, Y.; Haque, F.; Shu, D.; Li, W.; Zhu, Z.; Kotb, M.; Lyubchenko, Y.; Guo, P., Fabrication of 14 Different RNA Nanoparticles for Specific Tumor Targeting without Accumulation in Normal Organs. *RNA* 2013b, *19*, 766–777.

Shu, D.; Khisamutdinov, E.; Zhang, L.; Guo, P., Programmable folding of fusion RNA complex driven by the 3WJ motif of phi29 motor pRNA. *Nucleic Acids Research* 2013a, *42*, e10.

Shu, D.; Shu, Y.; Haque, F.; Abdelmawla, S.; Guo, P., Thermodynamically stable RNA three-way junctions for constructing multifuntional nanoparticles for delivery of therapeutics. *Nature Nanotechnology* 2011, *6*, 658–667.

Sparvath, S. L.; Geary, C. W.; Andersen, E. S., Computer-Aided Design of RNA Origami Structures. *Methods Mol Biol* 2017, *1500*, 51–80.

Tuerk, C.; Gold, L., Systematic evolution of ligands by exponential enrichment: RNA ligands to bacteriophage T4 DNA polymerase. *Science* 1990, *249* (4968), 505–510.

Westhof, E.; Masquida, B.; Jaeger, L., RNA tectonics: Towards RNA design. *Folding & Design* 1996, *1*, R78–R88.

Yingling, Y. G.; Shapiro, B. A., Computational design of an RNA hexagonal nanoring and an RNA nanotube. *Nano Letters* 2007, *7* (8), 2328–2334.

Zarzosa-Alvarez, A. L.; Sandoval-Cabrera, A.; Torres-Huerta, A. L.; Bermudez-Cruz, R. M., Electroeluting DNA fragments. *J Vis Exp* 2010, (43).

8 Purification, Characterization, and Structure Determination of RNA Nanoparticles

Satheesh Ellipilli, Nicolas D. Burns, and Peixuan Guo
The Ohio State University, Columbus, Ohio, USA

CONTENTS

The most distinguishable advantages of RNA nanoparticles are their defined size, shape, and stoichiometry, which is therefore controllable for various RNA nano-constructs fabrication such as RNA triangle, square, pentamer, hexamer, tetrahedron, etc. [1–4] Understanding the physical and structural properties of RNA nanoparticles is of great significance for the design and construction of functional nanoparticles by introducing different modules to the RNA scaffold while avoiding structural hindrance [5–7]. In this section, purification, physical characterization, and structural determination of RNA nanoparticles will be discussed.

PURIFICATION OF THE RNA NANOPARTICLES

Purification of RNA nanoparticles is not only a prerequisite for successful physical characterization but also an essential step before being applied to *in vivo* study. Several methods exist for the purification of RNA nanoparticles following *in vitro* assembly. Assembled RNA nanoparticles can be purified by HPLC, gel-electrophoresis, and gel-filtration methods.

High-performance liquid chromatography (HPLC) applied for the separation of the intact RNA nanoparticles from incomplete assemblies or single-stranded RNAs. Both reverse phase (RP) column and size exclusive column can be installed onto an HPLC system in order to separate the RNA nanoparticles by hydrophobicity and size, respectively. Buffer condition, mobile phase ratio, temperature, and flow speed need to be adjusted to optimize separation conditions.

Size-exclusion chromatography (SEC) is commonly used in RNA nanoparticle purification. SEC, also known as molecular sieve chromatography, is a chromatographic method in which macromolecules such as oligonucleotides, proteins, and polymers in solution are separated by their size. The chromatography column is packed with fine, porous beads which are composed of dextran polymers (Sephadex), agarose, or polyacrylamide. The pore sizes of these beads are used to estimate the dimensions of macromolecules.

DOI: 10.1201/9781003001560-9

Gel electrophoresis purification is another method used in RNA nanoparticles' separation. The advantages of purification on polyacrylamide gels are speed, simplicity, and relatively high resolution. The major band of RNA nanoparticles can be excised from the gel under UV light and two elution methods can be applied to purify the RNA nanoparticles. One is soaking the gel pieces into the RNA elution buffer (0.5 M ammonium acetate, 10 mM EDTA and 0.1% (wt/vol) SDS in 0.05% (vol/vol) DEPC-treated water) and incubation at 37 °C for at least 2 hours. After, incubation, the supernatant is collected and applied for ethanol perception to obtain purified RNA nanoparticles. Another method is the electro-separation system, which consists of an electrophoresis chamber and sample traps, driving the RNA samples to migrate out of the gel slices and retain in membrane traps under the driving force of an electric field [17]. Collected solution will be subjected to ethanol precipitation.

Gel filtration (size exclusive column) is another commonly used method for purification of RNA nanoparticles [18]. The chromatography column is packed with fine, porous beads, which are composed of polymers (Sephadex), agarose (Sepharose), or polyacrylamide (Sephacryl or BioGel P). The separation is mainly based on the particle size. The larger RNA nanoparticles are eluted first, which can pass through the pores quickly and the unassembled or single-stranded RNA strands are eluted later. There are different types of beads providing a different fractionation range. For example, commonly used Sephadex column includes Sephadex G25, G50, G75, G100, increasing in pore size, respectively. The choice of column is based on the difference of molecular weight between impurities and samples.

PHYSICAL CHARACTERIZATION OF RNA NANOPARTICLES

RNA nanoparticles are assembled through bottom-up self-assembly and are usually characterized by electrophoresis, size, and thermal melting using temperature-gradient gel electrophoresis (TGGE) and quantitative real-time PCR (qPCR).

Electrophoresis refers to the migration and separation of charged ions under the influence of electric current. Gel electrophoresis is one of the most commonly used methods for the analysis of RNA nanoparticles based on size and charge. An electric field is applied to move the generally negatively charged RNA nanoparticles through gel matrix soaked in a buffer. Assembly of RNA nanoparticles can be examined by the stepwise assembly using either native-polyacrylamide (PAGE) or agarose gel. After gel electrophoresis, dyes such as Ethidium bromide or SYBR Green can be used to stain the RNA nanoparticle to visualize via imaging equipment.

Dynamic Light Scattering (DLS) is a usually employed method to determine the size of the RNA nanoparticles. DLS is based on the Brownian motion of the particles dispersed in the solution. A single-frequency laser is directed to the sample present in a cuvette and the laser light gets scattered in all directions, if there are particles in the sample. The degree of light scattering depends upon the size of the particles present in the solution. The scattered light is detected at a certain angle over a period of time and the signal obtained is used to determine the diffusion coefficient and the particle size by the Stokes–Einstein equation [12].

Thermal meting (T_m) of the RNA nanoparticles is another important characterization to determine their thermal stability. TGGE is a commonly employed method to determine the T_m of the RNA nanoparticles. TGGE applies an increasing temperature gradient perpendicular to the electrical current on a polyacrylamide gel to determine melting temperature (T_m) of RNA nanoparticles [13]. RNA samples will be subjected to heating at a temperature-gradient gel, which may cause the assembled form of the RNA nanoparticles to denature. By quantifying the gel band intensity of each well, a melting curve can be generated to calculate T_m, which is defined by the temperature at which 50% of RNA samples are dissociated. Another approach for T_m estimation is real-time PCR. RNA samples mixed with intercalating dye (such as SYBR Green II) were put into a 96-well plate, which is subjected to a heating (denaturing) and cooling (assembling) processing. The melting curve can be attained by real-time monitoring of the fluorescence level of the dye, and T_m is estimated by the temperature where 50% of fluorescence intensity is observed.

STRUCTURE DETERMINATION OF RNA NANOPARTICLES

To visualize the shape of RNA nanoparticles, atomic force microscope (AFM) and cryo-electron microscope (cryo-EM) are favorable techniques to determine 3D structure.

AFM uses a cantilever with a sharp tip to scan over a sample surface [14]. The force between the surface and the tip causes the deflection of cantilever, which can be detected by the changes in the direction of the laser beam. A position-sensitive photo diode (PSPD) is used to track the changes and generates topographic map of the surface feature. Purified RNA nanoparticles can be dispersed onto a mica surface and subjected to AFM monitoring to decide the structural properties and size [3, 10]. However, detection limit of AFM is about 10 nm. Particles smaller than 10 nm cannot be visualized clearly due to the resolution problem.

Cryo-EM applied on samples cooled to cryogenic temperature to preserve the structure [15]. Electron microscopes use a beam of electrons to examine the structures of molecules and materials at the atomic scale. As the beam passes through a very thin sample, it interacts with the molecules, which projects an image of the sample onto the detector. By cryo-EM, the 3D structure of RNA nanoparticles can be re-constructed by calculation with very high resolution [3, 4, 16].

CONCLUSION

The assembled RNA nanoparticles can be purified efficiently using both HPLC and gel electrophoresis methods. The purified nanoparticles can be characterized for their size, thermal stability, and structure using DLS, TGGE/qPCR, and AFM/EM techniques, respectively. This allows for effective and subtle builds of nano-constructs useful to carry therapeutics and imaging markers to cells for therapy and imaging, respectively.

REFERENCES

1. Khisamutdinov, E. F., D. L. Jasinski, and P. Guo. 2014. RNA as a boiling-resistant anionic polymer material to build robust structures with defined shape and stoichiometry. ACS Nano. 8:4771–4781.
2. Jasinski, D., E. F. Khisamutdinov, Y. L. Lyubchenko, and P. Guo. 2014. Physicochemically Tunable Poly-Functionalized RNA Square Architecture with Fluorogenic and Ribozymatic Properties. ACS Nano 8:7620–7629.
3. Li, H., K. Zhang, F. Pi, S. Guo, L. Shlyakhtenko, W. Chiu, D. Shu, and P. Guo. 2016. Controllable Self-Assembly of RNA Tetrahedrons with Precise Shape and Size for Cancer Targeting. Adv. Mater. 28:7501–7507.
4. Khisamutdinov, E. F., D. L. Jasinski, H. Li, K. Zhang, W. Chiu, and P. Guo. 2016. Fabrication of RNA 3D Nanoprism for Loading and Protection of Small RNAs and Model Drugs. Advanced Materials 28:100079–100087.

5. Shu, D., H. Li, Y. Shu, G. Xiong, W. E. Carson, F. Haque, R. Xu, and P. Guo. 2015. Systemic delivery of anti-miRNA for suppression of triple negative breast cancer utilizing RNA nanotechnology. ACS Nano 9:9731–9740.

6. Binzel, D., Y. Shu, H. Li, M. Sun, Q. Zhang, D. Shu, B. Guo, and P. Guo. 2016. Specific Delivery of MiRNA for High Efficient Inhibition of Prostate Cancer by RNA Nanotechnology. Molecular Therapy 24:1267–1277.

7. Shu, Y., H. Yin, M. Rajabi, H. Li, M. Vieweger, S. Guo, D. Shu, and P. Guo. 2018. RNA-based micelles: A novel platform for paclitaxel loading and delivery. J Control Release 276:17–29.

8. Lee, P. Y., J. Costumbrado, C. Y. Hsu, and Y. H. Kim. 2012. Agarose gel electrophoresis for the separation of DNA fragments. J Vis.Exp.

9. Blundon, M., V. Ganesan, B. Redler, P. T. Van, and J. S. Minden. 2019. Two-Dimensional Difference Gel Electrophoresis. Methods Mol Biol 1855:229–247.

10. Shu, Y., F. Haque, D. Shu, W. Li, Z. Zhu, M. Kotb, Y. Lyubchenko, and P. Guo. 2013. Fabrication of 14 Different RNA Nanoparticles for Specific Tumor Targeting without Accumulation in Normal Organs. RNA 19:766–777.

11. Yin, H., G. Xiong, S. Guo, C. Xu, R. Xu, P. Guo, and D. Shu. 2019. Delivery of Anti-miRNA for Triple-Negative Breast Cancer Therapy Using RNA Nanoparticles Targeting Stem Cell Marker CD133. Mol Ther. 27:1252–1261. doi:S1525-0016(19)30184-4

12. Stetefeld, J., S. A. McKenna, and T. R. Patel. 2016. Dynamic light scattering: a practical guide and applications in biomedical sciences. Biophys Rev. 8:409–427.

13. Thatcher, D. R. and B. Hodson. 1981. Denaturation of proteins and nucleic acids by thermal-gradient electrophoresis. Biochem.J 197:105–109.

14. Binnig, G. and C. F. Quate. 1986. Atomic Force Microscope. Physical Review Letters 56:930–933.

15. Cheng, Y., N. Grigorieff, P. A. Penczek, and T. Walz. 2015. A primer to single-particle cryo-electron microscopy. Cell 161:438–449.

16. Xu, C., H. Li, K. Zhang, D. W. Binzel, H. Yin, W. Chiu, and P. Guo. 2019. Photo-controlled release of paclitaxel and model drugs from RNA pyramids. Nano Research 12:41–48.

17. Guo, S., X. Piao, H. Li, and P. Guo. 2018. Methods for construction and characterization of simple or special multifunctional RNA nanoparticles based on the 3WJ of phi29 DNA packaging motor. Methods 143:121–133.

18. Sun, T., R. R. Chance, W. W. Graessley, and D. J. Lohse. 2004. A Study of the Separation Principle in Size Exclusion Chromatography. Macromolecules 37:4304–4312.

9 SELEX

Chenxi Liang and Peixuan Guo
The Ohio State University, Columbus, Ohio, USA

Systematic Evolution of Ligands by Exponential (SELEX) enrichment is a molecular biochemistry technique to produce DNA and RNA aptamers. The DNA/RNA aptamers are oligonucleotides that could bind to a specific target, e.g., amino acids, proteins, bacteria, and cells. The aptamers are considered as artificial "nucleic acid antibodies". The idea of SELEX appeared in around 1990. According to Larry Gold (2015), this "in vitro mutagenesis on this scale" was first developed by Arnold Oliphant and Kevin Struhl in 1989. However, Craig Tuerk named this method as "SELEX" in his 1990 paper published in Science. The idea of SELEX is based on the fact that the RNA could have various kinds of secondary structures, which allows it to function as an oligonucleotide antibody. This idea indicates that SELEX can be used for only single-stranded DNA or RNA (ssDNA or ssRNA). The ssDNA and ssRNA can form various kinds of secondary structures that have different binding sites and binding abilities, while the DNA and RNA helices have a strict structure.

Aptamers are selected from a random nucleic acid library using the stringent SELEX method. Basically, the selection of the aptamers using SELEX is a repeated cyclic process that involves mainly three steps: i) incubation, ii) selection, and iii) amplification. Thus, the library is first incubated with the target molecules, which allows the oligonucleotides to bind to the target molecules, then, the target oligos will be separated from the unbound oligo library. Various methods are employed to separate the target-bound oligos such as immunoprecipitation. The separated oligos are further subjected to amplification, typically, using PCR or Reverse Transcription PCR methods. The amplified binding species become the new library and are used to go through the same selection cycle, and the ratio of the target-bound species increases after each cycle. Typically, after about 20 cycles, the majority of the library is enriched for the target molecule. The library of the selected oligo is cloned and sequenced. Although the SELEX procedure looks simple, it has some drawbacks, such as non-specific binding, requiring a longer time, and being labor-consuming. To overcome the limitations, especially, the non-specific binding, Ellington and Szostak developed a new SELEX method called negative SELEX. Normally, the target molecules are bound to a solid phase such as agarose. The negative SELEX is to incubate the library with only agarose after three selection cycles. This negative selection can remove the non-specific binding from the library pool. Also Jenison et al. developed counter SELEX similar to negative SELEX. Instead of using the purification support agarose, the counter SELEX uses structurally similar targets to remove the non-specific oligonucleotides. Another problem with the original procedure is that the repeated selection cycle is time-consuming and labor-intensive. To accelerate the selection procedure, Mendonsa and Bowser developed the Capillary Electrophoresis SELEX (CE-SELEX), which could reduce the selection rounds from 20 to 1-4. Depending upon the purpose, various other SELEX methods are developed in recent years. The SELEX method is widely used in different fields of studies, such as biomedical diagnostics and therapeutics. There are few other similar methods to select aptamers for a target molecule such as SAAB (selected and amplified binding site) and CASTing (cyclic amplification and selection of targets). The aptamers can form various structures compared to antibodies, thus, they possess some unique properties. Moreover, aptamers are easier and faster to produce, also cost-effective compared to antibodies. Since nucleic acids are more stable than proteins, aptamers are more capable of transportation and storage.

DOI: 10.1201/9781003001560-10

10 Uniqueness, Advantages, Challenges, Solutions, and Perspectives in Therapeutics Applying RNA Nanotechnology

Peixuan Guo
The Ohio State University, Columbus, OH, USA

Farzin Haque, Brent Hallahan, Randall Reif, and Hui Li
University of Kentucky, Lexington, Kentucky, USA

CONTENTS

DOI: 10.1201/9781003001560-11

INTRODUCTION

Nanotechnology is an energetic field that encompasses the fabrication and application of materials at the nanometer scale using either top-down approaches or bottom-up assembly. In the biological world, a large number of highly ordered structures and nanomachines made up of macromolecules have evolved to perform many diverse biological functions. Their intriguing configurations have inspired many biomimetic designs. DNA, RNA, and proteins have unique intrinsic characteristics at the nanometer scale and therefore can serve as the building blocks for the bottom-up design and construction of nanoscale structures and devices. Seeman (2010) pioneered the concept 30 years ago of using DNA as a material for creating nanostructures; this has led to an explosion of knowledge in the now well-established field of DNA nanotechnology. The potential of using peptides and proteins for nanotechnological applications has also been extensively explored (Moll et al., 2002; Rajagopal and Schneider, 2004; Vo-Dinh, 2005; Tsai et al., 2006; Banta et al., 2007). Recently, RNA molecules have become increasingly attractive (Guo, 2010; Abdelmawla et al., 2011) because of the amazing diversity of their structures and functions (Zuker, 1989; Pleij and Bosch, 1989; Westhof et al., 1996; Jaeger et al., 2001; McKinney et al., 2003; Leontis and Westhof, 2003; Guo, 2005, 2010; Leontis et al., 2006; Isambert, 2009; Guo et al., 2010). RNA can be manipulated with ease, much like DNA; yet it also has tremendous structural flexibility and functional diversity similar to the level of proteins. The unique properties in terms of free energy, folding, noncanonical base-pairing, base-stacking, in vivo transcription, and processing that distinguish RNA from DNA provide sufficient rationale to regard RNA nanotechnology as its own technological discipline (Guo, 2010). Several comprehensive reviews on RNA nanotechnology have been published previously (Guo, 2005, 2010; Jaeger and Chworos, 2006; Guo et al., 2010). This review will address several key advances, challenges, solutions, and future perspectives in the RNA nanotechnology field.

HISTORICAL EVOLUTION OF RNA NANOTECHNOLOGY

The studies on RNA structure and folding can be dated to decades ago. A rich source of literature on RNA structure and function are available (Privalov and Filiminov, 1978; Studnicka et al., 1978; Reid, 1981; Pleij et al., 1985; Freier et al., 1986; Ehresmann et al., 1987; Zuker, 1989). However, RNA nanotechnology is a unique field that is distinct from the classical studies on RNA structure and folding. RNA nanotechnology is the application of bottom-up approaches to assemble RNA architectures in nanometer scale. Besides intramolecular interaction and folding, the special knowledge of intermolecular interaction is necessary. RNA nanotechnology involves the characterization of the physical, chemical, biological, and pharmaceutical properties of nanoparticles that can be purified into homogeneity. In 1998, the pioneering work in Peixuan Guo's lab (Figure 10.1) demonstrated that RNA dimer, trimer, and hexamer nanoparticles can be assembled using re-engineered RNA fragments derived from the pRNA (packaging RNA), a vital component to gear the DNA packaging motor of the bacteriophage phi29. This finding was published in Molecular Cell (Guo et al., 1998) and was featured in Cell (Hendrix, 1998), proving the concept of RNA nanotechnology.

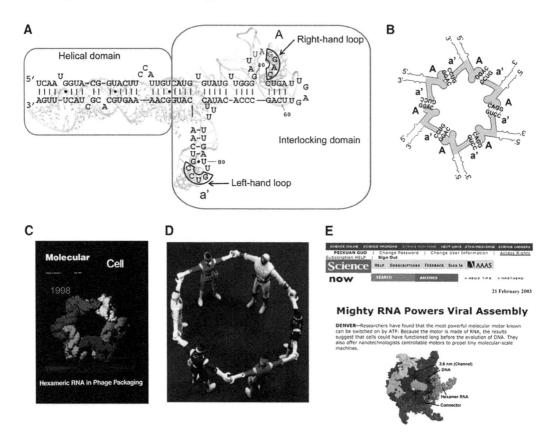

FIGURE 10.1 Structure of DNA-packaging RNA (pRNA) on the phi29 DNA-packaging motor and the hand-in-hand interaction used to build pRNA nanoparticles. (a) Sequence and secondary structure of phi29 pRNA. Superposition of the 2-dimensional and 3-dimensional structure of the phi29 pRNA Aa′. Uppercase letters represent the right-hand loop and lowercase letters the left-hand loop of pRNA. A pair of upper and lower case (e.g., Aa′) for same letters indicates a pair of complementary loops, whereas a pair of upper- and lowercase from different letters indicates noncomplementary loops (see Figure 10.2). The 4 bases in the right- and left-hand loops, which are responsible for inter-RNA interactions, are boxed. For example, pRNA Aa′ refers to a pRNA with complementary right-hand loop A and left-hand loop a′, which can form homo-hexamers (see also Figure 10.2 for homo-dimers and trimers). (b) Schematic of pRNA hexamer. (c) Packaging of phi29 DNA through the motor geared by 6 pRNA (Guo et al., 1998; Zhang et al., 1998). (d) Construction of hexameric pRNA nanoparticles using the hand-in-hand interaction approach (Chen et al., 1999). (e) Elucidation of phi29 pRNA hexamer on the motor. Figures adapted and reproduced with permission: (a) Liu et al., 2010, © 2010 ACS; (b) Shu et al. 2011b, © 2011 Elsevier Inc.; (c) © 1998 Cell Press; (d) Chen et al., 1999, © RNA Society; (e) © 2003 AAAS.

In 2004, Guo's group reported the systematic formation of pRNA nanoparticles using 2 technologies: hand-in-hand interactions and palindrome sequence-mediated self-annealing (Figures 10.1a–c, 10.2b,c, and 10.3) (Shu et al., 2004). In the succeeding years, through a series of papers, they showed that pRNA molecules could be conjugated with various therapeutic functionalities including aptamers, small interfering RNA (siRNA), ribozymes, and microRNA (miRNA) (Hoeprich et al., 2003; Guo et al., 2005, 2006; Khaled et al., 2005; Shu et al., 2009, 2011a, 2011b, 2011c; Abdelmawla et al., 2011; Ye et al., 2011; Zhang et al., 2009) (Figures 10.2a–d, 4). These findings have paved the way for RNA nanotechnology to develop into a novel area of therapeutics for the treatment of various diseases such as cancer, viral infections, and genetic diseases.

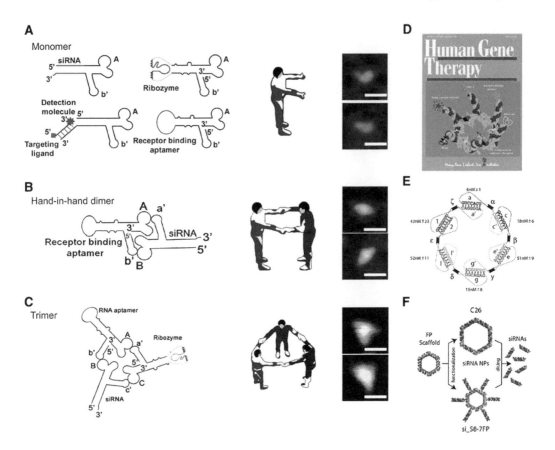

FIGURE 10.2 Construction of therapeutic pRNA nanoparticles via hand-in-hand interaction (see also Figure 10.1). Left to right column: Schematic, models, and atomic force microscopy (AFM) image showing the formation of different therapeutic nanoparticles containing small interfering RNA (siRNA), ribozymes, aptamers, and other moieties using bacteriophage phi29 pRNA that possess left- and right-hand interlocking loops; uppercase and lowercase letters represent right- and left-hand loops, respectively. Same letter pair (e.g., Aa′) indicates complementarity interlocking loops; different letter (e.g. Ab′) indicates noncomplementary loops (Hoeprich et al., 2003; Guo et al., 2005, 2006; Khaled et al., 2005; Shu et al., 2009, 2011a, 2011b, 2011c; Abdelmawla et al., 2011; Ye et al., 2012). (a) Construction of pRNA monomers bearing either siRNA, a ribozyme, a receptor-binding aptamer, a targeting ligand, or a detection molecule; scale bar = 15 nm. (b) Construction of pRNA dimers. Monomer Ab″, which contains a receptor-binding aptamer, and monomer Ba′, which contains an siRNA, assemble to form hand-in-hand dimers; scale bar = 30 nm. (c) Construction of pRNA trimers. Trimers are formed between monomer Ab′ (containing an RNA aptamer), Bc′ (containing an siRNA), and Ca′ (containing a ribozyme); scale bar = 30 nm. (d) Illustration of hexameric pRNA nanoparticles on the cover of Human Gene Therapy (Guo et al., 2005). (e–f) colE1 loop–loop interactions used to construct programmable a hexameric nanoring, via interlocking loops αα′, ββ′, γγ′, δδ′, εε′, and ζζ′. The siRNA sequences are attached to the vertices after the formation of the hexamer, instead of using the fusing approach, as in a–d. Figures adapted and reproduced with permission: (a–c) Shu et al., 2004, © 2004 ACS and Shu et al., 2003, © 2003 American Scientific Publishers (ASP); (d) Guo et al., 2005, © 2005 Mary Ann Liebert, Inc.; (e, f) Grabow et al., 2011, © 2011 American Chemical Society (ACS).

The development of multivalent pRNA nanoparticles in the Guo lab is just one facet of the rapidly emerging field of RNA nanotechnology and therapeutics. Investigations of the folding and structure of RNA motifs and junctions have laid a foundation for the further development of RNA nanotechnology. Significant contributions on the fundamental studies of RNA structural motifs were made by Eric Westhof (Leontis and Westhof, 2003; Lescoute and Westhof, 2006; Jossinet et al., 2007), Neocles Leontis (Jaeger et al., 2001; Leontis and Westhof., 2003; Leontis et al., 2006), David Lilley (Lilley, 1999; McKinney et al., 2003; Schroeder et al., 2010), and Luc Jaeger (Jaeger et al., 2001; Severcan et al., 2009, 2010; Afonin et al., 2010). Their fundamental work on RNA junctions (Leontis et al., 2006; Lescoute and Westhof, 2006; Schroeder et al., 2010) and RNA tectonics (Jaeger et al., 2001) have been used to construct diverse RNA nanoparticles, such as squares (Severcan et al., 2009), jigsaw puzzles (Chworos et al., 2004), filaments (Jaeger and Leontis, 2000; Nasalean et al., 2006; Geary et al., 2010), cubic scaffolds (Afonin et al., 2010), and polyhedrons (Severcan et al., 2010). Advances in RNA 3-dimensional computation expanding from the traditional intramolecular interactions to intermolecular interactions promoted by Bruce Sharpiro and others have brought new energy into the RNA nanotechnology field (Mathews and Turner, 2006; Shapiro et al., 2007, 2008; Yingling and Shapiro, 2007; Bindewald et al., 2008a; Afonin et al., 2010; Kasprzak et al., 2010; Laing and Schlick, 2010; Bindewald et al., 2011; Grabow et al., 2011). These newly developed inter-RNA computational programs will greatly facilitate RNA nanoparticle design and construction.

RNA nanotechnology is a vigorous and rapidly emerging new field of science, as evidenced by the burst of publications on RNA nanostructures over the last 5 years, indicating strong interest in RNA nanotechnologies in diverse fields such as chemistry, biophysics, biochemistry, structural biology, microbiology, cancer biology, pharmacy, cell biology, and nanomedicine. Currently, PubMed shows that 92% (1,002 of the total 1,090) of publications with the key words "RNA nanostructure" were published after 2005. With the continued development of RNA nanotechnology, many well-respected and prestigious journals have begun to include articles focused on RNA nanotechnology in their journals, including Science (Delebecque et al., 2011), Nature Nanotechnology (Afonin et al., 2010; Editorial Comment, 2011; Guo, 2010; Ohno et al., 2011; Shu et al., 2011a), PNAS (Dibrov et al., 2011), Nano Letters (Shu et al., 2004; Yingling and Shapiro, 2007; Grabow et al., 2011) Nano Today (Haque et al., 2012), and Nature Protocols (Afonin et al., 2011). In addition, new journals have been founded to cover topics on RNA nanotechnology, such as Nucleic Acid Therapeutics, WIREs RNA, and Molecular Therapy–Nucleic Acids. In 2009, the National Institutes of Health (NIH) launched the National Cancer Institute Alliance for Nanotechnology in Cancer to create and foster a community of scientists using novel nanotechnology approaches to diagnose, treat, and prevent cancers. As a result, a Cancer Nanotechnology Platform Partnership program entitled RNA Nanotechnology in Cancer Therapy directed by Dr. Peixuan Guo was established (http://nano.cancer.gov/action/programs/platforms/uc.asp). In 2010, the first International Conference of RNA Nanotechnology and Therapeutics (http://www.eng.uc.edu/nanomedicine/RNA2010) was held (Shukla et al., 2011) and a second conference is planned in April 3–5, 2013 at the University of Kentucky.

UNIQUENESS OF RNA NANOTECHNOLOGY

RNA has several unique attributes that make it a powerful biomaterial compared to DNA, such as high thermodynamic stability (Searle and Williams, 1993; Sugimoto et al., 1995; Freier et al., 1986), formation of canonical and noncanonical base pairs (Ikawa et al., 2004; Leontis et al., 2006; Li et al., 2006; Matsumura et al., 2009; Schroeder et al., 2010), base stacking properties (Searle and Williams, 1993; Sugimoto et al., 1995), and various in vivo attributes (Chang and Tinoco, 1994;

Guo et al., 1998; Zhang et al., 1998; Chen et al., 2000; Hoeprich et al., 2003; Wagner et al., 2004; Bindewald et al., 2008b; Laurenti et al., 2010). RNA molecules can fold into unique structural motifs mediated by canonical and noncanonical base pairings and further stabilized by tertiary interactions and complex 3-dimensional architectures exhibiting pseudoknots, single-stranded loops, bulges, hairpins, and base stacking. Currently, an RNA polymer up to 80 nt can be efficiently and commercially synthesized non-enzymatically. An 80-nt RNA can have up to 10^{48} (4^{80}) unique sequences with the sequence variation specifying for many individual possible structures. Such a huge pool is a great resource to identify diverse building blocks of RNA nanoparticles for the design, assembly, and manufacturing of therapeutic nanoparticles via intra- and intermolecular interactions. RNA–RNA interaction is the most stable with lowest free energy among the RNA–RNA, DNA–RNA, and DNA–DNA interactions (Lesnik and Freier, 1995; Gyi et al., 1996; Shu et al., 2011a; Binzel and Guo, unpublished results).

TECHNIQUES FOR THE CONSTRUCTION OF THERAPEUTIC

RNA NANOPARTICLES

RNA can fold into well-defined tertiary structures with specialized functionalities. The structural motifs and tertiary interactions have been examined in many RNA molecules and the information gleaned has been used to rationally design the building blocks that self-assemble into RNA nanoparticles (Figures 10.2–10.4) as discussed below.

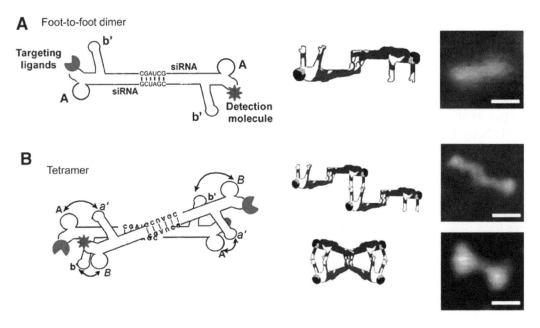

FIGURE 10.3 Construction of therapeutic pRNA nanoparticles via foot-to-foot interaction of palindrome sequences. Left to right column: Schematic, models, and AFM image showing the formation of different therapeutic nanoparticles containing siRNA, ribozymes, aptamers, and other moieties using bacteriophage phi29 pRNA containing a palindrome sequence (Shu et al., 2004). (a) Foot-to-foot dimers form through the palindrome sequence at the end of two Ab′ monomers, with one bearing a targeting ligand and the other a detection molecule; scale bar = 20 nm. (b) Tetramers assemble by the combination of hand-in-hand interlocking loops and foot-to-foot palindrome mechanism of 2 dimers (Ab′ and Ba′); scale bar = 20 nm. The models illustrate how the various structures are held together. Figures (a and b) reproduced with permission Shu et al., 2004, © 2004 ACS and Shu et al., 2003, © 2003 ASP.

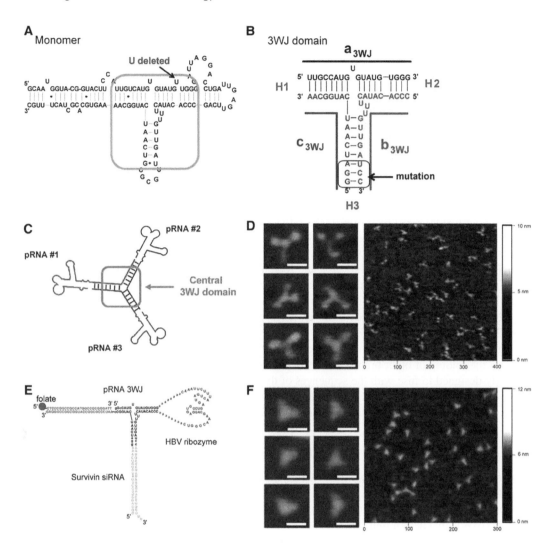

FIGURE 10.4 Construction of thermodynamically stable trivalent pRNA-based 3-way junction (3WJ) nanoparticles. (a) Sequence of pRNA monomer Ab′ (Guo et al., 1998). Green box: central 3WJ domain. In pRNA Ab′, A, and b′ represent right- and left-hand loops, respectively. (b) 3WJ domain composed of 3 RNA oligomers in black, red, and blue. Helical segments are represented as H1, H2, and H3. (c) Three pRNA molecules bound at the 3WJ-pRNA core sequence (black, red, and blue) and (d) its accompanying AFM images; scale bar = 30 nm. (e) Multi-module RNA nanoparticles harboring siRNA, ribozyme, and aptamer, and (f) its accompanying AFM images; scale bar = 20 nm. Figures reproduced with permission from Shu et al. 2011a, © 2011 Nature Publishing Group (NPG).

HAND-IN-HAND (LOOP–LOOP) INTERACTIONS

Bacteriophage phi29 pRNA has 2 defined domains (Figure 10.1): a 5′/3′-end helical domain (Zhang et al., 1994) and an interlocking loop region, which is located at the central part of the pRNA sequence (Reid et al., 1994; Zhang et al., 1994, 1995a; Chen et al., 2000). The central domain of each pRNA subunit contains 2 interlocking loops, known as the right- and left-hand loops, which can be re-engineered to form dimers, trimers, or hexamers via hand-in-hand interactions (Figure 10.2a–c) (Guo et al., 1987, 1998; Chen et al., 2000; Shu et al., 2003, 2004; Zhang et al., 1998). The 2 domains fold separately, and replacement of the helical domain with a siRNA does not affect pRNA structure, folding, or intermolecular interactions (Zhang et al., 1994; Trottier et al., 2000). This hand-in-hand interaction approach has recently been used by Bruce Shapiro and Luc Jaeger for the construction of RNA nanoparticles with different shapes (Figure 10.2e–f) (Yingling and Shapiro, 2007; Afonin et al., 2011; Grabow et al., 2011). The kissing loop of human immunodeficiency virus (HIV) RNA (Chang and Tinoco, 1994; Bindewald et al., 2008b) and the hand-in-arm interaction of Drosophila bicoid mRNA (Wagner et al., 2004) can be constructed utilizing a similar approach.

Robust RNA Motif as a Scaffold to Build Multivalent Nanoparticles

Mechanically constructing fusion complexes of DNA, RNA, or protein can be easily accomplished, but it is difficult to ensure that the individual modules within the complex will appropriately fold and function after fusion. Recently, it was reported that the 3-way junction (3WJ) is a motif of the phi29 pRNA that can be assembled from 3 small RNA oligos with unusually high affinity in the absence of metal salts. The resulting complex displays thermodynamically stable properties, resistant to denaturation even in the presence of 8 M urea and remains intact without dissociating at ultra-low concentrations. RNA nanoparticles harboring a variety of functionalities (siRNA, ribozyme, aptamer, riboswitch, miRNA, or folate) were constructed using the pRNA (Zhang et al., 1995b; Hoeprich et al., 2003), or its 3WJ core as a scaffold with perfect folding and function (Shu et al., 2011a; Haque et al., 2012) (Figure 10.4). The 3WJ-pRNA is tightly folded and serves as a driving force for the folding of other modules. As a result, individual functionalities can be placed at each branch without affecting the folding of other branches. The sequences for therapeutic and reporter moieties can be rationally designed to fuse with the sequences of the 3WJ strands a_{3WJ}, b_{3WJ}, and c_{3WJ}, respectively. The 3 RNA fragments can then be assembled into RNA nanoparticles and their folding evaluated by in vitro and in vivo functional assays (Figures 10.4–10.6).

Palindrome Sequence Mediated Formation of RNA Dimers

Palindrome sequences can promote the self-formation of pRNA dimers, tetramers, and arrays with high efficiency (Shu et al., 2004). In a similar manner, addition of self-complementary palindrome sequences to either the 5′ or 3′ end of one of the strands of the 3WJ-pRNA core results in the bridging of two 3WJs that harbor multiple functionalities via intermolecular interactions thereby generating a tetramer with 4 therapeutic and reporter moieties (Figure 10.3).

RNA Junctions as LEGO Pieces to Build Quaternary Structures

Large RNA constructs can be fabricated by non-templated assembly via modular design, such that the complex can be self-assembled from the basic building blocks without any external influence, assembled based on a modular design without any external template required (Chworos et al., 2004; Severcan et al., 2009, 2010). Examples include Tecto-RNA, 2-, 3-, and 4-way junctions (2-/3-/4WJ) (Figure 10.4), and self-assembly by colE1 kissing loop interactions or kissing loops engineered that mimic this type of interaction (Figure 10.2) and phi29 pRNA multimerization and quaternary architectures (Prats et al., 1990; Clever et al., 1996; Mujeeb et al., 1998; Jaeger and Leontis, 2000; Shu et al., 2003, 2004; Guo et al., 2005; Khaled et al., 2005; Grabow et al., 2011).

RNA Binding Proteins to Serve as Junctions for the Formation of Arrays

Ribosomal proteins have been shown to interact with RNA to form a nanostructure with a shape similar to an equilateral triangle (Ohno et al., 2011). In each triangle, three proteins are bound to an RNA scaffold containing kink-turn motifs (Schroeder et al., 2010) for protein binding. The kink-turn

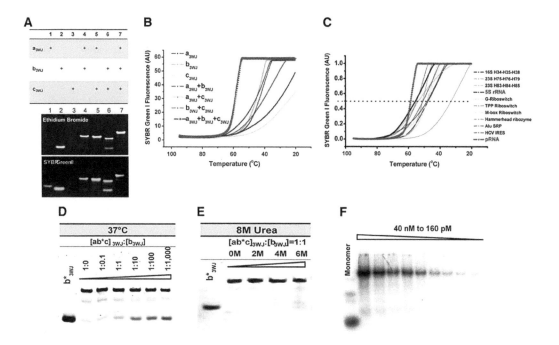

FIGURE 10.5 Assembly and stability studies of 3WJ-pRNA. In the tables, " + " indicates the presence of the strand in samples of the corresponding lanes. (a) 15% native polyacrylamide gel electrophoresis (PAGE) showing the assembly of the 3WJ core, stained by ethidium bromide (upper) and SYBR green 2 (lower). (b) Melting temperature curves for the assembly of the 3WJ core. Melting curves for the individual strands (brown, green, silver), the 2-strand combinations (blue, cyan, pink) and the 3-strand combination (red) are shown. (c) Melting curves for 11 different RNA 3WJ core motifs assembled from 3 oligos for each 3WJ motif under physiological buffer. (d–f) Competition and dissociation assays of 3WJ-pRNA. (d) Temperature effects on the stability of the 3WJ-pRNA core, denoted as [ab*c]$_{3WJ}$, evaluated by 16% native gel. A fixed concentration of Cy3-labeled [ab*c]$_{3WJ}$ was incubated with varying concentrations of unlabeled b$_{3WJ}$ at 37C. (e) Urea denaturing effects on the stability of [ab*c]$_{3WJ}$ evaluated by 16% native gel. A fixed concentration of labeled [ab*c]$_{3WJ}$ was incubated with unlabeled b$_{3WJ}$ at 1:1 ratio in the presence of 0–6 M urea at 25C. (f) Dissociation assay for the [^{32}P]-3WJ-pRNA complex harboring 3 monomeric pRNAs by 2-fold serial dilution (lanes 1–9). The monomer unit is shown on the left. Figures reproduced with permission from Shu et al., 2011a, © 2011 NPG.

allows the RNA to bend by *60 at three positions, thus forming a triangle. The resulting protein–RNA complex could have potential applications in medicine, biotechnology, and nanotechnology.

Combination of Rolling Circle Transcription of RNA and Self-assembly to Produce Giant Spherical RNA Particles

A method has been developed by using rolling circle transcription to form siRNA concatemers that self-assemble into sponge-like microspheres (Figure 10.7). The RNA interference (RNAi)-microsponges consisting of cleavable RNA strands can be processed by the cellular machinery to convert the stable hairpin RNA to siRNA after cellular uptake. This finding reveals that RNA, which is a special class of polymer, displays the intrinsic property of other chemical polymers that form lamellar spherulites, such as that from polyethylene when nucleated in the bulk state or in solution (Lee et al., 2012). Generally, pure RNA is negatively charged, and direct cellular uptake remains insignificant due to electrostatic repulsion from the negatively charged cell membrane. Hammond

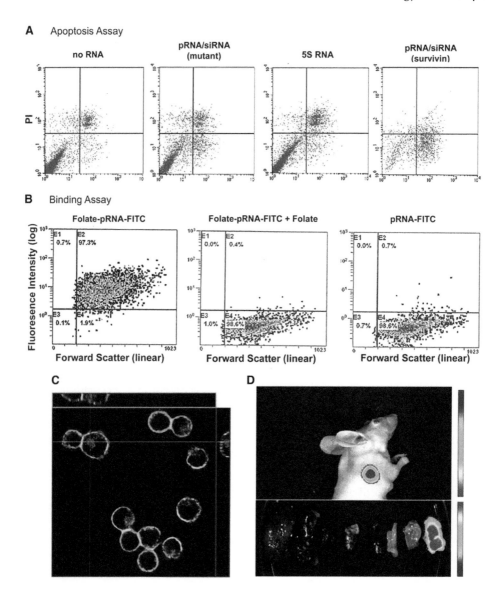

FIGURE 10.6 Apoptosis and binding assays of chimeric therapeutic pRNA. (a) Apoptosis induced by transfection of chimeric pRNA harboring siRNA targeting survivin using Lipofectamine 2000. Breast cancer MCF-7 cells were transfected with pRNA/siRNA (survivin) and apoptosis was monitored by propidium iodide–annexin A5 double labeling followed by flow cytometry. Cells in the bottom right quadrant represent apoptotic cells. The mutant pRNA/siRNA was transfected in parallel as a negative control. (b) Specific delivery of chimeric pRNA/siRNA by folate-pRNA. Flow cytometry analyses of the binding of fluorescein isothiocyanate (FITC)-labeled folate-pRNA to nasopharyngeal carcinoma (KB) cells. Left: Cells were incubated with folate-pRNA labeled with FITC. Middle: Cells were preincubated with free folate, which served as a blocking agent to compete with folate-pRNA for binding to the receptor. Right: Binding was also tested using folate-free pRNA labeled with FITC as a negative control. The percentages of FITC-positive cells are shown in the top-right quadrants. (c) Confocal images showed targeting of folate receptor positive (FR+)KB cells by co-localization (overlap, 4) of cytoplasm (green, 1) and RNA nanoparticles (red, 2). (d) 3WJ-pRNA nanoparticles target folate receptor positive (FR+) tumor xenografts on systemic administration in nude mice. Upper panel: whole body; lower panel: organ imaging (Lv, liver; K, kidney; H, heart; L, lung; S, spleen; I, intestine; M, muscle; T, tumor). Figures reproduced with permission: (a, b) Guo et al., 2005, © 2005 Mary Ann Liebert, Inc.; (c, d) Shu et al., 2011a, © 2011 NPG.

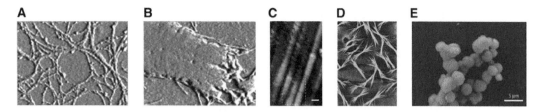

FIGURE 10.7 Self-assembled RNA nanoparticles as potential therapeutic agents. AFM images of (a, b) rationally designed RNA 1-dimensional and 2-dimensional arrays in vivo (Delebecque et al., 2011); (c) RNA bundles (scale bar = 50 nm) (Cayrol et al., 2009); (d) AFM images of pRNA arrays (Shu et al., 2004). (e) Transmission electron microscopy (TEM) images of RNA microsponges (Lee et al., 2012). Figures reproduced with permission: (a, b) Delebecque et al., 2011, © 2011 AAAS; (c) Cayrol et al., 2009, © 2009 ACS; (d) Shu et al., 2004, © 2004 ACS; (e) Lee et al., 2012, © 2012 NPG.

and colleagues (Lee et al., 2012) used synthetic poly(ethyleneimine) (PEI) to condense the RNAi-microsponge from 2 to 200 nm. By this approach, the net charge of microspheres were shifted from negative to positive and subsequently internalized into cells. It is commonly believed that at the lower pH environment within the endosome, protonation of amine residues of PEI can lower the osmotic potential and cause osmotic swelling, which can result in bursting of the endosome to release the siRNA. It would be interesting to investigate whether the feasibility of PEI/RNAi-microsponges reported by Lee et al. as therapeutic agents can be improved by including ligands for specific targeting. Variable mechanisms and routes such as phagocytosis, macropinocytosis, and clathrin- or caveolae-mediated endocytosis all can lead to the internalization of nanoparticles. It would be worthy to evaluate which route is involved in the cellular uptake of such large PEI/RNAi microsponges and to include targeting moieties to achieve specific delivery in vivo. In addition, extensive studies in xenograft models have revealed that delivery of nanomaterials requires a delicate balance between extravasation from the porous tumor vasculature and particles trapped by the monocyte phagocytic system or Kupffer cells in lung, spleen, and liver. It would be interesting to determine whether the PEI/RNAi-microsponge can be formulated in a way to evade the monocyte phagocytic system and organ accumulation.

CONSTRUCTION OF RNA-BASED NANOPARTICLES FOR THERAPEUTIC APPLICATIONS

There has been a heightened interest in RNA therapeutics since the discovery of siRNA (Fire et al., 1998; Hamilton and Baulcombe, 1999; Brummelkamp et al., 2002; Carmichael, 2002; Jacque et al., 2002; Li et al., 2002; Varambally et al., 2002), ribozymes (Guerrier-Takada et al., 1983; Zaug et al., 1983; Forster and Symons, 1987; Nava Sarver et al., 1990; Sarver et al., 1990; Chowrira et al., 1991) and anti-sense RNA (Coleman et al., 1985; Knecht and Loomis, 1987), since they have been shown to down-regulate the expression of specific genes in viral-infected or cancerous cells. However, the use of siRNA in gene therapy has been significantly limited due to the difficulty of targeting the siRNA to specific cells. The advantage of using phi29 pRNA as a delivery medium is based on its ability to form stable multimers, which can be manipulated and sequence-controlled (Guo et al., 1998; Chen et al., 2000; Shu et al., 2003). This particular system, by applying the hand-in-hand approach and the robust pRNA 3WJ motif, provides superior pliancy for constructing polyvalent delivery vehicles containing multiple components (Figures 10.2–10.4). For instance, one subunit of a dimeric, trimeric, or tetrameric RNA nanoparticle can be modified to contain a RNA aptamer that binds to a specific cell-surface receptor, thereby acting as a ligand for receptor-mediated endocytosis. A second subunit of the multimer can contain a reporter moiety such as a gold particle (Moll and Guo, 2007) or fluorescent dye for evaluating cell binding and entry. A third subunit can be designed to contain a component that enhances endosome disruption so that the therapeutic molecules are released. A fourth (or fifth or sixth, if necessary) subunit of the RNA nanoparticle can carry a therapeutic siRNA, ribozyme, riboswitch, miRNA, or another complementary drug. The incorporation of each of the functional modules to pRNA scaffold is discussed below.

siRNA

RNAi is a key post-transcriptional gene silencing mechanism that has evolved in plants and some animals. The siRNAs are typically 21–25-bp dsRNA strands with 2-nt overhangs at the 3′ ends. The siRNAs bind to a protein complex in the cytoplasm called the RNA- induced silencing complex (RISC). The siRNA/RISC complex then scans and intercepts intracellular mRNA containing a complementary sequence to the bound siRNA. The intercepted mRNA is cleaved and degraded, thereby silencing the expression of that gene (Fire et al., 1998; Brummelkamp et al., 2002; Carmichael, 2002; Jacque et al., 2002; Li et al., 2002; Varambally et al., 2002).

Since the siRNA is double stranded, the incorporation of siRNA into RNA nanoparticles is readily accomplished by simply fusing the siRNA sequences at one of the helical stems of the 3WJ (Figure 10.4) (Shu et al., 2011a) by replacing the end helical segment of the monomeric pRNA with siRNA sequences (Figure 10.2) (Liu et al., 2007) or by attaching the siRNA to the RNA assemblies, whereby the 5′ or 3′ ends of the sequences that constitute the assemblies are extended and either the sense or the antisense sequences are then hybridized to these extended sequences to form the siRNAs (Figure 10.2) (Afonin et al., 2011; Grabow et al., 2011). For increased stability, typically chemically modified 2′F nucleotides are used to modify the sense strand. Since chemical modifications might compromise the silencing potency of RNA nanoparticles containing siRNA components, the region of the Dicer processing site needs to be rationally designed and use of C or U nucleotides should be avoided if 2′-F C/U modifications were introduced.

miRNA

The miRNAs are typically short (*23 nt) RNA strands that are naturally found in plants and animals. They are part of noncoding RNA sequences and play an important role in gene regulation by binding onto specific messenger RNA sites responsible for protein coding (Bartel, 2009). Recently, it has been found that miRNA plays an important role in the control or development of cancers (He and Hannon, 2004), cardiac diseases (Chen et al., 2008), and regulation of the nervous system (Maes et al., 2009). Within each of the diseases, miRNA levels are either up- or down-regulated.

Recent work and discoveries have led to the idea of using miRNAs in therapy for gene regulation in cell mutations (Figure 10.9) (Bader et al., 2010). In diseases where miRNAs are seen to be down-regulated, levels can be synthetically increased through the delivery via the approach of the pRNA or its 3WJ core (Ye et al., 2011, 2012; Shu et al., 2011a). Similar to siRNA, the miRNA sequences can be conjugated onto each branch of the 3WJ and delivered to the diseased cells, which would then undergo normal Dicer processing in vivo through the RISC complex, returning normal gene regulation in the diseased cells. Different from siRNA, miRNA has a much broader target by regulating several genetic pathways (Kasinski and Slack, 2011).

RIBOZYMES

Ribozymes are RNA molecules that can catalyze chemical reactions (Kruger et al., 1982; Guerrier-Takada et al., 1983). Ribozymes have compact and specific structures that enable them to catalyze trans-esterification and hydrolysis reactions. They can intercept and cleave mRNA or the genome of RNA viruses and thus have a significant therapeutic impact.

1. chimeric pRNA monomer containing a hammerhead ribozyme was designed to cleave the poly(A) signal on hepatitis
2. virus (HBV) mRNA

Cleavage of the HBV mRNA was nearly complete in vitro and HBV replication was inhibited in vivo by this chimeric pRNA (Hoeprich et al., 2003). The antiapoptosis factor, survivin, regulates tumor development and progression. A chimeric pRNA containing a hammerhead ribozyme designed to target survivin mRNA was shown to suppress survivin gene expression and initiate apoptosis in cell cultures (Liu et al., 2007). It was shown that the HBV ribozyme can also cleave the poly(A) signal from HBV mRNA after being incorporated into the 3WJ nanoparticles (Figure 10.4) (Liu et al., 2007; Shu et al., 2011a).

Riboswitches

A riboswitch (Tucker and Breaker, 2005; Barrick and Breaker, 2007; Cheah et al., 2007; Breaker, 2008, 2012) is a component of some specific mRNAs that bind a small molecule and control the expression of that mRNA in response to the concentration of the small molecule. Riboswitches fold into intricate structures that typically recognize metabolites and have evolved as metabolic control mechanisms in bacteria. Riboswitches can regulate gene expression by several means, including premature termination of mRNA transcription, ribosome binding, and inhibition of mRNA translation, mRNA cleavage, and even mRNA degradation. There is substantial interest in engineering artificial riboswitches to create a new generation of regulators to control the expression level of targeted genes in response to interactions with small drug-like molecules.

Such RNA-based gene-control machines have the potential to supply nanoscale, cis-acting, modular systems, incapable of inducing antibody production for use in future gene therapies (Henkin, 2008; Ogawa and Maeda, 2008; Shahbabian et al., 2009). If RNA nanoparticles with riboswitch modules can be constructed in vivo, it would be possible to regulate biological functions in vivo.

Aptamers

A RNA aptamer is a RNA molecule that binds a specific ligand through the formation of a recognition structure (Ellington and Szostak, 1990; Tuerk and Gold, 1990; Mi et al., 2010). RNA aptamers with the ability to bind specific targets with high affinity can be extracted from a pool of random RNA oligonucleotides by in vitro SELEX (systematic evolution of ligands by exponential enrichment) (Ellington and Szostak, 1990; Tuerk and Gold, 1990).

Aptamers that specifically bind to target cancer receptors can be incorporated into the RNA nanoparticles as part of the functionality of the polyvalent therapeutics. The selected receptor-binding RNA aptamers can be rationally designed to link to the 5'/3' end of any helical region of the 3WJ. It is important to ensure that the aptamer folds correctly and that its binding affinity to the target cell-surface marker is maintained. Several chimeric pRNA containing aptamers have already been successfully used for binding to CD4 (Khaled et al., 2005), gp120 (Zhou et al., 2008, 2011) of HIV (Figure 10.8), or prostate cancer prostate-specific membrane antigen (McNamara et al., 2006; Dassie et al., 2009).

Advantages of RNA Nanotechnology for In Vivo Applications

Various types of therapeutic RNA have been developed and their applications for the treatment of diseases are just beginning to be fully realized. Although gene silencing with high efficacy and specificity by a variety of interference strategies and RNA molecules has been achieved in vitro, the effective delivery of therapeutic RNAs to specific cells in vivo remains challenging. Development of an efficient, specific, and nonpathogenic nanodevice for delivering multiple therapeutics in vivo is highly desirable. Application of RNA nanotechnology has significant advantages in this endeavor as outlined below.

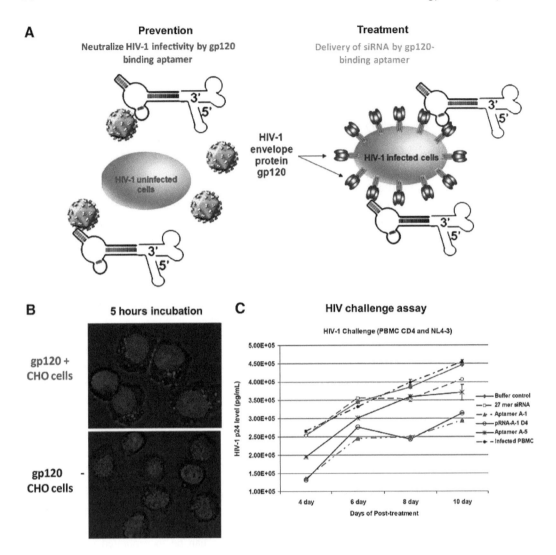

FIGURE 10.8 Chimeric pRNA-aptamer-siRNA nanoparticles for human immunodeficiency virus (HIV) therapy. (a) The pRNA-aptamer mediated targeted delivery of siRNA using chimeric pRNA–anti-gp120 aptamer. The anti-gp120 aptamer is responsible for binding to HIV-1 gp120 protein. (b) Cell-type specific binding studies of pRNA aptamer chimeras. Cy3-labeled pRNA aptamers were incubated with Chinese hamster ovary (CHO)-gp160 cells and CHO-EE control cells. Cell-surface binding of Cy3-labeled chimeras was assessed by confocal imaging. (c) The inhibition of HIV-1 infection mediated by pRNA-aptamer chimeras. Both antigp120 aptamer and pRNA-aptamer chimera neutralized HIV-1 infection in HIV-infected human peripheral blood mononuclear cells (PBMCs) (NL4-3 strain) culture. Data represent the average of triplicate measurements (Zhou et al., 2008, 2011). Figure (a) courtesy of Dr. Jiehua Zhou and Dr. John Rossi. Figures (b, c) reproduced with permission from Zhou et al., 2011, © 2011 Elseiver.

POLYVALENT DELIVERY FOR GENERATING SYNERGISTIC EFFECTS

The polyvalent RNA nanoparticles can deliver up to 6 kinds of molecules to specific cells including therapeutics, detection modules, drugs, or other functionalities (Guo, 2005; Guo et al., 2005; Nakashima et al., 2011; Chang et al., 2012a, 2012b). This particular system provides remarkable flexibility for the construction of polyvalent delivery vehicles since it is based on a modular design.

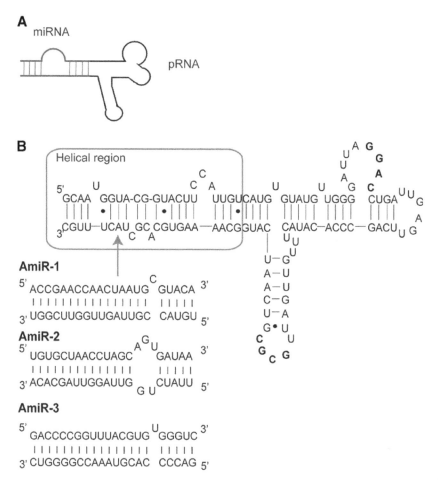

FIGURE 10.9 Schematic (a) and 2D structure (b) of chimeric pRNA–miRNA complexes for antiviral therapies. The helical region of pRNA is replaced by several artificial microRNA (AmiR) sequences. The AmiRs target the 3′ untranslated region (3′ UTR) of coxsackievirus B3 (CVB3) genome.

Hence, individual RNA subunits with various cargos can be constructed separately and assembled into the final quaternary complex by mixing them together in any desired combination (Shu et al., 2011a; Haque et al., 2012). For example, the deliverable RNA nanoparticle can be engineered to carry therapeutic siRNAs, ribozymes, or antisense RNAs against multiple genes or different regions of one target gene, and RNA aptamers or folic acid for targeted delivery (Figure 10.6). The other subunits of the RNA nanoparticle may carry anti-cancer drugs to enhance the therapeutic effect or to overcome the drug resistance by combination therapy. The therapeutics, detection molecules, or drug may also be combined into one nanoparticle, making the concomitant therapy and detection of the therapeutics possible with a single administration.

DEFINED SIZE, STRUCTURE, AND STOICHIOMETRY

Currently, the use of polymer for siRNA or drug delivery has been reported extensively (Nimesh et al., 2011; Singha et al., 2011; Troiber and Wagner, 2011; Duncan, 2011). RNA is a polymer (polynucleic acid). Different from other polymers such as polyethylene glycol, the homogeneity in size of the pRNA nanoparticles is of extreme importance. Highly efficient and controlled bottom-up self-assembly yields nanoparticles with well-defined structures and stoichiometry. This characteristic is highly valuable for the reproducible manufacturing of drugs and increased safety. The clearly defined structure and stoichiometry might facilitate FDA approval of RNA nanoparticles as therapeutic agents.

NANOSCALE SIZE FOR ENHANCED PERMEABILITY AND RETENTION EFFECTS

The size of a nanoparticle is commonly thought to be the fundamental factor for effective delivery to diseased tissues. Many studies suggest that nanoparticles ranging from 10 to 100 nm (Gao et al., 2005; Jain, 2005; Li and Szoka, 2007) are the optimal size because they are large enough to avoid excretion through the urine, yet small enough to bind to cell-surface receptors and enter the cells via receptor-mediated endocytosis (Li and Szoka, 2007). During the development of solid tumors, angiogenesis occurs to supply enough oxygen and nutrients to the fast-growing tumor cells. These newly formed blood vessels, unlike the tight blood vessels in most normal tissues, are leaky because of gaps between them and adjacent endothelial cells. This allows the particles that are usually excluded from the normal tissue to navigate through these gaps into the tumor interstitial space and concentrate in the tumor in a size-dependent manner. The pRNA nanoparticles (dimers, trimers, or tetramers) have sizes ranging between 20 and 40 nm (Liu et al., 2010; Shu et al., 2004, 2011a; Abdelmawla et al., 2011), which improves the biodistribution of the therapeutic pRNA nanoparticles in the blood circulation system, while the average size of a normal single siRNA molecule is well below 10 nm, which represents a major challenge for the siRNA delivery in vivo. In addition, the polyanionic nature of RNA makes it difficult to cross cell membranes, and non-formulated siRNAs have been reported to be easily excreted by the body (de Fougerolles et al., 2007; Kim and Rossi, 2007; Rozema et al., 2007). Nanoparticle delivery of siRNAs or other therapeutics has the potential to improve the pharmacokinetics (PK), pharmacodynamics (PD), and biodistribution, as well as reduce potential toxicity (Shu et al., 2004; Guo et al., 2005; Khaled et al., 2005; Abdelmawla et al., 2011). Furthermore, the PK and PD of pRNA nanoparticles can be improved by introducing chemical modifications to the RNA backbone. Chemically modified RNA is resistant to RNase degradation, which makes RNA nanoparticles more stable and increases their retention time during blood circulation. Specific delivery and longer retention time including the enhanced permeability and retention effect also reduces the dosage necessary for effective therapy.

Antiviral evaluation showed that the AmiRs displayed strong reduction of CVB3 replication (Ye et al., 2011). Figures adapted with permission from Guo et al., 2005; Ye et al., 2011, © 2011 Public Library of Science.

TARGETED DELIVERY TO CANCER CELLS

The pRNA nanoparticles can carry both a therapeutic agent and a ligand for the targeted delivery of the nanoparticles to specific tissues and cell types. Incorporation of a receptor-binding aptamer, folate, or other ligands to the pRNA complex with simple procedures ensures the specific binding and targeted delivery to cells. In combination with the advantage of nanoscale size, the pRNA system provides for both higher delivery efficiency and reduced off-target toxicity (Abdelmawla et al., 2011).

NON-INDUCTION OF AN ANTIBODY RESPONSE TO ENSURE REPEATED TREATMENTS

Protein-free RNA nanoparticles, such as the pRNA system, contain RNA aptamers designed to act as receptor antagonists with similar binding specificities as protein antagonists. However, RNA nanoparticles have a much lower antibody-inducing activity (Abdelmawla et al., 2011). Thus, the repeated administration of RNA nanoparticles during the treatment of chronic diseases is less likely to result in complications as a result of immune responses.

CHALLENGES, SOLUTIONS, AND PERSPECTIVES IN RNA NANOTECHNOLOGY

Although great progress has been achieved by applying RNA nanotechnology in medical applications, many challenges still remain. Herein we provide some solutions and perspectives on the chemical and thermodynamic instability, short in vivo half-life and biodistribution, low yield and high production cost, in vivo toxicity and side effects, and specific delivery and targeting, as well as endosomal escape.

CHEMICAL INSTABILITY

One of the major concerns on the use of RNA nanoparticles in therapeutics is the chemical stability of RNA itself. Natural RNA is extremely sensitive to degradation by RNases and is especially unstable in the body or serum. The stability of RNA has long been an obstacle to its application as a construction material. Over the last few years, rapid progress has been made in improving the stability of RNA, which includes chemical modifications of the bases (e.g., 5-Br-Ura and 5-I-Ura), modifications of the phosphate linkage (e.g., phosphothioate, boranophosphate), alteration of the 2′ carbon (e.g., 2′-F, 2′OMe or 2′-NH$_2$) (Watts et al., 2008; Singh et al., 2010); synthesis of peptide nucleic acids, locked nucleic acids, and their respective derivatives; polycarbamate nucleic acids (Madhuri and Kumar, 2010) or locked nucleic acids with a bridge at different positions (2′-4′, 1′-3′) (Mathe and Perigaud, 2008), and capping of the 3′-end (Patra and Richert, 2009). All these methods are very efficient in increasing RNase resistance in vitro and in vivo. However, the challenge is that after chemical modification, the folding properties and biological function of a RNA molecule change. The development of a method that confers resistance to RNase degradation while not changing the characteristic structure, self-assembly, and biological function of a RNA nanoparticle is critical. It was recently found that for all the aforementioned methods, the 2′F have minimal detrimental effect on folding, assembly, and function (Liu et al., 2010). While in some special cases, finetuning is necessary to find a location that can be modified with minimal detrimental effect, RNase degradation in vivo is no longer a concern. It has been shown that RNA degradation in serum occurred more frequently at vulnerable sites. Finetuning of these sites by mutation or alteration protect siRNAs from degradation in serum (Hong et al., 2010).

THERMODYNAMIC INSTABILITY

The thermodynamic stability of the RNA nanoparticles is of paramount importance with regard to the use of RNA nanoparticles as therapeutics. Injection of several microliters or milliliters of RNA solution into the body will result in several hundred thousand-fold dilutions. Dissociation of bottom-up assembled RNA nanoparticles at extremely low concentrations after in vivo dilution will be a serious concern. In a recent paper, thermodynamically stable pRNA 3WJ core scaffold was assembled from 3 to 6 pieces of RNA in the absence of metal salts. The 3WJ complex was stable in serum, remained intact at ultra-low concentrations and was even resistant to denaturation in 8 M urea (Figure 10.5) (Shu et al., 2011a; Haque et al., 2012). More importantly, various functionalities such siRNA, ribozyme, or receptor-binding aptamer incorporated into the 3WJ core resulted in the formation of polyvalent particles displaying all the authentic functionalities in vitro and in vivo (Figure 10.6). Therefore, the thermodynamic stability and in vivo dissociation are no longer a concern for pRNA-based nanoparticles.

SHORT IN VIVO HALF-LIFE

The other important factor in therapeutics is the PK of the drug. In order to improve the stability of RNA in vivo, a variety of chemical modifications have been introduced into RNA, as discussed previously. Chemically modified siRNA are RNase resistant, while retaining their biological activity

(Liu et al., 2010). However, the half-life of modified siRNA in vivo is only 15–45 min (Morrissey et al., 2005; Behlke, 2006).

Another critical factor that determines the in vivo retention time is the size of the RNA nanoparticles. Many studies suggest that particles ranging from 10 to 100 nm are the optimal size for a non-viral vector—large enough to be retained by the body, yet small enough to bind to cell-surface receptors and pass through cell membranes (Prabha et al., 2002). RNA nanoparticles designed in the range between 20 and 40 nm are usually excluded from the normal tissue that has tighter blood vessels, but enter into the tumor interstitial space and concentrate in the tumor via enhanced permeability and retention effects, since the angiogenic blood vessels have larger gaps. This optimal size range for RNA nanoparticles also improves the biodistribution and ensures the longer retention time for in vivo delivery. It has been reported that the half-life of chemically modified pRNA nanoparticles was 5–10 hours, in comparison to 0.8 hours for the siRNA counterparts (Abdelmawla et al., 2011). So the concerns about the in vivo retention and half-life of RNA nanoparticles have been significantly reduced by the application of chemical modifications.

Low Yield and High Production Costs

A major limiting factor of RNA nanotechnology in therapeutic applications is the cost of the nanoparticle construction, especially for RNA nanoparticles that require larger RNAs. RNA oligonucleotides can be prepared by enzymatic transcription or automated solid-phase synthesis. Enzymatic synthesis can produce relatively long transcripts in significant quantities, while commercial non-enzymatic RNA chemical synthesis can only produce RNAs that are 40–80 nt long. The longest chemically synthetic RNA with biological activity is 117-nt long (Guo et al., unpublished data). When it comes to the synthesis of relatively long RNA oligonucleotides, the yield of a RNA oligo decreases greatly as the length of the oligo increases (Reese, 2002; Marshall and Kaiser, 2004).

Classical approaches based on the t-Butyldimethylsiloxy protecting group for the 2′-hydroxyl are limited to short sequences, whereas more recent approaches based on 5′-O-DMT-2′-O-[(trisisopropylsilyl)-oxy]methyl (2′-O-TOM) protecting scheme (Pitsch et al., 2001), and 5′-O-Silyl-2′-Oorthoester (2′-ACE) protecting group combination (Scaringe et al., 1998) have provided a more effective tool for the chemical synthesis of longer RNAs. The cost of RNA synthesis is expected to gradually drop with the development of industrial-scale RNA production techniques. For example, the cost of synthesizing DNA oligos was 100-fold higher 20 years ago compared to the cost today.

RNA can also be produced by enzymatic synthesis with in vitro transcription, but the heterogeneity of the 3′-end of the RNA products presents a problem. To work around the heterogeneity issue, the transcribed sequence can be extended beyond the intended end and then cleaved at the desired site using small ribozymes, DNAzyme, RNase H, or a cis-cleaving hammerhead or ribozyme (Feng et al., 2001; Hoeprich et al., 2003). RNase ligase 2 has also been shown to be a good alternative over the traditional T4 DNA ligase to obtain longer RNAs by the ligation of 2 shorter synthetic RNA fragments.

To circumvent this issue of yield and cost, a clever approach is to employ the bottom-up assembly (1 of the 2 basic approaches of nanotechnology) of the RNA nanoparticles. The production of oligonucleotides with functional moieties then becomes a scalable process (chemically) and with a modular design the complex can be self-assembled from the basic building blocks. Using this methodology, thermodynamically and chemically stable pRNA-based nanoparticles with functional modules were successfully fabricated using a bipartite, tripartite, and tetrapartite approach with various modifications (Shu et al., 2011a; Shu et al., 2011b).

The most economic way for industrial scale production of RNA is by fermentation in bacteria. Cloning and production of RNA in bacteria with high yield has been reported (Wichitwechkarn et al., 1992; Ponchon and Dardel, 2007, 2011; Ponchon et al., 2009; Delebecque et al., 2011). Bacteria fermentation is the direction for industry production, but currently, the bacteria high yield production of RNA nanoparticles with therapeutic functionality has not been reported.

TOXICITY, IN VIVO SAFETY, AND SIDE EFFECTS

From an in vivo delivery and therapeutic point of view, it is essential that the nanoparticles have favorable pharmacological profiles concerning biodistribution, pharmacokinetics (stability, half-life, and clearance rate), immune response (antibody induction, a and b interferon, toll-like and innate immunity, PKR effect, and cytokine induction), specific targeting, and efficiency of gene silencing. The induction of innate immunity and certain organ toxicity has been a major concern in using RNA nanoparticles for therapeutic applications. If the RNA is single stranded then type-II interferons (IFN-g) should be used as a marker for toxicity assay; if the RNA is double stranded it is type-I interferons (INF-a/IFN-b) that are important. Other immunotoxicity issues, such as hypersensitivity, complement activation, and fever-like reactions can all be dose-limiting factors. In addition, it has been reported that the immunotoxicity of siRNA is sequence specific. The potential toxic effects of delivery vehicles should also be explored.

Some of the favorable pharmacological profiles of the pRNA-based nanoparticles have been reported recently (Abdelmawla et al., 2011; Shu et al., 2011b). It is exciting to find that the half-life of the pRNA nanoparticles has been extended 10-fold (5–10 hours) in comparison with their regular siRNA counterparts (15–45 minutes) with a clearance rate of < 0.13 L/kg/hour and a volume of distribution of 1.2 L/kg. The pRNA nanoparticles induced neither an interferon response (OAS1, MX1, or IFITM1) nor cytokine production in mice, even after repeated intravenous administrations in mice, up to 30 mg/kg. Fluorescent folate-pRNA nanoparticles efficiently and specifically bound to and were internalized by folate receptor-bearing cancer cells in vitro. Systemic injection of pRNA nanoparticles specifically and dose-dependently targeted folate receptor [FR(+)] xenograft tumor in mice with minimal or no accumulation in normal tissues (Figure 10.6) (Abdelmawla et al., 2011; Shu et al., 2011b). However, rodents might not necessarily present the same effects of nucleic acid toxicity as humans. This type of toxicity should be further tested in preclinical studies using non-human primates.

SPECIFIC DELIVERY AND TARGETING PROBLEMS

In order for RNA nanoparticles to be useful as therapeutic agents, they must be capable of targeting specific cells. The pRNA nanoparticles in the 15–50 nm size range will not enter cells randomly or nonspecifically. Hence, a nanotechnological approach was adopted to construct pRNA nanoparticles harboring various targeting molecules, such as RNA aptamers or folate. One example is the specific delivery of pRNA nanoparticles to the folate receptor, overexpressed on the surface of nasopharyngeal carcinoma cells (KB cells). The folate moiety readily binds to the folate receptor and the nanoparticles rapidly enter the cells via receptor-mediated endocytosis (Figure 10.6) (Guo et al., 2005, 2006; Khaled et al., 2005; Shu et al., 2011a).

Although ligand-mediated specific delivery is an exciting approach, RNA aptamers and ligands currently available are limited. Each cancer type requires a specific ligand. As a result, development of more RNA aptamers for specific targeting is imperative. Screening methods derived from SELEX have shown considerable promise and are compatible with RNA nanotechnology approaches. Once an aptamer has been selected, it could easily be incorporated into RNA nanoparticles and retain their binding function in vivo. The development of such a screening system is being actively pursued.

ENDOSOME TRAPPING

One of the major problems encountered in DNA or protein delivery is degradation of the therapeutic molecules in the endocytic pathway. Similarly, the greatest challenge in ligand-mediated endocytosis for specific delivery of siRNA or therapeutic RNA nanoparticles is endosomal escape. After receptor-mediated endocytosis, the nanoparticles are trapped in the endosomes within the cells. This keeps the siRNA from being able to be processed by the Dicer machinery, and therefore the siRNA is not able to knock down a specifically targeted gene.

By taking advantage of the oligomerization properties of pRNA, we can deliver therapeutic molecules in a complex with endosome-disrupting agents. A number of substances that disrupt endosomes and mediate endosome escape of therapeutic molecules have been described in the literature. Defective or psoralen-inactivated adenovirus particles have shown promise since they have considerable endosomolytic activity (Cotten et al., 1992). Synthetic peptides that mimic the membrane-fusing region of the hemaglutinin of influenza virus have also been successfully used in gene delivery systems to facilitate endosomal escape (Plank et al., 1994; Mastrobattista et al., 2002; Van Rossenberg et al., 2002). Polymeric endosome-disrupting gene delivery vectors, such as poly(amino ester) (Lim et al., 2002) or poly(DL-lactide-co-glycolide) (Panyam et al., 2002) have also been reported. Polymers harboring varieties of chemical moieties have been reported to enhance disrupture of the endosome. Using the multivalent property of RNA nanoparticles, the endosome-escaping reagents can be incorporated into the RNA nanoparticles. One subunit of the deliverable RNA complex (dimer, trimer, or hexamer) can be altered to contain an RNA aptamer that acts as a ligand for a cell-surface receptor and induces uptake by receptor-mediated endocytosis upon binding. Another 1 or 2 subunits of the RNA complex can be altered to contain components that facilitate endosome disruption for the release of the delivered therapeutic molecules from the endosome. The other subunits of the RNA complex can be used to carry a therapeutic siRNA, a ribozyme, miRNA, a riboswitch, or a chemical drug.

Methods for assisting endosome escape include the use of synthetic polymers to form siRNA/polymer polyplexes, the complexation of siRNA with lipids to form lipid nanoparticles, or the siRNA association with cell-penetrating peptides (CPPs) or endosome-disrupting chemicals. Design considerations for the formation of siRNA/polymer polyplexes for endosomal release and gene-silencing efficiency have been recently reviewed (Kwon, 2011). For increased endosomal escape, a common technique is to create polymers with various types of acid-responsive chemical functional groups such as acid-cleavable linkers including acetal, hydrazone, and maleic amides or acid protonating groups such as b-amino esters, imidazole, and sulfonamide (Kwon, 2011). In the case of acid-cleavable linkers, the cationic termini of the polymer complexes become protonated within the acidic pH of the endosome, which can destabilize the endosome via the proton sponge effect. In addition, the loss of the cationic branches after acid hydrolysis reduces the interactions between the siRNA and polymer, resulting in release of the siRNA for Dicer processing (Kwon, 2011).

Another approach to siRNA delivery and endosomal disruption is to use CPPs. In most ways, CPPs deliver siRNA into cells in the same way as siRNA/polymer polyplexes and lipid nanoparticles. The anionic siRNA interact with cationic peptides, which enter the cell via endocytosis and carry the siRNA in along with them. Recently, amphipathic peptides have received a great deal of attention for nucleic acid delivery. These peptides are usually short and contain a high number of histidine and leucine residues. In a recent paper (Langlet-Bertin et al., 2010), the siRNA delivery efficiency of a cationic amphipathic peptide, LAH4, and several of its derivatives were compared. Each of these peptides was able to deliver siRNA for the luciferase gene to mammalian cells and was shown to be at least as effective as common lipid based transfection reagents such as lipofectamine. The exact mechanism of endosomal disruption by amphipathic peptides is still unknown; however, it is likely that that acidic pH of the endosome leads to protonation of the histidine residues, which frees the peptides from the siRNA they were carrying and allows them to disrupt the endosome via the proton sponge effect (Langlet-Bertin et al., 2010). The proton sponge effect occurs after accumulation of a weak base in the endosome, which neutralizes the lumen of the endosome and results in an increase in the endosome's osmolarity (Midoux et al., 2009). As a result, the endosome swells and is no longer able to hold its contents leading to siRNA being released into the cytosol.

The pH dependence of endosome escape was investigated in a recent report using a commercially available amphipathic peptide called Endo-Porter that has been used to deliver various nucleic acid cargos to mammalian cells (Bartz et al., 2011). It was shown that Endo-Porter requires the acidification of the endosome since the activity of the peptide was blocked by bafilomycin A. At physiological pH, Endo-Porter does not appear to form secondary structures such as the a-helix that it forms at pH between 5.0 and 6.0. The exact mechanism for how the a-helical structure is able to disrupt the endosome is unknown, but it may be a result of interaction between the endosome membrane and the a-helix resulting in the formation of a large pore in the endosome or disruption of its membrane (Bartz et al., 2011).

Although there are numerous types of nanoparticles capable of endosomal disruption, there is still a major challenge associated with their future use for widespread therapeutic use: specific targeting. Currently, synthetic polymer nanoparticles, lipid nanoparticles, and amphipathic peptides are all non-specific, which greatly limits their usefulness in vivo. One of the advantages of RNA nanoparticles is the use of targeting moieties such as RNA aptamers or receptor-targeting ligands, such as folate, but endosome escape is still an issue. Both types of delivery have their associated challenges to overcome. If combined into one particle, it may be possible to create a specific targeting RNA nanoparticle capable of escaping the endosome and paving the way toward a new and powerful form of RNA therapeutics.

CONCLUSIONS

The self-assembly property of RNA can be utilized as a powerful bottom-up approach to rationally design and create nanostructures through the integration of biological, chemical, physical, and computational techniques. This approach relies upon the cooperative interactions of individual RNA subunits to spontaneously self-assemble into larger multimeric structures in a predefined manner. The feasibility and utility of RNA nanotechnology in therapeutics is beginning to be realized. Currently, the RNA therapeutics industry faces the following challenges: (1) chemical instability, (2) thermodynamic instability, (3) short in vivo half-life, (4) low yield and high production costs, (5) in vivo safety and side effect issues, (6) difficulty in specific delivery and targeting in vivo, and (7) endosome trapping after delivery into the cell. The first 5 challenges have been overcome to a significant extent. Specific targeting and endosome escape are still a major issue that remains hurdles and make many companies cautious about aggressively pursuing the use of RNA as therapeutics. Cancer targeting has progressed with the approach of RNA nanotechnology, but greater efforts are needed to tailor the delivery efficiency and specificity to individual cancers. Combination of RNA with other chemical polymers to improve endosome escape has faced the challenges of undefined structure and stoichiometry as well as nanoparticle accumulation in normal organs. Many research groups, including our own, are working to solve the endosome escape issues. Once these challenges are overcome, the field of RNA nanotechnology and therapeutics will surely become a clinical reality.

ACKNOWLEDGMENTS

We thank Yi Shu, Daniel Binzel, and Zhanxi Hao for their assistance in the preparation of this manuscript. We thank Dr. Markos Leggas for helpful discussions. The research was supported by NIH R01 EB003730, R01 EB012135, U01 CA151648, R01 GM059944, and NIH Nanomedicine Development Center: Phi29 DNA Packaging Motor for Nanomedicine, through the NIH Roadmap for Medical Research (PN2 EY 018230) directed by P. Guo.

AUTHOR DISCLOSURE STATEMENT

Guo is a cofounder of Kylin Therapeutics, Inc, and Biomotor and Nucleic Acids Nanotech Development, Ltd.

REFERENCES

Abdelmawla, S., Guo S., Zhang, L., Pulukuri, S., Patankar P., Conley, P., Trebley, J., Guo, P., and Li, Q.X. (2011). Pharmacological characterization of chemically synthesized monomeric pRNA nanoparticles for systemic delivery. *Mol. Ther.* 19, 1312–1322.

Afonin, K.A., Bindewald, E., Yaghoubian, A.J., Voss, N., Jacovetty, E., Shapiro, B.A., and Jaeger, L. (2010). In vitro assembly of cubic RNA-based scaffolds designed in silico. *Nat. Nanotechnol.* 5, 676–682.

Afonin, K.A., Grabow, W.W., Walker, F.M., Bindewald, E., Dobrovolskaia, M.A., Shapiro, B.A., and Jaeger, L. (2011). Design and self-assembly of siRNA-functionalized RNA nanoparticles for use in automated nanomedicine. *Nat. Protoc.* 6, 2022–2034.

Bader, A.G., Brown, D., and Winkler, M. (2010). The promise of microRNA replacement therapy. *Cancer Res.* 70, 7027–7030.

Banta, S., Megeed, Z., Casali, M., Rege, K., and Yarmush, M.L. (2007). Engineering protein and peptide building blocks for nanotechnology. *J. Nanosci. Nanotechnol.* 7, 387–401.

Barrick, J.E., and Breaker, R.R. (2007). The distributions, mechanisms, and structures of metabolite-binding riboswitches. *Genome Biol.* 8, R239.

Bartel, D.P. (2009). MicroRNAs: target recognition and regulatory functions. *Cell* 136, 215–233.

Bartz, R., Fan, H., Zhang, J., Innocent, N., Cherrin, C., Beck, S.C., Pei, Y., Momose, A., Jadhav, V., Tellers, D.M., Meng, F., Crocker, L.S., Sepp-Lorenzino, L., and Barnett, S.F. (2011). Effective siRNA delivery and target mRNA degradation using an amphipathic peptide to facilitate pH-dependent endosomal escape. *Biochem. J.* 435, 475–487.

Behlke, M.A. (2006). Progress towards in vivo use of siRNAs. *Mol. Ther.* 13, 644–670.

Bindewald, E., Afonin, K., Jaeger, L., and Shapiro, B.A. (2011). Multistrand RNA secondary structure prediction and nanostructure design including pseudoknots. *ACS Nano.* 5, 9542–9551.

Bindewald, E., Grunewald, C., Boyle, B., O'Connor, M., and Shapiro, B. A. (2008a). Computational strategies for the automated design of RNA nanoscale structures from building blocks using NanoTiler. *J. Mol. Graph. Model.* 27, 299–308.

Bindewald, E., Hayes, R., Yingling, Y.G., Kasprzak, W., and Shapiro, B.A. (2008b). RNAJunction: a database of RNA junctions and kissing loops for three-dimensional structural analysis and nanodesign. *Nucleic Acids Res.* 36, D392–D397.

Breaker, R.R. (2012). Riboswitches and the RNA world. *Cold Spring Harb. Perspect. Biol.* 4, a003566.

Breaker, R.R. (2008). Complex riboswitches. *Science* 319, 1795–1797.

Brummelkamp, T.R., Bernards, R., and Agami, R. (2002). A system for stable expression of short interfering RNAs in mammalian cells. *Science* 296, 550–553.

Carmichael, G.G. (2002). Medicine: silencing viruses with RNA. *Nature* 418, 379–380.

Cayrol, B., Nogues, C., Dawid, A., Sagi, I., Silberzan, P., and Isambert, H. (2009). A nanostructure made of a bacterial noncoding RNA. *J. Am. Chem. Soc.* 131, 17270–17276.

Chang, C.I., Lee, T.Y., Kim, S., Sun, X., Hong, S.W., Yoo, J.W., Dua, P., Kang, H.S., Kim, S., Li, C.J., and Lee, D.K. (2012a). Enhanced intracellular delivery and multi-target gene silencing triggered by tripodal RNA structures. *J. Gene Med.* 14, 138–146.

Chang, C.I., Lee, T.Y., Yoo, J.W., Shin, D., Kim, M., Kim, S., and Lee, D.K. (2012b). Branched, tripartite-interfering RNAs silence multiple target genes with long guide strands. *Nucleic Acid Ther.* 22, 30–39.

Chang, K.Y. and Tinoco, I. (1994). Characterization of a "kissing" hairpin complex derived from the human immunodeficiency virus genome. *Proc. Natl. Acad. Sci. U. S. A.* 91, 8705–8709.

Cheah, M.T., Wachter, A., Sudarsan, N., and Breaker, R.R. (2007). Control of alternative RNA splicing and gene expression by eukaryotic riboswitches. *Nature* 447, 497–500.

Chen, C., Sheng, S., Shao, Z., and Guo, P. (2000). A dimer as a building block in assembling RNA: A hexamer that gears bacterial virus phi29 DNA-translocating machinery. *J. Biol. Chem.* 275, 17510–17516.

Chen, C., Zhang, C., and Guo, P. (1999). Sequence requirement for hand-in-hand interaction in formation of pRNA dimers and hexamers to gear phi29 DNA translocation motor. *RNA* 5, 805–818.

Chen, J.F., Murchison, E.P., Tang, R., Callis, T.E., Tatsuguchi, M., Deng, Z., Rojas, M., Hammond, S.M., Schneider, M.D., Selzman, C.H., Meissner, G., Patterson, C., Hannon, G.J., and Wang, D.Z. (2008). Targeted deletion of Dicer in the heart leads to dilated cardiomyopathy and heart failure. *Proc. Natl. Acad. Sci. U. S. A.* 105, 2111–2116.

Chowrira, B.M., Berzal-Herranz, A., and Burke, J.M. (1991). Novel guanosine requirement for catalysis by the hairpin ribozyme. *Nature* 354, 320–322.

Chworos, A., Severcan, I., Koyfman, A.Y., Weinkam, P., Oroudjev, E., Hansma, H.G., and Jaeger, L. (2004). Building programmable jigsaw puzzles with RNA. *Science* 306, 2068–2072.

Clever, J.L., Wong, M.L., and Parslow, T.G. (1996). Requirements for kissing-loop-mediated dimerization of human immunodeficiency virus RNA. *J. Virol.* 70, 5902–5908.

Coleman, J., Hirashima, A., Inocuchi, Y., Green, P.J., and Inouye, M. (1985). A novel immune system against bacteriophage infection using complementary RNA (micRNA). *Nature* 315, 601–603.

Cotten, M., Wagner, E., Zatloukal, K., Phillips, S., Curiel, D.T., and Birnstiel, M.L. (1992). High-efficiency receptor-mediated delivery of small and large 48 kilobase gene constructs using the endosome-disruption activity of defective or chemically inactivated adenovirus particles. *Proc. Natl. Acad. Sci. U. S. A.* 89, 6094–6098.

Dassie, J.P., Liu, X.Y., Thomas, G.S., Whitaker, R.M., Thiel, K.W., Stockdale, K.R., Meyerholz, D.K., McCaffrey, A.P., McNamara, J.O., and Giangrande, P.H. (2009). Systemic administration of optimized aptamersiRNA chimeras promotes regression of PSMA-expressing tumors. *Nat. Biotechnol.* 27, 839–849.

de Fougerolles, A., Vornlocher, H.P., Maraganore, J., and Lieberman, J. (2007). Interfering with disease: a progress report on siRNA-based therapeutics. *Nat. Rev. Drug Discov.* 6, 443–453.

Delebecque, C.J., Lindner, A.B., Silver, P.A., and Aldaye, F.A. (2011). Organization of intracellular reactions with rationally designed RNA assemblies. *Science* 333, 470–474.

Dibrov, S.M., McLean, J., Parsons, J., and Hermann, T. (2011). Self-assembling RNA square. *Proc. Natl. Acad. Sci. U. S. A.* 108, 6405–6408.

Doermann, A.H. (1973). T4 and the rolling circle model of replication. *Annu. Rev. Genet.* 7, 325–341.

Duncan, R. (2011). Polymer therapeutics as nanomedicines: new perspectives. *Curr. Opin. Biotechnol.* 22, 492–501.

Editorial Comment (2011). The story so far: basic research in nanoscience and technology is flourishing, but obstacles to real-world applications remain. *Nat. Nanotechnol.* 6, 603.

Ehresmann, C., Baudin, F., Mougel, M., Romby, P., Ebel, J.-P., and Ehresmann, B. (1987). Probing the structure of RNAs in solution. *Nucleic Acids Res.* 15, 9109–9128.

Ellington, A.D. and Szostak, J.W. (1990). In vitro selection of RNA molecules that bind specific ligands. *Nature* 346, 818– 822.

Feng, Y., Kong, Y.Y., Wang, Y., and Qi, G.R. (2001). Inhibition of hepatitis B virus by hammerhead ribozyme targeted to the poly(A) signal sequence in cultured cells. *Biol. Chem.* 382, 655–660.

Fire, A., Xu, S., Montgomery, M.K., Kostas, S.A., Driver, S.E., and Mello, C.C. (1998). Potent and specific genetic interference by double-stranded RNA in Caenorhabditis elegans. *Nature* 391, 806–811.

Forster, A.C., and Symons, R.H. (1987). Self-cleavage of virusoid RNA is performed by the proposed 55-nucleotide active site. *Cell* 50, 9–16.

Freier, S.M., Kierzek, R., Jaeger, J.A., Sugimoto, N., Caruthers, M.H., Neilson, T., and Turner, D.H. (1986). Improved free-energy parameters for predictions of RNA duplex stability. *Proc. Natl. Acad. Sci. U. S. A.* 83, 9373–9377.

Gao, H., Shi, W., and Freund, L.B. (2005). Mechanics of receptor-mediated endocytosis. *Proc. Natl. Acad. Sci. U. S. A.* 102, 9469–9474.

Geary, C., Chworos, A., and Jaeger, L. (2010). Promoting RNA helical stacking via A-minor junctions. *Nucleic Acids Res.* 39, 1066–1080.

Grabow, W.W., Zakrevsky, P., Afonin, K.A., Chworos, A., Shapiro, B.A., and Jaeger, L. (2011). Selfassembling RNA nanorings based on RNAI/II inverse kissing complexes. *Nano Lett.* 11, 878–887.

Guerrier-Takada, C., Gardiner, K., Marsh, T., Pace, N., and Altman, S. (1983). The RNA moiety of ribonuclease P is the catalytic subunit of the enzyme. *Cell* 35, 849–857.

Guo, P. (2005). RNA nanotechnology: engineering, assembly, and applications in detection, gene delivery and therapy. *J. Nanosci. Nanotech* 5, 1964–1982.

Guo, P. (2010). The emerging field of RNA nanotechnology. *Nat. Nanotechnol.* 5, 833–842.

Guo, P., Coban, O., Snead, N.M., Trebley, J., Hoeprich, S., Guo, S., and Shu, Y. (2010). Engineering RNA for targeted siRNA delivery and medical application. *Adv. Drug Deliv. Rev.* 62, 650–666.

Guo, P., Erickson, S., and Anderson, D. (1987). A small viral RNA is required for in vitro packaging of bacteriophage phi29 DNA. *Science* 236, 690–694.

Guo, P., Zhang, C., Chen, C., Trottier, M., and Garver, K. (1998). Inter-RNA interaction of phage phi29 pRNA to form a hexameric complex for viral DNA transportation. *Mol. Cell* 2, 149–155.

Guo, S., Tschammer, N., Mohammed, S., and Guo, P. (2005). Specific delivery of therapeutic RNAs to cancer cells via the dimerization mechanism of phi29 motor pRNA. *Hum. Gene Ther.* 16, 1097–1109.

Guo, S., Huang, F., and Guo, P. (2006). Construction of folate-conjugated pRNA of bacteriophage phi29 DNA packaging motor for delivery of chimeric siRNA to nasopharyngeal carcinoma cells. *Gene Ther.* 13, 814–820.

Gyi, J.I., Conn, G.L., Lane, A.N., and Brown, T. (1996). Comparison of the thermodynamic stabilities and solution conformations of DNA center dot RNA hybrids containing purine-rich and pyrimidine-rich strands with DNA and RNA duplexes. *Biochemistry* 35, 12538–12548.

Hamilton, A.J., and Baulcombe, D.C. (1999). A species of small antisense RNA in posttranscriptional gene silencing in plants. *Science* 286, 950–952.

Haque, F., Shu, D., Shu, Y., Shlyakhtenko, L., Rychahou, P., Evers, M., and Guo, P. (2012). Ultrastable synergistic tetravalent RNA nanoparticles for targeting to cancers. *Nano Today* 2012. DOI: 10.1016/j.nantod.2012.06.010.

He, L., and Hannon, G.J. (2004). MicroRNAs: small RNAs with a big role in gene regulation. *Nat. Rev. Genet.* 5, 522–531.

Hendrix, R.W. (1998). Bacteriophage DNA packaging: RNA gears in a DNA transport machine (Minireview). *Cell* 94, 147–150.

Henkin, T.M. (2008). Riboswitch RNAs: using RNA to sense cellular metabolism. *Genes Dev.* 22, 3383–3390.

Hoeprich, S., Zhou, Q., Guo, S., Qi, G., Wang, Y., and Guo, P. (2003). Bacterial virus phi29 pRNA as a hammerhead ribozyme escort to destroy hepatitis B virus. *Gene Ther.* 10, 1258–1267.

Hong, J., Huang, Y., Li, J., Yi, F., Zheng, J., Huang, H., Wei, N., Shan, Y., An, M., Zhang, H., Ji, J., Zhang, P., Xi, Z., Du, Q., and Liang, Z. (2010). Comprehensive analysis of sequence-specific stability of siRNA. *FASEB J.* 24, 4844–4855.

Ikawa, Y., Tsuda, K., Matsumura, S., and Inoue, T. (2004). De novo synthesis and development of an RNA enzyme. *Proc. Natl. Acad. Sci. U. S. A.* 101, 13750–13755.

Isambert, H. (2009). The jerky and knotty dynamics of RNA. *Methods* 49, 189–196.

Jacque, J.M., Triques, K., and Stevenson, M. (2002). Modulation of HIV-1 replication by RNA interference. *Nature* 418, 435–438.

Jaeger, L., and Chworos, A. (2006). The architectonics of programmable RNA and DNA nanostructures. *Curr. Opin. Struct. Biol.* 16, 531–543.

Jaeger, L. and Leontis, N.B. (2000). Tecto-RNA: one dimensional self-assembly through tertiary interactions. *Angew Chem. Int. Ed. Engl.* 39, 2521–2524.

Jaeger, L., Westhof, E., and Leontis, N.B. (2001). TectoRNA: modular assembly units for the construction of RNA nano-objects. *Nucleic Acids Res.* 29, 455–463.

Jain, K.K. (2005). The role of nanobiotechnology in drug discovery. *Drug Discov. Today* 10, 1435–1442.

Jossinet, F., Ludwig, T.E., and Westhof, E. (2007). RNA structure: bioinformatic analysis. *Curr. Opin. Microbiol.* 10, 279–285.

Kasinski, A.L., and Slack, F.J. (2011). Epigenetics and genetics. MicroRNAs en route to the clinic: progress in validating and targeting microRNAs for cancer therapy. *Nat. Rev. Cancer* 11, 849–864.

Kasprzak, W., Bindewald, E., Kim, T.J., Jaeger, L., and Shapiro, B.A. (2010). Use of RNA structure flexibility data in nanostructure modeling. *Methods* 54, 239–250.

Khaled, A., Guo, S., Li, F., and Guo, P. (2005). Controllable self-assembly of nanoparticles for specific delivery of multiple therapeutic molecules to cancer cells using RNA nanotechnology. *Nano Lett.* 5, 1797–1808.

Kim, D.H., and Rossi, J.J. (2007). Strategies for silencing human disease using RNA interference. *Nat. Rev. Genet.* 8, 173–184.

Knecht, D.A., and Loomis, W.F. (1987). Antisense RNA inactivation of myosin heavy chain gene expression in *Dictyostelium discoideum*. *Science* 236, 1081–1086.

Kruger, K., Grabowski, P.J., Zaug, A.J., Sands, J., Gottschling, D.E., and Cech, T.R. (1982). Self-splicing RNA: autoexcision and autocyclization of the ribosomal RNA intervening sequence of Tetrahymena. *Cell* 31, 147–157.

Kwon, Y.J. (2011). Before and after endosomal escape: roles of stimuli-converting siRNA/Polymer interactions in determining gene silencing efficiency. *Acc. Chem. Res.* 45, 1077–1088.

Laing, C., and Schlick, T. (2010). Computational approaches to 3D modeling of RNA. *J. Phys. Condens. Matter* 22, 283101.

Langlet-Bertin, B., Leborgne, C., Scherman, D., Bechinger, B., Mason, A.J., and Kichler, A. (2010). Design and evaluation of histidine-rich amphipathic peptides for siRNA delivery. *Pharm. Res.* 27, 1426–1436.

Laurenti, E., Barde, I., Verp, S., Offner, S., Wilson, A., Quenneville, S., Wiznerowicz, M., Macdonald, H.R., Trono, D., and Trumpp, A. (2010). Inducible gene and shRNA expression in resident hematopoietic stem cells in vivo. *Stem Cells* 28, 1390–1398.

Lee, J.B., Hong, J., Bonner, D.K., Poon, Z., and Hammond, P.T. (2012). Self-assembled RNA interference microsponges for efficient siRNA delivery. *Nat. Mater.* 11, 316–322.

Leontis, N.B., Lescoute, A., and Westhof, E. (2006). The building blocks and motifs of RNA architecture. *Curr. Opin. Struct. Biol.* 16, 279–287.

Leontis, N.B., and Westhof, E. (2003). Analysis of RNA motifs. *Curr. Opin. Struct. Biol.* 13, 300–308.

Lescoute, A., and Westhof, E. (2006). Topology of threeway junctions in folded RNAs. *RNA* 12, 83–93.

Lesnik, E.A., and Freier, S.M. (1995). Relative thermodynamic stability of DNA, RNA, and DNA-RNA hybrid duplexes: relationship with base composition and structure. *Biochemistry* 34, 10807–10815.

Li, H., Li, W.X., and Ding, S.W. (2002). Induction and suppression of RNA silencing by an animal virus. *Science* 296, 1319–1321.

Li, W., and Szoka, F. (2007). Lipid-based nanoparticles for nucleic acid delivery. *Pharm Res*, 24, 438–449.

Li, X., Horiya, S., and Harada, K. (2006). An efficient thermally induced RNA conformational switch as a framework for the functionalization of RNA nanostructures. *J. Am. Chem. Soc.*, 128, 4035–4040.

Lilley, D.M. (1999). Structure, folding, and catalysis of the small nucleolytic ribozymes. *Curr. Opin. Struct. Biol.* 9, 330–338.

Lim, Y.B., Kim, S.M., Suh, H., and Park, J.S. (2002). Biodegradable, endosome disruptive, and cationic network-type polymer as a highly efficient and nontoxic gene delivery carrier. *Bioconjug. Chem.* 13, 952–957.

Liu, J., Guo, S., Cinier, M., Shlyakhtenko, L., Shu, Y., Chen, C., Shen, G., and Guo, P. (2010). Fabrication of stable and RNase-resistant RNA nanoparticles active in gearing the nanomotors for viral DNA packaging. *ACS Nano.* 5, 237–246.

Liu, H., Guo, S., Roll, R., Li, J., Diao, Z., Shao, N., Riley, M.R., Cole, A.M., Robinson, J.P., Snead, N.M., Shen, G., and Guo, P. (2007). Phi29 pRNA vector for efficient escort of hammerhead ribozyme targeting survivin in multiple cancer cells. *Cancer Biol. Ther.* 6, 697–704.

Madhuri, V., and Kumar, V.A. (2010). Design, synthesis, and DNA/RNA binding studies of nucleic acids comprising stereoregular and acyclic polycarbamate backbone: polycarbamate nucleic acids (PCNA). *Org. Biomol. Chem.* 8, 3734–3741.

Maes, O.C., Chertkow, H.M., Wang, E., and Schipper, H.M. (2009). MicroRNA: Implications for Alzheimer disease and other human CNS disorders. *Curr Genomics*, 10, 154–168.

Marshall, W.S., and Kaiser, R.J. (2004). Recent advances in the high-speed solid phase synthesis of RNA. *Curr. Opin. Chem. Biol.* 8, 222–229.

Mastrobattista, E., Koning, G.A., Van Bloois, L., Filipe, A.C., Jiskoot, W., and Storm, G. (2002). Functional characterization of an endosome-disruptive peptide and its application in cytosolic delivery of immuno-liposome-entrapped proteins. *J. Biol. Chem.* 277, 27135–27143.

Mathe, C., and Perigaud, C. (2008). Recent approaches in the synthesis of conformationally restricted nucleoside analogues. *European J. Org. Chem.* 2008, 1489–1505.

Mathews, D.H., and Turner, D.H. (2006). Prediction of RNA secondary structure by free energy minimization. *Curr. Opin. Struct. Biol.* 16, 270–278.

Matsumura, S., Ohmori, R., Saito, H., Ikawa, Y., and Inoue, T. (2009). Coordinated control of a designed trans-acting ligase ribozyme by a loop-receptor interaction. *FEBS Lett.* 583, 2819–2826.

McKinney, S.A., Declais, A.C., Lilley, D.M.J., and Ha, T. (2003). Structural dynamics of individual Holliday junctions. *Nat. Struct. Biol.* 10, 93–97.

McNamara, J.O., Andrechek, E.R., Wang, Y., Viles, K.D., Rempel, R.E., Gilboa, E., Sullenger, B.A., and Giangrande, P.H. (2006). Cell type-specific delivery of siRNAs with aptamer-siRNA chimeras. *Nat. Biotech.* 24, 1005–1015.

Mi, J., Liu, Y., Rabbani, Z.N., Yang, Z., Urban, J.H., Sullenger, B.A., and Clary, B.M. (2010). In vivo selection of tumor-targeting RNA motifs. *Nat. Chem. Biol.* 6, 22–24.

Midoux, P., Pichon, C., Yaouanc, J.J., and Jaffres, P.A. (2009). Chemical vectors for gene delivery: a current review on polymers, peptides and lipids containing histidine or imidazole as nucleic acids carriers. *Br J Pharmacol*, 157, 166–178.

Moll, D., and Guo, P. (2007). Grouping of ferritin and gold nanoparticles conjugated to pRNA of the phage phi29 DNApackaging motor. *J. Nanosci. Nanotech.* 7, 3257–3267.

Moll, D., Huber, C., Schlegel, B., Pum, D., Sleytr, U.B., and Sara, M. (2002). S-layer-streptavidin fusion proteins as template for nanopatterned molecular arrays. *Proc. Natl. Acad. Sci. U. S. A.* 99, 14646–14651.

Morrissey, D.V., Lockridge, J.A., Shaw, L., Blanchard, K., Jensen, K., Breen, W., Hartsough, K., Machemer, L., Radka, S., et al. (2005). Potent and persistent in vivo anti-HBV activity of chemically modified siRNAs. *Nat Biotechnol*, 23, 1002–1007.

Mujeeb, A., Clever, J.L., Billeci, T.M., James, T.L., and Parslow, T.G. (1998). Structure of the dimer initiation complex of HIV-1 genomic RNA. *Nat. Struct. Biol.* 5, 432–436.

Nakashima, Y., Abe, H., Abe, N., Aikawa, K., and Ito, Y. (2011). Branched RNA nanostructures for RNA interference. *Chem. Commun.* (Camb.) 47, 8367–8369.

Nasalean, L., Baudrey, S., Leontis, N.B., and Jaeger, L. (2006). Controlling RNA self-assembly to form filaments. *Nucleic Acids Res.* 34, 1381–1392.

Nimesh, S., Gupta, N., and Chandra, R. (2011). Cationic polymer based nanocarriers for delivery of therapeutic nucleic acids. *J. Biomed. Nanotechnol.* 7, 504–520.

Ogawa, A. and Maeda, M. (2008). An artificial aptazyme-based riboswitch and its cascading system in *E. coli*. *Chembiochem*. 9, 206–209.

Ohno, H., Kobayashi, T., Kabata, R., Endo, K., Iwasa, T., Yoshimura, S.H., Takeyasu, K., Inoue, T., and Saito, H. (2011). Synthetic RNA-protein complex shaped like an equilateral triangle. *Nat. Nanotechnol*. 6, 116–120.

Panyam, J., Zhou, W.Z., Prabha, S., Sahoo, S.K., and Labhasetwar, V. (2002). Rapid endo-lysosomal escape of poly(DL-lactide-co-glycolide) nanoparticles: implications for drug and gene delivery. *FASEB J*. 16, 1217–1226.

Patra, A., and Richert, C. (2009). High fidelity base pairing at the 3'-terminus. *J. Am. Chem. Soc*. 131, 12671–12681.

Pitsch, S., Weiss, P.A., Jenny, L., Stutz, A., and Wu, X. (2001). Reliable chemical synthesis of oligoribonucleotides (RNA) with 2'-O-[(trisisopropylsilyl)oxy]methyl (2'-O-tom)protected phosphoramidites. *Helv. Chim. Acta* 84, 3773–3795.

Plank, C., Oberhauser, B., Mechtler, K., Koch, C., and Wagner, E. (1994). The influence of endosome-disruptive peptides on gene transfer using synthetic virus-like gene transfer systems. *J. Biol. Chem*. 269, 12918–12924.

Pleij, C.W., and Bosch, L. (1989). RNA pseudoknots: structure, detection, and prediction. *Meth. Enzymol*. 180, 289–303.

Pleij, C.W.A., Rietveld, K., and Bosch, L. (1985). A new principle of RNA folding based on pseudoknotting. *Nucleic Acids Res*. 13, 1717.

Ponchon, L., and Dardel, F. (2007). Recombinant RNA technology; the tRNA scaffold. *Nat. Methods* 4, 571–576.

Ponchon, L., Beauvais, G., Nonin-Lecomte, S., and Dardel, F. (2009). A generic protocol for the expression and purification of recombinant RNA in *Escherichia coli* using a tRNA scaffold. *Nat. Protoc*., 4, 947–959.

Ponchon, L., and Dardel, F. (2011). Large scale expression and purification of recombinant RNA in *Escherichia coli*. *Methods* 54, 267–273.

Prabha, S., Zhou, W.Z., Panyam, J., and Labhasetwar, V. (2002). Size-dependency of nanoparticle-mediated gene transfection: studies with fractionated nanoparticles. *Int. J. Pharm*. 244, 105–115.

Prats, A.C., Roy, C., Wang, P.A., Erard, M., Housset, V., Gabus, C., Paoletti, C., and Darlix, J.L. (1990). Cis elements and trans-acting factors involved in dimer formation of murine leukemia virus RNA. *J. Virol*. 64, 774–783.

Privalov, P.L., and Filiminov, V.V. (1978). Thermodynamic analysis of transfer RNA unfolding. *J. Mol. Biol*. 122, 447–464.

Rajagopal, K., and Schneider, J.P. (2004). Self-assembling peptides and proteins for nanotechnological applications. *Curr. Opin. Struct. Biol*. 14, 480–486.

Reese, C.B. (2002). The chemical synthesis of oligo- and polynucleotides: a personal commentary. *Tetrahedron* 58, 8893–8920.

Reid, B.R. (1981). NMR studies on RNA structure and dynamics. *Annu. Rev. Biochem*. 50, 969–996.

Reid, R.J.D., Zhang, F., Benson, S., and Anderson, D. (1994). Probing the structure of bacteriophage phi29 prohead RNA with specific mutations. *J. Biol. Chem*. 269, 18656–18661.

Rozema, D.B., Lewis, D.L., Wakefield, D.H., Wong, S.C., Klein, J.J., Roesch, P.L., Bertin, S.L., Reppen, T.W., Chu, Q., et al. (2007). Dynamic polyconjugates for targeted in vivo delivery of siRNA to hepatocytes. *Proc. Natl. Acad. Sci. U. S. A*. 104, 12982–12987.

Sarver, N.A., Cantin, E.M., Chang, P.S., Zaia, J.A., Ladne, P.A., Stephens, D.A., and Rossi, J.J. (1990). Ribozymes as potential anti-HIV-1 therapeutic agents. *Science* 247, 1222–1225.

Scaringe, S.A., Wincott, F.E., and Caruthers, M.H. (1998). Novel RNA synthesis method using 5'-O-silyl-2'-Oorthoester protecting groups. *J. Am. Chem. Soc*. 120, 11820–11821.

Schroeder, K.T., McPhee, S.A., Ouellet, J., and Lilley, D.M. (2010). A structural database for k-turn motifs in RNA. *RNA* 16, 1463–1468.

Searle, M.S., and Williams, D.H. (1993). On the stability of nucleic acid structures in solution: enthalpy-entropy compensations, internal rotations, and reversibility. *Nucleic Acids Res*. 21, 2051–2056.

Seeman, N.C. (2010). Nanomaterials based on DNA. *Annu. Rev. Biochem*. 79, 65–87.

Severcan, I., Geary, C., Chworos, A., Voss, N., Jacovetty, E., and Jaeger, L. (2010). A polyhedron made of tRNAs. *Nat. Chem*. 2, 772–779.

Severcan, I., Geary, C., Verzemnieks, E., Chworos, A., and Jaeger, L. (2009). Square-shaped RNA particles from different RNA folds. *Nano Lett*. 9, 1270–1277.

Shahbabian, K., Jamalli, A., Zig, L., and Putzer, H. (2009). RNase Y, a novel endoribonuclease, initiates riboswitch turnover in *Bacillus subtilis*. *EMBO J*. 28, 3523–3533.

Shapiro, B.A., Bindewald, E., Kasprzak, W., and Yingling, Y. (2008). Protocols for the in silico design of RNA nanostructures. *Methods Mol. Biol.* 474, 93–115.

Shapiro, B.A., Yingling, Y.G., Kasprzak, W., and Bindewald, E. (2007). Bridging the gap in RNA structure prediction. *Curr. Opin. Struct. Biol.* 17, 157–165.

Shu, D., Huang, L., Hoeprich, S., and Guo, P. (2003). Construction of phi29 DNA-packaging RNA (pRNA) monomers, dimers and trimers with variable sizes and shapes as potential parts for nano-devices. *J. Nanosci. Nanotechnol.* 3, 295–302.

Shu, D., Moll, W.D., Deng, Z., Mao, C., and Guo, P. (2004). Bottom-up assembly of RNA arrays and super-structures as potential parts in nanotechnology. *Nano Lett.* 4, 1717–1723.

Shu, D., Shu, Y., Haque, F., Abdelmawla, S., and Guo, P. (2011a). Thermodynamically stable RNA three-way junctions as platform for constructing multifunctional nanoparticles for delivery of therapeutics. *Nat. Nanotechnol.* 6, 658–667.

Shu, Y., Cinier, M., Fox, S.R., Ben-Johnathan, N., and Guo, P. (2011b). Assembly of therapeutic pRNA-siRNA nanoparticles using bipartite approach. *Mol. Ther.* 19, 1304–1311.

Shu, Y., Cinier, M., Shu, D., and Guo, P. (2011c). Assembly of multifunctional phi29 pRNA nanoparticles for specific delivery of siRNA and other therapeutics to targeted cells. *Methods* 54, 204–214.

Shu, Y., Shu, D., Diao, Z., Shen, G., and Guo, P. (2009). Fabrication of polyvalent therapeutic RNA nanoparticles for specific delivery of siRNA, ribozyme, and drugs to targeted cells for cancer therapy. *IEEE NIH Life Sci. Syst. Appl. Workshop* 2009, 9–12.

Shukla, G.C., Haque, F., Tor, Y., Wilhelmsson, L.M., Toulme, J.J., Isambert, H., Guo, P., Rossi, J.J., Tenenbaum, S.A., and Shapiro, B.A. (2011). A boost for the emerging field of RNA nanotechnology. *ACS Nano.* 5, 3405–3418.

Singh, Y., Murat, P., and Defrancq, E. (2010). Recent developments in oligonucleotide conjugation. *Chem. Soc. Rev.* 39, 2054–2070.

Singha, K., Namgung, R., and Kim, W.J. (2011). Polymers in small-interfering RNA delivery. *Nucleic Acid Ther.* 21, 133–147.

Studnicka, G.M., Rahn, G.M., Cummings, I.W., and Salser, W.A. (1978). Computer method for predicting the secondary structure of single-stranded RNA. *Nucleic Acids Res.* 5, 3365–3387.

Sugimoto, N., Nakano, S., Katoh, M., Matsumura, A., Nakamuta, H., Ohmichi, T., Yoneyama, M., and Sasaki, M. (1995). Thermodynamic parameters to predict stability of RNA/DNA hybrid duplexes. *Biochemistry* 34, 11211–11216.

Troiber, C., and Wagner, E. (2011). Nucleic acid carriers based on precise polymer conjugates. *Bioconjug. Chem.* 22, 1737–1752.

Trottier, M., Mat-Arip, Y., Zhang, C., Chen, C., Sheng, S., Shao, Z., and Guo, P. (2000). Probing the structure of monomers and dimers of the bacterial virus phi29 hexamer RNA complex by chemical modification. *RNA* 6, 1257–1266.

Tsai, C.J., Zheng, J., Aleman, C., and Nussinov, R. (2006). Structure by design: from single proteins and their building blocks to nanostructures. *Trends Biotechnol.* 24, 449–454.

Tucker, B.J., and Breaker, R.R. (2005). Riboswitches as versatile gene control elements. *Curr. Opin. Struct. Biol.* 15, 342–348.

Tuerk, C., and Gold, L. (1990). Systematic evolution of ligands by exponential enrichment: RNA ligands to bacteriophage T4 DNA polymerase. *Science* 249, 505–510.

Van Rossenberg, S.M., Sliedregt-Bol, K.M., Meeuwenoord, N.J., Van Berkel, T.J., Van Boom, J.H., Van Der Marel, G.A., and Biessen, E.A. (2002). Targeted lysosome disruptive elements for improvement of parenchymal liver cell-specific gene delivery. *J. Biol. Chem.* 277, 45803–45810.

Varambally, S., Dhanasekaran, S.M., Zhou, M., Barrette, T.R., Kumar-Sinha, C., Sanda, M.G., Ghosh, D., Pienta, K.J., Sewalt, R.G., Et al. (2002). The polycomb group protein EZH2 is involved in progression of prostate cancer. *Nature* 419, 624–629.

Vo-Dinh, T. (2005). Protein nanotechnology: the new frontier in biosciences. *Methods Mol. Biol.* 300, 1–13.

Wagner, C., Ehresmann, C., Ehresmann, B., and Brunel, C. (2004). Mechanism of dimerization of bicoid mRNA: initiation and stabilization. *J. Biol. Chem.* 279, 4560–4569.

Watts, J.K., Deleavey, G.F., and Damha, M.J. (2008). Chemically modified siRNA: tools and applications. *Drug Discov. Today* 13, 842–855.

Westhof, E., Masquida, B., and Jaeger, L. (1996). RNA tectonics: towards RNA design. *Fold Des.* 1, R78–R88.

Wichitwechkarn, J., Johnson, D., and Anderson, D. (1992). Mutant prohead RNAs in the in vitro packaging of bacteriophage phi 29 DNA-gp3. *J. Mol. Biol.* 223, 991–998.

Ye, X., Hemida, M., Zhang, H.M., Hanson, P., Ye, Q., and Yang, D. (2012). Current advances in Phi29 pRNA biology and its application in drug delivery. *Wiley Interdiscip. Rev. RNA* 3, 469–481.

Ye, X., Liu, Z., Hemida, M.G., and Yang, D. (2011). Targeted delivery of mutant tolerant anti-coxsackievirus artificial microRNAs using folate conjugated bacteriophage Phi29 pRNA. *PLoS One* 6, e21215.

Yingling, Y.G., and Shapiro, B.A. (2007). Computational design of an RNA hexagonal nanoring and an RNA nanotube. *Nano Lett.* 7, 2328–2334.

Zaug, A.J., Grabowski, P.J., and Cech, T.R. (1983). Autocatalytic cyclization of an excised intervening sequence RNA is a cleavage-ligation reaction. *Nature* 301, 578–583.

Zhang, C.L., Lee, C.-S., and Guo, P. (1994). The proximate 5′ and 3′ ends of the 120-base viral RNA (pRNA) are crucial for the packaging of bacteriophage f29 DNA. *Virology* 201, 77–85.

Zhang, C.L., Tellinghuisen, T., and Guo, P. (1995a). Confirmation of the helical structure of the 5′/3′ termini of the essential DNA packaging pRNA of phage f29. *RNA* 1, 1041–1050.

Zhang, C.L., Trottier, M., and Guo, P.X. (1995b). Circularly permuted viral pRNA active and specific in the packaging of bacteriophage Phi29 DNA. *Virology* 207, 442–451.

Zhang, F., Lemieux, S., Wu, X., St. Arnaud, S., McMurray, C.T., Major, F., and Anderson, D. (1998). Function of hexameric RNA in packaging of bacteriophage phi29 DNA in vitro. *Mol. Cell* 2, 141–147.

Zhang, H.M., Su, Y., Guo, S., Yuan, J., Lim, T., Liu, J., Guo, P., Yang, D. (2009). Targeted delivery of anti-coxsackievirus siRNAs using ligand-conjugated packaging RNAs. *Antiviral Res.* 83, 307–316.

Zhou, J., Shu, Y., Guo, P., Smith, D., and Rossi, J. (2011). Dual functional RNA nanoparticles containing phi29 motor pRNA and anti-gp120 aptamer for cell-type specific delivery and HIV-1 inhibition. *Methods* 54, 284–294.

Zhou, J., Li, H., Zaia, J., and Rossi, J.J. (2008). Novel dual inhibitory function aptamer-siRNA delivery system for HIV-1 therapy. *Mol. Ther.* 16, 1481–1489.

Zuker, M. (1989). On finding all suboptimal foldings of an RNA molecule. *Science* 244, 48–52.

Part II

Design, Synthesis, and
Characterization Methods
in RNA Nanotechnology

11 The Natural Versatility of RNA

Lewis A. Rolband, Oleg A. Shevchenko,
Caroline M. West, Ciara E. Conway, Caryn D. Striplin,
and Kirill A. Afonin
The University of North Carolina at Charlotte, Charlotte, NC, USA

CONTENTS

While being mostly composed of only four different monomeric units, ribonucleic acid (RNA) serves as a versatile biopolymer with a wide array of biological roles. These roles are so diverse that RNA has been proposed to have been the first functional biopolymer.[1,2] It is commonly posited that for life to evolve it would have required a self-replicating system of biomolecular polymers.[1–3] Over time, these proto-biomolecules would form more complex structures, resulting in the present diversity observed in nature. One popular theory for the evolution of life from primordial conditions is the RNA world hypothesis, which states that RNA would act as both the encoder of genetic information and the catalytically active unit. By fulfilling both roles, RNA would be able to self-replicate and eventually make the evolution of life possible.[2,4] Experimental and theoretical papers have shown how ribonucleotides could have been synthesized near underwater vents providing evidence of the RNA world-hypothesis possibility.[5] Eventually, a reverse transcriptase catalyzed the movement from RNA to DNA which was evolutionarily advantageous due to the lack of a single hydroxyl group on the 2'-carbon. This change led to the improved genetic stability of DNA by preventing the autocleavage of the ribose backbone. The lack of this 2' hydroxyl group also accounts for many major differences in functionality and structure between DNA and RNA. Despite these evolutionary changes, RNA remains a critical biopolymer in all living systems.

STRUCTURAL ELEMENTS OF RNA

The levels of structure found in RNA are similar to the well-known structural organization of proteins. There are three more levels of structure beyond the foundational primary sequence that are essential to the understanding of protein function. Determining RNA secondary, tertiary, and quaternary structure gives insight into the molecule's function, and understanding the structure cascade is vital to designing of functional RNA products.[6]

DOI: 10.1201/9781003001560-13

The Anatomy of Nucleotides

RNA nucleotides are comprised of three main components: an anionic phosphate, a ribofuranose, and a nitrogenous base. As a point of nomenclature, when lacking a phosphate group, the structure is referred to as a nucleoside.[7] The numbering scheme for denoting atoms (Figure 11.1(A)) begins on the base, then continues to the ribose; the sugar atoms are numbered as primes with the base attached to C-1′ and the phosphate-bound carbon designated C-5′.[7] This numbering scheme provides a directionality to the RNA polymer as the chain of nucleobases runs from the 5′ to the 3′ carbon with phosphodiester bonds between each ribose. There are two key differences between DNA and RNA. The ribose unit of DNA is lacking a hydroxyl group at the 2′ position, and where RNA utilizes uracil, DNA uses thymine. The presence of this 2′ hydroxyl unit contributes to the chemical instability of RNA, as this hydroxyl group can lead to a self-cyclization reaction that cleaves the backbone of the oligonucleotide. This reaction is catalyzed by divalent cations and alkaline solution conditions.[13,14]

Sugar Pucker of RNA

Due to steric constraints, instead of forming a planar ring, the ribose unit of a nucleobase experiences torsion that changes the geometry of each carbon in the ring relative to the plane of the molecule. These steric interactions result in several ribose conformations, giving the five-membered ring a puckered shape.[7] As ribose units of nucleobases are asymmetrically substituted, the number of favorable sugar pucker conformers is limited. Bending or bulging the 2′ or 3′ carbons out of the plane formed by the C_{1}', O_{4}', and C_{4}' atoms of ribose is denoted as either *endo-* (bulging toward C_{5}') or *exo-* (away from the C_{5}').[7,8,15] Among many conformations, C_{2}'-*endo* and C_{3}'-*endo* pucker conformations are energetically preferred and are the most commonly found conformers in crystal structures.[8] Data from NMR experiments have shown that substituents on C_{2}' and C_{3}' influence the sugar pucker mode; larger and more electronegative groups increase the likelihood of the substituted carbon atom being found in the *exo* position.[8,16] As DNA nucleotides only have a slightly electronegative hydrogen at the C_{2}', DNA most commonly adopts a C_{2}'-*endo* pucker.[8,15] In contrast, RNA nucleotides, due to the 2′-hydroxyl, adopt the C_{3}'-*endo* pucker.[8,16] The inter-phosphorous distance along the oligonucleotide backbone changes from 5.9 Å to 7.0 Å when the ribose shifts from C_{3}'-*endo* to C_{2}'-*endo*.[8,16] These conformational shifts lead to large differences in oligonucleotide structure. A classic example of this is the difference seen in A-form and B-form double-stranded helices (Figure 11.1(B)). RNA tends to assemble into A-form helices with the C_{3}'-*endo* ribose conformer. Conversely, DNA arranges into B-form helices with its ribose in the C_{2}'-*endo* pucker.[7–9,15]

The Need for Thymine in DNA

Nitrogenous bases that have exocyclic amines can undergo spontaneous deamination in aqueous solutions.[17] The product of cytosine deamination is uracil. Within 24 hours, one of every 10^{7} cytosine residues are converted to uracil *via* spontaneous deamination.[18] In the context of the average mammalian cell, about 100 spontaneous cytosine deamination events occur each day.[18] Compared to cytosine, adenine and guanine bases undergo spontaneous deamination about 100 times more slowly.[18] In genomic DNA, uracil is recognized as foreign entity and is removed by the cell repair system, where it is replaced with cytosine when detected.[18–20] If DNA utilized uracil bases in the sequence, rather than thymine, there would be no means of determining which uracil residues were properly placed and which should be repaired to cytosines.[18–20] This process would result in the GC content of the genome being greatly reduced over time. Instead, the C_{5}-methyl of thymine provides an easy means of recognition, while maintaining the base pairing with adenine.[18–20]

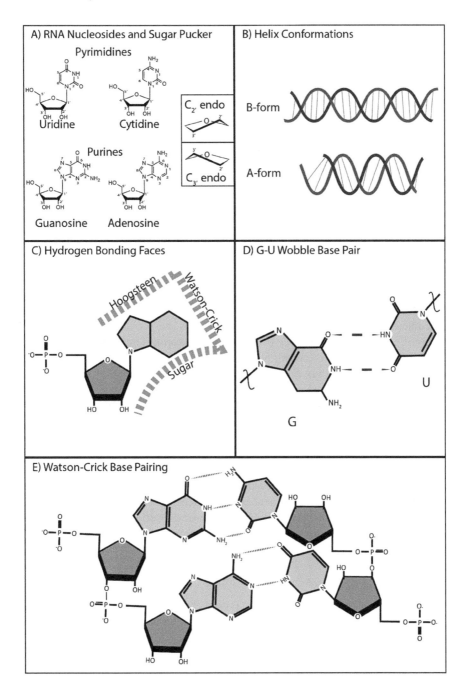

FIGURE 11.1 **A)** The structures of the four canonical RNA nucleosides are shown with numbering. The inset shows the ribofuranose 2′-*endo* and 3′-*endo* conformations with substituents out of the ring omitted for clarity.[7] **B)** The conformation of A- and B-form helices are shown in a cartoon representation. RNA tends to assemble into A-form helices due to the C_3'-*endo* ribose conformer preference. The B-form helix is typically adopted by DNA, as the C_2'-*endo* conformer is more favorable. The A-form helix is wider, but with more base pairs per unit of length than B-form helices.[8,9] **C)** The hydrogen bonding faces of purine bases are illustrated. For pyrimidine bases, the Hoogsteen face is also referred to as the C-H face. **D)** A wobble base pair is shown for a guanosine–uridine pair with the ribose units positioned in the *trans* orientation.[10–12] **E)** Canonical Watson-Crick base pairing is shown between two dinucleotides, 5′-GA-3′ and 5′-UC-3′, running anti-parallel to each other, with hydrogen bonds shown as dashed lines.[9]

PHYSICOCHEMICAL PROPERTIES OF RNA AND DNA

At physiological pH (generally considered to be 7.4), the phosphate backbone contributes a negative charge for each nucleotide in the sequence, while the nitrogenous bases are generally charge neutral, though the local environment can greatly impact the pK_a of the nitrogenous bases.[7,10] Protonation of the bases can lead to additional possibilities for non-canonical/non-Watson-Crick base pairing interactions.[10,21] In acidic conditions, the first locations to be protonated are the N_3 of cytosine and the N_1 of adenine, as they have pK_a values of 4.17 and 3.52, respectivley.[22] Under basic conditions, the N_1 nitrogen of guanine and the N_3 of thymine or uracil are deprotonated with pK_a values of 9.42, 9.93, and 9.38, respectively.[22] Additionally, in solution, the nucleotide triphosphates and oligonucleotides tend to chelate Mg^{2+} cations.[23] The Mg^{2+} ions have also been found to be important for proper RNA folding and function.[24,25] Below a pH of 3 and above a pH of 10, other protonation and deprotonation events occur, which can lead to the degradation of the oligonucleotide.[22] One of the most critical pH-dependent events is the base-catalyzed cleavage of the phosphodiester bond by a deprotonated 2′-hydroxyl.[13,22]

For the oligonucleotides to assemble through base pairing, hydrogen bonding needs to occur between nucleobases. Due to proximity of other bases on the sugar-phosphate backbone, a second base pair can form. Similarly, the first base pair can also fall apart.[22,26] If three consecutive base pairs form, duplex formation is seeded due to the attractive contributions from hydrogen bonding and stacking interactions.[22,26] This process is referred to as the cooperative zipper mechanism of helix formation.[22,26] In order for the three base pair seed to form, it must overcome an unfavorable positive free energy barrier. Once seeded, however, the free energy of further association between nucleobases becomes negative, thus, causing the assembly to grow spontaneously.[22,26] The stability of a folded oligonucleotide structure can be measured by following the denaturation of the structure.

A common means of assessing the stability of a nucleic acid assembly is by heating a solution of the structure and following the change in the absorbance of 260 nm light.[22] The temperature at which 50% of the structure has been denatured is termed the melting temperature and denoted as T_m.[22] Upon assembly, the absorbance of UV light by nucleobases is reduced due to base stacking interactions.[22,27] As the oligonucleotide denatures into a linear polymer, the absorbance of UV light by the nucleobases increases.[27,28] As adenine-uracil pairs (AU) share one less hydrogen bond than guanine-cytosine pairs (GC), AU rich sequences tend to have a lower T_m than GC rich oligonucleotides.[22,29,30] It is important to note that T_m is not an intrinsic property of an oligonucleotide and can be influenced by solution conditions such as pH or ionic strength.[31]

BASE-PAIR GEOMETRIES AND STABILIZING INTERACTIONS

The initial discovery of nucleic acid base pairing provided the Watson-Crick, also known as canonical, base pairing model (Figure 11.1(E)). In this model, adenine and thymine/uracil (A-T/AU) share two hydrogen bonds while guanine and cytosine (GC) share three hydrogen bonds.[26] The sugar moieties are close to each other, in a *cis* configuration, as the strands run anti-parallel to each other, in regard to the 5′–3′ directionality of oligonucleotides.[26,32,33] A variety of characterization methods have since demonstrated that there are a total of 12 geometric families for nucleotide base pairing, and the canonical Watson-Crick pairs are only one of these. Two additional faces on RNA nucleotides are available for base pairing interactions, the Hoogsteen face and the sugar edge (Figure 11.1(C)).[21,32,33] It has also been demonstrated that non-canonical "wobble" pairs can form (*e.g.* G-U).[10,11] Hydrogen bonding at the sugar edge occurs with the 2′-hydroxyl of the ribose and the functional groups of the nitrogenous base closer to the sugar unit. The combination of these three hydrogen bonding faces, with the additional degrees of freedom from having the oligonucleotides run parallel or antiparallel (with respect to the C-N glycosidic bond), results in the 12 possible geometric families of base pairing interactions. This also opens the possibility for base pairing to occur between more than the canonical AU or GC pairs.[21,26,32,33]

Just as nucleobases can interact horizontally within a plane, with some torsion, as base pairs, there are also interactions vertically between adjacent bases. Nitrogenous bases have aromatic systems that can interact through π-π stacking.[26] Crystal structures have shown that, within such a stack, the polarized regions ($-NH_2$, $=N-$, or $=O$) of one base superimpose over the π-electronic systems of adjacent bases, leading to an attractive interaction between the two nucleotides. This contributes dipole-induced dipole attractive forces to the stacking of bases along an oligonucleotide.[22,26,34] The strength of these π-π stacking attractions is greater than that of hydrogen bonds.[26] Furthermore, NMR studies suggested that the six-membered pyrimidine rings of adenine and guanine participate more in base stacking rather than the five-membered imidazole section of the nitrogenous bases and that base orientation within the stack depends on its amino- or keto- substituents.[26,35–37] Solubility experiments and NMR data have demonstrated that nucleobase stacking interactions between purine and pyrimidine base combinations are most stable in the order of: purine-purine, purine-pyrimidine, and pyrimidine-pyrimidine.[26,38–42]

The primary structure of an oligonucleotide is the order of the monomeric nucleobases. This ordering is generally denoted from 5′ to 3′, as this is the direction of enzymatic oligonucleotide synthesis. Secondary and tertiary structures of RNA arise from hydrogen bonding between nucleotides in 1 of the 12 previously described geometric orientations.[21,32,33,43] The specific geometry adopted during intramolecular base pairing results in diverse structural elements. RNA folding, in living systems, occurs sequentially as transcription occurs.[43] The common structural motifs (Figure 11.2)

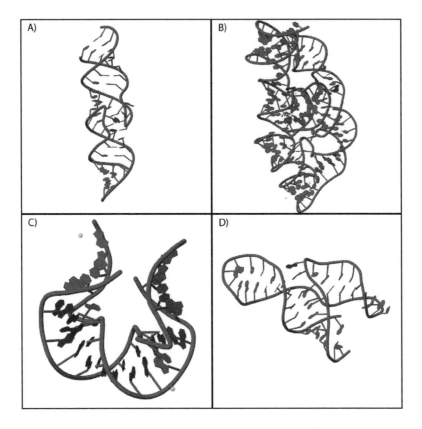

FIGURE 11.2 The structures are shown in a cartoon representation with Mg^{2+} ions shown as green balls of **A)** a kissing-loop RNA dimer originally discovered in HIV (PDB accession number 1ZCI)[44], **B)** the glmS ribozyme from *T. tengcongensis* bound to an inhibitory glucose-6-phosphate in ball-and-stick representation (PDB accession number 2Z74)[45], **C)** a double-stranded RNA kink-turn motif (PDB accession number 5FJO)[46], and **D)** an RNA three-way junction motif found in the Varkud satellite ribozyme (PDB accession number 2MTJ).[47]

are bulges, hairpin-loops, internal loops, and kink-turns.[46, 48,49] Each of these motifs can support significant amounts of variation. Helices of two anti-parallel regions with canonical Watson-Crick base pairing, for example, can form as A- or B-form helices, each with unique structural parameters.[9] Tertiary structure in RNA occurs when two or more elements of secondary structure interact to form a three-dimensional conformation.[50] The larger number of accessible tertiary structures of RNA, compared to DNA, can be partly explained by the 3′-*endo* sugar pucker being more accessible, causing increased flexibility and complexity in RNA's folding.[16] A variety of interactions between secondary elements work to stabilize these complex folding patterns. Cations in solution are frequently found to occupy distinct sites necessary for folding.[25] Single base intercalation from adjacent structural elements can ensure motifs remain in close proximity.[51,52] Commonly encountered tertiary structural motifs include kissing loops, tetraloops, multi-stranded junctions, and pseudoknots, though many more have been cataloged.[24,49,51–54] The adoption of folded conformations allow for conformation changes to be transmitted through long-range interacting motifs.[55,56] Tertiary motifs are also frequently involved in the function of naturally occurring RNAs. For example, junction structure is crucial for the function of many RNA molecules such as ribozymes, riboswitch binding pocket, and tRNA.[47,57,58] Quaternary structure occurs in RNAs when folded oligonucleotides interact to form complexes. Many functional RNAs possess this level of structural complexity with the most prominent example being the ribosome.[59] The same interactions affecting tertiary structure stability are used to promote the formation of quaternary assemblies.

RIBOZYMES

Ribozymes serve important roles as catalytic RNA oligonucleotides. These molecules have essential roles in the function of a cell and are comprised of tertiary RNA motifs that coordinate with metal ions for further structural reinforcement.[60] The tertiary motifs, discussed above, frequently form the active sites of ribozymes, and the variety of different structures encourages catalytic diversity. One of the most common reactions catalyzed by ribozymes is the cleavage of other RNA oligonucleotides.[61] These phosphodiester bond cleaving ribozymes can either act on other oligonucleotides, or they can serve as their own substrates through self-cleavage reactions. RNase ribozymes are thought to catalyze these reactions through concerted acid–base mechanisms.[58,61–63] Similar acid–base catalysis is found in the ribosome, where the nucleic acid active site catalyzes the formation of amide bonds. Ribozymes are also capable of utilizing divalent cations to form metalloribozymes, which can be useful for phosphoryl transfer reactions.[14,62,64]

RIBOSWITCHES

Riboswitches were discovered as a novel type of functional RNAs, having been first reported in 2002.[65,66] They are untranslated segments of mRNA that induce changes to RNA structure to modulate gene expression. Riboswitches consist of two parts: a ligand-binding region and an expression platform. The ligand can range greatly in size and chemical makeup from a single ion to larger cofactors or peptides with high affinity. The ligand-binding region determines which class the riboswitch is categorized into, and there are currently around 40 known classes of riboswitches.[65–68] The expression platform undergoes a structural change upon aptamer-target binding. This conformational shift in the expression platform causes alteration of gene expression by modulating transcription or translation.[66–69] This modulation can be accomplished through several different mechanisms. Self-cleavage of the mRNA containing an internal riboswitch-ribozyme is an effective means of preventing a functional peptide from being translated.[66–68] In some cases, a conformational shift will lead to the release or sequestration of the start codon or other ribosome binding sites, subsequently allowing or disrupting translation, respectively.[68,70,71]

CONCLUSION

Through the discoveries of structural biologists, the underlying structural features of complex biomolecules have been uncovered. Structural understanding of biopolymers is critical, as structure informs the function of biological macromolecules. As it relates to RNA nanotechnology, the structural motifs that have thus far been characterized have become a toolbox from which RNA nanoparticles (RNANPs) can be designed and built. Additionally, the mechanism by which ribozymes and riboswitches perform natural functions have served as inspiration for an array of synthetic RNANPs programmed to act similarly.[72,73] The variety of structural and functional RNAs present in nature serves as a source from which design elements can be drawn.

ACKNOWLEDGMENTS

Research reported in this publication was also supported by the National Institute of General Medical Sciences of the National Institutes of Health under Award Number R01GM120487 (to K.A.A.). The content is solely the responsibility of the authors and does not necessarily represent the official views of the National Institutes of Health.

REFERENCES

1. Bernhardt, H. S., The RNA world hypothesis: the worst theory of the early evolution of life (except for all the others). *Biol. Direct* 2012, *7*, 10.
2. Poole, A. M.; Jeffares, D. C.; Penny, D., The path from the RNA World 1998, *46* (1), 1–17.
3. Neveu, M.; Kim, H. J.; Benner, S. A., The "Strong" RNA world hypothesis: fifty years old. *Astrobiology* 2013, *13* (4), 391–403.
4. Kun, A.; Szilagyi, A.; Konnyu, B.; Boza, G.; Zachar, I.; Szathmary, E., The dynamics of the RNA world: insights and challenges. *DNA Habitats RNA Inhabitants* 2015, *1341*, 75–95.
5. Hud, N. V.; Fialho, D. M., RNA nucleosides built in one prebiotic pot. *Science* 2019, *366* (6461), 32–33.
6. Matveeva, O. V.; Kang, Y.; Spiridonov, A. N.; Saetrom, P.; Nemtsov, V. A.; Ogurtsov, A. Y.; Nechipurenko, Y. D.; Shabalina, S. A., Optimization of duplex stability and terminal asymmetry for shRNA design. *PLoS ONE* 2010, *5* (4), e10180.
7. Saenger, W., Defining Terms for the Nucleic Acids. In *Principles of Nucleic Acid Structure*, Saenger, W., Ed. Springer New York: New York, NY, 1984; pp 9–28.
8. Saenger, W., Structures and Conformational Properties of Bases, Furanose Sugars, and Phosphate Groups. In *Principles of Nucleic Acid Structure*, Saenger, W., Ed. Springer New York: New York, NY, 1984; pp 51–104.
9. Saenger, W., Polymorphism of DNA versus Structural Conservatism of RNA: Classification of A-, B-, and Z-TYPe Double Helices. In *Principles of Nucleic Acid Structure*, Saenger, W., Ed. Springer New York: New York, NY, 1984; pp 220–241.
10. Siegfried, N. A.; O'Hare, B.; Bevilacqua, P. C., Driving forces for nucleic acid pK(a) shifting in an A(+).C wobble: effects of helix position, temperature, and ionic strength. *Biochemistry* 2010, *49* (15), 3225–3236.
11. Garg, A.; Heinemann, U., A novel form of RNA double helix based on G·U and C·A. *RNA* 2018, *24* (2), 209–218.
12. Ananth, P.; Goldsmith, G.; Yathindra, N., An innate twist between Crick's wobble and Watson-Crick base pairs. *RNA* 2013, *19* (8), 1038–1053.
13. Oivanen, M.; Kuusela, S.; Lonnberg, H., Kinetics and mechanisms for the cleavage and isomerization of the phosphodiester bonds of RNA by Bronsted acids and bases. *Chem. Rev.* 1998, *98* (3), 961–990.
14. Steitz, T. A.; Steitz, J. A., A general 2-metal-ion mechanism for catalytic RNA. *Proc. Natl. Acad. Sci. U. S. A.* 1993, *90* (14), 6498–6502.
15. Freier, S., The ups and downs of nucleic acid duplex stability: structure-stability studies on chemically-modified DNA: RNA duplexes. *Nucleic Acids Res.* 1997, *25* (22), 4429–4443.
16. Kowiel, M.; Brzezinski, D.; Gilski, M.; Jaskolski, M., Conformation-dependent restraints for polynucleotides: the sugar moiety. *Nucleic Acids Res.* 2020, *48* (2), 962–973.

17. Shen, J. C.; Rideout, W. M.; Jones, P. A., The rate of hydrolytic deamination of 5-methylcytosine in double-stranded DNA. *Nucleic Acids Res.* 1994, *22* (6), 972–976.

18. Nelson, D. L., *Lehninger principles of biochemistry*. Fourth edition. W.H. Freeman: New York, 2005.

19. Rada, C.; Di Noia, J. M., Mismatch recognition and uracil excision provide complementary paths to both Ig switching and the A/T-focused phase of somatic mutation. *Mol. Cell* 2004, *16* (2), 163–171.

20. Savva, R.; McAuleyhecht, K.; Brown, T.; Pearl, L., The structural basis of specific base-excision repair by Uracil-DNA glycosylase. *Nature* 1995, *373* (6514), 487–493.

21. Leontis, N. B.; Stombaugh, J.; Westhof, E., The non-Watson-Crick base pairs and their associated isostericity matrices. *Nucleic Acids Res.* 2002, *30* (16), 3497–3531.

22. Saenger, W., Physical Properties of Nucleotides: Charge Densities, pK Values, Spectra, and Tautomerism. In *Principles of Nucleic Acid Structure*, Saenger, W., Ed. Springer New York: New York, NY, 1984; pp 105–115.

23. Sigel, H.; Griesser, R., Nucleoside 5′-triphosphates: self-association, acid–base, and metal ion-binding properties in solution. *Chem. Soc. Rev.* 2005, *34* (10), 875.

24. Brion, P.; Westhof, E., Hierarchy and dynamics of RNA folding. *Annu. Rev. Biophys. Biomol. Struct.* 1997, *26*, 113–137.

25. Pyle, A. M., Metal ions in the structure and function of RNA. *J. Biol. Inorg. Chem.* 2002, *7* (7–8), 679–690.

26. Saenger, W., Forces Stabilizing Associations Between Bases: Hydrogen Bonding and Base Stacking. In *Principles of Nucleic Acid Structure*, Saenger, W., Ed. Springer New York: New York, NY, 1984; pp 116–158.

27. Ackerman, M. M.; Ricciardi, C.; Weiss, D.; Chant, A.; Kraemer-Chant, C. M., Analyzing exonuclease-induced hyperchromicity by UV spectroscopy: an undergraduate biochemistry laboratory experiment. *J. Chem. Educ.* 2016, *93* (12), 2089–2095.

28. Cantor, C. R.; Tinoco, I., Absorption and optical rotatory dispersion of 7 trinucleoside diphosphates. *J. Mol. Biol.* 1965, *13* (1), 65–77.

29. Altan-Bonnet, G.; Libchaber, A.; Krichevsky, O., Bubble dynamics in double-stranded DNA. *Phys. Rev. Lett.* 2003, *90* (13), 4.

30. Putnam, B. F.; Vanzandt, L. L.; Prohofsky, E. W.; Mei, W. N., Resonant and localized breathing modes in terminal regions of the DNA double helix. *Biophys. J.* 1981, *35* (2), 271–287.

31. Owczarzy, R.; You, Y.; Moreira, B. G.; Manthey, J. A.; Huang, L. Y.; Behlke, M. A.; Walder, J. A., Effects of sodium ions on DNA duplex oligomers: improved predictions of melting temperatures. *Biochemistry* 2004, *43* (12), 3537–3554.

32. Leontis, N. B.; Lescoute, A.; Westhof, E., The building blocks and motifs of RNA architecture. *Curr. Opin. Struct. Biol.* 2006, *16* (3), 279–287.

33. Leontis, N. B.; Westhof, E., Geometric nomenclature and classification of RNA base Pairs. *RNA* 2001, *7* (4), 499–512.

34. Bugg, C. E.; Thomas, J. M.; Rao, S. T.; Sundaralingam, M., Stereochemistry of nucleic acids and their constituents. 10. Solid-state base-stacking patterns in nucleic acid constituents and polynucleotides. *Biopolymers* 1971, *10* (1), 175–219.

35. Broom, A. D.; Schweize, M. P.; Tso, P. O. P., Interaction and association of bases and nucleosides in aqueous solutions. V. Studies of association of purine nucleosides by vapor pressure osmometry and by proton magnetic resonance. *J. Am. Chem. Soc.* 1967, *89* (14), 3612–3622.

36. Tso, P. O. P., Hydrophobic-stacking properties of bases in nucleic acids. *Ann. NY Acad. Sci.* 1969, *153* (A3), 785–804.

37. Tso, P. O. P.; Melvin, I. S.; Olson, A. C., Interaction and association of bases and nucleosides in aqueous solutions. *J. Am. Chem. Soc.* 1963, *85* (9), 1289–1296.

38. Gratzer, W. B., Association of nucleic-acid bases in aqueous solution - a solvent partition study. *Eur. J. Biochem.* 1969, *10* (1), 184–187.

39. Mitchell, P. R.; Sigel, H., Proton NMR-study of self-stacking in purine and pyrimidine nucleosides and nucleotides. *Eur. J. Biochem.* 1978, *88* (1), 149–154.

40. Topal, M. D.; Warshaw, M. M., Dinucleoside monophosphates. 2. Nearest neighbor interactions. *Biopolymers* 1976, *15* (9), 1775–1793.

41. Davies, D. B. Co-Operative Conformational Properties of Nucleosides, Nucleotides and Nucleotidyl Units in Solution. In Nuclear Magnetic Resonance Spectroscopy in Molecular Biology, Pullman, B., Ed. Springer Netherlands: Dordrecht, 1978, pp 71–85.

42. Solie, T. N.; Schellman, J. A., The interaction of nucleosides in aqueous solution. *J. Mol. Biol.* 1968, *33* (1), 61–77.

43. Vandivier, L. E.; Anderson, S. J.; Foley, S. W.; Gregory, B. D., The conservation and function of RNA secondary structure in plants. *Annu. Rev. Plant Biol.* 2016, *67*, 463–488.

44. Ennifar, E.; Dumas, P., Polymorphism of bulged-out residues in HIV-1 RNA DIS kissing complex and structure comparison with solution studies. *J. Mol. Biol.* 2006, *356* (3), 771–782.

45. Klein, D. J.; Wilkinson, S. R.; Been, M. D.; Ferre-D'Amare, A. R., Requirement of helix p2.2 and nucleotide g1 for positioning the cleavage site and cofactor of the glmS ribozyme. *J. Mol. Biol.* 2007, *373* (1), 178–189.

46. Huang, L.; Wang, J.; Lilley, David M. J., A critical base pair in k-turns determines the conformational class adopted, and correlates with biological function. *Nucleic Acids Res.* 2016, *44* (11), 5390–5398.

47. Bonneau, E.; Legault, P., Nuclear magnetic resonance structure of the III–IV–V three-way junction from the Varkud satellite ribozyme and identification of magnesium-binding sites using paramagnetic relaxation enhancement. *Biochemistry* 2014, *53* (39), 6264–6275.

48. Hendrix, D. K.; Brenner, S. E.; Holbrook, S. R., RNA structural motifs: building blocks of a modular biomolecule. Q. Rev. Biophys. 2005, *38* (03), 221.

49. Liu, J.; Lilley, D. M. J., The role of specific 2'-hydroxyl groups in the stabilization of the folded conformation of kink-turn RNA. *RNA* 2007, *13* (2), 200–210.

50. Ganser, L. R.; Mustoe, A. M.; Al-Hashimi, H. M., An RNA tertiary switch by modifying how helices are tethered. *Genome Biol.* 2014, *15* (7), 425.

51. Batey, R. T.; Rambo, R. P.; Doudna, J. A., Tertiary motifs in RNA structure and folding. *Angew. Chem. Int. Edit.* 1999, *38* (16), 2327–2343.

52. Saenger, W., tRNA—A Treasury of Stereochemical Information. In *Principles of Nucleic Acid Structure*, Saenger, W., Ed. Springer New York: New York, NY, 1984; pp 331–349.

53. Butcher, S. E.; Pyle, A. M., The molecular interactions that stabilize RNA tertiary structure: RNA motifs, patterns, and networks. *Acc. Chem. Res.* 2011, *44* (12), 1302–1311.

54. Fiore, J. L.; Nesbitt, D. J., An RNA folding motif: GNRA tetraloop-receptor interactions. *Q. Rev. Biophys.* 2013, *46* (3), 223–264.

55. Mateos-Gomez, P. A.; Morales, L.; Zuniga, S.; Enjuanes, L.; Sola, I., Long-distance RNA-RNA Interactions in the coronavirus genome form high-order structures promoting discontinuous RNA synthesis during transcription. J. Virol. 2013, *87* (1), 177–186.

56. Serrano, P.; Pulido, M. R.; Saiz, M.; Martinez-Salas, E., The 3' end of the foot-and-mouth disease virus genome establishes two distinct long-range RNA-RNA interactions with the 5' end region. *J. Gen. Virol.* 2006, *87*, 3013–3022.

57. Laing, C.; Wen, D.; Wang, J. T.; Schlick, T., Predicting coaxial helical stacking in RNA junctions. *Nucleic Acids Res.* 2012, *40* (2), 487–498.

58. Birikh, K. R.; Heaton, P. A.; Eckstein, F., The structure, function and application of the hammerhead ribozyme. *Eur. J. Biochem.* 1997, *245* (1), 1–16.

59. Jones, C. P.; Ferré-D'Amaré, A. R., RNA quaternary structure and global symmetry. *Trends Biochem. Sci.* 2015, *40* (4), 211–220.

60. Gaines, C.; York, D., Computational RNA enzymology: new tools to unravelling the catalytic mechanism of nucleolytic ribozymes. *Abstr. Pap. Am. Chem. Soc.* 2018, *255*.

61. Zhuang, X. W.; Kim, H.; Pereira, M. J. B.; Babcock, H. P.; Walter, N. G.; Chu, S., Correlating structural dynamics and function in single ribozyme molecules. *Science* 2002, *296* (5572), 1473–1476.

62. Fedor, M. J.; Williamson, J. R., The catalytic diversity of RNAS. *Nat. Rev. Mol. Cell Biol.* 2005, *6* (5), 399–412.

63. Franzen, S., Expanding the catalytic repertoire of ribozymes and deoxyribozymes beyond RNA substrates. *Curr. Opin. Mol. Ther.* 2010, *12* (2), 223–232.

64. Shi, Y. G., The spliceosome: a protein-directed metalloribozyme. *J. Mol. Biol.* 2017, *429* (17), 2640–2653.

65. Subbaiah, K. C. V.; Hedaya, O.; Wu, J. B.; Jiang, F.; Yao, P., Mammalian RNA switches: molecular rheostats in gene regulation, disease, and medicine. *Comp. Struct. Biotechnol. J..* 2019, *17*, 1326–1338.

66. Serganov, A.; Nudler, E., A decade of riboswitches. *Cell* 2013, *152* (1–2), 17–24.

67. McCowen, P.; Corbino, K.; Stav, S.; Sherlock, M., Riboswitch diversity and distribution. *RNA* 2017, *23* (7).

68. Weinberg, Z.; Nelson, J. W.; Lünse, C. E.; Sherlock, M. E.; Breaker, R. R., Bioinformatic analysis of riboswitch structures uncovers variant classes with altered ligand specificity. *Proc. Natl Acad. Sci. U. S. A.* 2017, *114* (11), E2077–E2085.

69. Abduljalil, J. M., Bacterial riboswitches and RNA thermometers: nature and contributions to pathogenesis. *Noncoding RNA Res.* 2018, *3* (2), 54–63.

70. Bédard, A. V.; Hien, E. D. M.; Lafontaine, D. A., Riboswitch regulation mechanisms: RNA, metabolites and regulatory proteins. *Biochim. Biophys. Acta Gene. Regul. Mech.* 2020, *1863* (3), 194501.

71. Mehdizadeh Aghdam, E.; Hejazi, M. S.; Barzegar, A., Riboswitches: from living biosensors to novel targets of antibiotics. *Gene* 2016, *592* (2), 244–259.

72. Halman, J. R.; Satterwhite, E.; Roark, B.; Chandler, M.; Viard, M.; Ivanina, A.; Bindewald, E.; Kasprzak, W. K.; Panigaj, M.; Bui, M. N.; Lu, J. S.; Miller, J.; Khisamutdinov, E. F.; Shapiro, B. A.; Dobrovolskaia, M. A.; Afonin, K. A., Functionally-interdependent shape-switching nanoparticles with controllable properties. *Nucleic Acids Res.* 2017, *45* (4), 2210–2220.

73. Lincoln, T. A.; Joyce, G. F., Self-sustained replication of an RNA enzyme. *Science* 2009, *323* (5918), 1229–1232.

12 Nucleic Acids as a Building Material in Nanotechnology

Justin Halman and Kirill A. Afonin
The University of North Carolina at Charlotte, Charlotte, USA

CONTENTS

The combination of biocompatibility, amenability to bottom-up self-assembly, and wealth of functionalities in DNA and RNA has led to the discovery of a new field called nucleic acid nanotechnology. This field uses the folding and bonding principles of DNA and RNA to fold into rationally designed, unique, nanoscale architectures which are capable of performing a designated task.[1] Over the years, the field has evolved from simple assemblies with no function to complex, three-dimensional architectures that can be used as molecular machines, therapies, or scaffolds. These structures have shown advantages over traditional therapeutics, such as biocompatibility, multivalence, tunable physicochemical properties, and have been trending toward relatively low cost.

RATIONAL DESIGN AND COMPUTER-ASSISTED STRATEGIES

Early designs of nucleic acid nanoassemblies focused mainly on organizing DNA into controlled structures such as branched junctions and lattices.[2–4] In its nascence, these assemblies demonstrated some of the first uses of natural biomolecules in bottom-up assembly of nanoscale materials.[2,3] As strategies advanced, simple functional DNA-based materials were designed, including nanomechanical devices and rudimentary DNA walkers.[5–7] In the past decades, new classes and design methods of nucleic acid nanotechnology began to emerge.[8,9] Novel architectures wowed the scientific community with familiar images of smiling faces, stars, and triangles self-assembled from DNA strands in a technique now known as DNA origami.[10,11] At the same time, early programmable RNA nanodesigns began making their debut.[12,13]

DNA and RNA polymers demonstrated early abilities for precise bottom-up nucleic acid nanodesign.[10,14] Since then, the two fields, DNA and RNA nanotechnology, have move largely in tandem, with ever-increasingly complex designs being developed to answer critical questions in medicine, materials, and molecular biology.[1,15]

Analogous to DNA, RNA nanotechnology began with two-dimensional lattices and expanded thereafter.[16] Despite their molecular similarities, consequences of the 2′ hydroxyl group cause vast differences in the resulting molecule, demonstrated simply by their roles in molecular biology, and that RNA tertiary structures are relatively complex which makes their prediction even more challenging. As such, researchers take inspiration from natural biology, allowing natural evolution to guide their basic design. This idea was first brought forth through the concept of "RNA tectonics", which aims at constructing libraries using RNA mosaic units to create molecules with designed shapes and properties.[12] In fact, a "Jigsaw Puzzle" RNA nanostructure, among the first reported RNA nanodesigns, incorporated motifs taken from HIV genomes and ribosomal derived structures.[16] In another approach, a library of RNA nanostructures was designed by simply orienting the bacteriophage phi29 motif in different orientations to produce dimers, trimers, tetramers, and beyond in different orientations.[17–20] Despite the rigor of prediction, approaches used to construct DNA nanostructures can often times be applied to RNA nanostructures, such as junctions, branches, bundles, and twisted bundles.[21,22] Furthermore, examples of RNA origami have been demonstrated, expanding the domain of RNA nanotechnology.[23,24]

RNA nanotechnology made the generation of complex nanostructures a reality, but limitations still existed beyond using evolutionarily prepared motifs. As computational power has increased, the ability to predict intricacies in RNA structure has too. The simple nature of Watson-Crick base pairing makes predicting DNA structures a facile task; however, pseudoknots, non-canonical pairing, loop-loop, and other extraneous interactions make predicting RNA structures a challenging feat. Furthermore, the contributions of metal ions to RNA stabilization need to be accounted for and can play a major role in structure formation. To address these challenges, many programmers and mathematicians have aided the field with programs to predict RNA secondary and tertiary structure. These programs, including Mfold[25–27], Hyperfold[28], NUPACK[29–31], and Nanotiler[32] have aided RNA nanotechnologists in streamlining rational design for various facets of research. Despite advances, perfecting RNA structure prediction eludes researchers.

With these advances, numerous DNA, RNA, and DNA/RNA hybrid nanostructures have been designed and assembled including two-dimensional long-range structures such as tiles, sheets, and bundles.[4,33–35] Additionally, RNA has the advantage of being assembled co-transcriptionally, when RNA nanostructures fold and interact while simultaneously being transcribed.[36] Three-dimensional designs have been demonstrated with a variety of shapes and structures incorporating several RNA exclusive motifs or structures. For example, cubic RNA nanoparticles have been designed which use exclusively Watson-Crick base pairing, incorporating six individual RNA strands into a nanostructure of only 10 nm.[36] Furthermore, a ring-like structure was designed incorporating HIV-like kissing loops which assume near-perfect 120° geometry.[37,38]

In another approach, the bacteriophage phi29 motif was used in numerous orientations to allow for the generation of dozens of shapes with well-defined and predicted properties and sizes via one-pot assembly.[39] Several RNA structural motifs have been incorporated into nano-assemblies, allowing for a myriad of shapes including polygons, hearts, stars, and beyond.[40–42] In fact, using exclusively DNA and/or RNA, the repertoire of nanostructures that can be assembled is nearly infinite.[43] The scheme of designing RNA structures and their resulting architectures is shown in Figure 12.1.

Beyond static structures, DNA and RNA can be used to design dynamic machinery. The Watson-Crick base pairing predictability of nucleic acids allows for the generation of logic gates, switches, and various molecular machines.[44] Early designs used isothermal strand displacement to fuel reactions, generating desirable products downstream for biosensing or therapeutics, or to cause a physical or chemical change in a product.[45–47] This has been achieved for simple hybridization reactions to generate new structures based on strand input.[48,49] For example, DNA tweezers have been designed, using FRET as an output for their completed action.[45] This approach has been used to generate a series of functional designs, wherein the generation of a therapeutic product is achieved through strand displacement caused by a related input.[50–53] Furthermore, researchers have used strand displacement to cause shape-switching in nucleic acid nanostructures.[51,54–57] In addition to strand displacement, other strategies such as pH responsive,[58] biomarker-dependent,[59] and light-activated[60–62] dynamic nucleic acid-based structures have also been designed.

Outside of direct therapeutics, walkers and molecular machines have also been designed, using strand displacement to physically move cargo across a surface.[63,64] Both single-step and mutlistep walkers have been achieved with increasingly complex strand displacement mechanisms to achieve a greater depth of motion.[65] Expanding on this strategy and taking advantage of advances in folding prediction, complex logic gates and cascade pathways have been designed, activating recoverable products for biosensing purposes.[48,66] These strategies have since been incorporated into nucleic acid nanostructures for further advancements.

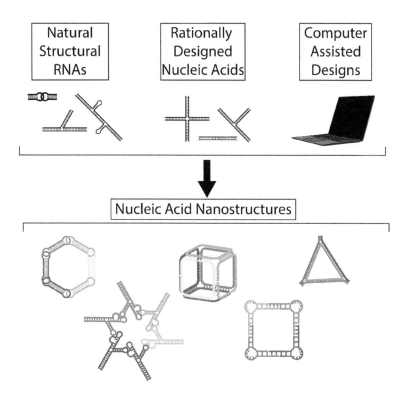

FIGURE 12.1 The combination of natural RNA structures, rational design, and computer-assisted folding prediction can result nearly unlimited RNA nanostructures.

Advancements in nucleic acid structure resolution and prediction have expanded the breadth of nucleic acid nanostructures. Beginning from simple junctions and branches, improvements have made the generation of complex three-dimensional structures a reality, using DNA origami, RNA tectonics, and rational design complemented by computer-assisted structure prediction. Furthermore, incorporating dynamic activity into these once static designs greatly expands the scope of their utilization, and further functionalization with therapeutics will aid in realizing their potential.

ACKNOWLEDGMENTS

Research reported in this publication was supported by the National Institute of General Medical Sciences of the National Institutes of Health under Award Numbers R01GM120487 and R35GM139587 (to K.A.A.). The content is solely the responsibility of the authors and does not necessarily represent the official views of the National Institutes of Health.

REFERENCES

1. Guo, P. X., The emerging field of RNA nanotechnology. *Nature Nanotechnology* 2010, *5* (12), 833–842.
2. Kallenbach, N. R.; Ma, R. I.; Seeman, N. C., An immobile nucleic-acid junction constructed from oligonucleotides. *Nature* 1983, *305* (5937), 829–831.
3. Seeman, N. C., Nucleic-acid junctions and lattices. *Journal of Theoretical Biology* 1982, *99* (2), 237–247.
4. Winfree, E.; Liu, F. R.; Wenzler, L. A.; Seeman, N. C., Design and self-assembly of two-dimensional DNA crystals. *Nature* 1998, *394* (6693), 539–544.
5. Reif, J. H., The design of autonomous DNA nano-mechanical devices: Walking and rolling DNA. *Natural Computing* 2003, *2* (4), 439.

6. Chen, Y.; Wang, M. S.; Mao, C. D., An autonomous DNA nanomotor powered by a DNA enzyme. *Angewandte Chemie-International Edition* 2004, *43* (27), 3554–3557.

7. Shin, J. S.; Pierce, N. A., A synthetic DNA walker for molecular transport. *Journal of the American Chemical Society* 2004, *126* (35), 10834–10835.

8. Mao, C. D.; LaBean, T. H.; Reif, J. H.; Seeman, N. C., Logical computation using algorithmic self-assembly of DNA triple-crossover molecules. *Nature* 2000, *407* (6803), 493–496.

9. Adleman, L. M., Molecular computation of solutions to combinatorial problems. *Science* 1994, *266* (5187), 1021–1024.

10. Rothemund, P. W. K., Folding DNA to create nanoscale shapes and patterns. *Nature* 2006, *440* (7082), 297–302.

11. Shih, W. M.; Quispe, J. D.; Joyce, G. F., A 1.7-kilobase single-stranded DNA that folds into a nanoscale octahedron. *Nature* 2004, *427* (6975), 618–621.

12. Westhof, E.; Masquida, B.; Jaeger, L., RNA tectonics: Towards RNA design. *Folding & Design* 1996, *1* (4), R78–R88.

13. Guo, P.; Zhang, C.; Chen, C.; Garver, K.; Trottier, M., Inter-RNA interaction of phage φ29 pRNA to form a hexameric complex for viral DNA transportation. *Molecular cell* 1998, *2* (1), 149–155.

14. Ishikawa, J.; Furuta, H.; Ikawa, Y., RNA tectonics (tectoRNA) for RNA nanostructure design and its application in synthetic biology. *Wiley Interdisciplinary Reviews: RNA* 2013, *4* (6), 651–664.

15. Seeman, N. C.; Sleiman, H. F., DNA nanotechnology. *Nature Reviews Materials* 2017, *3* (1), 1–23.

16. Chworos, A.; Severcan, I.; Koyfman, A. Y.; Weinkam, P.; Oroudjev, E.; Hansma, H. G.; Jaeger, L., Building programmable jigsaw puzzles with RNA. *Science* 2004, *306* (5704), 2068–2072.

17. Fang, Y.; Shu, D.; Xiao, F.; Guo, P. X.; Qin, P. Z., Modular assembly of chimeric phi29 packaging RNAs that support DNA packaging. *Biochemical and Biophysical Research Communications* 2008, *372* (4), 589–594.

18. Xiao, F.; Zhang, H.; Guo, P. X., Novel mechanism of hexamer ring assembly in protein/RNA interactions revealed by single molecule imaging. *Nucleic Acids Research* 2008, *36* (20), 6620–6632.

19. Guo, P.; Erickson, S.; Anderson, D., A small viral RNA is required for in vitro packaging of bacteriophage phi 29 DNA. *Science* 1987, *236* (4802), 690–694.

20. Guo, P.; Zhang, C.; Chen, C.; Garver, K.; Trottier, M., Inter-RNA interaction of phage φ29 pRNA to form a hexameric complex for viral DNA transportation. *Molecular cell* 1998, *2* (1), 149–155.

21. Cayrol, B.; Nogues, C.; Dawid, A.; Sagi, I.; Silberzan, P.; Isambert, H., A nanostructure made of a bacterial noncoding RNA. *Journal of the American Chemical Society* 2009, *131* (47), 17270–17276.

22. Sharma, A.; Haque, F.; Pi, F.; Shlyakhtenko, L. S.; Evers, B. M.; Guo, P., Controllable self-assembly of RNA dendrimers. *Nanomedicine: Nanotechnology, Biology and Medicine* 2016, *12* (3), 835–844.

23. Geary, C.; Rothemund, P. W. K.; Andersen, E. S., A single-stranded architecture for cotranscriptional folding of RNA nanostructures. *Science* 2014, *345* (6198), 799–804.

24. Geary, C.; Meunier, P. E.; Schabanel, N.; Seki, S., Oritatami: A computational model for molecular cotranscriptional folding. *International Journal of Molecular Sciences* 2019, *20* (9), 2259.

25. Zuker, M., Mfold web server for nucleic acid folding and hybridization prediction. *Nucleic Acids Research* 2003, *31* (13), 3406–3415.

26. Waugh, A.; Gendron, P.; Altman, R.; Brown, J. W.; Case, D.; Gautheret, D.; Harvey, S. C.; Leontis, N.; Westbrook, J.; Westhof, E.; Zuker, M.; Major, F., RNAML: A standard syntax for exchanging RNA information. *Rna-a Publication of the Rna Society* 2002, *8* (6), 707–717.

27. Zuker, M.; Jacobson, A. B., Using reliability information to annotate RNA secondary structures. *RNA* 1998, *4* (6), 669–679.

28. Bindewald, E.; Afonin, K. A.; Viard, M.; Zakrevsky, P.; Kim, T.; Shapiro, B. A., Multistrand structure prediction of nucleic acid assemblies and design of RNA switches. *Nano Letters* 2016, *16* (3), 1726–1735.

29. Zadeh, J. N.; Steenberg, C. D.; Bois, J. S.; Wolfe, B. R.; Pierce, M. B.; Khan, A. R.; Dirks, R. M.; Pierce, N. A., NUPACK: Analysis and design of nucleic acid systems. *Journal of Computational Chemistry* 2011, *32* (1), 170–173.

30. Dirks, R. M.; Pierce, N. A., An algorithm for computing nucleic acid base-pairing probabilities including pseudoknots. *Journal of Computational Chemistry* 2004, *25* (10), 1295–1304.

31. Dirks, R. M.; Bois, J. S.; Schaeffer, J. M.; Winfree, E.; Pierce, N. A., Thermodynamic analysis of interacting nucleic acid strands. *Siam Review* 2007, *49* (1), 65–88.

32. Bindewald, E.; Grunewald, C.; Boyle, B.; O'Connor, M.; Shapiro, B. A. J. J.; Modelling, computational strategies for the automated design of RNA nanoscale structures from building blocks using NanoTiler. *Journal of Molecular Graphics and Modelling* 2008, *27* (3), 299–308.

33. Lin, C.; Liu, Y.; Rinker, S.; Yan, H. J. C., DNA tile based self-assembly: Building complex nanoarchitectures. *Chem Phys Chem* 2006, *7* (8), 1641–1647.

34. Kosuri, S.; Church, G. M. J. N. m., Large-scale de novo DNA synthesis: Technologies and applications. *Nature Methods* 2014, *11* (5), 499.

35. Afonin, K. A.; Cieply, D. J.; Leontis, N. B. J. J., Specific RNA self-assembly with minimal paranemic motifs. *Journal of the American Chemical Society* 2008, *130* (1), 93–102.

36. Afonin, K. A.; Bindewald, E.; Yaghoubian, A. J.; Voss, N.; Jacovetty, E.; Shapiro, B. A.; Jaeger, L. J. N., In vitro assembly of cubic RNA-based scaffolds designed in silico. *Nature Nanotechnology* 2010, *5* (9), 676.

37. Grabow, W. W.; Zakrevsky, P.; Afonin, K. A.; Chworos, A.; Shapiro, B. A.; Jaeger, L. J. N. I. Self-assembling RNA nanorings based on RNAI/II inverse kissing complexes. *Nano Letters* 2011, *11* (2), 878–887.

38. Parlea, L.; Bindewald, E.; Sharan, R.; Bartlett, N.; Moriarty, D.; Oliver, J.; Afonin, K. A.; Shapiro, B. A. J. M., Ring Catalog: A resource for designing self-assembling RNA nanostructures. *Methods* 2016, *103*, 128–137.

39. Shu, D.; Shu, Y.; Haque, F.; Abdelmawla, S.; Guo, P. J. N., Thermodynamically stable RNA three-way junction for constructing multifunctional nanoparticles for delivery of therapeutics. *Nature Nanotechnology* 2011, *6* (10), 658.

40. Geary, C.; Chworos, A.; Verzemnieks, E.; Voss, N. R.; Jaeger, L. J. N. I., Composing RNA nanostructures from a syntax of RNA structural modules. *Nano Letters* 2017, *17* (11), 7095–7101.

41. Chworos, A. J. R. N., Rational design of RNA nanoparticles and nanoarrays. *RNA Nanotechnology: Pan Stanford Publishing Pte. Ltd* 2014, 213–234.

42. Bui, M. N.; Johnson, M. B.; Viard, M.; Satterwhite, E.; Martins, A. N.; Li, Z.; Marriott, I.; Afonin, K. A.; Khisamutdinov, E. F. J. N. N., Biology; Medicine, Versatile RNA tetra-U helix linking motif as a toolkit for nucleic acid nanotechnology. *Nanomedicine: Nanotechnology, Biology and Medicine* 2017, *13* (3), 1137–1146.

43. Veneziano, R.; Ratanalert, S.; Zhang, K.; Zhang, F.; Yan, H.; Chiu, W.; Bathe, M. J. S., Designer nanoscale DNA assemblies programmed from the top down. *Science* 2016, *352* (6293), 1534–1534.

44. Jahanban-Esfahlan, A.; Seidi, K.; Jaymand, M.; Schmidt, T. L.; Zare, P.; Javaheri, T.; Jahanban-Esfahlan, R. J. J., Dynamic DNA nanostructures in biomedicine: Beauty, utility and limits. *Journal of Controlled Release* 2019, *315*, 166–185.

45. Yurke, B.; Turberfield, A. J.; Mills, A. P.; Simmel, F. C.; Neumann, J. L. J. N., A DNA-fuelled molecular machine made of DNA. Nature 2000, *406* (6796), 605–608.

46. Yan, H.; Zhang, X.; Shen, Z.; Seeman, N. C. J. N., A robust DNA mechanical device controlled by hybridization topology. Nature 2002, *415* (6867), 62–65.

47. Feng, L.; Park, S. H.; Reif, J. H.; Yan, H., A two-state DNA lattice switched by DNA nanoactuator. Angewandte Chemie 2003, *42* (36), 4342–4346.

48. Yin, P.; Choi, H. M.; Calvert, C. R.; Pierce, N. A. J. N., Programming biomolecular self-assembly pathways. Nature 2008, *451* (7176), 318–322.

49. Dirks, R. M.; Pierce, N. A. J. P., Triggered amplification by hybridization chain reaction. *Proceedings of the National Academy of Sciences* 2004, *101* (43), 15275–15278.

50. Hochrein, L. M.; Schwarzkopf, M.; Shahgholi, M.; Yin, P.; Pierce, N. A. J. J., Conditional Dicer substrate formation via shape and sequence transduction with small conditional RNAs. *Journal of the American Chemical Society* 2013, *135* (46), 17322–17330.

51. Zakrevsky, P.; Parlea, L.; Viard, M.; Bindewald, E.; Afonin, K. A.; Shapiro, B. A., Preparation of a conditional RNA switch. In *RNA Nanostructures*, Springer: 2017; 303–324.

52. Martins, A. N.; Ke, W.; Jawahar, V.; Striplin, M.; Striplin, C.; Freed, E. O.; Afonin, K. A., Intracellular Reassociation of RNA–DNA Hybrids that Activates RNAi in HIV-Infected Cells. In *RNA Nanostructures*, Springer: 2017; 269–283.

53. Afonin, K. A.; Viard, M.; Tedbury, P.; Bindewald, E.; Parlea, L.; Howington, M.; Valdman, M.; Johns-Boehme, A.; Brainerd, C.; Freed, E. O. J. N. I., The use of minimal RNA toeholds to trigger the activation of multiple functionalities. *Nano Letters* 2016, *16* (3), 1746–1753.

54. Halman, J. R.; Satterwhite, E.; Roark, B.; Chandler, M.; Viard, M.; Ivanina, A.; Bindewald, E.; Kasprzak, W. K.; Panigaj, M.; Bui, M. N. J. N. Functionally-interdependent shape-switching nanoparticles with controllable properties. *Nucleic Acids Research* 2017, *45* (4), 2210–2220.

55. Chandler, M.; Ke, W.; Halman, J. R.; Panigaj, M.; Afonin, K. A., Reconfigurable nucleic acid materials for cancer therapy. In *Nanooncology*, Springer: 2018; pp. 365–385.

56. Ke, W.; Hong, E.; Saito, R. F.; Rangel, M. C.; Wang, J.; Viard, M.; Richardson, M.; Khisamutdinov, E. F.; Panigaj, M.; Dokholyan, N. V. J. N., RNA–DNA fibers and polygons with controlled immunorecognition activate RNAi, FRET and transcriptional regulation of NF-κB in human cells. *Nucleic Acids Research* 2019, *47* (3), 1350–1361.

57. Chandler, M.; Lyalina, T.; Halman, J.; Rackley, L.; Lee, L.; Dang, D.; Ke, W.; Sajja, S.; Woods, S.; Acharya, S. J. M., Broccoli fluorets: Split aptamers as a user-friendly fluorescent toolkit for dynamic RNA nanotechnology. *Molecules* 2018, *23* (12), 3178.

58. Idili, A.; Vallée-Bélisle, A.; Ricci, F., Programmable pH-triggered DNA nanoswitches. *Journal of the American Chemical Society* 2014, *136* (16), 5836–5839.

59. Zhang, X.; Battig, M. R.; Wang, Y. J. C. C., Programmable hydrogels for the controlled release of therapeutic nucleic acid aptamers via reversible DNA hybridization. *Chemical Communications* 2013, *49* (83), 9600–9602.

60. Schmidt, T. L.; Koeppel, M. B.; Thevarpadam, J.; Gonçalves, D. P.; Heckel, A. J. S., A light trigger for DNA nanotechnology. *Small* 2011, *7* (15), 2163–2167.

61. Yuan, Q.; Zhang, Y.; Chen, T.; Lu, D.; Zhao, Z.; Zhang, X.; Li, Z.; Yan, C.-H.; Tan, W. Photon-manipulated drug release from a mesoporous nanocontainer controlled by azobenzene-modified nucleic acid. *ACS Nano* 2012, *6* (7), 6337–6344.

62. Yang, Y.; Tashiro, R.; Suzuki, Y.; Emura, T.; Hidaka, K.; Sugiyama, H.; Endo, M., A photoregulated DNA-based rotary system and direct observation of its rotational movement. *Chemistry - A European Journal* 2017, *23* (16), 3979–3985.

63. Sherman, W. B.; Seeman, N. C., A precisely controlled DNA biped walking device. *Nano Letters* 2004, *4* (7), 1203–1207.

64. Omabegho, T.; Sha, R.; Seeman, N. C. J. S., A bipedal DNA Brownian motor with coordinated legs. *Science* 2009, *324* (5923), 67–71.

65. Chen, J.; Luo, Z.; Sun, C.; Huang, Z.; Zhou, C.; Yin, S.; Duan, Y.; Li, Y., Research progress of DNA walker and its recent applications in biosensor. *TrAC Trends in Analytical Chemistry* 2019, 115626.

66. Goldsworthy, V.; LaForce, G.; Abels, S.; Khisamutdinov, E. F. J. N., Fluorogenic RNA aptamers: A nanoplatform for fabrication of simple and combinatorial logic gates. *Nanomaterials* 2018, *8* (12), 984.

13 Multiple Functionalities for RNA Nanoparticles

Yelixza I. Avila
9201 University City Boulevard, Charlotte, USA

Martin Panigaj
Pavol Jozef Safarik University in Kosice, Kosice, Slovak Republic

CONTENTS

INTRODUCTION

The relatively simple rules that govern the interactions within one or between various RNA strands enable us to predict, design, and construct versatile RNA nanoassemblies composed from single or multiple RNA sequences. Over the last three decades, our growing insight into RNA biology has inspired the development of RNA structures that carry functional RNA moieties for the use in biotechnology or medicine.[1] Besides the functional RNA types such as siRNAs and aptamers, another therapeutically or diagnostically relevant molecules (*e.g.*, fluorescent dyes, chemotherapeutics, or peptides) extend universality of RNA nanoparticles (NANPs).[2,3] In addition to carried functional RNAs and linked chemistry, the shape, 3D structure, and composition of nucleic acid nanoparticles by itself have important bioactive roles.[4]

THE ASSEMBLY OF RNA NANOPARTICLES

RNA NANPs are rationally designed by reorganizing and modifying existing naturally occurring structural RNA motifs such as three- and four-way junctions, bulges, or internal and terminal loops (Figure 13.1(A)). The structure of RNA NANPs is based on the canonical and non-canonical Watson-Crick interactions that form secondary and tertiary structures. The RNA is highly modular; structural motifs may be recombined, increasing the structural and functional diversity of NANPs. However, direct incorporation of functional RNAs in NANPs must be optimized to not interfere with the local or overall assembly that would lead to disruption of either the bioactive RNA or the RNA scaffold. The overall engineering principles of RNA NANPs can be described by two main designing strategies.

RNA architectonics is based on the understanding of the structural parts typical for a given 3D shape. By studying the architecture of natural RNAs, individual structural motifs (tectoRNAs) are recognized and characterized. The modular tectoRNAs formed by folded single-stranded RNA

DOI: 10.1201/9781003001560-15

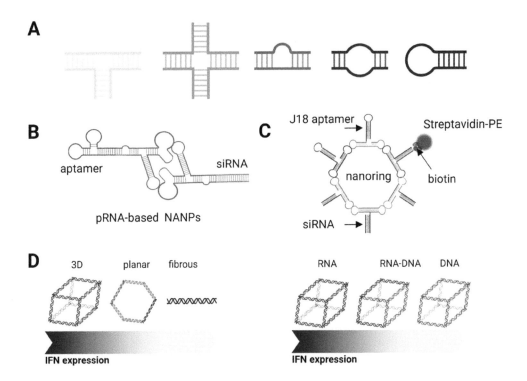

FIGURE 13.1 Multifunctional RNA NANPs. A) Schematic depiction of several structural motifs. From left to right- 3-way junction, 4-way junction, bulge, internal loop, and terminal loop. B) Dimeric NANP based on pRNA with functionalization of an aptamer and siRNA in the double helical region of each monomer. C) A hexameric nanoring with the capacity to display a combination of aptamers, siRNAs, fluorescent marker on streptavidin linked to NANP through biotinylated oligonucleotide. D) A pattern of selected immunogenic properties illustrating that as particles become less globular or more DNA-based, IFN expression decreases.

(ssRNA) then assemble into the wide range of RNA nanostructures held by tertiary interactions. In the second strategy, RNA NANPs self-assembles upon intermolecular base-pairing between ssRNAs lacking secondary structure.[5]

Currently, the individual strands for RNA NANPs are usually *in vitro* transcribed by T7 RNA polymerase using PCR-amplified DNA templates. The *in vitro* transcription, furthermore, allows in some instances, to produce RNA NANPs by cotranscriptional self-assembly. Purified RNA strands are thermally denatured and annealed together in one-pot or stepwise assembly. When using tectoR-NAs, the individual monomers must create secondary structure first. Therefore, immediate cooling on ice facilitates intramolecular interactions, while structures created by the strategy of ssRNA assembly do not require pre-folding step.[6]

In comparison to *in vitro* transcription and subsequent self-assembly of RNA NANPs, intracellular transcription and self-assembly of NANP strands is an almost unexplored field with vast potential. To mention a few, the *in vivo* synthesis of RNA NANPs can provide nanoscaffolds for the rational organization of metabolic pathways, cost-effective large-scale production of RNA, or NANPs production for further *in vivo* delivery inside the exosomes or virus-like particles.[7–9] The stoichiometry is vital for the arrangement of multi-stranded NANPs, but it is challenging to ensure it inside the cells. Encoding the whole NANP structure into one long ssRNA has been recently shown as a viable idea to avoid problems with stoichiometry.[10] How flexible is this *in vivo* ssRNA folding strategy to incorporate multiple functionalities is open for demonstration.

TOOLS FOR REAL-TIME ASSEMBLY VISUALIZATION

As RNA nanoparticle designs become more complex, the methods for real-time monitoring of the NANP assembly, tracking, reaction to the microenvironment, or other molecules are very instrumental. Fluorescent labeling by organic dyes is the method of choice, *e.g.*, to visualize the binding of RNA NANPs to specific receptors, folding, and degradation of RNA particles inside the cells.[3,11] The Förster resonance energy transfer (FRET) from an excited donor fluorophore to acceptor fluorophore depends on the close distance between both fluorophores. The conjugation of FRET pair to separated RNA strands thus allows observing dynamics of RNA strands base-pairing or separation *in vitro* and *in vivo*.[12–14]

We do not know any type of RNA with intrinsic fluorescent properties analogous to GFP. Therefore, any fluorescent RNA model that does not require the addition of oligonucleotides linked to dye would boost the field of RNA nanotechnology, especially in the research of intracellular NANP assembly. The closest system that answers this demand involves RNA aptamers that emit a fluorescent signal upon binding a fluorophore.[15] The last ten years witnessed the fast development of various fluorescent RNA aptamers such as Spinach, Broccoli, Mango, or Corn that in proper conformation bind to corresponding fluorophores and emit the light in diverse spectra.[16–19] The fusion of light-up RNA aptamers with gene constructs for *in vivo* transcription is the main advantage of fluorescent aptamers. Recently, the combination of Spinach and Mango aptamers within the sequence of the ssRNA origami construct demonstrated the functional feasibility of genetically encoded FRET systems to report conformational changes upon induction in bacterial cells.[20] Alternatively, splitting RNA aptamers into two separate strands renders them functional/fluorescent only upon successful reassembly of both parts, which makes split aptamer strategy an effective reporter system for the assembly of RNA NANPs.[21,22] Specifically, the fluorescent split aptamers have the potential in the development of protocols for *in vivo* assembly of RNA NANPs composed from multiple strands.

MULTIFUNCTIONAL RNA NANPs

The spectrum of RNA functions is sufficiently broad to cover almost all cellular processes. RNA nanoconstructs are modular and programmable with an option to combine and interchange various RNA species by their integration to RNA scaffold. The repertoire of functionalities mostly includes aptamers, RNAi activators (small interfering RNAs - siRNAs, small hairpin RNAs- shRNAs, micro RNAs- miRNAs), and ribozymes. In addition to RNA, DNA oligonucleotides may be used as anti-micro RNAs (antimiRs) or decoys. The DNA/RNA aptamers are selected by directed evolution to bind its target (*e.g.*, protein or metabolite) with an affinity comparable to monoclonal antibodies. The functionality of an aptamer is defined by its secondary and tertiary structure. In the last decade, aptamers have been shown to function not only as delivery agents for therapeutic cargo, but many aptamers upon receptor recognition elicit antagonistic or agonistic responses that, in combination with transported therapeutics, have the potential of synergism.[23] On the other side, RNAi activators exert their function by fully or partial complementary base-pairing to its target mRNA (siRNA, shRNA, and miRNA) downregulating the expression of a specific gene or set of genes while application of antimiR oligonucleotides inhibits endogenous miRNAs, consequently resulting in the upregulation of genes silenced by miRNA. The decoys are DNA oligonucleotides with sequences equal to transcription factor binding sites in genomic DNA. The presence of decoys in the cell prevents transcription factor binding to genomic DNA.

During the 20 years since the first description of the RNA nanoparticle, several concepts for the spatial organization of functional RNAs have been introduced. In the pioneering study, bottom-up design of the RNA nanoparticles was inspired by the DNA-packaging motor of bacteriophage phi29 as a building block to assemble RNA nanostructures.[24] Two domains folding independently are the main hallmark of packaging RNA (pRNA) monomers. The intermolecular binding domain composed of the left- and right-hand loops pair with other monomers to form multimeric pRNAs.[25]

The 5′ and 3′ ends of pRNA located in the double-stranded helical domain can be replaced for an aptamer or siRNA without affecting the pRNA structure. Generally, the 3WJ motif of phi29 DNA packaging motor offers many possibilities for bottom-up self-assembly of simple or more complex multifunctional RNA nanoparticles (Figure 13.1(B)).

In silico designed/programmed RNA nanoparticles are the alternative approach to nature-derived RNA particles. The complex hexameric RNA nanorings designed by Afonin *et al.* have the capacity to simultaneously carry on up to six aptamers to increase valency accompanying six siRNAs silencing the same or six different genes. Further, at least one aptamer can be exchanged for one biotinylated oligonucleotide that allows fluorescent visualization via coupling through streptavidin conjugated to phycoerythrin. This biotin-streptavidin connection represents the universal approach for displaying engineered proteins to nucleic acid nanoassembly (Figure 13.1(C)).[3]

The conditional activation of embedded functionalities only in the presence of a specific stimulus increases the level of NANP regulation. In this concept, each subunit is split and therefore nonfunctional but can be conditionally reassembled with its complementary cognate partner through the interactions of short single-stranded toeholds and become activated. As a proof of the concept, we initially studied functionally interdependent RNA-DNA hybrid cubes or fibers and polygons that can carry and activate different split functionalities, including RNAi, optical response, transcription, split aptamer reassembly, and NF-κB decoy DNA oligonucleotides.[13,14]

We also noted that simple optimization of the ratio between RNA and DNA strands that constitute assembly could adjust the immunomodulatory properties of nanoparticles depending on the goal of the application. Nonimmunogenic nanoparticles can carry on therapeutic oligonucleotides, while nanoconstructs with potent immunostimulatory properties may serve as vaccine adjuvants.

The later systematic research of NANP characteristics revealed patterns that contribute to immunostimulation. It appears that rather than the sequence, it is the structure itself that confers to immune activation where immunogenicity increases from fibrous through planar to globular NANPs. As for chemical composition, RNA nanoparticles are the most immunogenic, and the immune response weakens with an increasing ratio of DNA strands to RNAs (Figure 13.1(D)). Similarly, more functionalized NANPs with a higher number of Dicer substrate RNAs trigger more elevated interferon (IFN) production. From a therapeutic point of view, a delivery of naked NANPs leaves cell defense unresponsive, but the administration of NANPs in formulation with transfection agent activates IFN production.[4,26–29]

CONCLUSION AND PERSPECTIVE

RNA is a modular and programmable material for the synthesis of targetable drugs and a target itself. RNA being the evolutional and biochemical predecessor of proteins, serves as an ideal therapeutic target for precise intervention. When combined with the binding potential of aptamers, RNA has the capacity to be aimed not only at malfunctioned RNAs, but also at proteins, metabolites, or other relevant compounds. The fusion of structural motifs with functional RNA types leads to homogenous and biocompatible structures, thus, reducing chemical complexity and possible toxicity of medicament. Increasing knowledge of the immunostimulatory properties of nucleic acids enables the synthesis of assemblies with desired immunomodulatory attributes. Future research of conditional NANPs will inevitably lead to the development of smart single or functionally interdependent pair of NANPs that would require specific triggers from disease-related molecules. The activation of smart NANPs would begin upon toehold interaction with specific transcript (mRNA, miRNA, or non-coding RNA) or through structural switch upon binding of protein or metabolite to NANP-embedded aptamer. Additionally, if we knew a reliable way for efficient trafficking of NANPs to the cytoplasm upon cell-specific delivery mediated by aptamers, we could design therapeutic NANPs solely composed of nucleic acids. Parallelly with the expanding universe of RNA functions, options, and possibilities of multifunctional RNA nanoparticles inflate, too.

REFERENCES

1. Guo, P., The emerging field of RNA nanotechnology. *Nat. Nanotechnol.* 2010, *5* (12), 833–842.
2. Dao, B. N.; Viard, M.; Martins, A. N.; Kasprzak, W. K.; Shapiro, B. A.; Afonin, K. A., Triggering RNAi with multifunctional RNA nanoparticles and their delivery. *DNA RNA Nanotechnol.* 2015, *1* (1), 27–38.
3. Afonin, K. A.; Viard, M.; Koyfman, A. Y.; Martins, A. N.; Kasprzak, W. K.; Panigaj, M.; Desai, R.; Santhanam, A.; Grabow, W. W.; Jaeger, L.; Heldman, E.; Reiser, J.; Chiu, W.; Freed, E. O.; Shapiro, B. A., Multifunctional RNA nanoparticles. *Nano Lett.* 2014, *14* (10), 5662–5671.
4. Chandler, M.; Johnson, M. B.; Panigaj, M.; Afonin, K. A., Innate immune responses triggered by nucleic acids inspire the design of immunomodulatory nucleic acid nanoparticles (NANPs). *Curr. Opin. Biotechnol.* 2019, *63*, 8–15.
5. Grabow, W. W.; Jaeger, L., RNA self-assembly and RNA nanotechnology. *Acc. Chem. Res.* 2014, *47* (6), 1871–1880.
6. Afonin, K. A.; Grabow, W. W.; Walker, F. M.; Bindewald, E.; Dobrovolskaia, M. A.; Shapiro, B. A.; Jaeger, L., Design and self-assembly of siRNA-functionalized RNA nanoparticles for use in automated nanomedicine. *Nat. Protoc.* 2011, *6* (12), 2022–2034.
7. Delebecque, C. J.; Lindner, A. B.; Silver, P. A.; Aldaye, F. A., Organization of intracellular reactions with rationally designed RNA assemblies. *Science* 2011, *333* (6041), 470–474.
8. Ponchon, L.; Dardel, F., Large scale expression and purification of recombinant RNA in Escherichia coli. *Methods* 2011, *54* (2), 267–273.
9. Panigaj, M.; Reiser, J., Aptamer guided delivery of nucleic acid-based nanoparticles. *DNA and RNA Nanotechnol.* 2016, *2* (1), 42.
10. Li, M.; Zheng, M.; Wu, S.; Tian, C.; Liu, D.; Weizmann, Y.; Jiang, W.; Wang, G.; Mao, C., In Vivo production of RNA nanostructures via programmed folding of single-stranded RNAs. *Nat. Commun.* 2018, *9* (1), 2196.
11. Reif, R.; Haque, F.; Guo, P., Fluorogenic RNA nanoparticles for monitoring RNA folding and degradation in real time in living cells. *Nucleic Acid Ther.* 2012, *22* (6), 428–437.
12. Afonin, K. A.; Viard, M.; Martins, A. N.; Lockett, S. J.; Maciag, A. E.; Freed, E. O.; Heldman, E.; Jaeger, L.; Blumenthal, R.; Shapiro, B. A., Activation of different split functionalities on re-association of RNA-DNA hybrids. *Nat. Nanotechnol.* 2013, *8* (4), 296–304.
13. Halman, J. R.; Satterwhite, E.; Roark, B.; Chandler, M.; Viard, M.; Ivanina, A.; Bindewald, E.; Kasprzak, W. K.; Panigaj, M.; Bui, M. N.; Lu, J. S.; Miller, J.; Khisamutdinov, E. F.; Shapiro, B. A.; Dobrovolskaia, M. A.; Afonin, K. A., Functionally-interdependent shape-switching nanoparticles with controllable properties. *Nucleic Acids Res.* 2017, *45* (4), 2210–2220.
14. Ke, W.; Hong, E.; Saito, R. F.; Rangel, M. C.; Wang, J.; Viard, M.; Richardson, M.; Khisamutdinov, E. F.; Panigaj, M.; Dokholyan, N. V.; Chammas, R.; Dobrovolskaia, M. A.; Afonin, K. A., RNA-DNA fibers and polygons with controlled immunorecognition activate RNAi, FRET and transcriptional regulation of NF-kappa B in human cells. *Nucleic Acids Res.* 2019, *47* (3), 1350–1361.
15. Bouhedda, F.; Autour, A.; Ryckelynck, M., Light-Up RNA Aptamers and Their Cognate Fluorogens: From Their Development to Their Applications. *Int. J. Mol. Sci.* 2017, *19* (1), 44.
16. Paige, J. S.; Wu, K. Y.; Jaffrey, S. R., RNA mimics of green fluorescent protein. *Science* 2011, *333* (6042), 642–646.
17. Filonov, G. S.; Moon, J. D.; Svensen, N.; Jaffrey, S. R., Broccoli: rapid selection of an RNA mimic of green fluorescent protein by fluorescence-based selection and directed evolution. *J. Am. Chem. Soc.* 2014, *136* (46), 16299–16308.
18. Dolgosheina, E. V.; Jeng, S. C.; Panchapakesan, S. S.; Cojocaru, R.; Chen, P. S.; Wilson, P. D.; Hawkins, N.; Wiggins, P. A.; Unrau, P. J., RNA mango aptamer-fluorophore: a bright, high-affinity complex for RNA labeling and tracking. *ACS Chem. Biol.* 2014, *9* (10), 2412–2420.
19. Song, W.; Filonov, G. S.; Kim, H.; Hirsch, M.; Li, X.; Moon, J. D.; Jaffrey, S. R., Imaging RNA polymerase III transcription using a photostable RNA–fluorophore complex. *Nat. Chem. Biol.* 2017, *13* (11), 1187–1194.
20. Jepsen, M. D. E.; Sparvath, S. M.; Nielsen, T. B.; Langvad, A. H.; Grossi, G.; Gothelf, K. V.; Andersen, E. S., Development of a genetically encodable FRET system using fluorescent RNA aptamers. *Nat. Commun.* 2018, *9* (1), 18.
21. O'Hara, J.; Marashi, D.; Morton, S.; Jaeger, L.; Grabow, W.; O'Hara, J. M.; Marashi, D.; Morton, S.; Jaeger, L.; Grabow, W. W., Optimization of the split-spinach aptamer for monitoring nanoparticle assembly involving multiple contiguous RNAs. *Nanomaterials* 2019, *9* (3), 378.

22. Chandler, M.; Lyalina, T.; Halman, J.; Rackley, L.; Lee, L.; Dang, D.; Ke, W.; Sajja, S.; Woods, S.; Acharya, S.; Baumgarten, E.; Christopher, J.; Elshalia, E.; Hrebien, G.; Kublank, K.; Saleh, S.; Stallings, B.; Tafere, M.; Striplin, C.; Afonin, K. A., Broccoli fluorets: split aptamers as a user-friendly fluorescent toolkit for dynamic RNA nanotechnology. *Molecules* 2018, *23* (12).

23. Panigaj, M.; Johnson, M. B.; Ke, W.; McMillan, J.; Goncharova, E. A.; Chandler, M.; Afonin, K. A., Aptamers as modular components of therapeutic nucleic acid nanotechnology. *ACS Nano* 2019, *13* (11), 12301–12321.

24. Guo, P.; Zhang, C.; Chen, C.; Garver, K.; Trottier, M., Inter-RNA interaction of phage phi29 pRNA to form a hexameric complex for viral DNA transportation. *Mol. Cell* 1998, *2* (1), 149–155.

25. Shu, Y.; Haque, F.; Shu, D.; Li, W.; Zhu, Z.; Kotb, M.; Lyubchenko, Y.; Guo, P., Fabrication of 14 different RNA nanoparticles for specific tumor targeting without accumulation in normal organs. *RNA* 2013, *19* (6), 767–777.

26. Chandler, M.; Afonin, K. A., Smart-responsive Nucleic Acid Nanoparticles (NANPs) with the potential to modulate immune behavior. *Nanomaterials (Basel)* 2019, *9* (4), 611.

27. Hong, E.; Halman, J. R.; Shah, A.; Cedrone, E.; Truong, N.; Afonin, K. A.; Dobrovolskaia, M. A., Toll-like receptor-mediated recognition of Nucleic Acid Nanoparticles (NANPs) in human primary blood cells. *Molecules* 2019, *24* (6), 1094.

28. Rackley, L.; Stewart, J. M.; Salotti, J.; Krokhotin, A.; Shah, A.; Halman, J.; Juneja, R.; Smollett, J.; Roark, B.; Viard, M.; Tarannum, M.; Vivero-Escoto, J. L.; Johnson, P.; Dobrovolskaia, M. A.; Dokholyan, N. V.; Franco, E.; Afonin, K. A., RNA fibers as optimized nanoscaffolds for siRNA coordination and reduced immunological recognition. *Adv. Funct. Mater.* 2018, *28* (48): 1805959.

29. Hong, E.; Halman, J. R.; Shah, A. B.; Khisamutdinov, E. F.; Dobrovolskaia, M. A.; Afonin, K. A., Structure and composition define immunorecognition of nucleic acid nanoparticles. *Nano Lett.* 2018, *18* (7), 4309–4321.

14 From Computational RNA Structure Prediction to the Design of Biologically Active RNA-Based Nanostructures

W.K. Kasprzak
Frederick National Laboratory for Cancer Research, Frederick, MD, USA

Bruce A. Shapiro
National Cancer Institute, Frederick, MD, USA

CONTENTS

As many genetic and catalytic functions of RNA have been elucidated in recent years, most notably the RNA interference mechanism, its biological importance and ways of utilizing it to disrupt cancer or other disease pathways have been recognized. RNA is also a flexible structural material, capable of self-assembling into complicated three-dimensional objects, and, as such, it can be employed in the design of RNA nanostructures in which one kind of material provides biologically active elements as well as a scaffold to which these functional entities can be attached, allowing for multiple targeting and control of stoichiometry of the targeting agents. Discovery of common structural motifs in natural RNAs that are maintained outside the original context and that can be employed as building blocks outside the original structural contexts in a "mix-and-match" fashion have also been incorporated in the software for the prediction of RNA structures and for designing artificial nano-scale assemblies. Structure engineering reverses the structure prediction problem and aims to design self-folding and/or self-assembling sequences that can produce the desired 3D structure. Incorporation of RNA dynamics adds another level of information to employ in the modeling and characterization of the RNA structures. In addition, ligands docking to RNA targets and RNA interactions with proteins should be considered in the prediction and design processes. This short chapter covers multiple vast subjects and cannot be treated as a comprehensive review of all the tools and methods. Selected references are meant only as footholds to a rich literature that the readers are expected to pursue on their own.

From the structural point of view RNA is a polymer chain containing a combination of four nucleotides with a directionality corresponding to the natural order of RNA chain synthesis (transcription), from the 5' to 3' end. The bases that give nucleotides their names, usually referred to by the single letters A, G, C, and U, are Adenine, Guanine, Cytosine, and Uracil. A primary structure of RNA is defined by a sequence of nucleotides. Conceptually single-stranded, RNA chains are sufficiently flexible to fold onto themselves and base pair Gs with Cs and As with Us in the canonical, Watson-Crick scenario. An additional, so-called "wobble" base pair G-U is also frequent. In general, all kinds of non-canonical interactions are found in natural RNAs. The pattern of paired and unpaired nucleotides defines the secondary structure of a sequence. Pairing interactions stabilize the

DOI: 10.1201/9781003001560-16

folded RNA. Unpaired regions, out of which emanate two or more helices (or branches), are called loops, and they tend to destabilize the secondary structure.

In 3D terms, secondary structure folding leads to formation of helices between anti-parallel complementary fragments of the sequence. In addition, in 3D, nucleotides in the loop regions can form non-canonical pairs, driven in part by stabilizing stacking interactions with the flanking helices further reducing truly single-stranded regions. Beyond the secondary structure and its 3D equivalents, there are tertiary interactions between the single-stranded regions resulting in a variety of pseudoknots. Despite a relatively low number of RNA structures (~1,500 as of 2019) deposited in the PDB database (rcsb.org [1], analysis of them made it possible to identify and classify nucleotide pairing types based on the Leontis-Westhof combinations of three interacting nucleotide "edges," Watson-Crick (canonical), Hoogsteen and Sugar (with some extensions) [2–4]. From the modeling and RNA nanostructure engineering perspective, one of the most valuable principles gleaned from the analysis of the available structures is that of structural isostericity or preservation of the geometric features in functionally equivalent RNAs [5–7], which was shown to be evolutionarily preserved more strongly than the primary structure (sequence) [6].

Numerous programs have been developed for the prediction of the secondary structure of RNA given an input sequence. See [8, 9] and the website en. wikipedia.org/wiki/List_of_RNA_structure_prediction_software for a list of secondary structure prediction programs. They all use either experimentally obtained free energy parameters for nucleotide interactions or statistical potential parameters based on the distributions of the observed features in the known RNA structures. Various deterministic and non-deterministic algorithms are applied to produce the optimal free energy solutions and a sampling of suboptimal conformations for a full length or a co-transcriptional folding, i.e. folding with an elongating sequence. Some provide options to add tertiary interaction predictions to the secondary structure, based on partial free energy approximations of such features or other criteria. Most 3D structure prediction programs take primary and secondary structure information as input and build the model based on differently defined motifs giving premium to sequence homology or the closest equivalence. Other methods allow for de novo coarse-grained dynamics simulations to guide folding in 3D space. Again, refer to the site en. wikipedia.org/wiki/List_of_RNA_structure_prediction_software for pointers to further exploration. For comprehensive descriptions of the multiple tools and strategies for computational 3D RNA structure prediction refer to the publications on the RNA-Puzzle competitions [10–14]. The improving, but still not perfect, accuracy of these programs leaves room for further molecular dynamics (MD) exploration of the optimality of the models, their stability, and conformation alternatives based on the sampling of the free energy landscape around the initial states (models). All-atom MD simulations approximate dynamics of molecular interactions (for solute and solvent) with Newtonian equations of motion calculated at small times intervals (1–4 fs) based on a potential energy function and an associated force field. For RNA molecules, the AMBER force field has become the "industry standard" after generations of parameter tuning to best reproduce experimental results and match quantum mechanics (QM) simulations [15–20]. MD is an enormous subject that has been comprehensively reviewed by Sponer et al. [21]. For examples of MD applications in nanostructure characterization, see [22, 23].

The hierarchical and modular nature of RNA structures lends itself to nano-scale structure engineering. Compared to the structure prediction problem, the order of steps is reversed, with geometric validation of a desired 3D shape performed first, followed by the design of a sequence or sequences that could fold into the designed shape. Building on the experience with secondary/tertiary structure prediction and 3D modeling, we have first created a database of structural junctions (N-way junctions/loops and special hairpin-to-hairpin "kissing loops") [24]. These can be used as "nodes" and module interfaces in self-assembling nanostructures (e.g., see [25–31]). A good example of this design philosophy is a hexameric ring structure that was recently developed into a truncated tetrahedron nanostructure as illustrated in Figure 14.1 [31–33]. A kissing loop from the ColE1 plasmid (PDB: 2BJ2) was used as an interface between six "dumbbells" of the original ring design, i.e. each dumbbell consists of 15 base-pair-long helices with two hairpin loops at both ends with each chain's

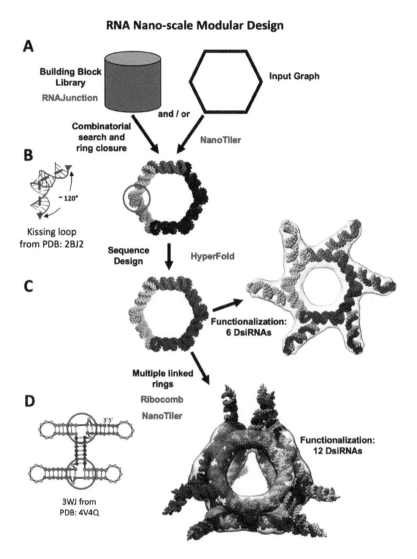

RNA Nano-scale Modular Design

FIGURE 14.1 An illustration of the modular nature or RNA nanostructure design. (A) The design process starts with a selection of a junction from a building block library, such as our RNAJunction database and, optionally, a target nano-shape graph to be translated into an actual molecule. In this case, six copies of the selected kissing loop (one shown inside a red circle) and linking helices are employed. (B) Combinatorial searches among multiple linker helix lengths with minor helix distortions to achieve a clean closure of the ring are performed by the program NanoTiler (or its "wrapper" script – Ribocomb). (C) Program HyperFold can be used to design a sequence or (six in this case) sequences that would most efficiently self-assemble into the desired shape. If the fully functionalized ring with six dicer substrate siRNA arms is being designed, sequence optimization is applied to the extended chains and their complements forming arms, preserving the immutable functional sequence fragments. Model of the functionalized nanoring inside a cryo-EM envelope is shown. (D) Multiple rings can be linked together by bridging two sides of the original hexamer with two copies of a three-way junction (3WJ) (shown inside red circles) and additional helix fragment yielding and "H"-shaped block. Optimization of the geometry and sequences can be achieved, as before, with the aid of Ribocomb/NanoTiler and HyperFold, respectively. A model of the fully functionalized tetrahedral nanoparticle is shown with its core perfectly fitting inside a cryo-EM reconstruction of the scaffold without the functional arms.

5′ and 3′ bases meeting in the middle of the helix. This kissing loop pairs all seven nucleotides of each half and maintains an approximately 120° angle between the emanating helices, a natural fit to a hexagonal shape when used six times. The kissing loop (hairpins A and B) can be split into halves and used as the hairpins at the end of each dumbbell. In this case the natural flexibility of the design, combined with complementarity of each module, tends to produce a distribution of polygons of varying sizes (smaller and larger than the intended hexagons). Using two dumbbell types, AA and BB, decreases the variability of the self-assembling products to a few even-sided species. The best pure hexagon yield is obtained when the sequences of the interacting halves of the kissing loops are mutated to promote six unique pairs [34]. The nanoring model can be built with the aid of our program NanoTiler [35, 36] or other 3D modeling tools, while the intra-strand folding of each dumbbell with inter-strand interactions limited to the hairpins can be optimized for six sequences using programs such as NanoFolder or newer HyperFold [37, 38]. Recently, we have added a bridging linker between every other dumbbell of four hexameric rings to achieve a truncated tetrahedral scaffold, which increased the number of possible functional elements from six to twelve. The new nanoparticle showed improved RNAi efficacy of its Dicer substrate siRNAs (DsiRNAs), even when normalized by the number of functional units, hinting at the importance of the full nanoparticle shape [33]. This linking of four rings was engineered by adding three-way junctions from *E coli's* 23S rRNA (PDB: 4V4Q) to the middle of the two bridged dumbbells. This junction maintains a straight stacking of the two halves of the ring dumbbell and a near right angle between them and bridging linker helix. In terms of a sequence design, the two linked dumbbells fold from one chain, effectively creating "H"-shaped building blocks. The tetrahedral structure maintained its integrity with low levels of distortions in MD simulations, which agreed well with experimental particle size measurements and cryo-EM reconstruction. In addition, we would like to briefly mention that another level of modeling and design involves potential influences of ligands on RNA. These include coordinating divalent ions, such as Mg^{2+}, that may stabilize particular conformations and various triggers of RNA aptamers that bind particular targets with high specificity and affinity, potentially changing their 3D geometry in the process. Finally, RNA–protein interactions should be considered. For a brief review of ligand docking programs appropriate for nucleic acid targets and references to studies that validated ligand-docking nanostructure designs in vitro and in vivo, refer to our review [39].

ACKNOWLEDGMENTS

This research was supported [in part] by the Intramural Research Program of the NIH, Center for Cancer Research, NCI-Frederick. This work has been funded in whole or in part with Federal funds from the Frederick National Laboratory for Cancer Research, National Institutes of Health, under contract no. HHSN261200800001E. The content of this publication does not necessarily reflect the views or policies of the Department of Health and Human Services, nor does mention of trade names, commercial products, or organizations implies endorsement by the U.S. Government.

REFERENCES

1. Berman HM, Westbrook J, Feng Z, Gilliland G, Bhat TN, Weissig H, Shindyalov IN, Bourne PE: The protein data bank. *Nucleic Acids Res* 2000, **28**:235–242.
2. Leontis NB, Westhof E: Geometric nomenclature and classification of RNA base pairs. *RNA* 2001, **7**:499–512.
3. Leontis NB, Westhof E: Analysis of RNA motifs. *Curr Opin Struct Biol* 2003, **13**:300–308.
4. Zirbel CL, Sponer JE, Sponer J, Stombaugh J, Leontis NB: Classification and energetics of the base-phosphate interactions in RNA. *Nucleic Acids Res* 2009, **37**:4898–4918.
5. Leontis NB, Stombaugh J, Westhof E: The non-Watson-Crick base pairs and their associated isostericity matrices. *Nucleic Acids Res* 2002, **30**:3497–3531.

6. Stombaugh J, Zirbel CL, Westhof E, Leontis NB: Frequency and isostericity of RNA base pairs. *Nucleic Acids Res* 2009, **37**:2294–2312.

7. Grabow W, Jaeger L: RNA modularity for synthetic biology. *F1000Prime Rep* 2013, **5**:46.

8. Shapiro BA, Wu JC, Bengali D, Potts MJ: The massively parallel genetic algorithm for RNA folding: MIMD implementation and population variation. *Bioinformatics* 2001, **17**:137–148.

9. Shapiro BA, Kasprzak W, Grunewald C, Aman J: Graphical exploratory data analysis of RNA secondary structure dynamics predicted by the massively parallel genetic algorithm. *J Mol Graph Model* 2006, **25**:514–531.

10. Cruz JA, Blanchet MF, Boniecki M, Bujnicki JM, Chen SJ, Cao S, Das R, Ding F, Dokholyan NV, Flores SC, et al.: RNA-Puzzles: a CASP-like evaluation of RNA three-dimensional structure prediction. *RNA* 2012, **18**:610–625.

11. Miao Z, Adamiak RW, Blanchet MF, Boniecki M, Bujnicki JM, Chen SJ, Cheng C, Chojnowski G, Chou FC, Cordero P, et al.: RNA-Puzzles Round II: assessment of RNA structure prediction programs applied to three large RNA structures. *RNA* 2015, **21**:1066–1084.

12. Miao Z, Adamiak RW, Antczak M, Batey RT, Becka AJ, Biesiada M, Boniecki MJ, Bujnicki JM, Chen SJ, Cheng CY, et al.: RNA-Puzzles Round III: 3D RNA structure prediction of five riboswitches and one ribozyme. *RNA* 2017, **23**:655–672.

13. Miao Z, Westhof E: RNA structure: Advances and assessment of 3D structure prediction. *Annu Rev Biophys* 2017, **46**:483–503.

14. Magnus M, Antczak M, Zok T, Wiedemann J, Lukasiak P, Cao Y, Bujnicki JM, Westhof E, Szachniuk M, Miao Z: RNA-Puzzles toolkit: A computational resource of RNA 3D structure benchmark datasets, structure manipulation, and evaluation tools. *Nucleic Acids Res* 2020, **48**:576–588.

15. Hopkins CW, Le Grand S, Walker RC, Roitberg AE: Long-Time-step molecular dynamics through hydrogen mass repartitioning. *J Chem Theory Comput* 2015, **11**:1864–1874.

16. Case DA, Cheatham TE, Darden T, Gohlke H, Luo R, Merz KM, Onufriev A, Simmerling C, Wang B, Woods RJ: The Amber biomolecular simulation programs. *Journal of Computational Chemistry* 2005, **26**:1668–1688.

17. Cornell WD, Cieplak P, Bayly CI, Gould IR, Merz KM, Ferguson DM, Spellmeyer DC, Fox T, Caldwell JW, Kollman PA: A 2nd generation force-field for the simulation of proteins, nucleic-acids, and organic-molecules. *Journal of the American Chemical Society* 1995, **117**:5179–5197.

18. Salomon-Ferrer R, Case DA, Walker RC: An overview of the Amber biomolecular simulation package. *Wiley Interdisciplinary Reviews-Computational Molecular Science* 2013, **3**:198–210.

19. Salomon-Ferrer R, Gotz AW, Poole D, Le Grand S, Walker RC: Routine Microsecond Molecular Dynamics Simulations with AMBER on GPUs. 2. Explicit Solvent Particle Mesh Ewald. *J Chem Theory Comput* 2013, **9**:3878–3888.

20. Kim T, Kasprzak WK, Shapiro BA: Protocols for molecular dynamics simulations of RNA nanostructures. *Methods Mol Biol* 2017, **1632**:33–64.

21. Sponer J, Bussi G, Krepl M, Banas P, Bottaro S, Cunha RA, Gil-Ley A, Pinamonti G, Poblete S, Jurecka P, et al.: RNA structural dynamics as captured by molecular simulations: A comprehensive overview. *Chem Rev* 2018, **118**:4177–4338.

22. Kasprzak W, Bindewald E, Kim TJ, Jaeger L, Shapiro BA: Use of RNA structure flexibility data in nanostructure modeling. *Methods* 2011, **54**:239–250.

23. Afonin KA, Kasprzak W, Bindewald E, Puppala PS, Diehl AR, Hall KT, Kim TJ, Zimmermann MT, Jernigan RL, Jaeger L, et al.: Computational and experimental characterization of RNA cubic nanoscaffolds. *Methods* 2014, **67**:256–265.

24. Bindewald E, Hayes R, Yingling YG, Kasprzak W, Shapiro BA: RNAJunction: A database of RNA junctions and kissing loops for three-dimensional structural analysis and nanodesign. *Nucleic Acids Res* 2008, **36**:D392–397.

25. Shu D, Moll WD, Deng Z, Mao C, Guo P: Bottom-up assembly of RNA arrays and superstructures as potential parts in nanotechnology. *Nano Lett* 2004, **4**:1717–1723.

26. Guo S, Tschammer N, Mohammed S, Guo P: Specific delivery of therapeutic RNAs to cancer cells via the dimerization mechanism of phi29 motor pRNA. *Hum Gene Ther* 2005, **16**:1097–1109.

27. Chworos A, Severcan I, Koyfman AY, Weinkam P, Oroudjev E, Hansma HG, Jaeger L: Building programmable jigsaw puzzles with RNA. *Science* 2004, **306**:2068–2072.

28. Severcan I, Geary C, Verzemnieks E, Chworos A, Jaeger L: Square-shaped RNA particles from different RNA folds. *Nano Lett* 2009, **9**:1270–1277.
29. Geary C, Chworos A, Verzemnieks E, Voss NR, Jaeger L: Composing RNA nanostructures from a syntax of RNA structural modules. *Nano Lett* 2017, **17**:7095–7101.
30. Afonin KA, Grabow WW, Walker FM, Bindewald E, Dobrovolskaia MA, Shapiro BA, Jaeger L: Design and self-assembly of siRNA-functionalized RNA nanoparticles for use in automated nanomedicine. *Nat Protoc* 2011, **6**:2022–2034.
31. Afonin KA, Viard M, Koyfman AY, Martins AN, Kasprzak WK, Panigaj M, Desai R, Santhanam A, Grabow WW, Jaeger L, et al.: Multifunctional RNA nanoparticles. *Nano Lett* 2014, **14**:5662–5671.
32. Yingling YG, Shapiro BA: Computational design of an RNA hexagonal nanoring and an RNA nanotube. *Nano Lett* 2007, **7**:2328–2334.
33. Zakrevsky P, Kasprzak WK, Heinz WF, Wu W, Khant H, Bindewald E, Dorjsuren N, Fields EA, de Val N, Jaeger L, et al.: Truncated tetrahedral RNA nanostructures exhibit enhanced features for delivery of RNAi substrates. *Nanoscale* 2020, **12**:2555–2568.
34. Grabow WW, Zakrevsky P, Afonin KA, Chworos A, Shapiro BA, Jaeger L: Self-assembling RNA nanorings based on RNAI/II inverse kissing complexes. *Nano Lett* 2011, **11**:878–887.
35. Bindewald E, Grunewald C, Boyle B, O'Connor M, Shapiro BA: Computational strategies for the automated design of RNA nanoscale structures from building blocks using NanoTiler. *J Mol Graph Model* 2008, **27**:299–308.
36. Sharan R, Bindewald E, Kasprzak WK, Shapiro BA: Computational generation of RNA nanorings. *Methods Mol Biol* 2017, **1632**:19–32.
37. Bindewald E, Afonin K, Jaeger L, Shapiro BA: Multistrand RNA secondary structure prediction and nanostructure design including pseudoknots. *ACS Nano* 2011, **5**:9542–9551.
38. Bindewald E, Afonin KA, Viard M, Zakrevsky P, Kim T, Shapiro BA: Multistrand structure prediction of nucleic acid assemblies and design of RNA switches. *Nano Lett* 2016, **16**:1726–1735.
39. Kasprzak WK, Ahmed NA, Shapiro BA: Modeling ligand docking to RNA in the design of RNA-based nanostructures. *Curr Opin Biotechnol* 2019, **63**:16–25.

15 Application of RNA Tertiary Structure Prediction Tool iFoldRNA in RNA Nanotechnology

Jian Wang and Nikolay V. Dokholyan
Penn State College of Medicine, Hershey, PA, USA

CONTENTS

INTRODUCTION

In addition to serving as genetic information carriers (mRNA), protein expression transporters (tRNA), and large biomolecule scaffolds (rRNA), RNA molecules play pivotal roles in nearly all aspects of cell life [1–4], and more RNA functions remain to be discovered. RNA functions are determined by the specific sequences and structures of RNAs. RNA structure is critical for the function of both mRNAs and non-coding RNAs (ncRNAs), including microRNAs [1] (miRNAs), riboswitches [2], ribozymes [3], and long non-coding RNAs [4] (lncRNAs). Like proteins, the functions of RNA are mostly decided by their specific tertiary structures, suggesting RNA molecules as promising potential drug targets. By understanding RNA functions, knowledge of RNA tertiary structure is not only essential for structure-based drug design but also beneficial to applications in the emerging RNA nanotechnology field [5–7]. RNAs are generally structurally versatile and prone to self-assembly, RNAs of different sequences fold readily, and RNA annealing needs low free energy; these properties provide the possibilities to control RNA structures [8]. These characteristics enable RNA as a potent biomaterial for applications in nanotechnology. More and more roles of RNA in nanomedicine applications have been found, and potential applications of RNA nanotechnology in disease therapies have been recently summarized [9, 10].

    RNA secondary structures are typically experimentally determined by DMS probing [11] and SHAPE [12], and computational RNA secondary structure prediction algorithms [13–16] have advanced significantly [16], especially when experimental data is integrated into the computational

DOI: 10.1201/9781003001560-17

prediction. The tertiary structure of RNA can be obtained by crystallography, nuclear magnetic resonance (NMR), and cryo-electron microscopy (cryo-EM). However, by 2018, only approximately 3,000 RNA structures are deposited in the RCSB PDB [17]. In contrast, more than 130,000 protein structures are deposited by that same year. The largest experimentally solved RNA is a ribosome subcomponent that is stabilized through interactions with proteins. Computational RNA tertiary structure prediction tools have promise to supplement experimental methods and facilitate elucidation of RNA structures and functions [18, 19]. Although the prediction of RNA tertiary structure remains challenging, new computational methodologies for RNA tertiary structure prediction are emerging [20–38] based on various strategies including fragment assembly (RNAComposer [21, 22], 3dRNA [23–26]), coarse-grained and atomic-level molecular dynamics or Monte Carlo simulation (FARFAR [33], iFoldRNA [27–32], SimRNA [34, 35]), and internal coordinate space dynamics (RNABuilder [36], Vfold [37, 38]). All of these methods have limitations, but the field of computational RNA tertiary structure prediction is rapidly evolving. In this chapter, we will present the workflow of utilizing iFoldRNA for RNA tertiary structure prediction and demonstrate its potential applications in nanotechnology. We apply iFoldRNA to build and compare the stability of tertiary structures of RNA triangles and rectangles [39]. We then conduct thermodynamic analysis of the tertiary structures of RNA cubes and rings [40, 41].

RNA 3D STRUCTURE PREDICTION USING iFoldRNA

THE OUTLINE OF iFoldRNA

The prediction of RNA structure is accomplished using a coarse-grained 3-bead RNA model. Each bead in the model represents a phosphate, sugar, or nucleobase. iFoldRNA performs folding simulations using the Discrete Molecular Dynamics (DMD) simulation engine [27, 42–45] and Medusa force field [46, 47]. Users can provide the secondary structure to facilitate RNA modeling. Base-pairing information in secondary structure is implemented as an additional potential promoting tertiary contacts between corresponding nucleotides. Given the sequence, with or without the secondary structure, iFoldRNA first performs an initial short simulation to preliminarily fold the RNA. The objective of the initial simulation is to provide a good starting point for the subsequent replica exchange molecular dynamics simulation aiming to thoroughly searching the conformational space. An ensemble of RNA molecules are generated from simulations at 8 different temperatures in 8 threads [48] (Figure 15.1). Subsequently, 1% of the lowest energy structures are selected and clustered by root mean squared deviation (RMSD). Finally, the centroids of the resulting clusters are subject to all-atom reconstruction. If more structural information is available [49, 50], such as inter-nucleotide proximity information [28], hydroxyl radical probing (HRP) data [29], correlated chemical probing data [51, 52], nuclear magnetic resonance (NMR) data [32], and fluorescence resonance energy transfer (FRET) data [53], an additional force field is applied [29], effectively biasing RNA conformations to satisfy this structural constraints. Particularly, for HRP data, in addition to low free energies, the structures selected for clustering are also required to have high structure reactivity correlations. Conversion to an all-atom representation is performed by replacing each 3-bead nucleotide by randomly selected rotamers of corresponding nucleotides in an all-atom representation. A 3-bead nucleotide consists of three atoms: P, C and O, which refer to the phosphate, sugar, and base group. The P, C, and O atoms in the 3-bead model are aligned with the P, C1', and N1/N3 atoms in the all-atom model through the Kabsch algorithm [54]. The initial all-atom structure is run through a short all-atom DMD simulation to connect bonds and remove clashes. This simulation is performed with a high heat exchange coefficient, which allows for rapid dissipation of excess heat generated.

Given a sequence, iFoldRNA first performs an initial short simulation to build a preliminary 3D structure. Secondary structure and constraints are accepted to facilitate the prediction. Next, iFoldRNA utilizes replica exchange simulations to thoroughly explore the conformational space. Subsequently, structures in the conformation ensemble are ranked by iFoldRNA force field.

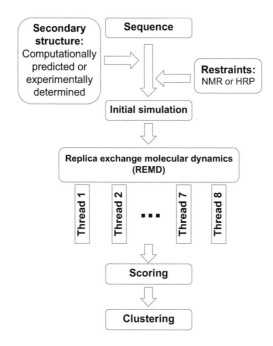

FIGURE 15.1   Workflow of iFoldRNA.

The 1% lowest free energy structures are selected and clustered. The centroid structures of all clusters are then subject to all-atom reconstruction in order to obtain the final candidate structures.

iFoldRNA WEBSERVER

iFoldRNA is an automated RNA structure prediction webserver (http://iFoldRNA.dokhlab.org) designed to predict three-dimensional RNA structure based on available sequence data, as well as optional secondary structure and distance constraints based on experiments [55] such as NMR [32], HRP [29]A, and SHAPE-JuMP [56]. As of September 2019, approximately 4,000 tasks have been submitted to the iFoldRNA server. The server operates in four different modes:

1. RNA modeling with sequence only. In this mode, the server predicts RNA tertiary structure based solely on the nucleotide sequence provided. This mode works for only short RNAs, which are less than 50 nucleotides long. The predicted structures are typically within 4 Å RMSD from experimentally determined structures. The limitation on RNA size is due to time requirements; increasing sequence size quickly becomes computationally prohibitive, since the RNA conformational space grows exponentially.
2. Thermodynamic analysis. In this mode, various thermodynamic properties of RNA folding are inferred using replica exchange DMD simulations.
3. RNA modeling using HRP restraints. In this mode, a user can provide nucleotide solvent accessibility data, which is measured as reactivities from HRP experiments.
4. RNA modeling using NMR constraints.

By use of experimentally derived constraints, the conformational space can be significantly reduced, allowing tertiary structure prediction of RNAs up to a few hundred nucleotides long. The webserver interface is self-explanatory. Examples are provided for the sequence and secondary structure. Restraint examples are provided as separate files.

RNA MODELING ACCURACY USING iFoldRNA

In our previous work [30, 31, 57], we thoroughly tested the accuracy and quality of iFoldRNA predictions using RNAs of varying complexities. We have also participated in the RNA tertiary structure prediction competition, RNA-Puzzles [58–60]. Targets in RNA-Puzzles contain sequences from large RNA structures with limited or no homology to previously solved RNA molecules. iFoldRNA with the aid of experimental data predicts one of the best models [60] in RNA-Puzzles (Target 12) and typically ranks top, suggesting that iFoldRNA can provide useful structural information for biological problems.

APPLICATIONS OF iFoldRNA IN RNA NANOTECHNOLOGY

TERTIARY STRUCTURE CONSTRUCTION AND COMPARISON OF THERMODYNAMIC STABILITY OF RNA TRIANGLE AND RECTANGLE

As a new class of reconfigurable materials, nucleic acid-based assemblies [61] overcome limitations of traditional biochemical approaches and improve the potential therapeutic utility of nucleic acids. Nucleic acid-based assemblies typically interact with each other and further communicate with the cellular machinery in a controlled manner. Ke and coworkers [39] designed RNA–DNA fibers and polygons that are able to cooperate in different human cell lines and that have defined immunostimulatory properties confirmed by *in vitro* experiments. Their work expands the possibilities of RNA nanotechnology by introducing simple design principles, potentially allowing for a simultaneous release of various siRNAs together with functional DNA sequences and providing controlled rates of reassociation, stabilities in human blood serum, and immunorecognition. They designed a set of dynamic RNA–DNA fibers and polygons (triangles and rectangles) that can interact inside the cells to release a large number of DS RNAs. They found that triangle shapes are the most common among polygon structures observed by AFM (ref. 38 Figure 15.2).

In order to compare the thermodynamic stabilities of triangle and rectangle assemblies, we utilize iFoldRNA to perform coarse-grained molecular dynamics simulations for both shapes. Based on the shape of the assemblies, we first derive their specific secondary structures. Next, we perform a short iFoldRNA simulation to build an initial tertiary structure of the assemblies. Next, we utilize Replica Exchange Molecular Dynamics (REMD) to thoroughly sample the conformational space. The temperatures of the 8 threads of REMD are distributed from 0.2 kT to 0.4 kT. From the final conformational ensemble of RNA triangles and rectangles, we select two conformations that feature ideal triangle and rectangle shapes. The two conformations are then used as initial structures for subsequent MD simulations. The structures undergo drastic conformational variations during the 3 million steps of simulations (Figure 15.2). The results suggest that the triangle shape is relatively stable, while the rectangle shape undergoes greater fluctuations. These dynamic differences explain why triangle shapes are the most common among polygon structures observed by AFM (ref. 36; Figure 15.2).

THERMODYNAMICS ANALYSIS OF RNA CUBE AND RING

Infusion reactions (IRs) hamper the translation of many novel therapeutics, including those utilizing nanotechnology. Nucleic acid nanoparticles (NANPs) hold tremendous potential to bring these reactions under control by tuning the particle's physicochemical properties to the desired type and magnitude of the immune response. NANPs are a novel class of therapeutics prepared by rational design of relatively short oligonucleotides to self-assemble into various programmable geometric shapes. Recently, Hong and coworkers [40, 41] reported the very first comprehensive study of the structure–activity relationship between NANP shape, size, composition, and immunorecognition in human cells. They chose a small representative group of NANPs to address the influence of their composition and shape (RNA rings and cubes) on the immunorecognition in human cells.

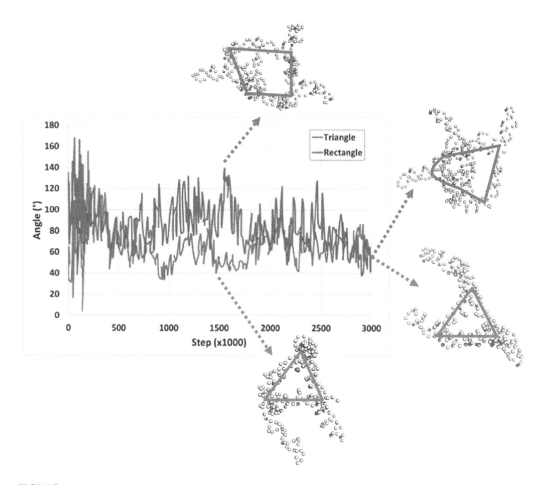

FIGURE 15.2 Molecular dynamics simulation of RNA triangle and rectangle. The simulation is performed by using iFoldRNA, and the model is coarse grained by using the "C3'" atom to represent each atom in the molecule. The triangle shape is always preserved during the simulation, while the rectangle shape is unstable. The angle between two edges in the triangle remains stable after the triangle shape is formed, while the angle between two edges in the rectangle decreases steadily after the rectangle shape is formed.

Similar to the thermodynamic analysis of RNA triangles and rectangles, we utilized iFoldRNA to compare the structural stabilities of RNA rings and cubes, which affects their immunorecognition in human cells. Given the tertiary structures of RNA ring and cube, we first manually extracted the secondary structures from the known tertiary structures. We directly utilize REMD to thoroughly sample the conformational space. The temperatures of 8 threads for REMD are distributed from 0.2 kT to 0.4 kT. Next, we calculate the free energy landscapes of RNA ring and cube on the basis of the conformational ensemble sampled by REMD. We derive the potential of mean force (PMF) as follows:

$$\text{PMF}\left(\text{RMSD, E}\right) = -k_b T ln\left(W\left(RMSD, E\right)\right) + C \tag{15.1}$$

where E is the iFoldRNA energy, k_b is the Boltzmann constant, T is the temperature (K), W is a function that defines the probability of a given pair of RMSD and the iFoldRNA energy, and the constant C sets the lowest PMF value at any given temperature to be zero.

The free energy landscape of the RNA ring is considerably smooth; it has only one free energy minimum (Figure 15.3). On the contrary, the free energy landscape of the RNA cube is relatively

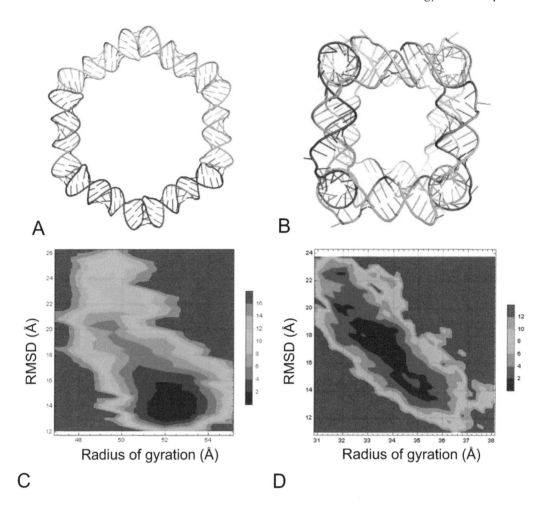

A

B

C

D

FIGURE 15.3 Tertiary structures and free energy landscapes of RNA ring and cube. (A) Tertiary structure of RNA ring. (B) Tertiary structure of RNA cube. (C) Free energy landscape of RNA ring. Values of the colors refer to free energies (units: kcal·mol⁻¹). (D) Free energy landscape of RNA cube.

rough; it has multiple local free energy minima. A smooth free energy landscape causes the conformation to readily stay in the global free energy minimum, reflecting that the tertiary structure is stable with only slight conformational change around that global minimum state. A rough free energy landscape causes the conformation to potentially become trapped in multiple local free energy minima, leading to multiple metastable states, which corresponds to multiple free energy minima, reflecting that the tertiary structure is not stable. The thermodynamic analysis results indicate that the tertiary structure of the RNA ring is more stable than that of RNA cube.

CONCLUSION

Similar to understanding protein structure [62–64], RNA structure has been studied for decades. However, contrary to proteins, methods for predicting the 3D structure of RNA have only recently emerged. iFoldRNA has successfully predicted [29–31] the three-dimensional structure of different RNA based on secondary structure. In this work, we briefly introduce the workflow of iFoldRNA and demonstrate the utilization of iFoldRNA in the field of nanotechnology [5–7]. In the future, we plan to expand the utilities of iFoldRNA to RNA nanotechnology field by 1) supporting the modeling of RNA/DNA complexes, 2) integrating fragment-assembly methods into iFoldRNA to enable motifs-based RNA design, and 3) developing RNA nanoparticles design interface and integrate it into iFoldRNA webserver.

ACKNOWLEDGMENTS

We acknowledge support from the National Institutes of Health (NIH) 1R35 GM134864 and 1R01AG071675, and the Passan Foundation. We gratefully appreciate Dr. Elizabeth A. Proctor for her proofreading of the chapter.

REFERENCES

1. Ambros, V. The functions of animal microRNAs. *Nature* **431**, 350 (2004).
2. Mandal, M. & Breaker, R. R. Gene regulation by riboswitches. *Nat. Rev. Mol. Cell Biol.* **5**, 451 (2004).
3. Doudna, J. A. & Cech, T. R. The chemical repertoire of natural ribozymes. *Nature* **418**, 222 (2002).
4. Prensner, J. R. & Chinnaiyan, A. M. The emergence of lncRNAs in cancer biology. *Cancer Discov.* **1**, 391–407 (2011).
5. Jasinski, D., Haque, F., Binzel, D. W. & Guo, P. Advancement of the emerging field of RNA nanotechnology. *ACS Nano* **11**, 1142–1164 (2017).
6. Grabow, W. W. & Jaeger, L. RNA self-assembly and RNA nanotechnology. *Acc. Chem. Res.* **47**, 1871–1880 (2014).
7. Guo, P. The emerging field of RNA nanotechnology. *Nat. Nanotechnol.* **5**, 833 (2010).
8. Guo, P., Haque, F., Hallahan, B., Reif, R. & Li, H. Uniqueness, advantages, challenges, solutions, and perspectives in therapeutics applying RNA nanotechnology. *Nucleic Acid Ther.* **22**, 226–245 (2012).
9. Afonin, K. A. et al. Multifunctional RNA nanoparticles. *Nano Lett.* **14**, 5662–5671 (2014).
10. Zhang, H. M. et al. Targeted delivery of anti-coxsackievirus siRNAs using ligand-conjugated packaging RNAs. *Antiviral Res.* **83**, 307–316 (2009).
11. Tijerina, P., Mohr, S. & Russell, R. DMS footprinting of structured RNAs and RNA–protein complexes. *Nat. Protoc.* **2**, 2608 (2007).
12. Wilkinson, K. A., Merino, E. J. & Weeks, K. M. Selective 2′-hydroxyl acylation analyzed by primer extension (SHAPE): quantitative RNA structure analysis at single nucleotide resolution. *Nat. Protoc.* **1**, 1610 (2006).
13. Zuker, M., Mathews, D. H. & Turner, D. H. Algorithms and thermodynamics for RNA secondary structure prediction: a practical guide. in *RNA Biochemistry and Biotechnology* 11–43 (Springer, 1999).
14. Sato, K., Hamada, M., Asai, K. & Mituyama, T. Centroid Fold: a web server for RNA secondary structure prediction. *Nucleic Acids Res.* **37**, 277–280 (2009).
15. Zakov, S., Goldberg, Y., Elhadad, M. & Ziv-Ukelson, M. Rich Parameterization Improves RNA Structure Prediction. *J. Comput. Biol.* **18**, 1525–1542 (2011).
16. Puton, T., Kozlowski, L. P., Rother, K. M. & Bujnicki, J. M. CompaRNA: a server for continuous benchmarking of automated methods for RNA secondary structure prediction. *Nucleic Acids Res.* **41**, 4307–4323 (2013).
17. Berman, H. M. et al. The Protein Data Bank and the challenge of structural genomics. *Nat. Struct. \& Mol. Biol.* **7**, 957–959 (2000).
18. Mendez, R. et al. Knowledge-Based Models RNA. *Coarse-Grained Modeling of Biomolecules* 50 (2017).
19. Krokhotin, A. & Dokholyan, N. V. Chapter Three - Computational Methods Toward Accurate RNA Structure Prediction Using Coarse-Grained and All-Atom Models. in *Computational Methods for Understanding Riboswitches* (eds. Chen, S.-J. & Burke-Aguero, D. H. B. T.-M.) **553**, 65–89 (Academic Press, 2015).
20. Westhof, E. RNA modeling, naturally. *Proc. Natl. Acad. Sci.* **109**, 2691–2692 (2012).
21. Popenda, M. et al. Automated 3D structure composition for large RNAs. *Nucleic Acids Res* **40**, e112 (2012).
22. Antczak, M. et al. New functionality of RNAComposer: an application to shape the axis of miR160 precursor structure. *Acta Biochim. Pol.* (2016).

23. Wang, J. et al. Optimization of RNA 3D structure prediction using evolutionary restraints of nucleotide–nucleotide interactions from direct coupling analysis. *Nucleic Acids Res.* **45**, 6299–6309 (2017).

24. Zhao, Y. et al. Automated and fast building of three-dimensional RNA structures. *Sci. Rep.* **2**, 734 (2012).

25. Wang, J. & Xiao, Y. Using 3dRNA for RNA 3-D structure prediction and evaluation. *Curr. Protoc. Bioinform.* 5.9.1–5.9.12 (2017). doi:10.1002bi.21

26. Wang, J., Zhao, Y., Zhu, C. & Xiao, Y. 3dRNAscore: a distance and torsion angle dependent evaluation function of 3D RNA structures. *Nucleic Acids Res.* **43**, e63–e63 (2015).

27. Ding, F. et al. Ab initio RNA folding by discrete molecular dynamics: from structure prediction to folding mechanisms. *RNA* **14**, 1164–1173 (2008).

28. Gherghe, C. M., Leonard, C. W., Ding, F., Dokholyan, N. V. & Weeks, K. M. Native-like RNA tertiary structures using a sequence-encoded cleavage agent and refinement by discrete molecular dynamics. *J Am Chem Soc* **131**, 2541–2546 (2009).

29. Ding, F., Lavender, C. A., Weeks, K. M. & Dokholyan, N. V. Three-dimensional RNA structure refinement by hydroxyl radical probing. *Nat. Methods* **9**, 603–608 (2012).

30. Sharma, S., Ding, F. & Dokholyan, N. V. iFoldRNA: three-dimensional RNA structure prediction and folding. *Bioinformatics* **24**, 1951–1952 (2008).

31. Krokhotin, A., Houlihan, K. & Dokholyan, N. V. iFoldRNA v2: Folding RNA with constraints. *Bioinformatics* **31**, 2891–2893 (2015).

32. Williams Benfeard, I. I. et al. Structure modeling of RNA using sparse NMR constraints. *Nucleic Acids Res.* **45**, 12638–12647 (2017).

33. Das, R., Karanicolas, J. & Baker, D. Atomic accuracy in predicting and designing noncanonical RNA structure. *Nat. Methods* **7**, 291–294 (2010).

34. Rother, K. et al. Template-based and template-free modeling of RNA 3D structure: Inspirations from protein structure modeling. RNA 3D Structure Analysis and Prediction **27**, 67–90 (2012).

35. Boniecki, M. J. et al. SimRNA: a coarse-grained method for RNA folding simulations and 3D structure prediction. *Nucleic Acids Res.* **44**, e63 (2016).

36. Flores, S. C. & Altman, R. B. Turning limited experimental information into 3D models of RNA. *RNA* **16**, 1769–1778 (2010).

37. Cao, S. & Chen, S.-J. Physics-Based De Novo Prediction of RNA 3D Structures. *J. Phys. Chem. B* **115**, 4216–4226 (2011).

38. Xu, X. J., Zhao, P. N. & Chen, S. J. Vfold: A Web Server for RNA Structure and Folding Thermodynamics Prediction. *PLoS One* **9**, e107504 (2014).

39. Ke, W. et al. RNA-DNA fibers and polygons with controlled immunorecognition activate RNAi, FRET and transcriptional regulation of NF-κB in human cells. *Nucleic Acids Res.* **47**, 1350–1361 (2018).

40. Hong, E. et al. Toll-Like Receptor-Mediated Recognition of Nucleic Acid Nanoparticles (NANPs) in Human Primary Blood Cells. *Molecules* **24**, 1094 (2019).

41. Hong, E. et al. Structure and composition define immunorecognition of nucleic acid nanoparticles. *Nano Lett.* **18**, 4309–4321 (2018).

42. Dokholyan, N. V., Buldyrev, S. V., Stanley, H. E. & Shakhnovich, E. I. Discrete molecular dynamics studies of the folding of a protein-like model. *Fold. Des.* **3**, 577–587 (1998).

43. Proctor, E. A., Ding, F. & Dokholyan, N. V. Discrete molecular dynamics. *Wiley Interdiscip. Rev. Comput. Mol. Sci.* **1**, 80–92 (2011).

44. Proctor, E. A. & Dokholyan, N. V. Applications of Discrete Molecular Dynamics in biology and medicine. *Curr. Opin. Struct. Biol.* **37**, 9–13 (2016).

45. Ding, F. & Dokholyan, N. V. Multiscale Modeling of RNA Structure and Dynamics BT - RNA 3D Structure Analysis and Prediction. in (eds. Leontis, N. & Westhof, E.) 167–184 (Springer Berlin Heidelberg, 2012). doi:10.1007/978-3-642-25740-7_9

46. Wang, J. & Dokholyan, N. V. MedusaDock 2.0: Efficient and Accurate Protein-Ligand Docking With Constraints. *J. Chem. Inf. Model.* **59**, 2509–2515 (2019).

47. Ding, F., Yin, S. & Dokholyan, N. V. Rapid flexible docking using a stochastic rotamer library of ligands. *J. Chem. Inf. Model.* **50**, 1623–1632 (2010).

48. Sugita, Y. & Okamoto, Y. Replica-exchange molecular dynamics method for protein folding. *Chem. Phys. Lett.* **314**, 141–151 (1999).

49. Ding, F. & Dokholyan, N. V. RNA three-dimensional structure determination using experimental constraints. *RNA Nanotechnol. Ther.* 159–175 (2013). doi:10.1201/b15152

50. Lavender, C. A., Ding, F., Dokholyan, N. V. & Weeks, K. M. Robust and Generic RNA Modeling Using Inferred Constraints: A Structure for the Hepatitis C Virus IRES Pseudoknot Domain. *Biochemistry* **49**, 4931–4933 (2010).

51. Homan, P. J. et al. Single-molecule correlated chemical probing of RNA. *Proc. Natl. Acad. Sci.* **111**, 13858–13863 (2014).

52. Krokhotin, A., Mustoe, A. M., Weeks, K. M. & Dokholyan, N. V. Direct identification of base-paired RNA nucleotides by correlated chemical probing. *RNA* **23**, 6–13 (2017).

53. Cole, D. I. et al. New Models of Tetrahymena Telomerase RNA from Experimentally Derived Constraints and Modeling. *J. Am. Chem. Soc.* **134**, 20070–20080 (2012).

54. Kabsch, W. A discussion of the solution for the best rotation to relate two sets of vectors. *Acta Crystallogr. Sect. A Cryst. Physics, Diffraction, Theor. Gen. Crystallogr.* **34**, 827–828 (1978).

55. Dokholyan, N.V. "Experimentally-driven protein structure modeling", *Journal of Proteomics*, **220**, 103777, (2020).

56. Christy, T. W., Giannetti, C. A., Houlihan, G., Smola, M. J., Rice, G. M., Wang, J., Dokholyan, N. V., Laederach, A., Holliger, P., and Weeks, K. M. "Direct mapping of higher-order RNA interactions by SHAPE-JuMP", *Biochemistry*, **60**, 1971–1982, (2021).

57. Wang, J. et al. Limits in accuracy and a strategy of RNA structure prediction using experimental information. *Nucleic Acids Research* **47**, (2019).

58. Cruz, J. A. et al. RNA-Puzzles: a CASP-like evaluation of RNA three-dimensional structure prediction. *RNA* **18**, 610–625 (2012).

59. Miao, Z. et al. RNA-Puzzles Round II: assessment of RNA structure prediction programs applied to three large RNA structures. *RNA* **21**, 1066–1084 (2015).

60. Miao, Z. et al. RNA-Puzzles Round III: 3D RNA structure prediction of five riboswitches and one ribozyme. *RNA* **23**, 655–672 (2017).

61. Yan, H. Nucleic acid nanotechnology. *Science (80-).* **306**, 2048–2049 (2004).

62. Ding, F. & Dokholyan, N. V. Simple but predictive protein models. *Trends Biotechnol.* **23**, 450–455 (2005).

63. Dokholyan, N. V. Studies of folding and misfolding using simplified models. *Curr. Opin. Struct. Biol.* **16**, 79–85 (2006).

64. Khatun, J., Khare, S. D. & Dokholyan, N. V. Can Contact Potentials Reliably Predict Stability of Proteins? *J. Mol. Biol.* **336**, 1223–1238 (2004).

16 HyperFold
A Web Server for Predicting Nucleic Acid Complexes

Eckart Bindewald
Basic Science Program, Frederick National Laboratory for Cancer
Research, Frederick, MD, USA

Bruce A. Shapiro
RNA Biology Laboratory, National Cancer Institute, Frederick, MD, USA

CONTENTS

INTRODUCTION

RNA and DNA nanostructures typically consist of several strands that form complexes with potentially intricate topologies (Afonin et al. 2010, Dibrov et al. 2011, Shu et al. 2011, Yingling and Shapiro 2007, Chworos et al. 2004). It is important for the characterization of such nucleic acid nanostructures to have a computational tool that is able to predict significant properties such as the secondary structure. Programs that predict base pairing of multiple RNA strands potentially forming complexes are Multifold and NUPACK (Zadeh et al. 2011, Andronescu, Zhang, and Condon 2005). Novel aspects of HyperFold are that the program allows predictions not only for RNA–RNA and DNA–DNA interactions but also for RNA–DNA hybrid interactions, potentially containing intra-strand and inter-strand pseudoknots.

An important approach for finding the minimum free energy of an RNA secondary structure is to exploit the potential nestedness of the base pairing interactions. If we imagine the RNA strands to be drawn in a circular fashion, and base pairs drawn as arcs that connect different bases, one can appreciate that each drawn arc (i.e. each placed base pair) splits the original secondary structure prediction diagram into parts that correspond to easier-to-solve substructures. This leads to an efficient recursion algorithm for finding the minimum free energy structure, albeit with the tradeoff that pseudoknotted RNA structures are not being considered (Nussinov et al. 1978).

RNA pseudoknots may have important biological functions. For example, pseudoknots in an mRNA may cause ribosomal frameshifting. Moreover, pseudoknots can be of even greater prevalence in designed RNA complexes. For example, in designed RNA squares that consist of L-shapes, each connecting kissing-loop interaction corresponds to a pseudoknot interaction. (Bindewald et al. 2016) An algorithm for predicting the secondary structure of such designed RNA complexes needs to be able to handle multiple RNA strands as well as pseudoknots.

DOI: 10.1201/9781003001560-18

The approach used in the HyperFold program is to perform a controlled version of a complete enumeration of all possible RNA secondary structures. The algorithm and how to use its web server will be described in more detail, but briefly the algorithm lists potential helices above a certain minimum length (called core helices) and considers all combinations of placed and not-placed core helices. For each combination of placed core helices, the remaining single-stranded regions are folded using shorter helices. It should be noted that strands can correspond to RNA and DNA, and the program is able to create predictions for RNA–RNA, DNA–DNA and RNA–DNA interactions. This approach has been used for the characterization of RNA switches and RNA–DNA hybrid duplexes (Afonin et al. 2016, Bindewald et al. 2016, Halman et al. 2017). Additional options for the further reduction of the search space are described below.

ALGORITHM

The main idea of the folding algorithm is that it approximates an exhaustive combinatorial search for the minimum free energy of the provided nucleic acid strands for a given temperature and concentration (Bindewald et al. 2016). This has the advantage that in principle complicated pseudoknotted structures can be predicted by the algorithm. Because this leads for non-trivially short sequences to a large number of potential secondary structures ("combinatorial explosion"); a heuristic is utilized in order to reduce the search space.

The algorithm starts by generating a complete list of all longer helices (called core helices) that consist of at least 6 base pairs. Folding energies of individual helices (that potentially can correspond to RNA, DNA and RNA–DNA hybrid helices) are computed according to established nearest neighbor energy models (SantaLucia et al. 1996, Sugimoto et al. 1995, Mathews et al. 1999). The algorithm utilizes two queues (called A and B) that each contain secondary structures models. The elements of each queue are at all times sorted by the free energy estimates of their corresponding secondary structure models (this is referred to as a priority queue).

The algorithm starts with a completely unfolded structure added to queue A. It then proceeds by removing the top ranked structure from queue A (which in the beginning corresponds to the completely unfolded structure) and generates a list of derived structures in which one additional core helix has been placed. The algorithm contains an adjustable limitation in regard to pseudoknot complexity, currently set to allowing for each placed core helix to intersect with at most one other core helix in a non-nested fashion. Also, helices shorter than 6 base pairs are added to each derived structure using a greedy algorithm. These derived structures (as well as the original structure) are added to queue B. We call the queue from which the top-ranking structures are taken the emitting queue, while the queue to which expanded structures are added the receiving queue. Next, the top-ranking structure from queue B is "expanded" in a similar fashion by adding on additional core helices as well as short helices, followed by adding those expanded structure models into the other queue (in this case queue A). Structures are removed from the emitting queue, expanded and added to the receiving queue until the emitting queue is empty. Now the roles of the emitting and receiving queues are swapped, in other words the receiving queue becomes the emitting queue and vice versa. In order to prevent a combinatorial explosion, both queues are "leaky", in other words the worst ranking structures are discarded once the queues exceed a specified maximum size. This approach is depicted in Figure 16.1. Note that in the limit of an extremely large maximum size, this search approach corresponds to a complete enumeration, while in the limit of a maximum queue size of 1 the algorithm corresponds to a greedy approach.

Entropy Estimation

Entropy terms are used in secondary structure evaluations in order to estimate the free energy contributions of the many energetically similar three-dimensional conformations that correspond to one particular set of base pairs (secondary structure). This is well established for cases of hairpin loops in the form of the Jacobson-Stockmayer formula (Jacobson, Beckmann, and Stockmayer 1950). But pseudoknotted structures may lead to 3D restrictions (and corresponding reduction in conformational entropy)

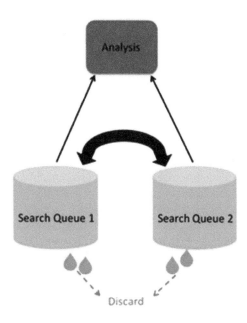

FIGURE 16.1 Visual depiction of the search algorithm. Structures are removed from one queue (the emitting queue), additional helices are placed and added to the other queue (called receiving queue). The roles of the emitting and receiving queues are swapped once the emitting queue is empty. Elements of each queue are at all times sorted by free energy. The worst-ranked structures are discarded from a queue if the maximum queue size is exceeded, leading to the notion of a "leaky queue". Reprinted (adapted) with permission from Bindewald *et al.*, Nano Letters (2016) (Bindewald et al. 2016). Copyright 2016 American Chemical Society

that is not captured by the Jacobson-Stockmayer approach. That is the reason why the HyperFold program utilizes an alternative method. In principle, entropic contributions could be estimated by having a 3D representation of the nucleic acid strands and counting how many 3D conformations are possible (including rotation and translational degrees of freedom as well as conformational changes of complexes) – this would however be impractical for all but trivially short sequences. Instead, an intermediate representation is chosen: as an alternative to 3D coordinates, two distance matrices corresponding to minimum and maximum distances between all residue pairs (each corresponding to one bead) are utilized. Minimum and maximum possible distances are easily obtained for nucleotides that are adjacent in sequence. For larger sequence separations, minimum and maximum distances can be updated so that adjacent nucleotides have the known minimum and maximum distances and the triangle inequality is fulfilled. For nucleotides that correspond to strands that are not part of the same complex, the maximum distance is estimated by computing the expected distance of nucleotides for unbound strands for a given concentration. The maximum and minimum distances correspond to a spherical space of possible confirmations and the logarithm of the number of available conformations is proportional to the logarithm of the spherical volume defined by the maximum and minimum distances. From these terms the entropy can be estimated. Each time a helix is placed, the two matrices of minimum and maximum distances are updated, leading to an updated value of the entropy and free energy of the complexes.

Partition Function

The computational search visits a lot of different conformations with corresponding energies. These visited energies can be utilized to compute a partition function (in other words a Boltzmann-weighted sum of the energies of all structures). We showed previously that it is legitimate to use Boltzmann-weighted sums of terms corresponding to free energies (in which each term represents an average over all 3D conformations of a particular secondary structure). (Afonin et al. 2015) The partition function is essentially one large normalization factor that allows one to estimate for a specific secondary structure the probability with which it can be expected to be found in an ensemble. Also, each visited secondary structure corresponds to a particular strand connectivity of associated and dissociated strands potentially forming complexes. While updating the partition function, also probabilities corresponding to each found set of strand complexes are updated. In this fashion, the final result can contain the important list of strand complex concentrations.

WEB PAGE USAGE

Step 1. *Prepare input sequences.*
A text file corresponding to nucleotide sequences has to be prepared in the common FASTA format. This means that in the text each sequence is preceded by a descriptor line consisting of a leading ">" character followed by a name or description of the sequence. Note that the description of a sequence will not be utilized by the web server. The sequence characters are case sensitive: RNA sequences are denoted as upper case (allowed characters are A, C, G, U) and DNA sequences are described by lower case characters (allowed characters are a, c, g, t). It is up to the user to adjust existing uppercase or lowercase nucleic acid characters according to this convention.

Step 2. *Enter input data into web server form*
The user needs to fill out the web form as shown in Figures 16.2 and 16.4. This involves the following steps:

a)   The prepared sequence information should be opened in a text editor. The sequences can be transferred from the text editor to the text area labeled "Sequences" in the browser form via "copy and paste".

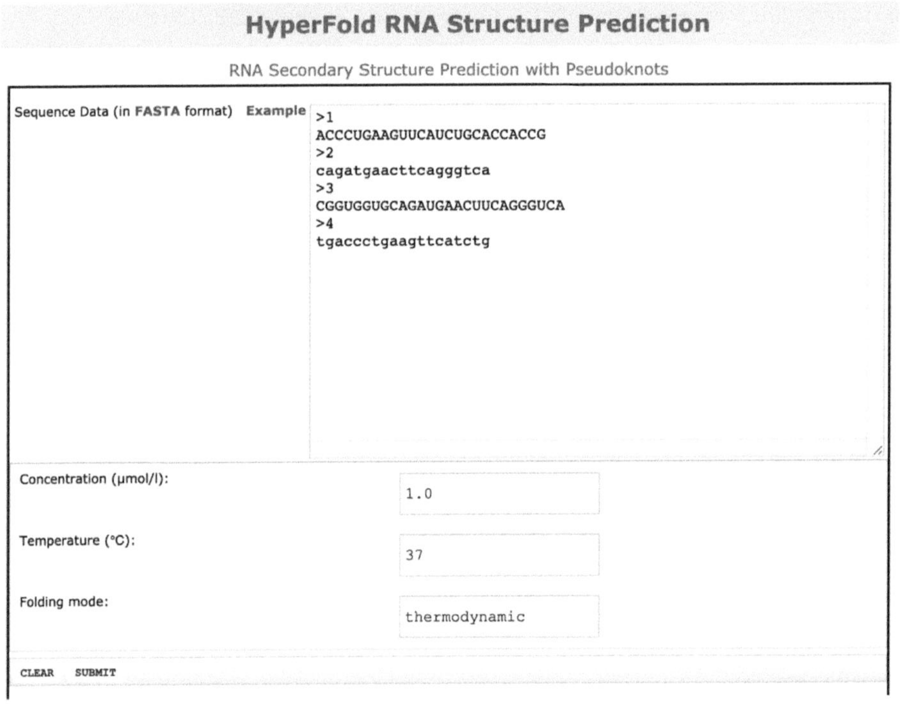

HyperFold RNA Structure Prediction

RNA Secondary Structure Prediction with Pseudoknots

Sequence Data (in **FASTA** format) Example	>1 ACCCUGAAGUUCAUCUGCACCACCG >2 cagatgaacttcagggtca >3 CGGUGGUGCAGAUGAACUUCAGGGUCA >4 tgaccctgaagttcatctg
Concentration (μmol/l):	1.0
Temperature (°C):	37
Folding mode:	thermodynamic
CLEAR SUBMIT	

FIGURE 16.2   Example of nucleotide sequences entered into the submission form for structure prediction using the FASTA format. RNA and DNA sequences are denoted as uppercase and lowercase characters, respectively. This example corresponds to the RNA and DNA sequences that form hybrid duplexes that re-associate into RNA and DNA duplexes as described in Afonin et al., 2016.

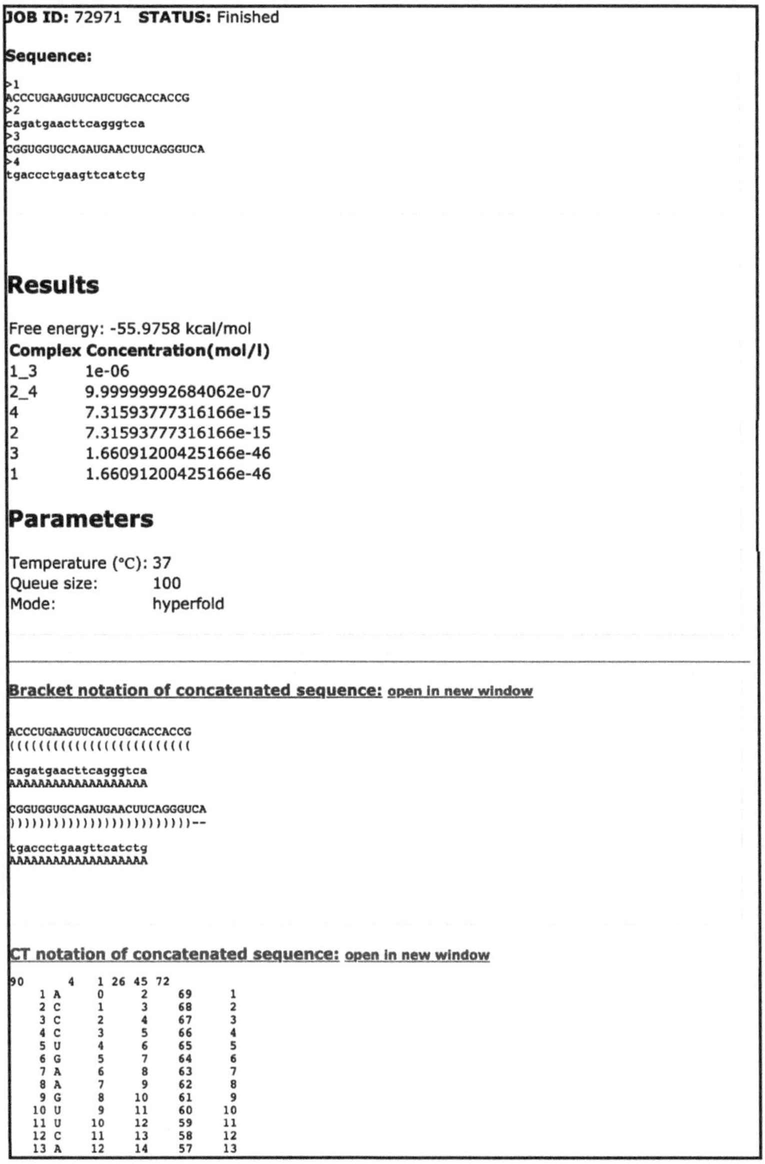

```
JOB ID: 72971   STATUS: Finished

Sequence:

>1
ACCCUGAAGUUCAUCUGCACCACCG
>2
cagatgaacttcagggtca
>3
CGGUGGUGCAGAUGAACUUCAGGGUCA
>4
tgaccctgaagttcatctg
```

Results

Free energy: -55.9758 kcal/mol

Complex Concentration(mol/l)

1_3	1e-06
2_4	9.99999992684062e-07
4	7.31593777316166e-15
2	7.31593777316166e-15
3	1.66091200425166e-46
1	1.66091200425166e-46

Parameters

Temperature (°C):	37
Queue size:	100
Mode:	hyperfold

Bracket notation of concatenated sequence: open in new window

```
ACCCUGAAGUUCAUCUGCACCACCG
(((((((((((((((((((((((((

cagatgaacttcagggtca
AAAAAAAAAAAAAAAAAAA

CGGUGGUGCAGAUGAACUUCAGGGUCA
)))))))))))))))))))))))))--

tgaccctgaagttcatctg
AAAAAAAAAAAAAAAAAAA
```

CT notation of concatenated sequence: open in new window

```
90      4   1 26 45 72
   1 A     0    2   69    1
   2 C     1    3   68    2
   3 C     2    4   67    3
   4 C     3    5   66    4
   5 U     4    6   65    5
   6 G     5    7   64    6
   7 A     6    8   63    7
   8 A     7    9   62    8
   9 G     8   10   61    9
  10 U     9   11   60   10
  11 U    10   12   59   11
  12 C    11   13   58   12
  13 A    12   14   57   13
```

FIGURE 16.3 Results returned by the HyperFold web server for the example of two RNA and two DNA strands that can form RNA–DNA hybrids that re-associate into RNA and DNA duplexes [10]. A list called "Complex concentration" shows formed complexes (named by their constituting strand ids) and predicted concentrations. The user can obtain results in a multi-strand version of the bracket notation and in the tabular ("CT") format (only the first rows of the resulting tabular format are shown in this figure). Further below in the screen output (not shown) is a list of suboptimal secondary structures provided in CT tabular text format. Note that RNA strands are denoted in upper case, while DNA strands are shown using lower case characters.

b) The desired concentration can be entered into the web form in units of micromol/l. Note that currently the concentration of all strands is assumed to be equal.

c) The folding temperature can be changed from the default of 37C to the desired value.

d) Decide on a folding "mode". The default is regular thermodynamic folding (correspond- ing to parameter "thermodynamic". It is also possible to emphasize kinetic effects in the

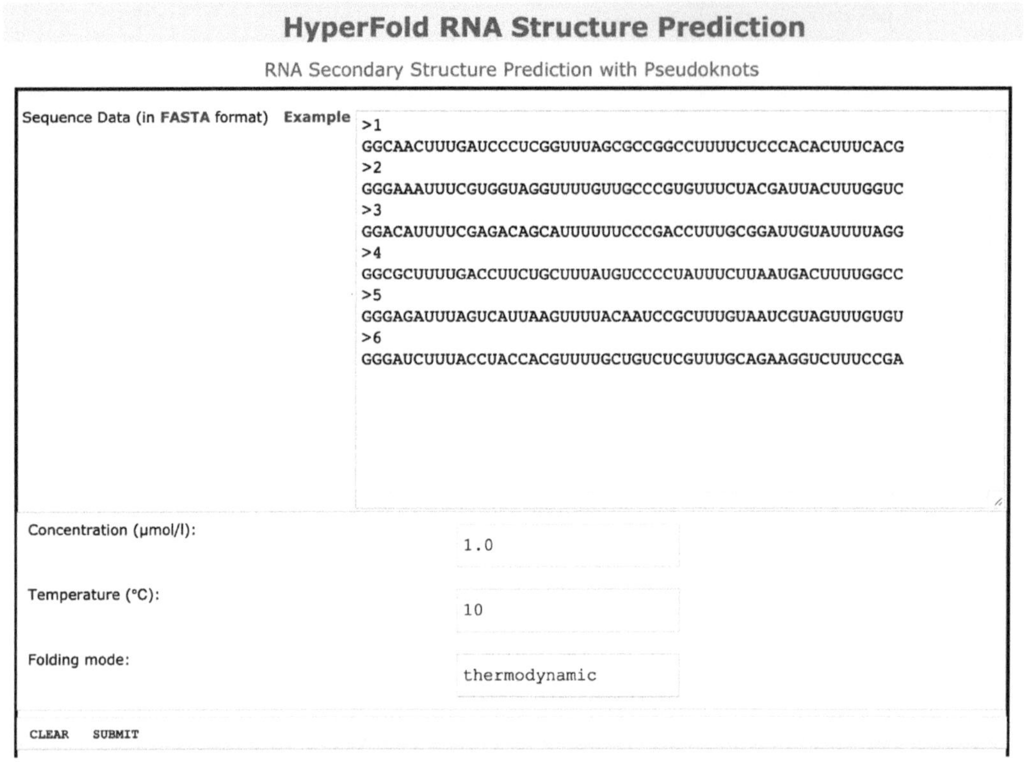

HyperFold RNA Structure Prediction

RNA Secondary Structure Prediction with Pseudoknots

Sequence Data (in **FASTA** format) Example	>1 GGCAACUUUGAUCCCUCGGUUUAGCGCCGGCCUUUUCUCCCACACUUUCACG >2 GGGAAAUUUCGUGGUAGGUUUUUGUUGCCCGUGUUUCUACGAUUACUUUGGUC >3 GGACAUUUUCGAGACAGCAUUUUUUCCCGACCUUUGCGGAUUGUAUUUUAGG >4 GGCGCUUUUGACCUUCUGCUUUAUGUCCCCUAUUUCUUAAUGACUUUUGGCC >5 GGGAGAUUUAGUCAUUAAGUUUUACAAUCCGCUUUGUAAUCGUAGUUUGUGU >6 GGGAUCUUUACCUACCACGUUUUGCUGUCUCGUUUGCAGAAGGUCUUUCCGA
Concentration (μmol/l):	1.0
Temperature (°C):	10
Folding mode:	thermodynamic
CLEAR SUBMIT	

FIGURE 16.4 Submission form for secondary structure prediction. The user specifies the sequences in FASTA format. Additionally, the user can specify the concentration of strands as well the folding temperature (set in this example to 10°C). The shown example corresponds to six sequences forming a cube-like nanostructure with linker regions consisting of 3U nucleotides (Afonin et al. 2010, Afonin et al. 2014).

folding algorithm by choosing parameter "kinetic". In that mode, intra-strand helices are folded before inter-strand helices and the search queue size is smaller compared to the thermodynamic folding mode.

Step 3. *Submit and store compute job*
Upon clicking the "submit" button in the web browser, the user is provided with a compute job id. Saving this id or bookmarking this page is important in the case of bigger structures for which the runtime can amount to several hours.

Step 4. *Retrieve results*
Results can be obtained with two approaches:

a) Reloading: By reloading the result web page presented to the user after job submission, it can be ascertained whether the compute job has finished. Note that due to the nature of enumerating structures, the run-time can be several hours.

b) Retrieve results of past compute jobs via Job ID. If the browser content that was created after job submission is not available (e.g. because the browser window has been closed or a different device is used for accessing the results), the results of the compute job can be obtained by entering the Job ID into the text field available at the home page of the web server.

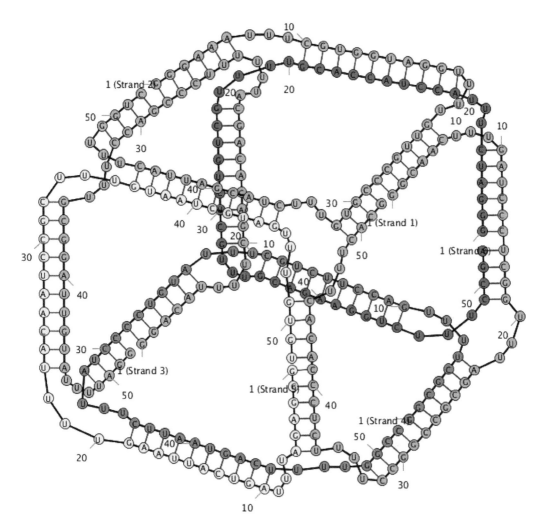

FIGURE 16.5 Visualization of the predicted secondary structure of the 6-stranded RNA cube corresponding to the sequences in Figure 16.4 (Afonin et al. 2010, Afonin et al. 2014) This image was created with the RiboSketch software (Lu et al. 2018).

Step 5. *Interpret Results*

Finally, the user is presented with a list of predicted base pairs of the RNA secondary structure in tabular format (called CT format) as well as an extended form of bracket notation. An example of the returned results is shown in Figure 16.3. The resulting secondary structure information can be post-processed with visualization programs such as RiboSketch (see Figure 16.5) [18].

LIMITATIONS

It should be noted that the approach is based on quasi-exhaustive enumeration of secondary structures. This can lead to the possibility of several hours of compute time for sequences that consist of more than 100 nucleotides in total. The user should therefore save the compute job id in order to retrieve a prediction result at a later time, even potentially from a different web browser window.

CONCLUSIONS

A web server for predicting folding properties of multi-strand RNA and DNA strands has been presented. It should find many uses, for example, in the field of RNA and DNA nanotechnology.

ACKNOWLEDGMENTS

This work was supported in whole or in part with Federal funds from the Frederick National Laboratory for Cancer Research, National Institutes of Health, under Contract No. HHSN261200800001E. This research was supported (in part) by the Intramural Research Program of the NIH, National Cancer Institute, Center for Cancer Research. The content of this publication does not necessarily reflect the views or policies of the Department of Health and Human Services, nor does mention of trade names, commercial products, or organizations imply endorsement by the U.S. Government.

REFERENCES

Afonin, K. A., E. Bindewald, M. Kireeva, and B. A. Shapiro. 2015. "Computational and experimental studies of reassociating RNA/DNA hybrids containing split functionalities." *Methods Enzymol* 553:313–34. doi:10.1016/bs.mie.2014.10.058.

Afonin, K. A., E. Bindewald, A. J. Yaghoubian, N. Voss, E. Jacovetty, B. A. Shapiro, and L. Jaeger. 2010. "In vitro assembly of cubic RNA-based scaffolds designed in silico." *Nat Nanotechnol* 5 (9):676–82. doi:10.1038/nnano.2010.160.

Afonin, K. A., W. Kasprzak, E. Bindewald, P. S. Puppala, A. R. Diehl, K. T. Hall, T. J. Kim, M. T. Zimmermann, R. L. Jernigan, L. Jaeger, and B. A. Shapiro. 2014. "Computational and experimental characterization of RNA cubic nanoscaffolds." *Methods* 67 (2):256–65. doi:10.1016/j.ymeth.2013.10.013.

Afonin, K. A., M. Viard, P. Tedbury, E. Bindewald, L. Parlea, M. Howington, M. Valdman, A. Johns-Boehme, C. Brainerd, E. O. Freed, and B. A. Shapiro. 2016. "The Use of Minimal RNA Toeholds to Trigger the Activation of Multiple Functionalities." *Nano Lett* 16 (3):1746–53. doi:10.1021/acs.nanolett.5b04676.

Andronescu, M., Z. C. Zhang, and A. Condon. 2005. "Secondary structure prediction of interacting RNA molecules." *J Mol Biol* 345 (5):987–1001. doi:10.1016/j.jmb.2004.10.082.

Bindewald, E., K. A. Afonin, M. Viard, P. Zakrevsky, T. Kim, and B. A. Shapiro. 2016. "Multistrand Structure Prediction of Nucleic Acid Assemblies and Design of RNA Switches." *Nano Lett* 16 (3):1726–35. doi:10.1021/acs.nanolett.5b04651.

Chworos, A., I. Severcan, A. Y. Koyfman, P. Weinkam, E. Oroudjev, H. G. Hansma, and L. Jaeger. 2004. "Building programmable jigsaw puzzles with RNA." *Science* 306 (5704):2068–72. doi:10.1126/science.1104686.

Dibrov, S. M., J. McLean, J. Parsons, and T. Hermann. 2011. "Self-assembling RNA square." *Proc Natl Acad Sci U S A* 108 (16):6405–8. doi:10.1073/pnas.1017999108.

Halman, J. R., E. Satterwhite, B. Roark, M. Chandler, M. Viard, A. Ivanina, E. Bindewald, W. K. Kasprzak, M. Panigaj, M. N. Bui, J. S. Lu, J. Miller, E. F. Khisamutdinov, B. A. Shapiro, M. A. Dobrovolskaia, and K. A. Afonin. 2017. "Functionally-interdependent shape-switching nanoparticles with controllable properties." *Nucleic Acids Res* 45 (4):2210–2220. doi:10.1093/nar/gkx008.

Jacobson, H., C.O. Beckmann, and W.H. Stockmayer. 1950. "Intramolecular Reaction in Polycondensations. II. Ring-Chain Equilibrium in Polydecamethy,en Adipate." *J. Chem. Phys.* 18:1607.

Lu, J. S., E. Bindewald, W. K. Kasprzak, and B. A. Shapiro. 2018. "RiboSketch: versatile visualization of multi-stranded RNA and DNA secondary structure." *Bioinformatics* 34 (24):4297–4299. doi:10.1093/bioinformatics/bty468.

Mathews, D. H., J. Sabina, M. Zuker, and D. H. Turner. 1999. "Expanded sequence dependence of thermodynamic parameters improves prediction of RNA secondary structure." *J Mol Biol* 288 (5):911–40. doi:10.1006/jmbi.1999.2700.

Nussinov, R., G. Pieczenik, J. R. Griggs, and D. J. Kleitman. 1978. "Algorithms for Loop Matchings." *SIAM Journal on Applied Mathematics* 35:68–82.

SantaLucia, J., Jr., H. T. Allawi, and P. A. Seneviratne. 1996. "Improved nearest-neighbor parameters for predicting DNA duplex stability." *Biochemistry* 35 (11):3555–62. doi:10.1021/bi951907q.

Shu, D., Y. Shu, F. Haque, S. Abdelmawla, and P. Guo. 2011. "Thermodynamically stable RNA three-way junction for constructing multifunctional nanoparticles for delivery of therapeutics." *Nat Nanotechnol* 6 (10):658–67. doi:10.1038/nnano.2011.105.

Sugimoto, N., S. Nakano, M. Katoh, A. Matsumura, H. Nakamuta, T. Ohmichi, M. Yoneyama, and M. Sasaki. 1995. "Thermodynamic parameters to predict stability of RNA/DNA hybrid duplexes." *Biochemistry* 34 (35):11211–6. doi:10.1021/bi00035a029.

Yingling, Y. G., and B. A. Shapiro. 2007. "Computational design of an RNA hexagonal nanoring and an RNA nanotube." *Nano Lett* 7 (8):2328–34. doi:10.1021/nl070984r.

Zadeh, J. N., C. D. Steenberg, J. S. Bois, B. R. Wolfe, M. B. Pierce, A. R. Khan, R. M. Dirks, and N. A. Pierce. 2011. "NUPACK: Analysis and design of nucleic acid systems." *J Comput Chem* 32 (1):170–3. doi:10.1002/jcc.21596.

17 RNA Switches
Towards Conditional Dynamic RNA-Based Constructs for Therapeutics and Bioassays

Paul Zakrevsky and Bruce A. Shapiro
National Cancer Institute, Frederick, USA

CONTENTS

Ribonucleic acid (RNA) plays a central role in the expression of genetically encoded information. The specificity afforded through complementary Watson–Crick base pairing allows information to be stored and read at the level of an RNA's primary sequence. Messenger RNA (mRNA) and transfer RNA (tRNA) are essential for transcribing and decoding the genomic information stored at the level of DNA for translation of this information into proteins. However, additional non-coding RNAs also play a crucial role in gene expression in the form of regulation. One such method of RNA-based regulation in eukaryotic cells is RNA interference (RNAi), a mechanism by which short duplexes known as small interfering RNAs (siRNA) downregulate the expression of a target gene via complementarity with a target mRNA [1]. While the RNAi pathway is used by the cell for endogenous regulation of gene expression, it also provides a potential platform for exogenous control over the expression of host genes with biological or therapeutic importance.

Harnessing this natural cellular mechanism holds great promise for the treatment of disease. In theory, an RNAi substrate can be delivered to cells to target any gene of interest. However, regulating the activity of this exogenous RNAi response is vital. Downregulating the expression of a target gene in unintended cells could have detrimental consequences and may ultimately lead to the death of otherwise healthy cells. One way to discern the state of a particular cell is by detection of an RNA biomarker [2]; a transcript that is highly expressed in the cells of interest relative to the greater population. Operating with the notion that an RNA biomarker can be used as a diagnostic indicator, dynamic nucleic acid systems have been developed that regulate their generation of an RNAi-substrate based on the presence or absence of an oligonucleotide trigger (RNA biomarker).

Many systems designed for conditional RNAi make use of strand displacement reactions to facilitate their structural dynamics. Just as genetic information can be encoded in a nucleic acid sequence, so too can structural information. Conditional constructs are designed to adopt an initial state that is defined by programmed base pairing. However, introduction of an additional RNA trigger strand with regions complementary to the initial construct can result in the possibility of forming a lower energy structure, ultimately inducing a conformational change. While the conformational change may be thermodynamically favorable, it may be kinetically slow due to the activation energy needed to first unfold portions of the stable initial structure. To accommodate this kinetic consideration, a single-stranded nucleation site, termed a toehold, can be incorporated in the initial structure [3]. Binding between a region of the trigger oligonucleotide and the single-stranded toehold facilitates the sequential displacement of adjacent complementary nucleotides, which leads to strand displacement and formation of a new lower energy structure. Thus, creative sequence design can produce systems in which dynamic conformational change of the system is regulated by interaction with a trigger oligonucleotide and used to generate functional structures on a conditional basis.

A common feature of many conditional RNAi systems is the generation of a ~21–27 bp double-stranded region of RNA (dsRNA) as their output product. This dsRNA is termed a Dicer substrate and could take the form of a free duplex or the stem of a short hairpin RNA (shRNA). In either case, it requires processing by the endonuclease Dicer in order to generate the siRNA that is used to produce an RNAi response. One such conditional RNAi system employs a two-stranded RNA switch design [4, 5]. Assembly of this all-RNA switch occurs between one strand that encodes an shRNA that acts as a Dicer substrate and a second that behaves as a diagnostic strand containing a single-stranded toehold region for detection of the trigger oligonucleotide (Figure 17.1a). This two-stranded RNA switch design exclusively utilizes conformational change to prevent premature Dicer processing. Initially, the two strands are assembled in such a way that the functional shRNA structure is sequestered in an inactive conformation, and in which the entire initial structure lacks any long continuous helices that could act as a typical Dicer substrate. However, pairing between the toehold and the trigger oligonucleotide induces a strand displacement reaction through extensive base pairing with the diagnostic strand. A subsequent conformational rearrangement generates the functional shRNA structure that can be processed by Dicer. The necessity to sequester the shRNA while avoiding long continuous helices in the initial structure complicates the folding strategy and sequence design, which may be a potential drawback of this all-RNA system. Additionally, an all-RNA system is particularly susceptible to nuclease degradation. However, this system has been shown to function in a conditional fashion after cellular delivery [4].

Other conditional RNAi systems have been constructed that initially impede formation of the functional RNA duplex by binding the RNAi sense and antisense sequences to regions of modified RNA or DNA. The inclusion of 2'-modified nucleotides can be used as a method to prevent premature Dicer processing of initial structures and intermediates, as well as increase a construct's resistance to degradation by other nucleases. Several schemes using small conditional RNAs (scRNAs) containing 2'-O-methyl modifications have been characterized *in vitro* [6]. A very simple single-component scRNA system can promote conditional shRNA Dicer substrate generation (Figure 17.1b, i) but leaves vital portions of the RNA unprotected in its initial conformation, rendering it vulnerable to nucleases. This can be rectified by using slightly more complex multi-construct scRNA systems (Figure 17.1b, ii). Despite its potential nuclease susceptibility, the single-component scRNA system has been demonstrated to conditionally generate RNAi substrates in a human cell lysate, including the processing of shRNA outputs by wildtype endogenous Dicer to produce short siRNA duplexes [7].

Conditional RNAi systems that make use of hybrid duplex formation between RNA and DNA strands have also been developed and show promise *in vitro* [8]. Whereas most conditional RNAi systems focus on activation of RNAi substrate release in response to a trigger oligonucleotide, researchers created RNA/DNA hybrid systems which can be designed to either activate or repress RNAi-substrate generation when a specific trigger is present (Figure 17.1c). Combining both the activation and repression mechanisms is also possible, resulting in RNA/DNA hybrid systems in which the degree of RNAi-substrate release can be attenuated based on the presence of two distinct trigger sequences [8]. Much like the use of 2'-O-methyl modification, the formation of RNA/DNA hybrid duplexes prevents processing by Dicer and non-specific degradation of the RNA strands by nucleases [9].

Each of these conditional RNAi systems have their own unique benefits and individual drawbacks. Aside from RNA interference based methods, other approaches and mechanisms for the conditional regulation of genes have been developed. Examples include insertion of toehold-driven riboregulators within the 5'-UTR of genes [10] and CRISPR/Cas systems that employ conditional, dynamic guide RNAs [11]. For regulation of an endogenous gene these systems would require the

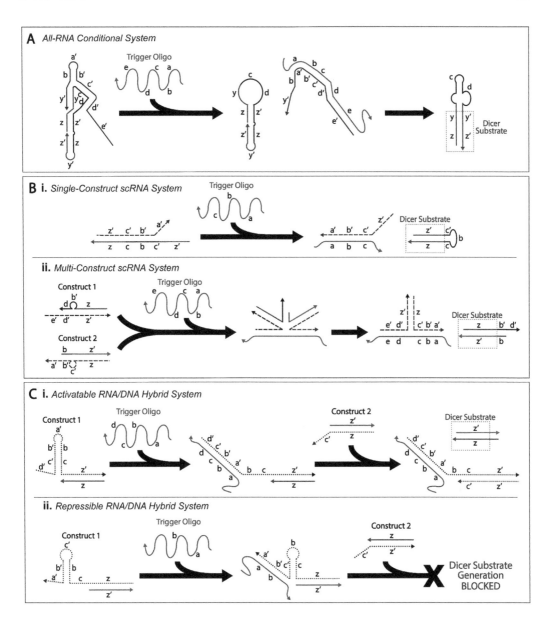

FIGURE 17.1 Examples of conditional RNAi systems that regulate their generation of a Dicer substrate in response to a specific trigger oligonucleotide. (a) An all-RNA two-stranded switch. (b) 2'-O-methyl containing small conditional RNA (scRNA) systems that make use of (i) one initial construct or (ii) multiple initial constructs. (c) Conditional RNA/DNA hybrid pairs designed to either (i) activate or (ii) repress generation of a Dicer substrate product in response to a trigger. Specific sequence regions within each system are indicated by lower case letters. Differences in backbone chemistries are defined by line depiction: RNA (solid line), 2'-O-methyl (long dash), DNA (short dash). Regions colored red indicate sense/antisense of Dicer substrate product. All sequence lengths are displayed in scale to one another. The schematics displayed in (a), (b), and (c) portray systems described in references [4, 5, 8], respectively.

editing of host genes or the expression of exogenous proteins, however, and would permanently alter the host genome. On the contrary, conditional RNAi methods require only the delivery of relatively small nucleic acid constructs, and their effect is temporary, limiting the probability of serious, permanent off-target effects. Continued development of many conditional RNAi systems is ongoing, with efforts being pursued to optimize their function for application in cellular systems.

ACKNOWLEDGMENTS

This research was supported [in part] by the Intramural Research Program of the NIH, Center for Cancer Research, NCI-Frederick. The content of this publication does not necessarily reflect the views or policies of the Department of Health and Human Services, nor does mention of trade names, commercial products, or organizations imply endorsement by the U.S. Government.

REFERENCES

1. R. C. Wilson and J. A. Doudna, *Annu Rev Biophys*, 2013, **42**, 217–239.
2. X. Xi, T. Li, Y. Huang, J. Sun, Y. Zhu, Y. Yang and Z. J. Lu, *Noncoding RNA*, 2017, **3**, 1–17.
3. K. A. Afonin, M. Viard, P. Tedbury, E. Bindewald, L. Parlea, M. Howington, M. Valdman, A. Johns-Boehme, C. Brainerd, E. O. Freed and B. A. Shapiro, *Nano Lett*, 2016, **16**, 1746–1753.
4. E. Bindewald, K. A. Afonin, M. Viard, P. Zakrevsky, T. Kim and B. A. Shapiro, *Nano Lett*, 2016, **16**, 1726–1735.
5. P. Zakrevsky, L. Parlea, M. Viard, E. Bindewald, K. A. Afonin and B. A. Shapiro, *Methods Mol Biol*, 2017, **1632**, 303–324.
6. L. M. Hochrein, M. Schwarzkopf, M. Shahgholi, P. Yin and N. A. Pierce, *J Am Chem Soc*, 2013, **135**, 17322–17330.
7. L. M. Hochrein, T. J. Ge, M. Schwarzkopf and N. A. Pierce, *ACS Synth Biol*, 2018, **7**, 2796–2802.
8. P. Zakrevsky, E. Bindewald, H. Humbertson, M. Viard, N. Dorjsuren and B. A. Shapiro, *Nanomaterials (Basel)*, 2019, **9**, 1–22.
9. K. A. Afonin, M. Viard, A. N. Martins, S. J. Lockett, A. E. Maciag, E. O. Freed, E. Heldman, L. Jaeger, R. Blumenthal and B. A. Shapiro, *Nat Nanotechnol*, 2013, **8**, 296–304.
10. A. A. Green, P. A. Silver, J. J. Collins and P. Yin, *Cell*, 2014, **159**, 925–939.
11. M. H. Hanewich-Hollatz, Z. Chen, L. M. Hochrein, J. Huang and N. A. Pierce, *ACS Cent Sci*, 2019, **5**, 1241–1249.

18 RNA Multiway Junction Motifs as Lego for the Construction of Multifunctional RNA Nanoparticles

Farzin Haque
University of Kentucky, Lexington, KY, USA

Xin Li
The Ohio State University, Columbus, OH, USA

Peixuan Guo
The Ohio State University, Columbus, OH, USA

CONTENTS

INTRODUCTION

Bottom-up approach in RNA nanotechnology has recently emerged as an important means to construct RNA nanoparticles *via* self-assembly with desired structure and stoichiometry. The approach relies on the intrinsic nanoscale attributes of RNA as a construction material (Guo, 2010). Structurally, RNA can fold into incredibly diverse structures, displaying single-stranded bulges, hairpins, internal loops, and pseudoknots, which makes it distinct from DNA (Zuker, 1989; Pleij and Bosch, 1989; Guo, 2005; Isambert, 2009). In addition to canonical Watson–Crick (W–C)

DOI: 10.1201/9781003001560-20

base pairing, RNA exhibits non-canonical W–C base pairing (such as, G-U wobble base pairs), base stacking, and tertiary interactions (Searle and Williams, 1993; Sugimoto *et al.*, 1995; Ikawa *et al.*, 2004; Leontis *et al.*, 2006; Li *et al.*, 2006; Matsumura *et al.*, 2009; Schroeder *et al.*, 2010). Furthermore, RNA/RNA helices are thermodynamically more stable than DNA/DNA equivalents (Searle and Williams, 1993; Sugimoto *et al.*, 1995), which makes it particularly attractive for *in vivo* delivery. Functionally, RNA is versatile, as evidenced by the existence of several functionally active molecules *in vivo*, such as short interfering RNA (siRNA) (Fire *et al.*, 1998; Li *et al.*, 2002), micro RNA (miRNA) (Fabian *et al.*, 2010), RNA aptamer (Ellington and Szostak, 1990; Tuerk and Gold, 1990), ribozyme (Kruger *et al.*, 1982; Guerrier-Takada *et al.*, 1983), and riboswitches (Sudarsan *et al.*, 2008).

Several assembly mechanisms of biologically active RNA nanoparticles can be used to construct synthetic RNA nanoparticles with defined structure and stoichiometry *via* intra- and/or intermolecular interactions. Examples include the following:

1. Loop–loop interactions, as observed in phi29 packaging RNA (pRNA), which can assemble into dimer, trimer, and hexamer *via* hand-in-hand interactions of right- and left-hand interlocking loops (Turner and Tijan, 1989; Guo *et al.*, 1998; Chen *et al.*, 1999; Chen *et al.*, 2000; Shu *et al.*, 2003; Shu *et al.*, 2004) and retrovirus kissing loops for designing tectoRNA (Chworos *et al.*, 2004; Severcan *et al.*, 2009);

2. RNA "architectonics" (Chworos *et al.*, 2004), defined as rational design of 3D RNA constructs, whereby structural RNA information is encoded within an artificial sequence to direct the self-assembly of supramolecular assemblies. Examples include RNA filaments (Jaeger and Leontis, 2000; Nasalean *et al.*, 2006; Geary *et al.*, 2010), tectosquares (molecular jigsaw puzzles) (Chworos *et al.*, 2004; Severcan *et al.*, 2009), and tRNA antiprisms (Woodson, 2010);

3. *In vitro* selection technique employing synthetic ribozyme ligase (Ikawa *et al.*, 2004; Matsumura *et al.*, 2009), as demonstrated by the construction of conformational switches of RNA nanostructures using peptide-binding RNA structural motifs (Li *et al.*, 2006);

4. Palindrome sequences, which is defined as the sequence of nucleotides, that read the same forwards (5′ → 3′) and backwards (3′ → 5′). By introducing a palindrome sequences at the 5′- or 3′-end, the molecule will spontaneously assemble *via* self-annealing (intermolecular interactions) after *in vitro* transcription or chemical synthesis. This method has proven to be useful for constructing RNA bundles with precise control over the angle or the direction for RNA fiber extension (Shu *et al.*, 2004);

5. RNA branched architectures employing multi-junction motifs (Figure 18.1) to serve as the scaffold in nanoparticle construction (Bindewald *et al.*, 2008a; Severcan *et al.*, 2009). Examples include kink-turn motif (Schroeder *et al.*, 2010) to direct the assembly of ribosomal proteins to form a nanostructure with a shape similar to an equilateral triangle (Ohno *et al.*, 2011); pRNA three-way junction motif for constructing trivalent therapeutic RNA nanoparticles (Shu *et al.*, 2011; Haque *et al.*, 2012); rRNA (ribosomal RNA) structural motif to direct the tetramer assembly of L-shaped tectoRNAs; 23S rRNA 3WJ-motif to build T-shaped architectures; and, tRNA four-way junction and five-way junction motifs to assemble L-shaped tertiary structures (Bindewald *et al.*, 2008b; Severcan *et al.*, 2009).

RNA junctions represent the branch-point between different double-stranded helical segments. These important branched architectural elements are abundant in many natural structured RNAs, such as ribosomal RNA (rRNA), transfer RNA (tRNA), ribozymes, and riboswitches (Chen *et al.*, 1999; Honda *et al.*, 1999; Lescoute and Westhof, 2006; Laing and Schlick, 2009; Wakeman *et al.*, 2009; Kulshina *et al.*, 2010). The focus of this chapter is on the use of biologically derived RNA junction elements, in particular three-way junction (3WJ) and four-way junction (4WJ) motifs

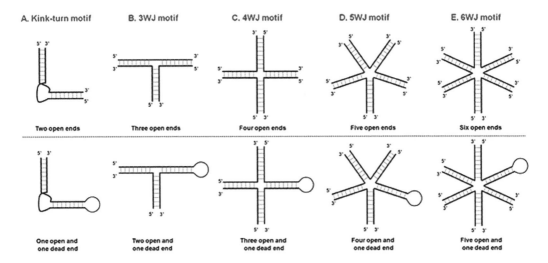

FIGURE 18.1   Illustration of branched RNA architectures for nanoparticle assembly: (a) kink-turn motif; (b) 3WJ motif; (c) 4WJ motif; (d) 5WJ motif; (e) 6WJ motif.

for constructing multivalent RNA nanoparticles for therapeutic and diagnostic applications. The approach relies on utilizing modular building blocks (two–six pieces of RNA oligos) to self-assemble into larger nanoparticles using the bottom-up approach of nanotechnology.

3WJ MOTIFS

CLASSIFICATIONS AND OCCURRENCES

A wide range of 3WJ motifs are prevalent in structured RNAs and their structures have been deduced from molecular modeling, NMR, and X-ray crystallography. Based on the available high-resolution structures of folded RNAs, Lescoute and Westhof (2006) classified 33 different 3WJ motifs into 3 families (A, B, and C) based on the number of nucleotides in the junction strands (Figure 18.2). In the nomenclature, H1, H2, and H3 represent the three helices: J1/2 (junction between H1 and H2), J3/1 (junction between H3 and H1), and J2/3 (junction between H2 and H3) refer to the junction strands.

1. *Family A*: coaxially stacked H1 and H2 helices; J3/1<J2/3; H3 helix is near perpendicular (but can have other angles) to the coaxially stacked helices; junction contacts are minimal. Family A occurs in 16S and 23S rRNAs.
2. *Family B*: coaxially stacked H1 and H2 helices; J3/1≈J23/; H3 helix is bent towards H2 helix; J2/3 faces the H1 helix, while J3/1 contacts H3 helix. Family B is relatively rare, and so far observed in 16S and 23S rRNAs.
3. *Family C*: coaxially stacked H1 and H2 helices; J3/1>J2/3; H3 helix is bent towards H1 helix; J3/1 contacts H2 helix extensively and adopts a hairpin structure. Family C is abundant and observed in various large RNA complexes, such as Alu domain, G-riboswitch, rRNA (16S, 23S, 5S, L11), Hammerhead ribozyme, RNase P B-type, Group I introns, etc.
4. *Unclassified:* pRNA from bacteriophages phi29, SF5, M2/NF, B103, and GA1 all contain a 3WJ motif (Shu *et al.*, 2011). The crystal structure of the phi29 pRNA core motif is currently being resolved (Zhang *et al.*, 2013); but as of now, pRNA 3WJs cannot be classified into the aforementioned three families.

A. 3WJ Nomenclature   **B. Classification of 3WJ in RNA structures**

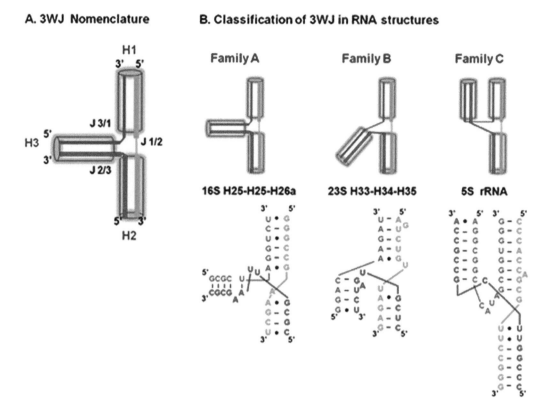

FIGURE 18.2   (A) Nomenclature used for characterizing 3WJ motifs. Helices: H1, H2, and H3. Junctions: J1/2, J2/3, and J1/3. (B) Classification of 3WJs in RNA structures into three families (A, B, and C). Schematic (top) and a representative example with secondary structure (bottom) (Lescoute and Westhof, 2006).

APPLICATION OF 3WJ IN NANOTECHNOLOGY

The feasibility of using 3WJ motifs as scaffold for therapeutic RNA nanoparticle assembly was elegantly demonstrated by Guo and colleagues (Shu *et al.*, 2011). The authors systematically evaluated the assembly and thermodynamic properties of 25 different 3WJ motifs present in different biological RNA structures. Fourteen of the twenty-five 3WJ motifs were not feasible to be studied using just the core sequences, since some of the RNA oligos were less than 10-nt long (lower limit for chemical synthesis).

For 11 3WJ motifs (Figure 18.3), the 3 strands comprising the core 3WJ structural motifs were chemically synthesized. The strands were mixed in 1: 1:1 molar ratio and the resulting complex formation was assayed in native and denaturing gels (Figure 18.3b). Of the 11 3WJ motifs, 6 of the core structures (in their published form (Chen *et al.*, 1999; Honda *et al.*, 1999; Lescoute and Westhof, 2006; Wakeman *et al.*, 2009; Kulshina *et al.*, 2010)) were observed to assemble in native gels, but only 2 (3WJ-pRNA and 3WJ-5S rRNA) were stable under strongly denaturing conditions (8M Urea).

The thermodynamic stabilities of the 11 3WJ motifs were extracted from T_M measurements. Among the 11 motifs, pRNA-3WJ displayed the highest T_M (~58°C) with the steepest slope, indicating simultaneous cooperative assembly of all 3 strands, followed closely by 5S-rRNA (T_M ~54°C). The melting curves for the other 3WJ motifs displayed temperature-dependent curves with gradual slopes, indicating multiple folding landscapes (Figure 18.3c).

The 3WJ domain of the pRNA of bacteriophage phi29 DNA packaging motor was assembled from three pieces of small RNA oligos with unusually stable properties (Figure 18.4b). Self-assembled

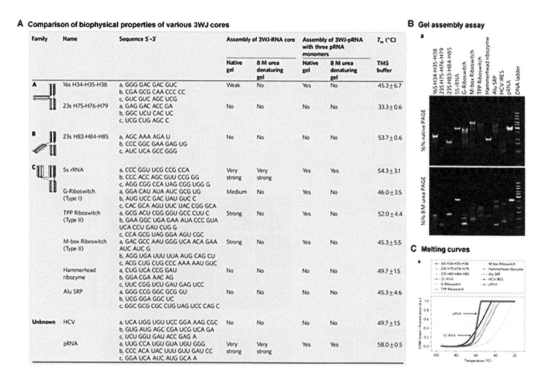

FIGURE 18.3 (a) Comparison of biophysical properties of various 3WJ structures. The sequences of the 3WJ cores were obtained from references (Chen *et al.*, 1999; Honda *et al.*, 1999; Lescoute and Westhof, 2006; Wakeman *et al.*, 2009; Kulshina *et al.*, 2010). Families A, B, and C are based on Lescoute and Westhof classification (Lescoute and Westhof, 2006). (b) Assembly and stability of 11 3WJ motifs assayed by 16% native (top) and 16% 8M urea (bottom) PAGE gel. (c) Melting curves for the 11 3WJ motifs under physiological buffer TMS. Refer to A for respective T_M values. Figures reproduced with permissions from Ref. (Shu *et al.*, 2011), © Nature publishing group.

RNA nanoparticles with three or six pieces of RNA guided by the 3WJ domain were resistant to 8 M urea denaturation and remained intact at extremely low concentrations, as demonstrated by competition and dilution assays. In addition, magnesium was not required for nanoparticle assembly. We further demonstrated that the centerfold domain of the pRNA could be reengineered to form X-shaped motif (Figure 18.4c, f), which also displayed thermodynamically stable properties.

THE CONSTRUCTION OF 3WJ-BASED RNA NANOPARTICLES HARBORING FUNCTIONAL MODULES

To assess the scaffolding abilities of the 25 different 3WJ core motifs, monomeric pRNA was used as a functional module (Zhang *et al.*, 1994; Guo *et al.*, 1998; Shu *et al.*, 2004; Xiao *et al.*, 2005; Shu *et al.*, 2007). The three strands comprising the core motif were placed at the 3′-end of pRNA monomer (117-nt) to serve as "sticky ends". Upon co-transcription (or annealing of individual pRNA with the 3′-end overhang strands), 9 of the 25 3WJ motifs were observed to self-assemble into trivalent RNA nanoparticles harboring one pRNA on each helical arm. AFM images strongly indicated the formation of the trivalent RNA nanoparticles, which were consistent with the designs (Figure 18.4d). However, only the 3WJ-pRNA and 3WJ-5S rRNA remained stable in the presence of 8 M urea (Shu *et al.*, 2011).

Since the 3WJ-pRNA and 3WJ-5S rRNA were the most thermodynamically stable cores, they could serve as scaffolds for constructing trivalent RNA nanoparticles harboring therapeutically

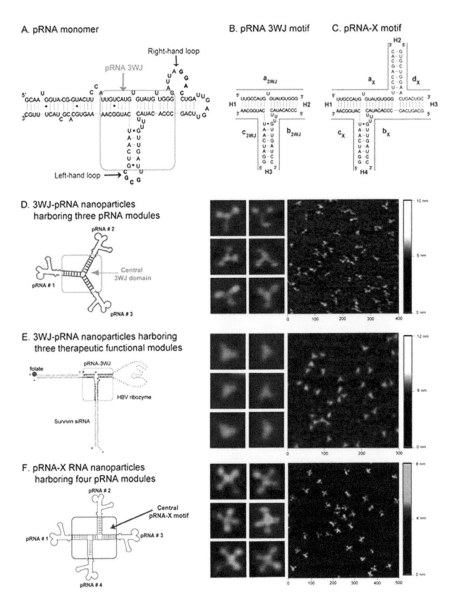

FIGURE 18.4 (a) Secondary structure of monomeric pRNA (packaging RNA) of phi29 DNA packaging motor. (b) 3WJ domain of pRNA comprising of three strands (a_{3WJ}, b_{3WJ}, and c_{3WJ}); Helices: H1, H2, and H3. (c) The core of the pRNA-X motif comprising four strands (a_X, b_X, c_X, and d_X); Helices: H1, H2, H3, and H4. (d) Trivalent RNA nanoparticles harboring one monomer pRNA on each arm of the 3WJ domain in B; schematic (left) and corresponding AFM images (right). (e) The construction of therapeutic 3WJ RNA nanoparticles harboring siRNA, ribozyme, and folate; schematic (left) and corresponding AFM images (right). (f) Tetravalent RNA nanoparticles harboring one monomer pRNA on each arm of the pRNA-X motif in C; schematic (left) and corresponding AFM images (right). Figures reproduced with permissions from: (a, b, d, and e) Ref. (Shu *et al.*, 2011), © Nature publishing group; (c, f) Ref. (Haque *et al.*, 2012) © Elsevier.

relevant RNAs, such as siRNA, miRNA, aptamer, and ribozyme. Accordingly, sequences for the desired siRNA, aptamer, and/or ribozyme were rationally designed with the three strands comprising the core. The purified strands were then mixed in stoichiometric molar ratio and annealed to assemble the final complex and assayed in native and denaturing gels. AFM images strongly indicated the formation of the trivalent RNA nanoparticles, which were consistent with the designs (Shu *et al.*, 2011) (Figure 18.4e).

EVALUATION OF THE FUNCTIONAL MODULES INCORPORATED IN 3WJ RNA NANOPARTICLES

One significant advantage of utilizing the bottom-up approach in RNA nanotechnology is the controlled step-by-step assembly of nanoparticles with desired structure and stoichiometry. It is important to ensure that the modules incorporated into the 3WJ scaffold retain their original folding and authentic functionality without affecting the assembly of the core scaffold. The aforementioned 3WJ-pRNA (or pRNA-X) and 3WJ-5S rRNA scaffold were fused with functional modules (siRNA, ribozyme, or aptamer) at each of the three helical arms and the functionalities were assayed by *in vitro* and *in vivo* assays (Figure 18.5) (Shu *et al.*, 2011; Haque *et al.*, 2012).

Hepatitis B Virus (HBV) ribozyme was able to cleave its substrate (Shu *et al.*, 2011) (Figure 18.5a); Malachite Green (MG) binding aptamer retained its ability to bind MG dye (Malachite Green dye, triphenylmethane), as demonstrated by fluorescence increase (Shu *et al.*, 2011) (Figure 18.5b); Folate, a targeting ligand was conjugated to one of the RNA strands and the resulting nanoparticle efficiently targeted folate receptor positive (FR+) KB cancer cells, as demonstrated by flow

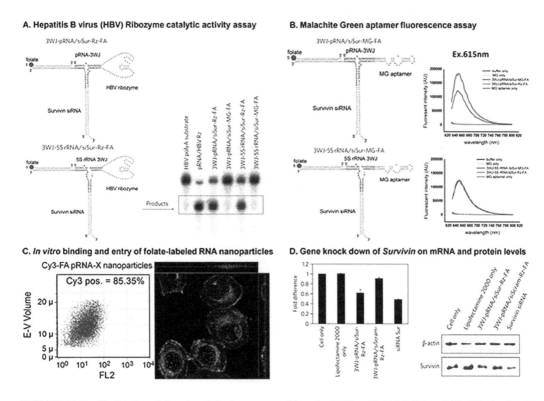

FIGURE 18.5 Evaluation of functionalities incorporated into the 3WJ scaffold (pRNA and 5S rRNA) using *in vitro* and *in vivo* functional assays. (a) Hepatitis B Virus (HBV) ribozyme catalytic assay. (b) Malachite Green (MG) aptamer fluorescence assay. (c) *In vitro* binding and entry of fluorescent and folate labeled pRNA-X RNA nanoparticles into folate receptor positive cells, as demonstrated by flow cytometry (left) and confocal imaging (right). Co-localization of cytoplasma (green) and RNA nanoparticles (red) are shown. (d) Target gene knockdown of survivin on mRNA (left) and protein (right) levels (Nakashima *et al.*, 2011). Figures reproduced with permissions: (a, b, and d) from Ref. (Shu *et al.*, 2011), © Nature publishing group; (c) Ref. (Haque *et al.*, 2012) © Elsevier.

cytometry and confocal imaging (Haque *et al.*, 2012) (Figure 18.5c); Survivin siRNA was able to silence the *Survivin* gene on both mRNA level (assayed by reverse transcription-PCR) and on protein level (assayed by Western blot) compared to the scramble controls (Shu *et al.*, 2011) (Figure 18.5d).

For the past decades, 3WJs have been increasingly used as drug-delivery platform for cancer treatment. Different therapeutic groups such as siRNA, anti-miRNA, and anticancer drugs have been incorporated into 3WJs for therapeutic effect. 3WJs with two siRNAs targeting Mediator Subunit 1 (MED1) have been used for breast cancer treatment in combination with tamoxifen, which has shown significant tumor inhibition effect (Zhang *et al.*, 2017a) (Figure 18.6a). 3WJs carrying different anti-miRNAs, anti-miR17, and anti-miR21 have been used prostate cancer treatment. The tumor growth rate of tumor-bearing mice treated with 3WJs with anti-miRNA is significantly slower compared to control groups (Binzel *et al.*, 2016) (Figure 18.6b). Besides gene silencing groups, 3WJs have also been applied for anticancer drug delivery both non-covalently and covalently. 3WJs with doxorubicin chelating in the extended helices demonstrated drug delivery to cell as shown in confocal microscopy images (Pi *et al.*, 2016) (Figure 18.6c). Camptothecin conjugated 3WJs have been used to treat cancer in KB tumor bearing mouse model with promising tumor suppression effect (Piao *et al.*, 2019) (Figure 18.6c).

For *in vivo* targeting, nude mice were injected with FR+ KB cancer cells to generate a xenograft. Systemic injection *via* the tail vein of fluorescently labeled 2′-F modified pRNA nanoparticles harboring the folate ligand efficiently targeted and strongly bound to xenograft tumors without accumulating in any other vital organs (liver, kidney, heart, lungs, spleen) (Shu *et al.*, 2011; Haque *et al.*, 2012) (Figure 18.6d). In addition to targeting ligand, 3WJs with RNA aptamers have also been constructed and investigated for in vivo targeting effect. Targeting effect of 3WJs harboring CD133 aptamer is demonstrated by the strong fluorescent signal observed in tumors instead of vital organs for mice injected with fluorescent labeled 3WJ/CD133apt (Yin *et al.*, 2019) (Figure 18.6d). Pharmacokinetic (PK) and pharmacodynamic (PD) studies in mice revealed favorable pharmacological and biodistribution profiles for pRNA-based nanoparticles: enhanced half-life (5–10 hours as opposed to 0.25–0.75 hours for bare siRNA counterparts); non-induction of cytokine or interferon I (α and β) responses even at very high doses of 30 mg/kg (Abdelmawla *et al.*, 2011).

THE CONSTRUCTION OF 3WJ-BASED SQUARE-SHAPED AND TRIANGULAR RNA NANOPARTICLES

Besides the usage of 3WJ as the targeting delivery platform for therapeutic groups, 3WJs have also been used to construct more complex RNA nanoparticles. Shapiro and colleagues reported the construction of a computationally designed (using NanoTiler program (Bindewald *et al.*, 2008a) and RNAJunction database (Bindewald *et al.*, 2008b)), experimentally self-assembled, triangle consisting of four strands using three copies of a 16S rRNA 3WJ motif (Bindewald *et al.*, 2011) (Figure 18.7a). Jeager and colleagues reported the construction of square-shaped RNA particles (tectosquares), utilizing the 3WJ motif (UA_h_3WJ) from 23S rRNA (Severcan *et al.*, 2009). The 3WJ adopts a T-shape with two helices coaxial stacked, and the third helix protruding out at a right angle. The 90° bend angle stabilized the corners to form a monodispersed closed square shape with a central cavity, as demonstrated by AFM images (Figure 18.7a). The 3WJ tectosquares can further self-assemble into a 1D ladder and 2D planar arrays *via* programmed tail–tail interactions. 3D RNA nanoparticles such as tetrahedron, prism, pyramid have also been constructed with the potential for drug delivery and protection utilizing 3WJ as the building block (Li *et al.*, 2016; Khisamutdinov *et al.*, 2016; Xu *et al.*, 2019) (Figure 18.7b). In addition to RNA nanoparticles with 3WJ as the single component, more complicated RNA nanoparticles utilizing versatile components such as micelle and exosome have been constructed and used for cancer treatment in animal models (micelles, exosomes) (Figure 18.7c). Exosomes with 3WJ decoration on the surface and siRNA encapsulation have shown therapeutic effect in breast cancer, prostate cancer, and colorectal cancer models

A. siRNA delivery

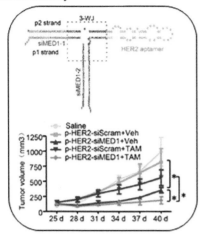

B. anti-miRNA delivery

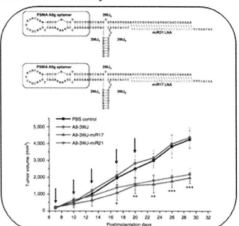

C. Drug delivery

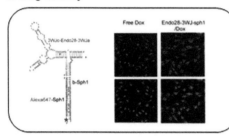

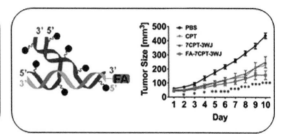

D. Target ligand and aptamer for tumor accumulation

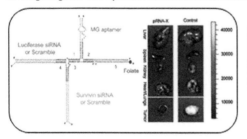

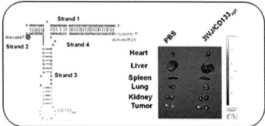

FIGURE 18.6 Evaluation of 3WJ nanoparticles functionalized with therapeutic groups and targeting ligands (a) 3WJs with two siRNAs (siMED1-1 and siMED1-2) and HER2 aptamer showed tumor inhibition effect in combination with tamoxifen. (b) 3WJs with anti-miRNA (anti-miR21 or anti-miR17) and PSMA aptamer showed tumor suppression effect in mice model (c) 3WJs with Endo28 aptamer and doxorubicin chelating into the extended helices showed successful drug delivery to cells as demonstrated by confocal imaging (left), 3WJs with folate ligand and camptothecin conjugation inhibited tumor growth rate (right) (d) Targeting of tumor xenografts in mice after systematic injection utilizing pRNA-X nanoparticles harboring folate (left). 3WJs with CD133 aptamer showed targeting accumulation in tumor instead of vital organs (right). Figures reproduced with permission from (a) Ref. (Zhang *et al.*, 2017b), © American Chemical Society; (b) Ref. (Binzel *et al.*, 2016), © Elsevier; (c) Ref. (Pi *et al.*, 2016), Ref. (Piao *et al.*, 2019); (d) Ref. (Haque *et al.*, 2012), Ref. (Yin *et al.*, 2019).

A. 3WJ based 2D RNA nanostructures

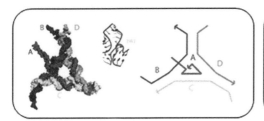
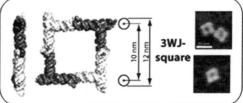

B. 3WJ based 3D RNA nanostructures

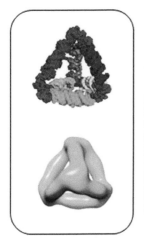
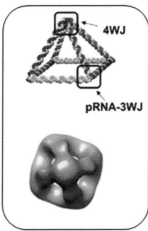
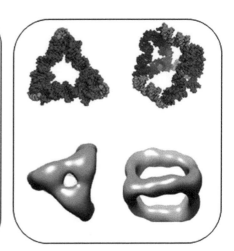

C. RNA nanostructures with RNA and other components

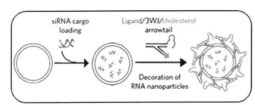
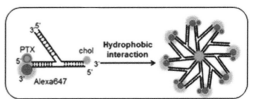

FIGURE 18.7 The construction of complicated RNA nanoparticles based on 3WJ motif. (a) The construction of triangular RNA nanostructures. Each of the three vertices contain a 3WJ motif (magnified); molecular model (left); schematic. Square-shaped RNA nanostructures built using four 3WJ motifs; schematic (left); AFM images of square (right). (b) Tetrahedron (left), pyramid (middle), and prism (right) using 3WJs as building blocks; schematic (top) and Cryo-EM reconstruction (bottom). (c) The construction of exosome as delivery platform with 3WJ decorated on the surface and siRNA encapsulated inside (left). RNA micelles built from 3WJ motifs with cholesterol, folate, and paclitaxel conjugation. Figures reproduced with permissions from: (a) Ref. (Bindewald *et al.*, 2011), Ref. (Severcan *et al.*, 2009), © American Chemical Society; (b); Ref. (Li *et al.*, 2016), Ref. (Xu *et al.*, 2019), Ref. (Khisamutdinov *et al.*, 2016); (c) Ref. (Pi *et al.*, 2018), Ref. (Shu *et al.*, 2018a), © Elsevier.

(Pi *et al.*, 2018) (Figure 18.7c). Micelles formed from 3WJ with three functional groups, folate for target delivery, paclitaxel for therapeutic effect and cholesterol for micelle formation, have been constructed (Shu *et al.*, 2018b) (Figure 18.7c). Such rationally designed supramolecular assemblies have the potential to be utilized in nanomedicine applications by conjugating functional siRNA and aptamers at each of the corners of the square and triangular RNA nanoparticles.

4WJ MOTIFS

CLASSIFICATIONS AND OCCURRENCES

Four-way junctions (4WJ) represent the second most abundant junction motifs in biological RNAs after 3WJ. Laing and Schlick (2009) classified 62 4WJs obtained from high- resolution 3D structures of folded RNAs containing the 4WJ into nine major families (Figure 18.8). The 4WJ classification

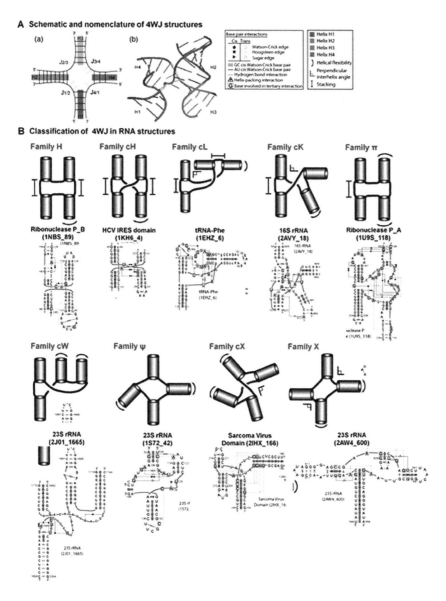

FIGURE 18.8 (a) Nomenclature used for characterizing 4WJ motifs. Helices: H1, H2, H3, and H4. Junctions: J1/2, J2/3, J3/4, and J4/1; 2D (left) and 3D (right) images are shown. Box: Symbols of various base pairs and notations observed in RNA structures shown in C. (b) Classification of 4WJs in RNA structures into nine families (H, cH, cL, cK, π, cW, ψ, X, and cX). Schematic (top) and a representative example with secondary structure (bottom). Symbols represent tertiary contacts and base pairing as shown in (a). Figures reproduced with permissions from Ref. (Laing and Schlick, 2009), © Elsevier.

is based on coaxial stacking signatures, helical arrangements (parallel and perpendicular), and helical arm configurations. Common tertiary motifs observed are coaxial stacking, non-canonical base pairings, internal loops, pseudoknots, kissing hairpins, and A-minor interactions, all of which play important roles in folding and stabilization of these 4WJ RNA structures. In the nomenclature, H1, H2, H3, and H4 represent the four helices; J1/2, J2/3, J3/4, and J4/1 represent the junctions (J) between the corresponding numbered helices (H1-4). For example, J1/2 is the junction between H1 and H2 helices.

1. *Family H*: two coaxially stacked helical configuration (roughly aligned); in each coaxial helix, continuous strands are anti-parallel to each other; coaxial helix arrangements stabilized by long-range interactions; Family H occurs in Ribonuclease P (A and B types), hairpin ribozymes, and 23S rRNA.
2. *Family cH*: two coaxially stacked helical configuration (roughly aligned); two types of coaxial stacking: H1H4 with H2H3 and H1H2 with H3H4; in each coaxial helix, continuous strands run in the same direction; the strands cross at the center stabilized by A-minor interactions at the cross-point. Family cH occurs in HCV IRES (Hepatitis C Virus Internal Ribosome Entry Site), 16S rRNA, 23S rRNA, and Flavin Mono-Nucleotide (FMN) riboswitches.
3. *Family cL*: two coaxially stacked helical configuration; "L"-shaped with coaxial stacks H1H4 and H2H3 perpendicular to each other; stabilized by long-range interactions, for example loop-helix, loop–loop, helix packing P-interactions, as well as ion concentrations; A-minor interactions stabilize strand exchange cross-points. Family cL occurs in 16S rRNA, 23S rRNA, tRNA, and S-adenosyl-methionine I (SAM I) riboswitches.
4. *Family cK*: one coaxially stacked helical configuration; two helices stacked with the third helix perpendicular to the coaxial helix, while the fourth helix projects out at an angle; perpendicular helical arrangements stabilized by long-range interactions; strand exchange cross-point involves A-minor interactions; helical arms rotate depending on pseudoknot formations, non-canonical base pairing and helix packing arrangements. Family cK occurs in 16S rRNA and 23S rRNA.
5. *Family π*: one coaxially stacked helical configuration; second pair of helices are aligned; helical arrangements stabilized by non-canonical base pairing and A-minor interactions. Family π occurs in Ribonuclease P_A.
6. *Family cW*: non-stacked helical configuration; long single-stranded segments; helical alignment between consecutive helices H1 and H4; contains high degree of junction symmetry; specific adopted conformations depend on tertiary interactions and/or upon protein binding. Family cW is uncommon and so far observed in 23S rRNA.
7. *Family ψ*: non-stacked helical configuration; long single-stranded segments; helical alignment between consecutive helices H2 and H4; contains high degree of junction symmetry; specific adopted conformations depend on tertiary interactions and/or upon protein binding. Family ψ is uncommon and so far observed in 23S rRNA.
8. *Family cX*: non-stacked helical configuration; long single-stranded segments; helical arms are arranged perpendicularly to each other; contains high degree of junction symmetry; specific adopted conformations depend on tertiary interactions and/or upon protein binding. Family cX is uncommon and so far observed in 23S rRNA.
9. *Family X*: non-stacked helical configuration; long single-stranded segments; helical arms are arranged perpendicularly to each other; contains high degree of junction symmetry; specific adopted conformations depend on tertiary interactions and/or upon protein binding. Family X is uncommon and so far observed in 16S rRNA and Sarcoma virus.

APPLICATION OF 4WJ MOTIFS IN NANOTECHNOLOGY

Several H-shaped tecto-RNA complexes were designed using GNRA loop/loop-receptor interaction motifs fused to a 4WJ motif derived from hairpin ribozyme (Nasalean *et al.*, 2006; Novikova *et al.*,

2010). Dimers, trimers, and tetramers in open and closed-ring form, as well as polymeric arrays were constructed by directed assembly of individual tecto-RNA subunits. The multimeric complexes were assembled in a defined and predictable stoichiometry by adjusting the 4WJ cross-over point. Interestingly, the cooperatively assembled closed dimers and trimers were resistant to RNase degradation and therefore suitable for therapeutic and diagnostic applications. Furthermore, each of the tecto-RNA subunits can potentially be derivatized at multiple locations to attach functional therapeutic and diagnostic modules (Novikova *et al.*, 2010).

FUTURE OUTLOOK AND PERSPECTIVES

RNA junctions are important structural elements in many folded RNAs performing diverse functions (Chen *et al.*, 1999; Honda *et al.*, 1999; Lescoute and Westhof, 2006; Laing and Schlick, 2009; Wakeman *et al.*, 2009; Kulshina *et al.*, 2010). They comprise of both rigid and flexible elements which contribute to the proper folding and functioning of the RNA molecule. The rigid components are stabilized by pseudoknots, coaxial stacking, and long-range interactions, whereas the flexible parts appear on longer helical segments and loop regions and the adopted conformation is dependent on ion concentrations and/or protein binding. A wide assortment of RNA 3WJ and 4WJ junction motifs are available for constructing diverse RNA nanoparticles with desired structure and function (Lescoute and Westhof, 2006; Laing and Schlick, 2009).

The 3WJ-pRNA-based therapeutic RNA nanoparticles have several attributes that make them particularly attractive for constructing nanoparticles for therapeutic and diagnostic applications. The key features include: (1) modular design, such that the larger nanoparticles can be assembled from individual strands that are within the limits of chemical synthesis and/or *in vitro* transcription (Shu *et al.*, 2011; Haque *et al.*, 2012); (2) homogenous distribution of nanoparticles *via* bottom-up self-assembly, which will significantly reduce off-site targeting and non-specific toxicity induced by heterogeneous nanoparticles; (3) chemically and thermodynamically stable nanoparticles such that they remain intact at ultra-low concentrations in the circulating blood without dissociating (Shu *et al.*, 2011; Liu *et al.*, 2011); (4) retains the authentic folding and function of therapeutic and diagnostic modules after incorporation into the scaffold without disrupting the folding of the core structure (Shu *et al.*, 2011; Haque *et al.*, 2012); (5) nanoscale size within 10–100 nm, which is small enough for cell entry *via* receptor- mediated endocytosis and large enough to avoid rapid kidney clearance; (6) multivalent nature, which facilitates specific targeting (*via* aptamers) and enhanced silencing effects (multiple copies of siRNA targeting one gene at same and/or different loci; or two or three genes simultaneously), and detection (using fluorophores), all-in-one nanoparticle (Shu *et al.*, 2011; Haque *et al.*, 2012); and (7) displays favorable pharmacokinetics, pharmacodynamics, biodistribution, and toxicology profiles in mice (Abdelmawla *et al.*, 2011).

To advance the development and utility of RNA junction scaffolds for nanotechnological applications, we require deeper knowledge of RNA folding, intermolecular assembly and control of stoichiometry. Several approaches are outlined below.

MUTATIONS TO INCREASE THERMODYNAMIC STABILITY

It is critically important to ensure that the RNA nanoparticles are thermodynamically stable and remain intact at ultra-low concentrations upon systemic injection. In case of 3WJ-pRNA, the length of the three helices was 8–9-bp (Shu *et al.*, 2011). Truncations and mutation analysis revealed that a minimum of 6-bp was necessary to assemble the 3WJ complex, while 8-bp was necessary to keep the junction stable in strongly denaturing conditions (Shu *et al.*, 2011). Furthermore, among 11 3WJ motifs found in biological RNA, 3WJ-pRNA displayed the highest T_M with the steepest slope (Shu *et al.*, 2011). Mutation experiments can be carried out in other 3WJ and 4WJ core motifs found in biological RNAs to find several alternative scaffolds with various shapes for diverse applications.

CHEMICAL MODIFICATIONS TO INCREASE SERUM STABILITY

Over the years, chemical modifications such as 2'-F have proven to have minimal detrimental effects on RNA folding, assembly, and function (Shu *et al.*, 2011; Liu *et al.*, 2011; Haque *et al.*, 2012). However, challenges still remain and improvements are necessary to optimize the conditions necessary to retain functionality while conferring resistance to nuclease degradation.

COMPUTATIONAL APPROACHES TO GUIDE NANOPARTICLE ASSEMBLY

Based on the sequences or predicted secondary structures of 3WJ and 4WJ, it is extremely challenging to predict the coaxial stacking signatures or helical configurations of folded RNA motifs. A thorough evaluation of the folding, energetic, and thermodynamic attributes of the junction motifs using computational approaches is necessary to construct RNA nanoparticles with desired structure and stoichiometry for diverse applications.

EVALUATION OF FUNCTIONALITY OF MODULES INCORPORATED INTO 3WJ OR 4WJ SCAFFOLD

It is important to ensure that the fusion of functional modules (such as siRNA, miRNA, aptamer, ribozyme, and riboswitch) does not interfere with the folding of the central 3WJ or 4WJ scaffold. Furthermore, the functional moieties must retain their folding properties and authentic function independently after incorporation in the 3WJ or 4WJ scaffolds.

ACKNOWLEDGMENTS

The research was supported by NIH grants EB003730 and CA151648 to P.G. P.G. is a cofounder of Kylin Therapeutics, Inc., and Biomotor and Nucleic Acid Nanotechnology Development Corp. Ltd.

REFERENCES

Abdelmawla S, Guo S, Zhang L, Pulukuri S, Patankar P, Conley P, Trebley J, Guo P, and Li QX (2011) Pharmacological characterization of chemically synthesized monomeric pRNA nanoparticles for systemic delivery. *Molecular Therapy*, **19**, 1312–1322.

Bindewald E, Afonin K, Jaeger L, and Shapiro BA (2011) Multistrand RNA secondary structure prediction and nanostructure design including pseudoknots. *ACS Nano*, **5**, 9542–9551.

Bindewald E, Grunewald C, Boyle B, O'Connor M, and Shapiro BA (2008a) Computational strategies for the automated design of RNA nanoscale structures from building blocks using NanoTiler. *Journal of Molecular Graphics & Modelling*, **27**, 299–308.

Bindewald E, Hayes R, Yingling YG, Kasprzak W, and Shapiro BA (2008b) RNAJunction: a database of RNA junctions and kissing loops for three-dimensional structural analysis and nanodesign. *Nucleic Acids Res*, **36**, D392–D397.

Binzel D, Shu Y, Li H, Sun M, Zhang Q, Shu D, Guo B, and Guo P (2016) Specific Delivery of MiRNA for High Efficient Inhibition of Prostate Cancer by RNA Nanotechnology. *Molecular Therapy*, **24**, 1267–1277.

Chen C, Sheng S, Shao Z, and Guo P (2000) A dimer as a building block in assembling RNA: A hexamer that gears bacterial virus phi29 DNA-translocating machinery. *J Biol Chem*, **275(23)**, 17510–17516.

Chen C, Zhang C, and Guo P (1999) Sequence requirement for hand-in-hand interaction in formation of pRNA dimers and hexamers to gear phi29 DNA translocation motor. *RNA*, **5**, 805–818.

Chworos A, Severcan I, Koyfman AY, Weinkam P, Oroudjev E, Hansma HG, and Jaeger L (2004) Building programmable jigsaw puzzles with RNA. *Science*, **306**, 2068–2072.

Ellington AD and Szostak JW (1990) *In vitro* selection of RNA molecules that bind specific ligands. *Nature*, **346**, 818–822.

Fabian MR, Sonenberg N, and Filipowicz W (2010) Regulation of mRNA translation and stability by microRNAs. *Annu Rev Biochem*, **79**, 351–379.

Fire A, Xu S, Montgomery MK, Kostas SA, Driver SE, and Mello CC (1998) Potent and specific genetic interference by double-stranded RNA in Caenorhabditis elegans. *Nature*, **391**, 806–811.

Geary C, Chworos A, and Jaeger L (2010) Promoting RNA helical stacking via A-minor junctions. *Nucleic Acids Res*, **39**, 1066–1080.

Guerrier-Takada C, Gardiner K, Marsh T, Pace N, and Altman S (1983) The RNA moiety of ribonuclease P is the catalytic subunit of the enzyme. *Cell*, **35**, 849–857.

Guo P (2005) RNA Nanotechnology: Engineering, Assembly and Applications in Detection, Gene Delivery and Therapy. *Journal of Nanoscience and Nanotechnology*, **5(12)**, 1964–1982.

Guo P (2010) The emerging field of RNA nanotechnology. *Nature Nanotechnology*, **5**, 833–842.

Guo P, Zhang C, Chen C, Trottier M, and Garver K (1998) Inter-RNA interaction of phage phi29 pRNA to form a hexameric complex for viral DNA transportation. *Mol Cell*, **2**, 149–155.

Haque F, Shu D, Shu Y, Shlyakhtenko L, Rychahou P, Evers M, and Guo P (2012) Ultrastable synergistic tetravalent RNA nanoparticles for targeting to cancers. *Nano Today*, **7**, 245–257.

Honda M, Beard MR, Ping LH, and Lemon SM (1999) A phylogenetically conserved stem-loop structure at the 5' border of the internal ribosome entry site of hepatitis C virus is required for cap-independent viral translation. *J Virol*, **73**, 1165–1174.

Ikawa Y, Tsuda K, Matsumura S, and Inoue T (2004) De novo synthesis and development of an RNA enzyme. *Proc Natl Acad Sci U S A*, **101**, 13750–13755.

Isambert H (2009) The jerky and knotty dynamics of RNA. *Methods*, **49**, 189–196.

Jaeger L and Leontis NB (2000) Tecto-RNA: One dimensional self-assembly through tertiary interactions. *Angew Chem Int Ed Engl*, **39**, 2521–2524.

Khisamutdinov EF, Jasinski DL, Li H, Zhang K, Chiu W, and Guo P (2016) Fabrication of RNA 3D Nanoprism for Loading and Protection of Small RNAs and Model Drugs. *Advanced Materials*, **28**, 100079–100087.

Kruger K, Grabowski PJ, Zaug AJ, Sands J, Gottschling DE, and Cech TR (1982) Self-splicing RNA: autoexcision and autocyclization of the ribosomal RNA intervening sequence of Tetrahymena. *Cell*, **31**, 147–157.

Kulshina N, Edwards TE, and Ferre-D'Amare AR (2010) Thermodynamic analysis of ligand binding and ligand binding-induced tertiary structure formation by the thiamine pyrophosphate riboswitch. *RNA*, **16**, 186–196.

Laing C and Schlick T (2009) Analysis of four-way junctions in RNA structures. *J Mol Biol*, **390**, 547–559.

Leontis NB, Lescoute A, and Westhof E (2006) The building blocks and motifs of RNA architecture. *Curr Opin Struct Biol*, **16**, 279–287.

Lescoute A and Westhof E (2006) Topology of three-way junctions in folded RNAs. *RNA*, **12**, 83–93.

Li H, Li WX, and Ding SW (2002) Induction and suppression of RNA silencing by an animal virus. *Science*, **296**, 1319–1321.

Li H, Zhang K, Pi F, Guo S, Shlyakhtenko L, Chiu W, Shu D, and Guo P (2016) Controllable Self-Assembly of RNA Tetrahedrons with Precise Shape and Size for Cancer Targeting. *Adv Mater*, **28**, 7501–7507.

Li X, Horiya S, and Harada K (2006) An efficient thermally induced RNA conformational switch as a framework for the functionalization of RNA nanostructures. *J Am Chem Soc*, **128**, 4035–4040.

Liu J, Guo S, Cinier M, Shlyakhtenko LS, Shu Y, Chen C, Shen G, and Guo P (2011) Fabrication of stable and RNase-resistant RNA nanoparticles active in gearing the nanomotors for viral DNA packaging. *ACS Nano*, **5**, 237–246.

Matsumura S, Ohmori R, Saito H, Ikawa Y, and Inoue T (2009) Coordinated control of a designed trans-acting ligase ribozyme by a loop-receptor interaction. *FEBS Lett*, **583**, 2819–2826.

Nakashima Y, Abe H, Abe N, Aikawa K, and Ito Y (2011) Branched RNA nanostructures for RNA interference. *Chem Commun (Camb)*, **47**, 8367–8369.

Nasalean L, Baudrey S, Leontis NB, and Jaeger L (2006) Controlling RNA self-assembly to form filaments. *Nucleic Acids Res*, **34**, 1381–1392.

Novikova IV, Hassan BH, Mirzoyan MG, and Leontis NB (2010) Engineering cooperative tecto-RNA complexes having programmable stoichiometries. *Nucleic Acids Res*, **39(7)**, 2903–2917.

Ohno H, Kobayashi T, Kabata R, Endo K, Iwasa T, Yoshimura SH, Takeyasu K, Inoue T, and Saito H (2011) Synthetic RNA-protein complex shaped like an equilateral triangle. *Nat Nanotechnol*, **6**, 116–120.

Pi F, Binzel DW, Lee TJ, Li Z, Sun M, Rychahou P, Li H, Haque F, Wang S, Croce CM, Guo B, Evers BM, and Guo P (2018) Nanoparticle orientation to control RNA loading and ligand display on extracellular vesicles for cancer regression. *Nat Nanotechnol*, **13**, 82–89.

Pi F, Zhang H, Li H, Thiviyanathan V, Gorenstein DG, Sood AK, and Guo P (2016) RNA nanoparticles harboring annexin A2 aptamer can target ovarian cancer for tumor-specific doxorubicin delivery. *Nanomedicine*, **13**, 1183–1193.

Piao X, Yin H, Guo S, Wang H, and Guo P (2019) RNA nanotechnology to solubilize hydrophobic antitumor drug for targeted delivery. *Advanced Science*, In Press.

Pleij CW and Bosch L (1989) RNA pseudoknots: Structure, Detection, and Prediction. *Meth Enzymol*, **180**, 289–303.

Schroeder KT, McPhee SA, Ouellet J, and Lilley DM (2010) A structural database for k-turn motifs in RNA. *RNA*, **16**, 1463–1468.

Searle MS and Williams DH (1993) On the stability of nucleic acid structures in solution: enthalpy-entropy compensations, internal rotations and reversibility. *Nucleic Acids Res*, **21**, 2051–2056.

Severcan I, Geary C, Verzemnieks E, Chworos A, and Jaeger L (2009) Square-shaped RNA particles from different RNA folds. *Nano Lett*, **9**, 1270–1277.

Shu D, Shu Y, Haque F, Abdelmawla S, and Guo P (2011) Thermodynamically stable RNA three-way junctions for constructing multifuntional nanoparticles for delivery of therapeutics. *Nature Nanotechnology*, **6**, 658–667.

Shu D, Huang L, Hoeprich S, and Guo P (2003) Construction of phi29 DNA-packaging RNA (pRNA) monomers, dimers and trimers with variable sizes and shapes as potential parts for nano-devices. *J Nanosci Nanotechnol*, **3**, 295–302.

Shu D, Moll WD, Deng Z, Mao C, and Guo P (2004) Bottom-up assembly of RNA arrays and superstructures as potential parts in nanotechnology. *Nano Lett*, **4**, 1717–1723.

Shu D, Zhang H, Jin J, and Guo P (2007) Counting of six pRNAs of phi29 DNA-packaging motor with customized single molecule dual-view system. *EMBO J*, **26**, 527–537.

Shu Y, Yin H, Rajabi M, Li H, Vieweger M, Guo S, Shu D, and Guo P (2018b) RNA-based micelles: A novel platform for paclitaxel loading and delivery. *J Control Release*, **276**, 17–29.

Shu Y, Yin H, Rajabi M, Li H, Vieweger M, Guo S, Shu D, and Guo P (2018a) RNA-based micelles: A novel platform for paclitaxel loading and delivery. *J Control Release*, **276**, 17–29.

Sudarsan N, Lee ER, Weinberg Z, Moy RH, Kim JN, Link KH, and Breaker RR (2008) Riboswitches in eubacteria sense the second messenger cyclic di-GMP. *Science*, **321**, 411–413.

Sugimoto N, Nakano S, Katoh M, Matsumura A, Nakamuta H, Ohmichi T, Yoneyama M, and Sasaki M (1995) Thermodynamic parameters to predict stability of RNA/DNA hybrid duplexes. *Biochemistry*, **34**, 11211–11216.

Tuerk C and Gold L (1990) Systematic evolution of ligands by exponential enrichment: RNA ligands to bacteriophage T4 DNA ploymerase. *Science*, **249**, 505–510.

Turner R and Tijan R (1989) Leucine repeats and an adjacent DNA binding domain mediate the formation of functional c-Fos and c-Jun heterodimers. *Science*, **243**, 1689–1694.

Wakeman CA, Ramesh A, and Winkler WC (2009) Multiple metal-binding cores are required for metalloregulation by M-box riboswitch RNAs. *J Mol Biol*, **392**, 723–735.

Woodson SA (2010) Compact intermediates in RNA folding. *Annu Rev Biophys*, **39**, 61–77.

Xiao F, Moll D, Guo S, and Guo P (2005) Binding of pRNA to the N-terminal 14 amino acids of connector protein of bacterial phage phi29. *Nucleic Acids Res*, **33**, 2640–2649.

Xu C, Li H, Zhang K, Binzel DW, Yin H, Chiu W, and Guo P (2019) Photo-controlled release of paclitaxel and model drugs from RNA pyramids. *Nano Research*, **12**, 41–48.

Yin H, Xiong G, Guo S, Xu C, Xu R, Guo P, and Shu D (2019) Delivery of Anti-miRNA for Triple-Negative Breast Cancer Therapy Using RNA Nanoparticles Targeting Stem Cell Marker CD133. *Mol Ther*.

Zhang CL, Lee C-S, and Guo P (1994) The proximate 5′ and 3′ ends of the 120-base viral RNA (pRNA) are crucial for the packaging of bacteriophage f29 DNA. *Virology*, **201**, 77–85.

Zhang H, Endrizzi JA, Shu Y, Haque F, Sauter C, Shlyakhtenko LS, Lyubchenko Y, Guo P, and Chi YI (2013) Crystal Structure of 3WJ Core Revealing Divalent Ion-promoted Thermostability and Assembly of the Phi29 Hexameric Motor pRNA. *RNA*, **19**, 1226–1237.

Zhang Y, Leonard M, Shu Y, Yang Y, Shu D, Guo P, and Zhang X (2017a) Overcoming Tamoxifen Resistance of Human Breast Cancer by Targeted Gene Silencing Using Multifunctional pRNA Nanoparticles. *ACS Nano*, **11**, 335–346.

Zhang Y, Leonard M, Shu Y, Yang Y, Shu D, Guo P, and Zhang X (2017b) Overcoming Tamoxifen Resistance of Human Breast Cancer by Targeted Gene Silencing Using Multifunctional pRNA Nanoparticles. *ACS Nano*, **11**, 335–346.

Zuker M (1989) On finding all suboptimal foldings of an RNA molecule. *Science*, **244**, 48–52.

19 Fabrication Methods for RNA Nanoparticle Assembly Based on Bacteriophage Phi29 pRNA Structural Features

Yi Shu and Hongzhi Wang
The Ohio State University, Columbus, OH, USA

Bahar Seremi
Texas Tech University, Lubbock, TX, USA

Peixuan Guo
The Ohio State University, Columbus, OH, USA

CONTENTS

RNA NANOTECHNOLOGY

Ribonucleic acid (RNA) molecules can be easily manipulated with the simplicity of DNA, while possessing versatile structures and functions similar to proteins. This property makes RNA a suitable candidate for applications in nanotechnology and nanomedicine (Shu *et al.*, 2003; Hansma *et al.*, 2003; Shu *et al.*, 2004; Guo, 2010; Jasinski *et al.*, 2017; Xu *et al.*, 2018a). RNA is made up of a sugar-phosphate backbone chain with a combination of four ribonucleobases: adenine (A), guanine (G), cytosine (C), and uracil (U). Most RNA molecules are single-stranded. Like DNA, RNA

DOI: 10.1201/9781003001560-21

can be synthesized both enzymatically through *in vitro* transcription and chemically via solid-phase synthesis. Unlike DNA, RNA contains complicated three-dimensional secondary structures which can play more versatile and specific functions.

The primary sequences of the RNA determine the folding of RNA structures. Secondary structure elements such as bulges, hairpin loops, internal loops, and junctions are formed through Watson–Crick and/or non-canonical base pairing. The inter- and intramolecular interactions such as loop–loop interaction and base stacking ensure the formation of RNA tertiary structures by hydrogen bonds within the RNA molecules. In addition, since RNA backbone is negatively charged, metal ions such as Mg^{2+} are needed to stabilize many secondary and tertiary structures (Pan *et al.*, 1993).

In addition to encoding genetic information, RNA molecules also play active roles in cells by catalyzing biological reactions (Strobel and Cochrane, 2007; Westhof, 2012; Fukuda *et al.*, 2014), regulating gene expression (Lee *et al.*, 2009; Zhang, 2009; Fabian *et al.*, 2010; Taft *et al.*, 2010; Lam *et al.*, 2015; Chakraborty *et al.*, 2017), or sensing and communicating responses to cellular signals (Sudarsan *et al.*, 2008; Henkin, 2008; Ogawa and Maeda, 2008). In recent years, more and more functional RNA molecules, naturally or artificially engineered, have been discovered. Like antibodies, RNA aptamers selected from systematic evolution of ligands by exponential enrichment (SELEX) (Ciesiolka *et al.*, 1995; Kraus *et al.*, 1998; Bouvet, 2001; Clark and Remcho, 2002; Shu and Guo, 2003; Zhou and Rossi, 2016; Lyu *et al.*, 2016) are able to bind to specific targets, including proteins, organic compounds, and nucleic acids (Ellington and Szostak, 1990; Tuerk and Gold, 1990; Gold, 1995; Ye *et al.*, 2012). The ability to recognize specific cell surface markers through the formation of binding pockets and the capability of internalization by the targeted cells paves a new way for targeted delivery (McNamara *et al.*, 2006; Zhou *et al.*, 2008; Dassie *et al.*, 2009; Cerchia and de Franciscis, 2010; Leonard *et al.*, 2015; Miller-Kleinhenz *et al.*, 2015; Zhou and Rossi, 2017). In the early 1980s, Thomas Cech (Cech *et al.*, 1981) and Sydney Altman (Guerrier-Takada *et al.*, 1983) found RNA molecules had the ability to catalyze chemical reactions. In 1998, Andrew Fire and Craig Mello discovered RNAi (RNA interference), a mechanism which regulates gene expression on a post-transcriptional level (Fire *et al.*, 1998). The discovery of RNAi has heightened interests in RNA therapeutics (Fire *et al.*, 1998). Several RNA-based therapeutic approaches using small interfering RNAs (siRNAs) (Li *et al.*, 2002; Brummelkamp *et al.*, 2002; Varambally *et al.*, 2002; Carmichael, 2002; Jacque *et al.*, 2002; Ghildiyal and Zamore, 2009; Guo *et al.*, 2010; Tatiparti *et al.*, 2017; Xu *et al.*, 2018b), ribozymes (Forster and Symons, 1987; Sarver *et al.*, 1990; Chowrira *et al.*, 1991; Westhof, 2012; Fukuda *et al.*, 2014), and anti-sense RNAs (Coleman *et al.*, 1985; Knecht and Loomis, 1987; Lennox *et al.*, 2013; Shu *et al.*, 2015) have been shown to downregulate specific gene expression by intercepting and cleaving mRNA or the genome of RNA viruses in cancerous or viral-infected cells. In 2016, Fomivirsen became the first and antisense **RNA** therapeutic to receive FDA approval and enter the market, which is approved for cytomegalovirus retinitis. In the meanwhile, the CRISPR-Cas9 system as a genome editing method has generated a lot of excitement in the scientific community. Researchers create a small piece of RNA with a short "guide" sequence that attaches (binds) to a specific target sequence of DNA in a genome. The sgRNA directs the Cas9 endonuclease to cleave both DNA strands in a sequence-specific manner. Genome editing using CRISPR-Cas9 system is of great interest in the prevention and treatment of human diseases (Cong *et al.*, 2013; Jiang and Marraffini, 2015).

The concept of RNA nanotechnology has been proposed for more than a decade (Guo *et al.*, 1998; Zhang *et al.*, 1998; Jaeger and Leontis, 2000; Jaeger *et al.*, 2001; Shu *et al.*, 2004; Chworos *et al.*, 2004; Guo, 2005; Jaeger and Chworos, 2006; Guo, 2010; Jasinski *et al.*, 2017; Guo *et al.*, 2018; Xu *et al.*, 2018a). Elucidation of the structural and folding mechanism of RNA motifs and junctions has laid a foundation for the further development of RNA nanotechnology. The structure motifs and tertiary interactions can be extracted from RNA molecules as the building blocks to self-assemble nanoscaled scaffolds (Leontis and Westhof, 2003; Leontis *et al.*, 2006), and the intra- and intermolecular interaction can be computed and rationally designed. Meanwhile, functional RNA molecules such as siRNA, ribozyme, and riboswitch can be incorporated into the scaffold and further form functionalized RNA nanoparticles. The field of RNA nanotechnology is becoming popular because the potential of RNA nanoparticles in the treatment of cancer, viral infection, or genetic diseases is recognized (Guo, 2010). RNA nanotechnology holds great potential because of the following reasons:

1. Homogeneous RNA nanoparticles can be manufactured with high reproducibility and known stoichiometry, thus avoiding unpredictable side effects or nonspecific toxicity associated with heterogeneous structures.
2. Using the bottom-up approach, RNA nanoparticles can be assembled harboring multiple therapeutic, reporters and/or targeting payloads for synergetic effects (Khaled *et al.*, 2005; Tarapore *et al.*, 2010; Binzel *et al.*, 2016; Yin *et al.*, 2019).
3. Cell type-specific gene targeting can be achieved via simultaneous delivery and detection modules which reduces off-target toxicity and lowers the concentration of the drug adminis- tered, thus reducing the side effects of the therapeutics.
4. RNA nanoparticle size typically ranges from 10 to 50 nm, an optimal size for a non-viral vec- tor as they are large enough to be retained by the body yet small enough to pass through the cell membrane via the cell surface receptors mediated endocytosis. The advantageous size has the potential to greatly improve the pharmacokinetics, pharmacodynamics, biodistribution, and toxicology profiles by avoiding non-specific cell penetration (Abdelmawla *et al.*, 2011).
5. Protein-free RNA nanoparticles with RNA aptamers as anti-receptors can yield superior spec- ificity compared to protein anti-receptors while displaying lowest antibody-inducing activity, thus providing an opportunity for repeated administration and treatment of chronic diseases.
6. Nanoparticles process size, shape, and sequence-dependent immunogenicity. immunore- sponse of RNA nanoparticles is tunable to produce either a minimal immune response that can serve as safe therapeutic vectors, or a strong immune response for cancer immunotherapy or vaccine adjuvants (Guo *et al.*, 2017).
7. RNA nanoparticles are treated as chemical drugs rather than biological entities, which will facilitate FDA approval. In 2016, became the first and antisense RNA therapeutic to receive FDA approval and enter the market, which is approved for cytomegalovirus retinitis. Until now, several antisense drugs were approved for various diseases, such as homozygous familial hypercholesterolemia and spinal muscular atrophy.

PRINCIPLES AND TOOLS FOR RNA NANOPARTICLE FABRICATION

RNA nanoparticle's construction is generally following three steps: RNA building block extraction, rational computational design/modeling, and RNA nanoparticle fabrication. RNA structural motifs such as bulges, internal loops (Ferrandon *et al.*, 1997), kissing loops (Laughrea and Jette, 1996; Clever *et al.*, 1996; Paillart *et al.*, 1996), and junctions (Lilley, 2000) can be extracted from the known RNA structures as the building blocks and rational designed to apply for RNA nanoparticle construction. Significant contributions on the fundamental studies of RNA structural motifs were made by group of scientists from Eric Westhof (Westhof *et al.*, 1996; Leontis and Westhof, 2003; Lescoute and Westhof, 2006; Jossinet *et al.*, 2007), Neocles Leontis (Jaeger *et al.*, 2001; Leontis and Westhof, 2001; Leontis and Westhof, 2002; Leontis and Westhof, 2003; Leontis *et al.*, 2006), and David Lilley's laboratory (Cadd and Patterson, 1994; Lilley, 1999; McKinney *et al.*, 2003; Schroeder *et al.*, 2010).

RNA folding and structural computation is essential for using RNA motifs for RNA nanoparticle assembly. The computational design of nanostructures is relatively inexpensive and provides a faster

TABLE 19.1

Resource for RNA Nanoparticle Computation Design

	Name	Description	Reference
RNA structure database	Protein Data Bank (PDB)	Atomic structure information derived from X-ray crystallography or NMR (http://www.wwpdb.org/)	Berman *et al.* (2000), Berman *et al.* (2007)
	Nucleic Acid Database (NDB)	Nucleic acid structures derived from X-ray crystallography or NMR and classified by type (http://ndbserver.rutgers.edu/)	Berman *et al.* (1992)
	Structure Classification of RNA Database (SCOR)	Classification of RNA internal loop and hairpin loop structures (http://scor.lbl.gov.)	Klosterman *et al.* (2004), Tamura *et al.* (2004)
	Non-Canonical Interactions in RNA database (NCIR)	Structure information about noncanonical RNA base pairs (http://prion.bchs.uh.edu/bp_type/)	Nagaswamy *et al.* (2002)
	RNAjunction database	Contains "over 12,000 extracted three-dimensional junction and kissing loop structures as well as detailed annotations for each motif"(http://rnajunction.abcc.ncifcrf.gov/).	Bindewald *et al.* (2008b)
RNA structure predication	Mfold UNAfold RNAstructure Sfold RNAshapes MPGAfold	RNA secondary structure prediction and determination	Zuker (2003), Markham and Zuker (2008), Mathews *et al.* (2004), Ding *et al.* (2004), Steffen *et al.* (2006), Shapiro and Wu (1997)
	Nanofolder	multi-stranded RNA Secondary Structure Prediction and Design	Bindewald *et al.* (2011)
	Stem Trace	Comparative RNA structure analysis	Kasprzak *et al.* (2011)
	RNA2D3D MANIP NAB RNA-Puzzles ifoldRNA	RNA 3D structure prediction and modeling	Martinez *et al.* (2008), Massire and Westhof (1998), Macke and Case (1997), Cruz *et al.* (2012), Sharma *et al.* (2008), Ding *et al.* (2008)
	Nanotiler	Design of RNA nanoscale structures from building blocks	Bindewald *et al.* (2008a)
	RADAR	web server for RNA data analysis and research	Khaladkar *et al.* (2007)

examination of new structural designs. Advances of RNA 3D computation was promoted by Bruce Shapiro and co-workers, which brought new energy into the RNA nanotechnology field (Yingling and Shapiro, 2007; Shapiro *et al.*, 2008; Bindewald *et al.*, 2008a; Shapiro, 2009; Afonin *et al.*, 2010; Grabow *et al.*, 2011; Qiu *et al.*, 2013). For the available RNA structure information and computation program, please refer to Table 19.1.

Luc Jaeger's group has shown that RNA can be used as efficient scaffold which are built from designed building blocks. TectoRNAs (Jaeger *et al.*, 2001), which were designed from structure motif such as ribosome RNA or HIV kissing loops, can built up varieties of RNA nanostructures such as tectosquares, Jigsaw puzzles, nanocubes, and nanorings (Chworos *et al.*, 2004; Severcan *et al.*, 2009; Severcan *et al.*, 2010; Afonin *et al.*, 2010; Grabow *et al.*, 2011). Cayrol et al. reported, for the first time, that RNA filaments and bundles were formed by self-assembly of the DsrA, the noncoding RNA of *Escherishia coli* through antisense interactions of three contiguous, self-complementary regions (Cayrol *et al.*, 2009).

Bacteriophage phi29 packaging RNA (pRNA) is another system for assembly of RNA nanoparticles via interlocking loop–loop interactions. Specific delivery of siRNA to target cells has been achieved using pRNA nanodelivery system (Guo *et al.*, 2005; Khaled *et al.*, 2005; Guo *et al.*, 2006). Previous extensive studies showed that pRNA could escort the siRNA to silence genes and to destroy cancer cells of leukemia, lung, breast, ovarian, prostate, among others (Hoeprich *et al.*, 2003; Guo *et al.*, 2005; Khaled *et al.*, 2005; Guo *et al.*, 2006; Liu *et al.*, 2007; Li *et al.*, 2009; Zhang *et al.*, 2009; Tarapore *et al.*, 2010). The detailed introduction and methods for pRNA nanoparticle construction will be introduced in the next session.

ASSEMBLY OF pRNA-BASED RNA NANOPARTICLES

THE BIOLOGICAL FUNCTION OF pRNA ON PHI29 DNA PACKAGING MOTOR

All linear double-stranded DNA viruses including bacteriophage phi29 (Guo and Trottier, 1994) possess a common feature that their genome is packaged into a preformed procapsid during the maturation of the viron. This process is accomplished by an ATP-driven packaging motor (Figure 19.1A) (Earnshaw and Casjens, 1980). However, the most exciting aspect of the phi29 DNA packaging process is the discovery of a small viral RNA called pRNA (Figure 19.1B) (Guo *et al.*, 1987), which is 120 bases and transcribed from the left end of the phi29 genome. It has also been revealed that pRNA contains two functional domains (Reid *et al.*, 1994b; Zhang *et al.*, 1995a), one facilitates the formation of pRNA hexameric ring and binds to the connector while the other binds to the DNA-packaging enzyme gp16 (Lee and Guo, 2006). By using the energy from gp16 hydrolyzing ATP,

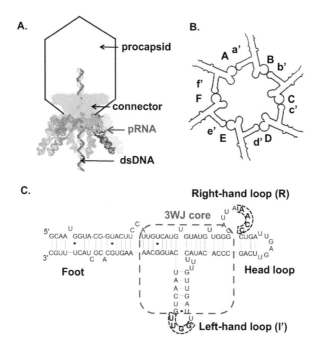

FIGURE 19.1 The illustration of constructing loop extended pRNA. (a) Bacteriophage phi29 DNA packaging motor. Six copies of wild-type pRNAs assemble a hexamer ring on the connector portal to gear the viral DNA into the prohead. (b) pRNA hexamer ring. (c) The primary sequence and secondary structure of wild-type pRNA. Two domains are connected by a 3WJ core.

this very powerful DNA packaging motor then gears the viral genome into the preformed procapsid. After the complete packaging, it is possible that pRNA leaves the connector before the collar protein gp11 and tail knob gp9 protein block the connector channel and assemble the intact viral particles (Chen *et al.*, 2000).

STRUCTURE FEATURES OF BACTERIOPHAGE PHI29 PRNA

pRNA is a crucial component in the phi29 DNA packaging motor. Each pRNA contains two functional domains. The helical domain is located at the 5'/3' paired ends (Cairns *et al.*, 1988; Zhang *et al.*, 1994; Reid *et al.*, 1994a; Reid *et al.*, 1994b; Reid *et al.*, 1994c; Garver and Guo, 1997) denoted as "foot". The central domain of each pRNA subunit (bases 23–97) embraces one head loop and two interlocking loops denoted as the right- and left-hand loops for intermolecular interaction (Reid *et al.*, 1994a; Garver and Guo, 1997; Chen *et al.*, 1999; Chen *et al.*, 2000) (Figure 19.1C). The two domains fold separately, and sequence alteration of the helical domain does not affect pRNA structure, folding or intermolecular interactions (Zhang *et al.*, 1994; Trottier *et al.*, 2000). Furthermore, the pRNA interlocking central domain is a nucleation core with very low free energy, which folds independently of the newly incorporated moieties (Chen *et al.*, 1999). Two interlocking loops can be reengineered to form dimers, trimers, or hexamers via hand-in-hand interactions (Guo *et al.*, 1987; Guo *et al.*, 1998; Chen *et al.*, 2000; Shu *et al.*, 2003; Shu *et al.*, 2004). In addition, the two domains are connected by a three-way junction (3WJ) motif (Figure 19.1C). The 3WJ motif of pRNA is thermodynamically stable, resistant to denaturation by 8 M urea, and remains intact at ultra-low concentrations (Shu *et al.*, 2011a). Recently, we demonstrated that the 3WJ motif of the pRNA can be used as a scaffold to assemble with high affinity trivalent RNA nanoparticles harboring therapeutic modules (Shu *et al.*, 2011a).

BOTTOM-UP ASSEMBLY OF PRNA NANOPARTICLES

The self-assembly of pRNA nanoparticles is a prominent bottom-up approach and represents an important idea that biomolecules can be successfully integrated into nanotechnology (Shu *et al.*, 2004; Glotzer, 2004; Gates *et al.*, 2005; Guo *et al.*, 2005; Khaled *et al.*, 2005; Shu *et al.*, 2015; Lee *et al.*, 2017; Shu *et al.*, 2018; Piao *et al.*, 2019). Such an approach relies on the cooperative interaction of individual RNA molecules that spontaneously assemble in a predefined manner and form larger 2D or 3D structures.

The unique structure of pRNA (Figure 19.2A) and its interlocking loops and 3WJ motif have motivated the studies on its application in RNA nanotechnology. There are variety of principles to construct pRNA-based nano-architectures with diverse shapes and angles. RNA loops, cores, junction motifs, and palindrome sequences derived from pRNA were gathered in a "toolkit" to fabricate dimers, trimers, and other higher order oligomers, as well as branched diverse architectures via hand-in-hand, foot-to-foot, and arm-on-arm interactions. 1) Hand-in-hand interactions: Assembly of pRNA dimers and trimers via the interlocking hand-in-hand loops has been reported previously (Figure 19.2B) (Guo *et al.*, 1998; Chen *et al.*, 1999; Shu *et al.*, 2003; Shu *et al.*, 2004; Guo *et al.*, 2005; Khaled *et al.*, 2005; Shu *et al.*, 2011c). We also tried to produce stable interlocking loops for nanoparticle construction by extending the loop sequences. Extensive investigations have resulted in stronger hand-in-hand interaction that can be used for the construction of higher ordered pRNA polygonal nanoparticles such as tetramer, pentamer, hexamer, heptamer, etc. 2) Foot-to-foot interactions: A palindrome sequence reads the same way either from 5' to 3' direction on one strand or from 5' to 3' direction on the complementary strand. The 6-nt palindrome sequences were designed and introduced at the 3'-end of the pRNA building blocks. These RNA building blocks were able to form foot-to-foot self-dimers *via* intermolecular interactions mediated by the palindrome sequences. In a similar fashion, palindrome sequences were utilized to bridge RNA nanostructures, motifs, or scaffolds for self-assembling RN foot-to-foot structures (Figure 19.2C). 3) Branched pRNA

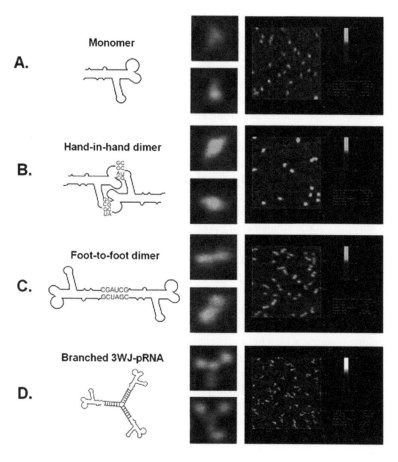

FIGURE 19.2   The AFM image of constructed pRNA nanoparticles via hand-in-hand, foot-to-foot, and arm-on-arm interactions. (a) pRNA monomer. (b) pRNA hand-in-hand dimer (adapted from (Liu *et al.*, 2011)). (c) pRNA foot-to-foot dimer (adapted from Liu *et al.*, 2011). (d) Branched 3WJ-pRNA (adapted from Shu *et al.*, 2011a).

nanostructures: larger macromolecular assemblies of pRNA nanoparticles were also constructed *via* intramolecular interactions utilizing junction motifs (Figure 19.2D) (Shu *et al.*, 2011a).

CONSTRUCTION OF PRNA NANOPARTICLES WITH BIOLOGICAL AND PHARMACEUTICAL FUNCTIONALITIES

Taking advantages of the folding independence of pRNA two functional domains, end conjugation of pRNA with chemical moiety, fusing pRNA with a receptor-binding RNA aptamer, siRNA, ribozyme, or other chemical groups generally do not disturb the interlocking interaction or interfere with the function of inserted moieties. Then, the engineered monomeric pRNA chimera can be further assembled as various through different strategies. The incorporation of all functionalities were achieved prior but not subsequent to the assembly of the RNA nanoparticles; thus ensure the production of homogeneous therapeutic nanoparticles for medical application to ensure appropriate quality control. These novel RNA nanostructures harbor resourceful functionalities for numerous applications in nanotechnology and medicine.

Incorporation of Small Interfering RNA (siRNA) into pRNA Nanoparticles

The pRNA double-stranded 5′/3′ end helical domain and intermolecular binding domain fold independently of each other. Complementary modification studies have revealed that altering the

primary sequences of any nucleotide of the helical region does not affect pRNA structure and folding as long as the two strands are paired (Zhang *et al.*, 1994). Extensive studies revealed that siRNA
was a ~21–23nt RNA duplex (Elbashir *et al.*, 2001; Li *et al.*, 2002; Brummelkamp *et al.*, 2002;
Carmichael, 2002). Thus, it was possible to replace the helical region in pRNA with double-stranded
siRNA (Figure 19.3A). A variety of chimeric pRNAs with different targets were constructed to
carry siRNA. Such a pRNA/siRNA chimera was proven to be successfully inhibited targeting gene
expression. A pRNA/siRNA chimera harboring siRNA targeting coxsackievirus B3 (CVB3) protease gene was constructed and was able to knock down gene expression specifically and inhibit viral
replication *in vitro* (Zhang *et al.*, 2009). We also selected a model involving the Metallothionein-IIA
(MT-IIA) gene as a proof of concept. The specific knockdown of the MT-IIA gene by constructed
chimera pRNA/siRNA (MT-IIA) can reduce ovarian cancer cells viability significantly (Tarapore
et al., 2010). Pre-clinical study also showed that pRNA/siRNA chimera targeting anti-apoptotic factor survivin gene can drive cancer cells undergo apoptosis and prevent tumorigenesis in xenograft
mice models (Guo *et al.*, 2005; Khaled *et al.*, 2005).

Incorporation of Ribozyme into pRNA Nanoparticles

Using the circular permutation approach (Nolan *et al.*, 1993; Pan and Uhlenbeck, 1993; Zhang *et al.*,
1997), almost any nucleotide of the entire pRNA can serve as either the new 5' or 3'-end of the RNA
monomer. Connecting the pRNA 5'/3' ends with variable sequences did not disturb its folding and
function (Hoeprich *et al.*, 2003; Guo *et al.*, 2005; Khaled *et al.*, 2005; Guo *et al.*, 2006; Liu *et al.*,
2007). These unique features, which help prevent two common problems, exonuclease degradation
and misfolding in the cell, make pRNA an ideal vector to carry therapeutic RNA such as ribozyme
(Figure 19.3B). A pRNA-based vector was designed to carry hammerhead ribozymes that cleave
the hepatitis B virus (HBV) polyA signal (Hoeprich *et al.*, 2003). The pRNA/ribozyme (survivin),
which targeted the anti-apoptosis factor survivin to downregulate genes involved in tumor development and progression, was also shown to suppress survivin expression and initiate apoptosis (Liu
et al., 2007).

Incorporation of RNA Aptamer into pRNA Nanoparticles

In vitro SELEX (Ellington and Szostak, 1990; Tuerk and Gold, 1990) of RNA aptamers which bind
to specific targets has become a powerful tool for selecting RNA molecules that target specific cell

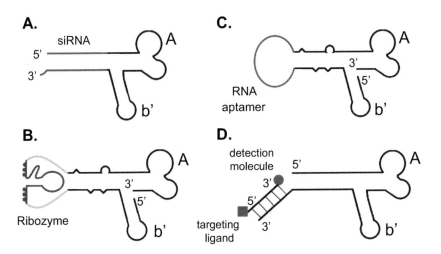

FIGURE 19.3 Illustration of incorporating functional modules into pRNA nanoparticles. (a) Incorporation
of siRNA. (b) Incorporation of ribozyme. (c) Incorporation of RNA aptamer. (d) Incorporation of chemical
ligands or fluorescent marker (adapted from Shu *et al.*, 2011c).

surface receptor. Aptamers were linked to the 3′ and 5′ end of pRNA. To facilitate independent folding, poly U or poly A linkers were placed between the pRNA and the aptamer. The nascent 5′/3′ end
of the pRNA was relocated to nt 71 and 75 using the circular permutation approach (Figure 19.3C)
(Zhang et al., 1995b; Zhang et al., 1997). The tightly folded nt 71 and 75 region protected the 5′/3′
end from exonuclease digestion. One of the pRNA/aptamer constructs harboring anti-HIV gp120
aptamer was proven to bind to and are internalized into cells expressing HIV gp120. Moreover, the
pRNA-aptamer chimeras alone also provide HIV inhibitory function by blocking viral infectivity in
an acute *in vitro* challenge assay (Zhou et al., 2011). Streptavidin (STV)-binding aptamer (Srisawat
and Engelke, 2001) can also be incorporated into pRNA nanoparticle. Upon incubation with STV
resin, pRNA nanoparticles harboring the STV-binding aptamer bound the STV resin and were then
eluted out by biotin. This STV-binding aptamer can be potentially utilized as the RNA nanoparticle
purification method.

Incorporation of Chemical Targeting Ligand such as Folate into pRNA Nanoparticles

Chemical ligands like folate which can recognize specific cell surface markers can also be conjugated to pRNA for specific targeting (Guo et al., 2006; Zhang et al., 2009; Tarapore et al., 2010). A
complementary DNA oligo can be annealed to the end of the pRNA which has the 3′-end extension
with 14–26 nucleotides (Figure 19.3D). The assembled folate-pRNA nanoparticles are able to bind
and internalize into cancer cells specifically and efficiently and were applied for systematic target
delivery of therapeutics *in vivo* (Abdelmawla et al., 2011; Shu et al., 2011a; Shu et al., 2011b).
Synthetic DNA oligos functionalized with different chemical groups, such as folate, were used for
therapeutic pRNA construction.

Unlike most of the common dyes and biotin, folic acid is not commercially available through its
NHS ester activated form. Reaction of folic acid in the presence of N, N′-dicyclohexylcarbodiimide
(DCC), NHS, and triethylamine (TEA) in dimethylsulfoxide (DMSO) lead to its conversion into
its NHS ester derivatives, which can be purified from the reaction mixture by precipitation (Zhang
et al., 2010). Alternatively, folate labeling of pRNA was also achieved through a two-step procedure involving primarily the coupling of the folate-NHS ester with a 5'-NH2-DNA oligo. HPLC
purification of the coupling product, followed by the annealing with pRNA has led to pRNA-folate
conjugates.

Incorporation of Fluorescent Marker into pRNA Nanoparticles

Many different strategies have been developed for post-transcriptional or co-transcriptional labeling
of RNA molecules with fluorescent dyes. These labeling strategies have been demonstrated to be
highly efficient for long-chain RNA, such as pRNA (120 bases) and can be distinguished following
single molecule labeling or random labeling of the whole chain of the RNA.

Whole chain labeling of pRNA: Bifunctional alkylating agents are known to promote crosslinking
of DNA or RNA molecules. Using a similar strategy, post-transcriptional fluorescent labeling of the
pRNA molecule was readily achieved using functionalized fluorophores, harboring a mono-alkyl-
ating reactive group, developed by Mirus (Label IT labeling reagent, Mirus). More recently, it was
demonstrated that T7 RNA polymerase can be used for *in vitro* enzymatic fluorescent labeling of
the RNA molecule using the new reagent tCTP (Stengel et al., 2010). Different T7 RNA polymerase
mutants were constructed to recognize and permit the incorporation of 2-modified triphosphate ribonucleotide such as 2′-OMe, 2′-F, 2′-NH$_2$, 2′-N$_3$ into the RNA chain (Sousa and Padilla, 1995; Huang
et al., 1997; Padilla and Sousa, 1999; Padilla and Sousa, 2002). Reactive functions (amino or azido)
can be imagined to further conjugate to the RNA molecules. Alternatively, 2′-F pRNA molecules
were constructed and shown to present an improved resistance to RNase degradation compared to
unmodified RNA, when both pyrimidines were substituted by their 2′-F counterparts (Kawasaki
et al., 1993; Shu et al., 2011a).

*Single and site-specific labeling of pRNA with fluorescent molecules or other chemical reactive
groups:* RNA solid-phase synthesis makes it possible to incorporate NH$_2$ group modified nucleotide

into specific site of RNA strands. NHS-activated fluorescent compounds can be directly coupled to the end NH_2 group of single-labeled RNA fragments as a post-transcriptional labeling approach with high labeling efficiency. However, solid phase synthesis of long RNAs greater than about 120 nts is hard to achieve with high synthesis efficiency and more rely on the enzymatic methods and the single labeling of the RNA. To overcome this challenge and label the longer RNA with a single functional group, AMP and GMP derivatives which have been demonstrated to be efficient initiators in RNA transcription by T7 RNA polymerase, but cannot be used in the elongation step, were designed. Efficient labeling of RNA molecules at the specific 5′position can be readily achieved by either one-step transcription initiation or a two-step procedure of transcription and post-transcriptional modification. Various kinds of AMP and GMP derivatives were synthesized by conjugating different chemical groups with AMP or GMP through linker molecules through established chemistry (Bruce and Uhlenbeck, 1978; Milligan et al., 1987; Zhang et al., 2001; Huang et al., 2003; Guo et al., 2006). The amino- and thiol- reactive derivatives, AMP-hexanediamine (HDA) (Huang et al., 2003; Huang et al., 2008) and 5′-deoxy-5′-thioguanosine-5′-monophosphorothioate (GSMP) (Zhang et al., 2001) respectively, present both reactive moieties for further conjugation with ligands or detection markers for the production of deliverable polyvalent therapeutic particles. Whereas the synthesis of GSMP requires a more complicated chemical process (Zhang et al., 2001), an AMP-HDA derivative was synthesized in one-step through the direct coupling of HDA to AMP at pH 6.5 in the presence of 1-Ethyl-3-[3-dimethylaminopropyl] carbodiimide hydrochloride (EDC) (Huang et al., 2008). AMP-HDA can then be purified either simply by removing the excess of HDA using ion exchange chromatography (Huang et al., 2008) or reverse-phase chromatography (Huang et al., 2003).

The reactive aliphatic amine of AMP-HDA can be used for further functionalization by common coupling reactions with N-hydroxysuccinimide (NHS) activated compounds. Fluorescent dyes (FITC, Cy5, Cy3) can be coupled to AMP-HDA, and the resulting derivatives are shown to be efficient for the construction of single-labeled pRNA molecule (Huang, 2003).

SH groups, provided by GSMP, can be used to link either a NH_2-group via a hetero-bifunctional crosslinker such as N-[β-Maleimidopropyloxy] succinimide ester (BMPS) or any common chemicals that contain maleimide, vinyl sulfone, or pyridyl disulfide. Using similar approaches, it was possible to efficiently synthesize RNA directly and incorporate traditional coupling reactive groups, such as amino –NH_2, -COOH, Maleimide, or NHS for constructing polyvalent RNA nanoparticles. The NH_2- group was used to link any particles with a COOH-group with the help of EDC. NH_2/NH_2 interactions can be achieved via heterobifunctional crosslinkers.

Labeling of pRNA with radioactive markers: In vitro transcription is commonly used in tritium labeling RNA molecules. T4 polynucleotide kinase transfers an organic phosphate from the [gamma-^{32}P] ATP to the 5′-hydroxyl group of DNA and RNA. Though the in vitro transcription method has high labeling efficiency, the transcription product is usually in small amount which is not suitable for pharmaceutical studies. Hereby, we will use hydrogen-tritium exchange method to intrinsically labeled RNA strands. Hydrogen exchange reactions are pseudomonomolecular reactions whose rates are described by a first-order rate equation. Exchange rates are also highly temperature dependent. The half times for exchange in single- stranded polyA are approximately 4 hours at 90°C, 2 months at 37°C, and extrapolated to be years at −20°C, which is ideal for efficient labeling at 90°C, pharmacokinetic experiments at 37°C and storage of unused material at −20°C.

CONCLUDING REMARKS

The use of small RNA in gene therapy was significantly hampered due to the difficulties of producing a safe and efficient system which is able to specifically recognize and target specific cells. The strength of using phi29 pRNA as a delivery vehicle relies on its ability to easily form stable oligomers which could be manipulated and sequence-controlled (Guo et al., 1998; Chen et al., 2000; Shu et al., 2003). This particular system provides unprecedented versatility in constructing polyvalent delivery vehicles by separately constructing individual pRNA subunits with various cargos and mixing them together in any desired combination (Guo, 2005).

The potential of the pRNA platform (Guo, 2005; Shu *et al.*, 2009) for the construction of RNA nanoparticles to carry receptor-binding ligands for specific delivery of siRNA to target and silence particular genes has been explored for a variety of cancer and viral-infected cells, including breast cancer (Li *et al.*, 2009), prostate cancer (Guo *et al.*, 2005), cervical cancer (Li *et al.*, 2009), nasopharyngeal carcinoma (Guo *et al.*, 2006), leukemia (Guo *et al.*, 2005; Khaled *et al.*, 2005), and ovarian cancer (Tarapore *et al.*, 2010), as well as coxsackievirus infected cells (Zhang *et al.*, 2009). The data demonstrated that the pRNA nanoparticle is a nanodelivery platform that can be applied broadly to diverse medical applications.

The phi29 pRNA system is a unique delivery system which takes advantage of the structural features of pRNA, including the presence of two domains which fold independently and the facility of re-engineering to form different kinds of desired particles through interlocking loop–loop interactions or end palindrome sequences. The assembled nanoscale particles harboring functional moieties offer many advantages and favorable pharmaceutical advantages over the numerous other anti-cancer delivery platforms under development such as polyvalent delivery, controllable structure and precise stoichiometry, nanoscale size, targeted delivery, long haft life *in vivo*, no induction of interferon, and toll-like immunity, low or no toxicity, and non-induction of an antibody response to ensure repeated treatments. Consequently, pRNA nanoparticles' delivery will potentially lead to an innovative therapeutic strategy for developing an effective and safe treatment for cancer and other diseases.

ACKNOWLEDGMENTS

This work was supported by NIH grant EB003730 and CA151648 to P.G. P.G. is a cofounder of Kylin Therapeutics, Inc., and Biomotor and Nucleic Acid Nanotechnology Development Corp. Ltd. P.G.'s Sylvan G. Frank Endowed Chair position in Pharmaceutics and Drug Delivery is funded by the CM Chen Foundation. P.G. is the consultant of Oxford Nanopore Technologies, as well as the cofounder of ExonanoRNA LLC & its subsidiary ExonanoRNA Biomedicine LTD Foshan.

REFERENCES

Abdelmawla S, Guo S, Zhang L, Pulukuri S, Patankar P, Conley P, Trebley J, Guo P, and Li QX (2011) Pharmacological characterization of chemically synthesized monomeric pRNA nanoparticles for systemic delivery. *Mol Ther*, **19**, 1312–1322.

Afonin KA, Bindewald E, Yaghoubian AJ, Voss N, Jacovetty E, Shapiro BA, and Jaeger L (2010) In vitro assembly of cubic RNA-based scaffolds designed in silico. *Nat Nanotechnol*, **5**, 676–682.

Berman HM, Henrick K, Nakamura H, Markley J, Bourne PE, and Westbrook J (2007) Realism about PDB. *Nat Biotechnol*, **25**, 845–846.

Berman HM, Olson WK, Beveridge DL, Westbrook J, Gelbin A, Demeny T, Hsieh SH, Srinivasan AR, and Schneider B (1992) The nucleic acid database. A comprehensive relational database of three-dimensional structures of nucleic acids. *Biophys J*, **63**, 751–759.

Berman HM, Westbrook J, Feng Z, Gilliland G, Bhat TN, Weissig H, Shindyalov IN, and Bourne PE (2000) The protein data bank. *Nucleic Acids Res*, **28**, 235–242.

Bindewald E, Afonin K, Jaeger L, and Shapiro BA (2011) Multistrand RNA secondary structure prediction and nanostructure design including pseudoknots. *ACS Nano*, **5**, 9542–9551.

Bindewald E, Grunewald C, Boyle B, O'Connor M, and Shapiro BA (2008a) Computational strategies for the automated design of RNA nanoscale structures from building blocks using NanoTiler. *J Mol Graph Model*, **27**, 299–308.

Bindewald E, Hayes R, Yingling YG, Kasprzak W, and Shapiro BA (2008b) RNAJunction: a database of RNA junctions and kissing loops for three-dimensional structural analysis and nanodesign. *Nucleic Acids Res*, **36**, D392–D397.

Binzel D, Shu Y, Li H, Sun M, Zhang Q, Shu D, Guo B, and Guo P (2016) Specific delivery of MiRNA for high efficient inhibition of prostate cancer by RNA nanotechnology. *Mol Ther*, **24**, 1267–1277.

Bouvet P (2001) Determination of nucleic acid recognition sequences by SELEX. *Methods Mol Biol*, **148**, 603–610.

Bruce AG and Uhlenbeck OC (1978) Reactions at the termini of tRNA with T4 RNA ligase. *Nucleic Acids Res*, **5**, 3665–3677.

Brummelkamp TR, Bernards R, and Agami R (2002) A system for stable expression of short interfering RNAs in mammalian cells. *Science*, **296**, 550–553.

Cadd TL and Patterson JL (1994) Synthesis of viruslike particles by expression of the putative capsid protein of *Leishmania* RNA virus in a recombinant baculovirus expression system. *J Virol*, **68**, 358–365.

Cairns J, Overbaugh J, and Miller S (1988) The origin of mutants. *Nature*, **335**, 142–145.

Carmichael GG (2002) Medicine: silencing viruses with RNA. *Nature*, **418**, 379–380.

Cayrol B, Nogues C, Dawid A, Sagi I, Silberzan P, and Isambert H (2009) A nanostructure made of a bacterial noncoding RNA. *J Am Chem Soc*, **131**, 17270–17276.

Cech TR, Zaug AJ, and Grabowski PJ (1981) In vitro splicing of the ribosomal RNA precursor of Tetrahymena: involvement of a guanosine nucleotide in the excision of the intervening sequence. *Cell*, **27**, 487–496.

Cerchia L and de Franciscis, V (2010) Targeting cancer cells with nucleic acid aptamers. *Trends Biotechnol*, **28**, 517–525.

Chakraborty C, Sharma AR, Sharma G, Doss CGP, and Lee SS (2017) Therapeutic miRNA and siRNA: Moving from Bench to Clinic as Next Generation Medicine. *Mol Ther Nucleic Acids*, **8**, 132–143.

Chen C, Sheng S, Shao Z, and Guo P (2000) A dimer as a building block in assembling RNA: A hexamer that gears bacterial virus phi29 DNA-translocating machinery. *J Biol Chem*, **275(23)**, 17510–17516.

Chen C, Zhang C, and Guo P (1999) Sequence requirement for hand-in-hand interaction in formation of pRNA dimers and hexamers to gear phi29 DNA translocation motor. *RNA*, **5**, 805–818.

Chowrira BM, Berzal-Herranz A, and Burke JM (1991) Novel guanosine requirement for catalysis by the hairpin ribozyme. *Nature*, **354**, 320–322.

Chworos A, Severcan I, Koyfman AY, Weinkam P, Oroudjev E, Hansma HG, and Jaeger L (2004) Building programmable jigsaw puzzles with RNA. *Science*, **306**, 2068–2072.

Ciesiolka J, Gorski J, and Yarus M (1995) Selection of an RNA domain that binds Zn2+. *RNA*, **1**, 538–550.

Clark S and Remcho V (2002) Aptamers as analytical reagents. *Electrophoresis*, **23**, 1335–1340.

Clever JL, Wong ML, and Parslow TG (1996) Requirements for kissing-loop-mediated dimerization of human immunodeficency virus RNA. *J Virol*, **70(9)**, 5902–5908.

Coleman J, Hirashima A, Inocuchi Y, Green PJ, and Inouye M (1985) A novel immune system against bacteriophage infection using complementary RNA (micRNA). *Nature*, **315**, 601–603.

Cong L, Ran FA, Cox D, Lin S, Barretto R, Habib N, Hsu PD, Wu X, Jiang W, Marraffini LA, and Zhang F (2013) Multiplex genome engineering using CRISPR/Cas systems. *Science*, **339**, 819–823.

Cruz JA, Blanchet MF, Boniecki M, Bujnicki JM, Chen SJ, Cao S, Das R, Ding F, Dokholyan NV, Flores SC, Huang L, Lavender CA, Lisi V, Major F, Mikolajczak K, Patel DJ, Philips A, Puton T, Santalucia J, Sijenyi F, Hermann T, Rother K, Rother M, Serganov A, Skorupski M, Soltysinski T, Sripakdeevong P, Tuszynska I, Weeks KM, Waldsich C, Wildauer M, Leontis NB, and Westhof E (2012) RNA-Puzzles: A CASP-like evaluation of RNA three-dimensional structure prediction. *RNA*, **18**, 610–625.

Dassie JP, Liu XY, Thomas GS, Whitaker RM, Thiel KW, Stockdale KR, Meyerholz DK, McCaffrey AP, McNamara JO, and Giangrande PH (2009) Systemic administration of optimized aptamer-siRNA chimeras promotes regression of PSMA-expressing tumors. *Nat Biotechnol*, **27**, 839–849.

Ding F, Sharma S, Chalasani P, Demidov VV, Broude NE, and Dokholyan NV (2008) Ab initio RNA folding by discrete molecular dynamics: from structure prediction to folding mechanisms 1. *RNA*, **14**, 1164–1173.

Ding Y, Chan CY, and Lawrence CE (2004) Sfold web server for statistical folding and rational design of nucleic acids. *Nucleic Acids Res*, **32**, W135–W141.

Earnshaw WC and Casjens SR (1980) DNA packaging by the double-stranded DNA bacteriophages. *Cell*, **21**, 319–331.

Elbashir SM, Harborth J, Lendeckel W, Yalcin A, Weber K, and Tuschl T (2001) Duplexes of 21-nucleotide RNAs mediate RNA interference in cultured mammalian cells. *Nature*, **411**, 494–498.

Ellington AD and Szostak JW (1990) *In vitro* selection of RNA molecules that bind specific ligands. *Nature*, **346**, 818–822.

Fabian MR, Sonenberg N, and Filipowicz W (2010) Regulation of mRNA translation and stability by microRNAs. *Annu Rev Biochem*, **79**, 351–379.

Ferrandon D, Koch I, Westhof E, and Nusslein-Volhard C (1997) RNA-RNA interaction is required for the formation of specific bicoid mRNA 3′ UTR-STAUFEN ribonucleoprotein particles. *EMBO J*, **16**, 1751–1758.

Fire A, Xu S, Montgomery MK, Kostas SA, Driver SE, and Mello CC (1998) Potent and specific genetic interference by double-stranded RNA in Caenorhabditis elegans. *Nature*, **391**, 806–811.

Forster AC and Symons RH (1987) Self-cleavage of virusoid RNA is performed by the proposed 55- nucleotide active site. *Cell*, **50**, 9–16.

Fukuda M, Kurihara K, Yamaguchi S, Oyama Y, and Deshimaru M (2014) Improved design of hammerhead ribozyme for selective digestion of target RNA through recognition of site-specific adenosine-to-inosine RNA editing. *RNA*, **20**, 392–405.

Garver K and Guo P (1997) Boundary of pRNA functional domains and minimum pRNA sequence requirement for specific connector binding and DNA packaging of phage phi29. *RNA*, **3**, 1068–1079.

Gates BD, Xu Q, Stewart M, Ryan D, Willson CG, and Whitesides GM (2005) New approaches to nanofabrication: molding, printing, and other techniques. *Chem Rev*, **105**, 1171–1196.

Ghildiyal M and Zamore PD (2009) Small silencing RNAs: an expanding universe. *Nat Rev Genet*, **10**, 94–108.

Glotzer SC (2004) Materials science. Some Assembly Required. *Science*, **306**, 419–420.

Gold L (1995) The SELEX process: a surprising source of therapeutic and diagnostic compounds. *Harvey Lect*, **91**, 47–57.

Grabow WW, Zakrevsky P, Afonin KA, Chworos A, Shapiro BA, and Jaeger L (2011) Self-Assembling RNA Nanorings Based on RNAI/II Inverse Kissing Complexes. *Nano Lett*, **11**, 878–887.

Guerrier-Takada C, Gardiner K, Marsh T, Pace N, and Altman S (1983) The RNA moiety of ribonuclease P is the catalytic subunit of the enzyme. *Cell*, **35**, 849–857.

Guo P (2005) RNA Nanotechnology: Engineering, Assembly and Applications in Detection, Gene Delivery and Therapy. *J Nanosci Nanotechnol*, **5(12)**, 1964–1982.

Guo P (2010) The emerging field of RNA nanotechnology. *Nat Nanotechnol*, **5**, 833–842.

Guo P, Erickson S, and Anderson D (1987) A small viral RNA is required for *in vitro* packaging of bacteriophage phi29 DNA. *Science*, **236**, 690–694.

Guo P and Trottier M (1994) Biological and biochemical properties of the small viral RNA (pRNA) essential for the packaging of the double-stranded DNA of phage φ29. *Semin Virol*, **5**, 27–37.

Guo P, Zhang C, Chen C, Trottier M, and Garver K (1998) Inter-RNA interaction of phage phi29 pRNA to form a hexameric complex for viral DNA transportation. *Mol Cell*, **2**, 149–155.

Guo P, Coban O, Snead NM, Trebley J, Hoeprich S, Guo S, and Shu Y (2010) Engineering RNA for Targeted siRNA Delivery and Medical Application. *Adv Drug Deliv Rev*, **62**, 650–666.

Guo S, Tschammer N, Mohammed S, and Guo P (2005) Specific delivery of therapeutic RNAs to cancer cells via the dimerization mechanism of phi29 motor pRNA. *Hum Gene Ther*, **16**, 1097–1109.

Guo S, Huang F, and Guo P (2006) Construction of folate-conjugated pRNA of bacteriophage phi29 DNA packaging motor for delivery of chimeric siRNA to nasopharyngeal carcinoma cells. *Gene Ther*, **13**, 814–820.

Guo S, Li H, Ma M, Fu J, Dong Y, and Guo P (2017) Size, Shape, and Sequence-Dependent Immunogenicity of RNA Nanoparticles. *Mol Ther Nucleic Acids*, **9**, 399–408.

Guo S, Piao X, Li H, and Guo P (2018) Methods for construction and characterization of simple or special multifunctional RNA nanoparticles based on the 3WJ of phi29 DNA packaging motor. *Methods*, **143**, 121–133.

Hansma HG, Oroudjev E, Baudrey S, and Jaeger L (2003) TectoRNA and 'kissing-loop' RNA: atomic force microscopy of self-assembling RNA structures. *J Microsc*, **212**, 273–279.

Henkin TM (2008) Riboswitch RNAs: using RNA to sense cellular metabolism. *Genes Dev*, **22**, 3383–3390.

Hoeprich S, Zhou Q, Guo S, Qi G, Wang Y, and Guo P (2003) Bacterial virus phi29 pRNA as a hammerhead ribozyme escort to destroy hepatitis B virus. *Gene Ther*, **10**, 1258–1267.

Huang F (2003) Efficient incorporation of CoA, NAD and FAD into RNA by in vitro transcription. *Nucleic Acids Res*, **31**, e8.

Huang F, He J, Zhang Y, and Guo Y (2008) Synthesis of biotin-AMP conjugate for 5′ biotin labeling of RNA through one-step in vitro transcription. *Nat Protoc*, **3**, 1848–1861.

Huang F, Wang G, Coleman T, and Li N (2003) Synthesis of adenosine derivatives as transcription initiators and preparation of 5′ fluorescein- and biotin-labeled RNA through one-step *in vitro* transcription. *RNA*, **9**, 1562–1570.

Huang Y, Eckstein F, Padilla R, and Sousa R (1997) Mechanism of ribose 2′-group discrimination by an RNA polymerase. *Biochemistry*, **36**, 8231–8242.

Jacque JM, Triques K, and Stevenson M (2002) Modulation of HIV-1 replication by RNA interference. *Nature*, **418**, 435–438.

Jaeger L and Chworos A (2006) The architectonics of programmable RNA and DNA nanostructures. *Curr Opin Struct Biol*, **16**, 531–543.

Jaeger L and Leontis NB (2000) Tecto-RNA: One dimensional self-assembly through tertiary interactions. *Angew Chem Int Ed Engl*, **39**, 2521–2524.

Jaeger L, Westhof E, and Leontis NB (2001) TectoRNA: modular assembly units for the construction of RNA nano-objects. *Nucleic Acids Res*, **29**, 455–463.

Jasinski D, Haque F, Binzel DW, and Guo P (2017) Advancement of the Emerging Field of RNA Nanotechnology. *ACS Nano*, **11**, 1142–1164.

Jiang W and Marraffini LA (2015) CRISPR-Cas: New Tools for Genetic Manipulations from Bacterial Immunity Systems. *Annu Rev Microbiol*, **69**, 209–228.

Jossinet F, Ludwig TE, and Westhof E (2007) RNA structure: bioinformatic analysis. *Curr Opin Microbiol*, **10**, 279–285.

Kasprzak W, Bindewald E, Kim TJ, Jaeger L, and Shapiro BA (2011) Use of RNA structure flexibility data in nanostructure modeling. *Methods*, **54**, 239–250.

Kawasaki AM, Casper MD, Freier SM, Lesnik EA, Zounes MC, Cummins LL, Gonzalez C, and Cook PD (1993) Uniformly modified 2′-deoxy-2′-fluoro phosphorothioate oligonucleotides as nuclease-resistant antisense compounds with high affinity and specificity for RNA targets. *J Med Chem*, **36**, 831–841.

Khaladkar M, Bellofatto V, Wang JTL, Tian B, and Shapiro BA (2007) RADAR: a web server for RNA data analysis and research. *Nucleic Acids Res*, **35**, W300–W304.

Khaled A, Guo S, Li F, and Guo P (2005) Controllable Self-Assembly of Nanoparticles for Specific Delivery of Multiple Therapeutic Molecules to Cancer Cells Using RNA Nanotechnology. *Nano Lett*, **5**, 1797–1808.

Klosterman PS, Hendrix DK, Tamura M, Holbrook SR, and Brenner SE (2004) Three-dimensional motifs from the SCOR, structural classification of RNA database: extruded strands, base triples, tetraloops and U-turns. *Nucleic Acids Res*, **32**, 2342–2352.

Knecht DA and Loomis WF (1987) Antisense RNA inactivation of myosin heavy chain gene expression in Dictyostelium discoideum. *Science*, **236**, 1081–1086.

Kraus E, James W, and Barclay AN (1998) Cutting edge: novel RNA ligands able to bind CD4 antigen and inhibit CD4+ T lymphocyte function. *J Immunol*, **160**, 5209–5212.

Lam JK, Chow MY, Zhang Y, and Leung SW (2015) siRNA Versus miRNA as Therapeutics for Gene Silencing. *Mol Ther Nucleic Acids*, **4**, e252.

Laughrea M and Jette L (1996) Kissing-Loop Model of HIV-1 Genome Dimerization: HIV-1 RNAs Can Assume Alternative Dimeric Forms, and All Sequences Upstream of Downstream of Hairpin 248–271 Are Dispensable for Dimer Formation. *Biochemistry*, **35 No. 5**, 1589–1598.

Lee CYF, Rennie PS, and Jia WWG (2009) MicroRNA Regulation of Oncolytic Herpes Simplex Virus-1 for Selective Killing of Prostate Cancer Cells. *Clin Cancer Res*, **15**, 5126–5135.

Lee TJ and Guo P (2006) Interaction of gp16 with pRNA and DNA for genome packaging by the motor of bacterial virus phi29. *J Mol Biol*, **356**, 589–599.

Lee TJ, Yoo JY, Shu D, Li H, Zhang J, Yu JG, Jaime-Ramirez AC, Acunzo M, Romano G, Cui R, Sun HL, Luo Z, Old M, Kaur B, Guo P, and Croce CM (2017) RNA Nanoparticle-Based Targeted Therapy for Glioblastoma through Inhibition of Oncogenic miR-21. *Mol Ther.* 25(7), 1544–1555.

Lennox KA, Owczarzy R, Thomas DM, Walder JA, and Behlke MA (2013) Improved Performance of Anti-miRNA Oligonucleotides Using a Novel Non-Nucleotide Modifier. *Mol Ther Nucleic Acids*, **2**, e117.

Leonard M, Zhang Y, and Zhang X (2015) Small non-coding RNAs and aptamers in diagnostics and therapeutics. *Methods Mol Biol*, **1296**, 225–233.

Leontis NB, Lescoute A, and Westhof E (2006) The building blocks and motifs of RNA architecture. *Curr Opin Struct Biol*, **16**, 279–287.

Leontis NB and Westhof E (2001) Geometric nomenclature and classification of RNA base pairs. *RNA*, **7**, 499–512.

Leontis NB and Westhof E (2002) The annotation of RNA motifs. *Comp Funct Genomics*, **3**, 518–524.

Leontis NB and Westhof E (2003) Analysis of RNA motifs. *Curr Opin Struct Biol*, **13**, 300–308.

Lescoute A and Westhof E (2006) Topology of three-way junctions in folded RNAs. *RNA*, **12**, 83–93.

Li H, Li WX, and Ding SW (2002) Induction and suppression of RNA silencing by an animal virus. *Science*, **296**, 1319–1321.

Li L, Liu J, Diao Z, Guo P, and Shen G (2009) Evaluation of specific delivery of chimeric phi29 pRNA/siRNA nanoparticles to multiple tumor cells. *Mol BioSyst*, **5**, 1361–1368.

Lilley DM (1999) Structure, folding and catalysis of the small nucleolytic ribozymes. *Curr Opin Struct Biol*, **9**, 330–338.

Lilley DM (2000) Structures of helical junctions in nucleic acids. *Q Rev Biophys*, **33**, 109–159.

Liu H, Guo S, Roll R, Li J, Diao Z, Shao N, Riley MR, Cole AM, Robinson JP, Snead NM, Shen G, and Guo P (2007) Phi29 pRNA Vector for Efficient Escort of Hammerhead Ribozyme Targeting Survivin in Multiple Cancer Cells. *Cancer Biol Ther*, **6**, 697–704.

Liu J, Guo S, Cinier M, Shlyakhtenko LS, Shu Y, Chen C, Shen G, and Guo P (2011) Fabrication of stable and RNase-resistant RNA nanoparticles active in gearing the nanomotors for viral DNA packaging. *ACS Nano*, **5**, 237–246.

Lyu Y, Chen G, Shangguan D, Zhang L, Wan S, Wu Y, Zhang H, Duan L, Liu C, You M, Wang J, and Tan W (2016) Generating Cell Targeting Aptamers for Nanotheranostics Using Cell-SELEX. *Theranostics*, **6**, 1440–1452.

Macke T and Case DA (1997) Modeling unusual nucleic acids with NAB. *Abstr Pap Am Chem Soc*, **213**, 289-COMP.

Markham NR and Zuker M (2008) UNAFold: software for nucleic acid folding and hybridization. *Methods Mol Biol*, **453**, 3–31.

Martinez HM, Maizel JV, and Shapiro BA (2008) RNA2D3D: A program for generating, viewing, and comparing 3-dimensional models of RNA. *J Biomol Str Dyn*, **25**, 669–683.

Massire C and Westhof E (1998) MANIP: an interactive tool for modelling RNA. *J Mol Graph Model*, **16**, 197.

Mathews DH, Disney MD, Childs JL, Schroeder SJ, Zuker M, and Turner DH (2004) Incorporating chemical modification constraints into a dynamic programming algorithm for prediction of RNA secondary structure. *Proc Natl Acad Sci U S A*, **101**, 7287–7292.

McKinney SA, Declais AC, Lilley DMJ, and Ha T (2003) Structural dynamics of individual Holliday junctions. *Nat Struct Biol*, **10**, 93–97.

McNamara JO, Andrechek ER, Wang Y, Viles KD, Rempel RE, Gilboa E, Sullenger BA, and Giangrande PH (2006) Cell type-specific delivery of siRNAs with aptamer-siRNA chimeras. *Nat Biotechnol*, **24**, 1005–1015.

Miller-Kleinhenz JM, Bozeman EN, and Yang L (2015) Targeted nanoparticles for image-guided treatment of triple-negative breast cancer: clinical significance and technological advances. *Wiley Interdiscip Rev Nanomed Nanobiotechnol*, **7**, 797–816.

Milligan JF, Groebe DR, Witherell GW, and Uhlenbeck OC (1987) Oligoribonucleotide synthesis using T7 RNA polymerase and synthetic DNA templates. *Nucleic Acids Res*, **15**, 8783–8798.

Nagaswamy U, Larios-Sanz M, Hury J, Collins S, Zhang Z, Zhao Q, and Fox GE (2002) NCIR: a database of non-canonical interactions in known RNA structures. *Nucleic Acids Res*, **30**, 395–397.

Nolan JM, Burke DH, and Pace NR (1993) Circularly Permuted tRNAs as Specific Photoaffinity Probes of Ribonuclease P RNA Structure. *Science*, **261**, 762–765.

Ogawa A and Maeda M (2008) An artificial aptazyme-based riboswitch and its cascading system in E. coli. *ChemBioChem*, **9**, 206–209.

Padilla R and Sousa R (2002) A Y639F/H784A T7 RNA polymerase double mutant displays superior properties for synthesizing RNAs with non-canonical NTPs. *Nucelic Acids Res*, **30**, e138.

Padilla R and Sousa R (1999) Efficient synthesis of nucleic acids heavily modified with non-canonical ribose 2′-groups using a mutantT7 RNA polymerase (RNAP). *Nucleic Acids Res*, **27**, 1561–1563.

Paillart JC, Skripkin E, Ehresmann B, Ehresmann C, and Marquet R (1996) A loop-loop "kissing" complex is the essential part of the dimer linkage of genomic HIV-1 RNA. *Proc Natl Acad Sci U S A*, **93**, 5572–5577.

Pan T, Long DM, and Uhlenbeck OC (1993). Divalent metal ions in RNA folding and catalysis. In Gesteland, R.F. and Atkins, J.F. (Eds.), *RNA World*. Cold Spring Harbor Laboratory Press, Cold Spring Harbor, NY, pp. 271–302.

Pan T and Uhlenbeck OC (1993) Circularly permuted DNA, RNA and proteins-a review. *Gene*, **125**, 111–114.

Piao X, Yin H, Guo S, Wang H, and Guo P (2019) RNA nanotechnology to solubilize hydrophobic antitumor drug for targeted delivery. *Adv Sci*, 6(22), 1900951.

Qiu M, Khisamutdinov E, Zhao Z, Pan C, Choi J, Leontis N, and Guo P (2013) RNA nanotechnology for computer design and *in vivo* computation. *Phil Trans R Soc A*, **371(2000)**, 20120310.

Reid RJD, Bodley JW, and Anderson D (1994a) Characterization of the prohead-pRNA interaction of bacteriophage phi29. *J Biol Chem*, **269**, 5157–5162.

Reid RJD, Bodley JW, and Anderson D (1994b) Identification of bacteriophage phi29 prohead RNA (pRNA) domains necessary for *in vitro* DNA-gp3 packaging. *J Biol Chem*, **269**, 9084–9089.

Reid RJD, Zhang F, Benson S, and Anderson D (1994c) Probing the structure of bacteriophage phi29 prohead RNA with specific mutations. *J Biol Chem*, **269**, 18656–18661.

Sarver NA, Cantin EM, Chang PS, Zaia JA, Ladne PA, Stephens DA, and Rossi JJ (1990) Ribozymes as Potential Anti-HIV-1 Therapeutic Agents. *Science*, **24**, 1222–1225.

Schroeder KT, McPhee SA, Ouellet J, and Lilley DM (2010) A structural database for k-turn motifs in RNA. *RNA*, **16**, 1463–1468.

Severcan I, Geary C, Verzemnieks E, Chworos A, and Jaeger L (2009) Square-shaped RNA particles from different RNA folds. *Nano Lett*, **9**, 1270–1277.

Severcan I, Geary C, Chworos A, Voss N, Jacovetty E, and Jaeger L (2010) A polyhedron made of tRNAs. *Nat Chem*, **2**, 772–779.

Shapiro BA (2009) Computational Design Strategies for RNA Nanostructures. *J Biomol Str Dyn*, **26**, 820.

Shapiro BA, Bindewald E, Kasprzak W, and Yingling Y (2008) Protocols for the in silico design of RNA nanostructures. *Methods Mol Biol*, **474**, 93–115.

Shapiro BA and Wu JC (1997) Predicting RNA H-type pseudoknots with the massively parallel genetic algorithm. *Comput Appl Biosci*, **13**, 459–471.

Sharma S, Ding F, and Dokholyan NV (2008) iFoldRNA: three-dimensional RNA structure prediction and folding. *Bioinformatics*, **24**, 1951–1952.

Shu D, Shu Y, Haque F, Abdelmawla S, and Guo P (2011a) Thermodynamically stable RNA three-way junctions for constructing multifuntional nanoparticles for delivery of therapeutics. *Nat Nanotechnol*, **6**, 658–667.

Shu D and Guo P (2003) A Viral RNA that binds ATP and contains an motif similar to an ATP-binding aptamer from SELEX. *J Biol Chem*, **278(9)**, 7119–7125.

Shu D, Huang L, Hoeprich S, and Guo P (2003) Construction of phi29 DNA-packaging RNA (pRNA) monomers, dimers and trimers with variable sizes and shapes as potential parts for nano-devices. *J Nanosci Nanotechnol*, **3**, 295–302.

Shu D, Li H, Shu Y, Xiong G, Carson WE, Haque F, Xu R, and Guo P (2015) Systemic delivery of anti-miRNA for suppression of triple negative breast cancer utilizing RNA nanotechnology. *ACS Nano*, **9**, 9731–9740.

Shu D, Moll WD, Deng Z, Mao C, and Guo P (2004) Bottom-up assembly of RNA arrays and superstructures as potential parts in nanotechnology. *Nano Lett*, **4**, 1717–1723.

Shu Y, Cinier M, Fox SR, Ben-Johnathan N, and Guo P (2011b) Assembly of Therapeutic pRNA-siRNA Nanoparticles Using Bipartite Approach. *Mol Ther*, **19**, 1304–1311.

Shu Y, Cinier M, Shu D, and Guo P (2011c) Assembly of multifunctional phi29 pRNA nanoparticles for specific delivery of siRNA and other therapeutics to targeted cells. *Methods*, **54**, 204–214.

Shu Y, Shu D, Diao Z, Shen G, and Guo P (2009) Fabrication of Polyvalent Therapeutic RNA Nanoparticles for Specific Delivery of siRNA, Ribozyme and Drugs to Targeted Cells for Cancer Therapy. *IEEE/NIH Life Science Systems and Applications Workshop*, 9–12.

Shu Y, Yin H, Rajabi M, Li H, Vieweger M, Guo S, Shu D, and Guo P (2018) RNA-based micelles: A novel platform for paclitaxel loading and delivery. *J Control Release*, **276**, 17–29.

Sousa R and Padilla R (1995) A mutant T7 RNA polymerase as a DNA polymerase. *EMBO J*, **14**, 4609–4621.

Srisawat C and Engelke DR (2001) Streptavidin aptamers: affinity tags for the study of RNAs and ribonucleoproteins. *RNA*, **7**, 632–641.

Steffen P, Voss B, Rehmsmeier M, Reeder J, and Giegerich R (2006) RNAshapes: an integrated RNA analysis package based on abstract shapes. *Bioinformatics*, **22**, 500–503.

Stengel G, Urban M, Purse BW, and Kuchta RD (2010) Incorporation of the fluorescent ribonucleotide analogue tCTP by T7 RNA polymerase. *Anal Chem*, **82**, 1082–1089.

Strobel SA and Cochrane JC (2007) RNA catalysis: ribozymes, ribosomes, and riboswitches. *Curr Opin Chem Biol*, **11**, 636–643.

Sudarsan N, Lee ER, Weinberg Z, Moy RH, Kim JN, Link KH, and Breaker RR (2008) Riboswitches in eubacteria sense the second messenger cyclic di-GMP. *Science*, **321**, 411–413.

Taft RJ, Pang KC, Mercer TR, Dinger M, and Mattick JS (2010) Non-coding RNAs: regulators of disease. *J Pathol*, **220**, 126–139.

Tamura M, Hendrix DK, Klosterman PS, Schimmelman NR, Brenner SE, and Holbrook SR (2004) SCOR: Structural Classification of RNA, version 2.0. *Nucleic Acids Res*, **32**, D182–D184.

Tarapore P, Shu Y, Guo P, and Ho SM (2010) Application of Phi29 Motor pRNA for Targeted Therapeutic Delivery of siRNA Silencing Metallothionein-IIA and Survivin in Ovarian Cancers. *Mol Ther*, **19**, 386–394.

Tatiparti K, Sau S, Kashaw SK, and Iyer AK (2017) siRNA delivery strategies: A comprehensive Review of Recent Developments. *Nanomaterials (Basel)*, **7(4)**, 77.

Trottier M, Mat-Arip Y, Zhang C, Chen C, Sheng S, Shao Z, and Guo P (2000) Probing the structure of monomers and dimers of the bacterial virus phi29 hexamer RNA complex by chemical modification. *RNA*, **6**, 1257–1266.

Tuerk C and Gold L (1990) Systematic evolution of ligands by exponential enrichment: RNA ligands to bacteriophage T4 DNA ploymerase. *Science*, **249**, 505–510.

Varambally S, Dhanasekaran SM, Zhou M, Barrette TR, Kumar-Sinha C, Sanda MG, Ghosh D, Pienta KJ, Sewalt RG, Otte AP, Rubin MA, and Chinnaiyan AM (2002) The polycomb group protein EZH2 is involved in progression of prostate cancer. *Nature*, **419**, 624–629.

Westhof E (2012) Ribozymes, catalytically active RNA molecules. Introduction. *Methods Mol Biol*, **848**, 1–4.

Westhof E, Masquida B, and Jaeger L (1996) RNA tectonics: Towards RNA design. *Folding Design*, **1**, R78–R88.

Xu C, Haque F, Jasinski DL, Binzel DW, Shu D, and Guo P (2018a) Favorable biodistribution, specific targeting and conditional endosomal escape of RNA nanoparticles in cancer therapy. *Cancer Lett*, **414**, 57–70.

Xu Y, Pang L, Wang H, Xu C, Shah H, Guo P, Shu D, and Qian SY (2018b) Specific delivery of delta-5-desaturase siRNA via RNA nanoparticles supplemented with dihomo-gamma-linolenic acid for colon cancer suppression. *Redox Biol*, **21**, 101085.

Ye M, Hu J, Peng M, Liu J, Liu J, Liu H, Zhao X, and Tan W (2012) Generating Aptamers by Cell-SELEX for Applications in Molecular Medicine. *Int J Mol Sci*, **13**, 3341–3353.

Yin H, Xiong G, Guo S, Xu C, Xu R, Guo P, and Shu D (2019) Delivery of Anti-miRNA for Triple-Negative Breast Cancer Therapy Using RNA Nanoparticles Targeting Stem Cell Marker CD133. *Mol Ther*, **27**(7), 1252–1261.

Yingling YG and Shapiro BA (2007) Computational design of an RNA hexagonal nanoring and an RNA nanotube. *Nano Lett*, **7**, 2328–2334.

Zhang C (2009) Novel functions for small RNA molecules. *Curr Opin Mol Ther*, **11**, 641–651.

Zhang CL, Lee C-S, and Guo P (1994) The proximate 5′ and 3′ ends of the 120-base viral RNA (pRNA) are crucial for the packaging of bacteriophage φ29 DNA. *Virology*, **201**, 77–85.

Zhang CL, Tellinghuisen T, and Guo P (1995a) Confirmation of the helical structure of the 5′/3′ termini of the essential DNA packaging pRNA of phage φ29. *RNA*, **1**, 1041–1050.

Zhang CL, Tellinghuisen T, and Guo P (1997) Use of circular permutation to assess six bulges and four loops of DNA-Packaging pRNA of bacteriophage phi29. *RNA*, **3**, 315–322.

Zhang CL, Trottier M, and Guo PX (1995b) Circularly permuted viral pRNA active and specific in the packaging of bacteriophage Phi29 DNA. *Virology*, **207**, 442–451.

Zhang F, Lemieux S, Wu X, St.-Arnaud S, McMurray CT, Major F, and Anderson D (1998) Function of hexameric RNA in packaging of bacteriophage phi29 DNA in vitro. *Mol Cell*, **2**, 141–147.

Zhang HM, Su Y, Guo S, Yuan J, Lim T, Liu J, Guo P, and Yang D (2009) Targeted delivery of anti-coxsackievirus siRNAs using ligand-conjugated packaging RNAs. *Antivir Res*, **83**, 307–316.

Zhang L, Sun L, Cui Z, Gottlieb RL, and Zhang B (2001) 5′-sulfhydryl-modified RNA: initiator synthesis, in vitro transcription, and enzymatic incorporation. *Bioconjug Chem*, **12**, 939–948.

Zhang P, Zhang Z, Yang Y, and Li Y (2010) Folate-PEG modified poly(2-(2-aminoethoxy)ethoxy)phosphazene/DNA nanoparticles for gene delivery: synthesis, preparation and in vitro transfection efficiency. *Int J Pharm*, **392**, 241–248.

Zhou J, Shu Y, Guo P, Smith D, and Rossi J (2011) Dual functional RNA nanoparticles containing phi29 motor pRNA and anti-gp120 aptamer for cell-type specific delivery and HIV-1 Inhibition. *Methods*, **54**, 284–294.

Zhou J, Li H, Zaia J, and Rossi JJ (2008) Novel dual inhibitory function aptamer-siRNA delivery system for HIV-1 therapy. *Mol Ther*, **16**, 1481–1489.

Zhou J and Rossi JJ (2016) Evolution of Cell-Type-Specific RNA Aptamers Via Live Cell-Based SELEX. *Methods Mol Biol*, **1421**, 191–214.

Zhou J and Rossi J (2017) Aptamers as targeted therapeutics: current potential and challenges. *Nat Rev Drug Discov*, **16**, 181–202.

Zuker M (2003) Mfold web server for nucleic acid folding and hybridization prediction. *Nucleic Acids Res*, **31**, 3406–3415.

20 Purification of RNA, Modified Oligos, and RNA Nanoparticles

Jordan Hill, Mario Vieweger, and Fengmei Pi
ExonanoRNA LLC, Columbus, Ohio, USA

Peixuan Guo
The Ohio State University, Columbus, OH, USA

CONTENTS

INTRODUCTION

Over the last two decades, RNA nanotechnology has flourished through rapid advances of RNA-based applications in biomedical and nanotechnological applications. In particular, its biomedical applications have shown promise in using RNA nanoparticles as a therapeutic delivery system to target cancers, viral infections, and genetic diseases [1, 2]. This recent onset of RNA therapeutics in clinical and preclinical studies has led to the prediction of a third milestone in drug development with focus on RNA-based and targeting drugs, following small molecule chemicals and protein-based therapeutics [3].

This advancement is driven by RNA's inherent ability to combine the structural flexibility and diversity of proteins with the programmability and precision of DNA base pairing [4]. Diverse nanoparticles of varying sizes and structural features have been created in recent years allowing the study of structure–function relationships and their optimization for a broad range of applications [5–12]. Many of these nanoparticles are now investigated for their potential in imaging, disease diagnosis, and RNA therapeutics [1–3, 13–15]. Several companies have sprung up to head the call and pursue RNA nanoparticle-based therapeutics and advance the field to the clinical stage.

Such efforts require vast amounts of homogeneous and contamination-free RNA nanoparticles. In general, thermodynamically driven design and assembly allows RNA nanoparticles to self-assemble at high efficiency into homogeneous nanoparticles. Yet, preclinical applications pose stringent purity requirements and quality control on the oligonucleotide subunits as well as fully assembled nanoparticles. To ensure batch-to-batch reproducibility, efficient and reproducible purification techniques for assembled nanocomplexes are needed which has been a challenge as traditional methods are designed for single-stranded oligonucleotides.

Traditionally, purification has been performed by polyacrylamide gel electrophoresis (PAGE) or agarose gel electrophoresis (AGE). Both techniques use size, shape, and charge to separate fully assembled nanostructures from single-stranded RNA oligos and partially assembled oligo complexes that migrate slower and faster, respectively [12]. This method has proven to be effective in separating the desired nanostructures, yet, retrieval from the gel has routinely been observed at low yields due to the steric hindrance of the structured particles. Moreover, traditional gel electrophoresis carries a high cost of labor and lacks the scalability required to obtain large quantities of highly purified nanostructures that are needed for pre-clinical studies.

In this chapter, we will review alternate techniques that have been demonstrated to be capable of purifying RNA nano-assemblies. Section 2 covers High Performance Liquid Chromatography (HPLC) techniques; Section 3 covers Continuous Elution Gel Electrophoresis (CEGE) and section 4 covers Ultracentrifugation techniques and their application for oligonucleotide and nanoparticle purification. Our illustrations focus on their mode of separation, useful applications, and range of scale. Our goal is to provide a robust framework for the RNA nanotechnology scientist to refer to when choosing the most suitable means of RNA nanoparticle purification for their applications.

LIQUID CHROMATOGRAPHY

PRINCIPLES OF OLIGONUCLEOTIDE PURIFICATION

Chromatographic separation of large molecules, such as biopolymers, began in the 1950s [16]. Since then, many improvements have been made to increase separation efficiency and yield [17]. Oligonucleotide purification is primarily governed by net charge differences – and therefore length – due to the increasing anionic contribution of each additional phosphate of the backbone [18]. As such, the most effective methods of purification utilize these charge differences to separate full-length oligonucleotides from their failure sequences by either a direct (anion exchange) or indirect (ion-pairing reverse-phase) mechanism.

In anion exchange HPLC, oligonucleotides are directly separated by their charge differences. The anionic oligonucleotides are introduced in a low ionic strength buffer and bind cationic adsorbents directly through electrostatic interactions. As the number of charge–charge interactions increases, retention to stationary adsorbents is enhanced. Therefore, long oligonucleotides are more strongly retained. After loading, a salt gradient utilizing a high ionic strength buffer is slowly introduced, weakening the retention of bound samples and resulting in the migration of oligonucleotides through the column at varying rates, proportional to their binding affinity. This migratory behavior ultimately leads to oligonucleotide separation and elution by exchange of the solid phase bound anion.

The indirect method employs a nonpolar stationary phase that promotes interactions with hydrophobic compounds. Due to the inherent hydrophilicity of oligonucleotides, the pairing of a hydrophobic counter ion – effectively neutralizing the charge and adding hydrophobic groups in the form of alkyl chains – is required for adequate resolution. Samples are loaded in an aqueous mobile phase containing the ion-paired oligonucleotides and adhere to the nonpolar adsorbent in an oligonucleotide-to-hydrophobic ion adsorbent fashion. Like the direct method, the number of charge interactions increases with length, and therefore longer sequences have a higher affinity for the nonpolar stationary phase. Upon the addition of an organic solvent gradient, the affinity of ion-pairs for the stationary phase decreases, giving rise to elution in order of increasing hydrophobicity and resolving bound oligonucleotides by the ion-paired reverse-phase method. It is important to note that various components of oligonucleotides – their bases and other modifications – can exert retention effects independent of ion-pairing [17, 18].

In both methods, the efficiency of purification decreases with oligonucleotide length [18]. This phenomenon is due to the lower relative charge difference between long and short oligos. For example, the relative difference between a 19 and 20mer is 2.5-fold greater than a 49 and 50mer.

ANION EXCHANGE HPLC

Anion exchange typically gives a more consistent and predictable elution pattern than IPRP [17]. Additionally, its ability to separate longer oligos (~50) exemplifies its value as a primary purification technique.

Anion-exchange columns are typically composed of porous silica or polymer adsorbent functionalized with tertiary amines or quaternary ammonium ions. The polymer-based oligonucleotides have the advantage of tolerating crude samples loaded directly after AMA or TEA.3HF treatment – often used in deprotection protocols of synthetic oligos – as they can tolerate pH 1–14 [17]. Mobile phase buffers are generally between pH 7 and 12, with higher pH buffers enhancing oligonucleotide resolution. However, not all oligonucleotides are compatible with highly alkaline conditions, such as 2'Fluoro, 2'Hydorxyl, and 2'Omethyl, which necessitates near-neutral (pH 8) buffers and other methods of denaturation [17]. Various salts have been used (NaCl, NaBr, NaOCL4, NaPO4) with each having their own elution strength corresponding with the Hofmeister [19] series. However, the strongest displacer is not always the best and factors such as cost, capacity effects, and upscaling feasibility need to be considered when optimizing a method. For pH sensitive oligos, perchlorate is often used, as it is chaotropic (Figure 20.1) [17, 19].

Oligonucleotides – especially ribonucleotides designed with secondary structure (e.g. aptamers [20], ribozymes [21, 22]) – often exhibit stable intermolecular and intramolecular structures. These secondary interactions can mask the negative charge of phosphate groups, broaden peaks, and distort the overall length-based effects on separation. Accordingly, it is vital to use denaturing conditions for oligos exhibiting these properties in order to attain sufficiently pure products. High temperature, high pH, and the use of denaturants such as acetonitrile, formamide, and urea are often used to temporarily eliminate these conformations and can easily be incorporated into AEX methods (Figure 20.2) [17].

An advantage of this purification method is that a trityl group does not need to be maintained during workup. However, tritylated oligos are still capable of being purified [17]. Smart fraction collection practices are crucial since impurities tend to come both before and after full-length products; the truncated sequences on the left side of the target peak, and oligos not fully deprotected as a

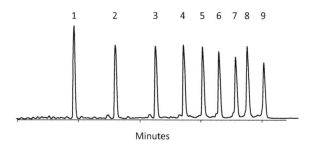

Minutes

FIGURE 20.1   Primary sequence and structure of wild-type pRNA. Figure is produced with permission from reference Liu et al. (2011), © American Chemical Society.

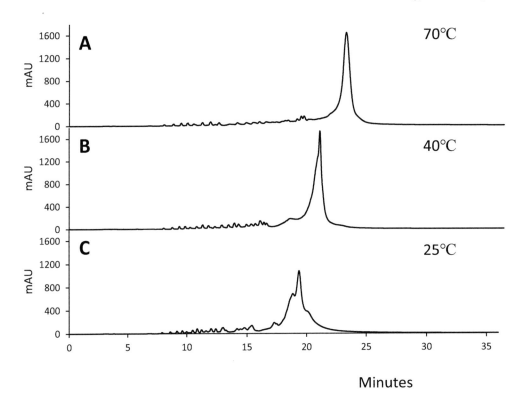

FIGURE 20.2 pRNA dimer formed through hand in hand complementary loop interactions. Figure is produced with permission from reference Guo et al. (2010), © Nature Publishing Group.

shoulder or to the right of the target peak. For applications requiring more than 90% purity, sacrificing yield is often a useful practice. A disadvantage of AEX is that sequences not fully deprotected do not display the significantly higher retention seen in IPRP methods, and therefore poor deprotection protocols may result in high amounts of impurities. Another disadvantage of this method is final products have substantial amounts of nonvolatile salts, therefore necessitating a desalting step for downstream applications [17].

(IP)-RP HPLC

Reversed phase HPLC is best suited for the purification of modified oligonucleotides. The upper limit of unmodified oligonucleotides tends to be 10–20 bases shorter than ion-exchange, as it is more difficult to control structural variations under reversed-phase constraints. However, this method is the only HPLC method capable of efficiently purifying large amounts of long (~100) oligonucleotides due to its applicability with DMT-on synthesis [18].

IPRP Info Combined

Reversed phase columns are typically composed of ODS silica packing or C^4–C^{18} alkylated polystyrene-divinyl benzene materials [17, 23]. Silica dissociates at high pH, and therefore, the use of polymer-based reversed phase columns has recently increased due to their enhanced temperature tolerance, wide pH tolerance, and increased column lifetime [17].

The ion-pairing agent triethylammonium acetate is the most commonly used mobile phase [23] and has been shown to be useful for tritylated and detritylated oligonucleotides. TEA+ and HFIP are also common, mainly due to their compatibility with the electrospray ionization process needed for MS analysis made possible by its high volatility [23, 24]. However, this system is not used for large-scale purifications due to the high cost of HFIP and toxicity of TEA+ [17]. Other ion-pairing agents have also been tested such as the acetate and bicarbonate salts of hexylamine (HA), dibutylamine (DB), diisopropylethylamine (DIPEA), and dimethylbutylamine (DBA) [24]. The organic eluents most often used are acetonitrile and methanol.

RP HPLC is particularly well-suited for the purification of oligonucleotides modified with hydrophobic moieties such as DMT, fluorophores [25], chemical drugs (ExonanoRNA unpublished), and various functional groups. During synthesis, the dimethoxytrityl (DMT) protective group can be left on the last base, and its hydrophobic nature produces an enormous retention enhancement under reverse phase conditions. As a result, high amounts of long (~100mer) unable to be purified by other chromatic methods can be feasibly resolved [18]. While theoretically only full-length oligos contain the DMT group, the imperfections of oligonucleotide synthesis can produce product mixtures containing up to 10% failure sequences with the DMT group [26]. These DMT-on truncates are difficult to separate and combining this method with anion-exchange can be advantageous for obtaining products with greater than 95% purity [18].

Resembling the DMT process is the purification of oligos conjugated with fluorophores. They impart significant hydrophobicity, and therefore are easily purified from unlabeled strands. However, crude 3'bioconjugates can present difficulties in separating truncated, labeled strands. In fact, some shorter oligos modified with fluorophores have been found to elute after the target strand [25]. This can be explained by the net hydrophobicity of the molecule; each additional oligo imparts hydrophilic character and decreases retention. Apparently, some hydrophobic modifications (DMT, fluorophores) overpower the ion-pairing effects on retention. Other modifications such as biotin [17] or azide functionalities have been purified by this method with success.

A disadvantage of IPRP for routine oligonucleotide purification is the existence of sequence-specific (i.e. base compositional) retention effects due to variability in hydrophobicity exhibited by individual bases [18]. This effect is not as prominent in AEX, making IPRP relatively less predictable. Buffer systems utilizing tetraalkylammonium pairing agents have been shown to limit sequence-specific effects and lead to purifications primarily based on charge; however, these mobile phases are not volatile and impractical for large-scale separations [26].

An advantage of IPRP over AEX is the separation of impurities unrelated to length [17]. DMT-on and not fully deprotected oligos are easily resolved from the full-length target sequences. Figure 20.3 shows a 22mer oligo that was purposely left partially deprotected. The IPRP method clearly resolves these impurities, while they are not visible in the AEX method (Figure 20.4).

Conclusions: For routine purifications of unmodified oligonucleotides, anion-exchange is the first choice. Most secondary structure effects can be avoided by a variety of techniques, and oligonucleotides up to ~50 nucleotides can be efficiently separated with 1–2 base resolution. The method is scalable, reliable, and avoids high amounts of toxic reagents. For purifications of many modified oligonucleotides, ion-paired reverse phase produces the most efficient separations. Additionally, it is a useful technique if long oligos are required in amounts larger than the capacity of PAGE. However, extra care must be taken with the DMT-on process.

CONTINUOUS ELUTION GEL-ELECTROPHORESIS (CEGE)

Purification of long full-length oligonucleotides (n>60) from n-1 terminated strands as well as purification of RNA nanoparticles from partially assembled particles is often a challenge. Ion-pair reverse-phase and anion-exchange HPLC suffer from low resolution for purification of long oligo nucleotides and require denaturing conditions which is not conducive for RNA nanoparticles. Polyacrylamide gel electrophoresis (PAGE) allows for better control over resolution by choice of polyacrylamide crosslinking percentages (typically 8–20%). Nonetheless, significant reduction of product yields is observed in the passive elution step following band excision when purifying long oligos or structured compounds. Consequently, a number of active elution procedures have been developed using electro-elution of oligos and nanoparticles from the excised gel pieces with varying degrees of success. The main obstacle in electro-elution is that the electric field and current flow follow the path of least

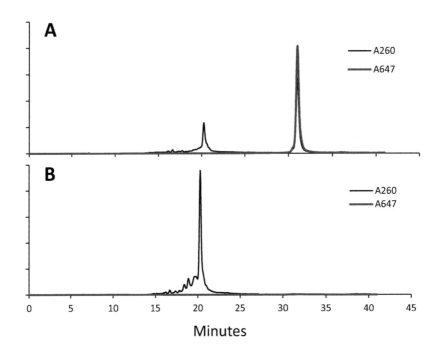

FIGURE 20.3 Schematic drawing of pRNA hexamer through complementary loop–loop interaction with six pieces of pRNA monomer. Figure is produced with permission from reference Chen et al. (2000), © American Society for Biochemistry and Molecular Biology.

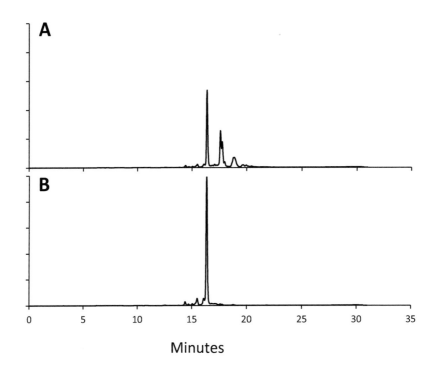

FIGURE 20.4 Three-way junction motif of pRNA, which can be used for RNA nanoparticles construction. Figure is produced with permission from reference Shu et al. (2011), © Nature Publishing Group.

resistance which leads around the gel piece thus causing elongated elution times and accumulation of the sample on the molecular weight cutoff membranes in the collection chamber.

An alternative that is especially useful for semi-preparative and preparative-scale oligo purification is CEGE which bypasses the shortcomings of passive elution and electroelution by continuously collecting samples as they run off the gel. This is achieved by casting the gel into a column rather than a sheet. The sample is then added in a thin layer on top of the gel bed and electrophoresed in the presence of dyes to indicate the progress of the gel. The samples running off the gel column are continuously collected by applying a slow buffer flow at the bottom of the column which is directed to a fraction collector. This process allows the purification of ~1 mg to 100 mg of oligonucleotides in a single run while bypassing tedious gel excision, passive elution, and ethanol precipitation steps. In addition, the type and percentage of the gel matrix can be adjusted based on application needs. In this section, we outline two application examples to demonstrate that good purification results can be obtained using denaturing Urea-PAGE matrices for long oligonucleotides and native PAGE matrices for RNA nanoparticles. Although not demonstrated in this section, Agarose gel matrices further allow purification of complexes and assemblies that are too large to run into polyacrylamide gels.

CEGE PURIFICATION OF TRANSCRIPTION PRODUCTS

One application of CEGE is the large-scale purification of transcription products. *In vitro* transcription provides a cost-effective and efficient means of producing microgram to milligram quantities of native or chemically modified RNA transcripts from a dsDNA template with a T7 promotor region. *In vivo* applications require enzymatically stable RNA which can be obtained by incorporating 2' Fluoro or 2'O-methyl NTPs into the runoff RNA transcript using commercially available mutant T7 polymerases, such as ExonanoRNA's ChemRNA™. While *in vitro* transcription is one of the most common techniques for laboratory-scale RNA production, scale-up can pose a significant challenge, especially when long or highly purified RNA is required for downstream applications. The Guo laboratory has recently demonstrated that use of CEGE can overcome this hurdle for the purification of milliliter-scale RNA transcription reactions.

In this mode, a commercially available CEGE devise, such as the BioRad model 491 Prep Cell, can be used to purify large-scale transcription reactions while continuously collecting eluting fractions during the gel electrophoresis run. To achieve purification of full-length RNA transcripts, the RNA is denatured and run over an 8% Urea-PAGE gel. A column of 5.5 centimeter height is sufficient to purify 1–2 milliliter-scale transcription reactions. The gel is prerun at 300–600 Volts to equilibrate the column and transcription reactions are denatured in gel loading buffer prior to loading onto the gel. Gel loading dyes can be used as indicator for gel run progression and start of fraction collection for eluting samples. Eluting fractions can be monitored online using a UV-VIS module or after elution by measuring the absorbance at 260 nm. Fractions are then analyzed on denaturing PAGE, concentrated and desalted to remove any remaining contamination from buffers and transcription reactions.

Using this technique, 2–4 ml sample volumes consisting of 1–50 mg of transcription reactions in denaturing gel loading buffer can be purified in 2–5 hours. Eluting fractions can be monitored online, concentrated and buffer exchanged within the same day, thus, significantly reducing otherwise labor-intensive workup processes.

CEGE PURIFICATION OF RNA NANOPARTICLES

RNA nanoparticles are assemblies of individual oligonucleotides into precisely designed three-dimensional nanostructures. Driven by base pairing thermodynamics, self-assembly of small RNA nanoparticles is an inherently efficient process that is often observed at larger than 90% assembly efficiencies. Applications that require assemblies of large numbers of individual oligonucleotides or a high degree of purity, such as therapeutic applications, often necessitate additional purifications steps. Purification of RNA nanoparticles poses a challenge due to their size and three-dimensional structure. While resolution in oligonucleotide purification is achieved by denaturation of oligonucleotides into homogenous samples, RNA nanoparticles need to remain in native conditions to stabilize their base-pairing interactions during purification.

On a laboratory scale this feat is typically achieved by separating the samples on a native TBE or TBM gel at 4 °C followed by gel excision and passive elution of the RNA nanoparticles from the gel. The three-dimensional structure and large size of the nanoparticles significantly hinders diffusion often leading to low recovery yields. Active elution techniques typically place the excised gel piece in an elution buffer under electrophoretic current and collect the run-off in a small chamber made of cut-off membranes that retain the large RNA nanoparticles. While electro-elution has been shown to increase the recovery yield, it is a very laborious technique and still suffers from inefficient elution causing long recovery times and losses from membrane aggregation of the samples.

CEGE can overcome the challenges in RNA nanoparticle purification by combining traditional native PAGE gels with fraction collection of the samples as they run off the gel. This approach bypasses the problematic workup steps of gel excision and elution completely and can be achieved in considerably shorter time periods which is beneficial for less stable complexes. Purification efficiencies approaching those of passively eluted of short oligos can be achieved while retaining the biological activity of functional moieties in the RNA nanoparticles. This feat has recently been demonstrated by Jasinski et al. using a 3WJ nanoparticle harboring a fluorescent MG aptamer produced by rolling circle replication.

As for purification of transcription products, a commercially available device, such as the BioRad Model 491 Gel Electrophoresis Prep Cell, can be used to purify large-scale RNA nanoparticle preparations. To ensure RNA nanoparticles retain their desired structure, the particles are separated on an 8% native TBM-PAGE gel at 4 °C in native loading buffer at 300V. This is demonstrated on MG-3WJ nanoparticles transcribed from circular dsDNA. Figure 20.5 illustrates fractions analyzed for both MG fluorescent signal and absorbance at 260 nm (Figure 20.5A). Three distinct populations containing MG fluorescence are separated and identified in order of migration speed as MG oligo, MG-3WJ nanoparticle, and MG-3WJ nanoparticle aggregates using PAGE analysis.

ULTRACENTRIFUGATION

In addition to chromatography and electrophoresis, Ultracentrifugation has been described as a scalable, cost-effective, and contamination-free method to purify RNA and RNA nanoparticles [22–24]. Centrifugation can be used to separate particles in solution according to their size, shape, and density. Centrifugal force, viscosity of the medium, and rotor speed allow adjustment of sedimentation speeds and thus resolution of particle separations from unwanted contaminations. In general, particles with densities higher than the solvent sediment while particles lighter than the solvent float to the top. The sedimentation speed depends on the difference in density between particle and solvent. Thus, choice of the gradient matrix allows separation based on density (isopycnic) or size and shape (rate zonal) of the sample. Exploiting these characteristics, isopycnic ultracentrifugation is being used in the purification of single- stranded (ss)RNA oligomers from proteins and nucleotides following large-scale *in vitro* transcription. In contrast, rate-zonal density gradient centrifugation is used to purify nanoparticles from single-stranded RNA and incompletely assembled complexes, based on their differences in size, shape, and mass causing difference in their sedimentation coefficients.

CsCl Density Gradient Ultracentrifugation for the Purification of Transcription Products

In vitro transcriptions contain several reagents that remain in solution at completion of the reaction, such as the DNA template, the RNA polymerase, unincorporated rNTPs, and the RNA transcript.

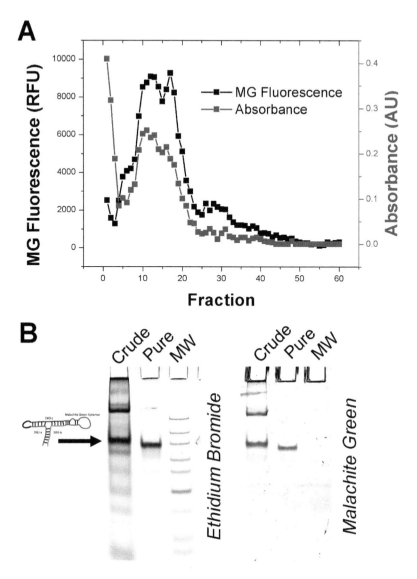

FIGURE 20.5 Diagram of RNA nanoparticle harboring malachite green (MG) aptamer, survivin siRNA, and folate-DNA/RNA sequence for targeting delivery, using 3WJ-pRNA as scaffolds. Figure is produced with permission from reference Shu et al. (2011) © Nature Publishing Group.

As outlined in section 3, PAGE purification is often used to separate the solution based on charge and shape of the components. An alternate approach is to separate the components based on their density using CsCl density gradient ultracentrifugation. Centrifugation can be scaled up for use in large-scale purification of transcription products when highly pure RNA transcripts are required. This method makes use of the differences in density for RNA, nucleotides and proteins and applies a CsCl isopycnic density gradient to purify the RNA transcripts via ultracentrifugation. A 1.65–1.95 CsCl density gradient is prepared using three density layers, 1.95 at the bottom, 1.65 at the top, and sample in 1.75 in the middle. The solutions are spun in a swinging bucket rotor at 45,000 RPM (246,078 x g) for 16 h followed by fractionation of the tubes to retrieve the separated samples. RNA equilibrates towards the bottom of the tube due to its high density. Figure XY shows a demonstration of this technique for the purification of phi29 pRNA monomers.

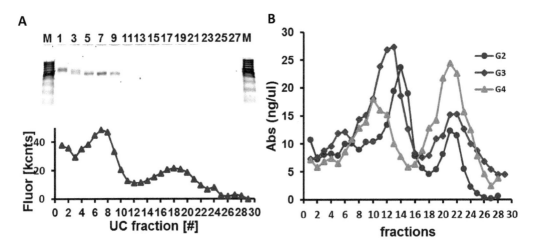

FIGURE 20.6   Schematic of pRNA nanoparticle harboring anti-HIV gp120 aptamer. (A) pRNA–A1-D3, the aptamer sequence was inserted into the 3'/5' double helical domain (23nt fragment) and loop domain (97nt fragment). (B) pRNA-A1-D4, the aptamer sequence was directly appended to the 5'end of pRNA 5'/3' double-stranded helical domain. Figure is produced with permission from reference Zhou et al. (2011) © Elsevier Inc.

Sucrose Gradient Rate Zonal Ultracentrifugation for the Purification of RNA Nanoparticles

The assembly of a large number of individual RNA oligos into a single RNA nanostructure tends to require additional purification to yield homogeneous products of large enough purity for biomedical applications. By means of example, a fourth generation RNA dendrimer (G4) consists of 11 unique sequences with a total of 61 individual oligos or 1872 nucleotides that assemble into a 609,675 Da nanoparticle. DLS and AFM analysis has shown that the particles are exceptionally homogeneous and remain thermodynamically and enzymatically stable but due to the sheer number of oligos, incomplete fragments are observed that were not able to find their counterpart during the assembly process. As assembly efficiency drops due to complexity and purification becomes more and more important, it also becomes more challenging due to the size of the particles. Particles of this size are too big to run into pores of even 4% PAGE gels and tend to get stuck in the wells of the gels leading to loss of yield. Agarose gel electrophoresis has a larger pore range but suffers in resolution and can often not separate partial assemblies from fully assembled particles. Rate zonal ultracentrifugation has proven extremely valuable for complex three-dimensional nanoparticles.

Under centrifugal force, unincorporated oligos, partial assemblies, and fully assembled nano-complexes exhibit different sedimentation rates in a viscous fluid. Larger particles sediment faster than smaller particles and at lower centrifugal forces. This allows the purification of nanoparticles from partial assemblies and unincorporated oligos based on their size and shape using sucrose gradient rate zonal ultracentrifugation. Using this technique, a 5–20% sucrose gradient is prepared and the sample is applied to the top of the gradient. The sample is centrifuged at 35krpm for 30 min and fractions are collected and examined on non-denaturing agarose gels stained with ethidium bromide (EtBr). Figure 20.6 demonstrates the separation capabilities for generation 2–4 dendrimers.

CHOOSING THE BEST METHOD FOR YOUR APPLICATION

On the laboratory scale, PAGE purification and semi-preparative HPLC techniques are among the most frequently used methods to purify oligonucleotides, modified oligonucleotides as well as dimers, duplexes and RNA nanoparticles. When larger scales of samples are required, these techniques quickly become labor- and time-intensive which has been a cause of hindrance in the advancement of the field of RNA nanotechnology. Continuous elution gel electrophoresis and ultra-centrifugation are but two alternate techniques that have been successfully used in the large-scale purification of structured RNA nanocomplexes. Nonetheless, each technique outlined in this chapter has a valuable contribution in its domain when employed appropriately. Based on our experiences, we outline our proposed uses for each in Table Z (Figure 20.7).

In laboratory-scale production of oligonucleotides of 15 – 60 nt in length, PAGE purification followed by band excision and passive elution is the most common approach. For long oligos or structured RNA nanoparticles, reduced yields are observed in passive elution. Consequently, active elution techniques such as CEGE can increase the yield of recovery for oligos with n > 60 nt. Anion exchange (AEX) HPLC has proven useful in the purification of synthetic single-stranded oligos of n ≤ 60 nt from n-1 terminations.

AEX purification can be scaled up to industrial cGMP scale and has become the industry standard for the purification of short therapeutic oligos. Purification of large-scale preparations of long (n>60 nt) oligos or RNA nanoparticles is not possible in AEX due to limitations in resolution and the need for denaturing conditions. CEGE is capable of covering this application range in the 1 to 100 mg scale using relatively affordable commercial elution devices. In addition, rate zonal ultra-centrifugation can be used to separate fully assembled RNA nanoparticles from monomers and incomplete fragments on the larger nanoparticle size.

When working with modified oligonucleotides, IPRP is extremely useful in separating conjugated oligos from unconjugated oligos, in particular when hydrophobic modifications such as fluorophores or cholesterol are used. In addition, IPRP is used in the purification of synthetic oligos. During deprotection, the dimethoxytrityl (DMT) protection group on the 5' oligo position can be selectively retained which gives a hydrophobic modifier for separation of full-length oligomers from prematurely terminated oligo fragments.

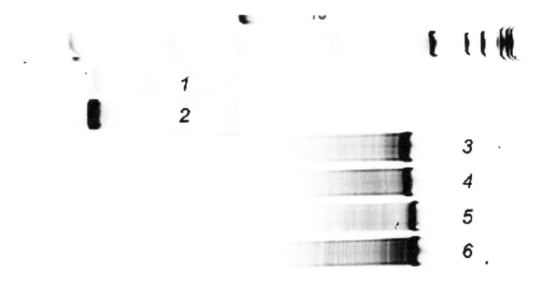

FIGURE 20.7   3WJ RNA nanoparticles harboring RNA aptamers for delivery of therapeutic RNAs in different cancer models. (A–D) Schematics and/or Atomic Force Microscopy (AFM) images of the 3WJ RNA nanoparticle scaffold, followed by in vitro and in vivo analyses following nanoparticle treatment in triple negative breast cancer (A), prostate cancer (B), HER2-positive breast cancer (C), and colon cancer (D). Figure is produced with permission from references Shu et al. (2015), Binzel et al. (2016), Zhang et al. (2016), and (Xu et al. 2019).

ACKNOWLEDGMENTS

P.G. is also a co-founder of ExonanoRNA LLC.

REFERENCES

1. Guo, P., Haque, F., Hallahan, B., Reif, R. & Li, H. Uniqueness, advantages, challenges, solutions, and perspectives in therapeutics applying RNA nanotechnology. *Nucleic Acid Therap* **22**, 226–245, doi:10.1089/nat.2012.0350 (2012).

2. Mitra, S., Shcherbakova, I. V., Altman, R. B., Brenowitz, M. & Laederach, A. High-throughput single-nucleotide structural mapping by capillary automated footprinting analysis. *Nucleic Acids Res* **36**, e63, doi:10.1093/nar/gkn267 (2008).

3. Shu, Y. et al. Stable RNA nanoparticles as potential new generation drugs for cancer therapy. *Adv Drug Deliv Rev* **66**, 74–89, doi:10.1016/j.addr.2013.11.006 (2014).

4. Guo, P. The emerging field of RNA nanotechnology. *Nat Nanotechnol* **5**, 833–842, doi:10.1038/nnano.2010.231 (2010).

5. Haque, F. et al. Ultrastable synergistic tetravalent RNA nanoparticles for targeting to cancers. *Nano Today* **7**, 245–257, doi:10.1016/j.nantod.2012.06.010 (2012).

6. Jasinski, D. L., Khisamutdinov, E. F., Lyubchenko, Y. L. & Guo, P. Physicochemically tunable polyfunctionalized RNA square architecture with fluorogenic and ribozymatic properties. *ACS Nano* **8**, 7620–7629, doi:10.1021/nn502160s (2014).

7. Khisamutdinov, E. F., Jasinski, D. L. & Guo, P. RNA as a boiling-resistant anionic polymer material to build robust structures with defined shape and stoichiometry. *ACS Nano* **8**, 4771–4781, doi:10.1021/nn5006254 (2014).

8. Khisamutdinov, E. F. et al. Enhancing immunomodulation on innate immunity by shape transition among RNA triangle, square and pentagon nanovehicles. *Nucleic Acids Res* **42**, 9996–10004, doi:10.1093/nar/gku516 (2014).

9. Liu, J. et al. Fabrication of stable and RNase-resistant RNA nanoparticles active in gearing the nanomotors for viral DNA packaging. *ACS Nano* **5**, 237–246, doi:10.1021/nn1024658 (2011).

10. Shu, Y., Cinier, M., Shu, D. & Guo, P. Assembly of multifunctional phi29 pRNA nanoparticles for specific delivery of siRNA and other therapeutics to targeted cells. *Methods* **54**, 204–214, doi:10.1016/j.ymeth.2011.01.008 (2011).

11. Shu, Y. et al. Fabrication of 14 different RNA nanoparticles for specific tumor targeting without accumulation in normal organs. *RNA* **19**, 767–777, doi:10.1261/rna.037002.112 (2013).

12. Shu, Y., Shu, D., Haque, F. & Guo, P. Fabrication of pRNA nanoparticles to deliver therapeutic RNAs and bioactive compounds into tumor cells. *Nat Protoc* **8**, 1635–1659, doi:10.1038/nprot.2013.097 (2013).

13. Leontis, N., Sweeney, B., Haque, F. & Guo, P. Conference scene: Advances in RNA nanotechnology promise to transform medicine. *Nanomedicine (London)* **8**, 1051–1054, doi:10.2217/nnm.13.105 (2013).

14. Shukla, G. C. et al. A boost for the emerging field of RNA nanotechnology. *ACS Nano* **5**, 3405–3418, doi:10.1021/nn200989r (2011).

15. Trautmann, L. et al. Upregulation of PD-1 expression on HIV-specific CD8+ T cells leads to reversible immune dysfunction. *Nat Med* **12**, 1198–1202, doi:10.1038/nm1482 (2006).

16. Snyder, L. R., Stadalius, M. A. & Quarry, M. A. Gradient elution in reversed-phase HPLC-separation of macromolecules. *Anal Chem* **55**, 1412A–1430A, doi:10.1021/ac00264a001 (1983).

17. Sinha, N. D. & Jung, K. E. Analysis and purification of synthetic nucleic acids using HPLC. *Current Protocols in Nucleic Acid Chemistry* **61**, 10.5.1–10.5.39, doi:10.1002/0471142700.nc1005s61 (2015).

18. Gilar, M. et al. Ion-pair reversed-phase high-performance liquid chromatography analysis of oligonucleotides: retention prediction. *J Chromatogr A* **958**, 167–182, doi:10.1016/s0021-9673(02)00306-0 (2002).

19. Zhang, Y. & Cremer, P. S. Interactions between macromolecules and ions: The Hofmeister series. *Curr Opin Chem Biol* **10**, 658–663, doi:10.1016/j.cbpa.2006.09.020 (2006).

20. Ellington, A. D. & Szostak, J. W. In vitro selection of RNA molecules that bind specific ligands. *Nature* **346**, 818–822, doi:10.1038/346818a0 (1990).

21. Symons, R. H. Small catalytic RNAs. *Annu Rev Biochem* **61**, 641–671, doi:10.1146/annurev.bi.61.070192.003233 (1992).

22. Noller, H. F. Structure of ribosomal RNA. *Annu Rev Biochem* **53**, 119–162, doi:10.1146/annurev. bi.53.070184.001003 (1984).

23. Apffel, A., Chakel, J. A., Fischer, S., Lichtenwalter, K. & Hancock, W. S. Analysis of oligonucleotides by HPLC-electrospray ionization mass spectrometry. *Anal Chem* **69**, 1320–1325, doi:10.1021/ac960916h (1997).

24. Gong, L. Comparing ion-pairing reagents and counter anions for ion-pair reversed-phase liquid chromatography/electrospray ionization mass spectrometry analysis of synthetic oligonucleotides. *Rapid Commun Mass Spectrom* **29**, 2402–2410, doi:10.1002/rcm.7409 (2015).

25. Fountain, K. J., Gilar, M., Budman, Y. & Gebler, J. C. Purification of dye-labeled oligonucleotides by ion-pair reversed-phase high-performance liquid chromatography. *J Chromatogr B Anal Technol Biomed Life Sci* **783**, 61–72, doi:10.1016/s1570-0232(02)00490-7 (2003).

26. Gilar, M. Analysis and purification of synthetic oligonucleotides by reversed-phase high-performance liquid chromatography with photodiode array and mass spectrometry detection. *Anal Biochem* **298**, 196–206, doi:10.1006/abio.2001.5386 (2001).

21 Physicochemical Characterization of Nucleic Acid Nanoparticles

Daniel Miller and Emil Khisamutdinov
Ball State University, Muncie, USA

CONTENTS

INTRODUCTION

Nucleic Acid Nanoparticles (NANPs) can be assembled into well-defined geometrical architectures from single or multiple oligonucleotides due to the nucleotides' unique ability to form Watson–Crick base pairing and base stacking interactions. Assessing physical and chemical properties of these nanostructures is an important task and is usually accomplished prior to their application *in vitro* and in animal models. The characterization includes standard biochemical and biophysical approaches. For example, the electrophoretic mobility shift assay is commonly used (using either agarose or polyacrylamide gel electrophoresis (PAGE) techniques) to assess assembly efficiency and evaluate remaining fractions after enzymatic stability assays of the nucleic acid assemblies [1–3]. The UV-melting technique is a convenient approach to extract melting temperature and sometimes thermodynamic properties of the nanoparticles [4]. The Dynamic Light Scattering (DLS) technique is helpful to assess the hydrodynamic diameter of a nanoparticle as well as information about homogeneity of the solution and average surface charge of the NANPs [5]. Microscopy tools, especially atomic force microscope, are vital instrumentation for assessing two-dimensional (2D) geometrical properties of nanoparticles [6,7]. Although more sophisticated and often more time-consuming, X-ray crystallography [8] and cryo-electron microscopy can provide absolute three-dimensional (3D) structural information [3]. Collectively, physicochemical properties and structural aspects of the NANPs can be used to further fine-tune those properties by adjusting the nucleotide composition of the assembling strands (e.g. increasing G-C content) or introduce chemically modified nucleic acid strands [9,10].

Noteworthy, there are well-known physiochemical trends for certain NANPs. Hybrid NANPs composed of higher percentages of DNA strands could be argued to have higher levels of *in vivo* stability because they withstand the presence of bovine serum for substantially longer periods than NANPs primarily composed of RNA [11,12]. However, NANPs with higher percent composition of

DOI: 10.1201/9781003001560-23

RNA could be argued to have higher levels of thermal stability, exhibiting higher melting temperatures than NANPs with higher percent composition DNA [12,13].

The focus of this chapter is to overview some of the most common approaches used in NANP characterization and briefly introduce principles behind these techniques.

ELECTROPHORETIC MOBILITY SHIFT ASSAYS (EMSA) TO TEST ASSEMBLY EFFICIENCY

The EMSA approach is a non-denaturing method used to separate NANPs using polyacrylamide (vertical) or agarose (horizontal) gel electrophoresis. The mobility shift (distance traveled) is influenced by several factors including ion concentrations, pH, temperature, the voltage applied to the matrix, total time of applied voltage, concentration of the gel matrix, and the number of nucleotides in the sample being tested. The principle of the gel electrophoresis is rather simple. The electrical current passes from the negative electrode (a.k.a. cathode often has a black color connecting port, and this is the region of the plate where samples are loaded) to the positive electrode (a.k.a. anode is the red-colored port and region of the plate where samples travel toward) through the gel matrix containing NANP samples (Figure 21.1).

The phosphate diester backbone of the oligonucleotides maintains their negative charge in buffer solutions of pH = 8.0. The common buffers are TBM (tris-borate magnesium used in PAGE) and TAE (tris-acetate EDTA used in agarose gel electrophoresis). Larger nanoparticles face more resistance traveling through the gel resulting in less distance traveled compared to smaller nanoparticles.

Mobility shift assays utilize agarose or polyacrylamide gels to confirm stepwise nanoparticle assemblies by loading strand combinations increasing the size of the nanoparticle fractions up to and including its projected shape. The step-wise base pairing of NANP strands increases the size of the nanoparticle being assessed. Increased NANP size results in reduced travel distance on the gel. Comparisons between the nucleic acid single strand, dimer, and trimer or more if applicable can be made with the fully assembled complex [14]. Additional confirmations including molecular weight (MW) ladders are often incorporated to confirm the MW of each of the previously mentioned structures.

Native PAGE utilizes a vertical plate with loading wells formed by a plastic well casting inserted into the top of the gel. Acrylamide and bis-acrylamide stock solution is diluted with TBM solution before initiating free radical polymerization with ammonium persulfate (APS) and N,N,N',N' Tetramethylethylene-1,2-diamine (TEMED). The bis-acrylamide allows for polymer cross-linking that results in a tight polymer gel for accurate separation of NANPs ranging from 20 to 2000 base pairs. Due to the specific polymer gel medium, NANPs with as little as a 1 base pair difference can

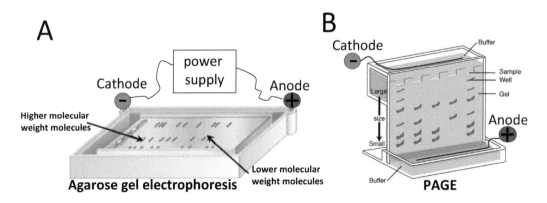

FIGURE 21.1 Representative images of (a) agarose gel electrophoresis and (b) polyacrylamide gel electrophoresis (PAGE) systems.

be detected on the polyacrylamide plate. Additional optimization is possible by adjusting the porosity of the gel. In our laboratory, we typically practice with acrylamide/bis-acrylamide ratios as 19:1 which is convenient for the relatively small NANP separation. Less bis-acrylamide (39,1) leads to less cross-linker and larger pore size which is more applicable for larger NANP samples. Native polyacrylamide gels require staining the plate with ethidium bromide after the run is complete. The native polyacrylamide gel is then washed with distilled or double deionized water to remove residual ethidium bromide before UV imaging.

Agarose gels are usually casted in a horizontal position (Figure 21.1a) where the plastic well forming comb is inserted into the cathode region of the gel. Agarose solution is often prepared with the stain (most often ethidium bromide) already pre-mixed into the polysaccharide dissolved in in TAE buffer. The interactions of polysaccharide strands are less quantifiable than the polymerization and cross-linking of polyacrylamide, but concentrations ranging from 0.5 % to 4 % agarose often result in adequate separation of NANPs.

Regardless of the gel, assembly efficiency can be visually assessed by comparing NANP assemblies as well as comparing the intensity of the band known to be the desired NANP with other individual stands or theoretical substructures.

EVALUATION THERMODYNAMIC STABILITY BY UV-THERMAL DENATURATION (UV-MELT) AND TEMPERATURE GRADIENT GEL ELECTROPHORESIS (TGGE)

NANPs are denatured (melted) with the application of heat. The point of 50% NANP dissociation from its native structural form is referred to as the melting temperature (Tm)[15]. Elevated temperature causes strand dissociation, breaking of hydrogen bonds between complimentary bases, as well as disrupting the stacking interactions between neighboring base pairs. Melting points correspond to the stability of NANPs as more stable nanoparticles have higher Tm values. The melting transitions of NANPs are influenced by sequence and structural variations, concentration, and external solvent variables including the presence of denaturing agents, monovalent and divalent metal ions, as well as the pH of the solution.

UV–Vis spectroscopy measures the increase UV light absorption at λ_{260} by the NANPs as a function of temperature, as the strands are denatured. Upon denaturation the individual aromatic nitrogenous bases became more exposed to absorb more UV light, this absorbance increase is known as hyperchromicity effect (Figure 21.2).

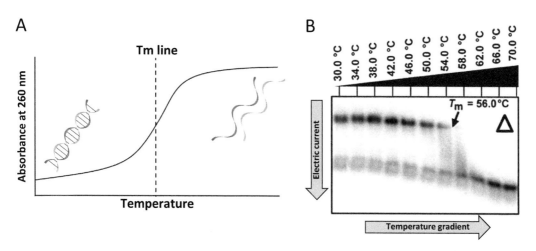

FIGURE 21.2   Examples of nucleic acid melting temperature determination by (a) UV-melting plot and (b) perpendicular Temperature Gradient Gel Electrophoresis. Figure 21.2b demonstrates melting temperature of RNA triangle nanoparticle. The image obtained from ref. #16, copyright 2014, Oxford University Press.

NANP samples are prepared in a range from 0.1 to 1.0 µM concentration depending on the complexity of the particles so the initial absorbance reading at 20° C falls below 1.0 absorbance unit for an accurate melting temperature (Tm) measurement. The sample is placed into a micro-cuvette (~100 µL) supplied with a PTFE stopper to avoid evaporation while exposed to a temperature gradient ranging from 20 to 100° C. In some cases, it is also beneficial to cover the sample solution with a thin layer of silicone oil to avoid evaporation. As temperature increases, the NANP samples slowly dissociate (melt). On an absorbance versus temperature graph, the point of 50% dissociation (Tm) is equal to the first derivative of the resulting sigmoidal curve. Additional conformational data sets can be obtained by reversing the temperature gradient starting at temperatures near 100°C and slowly decreasing the temperature to re-anneal the NANP sample. In this case, the point of 50% re-association (Tm) is also equal to the first derivative of the resulting sigmoidal curve (Figure 21.2a).

Temperature Gradient Gel Electrophoresis (TGGE) is another method providing visual confirmation of NANP thermal stability. NANP samples are exposed to increasing temperature gradients causing dissociation confirmed by electrophoretic analysis as demonstrated in Figure 21.2b [16]. Electrical current and linear temperature gradients can be applied perpendicular or parallel to each other with the appropriate electrophoresis equipment [17]. Computational analysis is used to assess 50% dissociation (or re-association) of the resulting bands on the gel plate. At low temperatures with complete assembly there is a strong band representing full assembly, and as the temperature increases the original band loses its intensity as the dissociated bands travel further distances on the gel plate. Upon complete dissociation, there is no longer a band present confirming full dissociation of the original NANP sample.

NANPs impose new challenges upon conventional methods of interpreting Tm values. UV-melt analysis of structures with three or more strands will often have several inflection points along the temperature gradient, and the standard application of the sigmoidal curve is no longer an accurate method to assess the Tm value of the NANP. Although bimodal sigmoidal curves can be utilized to enhance the accuracy of an additional Tm value, there are limitations to calculation accuracy beyond two regions of strand dissociation. TGGE allows visualization of strand displacement once the strand is fully dissociated; however, unlike UV-melt analysis, partial dissociation is undetectable using TGGE methods.

NUCLEASE RESISTANCE PROPERTIES OF NANP CAN BE ASSESSED BY SERUM STABILITY ASSAY

Serum stability assays are preformed to test resistance to the hydrolysis of phosphodiester bonds performed by nucleases that are abundant in a bloodstream. Exonucleases target external regions and endonucleases target internal regions to hydrolyze DNA and RNA strands. Certain levels of nuclease resistance are required for NANP applications focused on immunology and drug delivery. DNA strands are known to have greater stability than RNA strands in the presence of most nuclease testing protocols including fetal bovine serum (FBS) and human serum.

Control samples as well as experimental solutions of NANP containing 10–20% serum are prepared, and aliquots are taken from the experimental solutions at time intervals often ranging from one minute to 24 hours. Gel electrophoresis is again the method used to analyze the remaining fractions of the NANPs. High percent DNA strand composition extends the life-time of the NANPs and minimize nuclease recognition. NANP modifications including the replacement of an alcohol functional group on the ribose sugar with fluorine or ester substituents can increase nuclease stability up to and beyond 24 hours. Similar to analysis of the TGGE experiment, remaining fractions of the NANPs are calculated from the gel by quantification analysis using software such as Image J. Testing at higher serum concentrations is often more applicable for NANPs with higher levels of serum stability to avoid extremely long test run times. Although NANPs are often diluted to 1µM concentration with the serum stability assays, those concentrations can also be adjusted.

HYDRODYNAMIC DIAMETER ASSESSMENT BY DLS

The principle behind the DLS experiment relies on the Brownian motion of the NANPs in aqueous solution. The larger the particle, the slower the Brownian motion. This allows precise calculation of particle size by mathematical algorithms embedded within the DLS software (Figure 21.3.a). The DLS experiment directs a laser light beam through a polarizer and onto the NANP sample solution. This results in a speckle pattern of deflected laser beams interpreted by a fast photon detector placed at a 173° angle relative to the linear pathway between the laser and the sample. Additional calculations are derived from the Brownian motion of particles in solution, the Siegert relation of scattering volume, and the Stokes–Einstein equation to assess the hydrodynamic diameter, all of which are integrated within the DLS instrument program. Additional program features include auto-calculation of the Z-average; however, this is only applicable for monomodal test results.

DLS spectroscopic characterization does not disrupt the native particle configuration, allowing the same sample to be used for other experiments. In addition, testing sample volume sizes are relatively small, requiring about 100 μL of a 1 μM solution, as specified in various literature

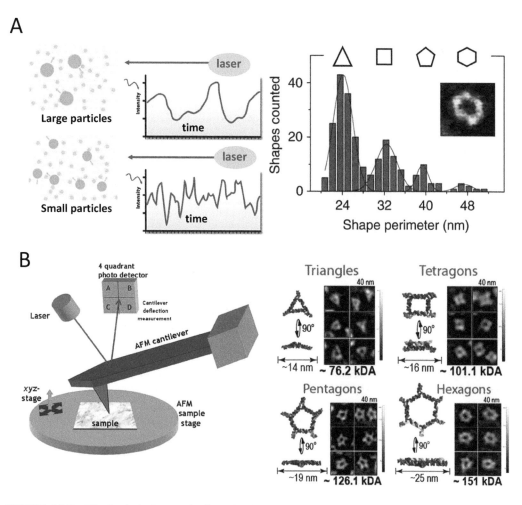

FIGURE 21.3 Biophysical characterization of NANPs. Hydrodynamic sizes can be obtained from DLS experiment (a) and 2D geometry of the particle shapes can be measured by AFM (b). DLS plot of the nucleic acid polygons in the Figure 21.3a were obtained from ref #7, copyright @ 2019 springer nature. AFM images in Figure 21.3b were adapted with permission from ref #13, copyright 2018, American Chemical Society.

reviews [1,7,12]. More recently, DLS became useful to extract information about NANP charges. Specifically, Guo's research group has reported several examples of implementing DLS for nanoparticle charge measurements [18–20].

NANP 2D IMAGING BY ATOMIC FORCE MICROSCOPY AND 3D STRUCTURE DETERMINATION BY CRYO-ELECTRON MICROSCOPY AND X-RAY CRYSTALLOGRAPHY

Microscopy imaging methods for NANPs often require sample preparation in solution while also maintaining appropriate conditions such as pH, temperature, divalent metal ion concentration to preserve native conformation of the constructs. NANP structural confirmations are assessed using atomic force, cryo-electron microscopies as well as X-ray crystallography. Atomic force microscopy (AFM) imaging reveals 2D structural confirmations, while cryo-EM and X-ray crystallography reveal 3D information about the investigated samples.

AMF is a scanning probe microscopy technique that results in imaging with resolution capabilities on the order of fractions of a nanometer. AFM is employed to determine the topography of NANPs (Figure 21.3b). Although optimization is required for the assessment of DNA samples, new modifications are evolving for RNA, hybrid, and modified NANPs. To obtain resolved images, samples are usually deposited to the atomically flat surface of a natural mineral muscovite mica. While the mica sheets are inexpensive materials with negative surface charge to hold cationic structures, this preparation is not helpful for most NANPs due to electrostatic repulsion. Prior to nucleic acid deposition, the mica surface requires pretreatment with positive charge carrying metal ions such as Ni^{2+} or Co^{2+} or polyamines such as poly-L-Lysine [21]. This provides a positively charged surface for better electrostatic attraction of negatively charged NANP. The mica surface also can be treated with 3-aminopropyltriethoxy silane (APTES) or other similar derivatives [22]. The APTES chemical modification links a silane triether with the exposed alcohol functional groups of the mica plate resulting in a primary amine capable of holding NANPs in place for imaging. Various intricate AFM images of NANPs have been obtained over the past two decades [11–13,23–25] and there are plenty valuable resources available specifying details of nucleic acid sample preparation for AFM [6,26–29].

The Cryo-EM working principle is similar to that of Transmission Electron Microscopy [30]. For each, the microscope fires a beam of electrons through the sample and the electrons are continuously captured to be sorted by a computer algorithm. A composite, high-resolution, and 3D image is produced within the program [31]. Most 3D NANPs are predicted in various programming models, and Cryo-EM is utilized to confirm positional accuracy of the predicted model. To attain the specific goals of various NANP applications including signal transduction, enzymatic catalysis, and molecular transport, it is vital to have an accurate confirmation of NANP structure predictions. Several examples of 3D NANPs structures resolved by Cryo-EM are available, including RNA triangular prisms [3,32], RNA cubes [33], and RNA tetrahedrons [34].

X-ray crystallography has recently made substantial gains to capture distinct 3D images for a diverse array of NANPs [8,35]. X-ray microscopy methods require crystallization of the NANPs often achieved by hanging drop vapor diffusion. Crystals are then soaked in a cryoprotectant buffer before flash freezing. Diffraction data is collected as X-ray beams diffract from the crystalline sample resulting in distinct patterns collected to determine the 3D structure of the crystal. Although focused on protein structures, the protein data bank also contains open source nucleic acid structural libraries.

ACKNOWLEDGMENT

This work was supported by the National Institute of Biomedical Imaging and Bioengineering of the National Institutes of Health under Award Number R03EB027910. The content is solely the responsibility of the authors and does not necessarily represent the official views of the National Institutes of Health.

REFERENCES

1. Goldsworthy, V., et al., *Fluorogenic RNA Aptamers: A nano-platform for fabrication of simple and combinatorial logic gates.* Nanomater, 2018. **8**(12).
2. Khisamutdinov, E.F., et al., *Simple method for constructing RNA triangle, square, pentagon by tuning interior RNA 3WJ angle from 60 degrees to 90 degrees or 108 degrees.* Methods Mol Biol, 2015. **1316**: p. 181–193.
3. Khisamutdinov, E.F., et al., *Fabrication of RNA 3D nanoprisms for loading and protection of small RNAs and model drugs.* Adv Mater, 2016. **28**(45): p. 10079–10087.
4. Schroeder, S.J. and D.H. Turner, *Optical melting measurements of nucleic acid thermodynamics.* Methods Enzymol, 2009. **468**: p. 371–387.
5. Stetefeld, J., S.A. McKenna, and T.R. Patel, *Dynamic light scattering: A practical guide and applications in biomedical sciences.* Biophys Rev, 2016. **8**(4): p. 409–427.
6. Sajja, S., et al., *Dynamic behavior of RNA nanoparticles analyzed by AFM on a mica/air interface.* Langmuir, 2018. **34**(49): p. 15099–15108.
7. Monferrer, A., et al., *Versatile kit of robust nanoshapes self-assembling from RNA and DNA modules.* Nat Commun, 2019. **10**(1): p. 608.
8. Dibrov, S.M., et al., *Self-assembling RNA square.* Proc Natl Acad Sci U S A, 2011. **108**(16): p. 6405–6408.
9. Jasinski, D.L., et al., *Physicochemically tunable polyfunctionalized RNA square architecture with fluorogenic and ribozymatic properties.* ACS Nano, 2014. **8**(8): p. 7620–7629.
10. Khisamutdinov, E.F., D.L. Jasinski, and P. Guo, *RNA as a boiling-resistant anionic polymer material to build robust structures with defined shape and stoichiometry.* ACS Nano, 2014. **8**(5): p. 4771–4781.
11. Johnson, M.B., et al., *Programmable nucleic acid based polygons with controlled neuroimmunomodulatory properties for predictive QSAR modeling.* Small, 2017. **13**(42).
12. Ke, W., et al., *RNA-DNA fibers and polygons with controlled immunorecognition activate RNAi, FRET and transcriptional regulation of NF-kappaB in human cells.* Nucleic Acids Res, 2019. **47**(3): p. 1350–1361.
13. Hong, E., et al., *Structure and composition define immunorecognition of nucleic acid nanoparticles.* Nano Lett, 2018. **18**(7): p. 4309–4321.
14. Bui, M.N., et al., *Versatile RNA tetra-U helix linking motif as a toolkit for nucleic acid nanotechnology.* Nanomedicine, 2017. **13**(3): p. 1137–1146.
15. Ansevin, A.T., et al., *High-resolution thermal denaturation of DNA. I. Theoretical and practical considerations for the resolution of thermal subtransitions.* Biopolymers, 1976. **15**(1): p. 153–174.
16. Khisamutdinov, E.F., et al., *Enhancing immunomodulation on innate immunity by shape transition among RNA triangle, square and pentagon nanovehicles.* Nucleic Acids Res, 2014. **42**(15): p. 9996–10004.
17. Benkato, K., et al., *Evaluation of thermal stability of RNA nanoparticles by temperature gradient gel electrophoresis (TGGE) in native condition.* Methods Mol Biol, 2017. **1632**: p. 123–133.
18. Yin, H., et al., *Delivery of anti-miRNA for triple-negative breast cancer therapy using RNA nanoparticles targeting stem cell marker CD133.* Mol Ther, 2019. **27**(7): p. 1252–1261.
19. Xu, C., et al., *Photo-controlled release of paclitaxel and model drugs from RNA pyramids.* Nano Res, 2019. **12**(1): p. 41–48.
20. Guo, P., *The emerging field of RNA nanotechnology.* Nat Nanotechnol, 2010. **5**(12): p. 833–842.
21. Thundat, T., et al., *Atomic force microscopy of DNA on mica and chemically modified mica.* Scanning Microsc, 1992. **6**(4): p. 911–918.
22. Shlyakhtenko, L.S., A.A. Gall, and Y.L. Lyubchenko, *Mica functionalization for imaging of DNA and protein-DNA complexes with atomic force microscopy.* Methods Mol Biol, 2013. **931**: p. 295–312.
23. Shu, Y., et al., *Fabrication of 14 different RNA nanoparticles for specific tumor targeting without accumulation in normal organs.* RNA, 2013. **19**(6): p. 767–777.
24. Chworos, A., et al., *Building programmable jigsaw puzzles with RNA.* Science, 2004. **306**(5704): p. 2068–2072.
25. Grabow, W.W., et al., *Self-assembling RNA nanorings based on RNAI/II inverse kissing complexes.* Nano Lett, 2011. **11**(2): p. 878–887.
26. Lyubchenko, Y.L., *Direct AFM visualization of the nanoscale dynamics of biomolecular complexes.* J Phys D Appl Phys, 2018. **51**(40).
27. Tong, Z., et al., *Novel polymer linkers for single molecule AFM force spectroscopy.* Methods, 2013. **60**(2): p. 161–168.

28. Ohno, H., S. Akamine, and H. Saito, *RNA nanostructures and scaffolds for biotechnology applications.* Curr Opin Biotechnol, 2019. **58**: p. 53–61.

29. Ohno, H., E. Osada, and H. Saito, *Design, assembly, and evaluation of RNA-protein nanostructures.* Methods Mol Biol, 2015. **1297**: p. 197–211.

30. Nogales, E. and S.H. Scheres, *Cryo-EM: A unique tool for the visualization of macromolecular complexity.* Mol Cell, 2015. **58**(4): p. 677–689.

31. Sigworth, F.J., *Principles of cryo-EM single-particle image processing.* Microsc, 2016. **65**(1): p. 57–67.

32. Hao, C., et al., *Construction of RNA nanocages by re-engineering the packaging RNA of Phi29 bacteriophage.* Nat Commun, 2014. **5**: p. 3890.

33. Afonin, K.A., et al., *In vitro assembly of cubic RNA-based scaffolds designed in silico.* Nat Nanotechnol, 2010. **5**(9): p. 676–682.

34. Li, H., et al., *Controllable self-assembly of RNA tetrahedrons with precise shape and size for cancer targeting.* Adv Mater, 2016. **28**(34): p. 7501–7507.

35. Boerneke, M.A., S.M. Dibrov, and T. Hermann, *Crystal-structure-guided design of self-assembling RNA nanotriangles.* Angew Chem Int Ed Eng, 2016. **55**(12): p. 4097–4100.

22 Light Scattering Techniques for Characterization of NANPs and Their Formulations

Lewis A. Rolband and Joanna K. Krueger
The University of North Carolina at Charlotte, Charlotte, USA

CONTENTS

Nucleic acid nanoparticles (NANPs) represent a class of nanotechnology that is designed primarily to be fully functional when in a solution environment. As such, the accurate characterization of native conformational states, interparticle interactions, size, molecular weight, and loading onto carrier molecules is necessary to gain proper insight into the functionality of these technologies. While significant characterization tools are available to the NANP researchers, it is important for the advantages and limitations of these tools to be well understood. Techniques that require the particle to be fixed to a surface, such as atomic force microscopy or transmission electron microscopy, or those that otherwise place the particle in the solid state, like cryogenic electron microscopy or X-ray crystallography for example, may unnecessarily bias conformational states and yield potentially misleading results [1,2]. Additionally, by analyzing NANPs in a dynamic solution environment, the response of the NANP to concentration, ionic strength, pH, temperature, and other changes can be readily determined in both a quantitative and qualitative manner [1,2]. The present work serves to discuss the utility of, and the key parameters provided by, various solution scattering techniques for the investigation and characterization of nucleic acid nanotechnologies.

DYNAMIC LIGHT SCATTERING

Dynamic light scattering (DLS) is an efficient method for initial characterization of particle sizes within a wide range. DLS is a technique that measures the change in the intensity of scattered light at a fixed angle, over microsecond intervals, due to the random Brownian motion of particles in solution as they diffuse [3–6]. As such, the translational diffusion coefficient, D_t, is directly calculated from the time-correlated scattering intensity changes and then related to several other important parameters. The two most relevant parameters to RNA nanotechnology derived from the D_t are the hydrodynamic radius, R_h, and the diffusion interaction parameter, k_D. The R_h is significant as an estimate of nanoparticle size in solution, as it represents the radius of a sphere that would diffuse at the same rate as the particle under investigation [7,8]. Additionally, R_h can be used to gain insights into complex formation or conformational changes [8]. To assess colloidal stability, k_D can be calculated from the scattering data and is directly related to the second virial coefficient, A_2 [6]. A positive value of k_D implies repulsive interactions between particles, while a negative value implies attractive interactions [6,9].

DOI: 10.1201/9781003001560-24

PHASE-ANALYSIS LIGHT SCATTERING

Often thought of as a partner technique of DLS, phase-analysis light scattering (PALS) or electrophoretic light scattering (ELS) serves to investigate the surface charges of macromolecules and nanoparticles. This is accomplished by applying an oscillating electric field to a solution of nanoparticles and observing the phase difference in the scattered light and of a reference beam. The difference in the phase is due to the Doppler shift that occurs as the light is scattered from a moving particle. The magnitude of this difference is proportional to the velocity of the particle as it moves in response to the electric field. In turn, the velocity of the particle as it translates through the solution is dependent on the solvent's dielectric constant, the solution's viscosity, and the zeta potential, ζ, of the particle, as given by the Smoluchowski equation [6, 10]. It is an important distinction that the ζ of a particle is not the same as its surface charge, as ζ is the charge at the slipping plane of the electrical double layer formed in solution. The zeta potential can serve as an important parameter for assessing colloidal stability, with values closer to neutral being more likely to aggregate [5–7, 10].

MULTI-ANGLE LIGHT SCATTERING

Multi-angle light scattering (MALS) is a static light scattering technique, meaning that the time-averaged intensity of light scattering from the particle is used for the characterization [6, 11]. The light scattering intensity is measured with an array of detectors placed around the sample. The intensity of scattered light is directly proportional to the molecular weight, M_w, of the particle [5, 6]. For particles with sizes less than 5% of the wavelength of incident light, the scattering will occur isotropically, and size information beyond the molecular weight will be minimal. Particles with sizes above this threshold will anisotropically scatter light, and the radius of gyration, R_g, can be calculated from the angular dependence of the scattering intensity. The R_g, also referred to as the root mean square radius, is the average distance of all point masses to the center of mass of a particle. While it is typically smaller than the hydrodynamic radius, R_g is dependent only on the morphology of the particle itself. MALS can also serve as a means of assessing the colloidal stability of a formulation by measuring the second virial coefficient, A_2, as it accounts for pairwise interactions between particles in solution [5, 6, 9, 11]. Similar to k_D, positive values of A_2 imply repulsive interactions between particles, while negative values imply interparticle attraction [6, 9, 12, 13]. MALS has also been used to analyze the loading of nucleic acid either into or onto carrier molecules by assessing changes in M_w and R_g as association occurs [14].

SMALL-ANGLE SCATTERING

Small-angle X-ray and neutron scattering (SAS) are structural characterization techniques that are ideal for examining the conformation of nucleic acid nanoparticles. The advantages of these techniques, over laser light scattering, are the increased resolution due to the shorter wavelength of X-rays and neutrons. As a result, there is a significantly greater angular dependence and greater structural information provided can result in accurate three-dimensional models of the particle under investigation through various *ab initio* computational methods [1, 2, 15–17]. SAS experiments provide information on the electron (atomic) density distribution within the scattering particle, thus the overall shapes and particle dimensions. SAS analyses provide valuable information regarding the M_w, R_g, and maximum linear dimension, d_{max} [18]. SAS experiments are ideal for following conformational changes of the molecular ensemble as it behaves in solution. If the flux is high enough, such as at a synchrotron source, the particles under investigation can be flowed through capillaries to avoid potential radiation damage of the sample. In addition, the inherent neutron scattering length differences of nucleic acid, lipid and protein, as well as between hydrogen and deuterium nuclei for specifically labeled components, allow one to characterize the conformations and relative dispositions of the individual molecules within an assembly of biomolecules using contrast variation techniques [1,19]. Contrast variation SANS experiments can then yield greater structural informa-tion about individual components in a complex in addition to the overall structure of the particle [19]. SAS has been used effectively to determine the structure of protein–aptamer complexes and as a basis for hypothesis on their mechanism of inhibition [1, 2, 15, 20]. Figure 22.1a used the Crysol program from the freely available ATSAS software package [21] to generate a SAS profile (relative intensity vs. the angular momentum vector, q) for the crystal structure of an RNA nanosquare [22] (pdb accession no. 3p59). Guinier analysis [23] of the low q region of this profile (see inset) gives an R_g of 24.0 Å. The inverse Fourier transform of this SAS data provides the P(r) profile, Figure 22.1b, which is the probability of finding a vector distance, r, between two atoms in the scattering particle. The P(r) profile was generated from these data using the GNOM program from ATSAS and selecting a D_{max} of 74 Å. Analysis of P(r) of 23.9 Å. Figure 22.1c shows the crystal structure of the RNA nanosquare (3p59) fit within an *ab initio* model of these SAS data generated using the program DAMMIN [24].

Solution-based scattering techniques all have unique roles in the realm of NANP characteriza-tion. Each of these provides complementary information and help to form a complete picture of NANPs as they exist in their native conformation. In addition to basic parameters of M_w and size, structural insights and colloidal stability assessments are readily accessible. Each of these tech-niques has been applied in a limited fashion to the world of nucleic acid nanotechnologies but will likely serve expanded roles in the future, particularly in the world of NANP therapeutics, where complexes between NANPs and other target molecules are frequently a requirement for their proper function. A clear understanding of the advantages and limitations of these techniques is required for the researcher to properly utilize them to their fullest potential. For more information regarding the underlying theory of these techniques, it is suggested that one of the referenced reviews be consulted [6, 10, 17, 19].

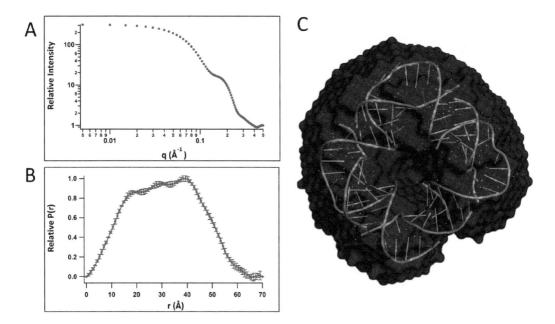

FIGURE 22.1   (a) calculated SAXS curve for the crystal structure (PDB accession no. 3p59) of an RNA nanosquare. Inset shows the Guinier fit to the low q region. (b) corresponding P(r) profile generated from the SAXS data using a D_{max} of 74 Å. (c) *ab initio* average of 15 DAMMIN- derived models of these SAXS data (semi-transparent beads) fit with a cartoon representation of the RNA nanosquare crystal structure.

REFERENCES

1. Oliver, R. C.; Rolband, L. A.; Hutchinson-Lundy, A. M.; Afonin, K. A.; Krueger, J. K., Small-angle scattering as a structural probe for nucleic acid nanoparticles (NANPs) in a dynamic solution environment. *Nanomaterials* 2019, *9* (5), 16.

2. Bruetzel, L. K.; Gerling, T.; Sedlak, S. M.; Walker, P. U.; Zheng, W. J.; Dietz, H.; Lipfert, J., Conformational changes and flexibility of DNA devices observed by small-angle x-ray scattering. *Nano Lett* 2016, *16* (8), 4871–4879.

3. Hassan, P. A.; Rana, S.; Verma, G., Making sense of Brownian motion: colloid characterization by dynamic light scattering. *Langmuir* 2015, *31* (1), 3–12.

4. Pecora, R., Dynamic light scattering measurement of nanometer particles in liquids. *J Nanopart Res* 2000, *2* (2), 123–131.

5. Xu, R. L., Light scattering: A review of particle characterization applications. *Particuology* 2015, *18*, 11–21.

6. Minton, A. P., Recent applications of light scattering measurement in the biological and biopharmaceutical sciences. *Anal Biochem* 2016, *501*, 4–22.

7. Bihari, P.; Vippola, M.; Schultes, S.; Praetner, M.; Khandoga, A. G.; Reichel, C. A.; Coester, C.; Tuomi, T.; Rehberg, M.; Krombach, F., Optimized dispersion of nanoparticles for biological in vitro and in vivo studies. *Part Fibre Toxicol* 2008, *5*, 14.

8. Scales, C. W.; Huang, F. Q.; Li, N.; Vasilieva, Y. A.; Ray, J.; Convertine, A. J.; McCormick, C. L., Corona-stabilized interpolyelectrolyte complexes of SiRNA with nonimmunogenic, hydrophilic/cationic block copolymers prepared by aqueous RAFT polymerization. *Macromolecules* 2006, *39* (20), 6871–6881.

9. Connolly, B. D.; Petry, C.; Yadav, S.; Demeule, B.; Ciaccio, N.; Moore, J. M. R.; Shire, S. J.; Gokarn, Y. R., Weak interactions govern the viscosity of concentrated antibody solutions: high-throughput analysis using the diffusion interaction parameter. *Biophys J* 2012, *103* (1), 69–78.

10. Doane, T. L.; Chuang, C. H.; Hill, R. J.; Burda, C., Nanoparticle zeta-Potentials. *Acc Chem Res* 2012, *45* (3), 317–326.

11. Wyatt, P. J., Light-scattering and the absolute characterization of macromolecules. *Anal Chim Acta* 1993, *272* (1), 1–40.

12. Araiso, Y.; Tsutsumi, A.; Qiu, J.; Imai, K.; Shiota, T.; Song, J.; Lindau, C.; Wenz, L. S.; Sakaue, H.; Yunoki, K.; Kawano, S.; Suzuki, J.; Wischnewski, M.; Schutze, C.; Ariyama, H.; Ando, T.; Becker, T.; Lithgow, T.; Wiedemann, N.; Pfanner, N.; Kikkawa, M.; Endo, T., Structure of the mitochondrial import gate reveals distinct preprotein paths. *Nature* 2019, *575* (7782), 395–+.

13. Koga, K.; Holten, V.; Widom, B., Deriving second osmotic virial coefficients from equations of state and from experiment. *J Phys Chem B* 2015, *119* (42), 13391–13397.

14. Jiang, X. L.; Chu, Y. F.; Liu, J.; Zhang, G. Y.; Zhuo, R. X., Aqueous sec analysis of cationic polymers as gene carriers. *Chin J Polym Sci* 2011, *29* (4), 421–426.

15. Yang, S. C., Methods for SAXS-based structure determination of biomolecular complexes. *Adv Mater* 2014, *26* (46), 7902–7910.

16. Lambert, D.; Leipply, D.; Draper, D. E., The osmolyte TMAO stabilizes native RNA tertiary structures in the absence of Mg2+: Evidence for a large barrier to folding from phosphate dehydration. *J Mol Biol* 2010, *404* (1), 138–157.

17. Putnam, C. D.; Hammel, M.; Hura, G. L.; Tainer, J. A., X-ray solution scattering (SAXS) combined with crystallography and computation: defining accurate macromolecular structures, conformations and assemblies in solution. *Q Rev Biophys* 2007, *40* (3), 191–285.

18. Grant, T. D.; Luft, J. R.; Carter, L. G.; Matsui, T.; Weiss, T. M.; Martel, A.; Snell, E. H., The accurate assessment of small-angle X-ray scattering data. *Acta Crystallogr Sect D-Biol Crystallogr* 2015, *71*, 45–56.

19. Mahieu, E.; Gabel, F., Biological small-angle neutron scattering: recent results and development. *Acta Crystallogr Sect D* 2018, *74* (8), 715–726.

20. Dupont, D. M.; Thuesen, C. K.; Botkjaer, K. A.; Behrens, M. A.; Dam, K.; Sorensen, H. P.; Staff, P. O., Protein-binding RNA aptamers affect molecular interactions distantly from their binding sites. *PLoS One* 2015, *10* (4), e0119207.

21. Franke, D.; Petoukhov, M. V.; Konarev, P. V.; Panjkovich, A.; Tuukkanen, A.; Mertens, H. D. T.; Kikhney, A. G.; Hajizadeh, N. R.; Franklin, J. M.; Jeffries, C. M.; Svergun, D. I., ATSAS 2.8: a comprehensive data analysis suite for small-angle scattering from macromolecular solutions. *J Appl Crystallogr* 2017, *50*, 1212–1225.

22. Chen, S.; Hermann, T., RNA-DNA hybrid nanoshapes that self-assemble dependent on ligand binding. *Nanoscale* 2020, *12* (5), 3302–3307.

23. Reyes, F. E.; Schwartz, C. R.; Tainer, J. A.; Rambo, R. P., Methods for Using New Conceptual Tools and Parameters to Assess RNA Structure by Small-Angle X-Ray Scattering. In *Riboswitch Discovery, Structure and Function*, BurkeAguero, D. H., Ed. Elsevier Academic Press Inc: San Diego, 2014; Vol. 549, pp 235–263.

24. Svergun, D. I., Restoring low resolution structure of biological macromolecules from solution scattering using simulated annealing (vol 76, pg 2879, 1999). *Biophys J* 1999, *77* (5), 2896–2896.

23 Electron Microscopy of Nucleic Acid Nanoparticles

Damian Beasock and Kirill A. Afonin
The University of North Carolina at Charlotte, Charlotte, USA

CONTENTS

Few techniques are currently available to achieve the resolution needed to visualize nucleic acid nanoparticles (NANPs). RNA and DNA helices only range from 2 to 2.5 nm in diameter and the dimensions of a typical NANP do not exceed tens of nanometers [1]. This size scale creates a challenge of visualizing NANPs when we need information regarding their precise structure and morphology. The maximum resolution of light microscopy depends on the wavelength of visible light and may resolve a few tenths of a micron at best [2]. Therefore, electron microscopy (EM) is one of the most viable methods to accurately assess NANP structures because of the resolving ability [3–25]. EM can also be useful as a method of characterization of NANPs encapsulation [26–42].

Similar to the light microscopy EM may be described by many of the same equations used in optics. However, the advantage of EM over the light microscopy is the wavelength of radiation used to analyze specimens [43]. While the wavelength of visible light is 400–700 nm, the typical wavelength of electrons is only a few picometers and depending on the accelerating voltage of microscope can range from 40 pm to 0.5 pm, thus enabling a much better resolution [44].

There are multiple types of EM that may become useful for analysis of biological samples. EM techniques that are suitable for visualizing nucleic acids are transmission electron microscopy (TEM), scanning transmission electron microscopy (STEM), scanning electron microscopy (SEM), field emission scanning electron microscopy (FESEM), and cryogenic electron microscopy (cryoEM). All techniques probe the sample with short wavelength electrons and create an image based on the interaction between the electron and the specimen. Considering the size scale and material properties of nucleic acids, aforementioned EM techniques have their own advantages and disadvantages when analyzing NANPs. Table 23.1 shows the relevant factors to consider between the available techniques. A low beam dose is ideal for nucleic acids as this prevents large increases in sample temperature. High beam dose may risk damaging the sample. Because some EM techniques require vacuum conditions, samples cannot contain water. Dehydrating samples may alter the structure relative to the typical structure in solution. Depending on the purpose, each technique offers different levels of resolution. Atomic resolution can determine structure down to the atomic level. At high resolution can distinguish individual strands of nucleic acids and structures of NANPs while the low resolution only allows to study the aggregations and supra-assemblies of NANPs [45].

The signal collected to produce electron micrographs is dependent on the interaction of the electron with the sample [46]. A challenge of resolution arises when observing nucleic acids because light elements (low atomic number) that they consist of are highly transparent to the electron radiation [47]. The most important factor in obtaining a good electron micrograph is the difference in signal across the image, a phenomenon known as contrast [48]. If the signal from the entire field of the image is the same, there will be no discernable image. Information conveyed from an image will be limited if there is little to no difference between the sample and the substrate or between separate

DOI: 10.1201/9781003001560-25

TABLE 23.1

EM Techniques and Important Factors to Consider When Visualizing NANPs

	TEM	STEM	SEM	FESEM	CryoEM
Beam dose	High	Low	High	Low	High
Dehydration	Yes	Yes	Yes	Yes	No
Resolution	Atomic	High	Low	High	Atomic

structures in the sample [45]. Ideally, contrast will be distinct enough to see the difference between the NANP, the substrate, and any substituents or structures that are part of the particle.

Another challenge during EM analysis is the potential of damage to NANPs incurred by electrons depositing energy into the material. Increased temperature during analysis may be destructive to samples, especially biological samples [49,50]. Biological samples are generally poor heat conductors and have low melting temperatures. Factors that affect temperature are the heat capacity, thermal conductivity, and sample morphology. Biological samples are prone to beam damage because of their weak bonds and because they are composed of light elements [51,52].

Sample preparation depends on the technique being used. Most EM techniques require biological particles to be dehydrated, dispersed, and fixed onto a substrate before analysis [53]. TEM requires transmission of electron beam, and therefore NANPs need to be dispersed and fixed onto a TEM gird [54,55]. A TEM grid is a substrate designed to allow as many electrons as possible to pass through the sample. As nucleic acids are made of light elements (i.e. H, C, N, O, P), maximizing contrast between the substrate and the sample is important. This can be done by choosing a more electron transparent grid coating such as a thin low atomic number material [56]. Sample staining is also a helpful step to maximize contrast of nucleic acids. Heavy metal salts can be added to nucleic acids to bind high atomic number elements to certain sites [54,56–59]. Figure 23.1a and 23.1b shows an example of contrast between the sample and substrate. This is achieved when a strong difference in signal can be determined between the sample and the substrate. Figure 23.1c and 23.1d shows examples of samples prepared with positive and negative staining. Here, high atomic number elements such as tungsten, uranium, osmium, or lead bond to the NANP with positive staining or surround the NANP with negative staining. This is to enhance the contrast between the sample structure and the surroundings [60]. Instead of dehydrating biological samples, cryoEM sample preparation requires vitrification of the sample in amorphous ice. This helps preserve the in solution structure of the sample during visualization [61]. Proper sample preparation is meant to enhance quality and contrast of images to reveal as much about the NANP structure as possible.

TEM involves focusing a beam with magnetic lenses and transmitting the electron beam through the sample. Magnification is varied by adjusting the strength of magnetic lenses [62]. The image is formed when electrons pass through the sample, losing energy and changing direction after transmission. This appears in the image as a different intensity signal than the electrons only interacting with the substrate. Care must be taken when analyzing biological samples under TEM to mitigate sample damage caused by the high beam dosage [63].

STEM focuses an electron beam on a small point and performs a raster scan of the sample area while detecting the transmitted beam. The lenses are not used to magnify the image but only to focus the electron beam into a small probe. The electron beam is pointed to a small point on the sample and the signal is taken after transmission. Each analyzed point on the sample represents a pixel in the final image. To image at a higher magnification, the beam is used to scan a smaller area [64]. Relative to traditional TEM, this method uses a much reduced beam dosage [65].

SEM scans samples in the same manner as STEM. However this technique detects secondary electrons scattered and emitted from the interaction volume [66]. SEM utilizing thermionic emission typically cannot achieve the resolution required to visualize individual nucleic acid structures

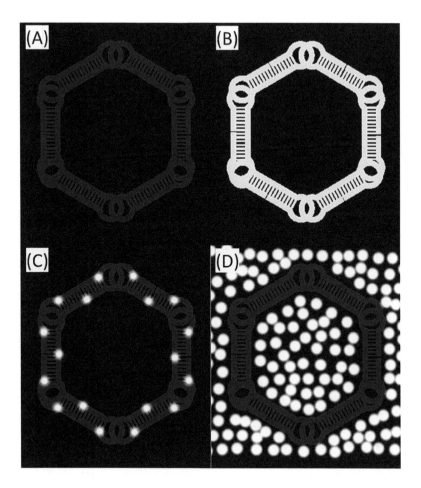

FIGURE 23.1   Illustrations of NANP contrast. (a) Poor contrast between substrate and sample; (b) good contrast between substrate and sample; (c) positive stained NANP where the high atomic number atoms are present as white dots; (d) negative stained NANP.

and is more commonly used for nanoparticle aggregations [67]. However, FESEM using a field emission source can resolve features small enough to characterize individual NANP structure [68]. FESEM is able to produce a narrow beam even at low accelerating voltage. This allows for better resolution than SEM and also mitigates energy deposited into the sample [69,70].

CryoEM is rapidly becoming a widely used and powerful technique for all biological samples. CryoEM transmits a beam through a section of a vitrified sample. With the recent development of direct electron detectors, atomic resolution can be achieved from single particles [71–77]. CryoEM is widely used to analyze nucleic acid structures as it provides good resolution and the low temperature conditions can prevent sample damage [78–81].

Each variation of EM has the capability to achieve the extremely high resolution needed to see nucleic acids. With the proper operating conditions and sample preparation, these techniques are well suited to observing and characterizing nucleic acid nanoparticles.

ACKNOWLEDGMENTS

Research reported in this publication was supported by the National Institute of General Medical Sciences of the National Institutes of Health under Award Numbers R01GM120487 and

R35GM139587 (to K.A.A.). The content is solely the responsibility of the authors and does not necessarily represent the official views of the National Institutes of Health.

REFERENCES

1. Afonin, K.A., et al., *Multifunctional RNA nanoparticles*. Nano Lett, 2014. **14**(10): p. 5662–71.
2. Davidson, M.W. *Resolution*. April, 4, 2020]; Available from: https://www.microscopyu.com/microscopy-basics/resolution.
3. Zuo, H., et al., *A case study of the likes and dislikes of DNA and RNA in self-assembly*. Angew Chem Int Ed Eng, 2015. **54**(50): p. 15118–21.
4. Zhang, C., et al., *Controlling the chirality of DNA nanocages*. Angew Chem Int Ed Eng, 2012. **51**(32): p. 7999–8002.
5. Zhang, C., et al., *Reversibly switching the surface porosity of a DNA tetrahedron*. J Am Chem Soc, 2012. **134**(29): p. 11998–2001.
6. Zhang, C., et al., *DNA-directed three-dimensional protein organization*. Angew Chem Int Ed Eng, 2012. **51**(14): p. 3382–5.
7. Zhang, C., et al., *Exterior modification of a DNA tetrahedron*. Chem Commun (Camb), 2010. **46**(36): p. 6792–4.
8. Zhang, C., et al., *DNA nanocages swallow gold nanoparticles (AuNPs) to form AuNP@DNA cage core-shell structures*. ACS Nano, 2014. **8**(2): p. 1130–5.
9. Zhang, C., et al., *Symmetry controls the face geometry of DNA polyhedra*. J Am Chem Soc, 2009. **131**(4): p. 1413–5.
10. Yu, J., et al., *De novo design of an RNA tile that self-assembles into a homo-octameric nanoprism*. Nat Commun, 2015. **6**: p. 5724.
11. Yu, G., et al., *Single-particle cryo-EM and 3D reconstruction of hybrid nanoparticles with electron-dense components*. Small, 2015. **11**(38): p. 5157–63.
12. Wu, X.-R., et al., *Binary self-assembly of highly symmetric DNA nanocages via sticky-end engineering*. Chin Chem Lett, 2017. **28**(4): p. 851–856.
13. Wang, P., et al., *Retrosynthetic analysis-guided breaking tile symmetry for the assembly of complex DNA nanostructures*. J Am Chem Soc, 2016. **138**(41): p. 13579–13585.
14. Tian, C., et al., *Directed self-assembly of DNA tiles into complex nanocages*. Angew Chem Int Ed Eng, 2014. **53**(31): p. 8041–4.
15. Liu, Z., et al., *Self-assembly of responsive multilayered DNA nanocages*. J Am Chem Soc, 2015. **137**(5): p. 1730–3.
16. Li, Y., et al., *Structural transformation: assembly of an otherwise inaccessible DNA nanocage*. Angew Chem Int Ed Eng, 2015. **54**(20): p. 5990–3.
17. Li, M., et al., *In vivo production of RNA nanostructures via programmed folding of single-stranded RNAs*. Nat Commun, 2018. **9**(1): p. 2196.
18. Ko, S.H., et al., *Synergistic self-assembly of RNA and DNA molecules*. Nat Chem, 2010. **2**(12): p. 1050–5.
19. Jones, R.A., *Challenges in soft nanotechnology*. Faraday Discuss, 2009. **143**: p. 9–14.
20. Huang, K., et al., *Self-assembly of wireframe DNA nanostructures from junction motifs*. Angew Chem Int Ed Eng, 2019. **58**(35): p. 12123–12127.
21. He, Y., et al., *Hierarchical self-assembly of DNA into symmetric supramolecular polyhedra*. Nature, 2008. **452**(7184): p. 198–201.
22. He, Y., et al., *On the chirality of self-assembled DNA octahedra*. Angew Chem Int Ed Eng, 2010. **49**(4): p. 748–51.
23. Hao, C., et al., *Construction of RNA nanocages by re-engineering the packaging RNA of Phi29 bacterio-phage*. Nat Commun, 2014. **5**: p. 3890.
24. Xiao, F., B. Demeler, and P. Guo, *Assembly mechanism of the sixty-subunit nanoparticles via interaction of RNA with the reengineered protein connector of phi29 DNA-packaging motor*. ACS Nano, 2010. **4**(6): p. 3293–301.
25. Afonin, K.A., et al., *In vitro assembly of cubic RNA-based scaffolds designed in silico*. Nat Nanotechnol, 2010. **5**(9): p. 676–82.
26. Liu, X., et al., *Inhibition of hypoxia-induced proliferation of pulmonary arterial smooth muscle cells by a mTOR siRNA-loaded cyclodextrin nanovector*. Biomaterials, 2014. **35**(14): p. 4401–16.

27. Chen, H., et al., *A pH-responsive cyclodextrin-based hybrid nanosystem as a nonviral vector for gene delivery.* Biomaterials, 2013. **34**(16): p. 4159–4172.

28. Zhang, H., et al., *"Push through one-way valve" mechanism of viral DNA packaging.* Adv Virus Res, 2012. **83**: p. 415–65.

29. Xiao, F., H. Zhang, and P. Guo, *Novel mechanism of hexamer ring assembly in protein/RNA interactions revealed by single molecule imaging.* Nucleic Acids Res, 2008. **36**(20): p. 6620–32.

30. Xiao, F., et al., *Binding of pRNA to the N-terminal 14 amino acids of connector protein of bacteriophage phi29.* Nucleic Acids Res, 2005. **33**(8): p. 2640–9.

31. Xiao, F., et al., *Adjustable ellipsoid nanoparticles assembled from re-engineered connectors of the bacteriophage phi29 DNA packaging motor.* ACS Nano, 2009. **3**(8): p. 2163–70.

32. Wendell, D., et al., *Translocation of double-stranded DNA through membrane-adapted phi29 motor protein nanopores.* Nat Nanotechnol, 2009. **4**(11): p. 765–72.

33. Sun, J., et al., *Controlling bacteriophage phi29 DNA-packaging motor by addition or discharge of a peptide at N-terminus of connector protein that interacts with pRNA.* Nucleic Acids Res, 2006. **34**(19): p. 5482–90.

34. Shu, D., et al., *Counting of six pRNAs of phi29 DNA-packaging motor with customized single-molecule dual-view system.* EMBO J, 2007. **26**(2): p. 527–37.

35. Shu, D. and P. Guo, *A viral RNA that binds ATP and contains a motif similar to an ATP-binding aptamer from SELEX.* J Biol Chem, 2003. **278**(9): p. 7119–25.

36. Moon, J.-M., et al., *Capture and alignment of phi29 viral particles in sub-40 nanometer porous alumina membranes.* Biomed Microdevices, 2008. **11**(1): p. 135–142.

37. Moll, W.D. and P. Guo, *Grouping of ferritin and gold nanoparticles conjugated to pRNA of the phage phi29 DNA-packaging motor.* J Nanosci Nanotechnol, 2007. **7**(9): p. 3257–67.

38. Lee, T.J., C. Schwartz, and P. Guo, *Construction of bacteriophage phi29 DNA packaging motor and its applications in nanotechnology and therapy.* Ann Biomed Eng, 2009. **37**(10): p. 2064–81.

39. Guo, P., C. Peterson, and D. Anderson, *Initiation events in in-vitro packaging of bacteriophage phi 29 DNA-gp3.* J Mol Biol, 1987. **197**(2): p. 219–28.

40. Green, D.J., et al., *Self-assembly of heptameric nanoparticles derived from tag-functionalized phi29 connectors.* ACS Nano, 2010. **4**(12): p. 7651–9.

41. Fang, H., et al., *Role of channel lysines and the "push through a one-way valve" mechanism of the viral DNA packaging motor.* Biophys J, 2012. **102**(1): p. 127–35.

42. Cai, Y., F. Xiao, and P. Guo, *The effect of N- or C-terminal alterations of the connector of bacteriophage phi29 DNA packaging motor on procapsid assembly, pRNA binding, and DNA packaging.* Nanomedicine, 2008. **4**(1): p. 8–18.

43. Slayter, E. M., and H.S. Slayter, *Light and Electron Microscopy* 1992: Cambridge University Press.

44. *Wavelength of electron.* April 4, 2020]; Available from: https://www.jeol.co.jp/en/words/emterms/search_result.html?keyword=wavelength%20of%20electron.

45. Williams, D.B. and C.B. Carter, *Transmission electron microscopy.* 2009: Springer US. LXII, 775.

46. Goldstein, J., et al., *Scanning electron microscopy and X-Ray microanalysis.* 1992: Springer US 840.

47. Okunishi, E., et al., *Visualization of light elements at ultrahigh resolution by STEM annular bright field microscopy.* Microsc Microanal, 2009. **15**(S2): p. 164–165.

48. Mehta, R., *Interactions, Imaging and Spectra in SEM.* 2012.

49. McMullan, G., et al., *Comparison of optimal performance at 300keV of three direct electron detectors for use in low dose electron microscopy.* Ultramicroscopy, 2014. **147**: p. 156–63.

50. Alfredsson, V., *Cryo-TEM studies of DNA and DNA–lipid structures.* Curr Opin Colloid Interface Sci, 2005. **10**(5-6): p. 269–273.

51. Egerton, R.F., *Control of radiation damage in the TEM.* Ultramicroscopy, 2013. **127**: p. 100–8.

52. Egerton, R.F., P. Li, and M. Malac, *Radiation damage in the TEM and SEM.* Micron, 2004. **35**(6): p. 399–409.

53. Al-Amoudi, A., L.P. Norlen, and J. Dubochet, *Cryo-electron microscopy of vitreous sections of native biological cells and tissues.* J Struct Biol, 2004. **148**(1): p. 131–5.

54. Hajibagheri, M.A., *Visualization of DNA and RNA molecules, and protein-DNA complexes for electron microscopy.* Mol Biotechnol, 2000. **15**(2): p. 167–84.

55. Weisman, S., et al., *Nanostructure of cationic lipid-oligonucleotide complexes.* Biophys J, 2004. **87**(1): p. 609–14.

56. Buckhout-White, S., et al., *TEM imaging of unstained DNA nanostructures using suspended graphene.* Soft Matter, 2013. **9**(5): p. 1414–1417.

57. Kabiri, Y., et al., *Distortion of DNA origami on graphene imaged with advanced TEM techniques.* Small, 2017. **13**(31).

58. Vasquez, C. and A.K. Kleinschmidt, *Electron microscopy of RNA strands released from individual Reovirus particles.* J Mol Biol, 1968. **34**(1): p. 137–47.

59. Montoliu, L., et al., *Visualization of large DNA molecules by electron microscopy with polyamines: application to the analysis of yeast endogenous and artificial chromosomes.* J Mol Biol, 1995. **246**(4): p. 486–92.

60. Harland, D.P., et al., *Transmission electron microscopy staining methods for the cortex of human hair: a modified osmium method and comparison with other stains.* J Microsc, 2011. **243**: p. 184–196.

61. Steere, R.L., *Electron microscopy of structural detail in frozen biological specimens.* J Biophys Biochem Cytol, 1957. **3**(1): p. 45–60.

62. *Transmission Electron Microscopy (TEM).* April 16, 2020]; Available from: https://www.nottingham.ac.uk/isac/facilities/tem.aspx.

63. Thach, R.E. and S.S. Thach, *Damage to biological samples caused by the electron beam during electron microscopy.* Biophys J, 1971. **11**(2): p. 204–10.

64. Pennycook, S.J. and P.D. Nellist, *Scanning Transmission Electron Microscopy Imaging and Analysis.* 2011, New York, NY: Springer New York,.

65. Dwyer, C., et al., *Sub-0.1 nm-resolution quantitative scanning transmission electron microscopy without adjustable parameters.* Appl Phys Lett, 2012. **100**(19).

66. Leng, Y., *Materials Characterization*, in *Scanning Electron Microscopy.* 2013, Wiley-VCH: Weinheim, Germany. p. 127–161.

67. Guo, P., *Rolling circle transcription of tandem siRNA to generate spherulitic RNA nanoparticles for cell entry.* Mol Ther Nucleic Acids, 2012. **1**: p. e36.

68. *Field Emission Scanning Electron Microscopy (FESEM) – PhotoMetrics.* April 16, 2020]; Available from: https://photometrics.net/field-emission-scanning-electron-microscopy-fesem/.

69. Murphy, E.L. and R.H. Good, *Thermionic emission, field emission, and the transition region.* Phys Rev, 1956. **102**(6): p. 1464–1473.

70. Christou, A., *Comparison of electron sources for high-resolution Auger spectroscopy in an SEM.* J Appl Phys, 1976. **47**(12): p. 5464–5466.

71. Stass, R., S.L. Ilca, and J.T. Huiskonen, *Beyond structures of highly symmetric purified viral capsids by cryo-EM.* Curr Opin Struct Biol, 2018. **52**: p. 25–31.

72. Kühlbrandt, W., *The Resolution Revolution*, in *SCIENCE.* 2014, AAAS.

73. Cheng, Y., et al., *A primer to single-particle cryo-electron microscopy.* Cell, 2015. **161**(3): p. 438–449.

74. Hanske, J., Y. Sadian, and C.W. Muller, *The cryo-EM resolution revolution and transcription complexes.* Curr Opin Struct Biol, 2018. **52**: p. 8–15.

75. Kuijper, M., et al., *FEI's direct electron detector developments: Embarking on a revolution in cryo-TEM.* J Struct Biol, 2015. **192**(2): p. 179–87.

76. Pospich, S. and S. Raunser, *Single particle cryo-EM-an optimal tool to study cytoskeletal proteins.* Curr Opin Struct Biol, 2018. **52**: p. 16–24.

77. DiMaio, F. and W. Chiu, *Tools for model building and optimization into near-atomic resolution electron cryo-microscopy density maps.* Methods Enzymol, 2016. **579**: p. 255–76.

78. Chen, W., et al., *Colorimetric detection of nucleic acids through triplex-hybridization chain reaction and DNA-controlled growth of platinum nanoparticles on graphene oxide.* Anal Chem, 2020. **92**(3): p. 2714–2721.

79. Rasoulianboroujeni, M., et al., *Development of a DNA-liposome complex for gene delivery applications.* Mater Sci Eng C Mater Biol Appl, 2017. **75**: p. 191–197.

80. Mobed, A., et al., *Immobilization of ssDNA on the surface of silver nanoparticles-graphene quantum dots modified by gold nanoparticles towards biosensing of microorganism.* Microchem J, 2020. **152**.

81. Jeffs, L.B., et al., *A scalable, extrusion-free method for efficient liposomal encapsulation of plasmid DNA.* Pharm Res, 2005. **22**(3): p. 362–72.

24 A Single-Molecule FRET Approach for Investigating the Binding Mechanisms of Anti-Viral Aptamers

Saptaswa Sen and Erik D. Holmstrom
University of Kansas, Lawrence, KS, USA

CONTENTS

INTRODUCTION

HEPATITIS C VIRUS BIOLOGY

Viruses are some of the smallest biological entities on earth. They are also extremely pathogenic and can infect organisms from all three domains of life. One such human viral pathogen is the hepatitis C virus (HCV), which causes progressive liver damage, resulting in cirrhosis and cancer. According to the World Health Organization (WHO), approximately 70 million people are currently infected with the virus and in 2016 alone approximately 400,000 people died from HCV-related complications (World Health Organization 2017).

DOI: 10.1201/9781003001560-26

The virus itself is a small 50 nm lipid-enveloped particle that contains a single-stranded positive-sense RNA genome that encodes for ten viral proteins, all of which are essential for viral infection, replication, and propagation (Kato 2000). The entire viral cycle can be divided into a series of key events, including particle entry, translation, post-translational processing, RNA replication, assembly, and particle release. There is no vaccine available for HCV infection and the current standard of care involves direct-acting antiviral (DAA) agents and pegylated interferon proteins. These DAAs target non-structural proteins and disrupt the viral cycle either via the inhibition of RNA-dependent RNA polymerases (e.g., NS5B) or by specifically binding to HCV proteases (e.g., NS3) and preventing essential cleavage/processing of the HCV polyprotein. However, the efficacy of these drugs is limited by resistance-acquired mutations associated with these non-structural proteins (Alves *et al.* 2013, Gaudieri *et al.* 2009, Halfon and Sarrazin 2012, Kuntzen *et al.* 2008). Therefore, the most effective treatments use combinations of DAAs to address this growing challenge, which unfortunately further increase the cost and limits availability of these therapies (Stepanova and Younossi 2017).

The HCV core protein (HCVcp) is the least variable of all ten HCV proteins and highly conserved among the six major viral genotypes, making it an attractive target for novel therapeutics (Strosberg *et al.* 2010). It is cleaved from the first ~191 amino acids of the HCV polyprotein and is known to interact with several host proteins during the viral cycle (Dolan *et al.* 2015). This 21 kDa intrinsically disordered protein also possesses lipid and RNA-binding activities (Counihan *et al.* 2011, Santolini *et al.* 1994, McLauchlan 2000). These biomolecular interactions permit the core protein to play a critical role in the formation of infectious virus particles (Gawlik and Gallay 2014). Specifically, HCVcp is the only structural protein component of the viral nucleocapsid – a large ribonucleoprotein complex formed via electrostatic interactions between HCVcp and the single-stranded genomic RNA (de Souza *et al.* 2016). The N-terminal nucleocapsid domain (NCD) of HCVcp (Figure 24.1a and b) is composed of highly polar and positively charged amino acids.

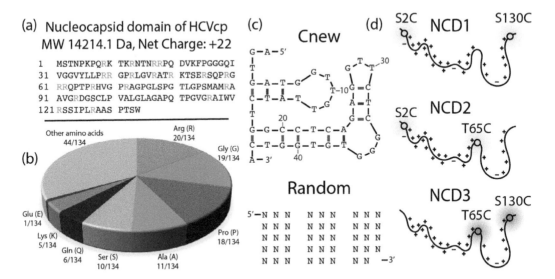

FIGURE 24.1   The primary sequence of the nucleocapsid domain and a putative anti-viral aptamer. (a) Amino acid sequence of the nucleocapsid domain (NCD) of the HCV core protein, highlighting positively charged arginine residues in orange. (b) Distribution of polar and disorder promoting amino acids associated with the NCD of HCVcp. (c) One of many mFold predicted (Zuker 2003) secondary structures of a SELEX generated 44-nt anti-viral aptamer termed Cnew. (d) Schematic representation of the three fluorescently labeled NCD variants discussed in this chapter. Relative positions of positively (+) and negatively (-) charged amino acids are shown along the polypeptide chains.

Importantly, the NCD is both necessary and sufficient for the formation of nucleocapsid-like particles (Majeau *et al.* 2004). Therefore, strategic inhibition of nucleocapsid assembly by disrupting the function of HCVcp, specifically the function of the NCD, represents a potentially promising approach for the development of novel HCV therapies (Strosberg *et al.* 2010). Additionally, the ability to disrupt nucleocapsid assembly with molecular therapeutics would also be useful for fundamental biochemical and biophysical studies of HCVcp, which would ultimately give rise to a more complete understanding of the series of events associated with the formation of infectious virus particles. Over the past 10 years, numerous molecules have been identified as potential HCVcp inhibitors, including low molecular weight compounds (Ni *et al.* 2011) and small peptides (Kota *et al.* 2009). In this chapter, we focus on the use of nucleic acid aptamers as prospective candidates for the inhibition of HCV nucleocapsid assembly.

Nucleic Acid Aptamer Therapies

Nucleic acid-based drugs (e.g., micro-RNAs, interfering RNAs, and antisense oligonucleotides) offer numerous opportunities for both academic and industrial endeavors. For example, aptamers are a promising class of molecular therapeutics, whose binding properties are comparable to that of antibodies and their various mimetics (Chen and Yang 2015, Keefe *et al.* 2010, Kong and Byun 2013, Sun *et al.* 2014). Particularly beneficial features of nucleic acid aptamers include high biocompatibility and solubility, numerous potential covalent modifications, and facile chemical synthesis. Most importantly, these oligonucleotides can also adopt a wide range of conformations that are stabilized via Van der Waals and hydrogen-bonding interactions, which lead to the formation of intricate secondary and tertiary structures. This structural diversity has made it possible to identify specific oligonucleotide sequences (i.e., aptamers) that bind to biomolecular targets (e.g., proteins, drugs, metal ions, antibiotics, amino acids) from random libraries of approximately 10^{15} sequences using a common *in vitro* selection method called SELEX (Ellington and Szostak 1990, Tuerk and Gold 1990).

The binding properties of aptamers have been exploited for a variety of therapeutic purposes. As an example, the first FDA-approved aptamer-based drug was Pegabtanib (Macugen), a small RNA aptamer that functions as an antagonist of the vascular endothelial growth factor (VEGF) and is used to treat age-related macular degeneration (Ng *et al.* 2006). Since then, aptamers have been developed for a wide range of medical conditions, including thrombosis, cancer, diabetes, and autoimmune diseases (Keefe *et al.* 2010). They are also being used to disrupt the function of viral proteins and could possibly be utilized as antiviral agents in the future (Zou *et al.* 2019). One such example is an anti-viral aptamer for HCV (Shi *et al.* 2014). Shi and colleagues used SELEX to identify Cnew, a 44-nt anti-viral DNA aptamer that preferentially binds to the HCVcp (Figure 24.1c) and inhibits production of HCV particles in Huh7.5 cells.

Over the past decades, there have been multiple studies on SELEX-derived anti-viral aptamers (Zou *et al.* 2019) and many of these are thought to bind to their targets via electrostatic and hydrogen-bonding interactions, as well as via shape complementarity. However, the molecular interactions governing the binding of Cnew to HCVcp are not well known. Additionally, it is not clear how aptamer binding affects the conformational properties (and thus function) of HCVcp. Therefore, we set out to address several unanswered questions related to the molecular interactions between the HCVcp and the Cnew anti-viral aptamer: Where does the aptamer bind? What types of interactions are important for bindings? Does aptamer binding change the conformation of the HCVcp? To address these questions, we used single-molecule fluorescence spectroscopy and Förster Resonance Energy Transfer (FRET) to monitor the structural and energetic properties of this nucleic acid–protein interaction. This spectroscopic approach is particularly well suited for this task because it enables quantitative biophysical investigations of heterogenous molecular ensembles at low nanomolar sample concentrations, which are not suitable for other approaches.

SINGLE-MOLECULE FLUORESCENCE SPECTROSCOPY

Classical structural approaches such as X-ray crystallography probe large molecular ensembles; and therefore, they only provide information about the average properties of the entire population. The properties of any one molecule cannot be identified in these ensemble approaches, and therefore it is challenging to study the various sub-populations of a heterogenous molecular ensemble (e.g., folded/unfolded) at the same time and under the same experimental conditions. Although the atomic information from these ensemble structural approaches is incredibly informative, they generally require high sample concentrations and/or specific solution conditions. Therefore, these classical approaches are not well suited for studies of conformationally heterogenous biomolecules, particularly those that are known to oligomerize at high concentrations. These drawbacks can be avoided by studying biomolecular interactions at the single-molecule level.

Single-molecule fluorescence spectroscopy is a well-established technique for probing the structural, energetic, and dynamic properties of biomolecules at the single-molecule level (Lerner *et al.* 2018). Replication, transcription, translation, and many other biomolecular processes accompanied by large conformational changes have been studied using this powerful approach. In most cases, a fluorophore is attached to a specific residue within the molecule of interest. Then, the fluorescence is monitored using a variety of experimental approaches to obtain structural, energetic, and dynamic properties of the fluorescently labeled molecules. The observables in single-molecule fluorescence studies can include spatial localization, fluorescence intensity, fluorescence lifetime, and fluorescence anisotropy.

The inclusion of a second fluorophore on the molecule of interest enables studies based on FRET. Briefly, FRET is a non-radiative photophysical process that permits energy transfer from an excited donor fluorophore to a spatially adjacent (e.g., 1–10 nm) ground state acceptor fluorophore. Perhaps most importantly, the efficiency of energy transfer between the two fluorophores has a strong inverse-distance dependence (Förster 1948). For this reason, the fluorophores in a typical FRET experiment form a nanoscale "spectroscopic ruler" that can be used to measure distances within and between biological macromolecules (Stryer 1978, Stryer and Haugland 1967).

However, one complication associated with ensemble FRET experiments is the presence of fluorescently labeled molecules that lack either an active donor or acceptor fluorophore, resulting in increased sample heterogeneity. These donor-only and acceptor-only impurities arise from imperfect fluorophore coupling reactions and/or unavoidable photochemistry. One of the many advantages associated with single-molecule measurements is that these FRET inactive species can be identified and excluded from data analysis. This can be accomplished using two-color excitation schemes to independently excite the donor and the acceptor fluorophores either via alternating excitation (ALEX) from continuous-wave lasers (Kapanidis *et al.* 2005) or interleaved excitation (PIE) from pulsed lasers (Müller *et al.* 2005). Importantly, this practice has helped to greatly advance single-molecule FRET-based research by enabling robust quantitative measurements of transfer efficiency (Hellenkamp *et al.* 2018, Holmstrom *et al.* 2018), which can then be quantitatively mapped back to molecular dimensions using the well-known distance dependence of FRET (Holmstrom *et al.* 2018).

In this chapter, we describe how single-molecule FRET can be used to study aptamer-induced conformational changes within the NCD of HCVcp. First, we measure the transfer efficiency of three FRET-labeled variants (Figure 24.1d) as a function of different solute concentrations to: (*i*) demonstrate our ability to detect conformational changes within NCD and (*ii*) characterize the molecular properties of the free form of this protein. Our single-molecule FRET results suggest that the dimensions of NCD are highly dependent on solution conditions with compact structural ensembles favored at high concentrations of salt and expanded ensembles favored at high concentrations of denaturant. Notably, the solute-dependent dimensions of these variants are similar to those of other highly charged intrinsically disordered proteins (Müller-Späth *et al.* 2010), indicating that repulsive electrostatic interactions modulate the dimensions of the nucleocapsid domain of HCVcp. Additionally, we show that a nucleic acid aptamer identified by Shi and colleagues binds to NCD with sub-nanomolar affinity and leads to a pronounced compaction of the protein. From these results, we propose a simple sequestration model for the inhibitory mechanism of these anti-viral aptamers that may one day be exploited as novel therapeutic agents for the treatment of HCV. Finally, our work provides a platform for future experiments that aim to further understand the molecular mechanisms associated with a wide range of nucleic acid aptamers and their biological targets.

EXPERIMENTAL APPROACH

BIOCONJUGATION TECHNIQUES

One of the key determinants of any single-molecule FRET experiment is sample quality. Therefore, the PFAM database and GenBank were used to identify a suitable variant (CAE46584.1) of the nucleocapsid domain (NCD) of the HCV core protein (Colina *et al.* 2004). The amino acid sequence associated with this clinical isolate (Figure 24.1a) was further modified to be maximally compatible with our single-molecule FRET experiments. First, the tryptophan residue at position 134 was removed to limit quenching of fluorophores proximal to the C-terminus of the protein. Then, in order to maximize our ability to incorporate fluorophores at specific locations within the protein via cysteine-maleimide conjugation chemistry, the less well-conserved cysteine residue at position 98 was converted to a methionine. Next, pairs of cysteine residues were systematically introduced at different locations (S2C, T65C, S130C) within the core protein for labeling. We expressed and purified three dual-cysteine variants of the NCD (Figure 24.1d): (*i*) S2C-S130C (NCD1); (*ii*) S2C-T65C (NCD2); and (*iii*) T65C-S130C (NCD3), following previously established protocols (Holmstrom *et al.* 2019, Holmstrom *et al.* 2018, Kunkel *et al.* 2001). For the NCD1 variant, the two cysteines are located close to the N- and C- termini so that the dimensions of the entire polypeptide chain could be probed via FRET. The NCD2 and NCD3 variants were used to independently study the molecular dimensions of the N- and C-terminal halves of the nucleocapsid domain within the context of the entire amino acid sequence. Notably, NCD2 and NCD3 exhibited markedly different values for the net charge per residue (+ 0.25 and + 0.09, respectively), particularly when compared to that of NCD1 (+ 0.16), which allowed us to assess how charged residues influence the behavior of the HCVcp. Importantly, the modification and purification of these proteins did not impact their biochemical function as they were still able to form nucleocapsid-like particles (Holmstrom *et al.* 2018) and chaperone viral genome dimerization (Holmstrom *et al.* 2019), which is another reported function of this protein.

The choice of donor and acceptor fluorophores for FRET can also greatly influence sample quality (Zosel *et al.* 2020). For the NCD variants in this chapter, fluorophores were chosen to minimally perturb their overall charge distribution of the protein. Additional factors influencing the final choice were also the photostability and quantum yields of the two fluorophores, as well as the absorption and emission spectra. Each dual-cysteine variant was simultaneously labeled with two molar equivalents of the maleimide-functionalized donor (Cy3b) and acceptor (CF660R) fluorophores. Reverse phase-HPLC was used to verify that each variant was coupled to donor and acceptor fluorophores (Holmstrom *et al.* 2018). Unfortunately, it was not possible to achieve complete chromatographic separation of the various labeling permutations – donor–donor (i.e., donor-only), acceptor–acceptor (i.e., acceptor-only), acceptor–donor (i.e., FRET-labeled), and donor–acceptor (i.e., FRET-labeled) – as they had very similar retention times. Nevertheless, after chromatographic enrichment, the FRET-labeled variants were lyophilized, resuspended to a final concentration of ~ 8 nM in buffered aqueous solution (25 mM HEPES pH 7.4; 0.01% (v/v) Tween 20; and 800 nM unlabeled HCV protein), and stored at −70 °C until use.

FLUORESCENCE EXCITATION SCHEME

The fluorescence excitation scheme is another important aspect of single-molecule FRET experiments. In our study, we used the alternating excitation method (ALEX) to independently excite the donor and acceptor fluorophores, thereby allowing us to: (*i*) accurately monitor the transfer efficiency of FRET-labeled NCD variants and (*ii*) rigorously study the molecular interactions between this domain and an unlabeled anti-viral aptamer. Direct excitation of the donor and acceptor fluorophores was achieved using dual-mode laser diodes (Omicron, Rodgau, Germany) operated in the continuous-wave configuration at 515 nm and 642 nm, respectively (Figure 24.2). To achieve

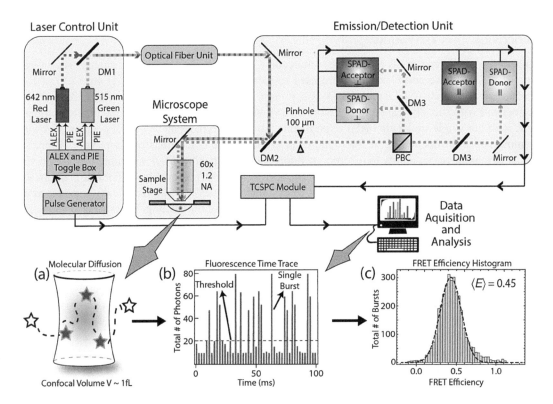

FIGURE 24.2 Schematic diagram of the confocal setup used for single-molecule FRET experiments. The alternating segments of collimated continuous-wave laser light (red and green) are coaxially aligned and directed towards a fiber-coupling unit using a dichroic mirror (DM1). A dual-band dichroic mirror (DM2) directs the laser light into the body of the microscope, where it is then focused into the sample by a high NA water objective. The fluorescence emitted from the sample is collected by the same objective, directed back through the major dichroic mirror (DM2), and focused by a lens through a 100 μm pinhole to reject out of plane fluorescence. The fluorescence that passes through the pinhole is split using a polarizing beam splitting cube (PBC). The polarized fluorescence is further split by two long-pass dichroic mirrors (DM3) to separate donor and acceptor fluorescence. Photons are detected by four single-photon avalanche diodes (SPAD), with the photon arrival times recorded by a time-correlated single photon counting (TCSPC) module. The arrival time, color, and polarization of each photon is used to calculate the transfer efficiency (E) of the fluorescently labeled biomolecules. Two key aspects to detecting single molecules include low concentrations and small detection volumes. (a) Schematic representation of confocal volume of the setup with a FRET-labeled molecule stochastically diffusing in the solution. (b) Diagram of a fluorescence time trace from a single-molecule FRET measurement where green and red bars represent the fraction of all detected donor and acceptor photons, respectively. (c) A representative FRET efficiency histogram resulting from the data analysis procedure.

alternating excitation, the continuous-wave output from each of the two lasers was modulated (i.e., ON/OFF) using a BNL Model 577-4C pulse generator (Berkley Nucleonics Corp., USA). Specifically, the two lasers were synchronized to a common 20 kHz signal using a fixed ON/OFF duty cycle for the green (88%) and red (10%) lasers. This excitation scheme generated alternating segments of continuous-wave light from the green (44 μs) and red (5 μs) diodes, with each segment followed by 0.5 μs where both lasers are OFF. The laser beams were coaxially aligned using a dichroic mirror (Figure 24.2 DM1) and then coupled into a single-mode optical fiber. After exiting the optical fiber, the average excitation powers were 85 μW for the green laser and 30 μW for the red laser. The coaxial beams were directed towards the MicroTime 200 confocal microscope system (PicoQuant, Berlin, Germany) using a dual-band dichroic mirror (Figure 24.2 DM2; ZT532/640rpc-UF3, Chroma Technology Corp., Bellows Falls, VT, USA). The 1.2 NA 60× water-immersion objective (UPLSAPO 60X, Olympus, Tokyo, Japan) associated with microscope system focused the light from the two excitation sources to a diffraction-limited spot approximately 50 μm inside the sample, which was held in place using the sample stage.

FLUORESCENCE EMISSION AND DETECTION

To facilitate observation of single molecules, the final concentration of fluorescently labeled NCD was maintained at ~ 80 pM. At these concentrations, the average number of fluorescently labeled molecules in the ~ 1 fL diffraction-limited excitation volume is significantly less than 1 (Figure 24.2a); and therefore, most of the time we only detect low levels of background fluorescence. However, stochastic diffusion of individual molecules through the focused alternating laser beams produces bursts of photons from the donor and acceptor fluorophores that are well above background levels. The high numerical aperture objective collects some of the donor and acceptor fluorescence and directs it back through the major dichroic mirror (Figure 24.2, DM2). This light is then focused through a 100 μm confocal pinhole using a lens. The pinhole rejects any out-of-plane fluorescence emitted from the sample solution and helps define the boundaries of the confocal volume (Figure 24.2a). After passing through the pinhole, the fluorescence photons are spatially separated first by a PBC and then split again by two long pass dichroic mirrors (DM3; 635LPXR, Chroma) to distinguish between donor and acceptor photons. The net result is four spatially separated photon streams (i.e., horizontally and vertically polarized light for both donor and acceptor emission). Bandpass filters are used to reject unwanted fluorescence from both donor (690/70 BrightLine, Semrock) and acceptor (582/64 BrightLine, Semrock) photon streams. Individual photons in the four streams are then detected by single-photon avalanche diodes (SPAD; SPCM-AQRH-14-TR, Excelitas Technologies, MA, USA). Finally, the arrival times of detected photons are processed and recorded with picosecond resolution using a TCSPC module (Figure 24.2; HydraHarp 400, PicoQuant) that is synchronized with the alternating excitation sources via the 20 kHz signal from the pulse generator.

DATA ANALYSIS

Fluorescence time traces (Figure 24.2b) were recorded for 5 to 10 minutes, which was sufficient for collection of at least 800 high-quality bursts of fluorescence from individual FRET-labeled molecules. The four steams of photon arrival times were analyzed in Mathematica (Version 12, Wolfram Research) using Fretica – a custom-built software package (Schuler Lab, University of Zurich, Switzerland). During data analysis, photon arrival times were divided into 500 μs time bins to construct a fluorescence time trace (Figure 24.2b). The average background signal was determined from all time bins with fewer than 20 total photons resulting from both donor (^{D,ex}) and acceptor (^{A,ex}) excitation. Any time bin with 20 or more photons was defined as a burst of fluorescence. Two ratiometric parameters were calculated for each burst: the stoichiometry (S) and the transfer efficiency (E). The fluorescence stoichiometry was calculated for each burst by dividing the total number of photons detected after donor excitation ($^{D,ex}n_{tot}$) by the sum of all photons detected after both donor and acceptor excitation: $S = {}^{D,ex}n_{tot}/({}^{A,ex}n_{tot} + {}^{D,ex}n_{tot})$. In this way, bursts arising from individual molecules with only one active fluorophore, referred to as donor-only or acceptor-only, will have stoichiometry values near 1 or 0, respectively. One of the many advantages of this dual-excitation technique is that it is possible to restrict data analysis to bursts with intermediate stoichiometry values near 0.5, which arise from FRET-labeled molecules containing both an active donor fluorophore and an active acceptor fluorophore. The transfer efficiency (E) for each of these FRET-labeled bursts can be calculated using the following equation: $E = {}^{D,ex}n_A/({}^{D,ex}n_A + {}^{D,ex}n_D)$. Here, the number of donor ($^{D,ex}n_D$) and acceptor ($^{D,ex}n_A$) photons resulting from donor excitation is corrected to account for: background, spectral crosstalk, and direct excitation of the acceptor fluorophore by the donor excitation source, as well as non-uniform excitation and detection efficiencies of the two fluorophores (Holmstrom *et al.* 2018). Finally, FRET efficiency histograms (Figure 24.2c) were constructed from all the high-quality bursts, which contain more than 50 total photons after applying the corrections described above. From here, the mean transfer efficiencies, $\langle E \rangle$, of the FRET-labeled molecules are obtained by fitting the FRET efficiency histograms to Gaussian and/or Log-Normal distributions.

MOLECULAR DIMENSION OF FREE HCV CORE PROTEIN

In order to further understand how various molecular therapeutics, specifically anti-viral aptamers, alter the conformation and thus function of HCVcp, it is important to establish a clear understanding of how this protein behaves in the absence of these compounds. To accomplish this task, we studied three fluorescently labeled variants of the nucleocapsid domain (NCD) of HCVcp (i.e., NCD1/2/3) under a wide range of solution conditions in the absence of aptamer. These studies also demonstrate that our single-molecule FRET approach can easily resolve changes in the molecular dimensions of the proteins caused by varying the solution conditions.

DEPENDENCE ON SOLUTION CONDITIONS

In the first set of experiments, we set out to determine how the molecular dimensions of the three NCD variants change in response to various chemical denaturants. For this, we choose to use guanidinium chloride (GdmCl), Urea, and KCl. GdmCl is generally considered one of the most effective chemical denaturants (Möglich *et al.* 2005). It is also a salt and therefore its ionic components can also modulate electrostatic interactions (e.g., salt bridges) within biomolecules. Figure 24.3 shows representative FRET efficiency histograms for the three NCD variants at increasing concentrations of GdmCl. At all concentrations of GdmCl (i.e., 0.01 M – 6.0 M), the histograms for each of the three variants contain a single dominant peak. This observation is consistent with the notion that HCVcp remains intrinsically disordered over wide range of experimental conditions and does not have distinct conformational ensembles separated by large free energy barriers (e.g., folded/unfold). These peaks were fit to a single Gaussian function to determine the mean transfer efficiency, $\langle E \rangle$, for the specified experimental conditions.

For the end-labeled variant, NCD1, the mean transfer efficiency increased from $\langle E \rangle = 0.27$ to $\langle E \rangle = 0.32$ with increasing concentrations from 0.05 M to 0.5 M GdmCl, indicative of molecular compaction (Figure 24.3a). However, further increasing the GdmCl concentrations resulted in lower values of $\langle E \rangle$ corresponding to an expansion of this disordered protein, which has previously been observed for the unfolded states of globular proteins and for many disordered proteins (Best 2020, Schuler and Eaton 2008, Stumpe and Grubmüller 2007, Uversky 2019, Ziv and Haran 2009). The location of the fluorophores in the NCD2 variant specifically probes the dimensions of the positively charged N-terminal amino acids. For this variant, we also observed a non-monotonic "rollover" of the $\langle E \rangle$ values. Specifically, the mean transfer efficiency increased from $\langle E \rangle = 0.49$ at 0.05 M GdmCl to $\langle E \rangle = 0.55$ at 0.5 M GdmCl with higher concentrations again leading to lower $\langle E \rangle$ values (Figure 24.3b). However, the compaction observed at low concentrations of GdmCl was mostly

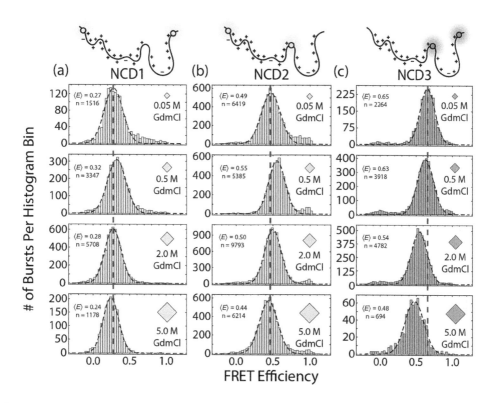

FIGURE 24.3 Representative single-molecule FRET efficiency histograms for the three NCD variants. Histograms for (a) NCD1 (gold), (b) NCD2 (teal), and (c) NCD3 (magenta) show how GdmCl (diamond) influences the molecular dimensions of the three variants. The black dashed lines represent Gaussian fits used to extract mean transfer efficiency, $\langle E \rangle$, values. The vertical red dashed lines indicate the $\langle E \rangle$ for each variant at 0.05 M GdmCl. Schematic representations of the labeling positions and charge densities for each of the fluorescently labeled proteins are depicted above the histograms.

absent in the experiments conducted with the C-terminally labeled variant, NCD3, which predominately displayed a denaturant-induced expansion at high concentrations (Figure 24.3c).

The non-monotonic "rollover" prominently observed in the high charge density segments of HCVcp is consistent with the notion that GdmCl can interact with proteins in at least two ways (Müller-Späth *et al.* 2010). First, the guanidinium and chloride ions can screen electrostatic interactions within highly charged sequences, which, for a positively charged segment of amino acids (e.g., NCD 2), would lead to a collapse of the conformational ensemble. Additionally, the guanidinium ion can also interact with and alter the solvation of the peptide backbone, leading to more expanded conformational ensembles. To further test the notion that the initial compaction observed at low concentrations of GdmCl was due largely to electrostatics, we also recorded FRET efficiency histograms for all three variants at increasing concentrations of KCl (Figure 24.4). Indeed, all three variants displayed a monotonic collapse to more compact conformational ensembles, with the extent of collapse governed by the number of charged residues between the two labeling site. Finally, we used urea as a non-ionic denaturant to further support the notion that high concentrations of denaturants give rise to expanded conformations of NCD. Indeed, higher concentrations of urea produce lower values of $\langle E \rangle$ regardless of where the fluorophores are located in the protein.

Our initial investigations of this highly positively charged intrinsically disordered protein reveal that its molecular dimensions are strongly dependent on the solution conditions. These studies also demonstrate our ability to precisely monitor conformational changes within HCVcp using

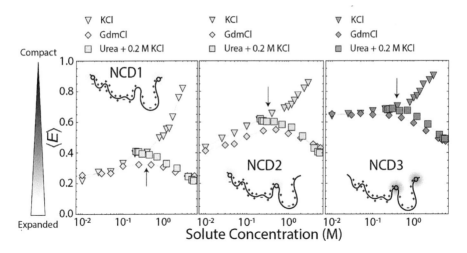

FIGURE 24.4 Solute-dependent expansion and compaction of the NCD variants. Dependence of the mean transfer efficiency, $\langle E \rangle$, on the concentrations of KCl (triangle), GdmCl (diamond), and urea (square) for the three variants: NCD1 (gold), NCD2 (teal), and NCD3 (magenta). The typical uncertainty in $\langle E \rangle$ for individual data point is ± 0.03. Schematic representations of the labeling positions and charge densities for each of the fluorescently labeled proteins are depicted in each of the data plots. The onset of denaturant-induced expansion is denoted with black arrows.

single-molecule fluorescence spectroscopy. Importantly, this approach provides us with a robust experimental foundation to begin characterizing the interactions between anti-viral aptamers and our three fluorescently labeled variants of the NCD of HCVcp.

MOLECULAR DIMENSION OF APTAMER-BOUND HCV CORE PROTEIN

The binding of ligands to macromolecules is an important aspect of most biochemical processes inside living cells. Such interactions are also central to the function of nearly all pharmaceuticals. With a rapidly growing population and newly emerging health concerns, there is a constant need for novel molecular therapeutics to combat a diverse range of medical conditions. Fortunately, molecular therapeutic agents derived from SELEX-generated aptamers open multiple avenues for treating such diseases. Single-stranded nucleic acid aptamers isolated via SELEX can fold into unique secondary and tertiary structures with high binding affinity and specificity towards various biological targets such as ATP, metal ions, and macromolecules (McKeague and Derosa 2012). Previously, Shi and colleagues used SELEX to identify a 44-nt aptamer (Cnew) that binds to HCVcp. Their Huh7.5 cell culture experiments showed that the presence of Cnew inhibits formation of infectious virus particles by disrupting the localization of HCVcp with lipid droplets and NS5A, as well as by preventing HCVcp from associating with the viral RNA and forming nucleocapsid particles. However, the molecular details and physical properties of the aptamer–HCVcp interaction remain unknown. For example, it is not known if/how aptamer binding alters the conformational properties (and thus function) of HCVcp. This lack of biophysical information has inspired us to begin characterizing this novel inhibitory interaction between the core protein and an anti-viral aptamer using a single-molecule FRET approach.

COMPACTION OF HCV CORE PROTEIN

FRET efficiency histograms were generated for each of the three NCD variants at increasing concentrations of Cnew (Figure 24.5). The FRET efficiency histograms for all three variants changed significantly upon addition of aptamer. In all three variants, addition of Cnew gave rise to an additional

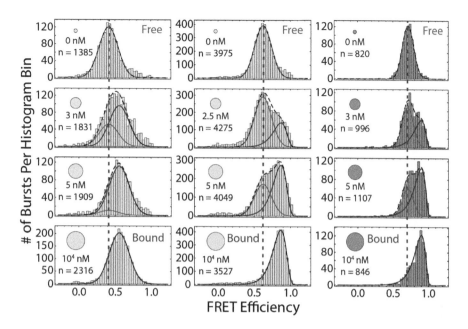

FIGURE 24.5 Anti-viral aptamer induces a global conformational change in HCVcp. FRET efficiency histograms of the NCD variants at various aptamer (circle) concentrations: NCD1 (gold); NCD2 (teal), and NCD3 (magenta). The red and blue solid lines represent fits used to extract the mean transfer efficiencies, $\langle E \rangle$, of the free or the bound form of the proteins. The increased $\langle E \rangle$ associated with the bound conformation suggests that aptamer binding causes a global compaction of NCD. The black dashed lines show the overall fit of the histograms for each protein. The vertical red dashed lines have been positioned to indicate the $\langle E \rangle$ associated with the unbound form of each NCD variant.

higher FRET efficiency population, presumably corresponding to the aptamer-bound conformation of NCD. This bound population was separated from the free population by $\Delta\langle E \rangle$ values of ~ 0.15, 0.25, and 0.16 for NCD1, NCD2, and NCD3, respectively. Given the polyanionic nature of the phosphodiester backbone, the compaction arising from aptamer binding is certainly consistent with the compaction observed with the addition of KCl and supports the notion that electrostatic interactions modulate the molecular dimensions of HCVcp. Specifically, the Cnew aptamer has a large negative net charge, which, when bound to the NCD of HCVcp, will screen the repulsive interactions between the numerous arginine residues within the core protein forming a macromolecular counterion complex (Holmstrom *et al.* 2019). In summary, the changes in the FRET efficiency histograms suggest that Cnew binding causes a global collapse of this nucleoprotein complex, a newly emerging theme for various disordered polyelectrolyte complexes (Schuler *et al.* 2019).

BINDING AFFINITY

To quantitatively assess the binding affinity between Cnew and the NCD of HCVcp, we returned to the FRET efficiency histograms of the three variants. The histograms from measurements at the highest (10^4 nM) and lowest (0 nM) concentrations of Cnew were fitted with a Log-Normal or Gaussian function to determine the location and width of the bound and free populations, respectively (Figure 24.5). All other histograms with intermediate aptamer concentrations were fitted with a sum of these two functions, with the positions and widths fixed to the values corresponding to the two limiting cases mentioned above. From this analysis, we were able to determine that the addition of the Cnew aptamer decreases the amplitude of the low FRET efficiency (e.g., free) population

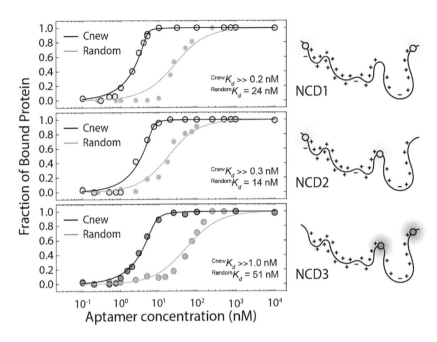

FIGURE 24.6 Binding isotherms for HCVcp variants as a function of oligonucleotide concentrations. Binding curves for the interaction of NCD1, NCD2, and NCD3 with the Cnew aptamer (dark circles) and the random library of 44-nt oligos (light circles). Schematic representations of the labeling positions and charge densities of each of the fluorescently labeled proteins are depicted next to each panel.

and increases the amplitude of the high FRET efficiency (e.g., bound) population. The widths and amplitudes of the fitted peak functions were used to calculate the relative area of the high FRET distribution, which reports on the fraction of bursts associated with molecules in the bound population. The fraction of bound HCVcp molecules was then plotted as a function of anti-viral aptamer concentration (Figure 24.6). These data were then fit to the quadratic form of the single-site binding equation (Equation 26.1) to determine the dissociation constant, K_d.

$$\text{Fraction bound}\left(L_{tot}\right) = \frac{\left(P_{tot} + L_{tot} + K_d\right) - \sqrt{-\left(4 \cdot L_{tot} \cdot P_{tot}\right) + \left(P_{tot} + L_{tot} + K_d\right)^2}}{\left(2 \cdot P_{tot}\right)} \quad (24.1)$$

In Equation 26.1, P_{tot} is the total core protein concentration, K_d is the dissociation constant, and L_{tot} is the total aptamer concentration. Surprisingly, the dissociation constants for binding of Cnew to all three NCD variants were well below P_{tot} (i.e., 8 nM). Therefore, we were only able to identify upper limits for the binding affinities between Cnew and NCD1 ($K_d \ll 0.2$ nM), NCD 2 ($K_d \ll 0.3$ nM) and NCD3 ($K_d \ll 1$ nM) based on the uncertainties associated with the fitting analysis.

BINDING SPECIFICITY

Given the large opposite charge associated with Cnew and the NCD of HCVcp, formation of the complex is likely to be governed by electrostatic interactions, and thus we wanted to determine if the NCD could bind to any 44-nt oligo or if there was a strong preference for the Cnew aptamer. To evaluate the specificity of this molecular interaction, we conducted a similar set of experiments using a random library of 44-nt single-stranded DNAs instead of the specific sequence associated with Cnew. Elevated concentrations of the random oligonucleotides also caused the unbound population to transition to a high FRET efficiency population similar to that of the NCD-Cnew complex, suggesting that each of the three variants can bind to a wide range of DNA sequences. For example, the oligonucleotide-induced change in $\langle E \rangle$ associated with the NCD3 variant was $\Delta\langle E \rangle = 0.16$ for the Cnew aptamer and $\Delta\langle E \rangle = 0.17$ for the random library, which suggests that the molecular dimensions of these two nucleoprotein complexes are quite similar.

A quantitative analysis of the fraction of bursts arising from NCD molecules in the bound conformation indicates that increasing concentrations of the random 44-nt oligo library shifts the conformational equilibrium of all three variants towards this high-FRET population. However, the data points in these binding curves depict a very different transition midpoint than what was observed for Cnew, which suggests that the dissociation constant for the random oligo library is significantly higher than the dissociation constant for the anti-viral aptamer. Indeed, after fitting the data we arrive at K_d values of 24 ± 3 nM, 14 ± 2 nM, and 51 ± 6 nM for the NCD1, NCD2, and NCD3 variants, respectively. From this analysis, we conclude that the sequence and structure of the Cnew aptamer allows it to preferentially bind to the NCD of HCVcp, but that any 44-nt sequence can compact this positively charged protein, resulting in macromolecular counterion complexes with similar dimensions.

DISCUSSION AND CONCLUSIONS

MODEL FOR INHIBITION

In this chapter, we described how a single-molecule fluorescence approach was used to begin characterizing the binding mechanism associated with a SELEX derived anti-viral aptamer and its biological target, the nucleocapsid domain (NCD) of the HCV core protein (HCVcp). Our results indicate that this 44-nt anti-viral aptamer binds to NCD with sub-nanomolar affinity under near physiological concentrations of monovalent cations. Upon binding, the negatively charged phosphate groups within the aptamer screen the numerous positive charges within the protein, allowing this intrinsically disordered polyelectrolyte to sample more compact conformations.

Previously, micromolar concentrations of tRNAs (Kunkel et al., 2001), short non-cognate DNAs (de Souza *et al*. 2016), and small structured RNA elements from the HCV genome (Holmstrom *et al*. 2018) have all been reported to induce formation of nucleocapsid-like particles. However, in our experiments, where the anti-viral aptamer concentration was as high as 10 µM, we never observed higher-order assemblies that would have been representative of nucleocapsid-like particles. Our findings suggest that perhaps the structure of Cnew prevents it from being incorporated into nucleocapsids-like particles. The above observations, together with the findings of Shi and colleagues, suggest that the mechanism of inhibition associated with the anti-viral aptamer, Cnew, likely involves sequestration of HCVcp in small compact nucleoprotein complexes that cannot readily assembly into infectious nucleocapsid particles (Figure 24.7).

Interestingly, in their cell culture experiments with Cnew, Shi and colleagues observed the emergence of a specific resistance associated mutation in HCVcp (V31A). The details of specific amino acid residues involved in the interaction between HCVcp and Cnew remain to be explored. Further investigation of residue-specific interactions and their role in aptamer resistance will be essential for future developments of anti-viral aptamers targeting the HCVcp. Such investigations will also likely unveil new information about the order of events driving the nucleocapsid assembly process. In this way, future mechanistic investigations of interactions between nucleic acid aptamers and their biological targets using quantitative single-molecule approaches will undoubtedly provide valuable insights into the use of nucleic acid aptamers as therapeutic agents against a wide range of ailments.

FUTURE PERSPECTIVES

Engineered and naturally occurring aptamers are an important class of functional nucleic acids with potential applications in many areas of scientific research. During the past decades, aptamer technology

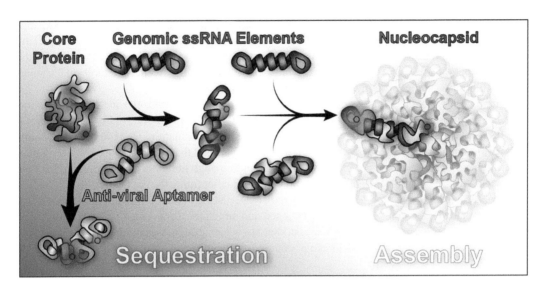

FIGURE 24.7 The proposed model for the inhibition of nucleocapsid assembly via sequestration of HCV core protein by an anti-viral aptamer. The HCV core protein (HCVcp; blue) is a positively charged intrinsically disordered polypeptide chain. HCVcp binds to structured single-stranded RNA elements within the HCV genome (red) forming compact ribonucleoprotein complexes (Holmstrom *et al.* 2018). At higher concentrations of genomic RNA, these ribonucleoprotein intermediates begin to assemble together forming much larger nucleocapsid-like particles, wherein the HCVcp adopts an expanded conformation (Holmstrom *et al.* 2018). As highlighted in this chapter, the core protein can also interact with an anti-viral aptamer forming a compact nucleoprotein complex that does not assembly into higher order structures. Thus, in the context of an infected host cell containing high concentrations of anti-viral aptamers, the newly expressed HCVcp will be sequestered by the aptamer in compact nucleoprotein complexes instead of being allowed to productively interact with the genomic RNA. The relative compactness of each of the various states of HCVcp is indicated by red and green circles corresponding to the fluorophores in our single-molecule FRET measurements.

has been used in a wide range of therapeutic applications associated with different diseases (Ulrich and Wrenger 2009, Keefe *et al.* 2010). However, these aptamers come with a unique set of pharmacological challenges (e.g., target delivery, aptamer stability) that have hampered their therapeutic utility and only a few select aptamers have entered the later phases of clinical trials. Importantly, a detailed mechanistic understanding of the molecular interactions associated with the therapeutic function of nucleic acid aptamers will almost certainly help to overcome some of these clinical challenges.

As we described in this chapter, single-molecule fluorescence methods are uniquely positioned to provide new insights into the molecular mechanisms underlying the formation of aptamer-target complexes. The insights from these types of biophysical and biochemical investigations will further our understanding of the inhibitory mechanisms and structure–activity relationships associated with aptamer-based therapeutics that target critical biological processes like viral nucleocapsid assembly. This information will not only aid in the development of new and potentially more-effective inhibitors, but it can also be used to enhance our fundamental understanding of the biological processes targeted by these therapeutic nucleic acids.

MATERIALS AND METHODS

The reagents used in this study were of the highest purity available from Sigma-Aldrich Corporation. The single-stranded oligonucleotides (i.e., Cnew and Random) were custom-synthesized by Integrated DNA Technology (Iowa, USA) The sequence for the Cnew anti-viral aptamer is 5′-AGT

GAT GGT TGT TAT CTG GCC TCA GAG GTT CTC GGG TGT GGT CA-3′. The sequence for the 44-nt random oligo library is 5′-N (NNN)$_{14}$ N-3′, where N represents equimolar incorporation of the four possible nucleobases (i.e., A, G, C and T). Oligonucleotide stocks were stored at −20 °C in 25 mM HEPES buffer pH 7.4 containing 1 mM EDTA to avoid degradation.

All single-molecule measurements were conducted in buffered aqueous solutions containing 25 mM HEPES (pH 7.4) and the specified concentrations of various solutes (i.e., GdmCl, Urea, KCl), as well as 0.01% *v/v* Tween20 and 8 nM unlabeled HCVcp. The unlabeled HCVcp and Tween20 served to inhibit the adsorption of the fluorescently labeled NCD variants onto the surface of the microscope slides (15 well Angiogenesis µ-slides, Ibidi, USA), which permitted longer measurement times. For the HCVcp-aptamer binding experiments, all measurements contained 200 mM KCl. Additionally, an incubation time of at least 5 minutes was necessary for sample equilibration prior to each measurement. All the experiments were carried out at 22 °C.

ACKNOWLEDGMENTS

Support for this work has been provided by the University of Kansas and the National Institutes of Health (P20GM103638). We would like to thank Zhaowei Liu for guidance with purification and bioconjugation.

Authors declared no conflict of interest.

REFERENCES

Alves R, Queiroz ATL, Pessoa MG, Da Silva EF, Mazo DFC, Carrilho FJ, … De Carvalho IMVG. 2013. "The presence of resistance mutations to protease and polymerase inhibitors in Hepatitis C virus sequences from the Los Alamos databank." *Journal of Viral Hepatitis*. 20(6):414–421. DOI: 10.1111/jvh.12051.

Best RB. 2020. "Emerging consensus on the collapse of unfolded and intrinsically disordered proteins in water." *Current Opinion in Structural Biology*. 60:27–38. DOI: 10.1016/j.sbi.2019.10.009.

Chen A and Yang S. 2015. "Replacing antibodies with aptamers in lateral flow immunoassay." *Biosensors & Bioelectronics*. 71:230–242. DOI: 10.1016/j.bios.2015.04.041.

Colina R, Casane D, Vasquez S, García-Aguirre L, Chunga A, Romero H, … Cristina J. 2004. "Evidence of intratypic recombination in natural populations of hepatitis C virus." *Journal of General Virology*. 85(1):31–37. DOI: 10.1099/vir.0.19472-0.

Counihan NA, Rawlinson SM, and Lindenbach BD. 2011. "Trafficking of hepatitis C virus core protein during virus particle assembly." *PLoS Pathogens*. 7(10):e1002302. DOI: 10.1371/journal.ppat.1002302.

Dolan PT, Roth AP, Xue B, Sun R, Dunker AK, Uversky VN, and Lacount DJ. 2015. "Intrinsic disorder mediates hepatitis C virus core-host cell protein interactions." *Protein Science*. 24(2):221–235. DOI: 10.1002/pro.2608.

Ellington AD and Szostak JW. 1990. "In vitro selection of RNA molecules that bind specific ligands." *Nature*. 346:818–822. DOI: 10.1038/346818a0.

Förster T. 1948. "Zwischenmolekulare Energiewanderung und Fluoreszenz." *Annalen der physik*. 437(1–2):55–75. DOI: 10.1002/andp.19484370105.

Gaudieri S, Rauch A, Pfafferott K, Barnes E, Cheng W, McCaughan G, … Lucas M. 2009. "Hepatitis C virus drug resistance and immune-driven adaptations: Relevance to new antiviral therapy." *Hepatology*. 49(4):1069–1082. DOI: 10.1002/hep.22773.

Gawlik K and Gallay PA. 2014. "HCV core protein and virus assembly: what we know without structures." *Immunologic Research*. 60:1–10. DOI: 10.1007/s12026-014-8494-3.

Halfon P and Sarrazin C. 2012. "Future treatment of chronic hepatitis C with direct acting antivirals: Is resistance important?" *Liver International*. 32(S1):79–87. DOI: 10.1111/j.1478-3231.2011.02716.x.

Hellenkamp B, Schmid S, Doroshenko O, Opanasyuk O, Kühnemuth R, Rezaei Adariani S, … Hugel T. 2018. "Precision and accuracy of single-molecule FRET measurements—a multi-laboratory benchmark study." *Nature Methods*. 15:669–676. DOI: 10.1038/s41592-018-0085-0.

Holmstrom ED, Holla A, Zheng W, Nettels D, Best RB, and Schuler B. 2018. "Accurate transfer efficiencies, distance distributions, and ensembles of unfolded and intrinsically disordered proteins from single-molecule FRET." *Methods in Enzymology*. 611:287–325. DOI: 10.1016/bs.mie.2018.09.030.

Holmstrom ED, Liu Z, Nettels D, Best RB, and Schuler B. 2019. "Disordered RNA chaperones can enhance nucleic acid folding via local charge screening." *Nature Communications*. 10:2453. DOI: 10.1038/s41467-019-10356-0.

Holmstrom ED, Nettels D, and Schuler B. 2018. "Conformational plasticity of hepatitis C virus core protein enables RNA-induced formation of nucleocapsid-like particles." *Journal of Molecular Biology*. 439(16):2453–2467. DOI: 10.1016/j.jmb.2017.10.010.

Kapanidis AN, Laurence TA, Nam KL, Margeat E, Kong X, and Weiss S. 2005. "Alternating-laser excitation of single molecules." *Accounts of Chemical Research*. 38(7):523–533. DOI: 10.1021/ar0401348.

Kato N. 2000. "Genome of human hepatitis C virus (HCV): gene organization, sequence diversity, and variation." *Microbial & Comparative Genomics*. 5(3):129–151. DOI: 10.1089/omi.1.2000.5.129.

Keefe AD, Pai S, and Ellington A. 2010. Aptamers as therapeutics. In *Nature Reviews Drug Discovery*.

Kong HY and Byun J. 2013. "Nucleic acid aptamers: New methods for selection, stabilization, and application in biomedical science." *Biomolecules & Therapeutics*. 21(6):423–434. DOI: 10.4062/biomolther.2013.085.

Kota S, Coito C, Mousseau G, Lavergne JP, and Strosberg AD. 2009. "Peptide inhibitors of hepatitis C virus core oligomerization and virus production." *Journal of General Virology*. 90(6):1319–1328. DOI: 10.1099/vir.0.008565-0.

Kunkel M, Lorinczi M, Rijnbrand R, Lemon SM, and Watowich SJ. 2001. "Self-Assembly of Nucleocapsid-Like Particles from Recombinant Hepatitis C Virus Core Protein." *Journal of Virology*. 75(5):2119–2129. DOI: 10.1128/jvi.75.5.2119-2129.2001.

Kuntzen T, Timm J, Berical A, Lennon N, Berlin AM, Young SK, … Allen TM. 2008. "Naturally occurring dominant resistance mutations to hepatitis C virus protease and polymerase inhibitors in treatment-naïve patients." *Hepatology*. 48(6):1769–1778. DOI: 10.1002/hep.22549.

Lerner E, Cordes T, Ingargiola A, Alhadid Y, Chung SY, Michalet X, and Weiss S. 2018. Toward dynamic structural biology: Two decades of single-molecule förster resonance energy transfer. In *Science*.

Majeau N, Gagné V, Boivin A, Bolduc M, Majeau JA, Ouellet D, and Leclerc D. 2004. "The N-terminal half of the core protein of hepatitis C virus is sufficient for nucleocapsid formation." *Journal of General Virology*. 85(4):971–981 DOI: 10.1099/vir.0.79775-0.

McKeague M and Derosa MC. 2012. "Challenges and opportunities for small molecule aptamer development." *Journal of Nucleic Acids*. 1-20. DOI: 10.1155/2012/748913.

McLauchlan J. 2000. "Properties of the hepatitis C virus core protein: A structural protein that modulates cellular processes." *Journal of Viral Hepatitis*. 7(1):2–14. DOI: 10.1046/j.1365-2893.2000.00201.x.

Möglich A, Krieger F, and Kiefhaber T. 2005. "Molecular basis for the effect of urea and guanidinium chloride on the dynamics of unfolded polypeptide chains." *Journal of Molecular Biology*. 345(1):153–162. DOI: 10.1016/j.jmb.2004.10.036.

Müller BK, Zaychikov E, Bräuchle C, and Lamb DC. 2005. "Pulsed interleaved excitation." *Biophysical Journal*. 89(11):3508–3522. DOI: 10.1529/biophysj.105.064766.

Müller-Späth S, Soranno A, Hirschfeld V, Hofmann H, Rüegger S, Reymond L, … Schuler B. 2010. "Charge interactions can dominate the dimensions of intrinsically disordered proteins." *Proceedings of the National Academy of Sciences of the United States of America*. 107(33):14609–14614. DOI: 10.1073/pnas.1001743107.

Ng EWM, Shima DT, Calias P, Cunningham ET, Guyer DR, and Adamis AP. 2006. "Pegaptanib, a targeted anti-VEGF aptamer for ocular vascular disease." *Nature Reviews Drug Discovery*. 5(2):123–132. DOI: 10.1038/nrd1955.

Ni F, Kota S, Takahashi V, Strosberg AD, and Snyder JK. 2011. "Potent inhibitors of hepatitis C core dimerization as new leads for anti-hepatitis C agents." *Bioorganic and Medicinal Chemistry Letters*. 21(8):2198–2202. DOI: 10.1016/j.bmcl.2011.03.014.

Santolini E, Migliaccio G, and La Monica N. 1994. "Biosynthesis and biochemical properties of the hepatitis C virus core protein." *Journal of Virology*. 68(6):3631–3641. DOI: 10.1128/jvi.68.6.3631-3641.

Schuler B, Borgia A, Borgia MB, Heidarsson PO, Holmstrom ED, Nettels D, and Sottini A. 2019. "Binding without folding - the biomolecular function of disordered polyelectrolyte complexes." *Current Opinion in Structural Biology*. 60:66–76. DOI: 10.1016/j.sbi.2019.12.006.

Schuler B and Eaton WA. 2008. "Protein folding studied by single-molecule FRET." *Current Opinion in Structural Biology*. 18(1):16–26. DOI: 10.1016/j.sbi.2007.12.003.

Shi S, Yu X, Gao Y, Xue B, Wu X, Wang X, … Zhu H. 2014. "Inhibition of hepatitis C virus production by aptamers against the Core Protein." *Journal of Virology*. 88(4):1990–1999. DOI: 10.1128/jvi.03312-13.

de Souza TLF, de Lima SMB, Braga VLA, Peabody DS, Ferreira DF, Bianconi ML, … de Oliveira AC. 2016. "Charge neutralization as the major factor for the assembly of nucleocapsid-like particles from C-terminal truncated hepatitis C virus core protein." *PeerJ*. 4:e2670. DOI: 10.7717/peerj.2670.

Stepanova M and Younossi ZM. 2017. "Economic burden of hepatitis C infection." *Clinical Liver Disease.* 21(3):579–594. DOI: 10.1016/j.cld.2017.03.012.

Strosberg AD, Kota S, Takahashi V, Snyder JK, and Mousseau G. 2010. "Core as a novel viral target for hepatitis C drugs." *Viruses.* 2(8):1734–1751. DOI: 10.3390/v2081734.

Stryer L. 1978. "Fluorescence energy transfer as a spectroscopic ruler." *Annual Review of Biochemistry.* 47:819–846. DOI: 10.1146/annurev.bi.47.070178.004131.

Stryer L and Haugland RP. 1967. "Energy transfer: a spectroscopic ruler." *Proceedings of the National Academy of Sciences of the United States of America.* 58(2):719–726. DOI: 10.1073/pnas.58.2.719.

Stumpe MC and Grubmüller H. 2007. "Interaction of urea with amino acids: Implications for urea-induced protein denaturation." *Journal of the American Chemical Society.* 129(51):16126–16131. DOI: 10.1021/ja076216j.

Sun H, Zhu X, Lu PY, Rosato RR, Tan W, and Zu Y. 2014. "Oligonucleotide aptamers: New tools for targeted cancer therapy." *Molecular Therapy--Nucleic Acids.* 3(8):e182. DOI: 10.1038/mtna.2014.32.

Tuerk C and Gold L. 1990. "Systematic evolution of ligands by exponential enrichment: RNA ligands to bacteriophage T4 DNA polymerase." *Science.* 249(4968):505–510. DOI: 10.1126/science.2200121.

Ulrich H and Wrenger C. 2009. "Disease-specific biomarker discovery by aptamers." *Cytometry Part A.* 75A(9):727–733. DOI: 10.1002/cyto.a.20766.

Uversky VN. 2019. "Intrinsically disordered proteins and their "Mysterious" (meta)physics." *Frontiers in Physics.* 7:10. DOI: 10.3389/fphy.2019.00010.

World Health Organization. 2017. *Global Hepatitis Report 2017.*

Ziv G and Haran G. 2009. "Protein folding, protein collapse, and Tanford's transfer model: Lessons from single-molecule FRET." *Journal of the American Chemical Society.* 131(8):2942–2947. DOI: 10.1021/ja808305u.

Zosel F, Holla A, and Schuler B. 2020. "Labeling of proteins for single-molecule fluorescence spectroscopy." *chemRxiv.*1–24. DOI: 10.26434/chemrxiv.11537913.v1.

Zou X, Wu J, Gu J, Shen L, and Mao L. 2019. "Application of aptamers in virus detection and antiviral therapy." *Frontiers in Microbiology.* 10:1462. DOI: 10.3389/fmicb.2019.01462.

Zuker M. 2003. "Mfold web server for nucleic acid folding and hybridization prediction." *Nucleic Acids Research.* 31(13):3406–3415. DOI: 10.1093/nar/gkg595.

25 Entropy and Enthalpy in RNA Nanoparticle Assembly and Thermodynamic Stability for Medical Application Using RNA Nanotechnology

Daniel W. Binzel, Xin Li, and Peixuan Guo
The Ohio State University, Columbus, United States

CONTENTS

INTRODUCTION

From the field of RNA biology over the past decades, it has come to be known that ribonucleic acid (RNA) plays an important role in cells for gene regulation, intracellular communication, gene splicing, and catalyzing; beyond the classic notion of RNA only acting as an intermediate for protein production (Cech, 1989; Cotten and Birnstiel, 1989; Lee *et al.*, 1993; Fire *et al.*, 1998; Lau *et al.*, 2001; Lagos-Quintana *et al.*, 2001; Bachellerie *et al.*, 2002; Mattick and Makunin, 2005; Claverie, 2005). Additionally, genome sequencing revealed 98.5% of the human genome does not code for proteins and is involved with short or long non-coding RNAs. Such discoveries have opened the door to study the structure and folding of RNA motifs involved in cells and viruses leading to the development of RNA nanotechnology (Hagenbuchle *et al.*, 1978; Westhof *et al.*, 1985; Pleij *et al.*, 1985; Guo *et al.*, 1987; Felden *et al.*, 1996; Westhof *et al.*, 1996; Paillart *et al.*, 1997; Shapiro and Wu, 1997; Guo *et al.*, 1998; Walter *et al.*, 1998; Zhou *et al.*, 2000; Zhuang *et al.*, 2000; Deindl, 2001; Jaeger *et al.*, 2001). RNA nanotechnology is defined as the bottom-up construction of materials of nanoscale size composed primarily of ribonucleic acids (Guo, 2010).

DOI: 10.1201/9781003001560-27

The field of RNA nanotechnology was first coined and proven in 1998, with the construction of controllable RNA dimers, trimer, and hexamers of the pRNA from the phi29 DNA packaging motor (Guo *et al.*, 1998). These nano-complexes demonstrated the ability to control the stoichiometry and structure of RNA through predictive design and sequence modifications. Since, the field of RNA nanotechnology has grown into a mature area of study through the construction of complex stable nano-platforms, including RNA multi-way junctions (including three-way junctions (3WJ), 4WJ, etc.) (Laing and Schlick, 2009; de la *et al.*, 2009; Shu *et al.*, 2011; Haque *et al.*, 2012), RNA planar shapes (triangle, square, pentagon, etc.) (Severcan *et al.*, 2009; Severcan *et al.*, 2010; Dibrov *et al.*, 2011; Jasinski *et al.*, 2014; Khisamutdinov *et al.*, 2014; Khisamutdinov *et al.*, 2014), RNA dendrimers (Sharma *et al.*, 2015), RNA prisms (Yu *et al.*, 2015; Li *et al.*, 2016; Khisamutdinov *et al.*, 2016; Xu *et al.*, 2019), and RNA micelles (Shu *et al.*, 2018; Yin *et al.*, 2019). As such, nucleic acid nanoparticles, including RNA nanoparticles, are unique over other nanoparticle platforms due to their controllable and defined size, shape, and stoichiometry (Shu *et al.*, 2013; Jasinski *et al.*, 2014; Khisamutdinov *et al.*, 2015; Jasinski *et al.*, 2018). RNA nanoparticles form into a stable complex and allows for the conjugation of functional RNA groups, such as receptor-binding aptamer (Ellington and Szostak, 1992; Gold, 1995), siRNA (Guo *et al.*, 2005; Khaled *et al.*, 2005; Guo *et al.*, 2006), ribozyme (Sarver *et al.*, 1990; Hoeprich 2003), miRNA (Chen *et al.*, 2010; Pegtel *et al.*, 2010; Ye *et al.*, 2011), or riboswitch (Winkler *et al.*, 2004; Mulhbacher *et al.*, 2010), while maintaining the natural folding and functionalities of the conjugated groups and the core motifs. Due to their versatility and robust characteristics, RNA nanoparticles have been applied in a broad range of applications for sensing operates, computer parts (Rinaudo *et al.*, 2007; Breaker, 2008; Win and Smolke, 2008; Benenson, 2009; Xie *et al.*, 2010; Xie *et al.*, 2011; Qiu *et al.*, 2013), imaging reagents (Chen *et al.*, 2012; Calzada *et al.*, 2012; Paige *et al.*, 2012; Kellenberger *et al.*, 2013), NEM devices (Noy, 2011), and therapeutics including the delivery, specific targeting, and treatment of cancer and viral infections (Guo *et al.*, 1998; Shu *et al.*, 2003; Yingling and Shapiro, 2007; Severcan *et al.*, 2009; Shu *et al.*, 2011; Afonin *et al.*, 2011; Lee *et al.*, 2012; Chang *et al.*, 2012; Haque *et al.*, 2012; Delebecque *et al.*, 2012; Tabernero *et al.*, 2013; Shu *et al.*, 2013; Lee *et al.*, 2015). Moreover, RNA nanoparticles have shown potential to revolutionize the pharmaceutical industry with the ability to specifically target and deliver therapeutics to cancers (Shu *et al.*, 2011; Shu *et al.*, 2013; Shu *et al.*, 2015; Cui *et al.*, 2015; Yin *et al.*, 2019), while maintaining a controlled immune response, that is size, shape, and sequence dependent on the nanoparticles (Guo *et al.*, 2017). RNA nanoparticles can be designed to avoid immune detection and elicit no response or create a large macrophage response to serve as an adjuvant in cancer therapies. They are designed to be within 10 and 50 nanometers in size, leading to favorable pharmacokinetic profiles; meaning they are retained in the human body for extended circulation, yet are small enough to pass through leaky blood vessels by the EPR effect and cell membranes through endocytosis (Maeda, 2001; Gao *et al.*, 2005; Jain, 2005; Li and Szoka, 2007; Maeda *et al.*, 2013; Jasinski *et al.*, 2014; Khisamutdinov *et al.*, 2014). Additionally the high thermodynamic stability of RNA nanoparticles allows them to remain intact at ultra-low concentrations and high temperatures between 70 and 100 °C, and their negative charge reduces non-specific interactions with negatively charged cell membranes. These characteristics allow for specific accumulation and targeting to cancer tumors by RNA nanoparticles at a level that has previously not been seen. As such Peixuan Guo predicted in 2014 that RNA drugs or RNA-targeting drugs will become the third milestone in drug development in addition to chemical and protein drugs (Shu *et al.*, 2014).

The Peixuan Guo lab discovered a three-way junction (3WJ) as the central core motif within packaging RNA (pRNA) on the phi29 bacteriophage DNA packaging motor in 2011 (Figure 25.1) (Shu *et al.*, 2011). This pRNA-3WJ motif is well studied and has developed into a therapeutic platform for the treatment of cancers and more complex nanoparticle. This core motif has been found to be ultra-stable, as it resists denaturation in the presence of urea, remains stable at ultralow concentrations during *in vivo* studies, and a high melting temperature (T_m) of near 60 °C (Figure 25.2a) (Shu *et al.*, 2011; Binzel *et al.*, 2014). The individual strands of the pRNA-3WJ assemble into the complex with very high affinity to each other in the absence of supporting metal ions or heating

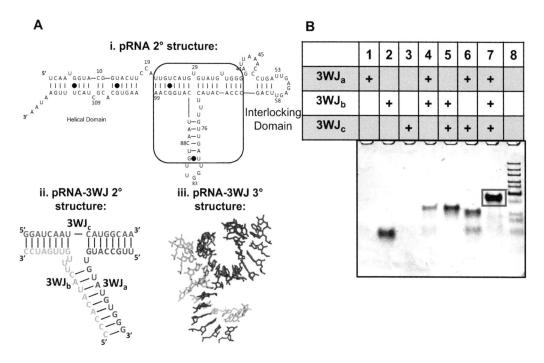

FIGURE 25.1 Overview of the Phi29 pRNA and the three-way junction (3WJ). (a) (i) Secondary structure of the phi29 pRNA monomer with the pRNA-3WJ outlined by the box, which connects the helical domain to the interlocking procapsid binding domains. (ii) Secondary structure and sequence of the pRNA-3WJ and the (iii) crystal structure of the pRNA-3WJ. (b) Assembly gel of the pRNA-3WJ from the three short RNA oligo strands showing efficiency in folding and particle homogeneity. Reprinted with permission from Binzel *et al.* (2016).

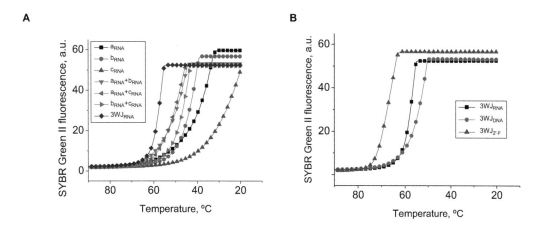

FIGURE 25.2 Melting temperatures of the pRNA-3WJ. (a) 3WJ$_{RNA}$ shows the highest melting temperature (T$_m$) over any of its components of dimer species. (b) Thermostability comparison of three 3WJ with chemical modifications. Reprinted with permission from Binzel *et al.* (2014). Further permissions related to the material excerpted should be directed to the ACS.

cycles that are typically needed for association of nucleic acid strands. The high yield of formation of the pRNA-3WJ and its stability makes it an ideal candidate to serve as a scaffold for a therapeutic platform and has been proven to deliver RNAi- based therapeutics to several cancer lines (Shu *et al.*, 2015; Cui *et al.*, 2015; Rychahou *et al.*, 2015; Lee *et al.*, 2015; Yin *et al.*, 2019).

Within the design and structuring of RNA nanoparticles, including the 3WJ, one must take into consideration the thermodynamics and kinetics to lead to a successful nanoparticle design (Jaeger *et al.*, 1993; Mathews and Turner, 2006; Tinoco, Jr. *et al.*, 2006; Fiore *et al.*, 2012; Rauzan *et al.*, 2013). Thermodynamics govern the assembly of multiple RNA oligo strands into a nanoparticle complex and their assembled stability. Furthermore, looking at the kinetics or speed of interactions between the components of RNA nanoparticles allows for information to be gained in the stability of RNA nano-complexes, especially while looking at dissociation rates (Zarrinkar and Williamson, 1994; Sclavi *et al.*, 1998; Xayaphoummine *et al.*, 2005; Chen, 2008; Binzel *et al.*, 2016). Here we focus and discuss the thermodynamics and kinetics behind the construction of RNA nanoparticles, specifically 3WJ-based nanoparticles and how they play into design considerations. The calculation of thermodynamic parameters will first be discussed and how this has led to computational methods to predict secondary and 3D structuring of nucleic acids of single and double strands. We will then look at how thermodynamic parameters play a role in three-way junction folding and how it's related to the kinetics for association of three-way junctions.

ENERGETICS IN THE FORMATION OF RNA NANOPARTICLES

RNA nanotechnology is defined by the bottom-up construction of complex RNA motifs that are composed primarily of RNA (Guo, 2010; Shukla *et al.*, 2011; Jasinski *et al.*, 2017). By nature RNA nanoparticles are typically composed of several single-stranded, short RNA oligos that self-assemble upon incubation of each of the component strands (Binzel *et al.*, 2014; Binzel *et al.*, 2016). In order to construct such RNA nanoparticles, during the design-process sequences must carefully be selected to ensure that association of the multiple RNA strands takes place in a spontaneous manner and results in the production of a stable RNA particle. The assembly of single-stranded RNAs (ssRNA) into nanoparticles is governed and described by thermodynamics, or the changes in heat and energies of the ssRNAs to nanoparticles. Thermodynamics allow for the prediction of spontaneous nanoparticle assembly and degree of stability the ssRNAs to stay assembled as nanoparticle complexes. Gibbs free energy, a central energy calculation in thermodynamics, known as available energy, is used to understand the complex's stability while looking at the change in energy over a reaction. Assembly of nanoparticle results in changes in energies and is composed of enthalpy and entropy (Binzel *et al.*, 2014).

Gibbs Free Energy

As stated above Gibbs free energy is the available work or available energy of a system and is comprised of enthalpy and entropy while being temperature dependent. Changes in Gibbs free energy (ΔG) combines the changes in enthalpy (ΔH) and entropy (ΔS) as described in the equation below:

$$\Delta G^\circ = \Delta H^\circ - T \Delta S^\circ \tag{25.1}$$

where T is the temperature of the reaction (Binzel *et al.*, 2014). Combining enthalpy and entropy at a given temperature of the reaction provides a ΔG° in which the spontaneity of the reaction is determined by its value. If a change in free energy is negative, the chemical reaction, or in this case the assembly of RNA nanoparticles is spontaneous while a positive value does not spontaneously occur and requires input of energy for the assembly. Thus, it is important to ensure during the design of RNA nanoparticles, they assemble in a spontaneous matter so that the nanoparticle remains folded and stable during future applications. Additionally the more negative the ΔG of the formation of nanoparticle the more readily it will fold and indicates a stable nanoparticle, as the dissociation

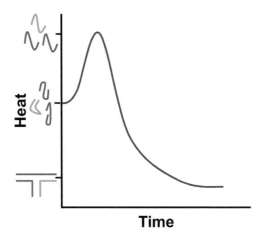

FIGURE 25.3 Example of heat energy required for RNA nanoparticle assembly. RNA strands preferentially folding into the lowest energy conformation (low enthalpy). RNA strands are initially heated to remove structuring and slowly cooled allowing for the formation of RNA nanoparticles. The folding of RNA is an exothermic process (−ΔH) releasing heat until reaching the most stable complex.

constant of the nanoparticle can be calculated from the change in Gibbs free energy. While we discuss the calculation and modeling of energy levels of RNA folding below, generally speaking an increase in Watson-Crick base pairing in a structure will lead to a more stable structure that folds more spontaneously. Addition stability can be gained by increasing the guanine and cytosine (G:C) contents of the strands.

It is important to note the inverse relationship of enthalpy and entropy in relationship to ΔG. This means that a release of heat energy (−ΔH) and an increase in disorder (+ΔS) are beneficial to a spontaneous reaction. Generally the formation of RNA structures goes from a disordered to ordered system resulting in a −ΔS. Thus spontaneity can be created if the change in enthalpy outweighs a negative change in entropy. The assembly of dsRNA is generally a spontaneous reaction that relies on the release of internal energy of ssRNA to form a much more energy conserving dsRNA structure (Figure 25.3). Unpaired RNA will fold upon itself forming short dsRNA regions with loops and bulges; however, base pairing with a second RNA strand conserves energy thus creating a negative change in enthalpy. The entropically unfavored process of going to a more structure system during RNA folding is typically far outweighed by the larger negative change in enthalpy during assembly (Tinoco *et al.*, 1971; Freier *et al.*, 1986; Turner *et al.*, 1988; Xia *et al.*, 1998; Mathews and Turner, 2002; Binzel *et al.*, 2014).

ENTHALPY IN NANOPARTICLE ASSEMBLY

Enthalpy, a state function, defined as the heat energy of a system, or in this case a RNA molecule or RNA structure. Heat is the combination of the internal energy and pressure and volume; when considering RNA nanoparticle assembly, the volume and pressure are held constant, thus we can consider enthalpy the internal energy of the RNA itself. Changes in enthalpy results in either the release or absorption of heat during a reaction, thus is often referred to as the heat of a reaction or heat of formation. The formation of RNA nanoparticles consists of ssRNAs base pairing into complex structures and motifs (Binzel *et al.*, 2014; Binzel *et al.*, 2016). The structuring of ssRNA generally is much less favorable and not nearly as stable as double- stranded RNA (dsRNA), as ssRNA has some of self-folding and base pairing but is incomplete in structuring compared to dsRNA (Figure 25.3) (Xia *et al.*, 1998; Martinez *et al.*, 2008; Bindewald *et al.*, 2011; Binzel *et al.*, 2014; Binzel *et al.*, 2016). These less favorable structures hold higher energy levels compared to complete Watson-Crick base pairing, therefore the assembly of RNA nanoparticles releases the internal energy as heat resulting in a negative enthalpy ($-\Delta H$). Furthermore, this release of heat during a reaction or process is known as an exothermic reaction (Xia *et al.*, 1998), while the denaturing of helical RNA or folded complexes requires energy input and is known as an endothermic process.

ENTROPY IN NANOPARTICLE ASSEMBLY

Entropy is more difficult to visualize and comprehend than previously discussed enthalpy, and is described as the level of disorder or randomness of a system or molecule. The entropy of a system can be calculated by looking at the number of the same species of particles in different conformations at any given time (Chen, 2008; Jacobson *et al.*, 2017); this is a difficult task, but looking at the change in structuring over time during a reaction or assembly of RNA nanoparticles is more feasible. A change in level of disorder of a system is described by ΔS, where a positive value is a result of an increase in the disorder or randomness, and a negative ΔS indicates a change to more ordered structuring. It is important to note that while the entropy of a system can be negative and return to being more ordered, the entropy of the universe is constantly increasing and moving to a more disordered setup. In looking at the assembly of RNA nanoparticles, moving from ssRNA to primarily dsRNA structuring results in a decrease in disorder, as ssRNAs are capable of self-folding into many unstable conformations, whereas RNA nanoparticles fold into predicted and stable structures. This results in a negative change in entropy.

PREDICTION AND CALCULATION OF THERMODYNAMIC PARAMETERS OF RNAs

Understanding and predicting thermodynamic parameters is a powerful tool in RNA nanoparticle design to ensure that resulting motifs and structures from multiple RNA strands fold easily while remaining stable. Significant research has been completed in understanding sequences of RNA strands and their base pairing to complement sequences in order to calculate the thermodynamic contributions to RNA folding for sequence predictions. Thermodynamic parameters can be calculated *via* a van't Hoff plot, in which the concentration of materials are varied while examining the difference in melting temperatures (T_m) of the duplexes (Figure 25.4) (Freier *et al.*, 1986; Lesnik and Freier, 1995; Binzel *et al.*, 2014). The melting temperature of a duplexed nucleic acid structure is defined as the temperature in which 50% of the species exists as a duplex, while 50% exist as single strands (Shu *et al.*, 2011; Binzel *et al.*, 2014). The T_m of nucleic acids are dependent on strand concentrations and can be created into a linear relationship by plotting the natural log (ln) of concentration versus the inverse of T_m ($1/T_m$). As a result the slope of the van't Hoff plot is related to enthalpy and the y-intercept is related to the entropy and enthalpy as described in the equation below:

$$\frac{1}{T_m} = \frac{2R}{\Delta H^\circ} \ln C_t + \frac{\Delta S^\circ - 2R \ln 6}{\Delta H^\circ} \tag{25.2}$$

where R is the universal gas constant and C_t is the concentration of RNAs.

Studies were originally completed on DNA sequences with strands including single nucleotide mutations, and it was found that thermodynamic parameters rely on what is known as nearest neighbor laws, in which the thermodynamics are affected by the two consecutive nucleotides in a sequence rather than a single nucleotide (Breslauer *et al.*, 1986; Santalucia *et al.*, 1996). These pioneering studies in DNA were then further extended to RNA (Tinoco *et al.*, 1971; Freier *et al.*, 1986; Xia *et al.*, 1998) and DNA/RNA hybrids (Sugimoto *et al.*, 1995; Lesnik and Freier, 1995). These studies allow for one to calculate by hand the thermodynamic parameters along with melting temperature for any given duplex structure and aid in the sequence design of RNA nanoparticles. However, the nature of folding of RNA with Watson-Crick base pairing, non-canonical base pairing,

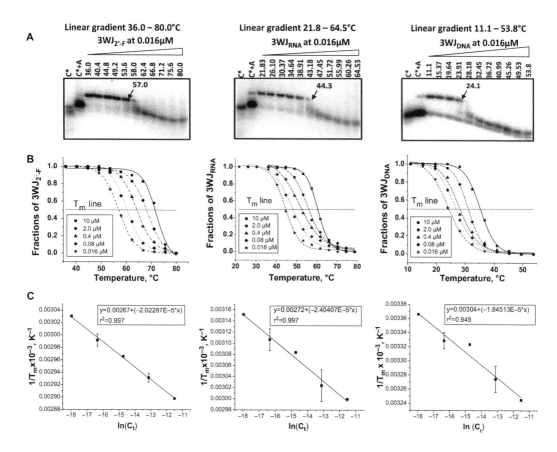

FIGURE 25.4 Calculation of thermodynamic parameter for 3WJs formation. (a) Representation of temperature gradient gels to calculate T_m of each 3WJ. (b) Melting temperature profiles of RNA, 2′-F RNA, and DNA 3WJs by various concentrations. (c) Plots of T_m vs. 3WJ concentrations (van't Hoff analysis) to calculate thermodynamic parameters. Reprinted with permission from Binzel *et al.* (2014). Further permissions related to the material excerpted should be directed to the ACS.

and base stacking creates multi-way junctions, bulges, and loops, all which play into the thermodynamic parameters. As such Mathews *et al.* was able to experimentally derive these parameters for three-way and four-way junctions (Mathews and Turner, 2002). More recently, several additional studies have been published on the folding energies behind RNA and DNA sequences (Gyi *et al.*, 1996; Conte *et al.*, 1997; Gyi *et al.*, 2003; Rauzan *et al.*, 2013).

Using the calculated parameters above, RNA and DNA secondary structure modeling software have become available to predict base pairing within single-stranded and double- stranded nucleic acids. These software allow researchers to predict folding and structuring of RNA nanoparticles for provided sequences as well as predict the thermodynamic stabilities. mFold, a public server, provides DNA and RNA secondary structure and base pairing modeling that allows users to see the most energy preferred folding of an RNA strand along with the predicted folding energies and stabilities (Zuker, 2003). Mathews and Turner have created predictions of secondary structures by minimizing the free energies following the methodology above (Mathews and Turner, 2006). Additionally UNAfold was developed to create and predict secondary structuring between two RNA strands building upon the development of mfold (Markham and Zuker, 2008). Several other prediction tools have been developed for the base pairing of RNA based off nucleic acids' thermodynamics (Bindewald *et al.*, 2011). More recently, 3D predictions of RNA folding and structuring have become available with software, including iFoldRNA, Nanotiler, RNA2D3D, and Cartaj (Lamiable *et al.*, 2012; Petrov *et al.*, 2013), to build models for given sequences.

THERMODYNAMIC STABILITY OF 3WJ RNA NANOPARTICLES

STABILITY OF THE pRNA-3WJ

RNA nanoparticles to be used for *in vivo* applications requires a high thermodynamic stability to ensure nanoparticle stability in harsh conditions and keep base pairing when diluted within the body during circulation. The previous studies described above looked at the folding of RNA duplexes and junctions based on single- and double-stranded systems. However, RNA nanoparticles are typically composed of several RNA oligos and bring in a more complex structure than dsRNA used in many of the studies above (Guo, 2010). Looking at the thermodynamics of a multi-stranded system brings a level of complexity that is difficult to study; however, Binzel *et al.* used the pRNA-3WJ from the phi29 DNA packaging motor as a model system in identifying the thermodynamic stability of three short RNA strands folding together (Binzel *et al.*, 2014). The pRNA-3WJ serves as the central folding domain of the pRNA that remains stable while undergoing strong forces of DNA packaging by the pg16 ATPase (Figure 25.1) (Guo *et al.*, 1987).

The pRNA-3WJ proved to be stable at high temperatures as the T_m was found to be nearly 60 °C at 10 **μM** (Figure 25.2) (Binzel *et al.*, 2014). This high melting temperature demonstrated high cooperation between the three short oligos to form three separate helical regions joined by the center junction. Additionally the assembly and melting profiles showed a single transition from ssRNA to pRNA-3WJ hinting at all three strands assembling together without any sign of dimer intermediate formation. The 3WJ formed with the highest stability over any of the dimer species with the highest melting temperature (Figure 25.2a). Thermodynamic parameters were calculated through a van't Hoff plot as described above, resulting in elucidating a significantly negative ΔG (Figure 25.3 & Table 25.1) (Binzel *et al.*, 2014). It is interesting to note that the RNA and 2′-Fluoro (2′-F) chemically modified 3WJs are more thermodynamically stable than the DNA 3WJ. This increased stability is seen in a more favorable (less negative) entropy, most likely due to the less rigid structuring of RNA over DNA to allow for flexibility in the RNA and slight structure variances. This results in a higher level of disorder. Finally, the pRNA-3WJ has been shown to remain stable at concentrations as low as 160 pM concentrations as assay by PAGE and radiolabeling (Shu *et al.*, 2011). These results indicate the pRNA-3WJ is able to form into stable complexes suitable for *in vivo* applications.

TABLE 25.1
Thermodynamic Parameters for 3WJs Formations[a]

3WJs	1/T_m vs Log (C_t)			
	ΔG°$_{37}$ (kcal/mol)	ΔH°$_{37}$ (kcal/mol)	ΔS°$_{37}$ (e.u.)	T_m[b] (°C)
2′-F RNA	−36 ± 0.45	−200 ± 5.7	−520 ± 17	72.1
RNA	−27 ± 0.58	−170 ± 13	−440 ± 39	60.4
DNA	−15 ± 0.71	−220 ± 25	−650 ± 83	35.2

[a] Thermodynamic parameters of pRNA-3WJ formation derived from van't Hoff plots of melting curves by temperature gradient gel electrophoresis. Reprinted with permission from Binzel *et al.* (2014). Further permissions related to the material excerpted should be directed to the ACS.
Parameters derived from 15% native TGGE.
[b] T_m values for 3WJ strand concentrations of 10^{-6} M

Effects of Chemical Modifications on the Stability of RNA and 3WJ Nanoparticles

There have long been concerns of thermodynamic and enzymatic stability of RNA and RNA nanoparticles preventing the translation to the clinic. As shown above, 3WJ nanoparticles have overcome the thermodynamic hurdle, as their high melting temperature and low ΔG indicated high stability to remain stable at ultra-low concentrations. While RNA is susceptible to RNases found within the body, it has been found modifying the backbone of nucleotides at the 2′ carbon location changes the conformation of the nucleotide and prevents RNase identification (Kawasaki et al., 1993; Pagratis et al., 1997; Layzer et al., 2004; Pallan et al., 2011; Khvorova and Watts, 2017; Stein and Castanotto, 2017). Commonly used modifications to add enzymatic stability to RNAs include 2′-fluoro (2′-F) and 2′-OMethyl and modified RNAs remain stability in serums for over 24 hours while natural RNA is degraded within the first 30 minutes (Shaw et al., 1991; Kawasaki et al., 1993).

Additionally, the 2′ chemical modifications to RNA have been found to further alter their thermodynamic stabilities. In looking at the pRNA-3WJ, chemical modifications could significantly increase the melting temperatures. 2′-F modifications on pyrimidine nucleotides raised the melting temperature of the 3WJ from ~60 °C to ~70 °C (Figure 25.2b) (Shu et al., 2011; Binzel et al., 2014; Piao et al., 2018), while substituting to locked nucleic acids (LNA) above 80 °C (Piao et al., 2018), the limit of detection using temperature gradient gel electrophoresis. These results were further confirmed by examining the thermodynamic stabilities of RNA/2′-F and RNA/LNA hybrids. As the chemical modification content increased from one to three strands the melting temperatures of the three-way junctions continued to increase. Additionally, the Gibbs free energy of formation for 2′-F pRNA-3WJ was found to be more negative than the native pRNA-3WJ, indicating a higher level of stability and more preferential folding (Binzel et al., 2014). The high stability of the LNA modifications and the pRNA-3WJ allowed for the nanoparticles to remain stable within the body, as complete 3WJ assemblies were purified from urine of mice after being intravenously administered (Piao et al., 2018). Alternatively, the thermodynamic stability was decreased as the DNA composition was increased within the pRNA-3WJs (Binzel et al., 2014; Piao et al., 2018).

ASSEMBLIES OF RNA THREE-WAY JUNCTION (3WJS) DRIVEN BY ENTROPY AND ENTHALPY

The pRNA-3WJ is composed of three short RNA oligos that co-assemble into a junction. Initial assembly studies of the pRNA-3WJ led to the belief that all three RNA strands co-associate at the same time, as there was no evidence of dimer species forming (Shu et al., 2011; Binzel et al., 2014). While, it is highly rare for three components to associate together at the same time, the high speed and ease of formation of the pRNA-3WJ with its strong stability further supported the idea that three component strands associated together at the same time. Through assembly gels and assembly profiles elucidating melting temperatures showed a high yield of three-way junction formation with no sign of dimer intermediates formation as shown in a single transition (Shu et al., 2011; Binzel et al., 2014).

With this hypothesis, Binzel et al. completed kinetic studies of the three component strands associating into the 3WJ by Surface Plasmon Resonance (SPR) (Binzel et al., 2016). It was found that the pRNA-3WJ follows a two-step reaction of association in which a dimer species is very temporarily formed. During the association of each of the three dimer associations, it was found that the $3WJ_{bc}$ dimer formed at a more rapid rate than the modeled 3WJ itself and the other two dimers (Figure 25.4). However, this dimer intermediate is normally not observed as immediately upon its formation the third strand, $3WJ_a$, interacts with the $3WJ_{bc}$ folding into the 3WJ. This is due to the dimer formation increasing the number of nucleotides available for the $3WJ_a$ to bind, thus significantly increasing its affinity to the forming 3WJ. This was evidenced by the fact that the second association step occurring at a rate too rapid to be observed by SPR due to a limit of detectable concentrations. This led to the development of a pseudo-one step association model of the pRNA-3WJ

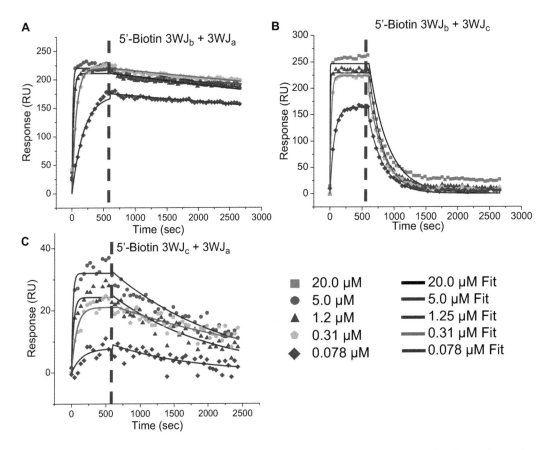

FIGURE 25.5 Surface Plasmon Resonance (SPR) of pRNA-3WJ dimers. Dimer species that make up the pRNA-3WJ were examined through SPR at concentrations ranging from 20 μM to 78 nM. Dimers were formed on the chip for the first 660 sec (association phase) followed by a dissociation phase until 2700 sec. The $3WJ_{ab}$ dimer (a) shows the slowest association and dissociation, while the $3WJ_{bc}$ (b) shows the fasted formation and dissociation and the $3WJ_{ac}$ a mix of strong formation and slow dissociation. Data was fit using a two-component pseudo-first order Langmuir model. Reprinted with permission from Binzel *et al.* (2016).

in which a three-way junction is immediately formed upon folding of the dimer species as described below and in Figure 25.5 (Binzel *et al.*, 2016).

$$3WJ_a + 3WJ_b + 3WJ_c \leftrightarrow 3WJ_a + 3WJ_{bc} \leftrightarrow 3WJ \qquad (25.3)$$

Interestingly from the SPR studies, it was found that the pRNA-3WJ stability and formation benefited from being composed of three strands. In kinetics, a fast association of a complex typically leads to a rapid dissociation; while the opposite is true in that a slow reaction results in a slow dissociation within RNA reactions. However, the pRNA-3WJ benefits from the fast association of the $3WJ_b$ and the $3WJ_c$ strands into a dimer, but the slow dissociation of the $3WJ_{ab}$ dimer. This results in a multi-stranded RNA motif that associated rapidly into a stable complex that is slow to dissociate. As one helical region of the 3WJ dissociates, the other two branches hold the structure together instead of fully unzipping and quickly refolds due to a high local concentration (Binzel *et al.*, 2016).

In looking at the structuring of the ssRNA components of the pRNA three-way junction, they are of high entropy or high levels of disorder, as the $3WJ_a$ and $3WJ_c$ strands generally do not stain by ethidium bromide intercalating dye. This shows the strands do not have defined structures and are in

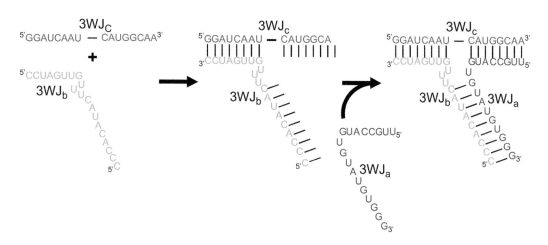

FIGURE 25.6 Assembly Mechanism of pRNA-3WJ. The pRNA-3WJ assembles through a two-step association mechanism in which the 3WJ$_{bc}$ dimer first forms as shown in the first panel. This association step greatly increases the attraction of the 3WJ$_a$ strand by doubling the nucleotides presented for binding resulting in a dropped ΔG°. This second reaction occurs at an unobservable rate and results in forming a stable nanostructure, in which each of the three strands are locked into the structure by two areas of base pairing. Reprinted with permission from Binzel *et al.* (2016).

a transient state. When the three strands assemble together into the 3WJ, a change is seen to a defined structure and distinct band on the gel indicating an unfavorable decrease in entropy. The formation of the 3WJ$_{bc}$ dimer presents a defined motif for the binding of the 3WJ$_a$ strand to bind to further demonstrating this decrease in entropy. However, the formation of the three-way junction produces a stable structure that has low internal energy; thus, the release of enthalpy is able to overcome the unfavorable change in entropy. However, as mentioned above the branched structuring of the pRNA-3WJ allows for flexibility between each of the helical regions, as well as strand breathing. These slight structure variations allow for a higher level of entropy over dsRNA further improving the stability (Figure 25.6).

APPLICATIONS OF STABLE 3WJ NANOPARTICLES

As proven in the thermodynamic and kinetic studies, the pRNA-3WJ serves as an ideal scaffold to translate into *in vivo* applications due to its strong stability upon rapid dilution within the bloodstream (Shu *et al.*, 2011). Additionally, the inclusion of 2′-F modifications to the 3WJ resulted in stability against RNases in serum for over 24 hours (Piao *et al.*, 2018). Here, we look at the application of pRNA-3WJ-based RNA nanoparticles that have translated into a therapeutic platform as well as look at the ability to utilize the pRNA-3WJ as building block for the construction of more complex nanoparticles and their applications.

With the high thermostability of the pRNA-3WJ, RNA nanoparticles of diverse shape, size, and stoichiometry are able to be constructed with the 3WJ serving as the core motif and guiding the folding. The pRNA-3WJ's branched structuring allows for versatility in nanoparticle design, creating a wide array of RNA nanoparticles. The phi29 pRNA-3WJ has been expanded into triangle, square, and pentagon polygons (Jasinski *et al.*, 2014; Khisamutdinov *et al.*, 2014; Khisamutdinov *et al.*, 2014; Khisamutdinov *et al.*, 2015). Through stoichiometry of the RNA strands, the shape of RNA nanoparticles can be precisely controlled (Khisamutdinov *et al.*, 2014) and RNA polygons have been constructed by several other groups (Chworos *et al.*, 2004; Yingling and Shapiro, 2007; Severcan *et al.*, 2009; Dibrov *et al.*, 2011; Bindewald *et al.*, 2011; Boerneke *et al.*, 2016; Huang and Lilley, 2016; Parlea *et al.*, 2016; Bui *et al.*, 2017). Furthermore, RNA nanoparticles have recently been constructed into three-dimensional prisms and tetrahedral geometry (Li *et al.*, 2016; Xu *et al.*, 2019).

Similarly other groups have constructed 3D RNA nanoparticles composed of numerous RNA strands self-assembling into thermostable complexes that resemble the pRNA-based prisms with RNA wire frames (Severcan *et al.*, 2010; Afonin *et al.*, 2010; Hao *et al.*, 2014; Yu *et al.*, 2015).

A major roadblock in cancer therapeutics is overcoming the toxicities to healthy tissues and organs. In order to revolutionize the cancer therapeutic field, targeted delivery is required in order to reduce these toxicities. RNA aptamers are oligonucleotides that bind to specific targets, and they have shown great therapeutic potential (Ellington and Szostak 1992; Gold, 1995). RNA aptamers have been added to helical regions of RNA nanoparticles including the pRNA-3WJ and other RNA nanoparticles to control targeted delivery to specific receptor displaying cells (Mamot *et al.*, 2005; Shigdar *et al.*, 2011; Esposito *et al.*, 2011; Thiel *et al.*, 2012; Sundaram *et al.*, 2013; Shu *et al.*, 2015; Li *et al.*, 2016; Binzel *et al.*, 2016; Pi *et al.*, 2018; Xu *et al.*, 2018; Porciani *et al.*, 2018; Panigaj *et al.*, 2019; Yin *et al.*, 2019; Yin *et al.*, 2019). Additionally, chemical ligands can be linked onto the RNA nanoparticles with folate used as a common cancer ligand (Shu *et al.*, 2011; Cui *et al.*, 2015; Rychahou *et al.*, 2015; Lee *et al.*, 2015). The modified RNA nanoparticles with tumor-targeting ligands are able to specifically target, bind, and enter tumor cells both in *in vitro* and *in vivo* experiments in murine models with no evidence of accumulation in healthy organs, especially the liver and lungs that are common accumulation sights of nanoparticles.

As RNA nanoparticles have proven their ability to specifically accumulate in and internalize in cancer tumors, they are an ideal platform for delivering therapeutics to cancers. RNA nanoparticles are perfect for the inclusion of RNAi-based therapeutics such as siRNA, miRNA, or anti-miRNAs (Fire *et al.*, 1998; Afonin *et al.*, 2012; Stewart *et al.*, 2016; Pastor *et al.*, 2018). Such inclusion into RNA nanoparticles only requires sequence extensions off each helix to include the desired siRNA or miRNA sequence. siRNA delivery has been realized by pRNA-3WJ nanoparticles against several different cancer lines by targeting different cell receptors for each tumor line (Cui *et al.*, 2015; Lee *et al.*, 2015; Zhang *et al.*, 2017; Xu *et al.*, 2018). An additional powerful therapeutic tool is the delivery of miRNA to cancers for tumor suppressor miRNA or anti-miRNA strands to knockdown oncogenic miRNA in tumors. pRNA-3WJ nanoparticles have proven to carry anti-miRNA sequences against miR21 and miR17, two well-studied and classified oncogenic miRNAs (Shu *et al.*, 2015; Binzel *et al.*, 2016; Lee *et al.*, 2017; Yin *et al.*, 2019). These short locked nucleic acid (LNA) strands were developed to bind to the seed region of specific miRNAs that then inhibit the miRNA from binding to mRNA, rendering it useless (Obad *et al.*, 2011). Most recently, RNA nanoparticles were realized for the conjugation and delivery of chemotherapeutic drugs (Pi *et al.*, 2016; Shu *et al.*, 2018; Piao *et al.*, 2019; Guo *et al.*, 2020). The hydrophobic and insoluble nature of many small molecule drugs hinders their bioavailability and resulting in toxicities. Conjugation of these chemotherapeutics to highly soluble RNA nanoparticles serves as a solution of overcoming bioavailability issues. As a result the RNA nanoparticle significantly improved the solubility of conjugated paclitaxel (Guo *et al.*, 2020). *In vivo* testing proved the specific delivery of paclitaxel conjugated pRNA-3WJs to tumors and had a more significant tumor inhibition effect over paclitaxel on its own (Piao *et al.*, 2019; Guo *et al.*, 2020). Finally, RNA nanoparticles can serve as a carrier of intercalating chemotherapeutics such as doxorubicin as the drug is bound within the double-stranded regions of the 3WJ (Pi *et al.*, 2016). These studies have further enhanced the ability of RNA nanoparticles and show significant promise for the treatment of cancers.

CONCLUSIONS

RNA nanotechnology applies the bottom-up self-assembly of multiple RNA strands that is governed by thermodynamics and related energy levels. Thermostability of RNA nanoparticles is crucial in relevance to particle stability, module dissociation, strand breathing, RNase accessibility, and capacity in drug loading. Their stabilities are related to the Gibbs free energy that is controlled by enthalpy and entropy. In depth thermodynamic and kinetic studies were completed on the three-way junction from the phi29 DNA packaging motor RNA. It was demonstrated the pRNA-3WJ composed of three short RNA strands folds rapidly and remains highly stable. The folding relies on changes in both enthalpy and entropy, through a heat energy releasing step during assembly to form a stable three-way junction with structure flexibility to allow a more favorable entropy over dsRNA structures. Such unique structural interaction of the three strands of RNA sequences results in the development of RNA 3WJ into a strong motif that has been used as therapeutic delivery platform for specific treatment of cancers.

ACKNOWLEDGMENTS

The research in P.G.'s lab was supported by NIH grants R01EB019036, U01CA151648 and U01CA207946. P.G.'s Sylvan G. Frank Endowed Chair position in Pharmaceutics and Drug Delivery is funded by the CM Chen Foundation. P.G. is the consultant of Oxford Nanopore Technologies, as well as the cofounder of ExoNanoRNA LLC.

REFERENCES

Afonin KA, Bindewald E, Yaghoubian AJ, Voss N, Jacovetty E, Shapiro BA, and Jaeger L (2010) In vitro assembly of cubic RNA-based scaffolds designed in silico. *Nat Nanotechnol*, **5**, 676–682.

Afonin KA, Grabow WW, Walker FM, Bindewald E, Dobrovolskaia MA, Shapiro BA, and Jaeger L (2011) Design and self-assembly of siRNA-functionalized RNA nanoparticles for use in automated nanomedicine. *Nat Protoc*, **6**, 2022–2034.

Afonin KA, Kireeva M, Grabow WW, Kashlev M, Jaeger L, and Shapiro BA (2012) Co-transcriptional assembly of chemically modified RNA nanoparticles functionalized with siRNAs. *Nano Lett*, **12**, 5192–5195.

Bachellerie JP, Cavaille J, and Huttenhofer A (2002) The expanding snoRNA world. *Biochimie*, **84**, 775–790.

Benenson Y (2009) RNA-based computation in live cells. *Curr Opin Biotechnol*, **20**, 471–478.

Bindewald E, Afonin K, Jaeger L, and Shapiro BA (2011) Multistrand RNA secondary structure prediction and nanostructure design including pseudoknots. *ACS Nano*, **5**, 9542–9551.

Binzel DW, Khisamutdinov EF, and Guo P (2014) Entropy-driven one-step formation of Phi29 pRNA 3WJ from three RNA fragments. *Biochemistry*, **53**, 2221–2231.

Binzel DW, Khisamutdinov E, Vieweger M, Ortega J, Li J, and Guo P (2016) Mechanism of three-component collision to produce ultrastable pRNA three-way junction of Phi29 DNA-packaging motor by kinetic assessment. *RNA*, **22**, 1710–1718.

Binzel D, Shu Y, Li H, Sun M, Zhang Q, Shu D, Guo B, and Guo P (2016) Specific Delivery of MiRNA for High Efficient Inhibition of Prostate Cancer by RNA Nanotechnology. *Mol Ther*, **24**, 1267–1277.

Boerneke MA, Dibrov SM, and Hermann T (2016) Crystal-structure-guided design of self-assembling RNA nanotriangles. *Angew Chem Int Ed Eng*, **55**, 4097–4100.

Breaker RR (2008) Complex riboswitches. *Science*, **319**, 1795–1797.

Breslauer KJ, Frank R, Blocker H, and Marky LA (1986) Predicting DNA duplex stability from the base sequence. *Proc Natl Acad Sci U S A*, **83**, 3746–3750.

Bui MN, Brittany JM, Viard M, Satterwhite E, Martins AN, Li Z, Marriott I, Afonin KA, and Khisamutdinov EF (2017) Versatile RNA tetra-U helix linking motif as a toolkit for nucleic acid nanotechnology. *Nanomedicine*, **13**, 1137–1146.

Calzada V, Zhang X, Fernandez M, z-Diaz-Miqueli A, Iznaga-Escobar N, Deutscher SL, Balter H, Quinn TP, and Cabral P (2012) A potencial theranostic agent for EGF-R expression tumors: (177)Lu-DOTA-nimotuzumab. *Curr Radiopharm*, **5**, 318–324.

Cech TR (1989) RNA chemistry. Ribozyme self-replication? *Nature*, **339**, 507–508.

Chang CI, Lee TY, Kim S, Sun X, Hong SW, Yoo JW, Dua P, Kang HS, Kim S, Li CJ, and Lee DK (2012) Enhanced intracellular delivery and multi-target gene silencing triggered by tripodal RNA structures. *J Gene Med*, **14**, 138–146.

Chen SJ (2008) RNA folding: conformational statistics, folding kinetics, and ion electrostatics. *Annu Rev Biophys*, **37**, 197–214.

Chen Z, Penet MF, Nimmagadda S, Li C, Banerjee SR, Winnard PT, Jr., Artemov D, Glunde K, Pomper MG, and Bhujwalla ZM (2012) PSMA-targeted theranostic nanoplex for prostate cancer therapy. *ACS Nano*, **6**, 7752–7762.

Chen Y, Zhu X, Zhang X, Liu B, and Huang L (2010) Nanoparticles modified with tumor-targeting scFv deliver siRNA and miRNA for cancer therapy. *Mol Ther*, **18**, 1650–1656.

Chworos A, Severcan I, Koyfman AY, Weinkam P, Oroudjev E, Hansma HG, and Jaeger L (2004) Building programmable jigsaw puzzles with RNA. *Science*, **306**, 2068–2072.

Claverie JM (2005) Fewer genes, more noncoding RNA. *Science*, **309**, 1529–1530.

Conte MR, Conn GL, Brown T, and Lane AN (1997) Conformational properties and thermodynamics of the RNA duplex r(CGCAAAUUUGCG)(2): Comparison with the DNA analogue d(CGCAAATTTGCG)(2). *Nucleic Acids Res*, **25**, 2627–2634.

Cotten M and Birnstiel M (1989) Ribozyme mediated destruction of RNA in vivo. *EMBO J*, **8**, 3861–3866.

Cui D, Zhang C, Liu B, Shu Y, Du T, Shu D, Wang K, Dai F, Liu Y, Li C, Pan F, Yang Y, Ni J, Li H, Brand-Saberi B, and Guo P (2015) Regression of gastric cancer by systemic injection of RNA nanoparticles carrying both ligand and siRNA. *Sci Rep*, **5**, 10726.

Deindl E (2001) 18S ribosomal RNA detection on Northern blot employing a specific oligonucleotide. *BioTechniques*, **31**, 1250, 1252.

Delebecque CJ, Silver PA, and Lindner AB (2012) Designing and using RNA scaffolds to assemble proteins in vivo. *Nat Protoc*, **7**, 1797–1807.

Dibrov SM, McLean J, Parsons J, and Hermann T (2011) Self-assembling RNA square. *Proc Natl Acad Sci U S A*, **108**, 6405–6408.

Ellington AD and Szostak JW (1992) Selection in vitro of single-stranded DNA molecules that fold into specific ligand-binding structures. *Nature*, **355**, 850–852.

Esposito CL, Passaro D, Longobardo I, Condorelli G, Marotta P, Affuso A, de F, V, and Cerchia L (2011) A neutralizing RNA aptamer against EGFR causes selective apoptotic cell death. *PLoS One*, **6**, e24071.

Felden B, Florentz C, Giegé R, and Westhof E (1996) A central pseudoknotted three-way junction imposes tRNA-like mimicry and the orientation of three 5' upstream pseudoknots in the 3' terminus of tobacco mosaic virus RNA. *RNA*, **2**, 201–212.

Fiore JL, Holmstrom ED, and Nesbitt DJ (2012) Entropic origin of Mg2+-facilitated RNA folding. *Proc Natl Acad Sci U S A*, **109**, 2902–2907.

Fire A, Xu S, Montgomery MK, Kostas SA, Driver SE, and Mello CC (1998) Potent and specific genetic interference by double-stranded RNA in Caenorhabditis elegans. *Nature*, **391**, 806–811.

Freier SM, Kierzek R, Jaeger JA, Sugimoto N, Caruthers MH, Neilson T, and Turner DH (1986) Improved free-energy parameters for predictions of RNA duplex stability. *Proc Natl Acad Sci U S A*, **83**, 9373–9377.

Gao H, Shi W, and Freund LB (2005) Mechanics of receptor-mediated endocytosis. *Proc Natl Acad Sci U S A*, **102**, 9469–9474.

Gold L (1995) The SELEX process: a surprising source of therapeutic and diagnostic compounds. *Harvey Lect*, **91**, 47–57.

Guo P (2010) The emerging field of RNA nanotechnology. *Nat Nanotechnol*, **5**, 833–842.

Guo P, Erickson S, and Anderson D (1987) A small viral RNA is required for *in vitro* packaging of bacteriophage phi29 DNA. *Science*, **236**, 690–694.

Guo S, Huang F, and Guo P (2006) Construction of folate-conjugated pRNA of bacteriophage phi29 DNA packaging motor for delivery of chimeric siRNA to nasopharyngeal carcinoma cells. *Gene Ther*, **13**, 814–820.

Guo S, Li H, Ma M, Fu J, Dong Y, and Guo P (2017) Size, shape, and sequence-dependent immunogenicity of rna nanoparticles. *Mol Ther Nucleic Acids*, **9**, 399–408.

Guo S, Tschammer N, Mohammed S, and Guo P (2005) Specific delivery of therapeutic RNAs to cancer cells via the dimerization mechanism of phi29 motor pRNA. *Hum Gene Ther*, **16**, 1097–1109.

Guo S, Vieweger M, Zhang K, Yin H, Wang H, Li X, Li S, Sparreboom A, Evers BM, Dong Y, Chui W, and Guo P (2020) Ultra-thermostable RNA nanoparticles for solubilizing and high yield loading of paclitaxel for breast cancer therapy with undetectable toxicity. *Nature Communications*

Guo P, Zhang C, Chen C, Trottier M, and Garver K (1998) Inter-RNA interaction of phage phi29 pRNA to form a hexameric complex for viral DNA transportation. *Mol Cell*, **2**, 149–155.

Gyi JI, Conn GL, Lane AN, and Brown T (1996) Comparison of the thermodynamic stabilities and solution conformations of DNA center dot RNA hybrids containing purine-rich and pyrimidine-rich strands with DNA and RNA duplexes. *Biochemistry*, **35**, 12538–12548.

Gyi JI, Gao DQ, Conn GL, Trent JO, Brown T, and Lane AN (2003) The solution structure of a DNA center dot RNA duplex containing 5-propynyl U and C; comparison with 5-Me modifications. *Nucleic Acids Res*, **31**, 2683–2693.

Hagenbuchle O, Santer M, and Steitz JA (1978) Conservation of primary structure at 3'-end of 18s RNA from eukaryotic cells. *Cell*, **13**, 551–563.

Hao C, Li X, Tian C, Jiang W, Wang G, and Mao C (2014) Construction of RNA nanocages by re-engineering the packaging RNA of Phi29 bacteriophage. *Nat Commun*, **5**, 3890.

Haque F, Shu D, Shu Y, Shlyakhtenko L, Rychahou P, Evers M, and Guo P (2012) Ultrastable synergistic tetra-valent RNA nanoparticles for targeting to cancers. *Nano Today*, **7**, 245–257.

Hoeprich S, ZHou Q, Guo S, Qi G, Wang Y, and Guo P (2003) Bacterial virus phi29 pRNA as a hammerhead ribozyme escort to destroy hepatitis B virus. *Gene Ther*, **10**, 1258–1267.

Huang L and Lilley DM (2016) A quasi-cyclic RNA nano-scale molecular object constructed using kink turns. *Nanoscale*, **8**, 15189–15195.

Jacobson DR, McIntosh DB, Stevens MJ, Rubinstein M, and Saleh OA (2017) Single-stranded nucleic acid elasticity arises from internal electrostatic tension. *Proc Natl Acad Sci U S A*, **114**, 5095–5100.

Jaeger JA, SantaLucia JJ, and Tinoco IJ (1993) Determination of RNA structure and thermodynamics. *Annu Rev Biochem*, **62**, 255–285.

Jaeger L, Westhof E, and Leontis NB (2001) TectoRNA: modular assembly units for the construction of RNA nano-objects. *Nucleic Acids Res*, **29**, 455–463.

Jain KK (2005) The role of nanobiotechnology in drug discovery. *Drug Discov Today*, **10**, 1435–1442.

Jasinski D, Haque F, Binzel DW, and Guo P (2017) Advancement of the emerging field of RNA nanotechnol-ogy. *ACS Nano*, **11**, 1142–1164.

Jasinski D, Khisamutdinov EF, Lyubchenko YL, and Guo P (2014) Physicochemically tunable poly-functionalized RNA square architecture with fluorogenic and ribozymatic properties. *ACS Nano*, **8**, 7620–7629.

Jasinski DL, Li H, and Guo P (2018) The effect of size and shape of RNA nanoparticles on biodistribution. *Mol Ther*, **26**, 784–792.

Kawasaki AM, Casper MD, Freier SM, Lesnik EA, Zounes MC, Cummins LL, Gonzalez C, and Cook PD (1993) Uniformly modified 2'-deoxy-2'-fluoro phosphorothioate oligonucleotides as nuclease-resistant antisense compounds with high affinity and specificity for RNA targets. *J Med Chem*, **36**, 831–841.

Kellenberger CA, Wilson SC, Sales-Lee J, and Hammond MC (2013) RNA-based fluorescent biosensors for live cell imaging of second messengers cyclic di-GMP and cyclic AMP-GMP. *J Am Chem Soc*, **135**, 4906–4909.

Khaled A, Guo S, Li F, and Guo P (2005) Controllable self-assembly of nanoparticles for specific delivery of multiple therapeutic molecules to cancer cells using RNA nanotechnology. *Nano Lett*, **5**, 1797–1808.

Khisamutdinov EF, Bui MN, Jasinski D, Zhao Z, Cui Z, and Guo P (2015) Simple method for constructing RNA triangle, square, pentagon by tuning interior RNA 3WJ angle from 60 degrees to 90 degrees or 108 degrees. *Methods Mol Biol*, **1316**, 181–193.

Khisamutdinov EF, Jasinski DL, and Guo P (2014) RNA as a boiling-resistant anionic polymer material to build robust structures with defined shape and stoichiometry. *ACS Nano*, **8**, 4771–4781.

Khisamutdinov EF, Jasinski DL, Li H, Zhang K, Chiu W, and Guo P (2016) Fabrication of RNA 3D nanoprism for loading and protection of small RNAs and model drugs. *Adv Mater*, **28**, 100079–100087.

Khisamutdinov E, Li H, Jasinski D, Chen J, Fu J, and Guo P (2014) Enhancing immunomodulation on innate immunity by shape transition among RNA triangle, square, and pentagon nanovehicles. *Nucleic Acids Res*, **42**, 9996–10004.

Khvorova A and Watts JK (2017) The chemical evolution of oligonucleotide therapies of clinical utility. *Nat Biotechnol*, **35**, 238–248.

Lagos-Quintana M, Rauhut R, Lendeckel W, and Tuschl T (2001) Identification of novel genes coding for small expressed RNAs. *Science*, **294**, 853–858.

Laing C and Schlick T (2009) Analysis of four-way junctions in RNA structures. *J Mol Biol*, **390**, 547–559.

Lamiable A, Barth D, Denise A, Quessette F, Vial S, and Westhof E (2012) Automated prediction of three-way junction topological families in RNA secondary structures. *Comput Biol Chem*, **37**, 1–5.

Lau NC, Lim LP, Weinstein EG, and Bartel DP (2001) An abundant class of tiny RNAs with probable regula-tory roles in Caenorhabditis elegans. *Science*, **294**, 858–862.

Layzer JM, McCaffrey AP, Tanner AK, Huang Z, Kay MA, and Sullenger BA (2004) In vivo activity of nucle-ase-resistant siRNAs. *RNA*, **10**, 766–771.

Lee RC, Feinbaum RL, and Ambros V (1993) The C. elegans heterochronic gene lin-4 encodes small RNAs with antisense complementarity to lin-14. *Cell*, **75**, 843–854.

Lee TJ, Haque F, Shu D, Yoo JY, Li H, Yokel RA, Horbinski C, Kim TH, Kim S-H, Nakano I, Kaur B, Croce CM, and Guo P (2015) RNA nanoparticles as a vector for targeted siRNA delivery into glioblastoma mouse model. *Oncotarget*, **6**, 14766–14776.

Lee JB, Hong J, Bonner DK, Poon Z, and Hammond PT (2012) Self-assembled RNA interference micro-sponges for efficient siRNA delivery. *Nat Mater*, **11**, 316–322.

Lee T, Yagati AK, Pi F, Sharma A, Choi J-W, and Guo P (2015) Construction of RNA-quantum dot chimera for nanoscale resistive biomemory application. *ACS Nano*, **9**, 6675–6682.

Lee TJ, Yoo JY, Shu D, Li H, Zhang J, Yu JG, Jaime-Ramirez AC, Acunzo M, Romano G, Cui R, Sun HL, Luo Z, Old M, Kaur B, Guo P, and Croce CM (2017) RNA nanoparticle-based targeted therapy for glioblastoma through inhibition of oncogenic miR-21. *Mol Ther*, **25**, 1544–1555.

Lesnik EA and Freier SM (1995) Relative thermodynamic stability of DNA, RNA, and DNA-RNA hybrid duplexes - relationship with base composition and structure. *Biochemistry*, **34**, 10807–10815.

Li W and Szoka F (2007) Lipid-based nanoparticles for nucleic acid delivery. *Pharm Res*, **24**, 438–449.

Li H, Zhang K, Pi F, Guo S, Shlyakhtenko L, Chiu W, Shu D, and Guo P (2016) Controllable self-assembly of RNA tetrahedrons with precise shape and size for cancer targeting. *Adv Mater*, **28**, 7501–7507.

Maeda H (2001) The enhanced permeability and retention (EPR) effect in tumor vasculature: the key role of tumor-selective macromolecular drug targeting. *Adv Enzym Regul*, **41**, 189–207.

Maeda H, Nakamura H, and Fang J (2013) The EPR effect for macromolecular drug delivery to solid tumors: Improvement of tumor uptake, lowering of systemic toxicity, and distinct tumor imaging in vivo. *Adv Drug Deliv Rev*, **65**, 71–79.

Mamot C, Drummond DC, Noble CO, Kallab V, Guo ZX, Hong KL, Kirpotin DB, and Park JW (2005) Epidermal growth factor receptor-targeted immunoliposomes significantly enhance the efficacy of multiple anticancer drugs in vivo. *Cancer Res*, **65**, 11631–11638.

Markham NR and Zuker M (2008) UNAFold: software for nucleic acid folding and hybridization. *Methods Mol Biol*, **453**, 3–31.

Martinez HM, Maizel JV, and Shapiro BA (2008) RNA2D3D: A program for generating, viewing, and comparing 3-dimensional models of RNA. *J Biomol Struct Dyn*, **25**, 669–683.

Mathews DH and Turner DH (2002) Experimentally derived nearest-neighbor parameters for the stability of RNA three- and four-way multibranch loops. *Biochemistry*, **41**, 869–880.

Mathews DH and Turner DH (2006) Prediction of RNA secondary structure by free energy minimization. *Curr Opin Struct Biol*, **16**, 270–278.

Mattick JS and Makunin IV (2005) Small regulatory RNAs in mammals. *Hum Mol Genet*, **14**, R121–R132.

Mulhbacher J, St-Pierre P, and Lafontaine DA (2010) Therapeutic applications of ribozymes and riboswitches. *Curr Opin Pharmacol*, **10**, 551–556.

Noy A (2011) Bionanoelectronics. *Adv Mater*, **23**, 807–820.

Obad S, dos Santos CO, Petri A, Heidenblad M, Broom O, Ruse C, Fu C, Lindow M, Stenvang J, Straarup EM, Hansen HF, Koch T, Pappin D, Hannon GJ, and Kauppinen S (2011) Silencing of microRNA families by seed-targeting tiny LNAs. *Nat Genet*, **43**, 371–378.

Pagratis NC, Bell C, Chang YF, Jennings S, Fitzwater T, Jellinek D, and Dang C (1997) Potent 2'-amino-, and 2'-fluoro-2'-deoxyribonucleotide RNA inhibitors of keratinocyte growth factor. *Nat Biotechnol*, **15**, 68–73.

Paige JS, Nguyen-Duc T, Song W, and Jaffrey SR (2012) Fluorescence imaging of cellular metabolites with RNA. *Science*, **335**, 1194.

Paillart JC, Westhof E, Ehresmann C, Ehresmann B, and Marquet R (1997) Non-canonical interactions in a kissing loop complex: The dimerization initiation site of HIV-1 genomic RNA. *J Mol Biol*, **270**, 36–49.

Pallan PS, Greene EM, Jicman PA, Pandey RK, Manoharan M, Rozners E, and Egli M (2011) Unexpected origins of the enhanced pairing affinity of 2'-fluoro-modified RNA7. *Nucleic Acids Res*, **39**, 3482–3495.

Panigaj M, Johnson MB, Ke W, McMillan J, Goncharova EA, Chandler M, and Afonin KA (2019) Aptamers as modular components of therapeutic nucleic acid nanotechnology. *ACS Nano*, **13**, 12301–12321.

Parlea L, Puri A, Kasprzak W, Bindewald E, Zakrevsky P, Satterwhite E, Joseph K, Afonin KA, and Shapiro BA (2016) Cellular delivery of RNA nanoparticles. *ACS Comb Sci*, **18**, 527–547.

Pastor F, Berraondo P, Etxeberria I, Frederick J, Sahin U, Gilboa E, and Melero I (2018) An RNA toolbox for cancer immunotherapy. *Nat Rev Drug Discov*, **17**, 751–767.

Pegtel DM, Cosmopoulos K, Thorley-Lawson DA, van Eijndhoven MA, Hopmans ES, Lindenberg JL, de Gruijl TD, Wurdinger T, and Middeldorp JM (2010) Functional delivery of viral miRNAs via exosomes. *Proc Natl Acad Sci U S A*, **107**, 6328–6333.

Petrov AI, Zirbel CL, and Leontis NB (2013) Automated classification of RNA 3D motifs and the RNA 3D Motif Atlas. *RNA*, **19**, 1327–1340.

Pi F, Binzel DW, Lee TJ, Li Z, Sun M, Rychahou P, Li H, Haque F, Wang S, Croce CM, Guo B, Evers BM, and Guo P (2018) Nanoparticle orientation to control RNA loading and ligand display on extracellular vesicles for cancer regression. *Nat Nanotechnol*, **13**, 82–89.

Pi F, Zhang H, Li H, Thiviyanathan V, Gorenstein DG, Sood AK, and Guo P (2016) RNA nanoparticles harboring annexin A2 aptamer can target ovarian cancer for tumor-specific doxorubicin delivery. *Nanomedicine*, **13**, 1183–1193.

Piao X, Wang H, Binzel DW, and Guo P (2018) Assessment and comparison of thermal stability of phosphorothioate-DNA, DNA, RNA, 2'-F RNA, and LNA in the context of Phi29 pRNA 3WJ. *RNA*, **24**, 67–76.

Piao X, Yin H, Guo S, Wang H, and Guo P (2019) RNA nanotechnology to solubilize hydrophobic antitumor drug for targeted delivery. *Adv Sci*, **6**, 1900951–1900958.

Pleij CWA, Rietveld K, and Bosch L. (1985) A new principle of RNA folding based on pseudonotting. *Nucleic Acids Res*, **13(5)**, 1717–1731.

Porciani D, Cardwell LN, Tawiah KD, Alam KK, Lange MJ, Daniels MA, and Burke DH (2018) Modular cell-internalizing aptamer nanostructure enables targeted delivery of large functional RNAs in cancer cell lines. *Nat Commun*, **9**, 2283.

Qiu M, Khisamutdinov E, Zhao Z, Pan C, Choi J, Leontis N, and Guo P (2013) RNA nanotechnology for computer design and *in vivo* computation. *Phil Trans R Soc A*, **371(2000)**, 20120310.

Rauzan B, McMichael E, Cave R, Sevcik LR, Ostrosky K, Whitman E, Stegemann R, Sinclair AL, Serra MJ, and Deckert AA (2013) Kinetics and thermodynamics of DNA, RNA, and hybrid duplex formation. *Biochemistry*, **52**, 765–772.

Rinaudo K, Bleris L, Maddamsetti R, Subramanian S, Weiss R, and Benenson Y (2007) A universal RNAi-based logic evaluator that operates in mammalian cells. *Nat Biotechnol*, **25**, 795–801.

Rychahou P, Haque F, Shu Y, Zaytseva Y, Weiss HL, Lee EY, Mustain W, Valentino J, Guo P, and Evers BM (2015) Delivery of RNA nanoparticles into colorectal cancer metastases following systemic administration. *ACS Nano*, **9**, 1108–1116.

Santalucia J, Allawi HT, and Seneviratne A (1996) Improved nearest-neighbor parameters for predicting DNA duplex stability. *Biochemistry*, **35**, 3555–3562.

Sarver NA, Cantin EM, Chang PS, Zaia JA, Ladne PA, Stephens DA, and Rossi JJ (1990) Ribozymes as potential anti-HIV-1 therapeutic agents. *Science*, **24**, 1222–1225.

Sclavi B, Sullivan M, Chance MR, Brenowitz M, and Woodson SA (1998) RNA folding at millisecond intervals by synchrotron hydroxyl radical footprinting. *Science*, **279**, 1940–1943.

Severcan I, Geary C, Chworos A, Voss N, Jacovetty E, and Jaeger L (2010) A polyhedron made of tRNAs. *Nat Chem*, **2**, 772–779.

Severcan I, Geary C, Jaeger L, Bindewald E, Kasprzak W, and Shapiro BA (2009). Computational and Experimental RNA Nanoparticle Design. In Alterovitz,G. and Ramoni,M. (Eds.), *Automation in Genomics and Proteomics: An Engineering Case-Based Approach*,. Wiley, Boston, MA, pp. 193–220.

Severcan I, Geary C, Verzemnieks E, Chworos A, and Jaeger L (2009) Square-shaped RNA particles from different RNA folds. *Nano Lett*, **9**, 1270–1277.

Shapiro BA and Wu JC (1997) Predicting RNA H-type pseudoknots with the massively parallel genetic algorithm. *Comput Appl Biosci*, **13**, 459–471.

Sharma A, Haque F, Pi F, Shlyakhtenko L, Evers BM, and Guo P (2015) Controllable self-assembly of RNA dendrimers. *Nanomed Nanotechnol Biol Med*, **12**, 835–844.

Shaw JP, Kent K, Bird J, Fishback J, and Froehler B (1991) Modified deoxyoligonucleotides stable to exonuclease degradation in serum. *Nucleic Acids Res*, **19**, 747–750.

Shigdar S, Lin J, Yu Y, Pastuovic M, Wei M, and Duan W (2011) RNA aptamer against a cancer stem cell marker epithelial cell adhesion molecule. *Cancer Sci*, **102**, 991–998.

Shu Y, Haque F, Shu D, Li W, Zhu Z, Kotb M, Lyubchenko Y, and Guo P (2013) Fabrication of 14 different RNA nanoparticles for specific tumor targeting without accumulation in normal organs. *RNA*, **19**, 766–777.

Shu D, Huang L, Hoeprich S, and Guo P (2003) Construction of phi29 DNA-packaging RNA (pRNA) monomers, dimers and trimers with variable sizes and shapes as potential parts for nano-devices. *J Nanosci Nanotechnol*, **3**, 295–302.

Shu D, Li H, Shu Y, Xiong G, Carson WE, Haque F, Xu R, and Guo P (2015) Systemic delivery of anti-miRNA for suppression of triple negative breast cancer utilizing RNA nanotechnology. *ACS Nano*, **9**, 9731–9740.

Shu Y, Pi F, Sharma A, Rajabi M, Haque F, Shu D, Leggas M, Evers BM, and Guo P (2014) Stable RNA nanoparticles as potential new generation drugs for cancer therapy. *Adv Drug Deliv Rev*, **66C**, 74–89.

Shu D, Shu Y, Haque F, Abdelmawla S, and Guo P (2011) Thermodynamically stable RNA three-way junctions for constructing multifuntional nanoparticles for delivery of therapeutics. *Nat Nanotechnol*, **6**, 658–667.

Shu Y, Yin H, Rajabi M, Li H, Vieweger M, Guo S, Shu D, and Guo P (2018) RNA-based micelles: A novel platform for paclitaxel loading and delivery. *J Control Release*, **276**, 17–29.

Shukla GC, Haque F, Tor Y, Wilhelmsson LM, Toulme JJ, Isambert H, Guo P, Rossi JJ, Tenenbaum SA, and Shapiro BA (2011) A boost for the emerging field of RNA nanotechnology. *ACS Nano*, **5**, 3405–3418.

Stein CA and Castanotto D (2017) FDA-approved oligonucleotide therapies in 2017. *Mol Ther*, **25**, 1069–1075.

Stewart JM, Viard M, Subramanian HK, Roark BK, Afonin KA, and Franco E (2016) Programmable RNA microstructures for coordinated delivery of siRNAs. *Nanoscale*, **8**, 17542–17550.

Sugimoto N, Nakano S, Katoh M, Matsumura A, Nakamuta H, Ohmichi T, Yoneyama M, and Sasaki M (1995) Thermodynamic parameters to predict stability of RNA/DNA hybrid duplexes. *Biochemistry*, **34**, 11211–11216.

Sundaram P, Kurniawan H, Byrne ME, and Wower J (2013) Therapeutic RNA aptamers in clinical trials. *Eur J Pharm Sci*, **48**, 259–271.

Tabernero J, Shapiro GI, Lorusso PM, Cervantes A, Schwartz GK, Weiss GJ, Paz-Ares L, Cho DC, Infante JR, Alsina M, Gounder MM, Falzone R, Harrop J, Seila White AC, Toudjarska I, Bumcrot D, Meyers RE, Hinkle G, Svrzikapa N, Hutabarat RM, Clausen VA, Cehelsky J, Nochur SV, Gamba-Vitalo C, Vaishnaw AK, Sah DW, Gollob JA, and Burris HA, III (2013) First-in-Man Trial of an RNA Interference Therapeutic Targeting VEGF and KSP in Cancer Patients with Liver Involvement. *Can Discov*, **3**, 406–417.

Thiel KW, Hernandez LI, Dassie JP, Thiel WH, Liu X, Stockdale KR, Rothman AM, Hernandez FJ, McNamara JO, and Giangrande PH (2012) Delivery of chemo-sensitizing siRNAs to HER2+-breast cancer cells using RNA aptamers. *Nucleic Acids Res*, **40**, 6319–6337.

Tinoco I, Jr., Li PT, and Bustamante C (2006) Determination of thermodynamics and kinetics of RNA reactions by force. *Q Rev Biophys*, **39**, 325–360.

Tinoco I, Uhlenbec O, and Levine MD (1971) Estimation of secondary structure in ribonucleic acids. *Nature*, **230**, 362.

Turner DH, Sugimoto N, and Freier SM (1988) RNA structure prediction. *Annu Rev Biophys Chem*, **17**, 167–192.

Walter F, Murchie AI, and Lilley DM (1998) Folding of the four-way RNA junction of the hairpin ribozyme. *Biochemistry*, **37**, 17629–17636.

Westhof E, Dumas P, and Moras D (1985) Crystallographic refinement of yeast aspartic acid transfer RNA. *J Mol Biol*, **184**, 119–145.

Westhof E, Masquida B, and Jaeger L (1996) RNA tectonics: Towards RNA design. *Fold Des*, **1**, R78–R88.

Win MN and Smolke CD (2008) Higher-order cellular information processing with synthetic RNA devices. *Science*, **322**, 456–460.

Winkler WC, Nahvi A, Roth A, Collins JA, and Breaker RR (2004) Control of gene expression by a natural metabolite-responsive ribozyme. *Nature*, **428**, 281–286.

Xayaphoummine A, Bucher T, and Isambert H (2005) Kinefold web server for RNA/DNA folding path and structure prediction including pseudoknots and knots. *Nucleic Acids Res*, **33**, W605–W610.

Xia TB, Santalucia J, Burkard ME, Kierzek R, Schroeder SJ, Jiao XQ, Cox C, and Turner DH (1998) Thermodynamic parameters for an expanded nearest-neighbor model for formation of RNA duplexes with Watson-Crick base pairs. *Biochemistry*, **37**, 14719–14735.

Xie Z, Liu SJ, Bleris L, and Benenson Y (2010) Logic integration of mRNA signals by an RNAi-based molecular computer. *Nucleic Acids Res*, **38**, 2692–2701.

Xie Z, Wroblewska L, Prochazka L, Weiss R, and Benenson Y (2011) Multi-input RNAi-based logic circuit for identification of specific cancer cells. *Science*, **333**, 1307–1311.

Xu C, Li H, Zhang K, Binzel DW, Yin H, Chiu W, and Guo P (2019) Photo-controlled release of paclitaxel and model drugs from RNA pyramids. *Nano Res*, **12**, 41–48.

Xu Y, Pang L, Wang H, Xu C, Shah H, Guo P, Shu D, and Qian SY (2018) Specific delivery of delta-5-desaturase siRNA via RNA nanoparticles supplemented with dihomo-gamma-linolenic acid for colon cancer suppression. *Redox Biol*, **21**, 101085.

Ye X, Liu Z, Hemida MG, and Yang D (2011) Targeted delivery of mutant tolerant anti-coxsackievirus artificial microRNAs using folate conjugated bacteriophage Phi29 pRNA. *PLoS One*, **6**, e21215.

Yin H, Wang H, Li Z, Shu D, and Guo P (2019) RNA micelles for systemic delivery of Anti-miRNA for cancer targeting and inhibition without ligand. *ACS Nano*, **13(1)**, 706–717.

Yin H, Xiong G, Guo S, Xu C, Xu R, Guo P, and Shu D (2019) Delivery of anti-miRNA for triple-negative breast cancer therapy using RNA nanoparticles targeting stem cell marker CD133. *Mol Ther*, **27**, 1252–1261.

Yingling YG and Shapiro BA (2007) Computational design of an RNA hexagonal nanoring and an RNA nanotube. *Nano Lett*, **7**, 2328–2334.

Yu JW, Liu ZY, Jiang W, Wang GS, and Mao CD (2015) De novo design of an RNA tile that self-assembles into a homo-octameric nanoprism. *Nat Commun*, **6**, 5724–5729.

Zarrinkar PP and Williamson JR (1994) Kinetic intermediates in RNA folding. *Science*, **265**, 918–924.

Zhang Y, Leonard M, Shu Y, Yang Y, Shu D, Guo P, and Zhang X (2017) Overcoming tamoxifen resistance of human breast cancer by targeted gene silencing using multifunctional pRNA nanoparticles. *ACS Nano*, **11**, 335–346.

Zhou ZH, Dougherty M, Jakana J, He J, Rixon FJ, and Chiu W (2000) Seeing the herpesvirus capsid at 8.5 A. *Science*, **288**, 877–880.

Zhuang X, Bartley LE, Babcock HP, Russell R, Ha T, Herschlag D, and Chu S (2000) A single-molecule study of RNA catalysis and folding. *Science*, **288**, 2048–2051.

Zuker M (2003) Mfold web server for nucleic acid folding and hybridization prediction. *Nucleic Acids Res*, **31**, 3406–3415.

Part III

Immunorecognition of RNA
Nanoparticles

26 Immunorecognition of Nucleic Acid Nanoparticles

Jessica McMillan and M. Brittany Johnson
University of North Carolina at Charlotte, Charlotte, USA

CONTENTS

Pattern recognition receptors (PRRs) are a double-edged sword that all nucleic acid nanoparticles (NANPs) must navigate. PRRs scan the extracellular, endosomal, and cytosolic environments for pathogen-associated molecular patterns (PAMPs) which are associated with microbial pathogens and damage-associated molecular patterns (DAMPs) associated with damage or death to the host cell. Ligand binding to PRRs promptly triggers a signaling cascade to drive tailored innate and adaptive immune responses (Takeuchi & Akira, 2010). Notably, off-target or detrimental immune responses restrict the therapeutic use of NANPs (Dobrovolskaia & McNeil, 2015a, b). However, therapeutics such as vaccine adjuvants and cancer immunotherapies are designed to stimulate advantageous immune responses (Goutagny, Estornes, Hasan, Lebecque, & Caux, 2012; Iurescia, Fioretti, & Rinaldi, 2017; Steinhagen, Kinjo, Bode, & Klinman, 2011; Temizoz, Kuroda, & Ishii, 2018). Understanding the characteristics of PRR ligands and the fine-tuneability of ligand parameters that can alter recognition provides valuable information to generate design principles for assembling therapeutic NANPs. This chapter will introduce PRRs known to identify nucleic acids and will explore parameters used to rationally design nanoparticles with specific immunostimulatory or immunoquiescent properties.

PRR RECOGNITION OF NUCLEIC ACIDS

Toll-like receptors (TLRs), are important PRRs that are expressed on the cell surface and endosomal compartment to survey the extracellular and endosomal space for PAMPs and DAMPs (Botos, Segal, & Davies, 2011; Takeuchi & Akira, 2010). TLRs possess three highly conserved structural domains: a ligand-binding domain composed of leucine-rich repeats, a transmembrane domain, and a cytoplasmic signaling domain. Notably, TLRs are differentially expressed among cell types. For instance, plasmacytoid dendritic cells (pDCs) and B-cells readily express TLR7, while phagocytic cells such as monocytes, macrophages, and myeloid dendritic cells express TLR8 (Eng, Hsu, & Lin, 2018). In this section, we will focus on the endosomal TLRs, TLR3, TLR 7/8, and TLR9 that have been documented to recognize nucleic acids. Importantly, the ligands for these TLRs are characterized by composition (DNA, RNA, hybrid), strand length (base pairs), and other modifiers (unmethylated, analogs). TLR3 preferentially recognizes dsRNA that is 40–50bp in length (Leonard et al., 2008; Wang, Liu, Davies, & Segal, 2010). In contrast, TLR7/8 binds poly U or G/U-rich ssRNA, nucleoside analogs, and imidazoquinolines (Barbalat, Ewald, Mouchess, & Barton, 2011; Forsbach et al., 2008; Hornung et al., 2005). Finally, unmethylated CpG DNA that is a minimum of

20 nucleotides is recognized by TLR9 (Hemmi et al., 2000). Activation of TLR7/8 and 9 stimulates MyD88-dependent pathways while TLR3 stimulates toll/interleukin-1-receptor (TIR)- domain-containing adapter-inducing interferon-β (TRIF)-dependent production of interferons (IFNs) and pro-inflammatory cytokines (Kawasaki & Kawai, 2014).

In addition to endosomal TLRs, the cytosolic nucleic acid sensors, retinoic acid inducible gene-I (RIG-I)-like receptors (RLRs) and DNA sensors identify nucleic acids. RIG-I is the founding member of the RLR family that also includes melanoma differentiation-associated protein-5 (MDA-5) and laboratory of genetics and physiology 2 (LGP2) (Loo & Gale, 2011; Yoneyama et al., 2004). These sensors possess common structural domains: All RLRs possess an ATPase containing DEAD box helicase and a C-terminal domain, RIG-I and LGP2 are autoregulated via a C-terminal repressor domain, RIG-I and MDA-5 additionally possess N-terminal signaling domains known as caspase activation and recruitment domains (CARDs) (T. Saito et al., 2007; Yoneyama et al., 2005, 2004). LGP2 serves as a regulator of RLR signaling as it lacks the required signaling domains (Satoh et al., 2010; Venkataraman et al., 2007; Yoneyama et al., 2005). Each of these receptors recognizes specific RNA signatures. RIG-I and MDA-5 recognize 19-300bp 5′ tri-phosphorylated (5′ppp) dsRNA and 1–4kb dsRNA, respectively (Hornung et al., 2006; Kato et al., 2008; Loo & Gale, 2011; Takeshi Saito, Owen, Jiang, Marcotrigiano, & Gale, 2008; Uzri & Gehrke, 2009). Additionally, AT-rich dsDNA can be transcribed by RNA polymerase III into a 5′ppp dsRNA ligand that can then be identified by RIG-I (Ablasser et al., 2009; Chiu, MacMillan, & Chen, 2009). Ligand recognition by RIG-I or MDA-5 triggers a mitochondrial antiviral signaling protein (MAVS) (also known as interferon-β promoter stimulator 1, IPS-1)-dependent signaling cascade driving nuclear translocation of the transcription factors, nuclear factor kappa-light-chain-enhancer of activated B cells (NF-κB), and interferon regulator factor (IRF) (Hiscott, Lacoste, & Lin, 2006; Öhman, Rintahaka, Kalkkinen, Matikainen, & Nyman, 2009). Sequentially, these transcription factors activate expression of pro-inflammatory cytokines and interferons (Loo & Gale, 2011). Finally, the DNA sensor cyclic GMP-AMP synthase (cGAS) monitors the cytosol for B or Y forming DNA and RNA–DNA hybrids longer than 40 bp. Ligand binding to cGAS stimulates the production of a second messenger, 2′-3′ cyclic GMP-AMP (cGAMP) which then binds to stimulator of interferon genes (STING). Similar to MAVS, STING activation triggers nuclear translocation of NF-κB and IRF to stimulate production of pro-inflammatory cytokines and interferons (Bhat & Fitzgerald, 2014; Gürtler & Bowie, 2013).

ENGINEERING NANP IMMUNORECOGNITION

Understanding the ligand characteristics for PRRs is important for the development of NANPs that are engineered to either stimulate or abrogate recognition thereby generating desired immune responses or delivering functional groups in an immunoquiescent manner. There are four primary factors we will discuss regarding the rational design of NANP immunostimulation: carrier, nucleic acid composition, sequence, and chemical modifications. First, as previously discussed, free exogenous RNA nanoparticles cannot enter cells without a carrier due to their negative charge caused by the phosphate backbone. Carriers including nanostructures, liposomes, polymers, dendrimers, silicon, carbon materials, and magnetic nanoparticles range in toxicity, biodistribution, and mechanism of delivery (Behzadi et al., 2017; Foroozandeh & Aziz, 2018; Wilczewska, Niemirowicz, Markiewicz, & Car, 2012). Importantly, PRRs are localized to specific subcellular compartments and carrier selection greatly impacts the PRRs that NANPs will encounter upon delivery. For example, gold nanoparticles coated with polyethylene glycol (PEG), magnetic nanoparticles coated with polyethylenimine (PEI), poly (lactide-co-glycolide)-graft-polyethylenimine (PgP) polymer, and nanoparticles with cell penetrating peptides traffic through the endosome and NANPs are delivered to the cytosol upon endosomal escape exposing NANPs to both endosomal and cytosolic nucleic acid sensors (Halman et al., 2020; Panyam, Zhou, Prabha, Sahoo, & Labhasetwar, 2002; Shenoy et al., 2006; Silva, Almeida, & Vale, 2019). In contrast, liposomes can undergo membrane fusion for direct delivery to the cytosol thereby avoiding endosomal PRRs (Wilczewska et al., 2012). Finally, dynamic RNA–DNA hybrid NANP platforms can be designed to avoid recognition by endosomal TLRs. These sense-antisense fibers, at 40 bp long, consist of a sense DNA strand that is 40 bp long with three sense short RNA strands complementarily bound. The anti-duplex consists of the reverse complement of the four sense strands. In cells, this pair of duplexes re-anneals with their cis NA complement (i.e. DNA pairs with DNA, RNA pairs with RNA). Thus, creating a dsDNA duplex that goes on to induce the cGAS-STING pathway and a dsRNA duplex that goes on to be sliced by a dicer substrate and partakes in RNA interference (RNAi) (Afonin et al., 2014, 2013). The diverse range of carriers and the ability to engineer dynamic NANP platforms provides the means to target or avoid specific PRRs through tightly controlled subcellular delivery.

Second, the nucleic acid composition of NANPs greatly impacts PRR recognition given most PRRs recognize only RNA or DNA. The exceptions include RIG-I which recognizes RNA directly and DNA indirectly via RNA polymerase III and cGAS that can recognize RNA–DNA hybrids. Our previous data indicates RNA NANPs are highly immunostimulatory and stimulate significant production of type I IFNs. In contrast, DNA NANPs stimulate minimal release of pro-inflammatory cytokines and interferons providing a more immunoquiescent platform (Bui et al., 2017; Hong et al., 2018; Johnson et al., 2017). Importantly, quantitative structure–activity relationships (QSAR) modeling demonstrates nanoparticle molecular weight, melting temperature, and half-life are predictive of NANP immunostimulation. Therefore, QSAR modeling can be applied to develop guidelines for therapeutic NANP design (Johnson et al., 2017).

Third, many PRRs have sequence-dependent recognition of nucleic acids (Pandey, Kawai, & Akira, 2015; Wu & Chen, 2014). For instance, TLR7/8 preferentially identifies poly U or G/U-rich ssRNA (Pandey et al., 2015; Sioud, 2006). Previous research indicates short interfering RNAs (siRNA) stimulate type I IFN responses. In pDCs, TLR7 mediates production of IFNs. Importantly, inclusion of a nine base pair motif (5′-GUCCUUCAA-3′) was required for immunostimulation denoting a sequence- dependent response (Hornung et al., 2005). Similarly, mesoporous silica nanoparticle or spherical nucleic acid delivery of sequences with CpG oligodeoxynucleotides stimulatesTLR9 due to inclusion of the known TLR9 ligand motif (Hanagata, 2012; Radovic-Moreno et al., 2015). Previously, NANPs have been demonstrated to trigger interferon responses (Bui et al., 2017; Halman et al., 2020; Hong et al., 2018, 2019; Johnson et al., 2017). Moreover, studies in peripheral blood mononuclear cells (PBMC) indicate TLR7 is important for the identification of RNA cubes and rings. However, TLR7 does not have a role in the recognition of RNA fibers, suggesting in addition to sequence there is a role for nanoparticle structure in determining immunostimulatory properties (Hong et al., 2019). Data also suggests TLR9 contributes to RNA cube driven cellular response, but the mechanism is currently unknown (Hong et al., 2019).

Finally, chemical modifications of nucleic acids affect immunorecognition by PRRs. Particularly, 2′-hydroxyl uridines are important for PRR recognition. Replacement of 2′-hydroxyl uridines with either 2′-O-methyl or 2′deoxy residues potently inhibit the production of inflammatory cytokines and interferons (Robbins et al., 2007; Schmitt et al., 2017; Sioud, 2006). Specifically, 2′-O-methyl residues abrogate TLR7 stimulation and may serve as an antagonist (Robbins et al. 2007, 2′O methyl modified RNAs act as TLR7 antagonist). Interestingly, incorporation of 2′-fluoro uridines abrogated TLR3 and TLR7 activation, but enhanced RIG-I mediated immune responses (Lee, Urban, Xu, Sullenger, & Lee, 2016). Collectively, these four parameters: carrier, nucleic acid composition, sequence, and chemically modified nucleic acids provide a means to fine-tune the activation or avoidance of specific PRRs. Generating design principles that incorporate these parameters will allow for the rational design of NANPs that harness the immune system for therapeutic applications (Figure 26.1).

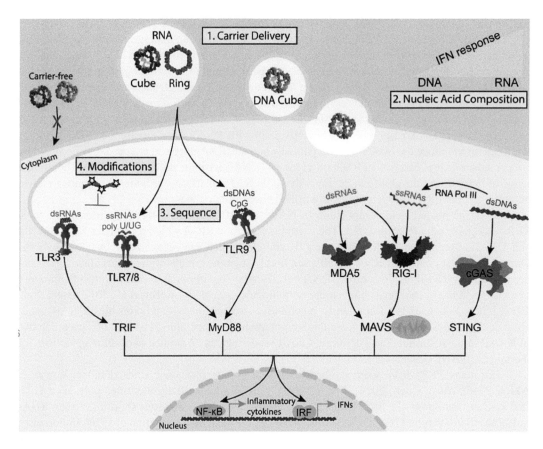

FIGURE 26.1 Immunorecognition of NANPs 1. In the absence of a carrier, NANPs are not internalized. Carrier determines NANP delivery to either the endosome or cytosol. Once inside the cell, NANPs can be recognized by endosomal and/or cytosolic nucleic acid sensors. 2. Nucleic acid composition affects production of IFNs with RNA-inducing potent IFN responses. For examples, RNA cubes and rings are identified by TLR7 and/or TLR9 and stimulate the production of IFNs. 3. PRRs identify specific nucleic acid motifs. TLR3, TLR7/8, and TLR9 recognize dsRNA, poly U/UG ssRNA, and CpG DNA, respectively. Within the cytosol RNA ligands are identified by the RLRs, MDA-5, and RIG-I and DNA is identified by DNA sensors such as cGAS. RIG-I is known to specifically identify polyuridine-rich 5' tri-phosphorylated RNA. Recognition by PRRs is affected by inclusion or exclusion of these specific ligand motifs. Recognition of nucleic acid ligands activates signaling cascades and drives nuclear translocation of the transcription factors NF-κB and IRF. Then, transcription factors initiate expression of inflammatory cytokines and IFNs. 4. Chemically modified nucleic acids can prevent TLR3 and TLR7 stimulation and reduce inflammatory cytokine and IFN responses.

REFERENCES

Ablasser, A., Bauernfeind, F., Hartmann, G., Latz, E., Fitzgerald, K. A., & Hornung, V. (2009). RIG-I-dependent sensing of poly(dA:dT) through the induction of an RNA polymerase III-transcribed RNA intermediate. *Nature Immunology, 10*(10), 1065–1072. doi:10.1038/ni.1779

Afonin, K. A., Desai, R., Viard, M., Kireeva, M. L., Bindewald, E., Case, C. L., ... Shapiro, B. A. (2014). Co-transcriptional production of RNA-DNA hybrids for simultaneous release of multiple split functionalities. *Nucleic Acids Research.* doi:10.1093/nar/gkt1001

Afonin, K. A., Viard, M., Martins, A. N., Lockett, S. J., MacIag, A. E., Freed, E. O., ... Shapiro, B. A. (2013). Activation of different split functionalities on re-association of RNA-DNA hybrids. *Nature Nanotechnology.* doi:10.1038/nnano.2013.44

Barbalat, R., Ewald, S. E., Mouchess, M. L., & Barton, G. M. (2011). Nucleic acid recognition by the innate immune system. *Annual Review of Immunology*. doi:10.1146/annurev-immunol-031210-101340

Behzadi, S., Serpooshan, V., Tao, W., Hamaly, M. A., Alkawareek, M. Y., Dreaden, E. C., ... Mahmoudi, M. (2017). Cellular uptake of nanoparticles: Journey inside the cell. *Chemical Society Reviews*. doi:10.1039/c6cs00636a

Bhat, N., & Fitzgerald, K. A. (2014). Recognition of cytosolic DNA by cGAS and other STING-dependent sensors. *European Journal of Immunology*. doi:10.1002/eji.201344127

Botos, I., Segal, D. M., & Davies, D. R. (2011). The structural biology of Toll-like receptors. *Structure*. doi:10.1016/j.str.2011.02.004

Bui, M. N., Brittany Johnson, M., Viard, M., Satterwhite, E., Martins, A. N., Li, Z., ... Khisamutdinov, E. F. (2017). Versatile RNA tetra-U helix linking motif as a toolkit for nucleic acid nanotechnology. *Nanomedicine: Nanotechnology, Biology and Medicine*, *13*(3), 1137–1146. doi:10.1016/j.nano.2016.12.018

Chiu, Y. H., MacMillan, J. B., & Chen, Z. J. (2009). RNA polymerase III detects cytosolic DNA and Induces Type I Interferons through the RIG-I Pathway. *Cell*, *138*(3), 576–591. doi:10.1016/j.cell.2009.06.015

Dobrovolskaia, M. A., & McNeil, S. E. (2015a). Strategy for selecting nanotechnology carriers to overcome immunological and hematological toxicities challenging clinical translation of nucleic acid-based therapeutics. *Expert Opinion on Drug Delivery*. doi:10.1517/17425247.2015.1042857

Dobrovolskaia, M. A., & McNeil, S. E. (2015b). Immunological and hematological toxicities challenging clinical translation of nucleic acid-based therapeutics. *Expert Opinion on Biological Therapy*. doi:10.1517/14712598.2015.1014794

Eng, H. L., Hsu, Y. Y., & Lin, T. M. (2018). Differences in TLR7/8 activation between monocytes and macrophages. *Biochemical and Biophysical Research Communications*. doi:10.1016/j.bbrc.2018.02.079

Foroozandeh, P., & Aziz, A. A. (2018). Insight into cellular uptake and intracellular trafficking of nanoparticles. *Nanoscale Research Letters*. doi:10.1186/s11671-018-2728-6

Forsbach, A., Nemorin, J.-G., Montino, C., Müller, C., Samulowitz, U., Vicari, A. P., ... Vollmer, J. (2008). Identification of RNA sequence motifs stimulating sequence-specific TLR8-dependent immune responses. *The Journal of Immunology*. doi:10.4049/jimmunol.180.6.3729

Goutagny, N., Estornes, Y., Hasan, U., Lebecque, S., & Caux, C. (2012). Targeting pattern recognition receptors in cancer immunotherapy. *Targeted Oncology*. doi:10.1007/s11523-012-0213-1

Gürtler, C., & Bowie, A. G. (2013). Innate immune detection of microbial nucleic acids. *Trends in Microbiology*. doi:10.1016/j.tim.2013.04.004

Halman, J. R., Kim, K. T., Gwak, S. J., Pace, R., Johnson, M. B., Chandler, M. R., ... Afonin, K. A. (2020). A cationic amphiphilic co-polymer as a carrier of nucleic acid nanoparticles (Nanps) for controlled gene silencing, immunostimulation, and biodistribution. *Nanomedicine: Nanotechnology, Biology, and Medicine*. doi:10.1016/j.nano.2019.102094

Hanagata, N. (2012). Structure-dependent immunostimulatory effect of CpG oligodeoxynucleotides and their delivery system. *International Journal of Nanomedicine*. doi:10.2147/IJN.S30197

Hemmi, H., Takeuchi, O., Kawai, T., Kaisho, T., Sato, S., Sanjo, H., ... Akira, S. (2000). A Toll-like receptor recognizes bacterial DNA. *Nature*. doi:10.1038/35047123

Hiscott, J., Lacoste, J., & Lin, R. (2006). Recruitment of an interferon molecular signaling complex to the mitochondrial membrane: Disruption by hepatitis C virus NS3-4A protease. *Biochemical Pharmacology*. doi:10.1016/j.bcp.2006.06.030

Hong, E., Halman, J. R., Shah, A. B., Khisamutdinov, E. F., Dobrovolskaia, M. A., & Afonin, K. A. (2018). Structure and composition define immunorecognition of nucleic acid nanoparticles. *Nano Letters*. doi:10.1021/acs.nanolett.8b01283

Hong, E., Halman, J. R., Shah, A., Cedrone, E., Truong, N., Afonin, K. A., & Dobrovolskaia, M. A. (2019). Toll-like receptor-mediated recognition of nucleic acid nanoparticles (NANPs) in human primary blood cells. *Molecules*. doi:10.3390/molecules24061094

Hornung, V., Ellegast, J., Kim, S., Brzózka, K., Jung, A., Kato, H., ... Hartmann, G. (2006). 5′-Triphosphate RNA is the ligand for RIG-I. *Science*. doi:10.1126/science.1132505

Hornung, V., Guenthner-Biller, M., Bourquin, C., Ablasser, A., Schlee, M., Uematsu, S., ... Hartmann, G. (2005). Sequence-specific potent induction of IFN-α by short interfering RNA in plasmacytoid dendritic cells through TLR7. *Nature Medicine*. doi:10.1038/nm1191

Iurescia, S., Fioretti, D., & Rinaldi, M. (2017). Nucleic acid sensing machinery: Targeting innate immune system for cancer therapy. *Recent Patents on Anti-Cancer Drug Discovery*. doi:10.2174/1574892812666171030163804

Johnson, M. B., Halman, J. R., Satterwhite, E., Zakharov, A. V., Bui, M. N., Benkato, K., ... Afonin, K. A. (2017). Programmable nucleic acid based polygons with controlled neuroimmunomodulatory properties for predictive QSAR modeling. *Small*, 1701255. doi:10.1002/smll.201701255

Kato, H., Takeuchi, O., Mikamo-Satoh, E., Hirai, R., Kawai, T., Matsushita, K., … Akira, S. (2008). Length-dependent recognition of double-stranded ribonucleic acids by retinoic acid–inducible gene-I and melanoma differentiation–associated gene 5. *The Journal of Experimental Medicine, 205*(7), 1601–1610. doi:10.1084/jem.20080091

Kawasaki, T., & Kawai, T. (2014). Toll-like receptor signaling pathways. *Frontiers in Immunology.* doi:10.3389/fimmu.2014.00461

Lee, Y., Urban, J. H., Xu, L., Sullenger, B. A., & Lee, J. (2016). 2′Fluoro modification differentially modulates the ability of RNAs to activate pattern recognition receptors. *Nucleic Acid Therapeutics.* doi:10.1089/nat.2015.0575

Leonard, J. N., Ghirlando, R., Askins, J., Bell, J. K., Margulies, D. H., Davies, D. R., & Segal, D. M. (2008). The TLR3 signaling complex forms by cooperative receptor dimerization. *Proceedings of the National Academy of Sciences of the United States of America.* doi:10.1073/pnas.0710779105

Loo, Y. M., & Gale, M. (2011). Immune Signaling by RIG-I-like Receptors. *Immunity.* doi:10.1016/j.immuni.2011.05.003

Öhman, T., Rintahaka, J., Kalkkinen, N., Matikainen, S., & Nyman, T. A. (2009). Actin and RIG-I/MAVS Signaling Components Translocate to Mitochondria upon Influenza A Virus Infection of Human Primary Macrophages. *The Journal of Immunology.* doi:10.4049/jimmunol.0803093

Pandey, S., Kawai, T., & Akira, S. (2015). Microbial sensing by toll-like receptors and intracellular nucleic acid sensors. *Cold Spring Harbor Perspectives in Biology.* doi:10.1101/cshperspect.a016246

Panyam, J., Zhou, W., Prabha, S., Sahoo, S. K., & Labhasetwar, V. (2002). Rapid endo-lysosomal escape of poly(DL-lactide- co glycolide) nanoparticles: implications for drug and gene delivery. *The FASEB Journal.* doi:10.1096/fj.02-0088com

Radovic-Moreno, A. F., Chernyak, N., Mader, C. C., Nallagatla, S., Kang, R. S., Hao, L., … Gryaznov, S. M. (2015). Immunomodulatory spherical nucleic acids. *Proceedings of the National Academy of Sciences of the United States of America.* doi:10.1073/pnas.1502850112

Robbins, M., Judge, A., Liang, L., McClintock, K., Yaworski, E., & MacLachlan, I. (2007). 2′-O-methyl-modified RNAs act as TLR7 antagonists. *Molecular Therapy.* doi:10.1038/sj.mt.6300240

Saito, T., Hirai, R., Loo, Y.-M., Owen, D., Johnson, C. L., Sinha, S. C., … Gale, M. (2007). Regulation of innate antiviral defenses through a shared repressor domain in RIG-I and LGP2. *Proceedings of the National Academy of Sciences, 104*(2), 582–587. doi:10.1073/pnas.0606699104

Saito, Takeshi, Owen, D. M., Jiang, F., Marcotrigiano, J., & Gale, M. (2008). Innate immunity induced by composition-dependent RIG-I recognition of hepatitis C virus RNA. *Nature, 454*(7203), 523–527. doi:10.1038/nature07106

Satoh, T., Kato, H., Kumagai, Y., Yoneyama, M., Sato, S., Matsushita, K., … Takeuchi, O. (2010). LGP2 is a positive regulator of RIG-I- and MDA5-mediated antiviral responses. *Proceedings of the National Academy of Sciences of the United States of America.* doi:10.1073/pnas.0912986107

Schmitt, F. C. F., Freund, I., Weigand, M. A., Helm, M., Dalpke, A. H., & Eigenbrod, T. (2017). Identification of an optimized 2′-O-methylated trinucleotide RNA motif inhibiting Toll-like receptors 7 and 8. *RNA.* doi:10.1261/rna.061952.117

Shenoy, D., Fu, W., Li, J., Crasto, C., Jones, G., DiMarzio, C., … Amiji, M. (2006). Surface functionalization of gold nanoparticles using hetero-bifunctional poly(ethylene glycol) spacer for intracellular tracking and delivery. *International Journal of Nanomedicine.* doi:10.2147/nano.2006.1.1.51

Silva, S., Almeida, A. J., & Vale, N. (2019). Combination of cell-penetrating peptides with nanoparticles for therapeutic application: A review. *Biomolecules.* doi:10.3390/biom9010022

Sioud, M. (2006). Single-stranded small interfering RNA are more immunostimulatory than their double-stranded counterparts: A central role for 2′-hyroxyl uridines in immune responses. *European Journal of Immunology.* doi:10.1002/eji.200535708

Steinhagen, F., Kinjo, T., Bode, C., & Klinman, D. M. (2011). TLR-based immune adjuvants. *Vaccine.* doi:10.1016/j.vaccine.2010.08.002

Takeuchi, O., & Akira, S. (2010). Pattern recognition receptors and inflammation. *Cell, 140*(6), 805–820. doi:S0092-8674(10)00023-1 [pii]\r10.1016/j.cell.2010.01.022

Temizoz, B., Kuroda, E., & Ishii, K. J. (2018). Combination and inducible adjuvants targeting nucleic acid sensors. *Current Opinion in Pharmacology.* doi:10.1016/j.coph.2018.05.003

Uzri, D., & Gehrke, L. (2009). Nucleotide sequences and modifications that determine RIG-I/RNA binding and signaling activities. *Journal of Virology, 83*(9), 4174–4184. doi:10.1128/JVI.02449-08

Venkataraman, T., Valdes, M., Elsby, R., Kakuta, S., Caceres, G., Saijo, S., … Barber, G. N. (2007). Loss of DExD/H Box RNA Helicase LGP2 Manifests Disparate Antiviral Responses. *The Journal of Immunology.* doi:10.4049/jimmunol.178.10.6444

Wang, Y., Liu, L., Davies, D. R., & Segal, D. M. (2010). Dimerization of Toll-like receptor 3 (TLR3) is required for ligand binding. *Journal of Biological Chemistry*. doi:10.1074/jbc.M110.167973

Wilczewska, A. Z., Niemirowicz, K., Markiewicz, K. H., & Car, H. (2012). Nanoparticles as drug delivery systems. *Pharmacological Reports*. doi:10.1016/S1734-1140(12)70901-5

Wu, J., & Chen, Z. J. (2014). Innate immune sensing and signaling of cytosolic nucleic acids. *Annual Review of Immunology*, *32*(1), 461–488. doi:10.1146/annurev-immunol-032713-120156

Yoneyama, M., Kikuchi, M., Matsumoto, K., Imaizumi, T., Miyagishi, M., Taira, K., … Fujita, T. (2005). Shared and unique functions of the DExD/H-Box Helicases RIG-I, MDA5, and LGP2 in antiviral innate immunity. *The Journal of Immunology*. doi:10.4049/jimmunol.175.5.2851

Yoneyama, M., Kikuchi, M., Natsukawa, T., Shinobu, N., Imaizumi, T., Miyagishi, M., … Fujita, T. (2004). The RNA helicase RIG-I has an essential function in double-stranded RNA-induced innate antiviral responses. *Nature Immunology*. doi:10.1038/ni1087

27 Viral Noncoding RNAs in Modulating Cellular Defense and Their Potential for RNA Nanotechnology

Martin Panigaj
Pavol Jozef Safarik University in Kosice, Kosice, Slovak Republic

Marina A. Dobrovolskaia
Frederick National Laboratory for Cancer Research, Frederick, MD, USA

Kirill A. Afonin
The University of North Carolina at Charlotte, Charlotte, NC, USA

CONTENTS

CELL DEFENSE AGAINST VIRUSES

Viruses are genetically and phenotypically diverse infectious agents providing profound selective pressure in the evolution of their host species with potential to significantly alter host population dynamics on global scale [1–4]. The entire virus life cycle depends on the hosts' intracellular environment and its molecular machinery. As a result, virus replication affects cell viability. To cope with infection, cells evolved diverse strategies suppressing the virus infection. Recognition of invading virus by sensing its genome, transcripts or replication intermediates plays a key role in the cell defense. Accordingly, viral pathogens have evolved intricate methods to circumvent or inhibit cell surveillance pathways [5–7].

The innate immune response provides a first line of defense against viral infection while avoiding reactions to its own nucleic acids (NA). Four determinants are considered in the recognition of self from non-self NA: (i) structural patterns (*e.g.*, composition (DNA vs. RNA), sequence, and/or chemical modification of NAs), (ii) intracellular localizations (e.g., occurrence in compartments unusual for NAs presence), (iii) relative quantities (e.g., when compared to physiological conditions), and (iv) threshold (e.g., expression of components for sensing of NAs and downstream signaling) [8]. The frontline of the innate immune response is represented by pattern recognition receptors (PRRs) that in the case of the viral infections detect pathogen-associated molecular patterns, or PAMPs. PRRs sensing PAMPs outside the cellular interior are located on the cell surface and, among others, include Toll-like receptors, or TLRs [9]. TLR9 sensing nucleic acids (TLR3, TLR7, TLR8, and

DOI: 10.1201/9781003001560-30

TLR9) are located in the endosomal compartment of a cell. Other groups of NA-sensing receptors reside in the cytoplasm and nucleus and include RNA sensing RIG-I-like receptors (RLRs), and DNA-sensing cyclic GMP-AMP synthase (cGAS) and interferon-γ-inducible protein 16 (IFI16), just to name a few [10]. The presence of individual PRRs differs among various tissues. While TLRs are mostly specific for cells of immune system, intracellular PRRs are expressed broadly.

Detection of viral PAMPs triggers intricate signaling cascades passing through the pathway's specific or common adaptor proteins depending on the NA trigger, finally merging to the transcription factors (Figure 27.1). The PRRs activation culminates into the expression of host defense genes such as chemokines, proinflammatory cytokines, and interferons (IFNs) that are the major players

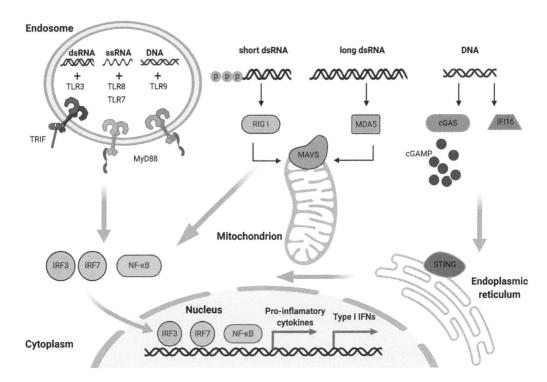

FIGURE 27.1   The simplified schematic pathways of virus sensing based on nucleic acids. TLR3, TLR7, TLR8, and TLR9 detect distinct forms of viral nucleic acids in endolysosomes (TLR3- dsRNA, TLR7/8- ssRNA, and TLR9- CpG DNA). After detection of foreign nucleic acids TLR7/8/9 draft Myeloid differentiation primary response 88 (MyD88) adaptor protein while TLR3 recruits TIR-domain-containing adapter-inducing interferon-β (TRIF). Depending on the adaptor protein, signaling cascade activates transcription factors NF-κB, regulatory factor 3 (IRF3), and IRF7 that results in the transcription of IFN genes, inflammatory cytokines, and chemokine. After entry to the cytoplasm, virus genetic material is detected depending on its nature by RNA or DNA sensors. The retinoic acid-inducible gene-I protein (RIG-I) and melanoma differentiation-associated protein 5 (MDA5) are activated by binding short double-stranded RNA (dsRNA) with a 5′-triphosphate and 5′-diphosphate or long dsRNA structures, respectively. Subsequently, RIG-I and MDA5 bind the mitochondrial antiviral signaling protein (MAVS) residing predominantly on mitochondrial membrane. Finally, kinase complexes activated by MAVS induce transcription through IRF3, IRF7, and NF-κB. The main cytoplasmic sensors of dsDNA are cGAS and IFNγ-inducible protein 16 (IFI16) that is located also in the nucleus where it probably detects naked viral DNA. After binding dsDNA, cGAS synthesizes the second messenger 2′3′-cyclic-GMP-AMP (cGAMP) that subsequently mobilizes stimulator of interferon (IFN) genes (STING) on the endoplasmic reticulum. The STING then induces transcription of antiviral genes through IRF3 and NF-κB [5,9]. Created with BioRender.com

of innate immunity against viral infection [11]. The IFN production in both autocrine and paracrine fashion subsequently stimulates further expression of IFN stimulated genes (ISGs), which restricts viral replication and spread, attracts innate immune cells to the site of infection, and induces the immune cells antiviral activity together with modulation of the adaptive immune responses to the viral infection in an effort to establish so-called antiviral state in adjacent healthy cells [12,13].

Replication and assembly of infectious particles strictly depend on the cellular metabolic and translational machinery that is accomplished by successful subversion of numerous host cell factors [5–7,14]. The intricacy of virus–host interactions reflects the complexity of virus replication strategies determined by its genome architecture, size, mutation rates, *etc.* [15,16]. The spectrum of diverse evasion mechanisms spans from passive to active approaches with the simplest strategy aiming to avoid PRRs by mimicking host NAs or by sequestering PRRs from the viral genome (*e.g.*, $2'$-*O* methylation of cap structure, internal N^6-methyladenosine (m^6A) modifications, or remodeling intracellular membranes to create protective replication compartments, respectively [17–23]).

Increasing experimental evidence has demonstrated that viral noncoding RNAs (ncRNAs) such as micro RNAs (miRNAs) and long ncRNAs (lncRNAs) are important players in virus counter-defense [24]. The involvement of virus encoded or genome-derived ncRNAs in virus infection is an attractive research field with potential for development of novel efficient antiviral therapies. Similarly, the host-encoded ncRNAs can be hijacked by virus or participate in regulating the innate responses (*e.g.*, against overreaction to minimize host tissue damage are relevant targets for therapy [25,26]). Furthermore, RNA motifs or RNA modifications involved in subverting cellular immunity can enrich the expanding field of RNA nanotechnology [27]. Embedding such motifs in nucleic acid nanoparticles (NANPs) would allow designing of assemblies with attenuated immunogenicity and enhanced stability for transfected or *in vivo* co-transcriptionally assembled NANPs [28].

VIRAL RNAs IN EVASION OF CELL DEFENSE

The virus evolution favors increasing functional density in viral genomes that are typical for overlapping reading frames, polycistronic mRNAs, multifunctional proteins, and functional RNAs within coding or noncoding regions. The secondary and tertiary structures of RNA transcripts expand the overall functional versatility with effects on diverse stages of virus replication to ensure successful infection. In addition, the plasticity of viral RNA structures presents a further level of regulation at various stages of the virus life cycle [29–32].

The simplest manipulation of viral RNAs with host defense system is silencing the expression of target genes by RNA interference (RNAi) [33]. RNAi is crucial post-transcriptional regulation of gene expression that is guided by miRNAs. Cellular miRNAs are imperfectly complementary to $3'$ untranslated regions (UTRs) of target mRNAs. Usually one miRNA sequence silences many different mRNAs. Not surprisingly, viruses generate their own miRNAs that help them to survive and replicate inside the cells [34]. Frequently, biogenesis of viral miRNAs differs from canonical cellular pathways and in RNA viruses can be even excised from the genome [35,36]. Interestingly, not only mRNA is subject for virus targeting, Herpesvirus saimiri (HVS) ncRNA transcripts, known as Herpesvirus saimiri U-rich RNA 1 (HSUR1), bind host miRNA (miR-27) and induce its decay [37]. Unusual targeting of mRNAs is achieved by HSUR2 that base pairs with two host miRNAs (miR-142-3p and miR-16) as well as to target mRNAs that code apoptotic proteins. HSUR2 recruited miRNAs draft Ago proteins that degrade or repress translation of bound mRNA. The final outcome is inhibition of apoptosis [38].

Viral lncRNAs are functionally heterogeneous group of RNAs longer than 200 nucleotides. Usually, lncRNAs are typical for dsDNA viruses with large genomes (*e.g.*, Herpes viruses). The most notable lncRNAs are one of the factors that orchestrate establishing and maintenance of latency and virus reactivation. Viral latency is a virus survival strategy to maintain its genome inside the infected host cell for long term either as an episome or integrated into the host chromosome. Virus dormancy is a sophisticated approach that allows avoiding the cell immune system [39].

The replication of herpes viruses in the nucleus allows them to exploit host cell transcription machinery RNA polymerase II and III (Pol II, Pol III) including subsequent processing (*e.g.*, capping and polyadenylation of mRNA). One of the most intensively studied lncRNA is ~1077 nts long with a polyadenylated nuclear (PAN) RNA that is present predominantly in the cell nucleus during the lytic phase of cells infected with Kaposi's sarcoma-associated herpesvirus (KSHV). PAN RNA is highly abundant due to the element for nuclear expression (ENE) that, by creating a triple helix with its poly(A) tail, blocks lncRNA degradation. Multifunctional PAN RNA is a scaffold for several viral and host proteins that epigenetically advance lytic reactivation by recruiting chromatin modifying factors or sequesters repressive proteins from the viral genome [40,41].

The latency associated transcript (LAT) of herpes simplex virus (HSV) is another viral lncRNA with an important role in latent infection. In infected neurons, the HSV transcription is largely restricted except the ~8.3kb long LAT. The primary LAT transcript is spliced into two stable introns (1.5 and 2 kb) the major LAT lncRNAs. The rest of LAT transcript is processed to miRNAs. The exact role of LAT and molecular interactions are unclear; however, the LAT lncRNA participates in establishment and maintenance of latency through suppression of viral genes transcription. Separately from the functions during latency, LAT has a direct role in reactivation from latency [42,43].

Similarly, HIV-1 encodes an antisense lncRNA to the viral 5'- long terminal repeat (LTR) that promotes the latent state of integrated HIV-1. This was experimentally supported by the observation of HIV activation upon targeting of antisense lncRNA by small RNAs. The antisense lncRNA mediates silencing of viral transcription by recruiting epigenetic complexes to the LTR that sets up changes in chromatin structure [44].

A sophisticated approach to sustain energy for virus replication and prevention of apoptosis during the infection is used by Human Cytomegalovirus (HCMV). At early stages of lytic phase of infection, HCMV transcribe 2.7 kb long β2.7 lncRNA that physically interacts with a subunit of mitochondrial multimeric enzyme Complex I- the genes associated with retinoid/interferon–induced mortality (GRIM-19). The complex I is a part of the mitochondrial respiratory chain that facilitates electron transport coupled with a proton gradient across the mitochondrial inner membrane to drive adenosine triphosphate (ATP) production. Interaction of β2.7 lncRNA with GRIM-19 prevents its relocation in response to apoptotic stimuli. The stabilized mitochondrial membrane potential then ensures steady production of ATP [45].

In comparison to direct interference with host gene expression by viral miRNAs or more complex indirect manipulation through viral lncRNAs, ncRNAs in several viruses evolved to directly impair PRR functions to avoid inhibition of translation, virus genome degradation, triggering interferon production, *etc*. So far, all known viruses are completely dependent on the host cell protein synthesis. Therefore, as a fundamental requirement for the viral replication, access to cell translation machinery is tightly controlled by the cellular defense apparatus. The viruses use various tactics to evade or subvert cellular protection components to exploit host translation [46].

Eukaryotic mRNA is protected at 5' end by cap structures with 2'-*O* methylation at the N-7 position as well as on the first and/or second nucleotide. The absence of RNA cap and/or modification results in recognition of viral RNA by IFIT1 (Interferon-Induced Protein with Tetratricopeptide Repeats) member of ISGs. IFIT1 is a cytosolic viral RNA sensor that upon detection of viral 5'-ppp RNA and absence of 2'-*O* methylation at 5'end shuts down host translation to halt virus replication. To avoid recognition some viruses encode enzymes for methylation and 2'-*O* capping or they can steal caps from the host mRNA. Alternatively, some viruses can use cap-independent translation. In contrast to that, positive-strand RNA alphaviruses, even their RNA genomic 5' end lacks 2'-*O* methylation, avoid IFIT1 recognition by antagonizing IFIT1 through their secondary structural motifs in the 5' UTR [47]. Interestingly, *in silico* analysis revealed that 5'UTR secondary structures in alphaviruses are diverse and are even missing in some members (Figure 27.2a) [48].

The ISGs-encoded RNA-dependent protein kinase R (PKR) and oligoadenylate synthase (OAS) together with RNase L are core enzymes guarding host translation. PKR and OAS are both activated

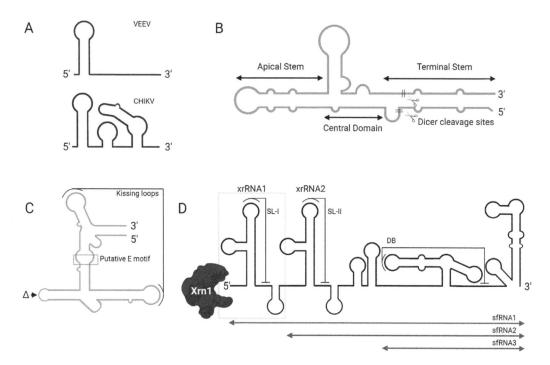

FIGURE 27.2 Schematic depictions of selected viral ncRNAs subverting cellular defense. (a) Secondary structures located in 5′UTR of [+] ssRNA genome of alphaviruses counteracting IFIT1. VEEV- Venezuelan equine encephalitis virus, CHIKV- Chikungunya virus [48]. (b) Adenovirus virus-associated RNA I (VA-I). Apical stem facilitates binding to PKR while CD creating a pseudoknot blocks PKR. Terminal stem processed by Dicer is involved in RNAi [53]. (c) Competitively inhibiting RNase L ncRNA (ciRNA) located in open reading frame of poliovirus [+] ssRNA genome. The two stem-loops form tertiary kissing loop, a putative E motif is essential for inhibition by unclear mechanism. Δ- the intermediate sequence between (5825-5905) is not necessary for ciRNA function and can be excised [54]. (d) Exonuclease Xrn1 produces sfRNAs by incomplete degradation of their [+] ssRNA genome. The sfRNAs originate from the 3′UTR containing xrRNAs that block cellular 5′→3′ Xrn1 at a specific site. Depending on the flavivirus, 3′UTRs consist of one or two SLs, one or two DB, and short hairpin followed by terminal 3′-stem loop. The number of sfRNAs may vary if Xrn1 proceeds through the first or second xrRNA [63]. Created with BioRender.com

by dsRNA (dsRNA genomes, intermediates in RNA virus genome replication, and secondary structures in ssRNA or bidirectional transcription). Activated PKR phosphorylates the subunit α of eukaryotic initiation factor 2 (eIF2α) leading to inactivation of eIF2 and subsequent overall inhibition of translation initiation. The OAS activity is triggered by dsRNA, too, but in turn it synthesizes second messenger 2′-5′ oligoadenylate (2′-5′ OA) from ATP. Next, 2′-5′ OA binds to an endogenous ribonuclease RNase L, activated monomeric RNase L dimerizes and then cleaves all RNA in the cell leading to apoptosis [49].

As countermeasures to immune restriction facilitated by PKR, almost all common viruses evolved strategies to bypass or antagonize PKR. The best described ncRNA interfering with PKR function that was also discovered as the first viral ncRNA is virus-associated RNA I (VA-I). The VA-I (~150- 200 nts long) is RNA that is transcribed by host RNA polymerase III from Adenoviral DNA in late phase infection. The secondary structure of VA-I ncRNA is composed of an apical stem (AS), a central domain (CD) containing a stem-loop forming a pseudoknot, and a terminal stem (TS) that is dispensable for PKR binding and inhibition (Figure 27.2b). The exact mechanism of PKR inhibition is not fully explained. Experimental evidence suggests that AS mediates binding of VA-I to PKR in a 1:1 ratio but AS itself does not inhibit PKR. Creating mutations in CD may change the inhibiting function of VA-I to activating, which suggest that stem in CD is responsible for PKR inhibition. In current model it is presumed that CD does not allow dimerization and thus autophosphorylation of PKR. Interestingly, experimentally introduced substitutions in VA-I sequence even increased inhibition of PKR in comparison to the wildtype VA-I. Frequent occurrence of wobble base pairs throughout the VA-I, similar to several other PKR inhibiting ncRNAs, is structural factor that may participate in inhibition of PKR, OAS1 or another dsRNA binding proteins such as ADAR. In addition, due to its high copy number (up to 10^8 copies/cell) VA-I together with less abundant VA-II transcripts disrupt miRNA biogenesis by saturating pre-miRNA export pathway by binding to the nuclear export protein Exportin 5 and the Dicer. Subsequently, Dicer processed TS strands are loaded into functional RISC complexes and target cellular genes [50–53].

A highly structured RNA sequence within the protein-coding region of poliovirus positive ssRNA genome is an illustrative example of evolutionary pressure on functional versatility of viral RNAs. The phylogenetically conserved but dynamic structure must maintain the correct protein-coding sequence, be structurally accessible for the translocation of the translation and replication machinery and yet to adopt a conformation that is able to competitively inhibit RNase L, a feature that conferred its name ciRNA. The proposed structure consists of two stem-loops forming tertiary kissing loop interaction and a putative loop E motif that is essential for the inhibition of RNase L (Figure 27.2c). However, NMR or crystallography will be required to confirm the existence of loop E within ciRNA. The mechanism of interaction between ciRNA and RNase L is unclear, too. Currently, it is hypothesized that ciRNA is binding to RNase L as a normal RNA substrate but owing to a specific conformation it inhibits RNase L probably by direct interaction with a cleavage domain [54–56].

The unusual way for generation of ncRNAs has evolved in flaviviruses including important arthropod-born human pathogens as dengue virus, West Nile virus, Zika virus, tick-borne encephalitis virus, and yellow fever virus. Recently, similar strategy of ncRNAs creation was described in plant infecting viruses. The hallmark of flavivirus infection is incomplete degradation of their positive ~10 kb long ssRNA genome that results in the accumulation of high levels of ~300–500 nts long subgenomic flaviviral RNAs (sfRNAs). Accumulated sfRNAs have profound pathogenic impact on infected cells, directly contributing to viral evasion of the type I interferon response, influences RNAi, and impedes 5′→3′ RNA decay, thus interfering with cellular turnover of RNA. The sfRNAs originate from the highly structured 3′ UTR that contain structures called XRN1-resistant RNAs (xrRNAs) that block cellular 5′→3′ exonuclease Xrn1 at a defined site. So far tested flavivirus 3′UTRs consist from one or duplicated stem loop (SL), one or two dumbbell structures (DB) and short hairpin followed by terminal 3′-stem loop (Figure 27.2d). Depending on the number of xrRNA elements, several types of sfRNA can arise suggesting that XRN1 can proceed through the first xrRNA and is stalled by the next resistant structure. According to the current model, the resistance is achieved by the formation of tertiary structures where xrRNAs participate in pseudoknot interactions. Proposed conformation resembles a ring-like structure with the 5′-end running through the ring. This 3D shape constitutes a barrier for XRN1, but permits viral polymerase to synthesize the opposite strand in the 3′→5′ direction [57–63].

NANPs AND MODULATION OF IMMUNE RESTRICTIONS

The functional nucleic acids transfer mediated by engineered viruses has been proven as an invaluable approach in gene therapy, yet the production of the necessary amount of recombinant vectors is not trivial and safety concerns may limit its use. Therefore, bottom-up creation of the nanoassemblies by using nucleic acids is an appealing idea for therapeutic applications [64–74]. Nucleic acids have been acknowledged as a convenient building material for nanotechnology due to their biocompatibility and programmability that benefit from the ability to form both canonical and non-canonical base pairings leading to a diverse set of structural motifs used as building blocks [75–85].

The modularity of DNA/RNA allows linking several functional DNA oligonucleotides or RNAs into one sequence. Subsequently, such functionally chimeric molecules based on the complementarity of its bases/monomers can be designed to self-assemble into complex multi-stranded nanoparticles. The programmable multitasking together with the ability to dynamically respond to the molecular clues makes nucleic acids an attractive material for tailor-made applications in biotechnology or personalized medicine [86–97].

Interestingly, delivery of NANPs mediated by transfection reagents demonstrates higher immunostimulatory potential in comparison to application of naked NANPs [98,99]. However, currently there is no reliable widespread method for the transport of naked NANPs to target cells, which highlights the need for the development of NANPs with ability to adjust cellular immune response. The relationship between NANPs and the cellular defense is a relatively new research field revealing that size, composition, connectivity and dimensionality of NANPs participate in the interaction of NANPs with cellular immune system. Globular rather than planar NANPs are more immunostimulatory, while fibrous structures are the least probable to trigger immune response, when NANPs are delivered inside the cell using carriers. As for the nucleic acid composition the reactivity decreases from RNA to RNA/DNA hybrid and DNA NANPs being almost immunoquiescent [28,86,87,90,98–101]. This trend was further observed by experiments on human peripheral blood mononuclear cells where TLR7 recognized RNA cubes and rings in contrast to RNA fibers and DNA cubes, which were not detected [98]. Additionally, based on the sequence and orientation in space, addition of functional extensions can alter immunorecognition of NANPs [102,103]. The immune-mediated effects of therapeutic NANPs delivering various functionalities, therefore, should be carefully evaluated when their medical application is considered.

This situation is an opportunity for employment of viral RNA motifs with ability to silence immune system into therapeutic NANPs. Various viral ncRNAs subvert cellular defense in different ways and their combination embedded on one assembly could theoretically increase immunosuppressive effect. To date none viral ncRNAs capable to restrict antiviral pathways were applied in therapeutic RNA nanotechnology, but some motifs or full length lncRNAs were functional as heterologous or as standalone transcript. The KSHV PAN RNA is highly abundant in nucleus of lytically infected cells. The reason for the PAN RNA accumulation is the presence of ENE motif at 3′end that interacts with poly(A). This *cis-acting* formation can increase the level of intronless RNA, but in comparison with similar enhancers it maintains transcript in nucleus as has been shown by the insertion of the 79-nt ENE motif near the 3′ end of a heterologous intronless globin gene [104]. The β2.7 lncRNA prevents apoptosis in cytomegalovirus infected cells by direct interaction with Complex I from the mitochondrial respiratory chain whose malfunction otherwise initiate production of reactive oxygen species (ROS) that damage mitochondria and promote apoptosis. Therapeutic potential of β2.7 RNA was subsequently indicated by study where the overtranscription of β2.7 led to the reduction of ROS. The lower ROS levels protected rat aortic endothelial cells from apoptosis induced by hypoxia/reperfusion and ischemia/reperfusion injury [105].

One of the major drawbacks for widespread testing of immunosuppressive RNA motifs in NANP technology is deficiency of defined minimal viral structures responsible for supposed function. In addition to local secondary and tertiary structures the situation is complicated also by involvement of long-range RNA-RNA interactions in folding of functional motifs. Today, the number of known viral ncRNAs implicated in host defense subversion is relatively small and their potential for modulation of immune system by NANPs is unexplored. The viruses evolved to pack and carry on as much functional elements as their minimalistic lifestyle allows; therefore, it is highly probable that each ncRNA provides certain functions. Furthermore, modifications of viral ncRNAs are likely another functional layer playing a role in virus survival. It will be intriguing to witness to what extent advances in molecular biology techniques reveal the world of viral ncRNAs and their interactions with host cells. Even more interesting would be to see how this knowledge will enrich the field of NANP technology.

ACKNOWLEDGMENTS

The study was supported in part (to M.A.D.) by federal funds from the National Cancer Institute, National Institutes of Health, under contract HHSN261200800001E and 75N91019D00024. The content of this publication does not necessarily reflect the views or policies of the Department of Health and Human Services, nor does mention of trade names, commercial products, or organizations imply endorsement by the U.S. Government. Research reported in this publication was also supported by the National Institute of General Medical Sciences of the National Institutes of Health under Award Numbers R01GM120487 and R35GM139587 (to K.A.A.). The content is solely the responsibility of the authors and does not necessarily represent the official views of the National Institutes of Health.

REFERENCES

1. Enard, D., Cai, L., Gwennap, C. and Petrov, D.A. (2016) Viruses are a dominant driver of protein adaptation in mammals. *eLife*, **5**.
2. Gilbert, C. and Cordaux, R. (2017) Viruses as vectors of horizontal transfer of genetic material in eukaryotes. *Current Opinion in Virology*, **25**, 16–22.
3. Gregory, A.C., Zayed, A.A., Conceição-Neto, N., Temperton, B., Bolduc, B., Alberti, A., Ardyna, M., Arkhipova, K., Carmichael, M., Cruaud, C. et al. (2019) Marine DNA viral macro- and microdiversity from pole to pole. *Cell*, **177**, 1109–1123.
4. Weitz, J.S. and Wilhelm, S.W. (2012) Ocean viruses and their effects on microbial communities and biogeochemical cycles. *F1000 Biology Reports*, **4**, 17.
5. Chan, Y.K. and Gack, M.U. (2016) Viral evasion of intracellular DNA and RNA sensing. *Nature Reviews. Microbiology*, **14**, 360–373.
6. Garcia-Sastre, A. (2017) Ten Strategies of Interferon Evasion by Viruses. *Cell Host & Microbe*, **22**, 176–184.
7. Unterholzner, L. and Almine, J.F. (2019) Camouflage and interception: how pathogens evade detection by intracellular nucleic acid sensors. *Immunology*, **156**, 217–227.
8. Roers, A., Hiller, B. and Hornung, V. (2016) Recognition of endogenous nucleic acids by the innate immune system. *Immunity*, **44**, 739–754.
9. Lester, S.N. and Li, K. (2014) Toll-like receptors in antiviral innate immunity. *Journal of Molecular Biology*, **426**, 1246–1264.
10. Ori, D., Murase, M. and Kawai, T. (2017) Cytosolic nucleic acid sensors and innate immune regulation. *International Reviews of Immunology*, **36**, 74–88.
11. Nan, Y., Nan, G. and Zhang, Y.J. (2014) Interferon induction by RNA viruses and antagonism by viral pathogens. *Viruses*, **6**, 4999–5027.
12. Kell, A.M. and Gale, M., Jr. (2015) RIG-I in RNA virus recognition. *Virology*, **479–480**, 110–121.
13. Pestka, S. (2007) The interferons: 50 years after their discovery, there is much more to learn. *The Journal of Biological Chemistry*, **282**, 20047–20051.
14. Beachboard, D.C. and Horner, S.M. (2016) Innate immune evasion strategies of DNA and RNA viruses. *Current Opinion in Microbiology*, **32**, 113–119.
15. Dolan, P.T., Whitfield, Z.J. and Andino, R. (2018) Mapping the evolutionary potential of RNA viruses. *Cell Host & Microbe*, **23**, 435–446.
16. Duffy, S., Shackelton, L.A. and Holmes, E.C. (2008) Rates of evolutionary change in viruses: patterns and determinants. *Nature Reviews Genetics*, **9**, 267–276.
17. Ariza-Mateos, A. and Gómez, J. (2017) Viral tRNA mimicry from a biocommunicative perspective. *Frontiers in Microbiology*, **8**.
18. Greenbaum, B.D., Levine, A.J., Bhanot, G. and Rabadan, R. (2008) Patterns of evolution and host gene mimicry in influenza and other RNA viruses. *PLoS Pathogens*, **4**, e1000079.
19. Shulla, A. and Randall, G. (2016) (+) RNA virus replication compartments: a safe home for (most) viral replication. *Current Opinion in Microbiology*, **32**, 82–88.
20. Bowie, A.G. and Unterholzner, L. (2008) Viral evasion and subversion of pattern-recognition receptor signalling. *Nature Reviews. Immunology*, **8**, 911–922.

21. den Boon, J.A. and Ahlquist, P. (2010) Organelle-like membrane compartmentalization of positive-strand RNA virus replication factories. *Annual Review of Microbiology*, **64**, 241–256.
22. Lu, M., Zhang, Z., Xue, M., Zhao, B.S., Harder, O., Li, A., Liang, X., Gao, T.Z., Xu, Y., Zhou, J. et al. (2020) N(6)-methyladenosine modification enables viral RNA to escape recognition by RNA sensor RIG-I. *Nature Microbiology*
23. Daffis, S., Szretter, K.J., Schriewer, J., Li, J., Youn, S., Errett, J., Lin, T.Y., Schneller, S., Zust, R., Dong, H. et al. (2010) 2'-O methylation of the viral mRNA cap evades host restriction by IFIT family members. *Nature*, **468**, 452–456.
24. Damas, N.D., Fossat, N. and Scheel, T.K.H. (2019) Functional Interplay between RNA Viruses and Non-Coding RNA in Mammals. *Noncoding RNA*, **5**.
25. Wang, P., Xu, J., Wang, Y. and Cao, X. (2017) An interferon-independent lncRNA promotes viral replication by modulating cellular metabolism. *Science*, **358**, 1051–1055.
26. Jiang, M., Zhang, S., Yang, Z., Lin, H., Zhu, J., Liu, L., Wang, W., Liu, S., Liu, W., Ma, Y. et al. (2018) Self-recognition of an inducible host lncRNA by RIG-I feedback restricts innate immune response. *Cell*, **173**, 906–919 e913.
27. Jasinski, D., Haque, F., Binzel, D.W. and Guo, P. (2017) Advancement of the emerging field of RNA nanotechnology. *ACS Nano*, **11**, 1142–1164.
28. Chandler, M., Johnson, M.B., Panigaj, M. and Afonin, K.A. (2019) Innate immune responses triggered by nucleic acids inspire the design of immunomodulatory nucleic acid nanoparticles (NANPs). *Current Opinion in Biotechnology*, **63**, 8–15.
29. Rausch, J.W., Sztuba-Solinska, J. and Le Grice, S.F.J. (2017) Probing the structures of viral RNA regulatory elements with SHAPE and related methodologies. *Frontiers in Microbiology*, **8**, 2634.
30. Smyth, R.P., Negroni, M., Lever, A.M., Mak, J. and Kenyon, J.C. (2018) RNA structure-A neglected puppet master for the evolution of virus and host immunity. *Frontiers in Immunology*, **9**, 2097.
31. Ganser, L.R., Kelly, M.L., Herschlag, D. and Al-Hashimi, H.M. (2019) The roles of structural dynamics in the cellular functions of RNAs. *Nature Reviews. Molecular Cell Biology*, **20**, 474–489.
32. Dethoff, E.A., Chugh, J., Mustoe, A.M. and Al-Hashimi, H.M. (2012) Functional complexity and regulation through RNA dynamics. *Nature*, **482**, 322–330.
33. Fire, A., Xu, S., Montgomery, M.K., Kostas, S.A., Driver, S.E. and Mello, C.C. (1998) Potent and specific genetic interference by double-stranded RNA in Caenorhabditis elegans. *Nature*, **391**, 806–811.
34. Bernier, A. and Sagan, S.M. (2018) The Diverse Roles of microRNAs at the Host(-)Virus Interface. *Viruses*, **10**.
35. Xie, M. and Steitz, J.A. (2014) Versatile microRNA biogenesis in animals and their viruses. *RNA Biology*, **11**, 673–681.
36. Hussain, M., Torres, S., Schnettler, E., Funk, A., Grundhoff, A., Pijlman, G.P., Khromykh, A.A. and Asgari, S. (2012) West Nile virus encodes a microRNA-like small RNA in the 3' untranslated region which up-regulates GATA4 mRNA and facilitates virus replication in mosquito cells. *Nucleic Acids Research*, **40**, 2210–2223.
37. Cazalla, D., Yario, T. and Steitz, J.A. (2010) Down-regulation of a host microRNA by a Herpesvirus saimiri noncoding RNA. *Science*, **328**, 1563–1566.
38. Gorbea, C., Mosbruger, T. and Cazalla, D. (2017) A viral Sm-class RNA base-pairs with mRNAs and recruits microRNAs to inhibit apoptosis. *Nature*, **550**, 275–279.
39. Lieberman, P.M. (2016) Epigenetics and genetics of viral latency. *Cell Host & Microbe*, **19**, 619–628.
40. Conrad, N.K. (2016) New insights into the expression and functions of the Kaposi's sarcoma-associated herpesvirus long noncoding PAN RNA. *Virus Research*, **212**, 53–63.
41. Sztuba-Solinska, J., Rausch, J.W., Smith, R., Miller, J.T., Whitby, D. and Le Grice, S.F.J. (2017) Kaposi's sarcoma-associated herpesvirus polyadenylated nuclear RNA: a structural scaffold for nuclear, cytoplasmic and viral proteins. *Nucleic Acids Research*, **45**, 6805–6821.
42. Watson, Z.L., Washington, S.D., Phelan, D.M., Lewin, A.S., Tuli, S.S., Schultz, G.S., Neumann, D.M. and Bloom, D.C. (2018) In vivo knockdown of the herpes simplex virus 1 latency-associated transcript reduces reactivation from latency. *Journal of Virology*, **92**.
43. Nicoll, M.P., Hann, W., Shivkumar, M., Harman, L.E., Connor, V., Coleman, H.M., Proenca, J.T. and Efstathiou, S. (2016) The HSV-1 latency-associated transcript functions to repress latent phase lytic gene expression and suppress virus reactivation from latently infected neurons. *PLoS Pathogens*, **12**, e1005539.

44. Saayman, S., Ackley, A., Turner, A.-M.W., Famiglietti, M., Bosque, A., Clemson, M., Planelles, V. and Morris, K.V. (2014) An HIV-encoded antisense long noncoding RNA epigenetically regulates viral transcription. *Molecular Therapy*, **22**, 1164–1175.

45. Reeves, M.B., Davies, A.A., McSharry, B.P., Wilkinson, G.W. and Sinclair, J.H. (2007) Complex I binding by a virally encoded RNA regulates mitochondria-induced cell death. *Science*, **316**, 1345–1348.

46. Stern-Ginossar, N., Thompson, S.R., Mathews, M.B. and Mohr, I. (2019) Translational control in virus-infected cells. *Cold Spring Harbor Perspectives in Biology*, **11**.

47. Hyde, J.L., Gardner, C.L., Kimura, T., White, J.P., Liu, G., Trobaugh, D.W., Huang, C., Tonelli, M., Paessler, S., Takeda, K. et al. (2014) A viral RNA structural element alters host recognition of nonself RNA. *Science*, **343**, 783–787.

48. Barik, S. (2019) In silico structure analysis of alphaviral RNA genomes shows diversity in the evasion of IFIT1-mediated innate immunity. *Journal of Biosciences*, **44**.

49. Drappier, M. and Michiels, T. (2015) Inhibition of the OAS/RNase L pathway by viruses. *Current Opinion in Virology*, **15**, 19–26.

50. Dzananovic, E., Astha, N., Chojnowski, G., Deo, S., Booy, E.P., Padilla-Meier, P., McEleney, K., Bujnicki, J.M., Patel, T.R. and McKenna, S.A. (2017) Impact of the structural integrity of the three-way junction of adenovirus VAI RNA on PKR inhibition. *PLoS One*, **12**, e0186849.

51. Hood, I.V., Gordon, J.M., Bou-Nader, C., Henderson, F.E., Bahmanjah, S. and Zhang, J. (2019) Crystal structure of an adenovirus virus-associated RNA. *Nature Communications*, **10**, 2871.

52. Andersson, M.G., Haasnoot, P.C.J., Xu, N., Berenjian, S., Berkhout, B. and Akusjärvi, G. (2005) Suppression of RNA interference by adenovirus virus-associated RNA. *Journal of Virology*, **79**, 9556–9565.

53. Vachon, V.K. and Conn, G.L. (2016) Adenovirus VA RNA: An essential pro-viral non-coding RNA. *Virus Research*, **212**, 39–52.

54. Townsend, H.L., Jha, B.K., Silverman, R.H. and Barton, D.J. (2008) A putative loop E motif and an H-H kissing loop interaction are conserved and functional features in a group C enterovirus RNA that inhibits ribonuclease L. *RNA Biology*, **5**, 263–272.

55. Keel, A.Y., Jha, B.K. and Kieft, J.S. (2012) Structural architecture of an RNA that competitively inhibits RNase L. *RNA*, **18**, 88–99.

56. Townsend, H.L., Jha, B.K., Han, J.Q., Maluf, N.K., Silverman, R.H. and Barton, D.J. (2008) A viral RNA competitively inhibits the antiviral endoribonuclease domain of RNase L. *RNA*, **14**, 1026–1036.

57. Chapman, E.G., Costantino, D.A., Rabe, J.L., Moon, S.L., Wilusz, J., Nix, J.C. and Kieft, J.S. (2014) The structural basis of pathogenic subgenomic flavivirus RNA (sfRNA) production. *Science*, **344**, 307–310.

58. Steckelberg, A.L., Akiyama, B.M., Costantino, D.A., Sit, T.L., Nix, J.C. and Kieft, J.S. (2018) A folded viral noncoding RNA blocks host cell exoribonucleases through a conformationally dynamic RNA structure. *Proceedings of the National Academy of Sciences of the United States of America*, **115**, 6404–6409.

59. Schuessler, A., Funk, A., Lazear, H.M., Cooper, D.A., Torres, S., Daffis, S., Jha, B.K., Kumagai, Y., Takeuchi, O., Hertzog, P. et al. (2012) West Nile virus noncoding subgenomic RNA contributes to viral evasion of the type I interferon-mediated antiviral response. *Journal of Virology*, **86**, 5708–5718.

60. Schnettler, E., Sterken, M.G., Leung, J.Y., Metz, S.W., Geertsema, C., Goldbach, R.W., Vlak, J.M., Kohl, A., Khromykh, A.A. and Pijlman, G.P. (2012) Noncoding flavivirus RNA displays RNA interference suppressor activity in insect and Mammalian cells. *Journal of Virology*, **86**, 13486–13500.

61. Moon, S.L., Anderson, J.R., Kumagai, Y., Wilusz, C.J., Akira, S., Khromykh, A.A. and Wilusz, J. (2012) A noncoding RNA produced by arthropod-borne flaviviruses inhibits the cellular exoribonuclease XRN1 and alters host mRNA stability. *RNA (New York, N.Y.)*, **18**, 2029–2040.

62. Slonchak, A. and Khromykh, A.A. (2018) Subgenomic flaviviral RNAs: What do we know after the first decade of research. *Antiviral Research*, **159**, 13–25.

63. MacFadden, A., O'Donoghue, Z., Silva, P., Chapman, E.G., Olsthoorn, R.C., Sterken, M.G., Pijlman, G.P., Bredenbeek, P.J. and Kieft, J.S. (2018) Mechanism and structural diversity of exoribonuclease-resistant RNA structures in flaviviral RNAs. *Nature Communications*, **9**, 119.

64. Binzel, D.W., Shu, Y., Li, H., Sun, M., Zhang, Q., Shu, D., Guo, B. and Guo, P. (2016) Specific delivery of MiRNA for high efficient inhibition of prostate cancer by RNA nanotechnology. *Molecular therapy: the journal of the American Society of Gene Therapy*, **24**, 1267–1277.

65. Zhang, H., Pi, F., Shu, D., Vieweger, M. and Guo, P. (2015) Using RNA nanoparticles with thermostable motifs and fluorogenic modules for real-time detection of RNA folding and turnover in prokaryotic and eukaryotic cells. *Methods in Molecular Biology*, **1297**, 95–111.

66. Shu, D., Li, H., Shu, Y., Xiong, G., Carson, W.E., Haque, F., Xu, R. and Guo, P. (2015) Systemic delivery of anti-miRNA for suppression of triple negative breast cancer utilizing RNA nanotechnology. *ACS Nano*, **9**, 9731–9740.
67. Rychahou, P., Haque, F., Shu, Y., Zaytseva, Y., Weiss, H.L., Lee, E.Y., Mustain, W., Valentino, J., Guo, P. and Evers, B.M. (2015) Delivery of RNA nanoparticles into colorectal cancer metastases following systemic administration. *ACS Nano*, **9**, 1108–1116.
68. Li, H., Lee, T., Dziubla, T., Pi, F., Guo, S., Xu, J., Li, C., Haque, F., Liang, X.J. and Guo, P. (2015) RNA as a stable polymer to build controllable and defined nanostructures for material and biomedical applications. *Nano Today*, **10**, 631–655.
69. Shu, Y., Pi, F., Sharma, A., Rajabi, M., Haque, F., Shu, D., Leggas, M., Evers, B.M. and Guo, P. (2014) Stable RNA nanoparticles as potential new generation drugs for cancer therapy. *Advanced Drug Delivery Reviews*, **66**, 74–89.
70. Feng, L., Li, S.K., Liu, H., Liu, C.Y., LaSance, K., Haque, F., Shu, D. and Guo, P. (2014) Ocular delivery of pRNA nanoparticles: distribution and clearance after subconjunctival injection. *Pharmaceutical Research*, **31**, 1046–1058.
71. Shu, Y., Haque, F., Shu, D., Li, W., Zhu, Z., Kotb, M., Lyubchenko, Y. and Guo, P. (2013) Fabrication of 14 different RNA nanoparticles for specific tumor targeting without accumulation in normal organs. *RNA*, **19**, 767–777.
72. Hoiberg, H.C., Sparvath, S.M., Andersen, V.L., Kjems, J. and Andersen, E.S. (2018) An RNA origami octahedron with intrinsic siRNAs for potent gene knockdown. *Biotechnology Journal*, e1700634.
73. Lee, H., Lytton-Jean, A.K.R., Chen, Y., Love, K.T., Park, A.I., Karagiannis, E.D., Sehgal, A., Querbes, W., Zurenko, C.S., Jayaraman, M. et al. (2012) Molecularly self-assembled nucleic acid nanoparticles for targeted in vivo siRNA delivery. *Nature Nanotechnology*, **7**, 389–393.
74. Afonin, K.A., Viard, M., Koyfman, A.Y., Martins, A.N., Kasprzak, W.K., Panigaj, M., Desai, R., Santhanam, A., Grabow, W.W., Jaeger, L. et al. (2014) Multifunctional RNA nanoparticles. *Nano Letters*, **14**, 5662–5671.
75. Grabow, W.W., Zakrevsky, P., Afonin, K.A., Chworos, A., Shapiro, B.A. and Jaeger, L. (2011) Self-assembling RNA nanorings based on RNAI/II inverse kissing complexes. *Nano Letters*, **11**, 878–887.
76. Afonin, K.A., Kasprzak, W.K., Bindewald, E., Kireeva, M., Viard, M., Kashlev, M. and Shapiro, B.A. (2014) In silico design and enzymatic synthesis of functional RNA nanoparticles. *Accounts of Chemical Research*, **47**, 1731–1741.
77. Jaeger, L., Westhof, E. and Leontis, N.B. (2001) TectoRNA: modular assembly units for the construction of RNA nano-objects. *Nucleic Acids Research*, **29**, 455–463.
78. Guo, P. (2010) The emerging field of RNA nanotechnology. *Nature Nanotechnology*, **5**, 833–842.
79. Geary, C., Chworos, A., Verzemnieks, E., Voss, N.R. and Jaeger, L. (2017) Composing RNA nanostructures from a syntax of RNA structural modules. *Nano Letters*, **17**, 7095–7101.
80. Grabow, W.W. and Jaeger, L. (2014) RNA self-assembly and RNA nanotechnology. *Accounts of Chemical Research*, **47**, 1871–1880.
81. Grabow, W.W., Zhuang, Z., Shea, J.E. and Jaeger, L. (2013) The GA-minor submotif as a case study of RNA modularity, prediction, and design. *Wiley Interdisciplinary Reviews: RNA*, **4**, 181–203.
82. Severcan, I., Geary, C., Chworos, A., Voss, N., Jacovetty, E. and Jaeger, L. (2010) A polyhedron made of tRNAs. *Nature Chemistry*, **2**, 772–779.
83. Chworos, A., Severcan, I., Koyfman, A.Y., Weinkam, P., Oroudjev, E., Hansma, H.G. and Jaeger, L. (2004) Building programmable jigsaw puzzles with RNA. *Science*, **306**, 2068–2072.
84. Jaeger, L., Westhof, E. and Leontis, N.B. (2001) TectoRNA: modular assembly units for the construction of RNA nano-objects. *Nucleic Acids Research*, **29**, 455–463.
85. Jaeger, L. and Leontis, N.B. (2000) Tecto-RNA: One-dimensional self-assembly through tertiary interactions this work was carried out in Strasbourg with the support of grants to N.B.L. from the NIH (1R15 GM55898) and the NIH Fogarty Institute (1-F06-TW02251-01) and the support of the CNRS to L.J. The authors wish to thank Eric Westhof for his support and encouragement of this work. *Angewandte Chemie (International Ed. in English)*, **39**, 2521–2524.
86. Panigaj, M., Johnson, M.B., Ke, W., McMillan, J., Goncharova, E.A., Chandler, M. and Afonin, K.A. (2019) Aptamers as modular components of therapeutic nucleic acid nanotechnology. *ACS Nano*, **13**, 12301–12321.

87. Ke, W., Hong, E., Saito, R.F., Rangel, M.C., Wang, J., Viard, M., Richardson, M., Khisamutdinov, E.F., Panigaj, M., Dokholyan, N.V. et al. (2019) RNA-DNA fibers and polygons with controlled immuno-recognition activate RNAi, FRET and transcriptional regulation of NF-kappaB in human cells. *Nucleic Acids Research*, **47**, 1350–1361.

88. Chandler, M., Ke, W., Halman, J.R., Panigaj, M. and Afonin, K.A. (2018) In Gonçalves, G. and Tobias, G. (eds.), *Nanooncology: Engineering nanomaterials for cancer therapy and diagnosis.* Springer International Publishing, Cham, pp. 365–385.

89. Halman, J.R., Satterwhite, E., Roark, B., Chandler, M., Viard, M., Ivanina, A., Bindewald, E., Kasprzak, W.K., Panigaj, M., Bui, M.N. et al. (2017) Functionally-interdependent shape-switching nanoparticles with controllable properties. *Nucleic Acids Research*, **45**, 2210–2220.

90. Chandler, M. and Afonin, K.A. (2019) Smart-responsive nucleic acid nanoparticles (NANPs) with the potential to modulate immune behavior. *Nanomaterials (Basel)*, **9**.

91. Chandler, M., Lyalina, T., Halman, J., Rackley, L., Lee, L., Dang, D., Ke, W., Sajja, S., Woods, S., Acharya, S. et al. (2018) Broccoli fluorets: Split aptamers as a user-friendly fluorescent toolkit for dynamic RNA nanotechnology. *Molecules*, **23**.

92. Zakrevsky, P., Parlea, L., Viard, M., Bindewald, E., Afonin, K.A. and Shapiro, B.A. (2017) Preparation of a conditional RNA switch. *Methods in Molecular Biology*, **1632**, 303–324.

93. Bindewald, E., Afonin, K.A., Viard, M., Zakrevsky, P., Kim, T. and Shapiro, B.A. (2016) Multistrand structure prediction of nucleic acid assemblies and design of RNA switches. *Nano Letters*, **16**, 1726–1735.

94. Afonin, K.A., Viard, M., Tedbury, P., Bindewald, E., Parlea, L., Howington, M., Valdman, M., Johns-Boehme, A., Brainerd, C., Freed, E.O. et al. (2016) The use of minimal RNA toeholds to trigger the activation of multiple functionalities. *Nano Letters*, **16**, 1746–1753.

95. Afonin, K.A., Viard, M., Kagiampakis, I., Case, C.L., Dobrovolskaia, M.A., Hofmann, J., Vrzak, A., Kireeva, M., Kasprzak, W.K., KewalRamani, V.N. et al. (2015) Triggering of RNA interference with RNA-RNA, RNA-DNA, and DNA-RNA nanoparticles. *ACS Nano*, **9**, 251–259.

96. Afonin, K.A., Desai, R., Viard, M., Kireeva, M.L., Bindewald, E., Case, C.L., Maciag, A.E., Kasprzak, W.K., Kim, T., Sappe, A. et al. (2014) Co-transcriptional production of RNA-DNA hybrids for simultaneous release of multiple split functionalities. *Nucleic Acids Research*, **42**, 2085–2097.

97. Bindewald, E., Afonin, K., Jaeger, L. and Shapiro, B.A. (2011) Multistrand RNA secondary structure prediction and nanostructure design including pseudoknots. *ACS Nano*, **5**, 9542–9551.

98. Hong, E., Halman, J.R., Shah, A., Cedrone, E., Truong, N., Afonin, K.A. and Dobrovolskaia, M.A. (2019) Toll-Like receptor-mediated recognition of nucleic acid nanoparticles (NANPs) in human primary blood cells. *Molecules*, **24**.

99. Hong, E., Halman, J.R., Shah, A.B., Khisamutdinov, E.F., Dobrovolskaia, M.A. and Afonin, K.A. (2018) Structure and composition define immunorecognition of nucleic acid nanoparticles. *Nano Letters*, **18**, 4309–4321.

100. Rackley, L., Stewart, J.M., Salotti, J., Krokhotin, A., Shah, A., Halman, J.R., Juneja, R., Smollett, J., Lee, L., Roark, K. et al. (2018) RNA fibers as optimized nanoscaffolds for siRNA coordination and reduced immunological recognition. *Advanced Functional Materials*, **28**, 1805959

101. Johnson, M.B., Halman, J.R., Satterwhite, E., Zakharov, A.V., Bui, M.N., Benkato, K., Goldsworthy, V., Kim, T., Hong, E., Dobrovolskaia, M.A. et al. (2017) Programmable nucleic acid based polygons with controlled neuroimmunomodulatory properties for predictive QSAR modeling. *Small*, **13**.

102. Rackley, L., Stewart, J.M., Salotti, J., Krokhotin, A., Shah, A., Halman, J., Juneja, R., Smollett, J., Roark, B., Viard, M. et al. (2018) RNA fibers as optimized nanoscaffolds for siRNA coordination and reduced immunological recognition. *Advanced Functional Materials*, **28**, 1805959

103. Guo, S., Li, H., Ma, M., Fu, J., Dong, Y. and Guo, P. (2017) Size, shape, and sequence-dependent immunogenicity of RNA nanoparticles. *Molecular Therapy-Nucleic Acids*, **9**, 399–408.

104. Conrad, N.K. and Steitz, J.A. (2005) A Kaposi's sarcoma virus RNA element that increases the nuclear abundance of intronless transcripts. *The EMBO Journal*, **24**, 1831–1841.

105. Zhao, J., Sinclair, J., Houghton, J., Bolton, E., Bradley, A. and Lever, A. (2010) Cytomegalovirus beta2.7 RNA transcript protects endothelial cells against apoptosis during ischemia/reperfusion injury. *The Journal of Heart and Lung Transplantation*, **29**, 342–345.

28 RIG-I as a Therapeutic Target for Nucleic Acid Nanoparticles (NANPs)

M. Brittany Johnson
University of North Carolina at Charlotte, Charlotte, USA

CONTENT

Pattern recognition receptors (PRRs) are critical for the identification of pathogen motifs, pathogen-associated molecular patterns (PAMPs) and endogenous host molecules, damage-associated molecular patterns (DAMPs) that are released from stressed cells. Toll-like receptors (TLRs), nucleotide-biding oligomerization domain (NOD)-like receptors, retinoic acid-inducible gene-I (RIG-I)-like receptors (RLRs), and DNA sensors are classes of PRRs (Barbalat, Ewald, Mouchess, & Barton, 2011; Desmet & Ishii, 2012; Wu & Chen, 2014). Recognition of PAMPs and DAMPs initiates the expression and release of immune mediators to recruit and activate immune cells. Additionally, PRR signaling stimulates expression of costimulatory molecules, antigen uptake, processing, and presentation via the major histocompatibility complex (MHC) that contribute to the generation of antigen-specific adaptive immune responses (Desmet & Ishii, 2012). Due to the role of PRRs in generating innate and adaptive immune responses, PRRs remain a promising therapeutic target.

With the discovery of RLRs, cytosolic dsRNA sensors in 2004, there have been considerable research efforts to understand their structure, signaling, and function (Kato et al., 2006; Yoneyama et al., 2004; 2005). Due to our current knowledge of the broad cell-type expression of RLRs and their important role in the detection of PAMPs and DAMPs, more recent studies have investigated the development and application of RLR agonists as antivirals, vaccine adjuvants, and cancer immunotherapeutics (Elion & Cook, 2018; Iurescia, Fioretti, & Rinaldi, 2018; Yong & Luo, 2018). There are three RLR family members: RIG-I, melanoma differentiation-associated protein-5 (MDA-5), and laboratory of genetics and physiology 2 (LGP2) (Kato et al., 2006; Yoneyama et al., 2004; 2005). RLRs possess a helicase domain and a C-terminal domain (CTD) important for ligand recognition. RIG-I and MDA-5 additionally have two N-terminal caspase-recruitment domains (CARDs) required for signaling (Saito et al., 2007; Yoneyama et al., 2004; 2005). LGP2 lacks CARDs and has been demonstrated to regulate RIG-I and MDA-5 (Satoh et al., 2010; Venkataraman et al., 2007; Yoneyama et al., 2005). While both RIG-I and MDA-5 recognize dsRNA, RIG-I recognizes specific RNA motifs. RIG-I preferentially recognizes blunt-end 5′ triphosphorylated dsRNA or ssRNA. However, one 5′ phosphate is sufficient for signaling and the presence of polyuridine-rich motifs can enhance signaling to ssRNA (Hornung et al., 2006; Kato et al., 2008; Saito, Owen, Jiang, Marcotrigiano, & Gale, 2008; Uzri & Gehrke, 2009). Notably, RIG-I can indirectly identify DNA ligands via RNA polymerase III mediated transcription of DNA ligands such as BDNA (dA:dT) into 5′ppp AU-RNA ligands (Ablasser et al., 2009; Chiu, MacMillan, & Chen, 2009). Both RIG-I and MDA-5 associate with mitochondrial antiviral signaling protein (MAVS) (also known as interferon-β promoter stimulator 1, IPS-1) via CARD domains to activate the transcription factors, nuclear factor kappa-light-chain-enhancer of activated B cells (NF-κB), and interferon regulator factor (IRF) (Hiscott, Lacoste, & Lin, 2006; Öhman, Rintahaka, Kalkkinen, Matikainen, & Nyman, 2009; Wu et al., 2014). Then NF-κB and IRF transcription factors translocate to the nucleus and initiate transcription of proinflammatory cytokine and interferons (Barbalat et al., 2011; Loo, Gale, 2011; Wilkins & Gale, 2010). Subsequently, interferon production signals via interferon-α/β receptor to drive the expression of interferon-stimulated genes (ISGs) that inhibit the viral replication and spread to surrounding cells. Together RLR-mediated responses trigger a protective antiviral cellular state (Loo et al., 2008; Poeck et al., 2010).

Currently, 90 approved antiviral drugs are used to inhibit viral infection or replication (De Clercq & Li, 2016). However, broad-spectrum antivirals are still of great interest because most antivirals are limited to only certain viral strains, and emerging resistance can render these antivirals ineffective (Strasfeld & Chou, 2010). RIG-I is a promising target due to the antiviral cell state triggered by RIG-I activation (Goulet et al., 2013). Currently, small molecular compounds and 5′ triphosphory-lated short dsRNAs have been investigated as potential antivirals. SB9200, a small molecule compound and M8, a uridine-rich 5′ triphosphorylated RNA have displayed effective RIG-I-mediated antiviral activity against RNA viruses in both in vitro and in vivo studies (Chiang et al., 2015; Goulet et al., 2013; Jones et al., 2017; Korolowicz et al., 2016; Lee et al., 2018; Olagnier et al., 2014; Sato et al., 2015). Specifically, SB9200 has been demonstrated as an effective antiviral in clinical trials for patients infected with chronic Hepatitis C (HCV) (Jones et al., 2017).

Also, RIG-I agonists can be employed as vaccine adjuvants to enhance antigen-specific immune responses to a vaccine antigen. In human clinical trials, the use of the TLR7/8/RIG-I agonist, CV8102, as an adjuvant for the licensed rabies vaccine demonstrated two safe effective doses that significantly enhanced immunogenicity of the rabies vaccine (Doener et al., 2019). Early investigation of M8 as a vaccine adjuvant in conjugation the virus-like particle (VLP) expressing H5N1 influenza hemagglutinin and neuraminidase demonstrated M8-induced higher antibody titer and increased murine survival against the H5N1 influenza virus when compared to alum, an approved human vaccine adjuvant (Beljanski et al., 2015). Importantly, gold nanorods have been used as a biocompatible and site-specific carrier to effectively deliver 5′ppp ssRNA. Gold nanorod delivery of 5′ppp ssRNA stimulated antiviral responses against influenza virus in human respiratory bronchial epithelial cells. In this system, RIG-I activation stimulated IFN and ISGs production resulting in diminished influenza virus replication (Chakravarthy et al., 2010). Together these data indicated RIG-I is a promising target for the development of host direct antivirals and vaccine adjuvants to combat antiviral resistance and emerging viruses that are of great public health concern.

In addition, to the use as antiviral and vaccine adjuvants, RIG-I agonist can be employed as cancer immunotherapeutics. Current immunotherapy employs immune checkpoint inhibitors that relieve the negative regulation on immune activation (Ribas & Wolchok, 2018). However, the effectiveness of immune checkpoint inhibitors is limited due to many tumors lacking infiltration of important immune cells such as cytotoxic T cells (Smith et al., 2018). RIG-I activation stimulates pro-inflammatory cytokines and interferons to recruit and activate adaptive immune cells (Wu & Chen, 2014). Interferons also stimulate upregulated MHC class I and costimulatory molecule expression which is necessary for antigen presentation and antigen-specific immune responses (Gessani, Conti, Del Cornò, & Belardelli, 2014). RIG-I dependent interferon production has been demonstrated to be important for dendritic cell activation, migration to the tumor-draining lymph node, and MHC class I cross-presentation to CD8 T cells (Duewell et al., 2014; Ellermeier et al., 2013). Additionally, RIG-I signaling initiates programmed cell death in cancer cells (Besch et al., 2009; Bhoopathi et al., 2014; Kübler et al., 2010; Poeck et al., 2008). Hiltonol an agonist for TLR3, MDA-5, and RIG-I has shown success in preclinical and early clinical studies at increases patient lifespan when tested in combination with chemotherapy or radiation (Grossman et al., 2010; Rosenfeld et al., 2010; Salazar et al., 1996). Phase I and II clinical trials have also demonstrated it is a promising tumor vaccine adjuvant (Smith et al., 2018). Notably, there has been early investigation into the use of nanoparticles as delivery agents for RIG-I agonists. Liposome-silica hybrid (LSH) and lipid calcium phosphate nanoparticles have been used to successfully deliver poly(I:C) and a RIG-I agonist siRNA targeting Bcl2. Consequently, RIG-I activation triggered tumor cell killing in prostate, pancreatic, and breast cancer models (Colapicchioni et al., 2015; Das, Shen, Liu, Goodwin, & Huang, 2019; Poeck et al., 2008). Additionally, endosomolytic polymer nanoparticles which are pH-responsive and facilitate active endosomal escape effectively deliver 5'triphosphate ds RNA to cells. Endosomolytic polymer mediated delivery induced proinflammatory cytokines and interferons, increased infiltration of CD8 t cells, and cell death in a colon cancer model. Furthermore, it enhanced the anti-tumor activity of an anti-PD-1 immune checkpoint blockade (Jacobson, Wang-Bishop, Becker, & Wilson, 2019). This evidence supports RIG-I as an important therapeutic target to apply in combination with immune checkpoint inhibitors due to its wide cellular expression, ability to promote antigen presentation, stimulate cytotoxic immune cells, and initiate programmed cell death pathways.

The central biological role of RIG-I in monitoring the cytosol for nucleic acid PAMPs and DAMPs allows RIG-I agonists to be applied to broad therapeutic applications including antivirals, vaccine adjuvants, and cancer immunotherapeutics (Figure 28.1). Nucleic acid nanoparticles (NANPs) are an ideal platform for the development of RIG-I agonists due to their fine-tunable properties (Bui et al., 2017). NANPs can be composed of RNA, DNA, and RNA/DNA hybrids that self-assemble with high batch to batch consistency due to Watson–Crick base pairing (Afonin, Kasprzak, et al., 2014). Importantly, the RNA strands contain the necessary 5'ppp ligand for RIG-I. The RNA and DNA composition of nanoparticles impacts many of the physiochemical properties including melting temperature and serum stability (Bui et al., 2017; Hong et al., 2018; Johnson et al., 2017). Higher RNA composition generates nanoparticles with higher melting temperatures and lower resistance to degradative nuclease found in serum. In contrast, DNA composition generates nanoparticles with lower melting temperatures and higher resistance to serum degradation. Nuclease degradation often limits the application of RNA-based therapeutics. However, RNA/DNA hybrid NANPs and NANPs containing modified nucleic acids provide a platform to increase the stability of RNA functional groups (Bui et al., 2017; Hong et al., 2018; Johnson et al., 2017; Lee, Urban, Xu, Sullenger, & Lee, 2016; Robbins et al., 2007; Schmitt et al., 2017). Interestingly, QSAR modeling demonstrates NANP physicochemical properties are strong predictors of the immunostimulatory activity (Johnson et al., 2017). This provides a means to rationally design the nucleic acid composition and predict the immunostimulation of NANPs.

PRRs recognize specific motifs to differentiate self and non-self-nucleic acids (Anchisi, Guerra, & Garcin, 2015; Hornung et al., 2006). Therefore, the immunostimulation can further be engineered through the incorporation of modified nucleic acid strands (Lee et al., 2016; Robbins et al., 2007; Schmitt et al., 2017; Sioud, 2006). Incorporation of 2'F pyrimidine enhanced RIG-I-mediated interferon production but significantly reduced TLR3 and TLR7 stimulation. Enhanced RIG-I stimulation also resulted in increased cell death compared to unmodified RNAs (Y. Lee et al., 2016). In contrast, 2' O-methyl modifications blocked IFN-β production and prevented cell death (Lee et al., 2016). Modified nucleic acids offer a promising approach to stimulate or abrogate activation of specific receptors.

Additionally, this platform offers a way to successfully deliver multiple functional siRNAs to target cells and provide a combinatorial treatment option (Afonin et al., 2011; Afonin, Viard, et al., 2014; Stewart et al., 2016). For example, delivery of multifunctional siRNA designed to stimulate RIG-I activation and silence Bcl2 or MDR1 have previously been demonstrated to promote apoptosis in lung metastases and drug-resistant leukemia cell lines, respectively (Bhoopathi et al., 2014; Poeck et al., 2008). NANPs can be assembled into a variety of shapes allowing for the incorporation of multiple functional groups including siRNA, chemical drugs, imaging molecules for detection, and aptamers (Afonin, Viard, et al., 2014).

Finally, cube and complementary anti-cube structures can be designed to release functional groups upon reassociation. This allows for the conditional activation of nanoparticle functional groups including siRNA, RNA aptamers, and NF-κB decoys (Halman et al., 2017; Ke et al., 2019; Kim et al., 2009; Porciani et al., 2015). NF-κB decoys that bind and prevent translocation of NF-κB to the nucleus diminish NF-κB- dependent production of inflammatory cytokines. This provides

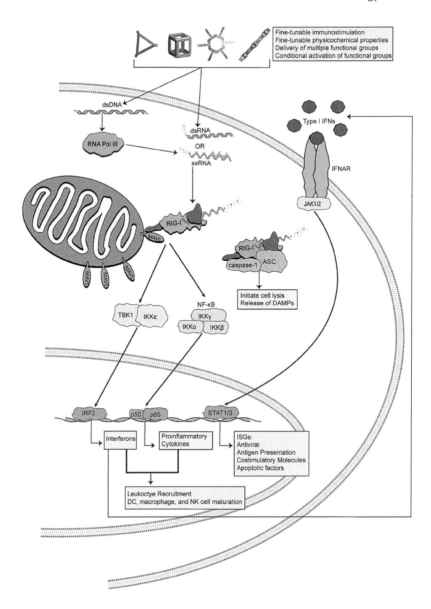

FIGURE 28.1 RIG-I as a target for NANPs stimulation of antiviral and antitumorigenic responses. RIG-I binding to 5'pppRNA drives RIG-I recruitment to the mitochondria outer membrane and association with MAVS. MAVS initiate a signaling cascade resulting in translocation of the transcription factors, IRF and NF-κB, to the nucleus where they induce expression of interferons and proinflammatory cytokines, respectively. Interferons and proinflammatory cytokines are important for leukocyte recruitment and maturation of dendritic cells, macrophages, and NK cells. Additionally, interferons signal in a paracrine or autocrine manner via IFNAR to stimulate expression of ISGs. ISGs include antiviral genes, antigen presentation and costimulatory molecules, and apoptotic factors. RIG-I activation can also recruit the inflammasome adaptor protein ASC which activates caspase-I thereby initiating cell lysis.

an additional mechanism to enhance desired IFN responses and diminish damaging inflammatory response. Collectively, the ability to rationally engineer the physicochemical and immunostimulatory properties and tightly control delivery and activation of functional groups provides a versatile platform for the future development of RIG-I therapeutic agonists (Figure 28.1).

REFERENCES

Ablasser, A., Bauernfeind, F., Hartmann, G., Latz, E., Fitzgerald, K. A., & Hornung, V. (2009). RIG-I-dependent sensing of poly(dA:dT) through the induction of an RNA polymerase III-transcribed RNA intermediate. *Nature Immunology*, *10*(10), 1065–1072. doi:10.1038/ni.1779

Afonin, K. A., Grabow, W. W., Walker, F. M., Bindewald, E., Dobrovolskaia, M. A., Shapiro, B. A., & Jaeger, L. (2011). Design and self-assembly of siRNA-functionalized RNA nanoparticles for use in automated nanomedicine. *Nature Protocols*, *6*(12), 2022–2034. doi:10.1038/nprot.2011.418

Afonin, K. A., Kasprzak, W., Bindewald, E., Puppala, P. S., Diehl, A. R., Hall, K. T., … Shapiro, B. A. (2014). Computational and experimental characterization of RNA cubic nanoscaffolds. *Methods*. doi:10.1016/j.ymeth.2013.10.013

Afonin, K. A., Viard, M., Koyfman, A. Y., Martins, A. N., Kasprzak, W. K., Panigaj, M., … Shapiro, B. A. (2014). Multifunctional RNA nanoparticles. *Nano Letters*, *14*(10), 5662–5671. doi:10.1021/nl502385k

Anchisi, S., Guerra, J., & Garcin, D. (2015). RIG-I atpase activity and discrimination of self-RNA versus non-self-RNA. *MBio*, *6*(2). doi:10.1128/mBio.02349-14

Barbalat, R., Ewald, S. E., Mouchess, M. L., & Barton, G. M. (2011). Nucleic acid recognition by the innate immune system. *Annual Review of Immunology*. doi:10.1146/annurev-immunol-031210-101340

Beljanski, V., Chiang, C., Kirchenbaum, G. A., Olagnier, D., Bloom, C. E., Wong, T., … Hiscott, J. (2015). Enhanced influenza virus-like particle vaccination with a structurally optimized RIG-I agonist as adjuvant. *Journal of Virology*, *89*(20), 10612–10624. doi:10.1128/JVI.01526-15

Besch, R., Poeck, H., Hohenauer, T., Senft, D., Häcker, G., Berking, C., … Hartmann, G. (2009). Proapoptotic signaling induced by RIG-I and MDA-5 results in type I interferon-independent apoptosis in human melanoma cells. *Journal of Clinical Investigation*. doi:10.1172/JCI37155

Bhoopathi, P., Quinn, B. A., Gui, Q., Shen, X. N., Grossman, S. R., Das, S. K., … Emdad, L. (2014). Pancreatic cancer-specific cell death induced in vivo by cytoplasmic-delivered polyinosine-polycytidylic acid. *Cancer Research*. doi:10.1158/0008-5472.CAN-14-0819

Bui, M. N., Brittany Johnson, M., Viard, M., Satterwhite, E., Martins, A. N., Li, Z., … Khisamutdinov, E. F. (2017). Versatile RNA tetra-U helix linking motif as a toolkit for nucleic acid nanotechnology. *Nanomedicine: Nanotechnology, Biology and Medicine*, *13*(3), 1137–1146. doi:10.1016/j.nano.2016.12.018

Chakravarthy, K. V., Bonoiu, A. C., Davis, W. G., Ranjan, P., Ding, H., Hu, R., … Prasad, P. N. (2010). Gold nanorod delivery of an ssRNA immune activator inhibits pandemic H1N1 influenza viral replication. *Proceedings of the National Academy of Sciences of the United States of America*. doi:10.1073/pnas.0914561107

Chiang, C., Beljanski, V., Yin, K., Olagnier, D., Ben Yebdri, F., Steel, C., … Hiscott, J. (2015). Sequence-specific modifications enhance the broad-spectrum antiviral response activated by RIG-I agonists. *Journal of Virology*. doi:10.1128/jvi.00845-15

Chiu, Y. H., MacMillan, J. B., & Chen, Z. J. (2009). RNA polymerase III detects cytosolic DNA and induces type I interferons through the RIG-I pathway. *Cell*, *138*(3), 576–591. doi:10.1016/j.cell.2009.06.015

Colapicchioni, V., Palchetti, S., Pozzi, D., Marini, E. S., Riccioli, A., Ziparo, E., … Caracciolo, G. (2015). Killing cancer cells using nanotechnology: Novel poly(I:C) loaded liposome-silica hybrid nanoparticles. *Journal of Materials Chemistry B*. doi:10.1039/c5tb01383f

Das, M., Shen, L., Liu, Q., Goodwin, T. J., & Huang, L. (2019). Nanoparticle delivery of RIG-I agonist enables effective and safe adjuvant therapy in pancreatic cancer. *Molecular Therapy*. doi:10.1016/j.ymthe.2018.11.012

De Clercq, E., & Li, G. (2016). Approved antiviral drugs over the past 50 years. *Clinical Microbiology Reviews*. doi:10.1128/CMR.00102-15

Desmet, C. J., & Ishii, K. J. (2012). Nucleic acid sensing at the interface between innate and adaptive immunity in vaccination. *Nature Reviews Immunology*. doi:10.1038/nri3247

Doener, F., Hong, H. S., Meyer, I., Tadjalli-Mehr, K., Daehling, A., Heidenreich, R., … Gnad-Vogt, U. (2019). RNA-based adjuvant CV8102 enhances the immunogenicity of a licensed rabies vaccine in a first-in-human trial. *Vaccine*. doi:10.1016/j.vaccine.2019.02.024

Duewell, P., Steger, A., Lohr, H., Bourhis, H., Hoelz, H., Kirchleitner, S. V., … Schnurr, M. (2014). RIG-I-like helicases induce immunogenic cell death of pancreatic cancer cells and sensitize tumors toward killing by CD8(+) T cells. *Cell Death and Differentiation*. doi:10.1038/cdd.2014.96

Elion, D. L., & Cook, R. S. (2018). Harnessing RIG-I and intrinsic immunity in the tumor microenvironment for therapeutic cancer treatment. *Oncotarget*. doi:10.18632/oncotarget.25626

Ellermeier, J., Wei, J., Duewell, P., Hoves, S., Stieg, M. R., Adunka, T., … Schnurr, M. (2013). Therapeutic efficacy of bifunctional siRNA combining TGF-β1 silencing with RIG-I activation in pancreatic cancer. *Cancer Research*. doi:10.1158/0008-5472.CAN-11-3850

Gessani, S., Conti, L., Del Cornò, M., & Belardelli, F. (2014). Type I interferons as regulators of human antigen presenting cell functions. *Toxins*. doi:10.3390/toxins6061696

Goulet, M. L., Olagnier, D., Xu, Z., Paz, S., Belgnaoui, S. M., Lafferty, E. I., ... Hiscott, J. (2013). Systems Analysis of a RIG-I Agonist Inducing Broad Spectrum Inhibition of Virus Infectivity. *PLoS Pathogens*, 9(4). doi:10.1371/journal.ppat.1003298

Grossman, S. A., Ye, X., Piantadosi, S., Desideri, S., Nabors, L. B., Rosenfeld, M., & Fisher, J. (2010). Survival of patients with newly diagnosed glioblastoma treated with radiation and temozolomide in research studies in the United States. *Clinical Cancer Research*. doi:10.1158/1078-0432.CCR-09-3106

Halman, J. R., Satterwhite, E., Roark, B., Chandler, M., Viard, M., Ivanina, A., ... Afonin, K. A. (2017). Functionally-interdependent shape-switching nanoparticles with controllable properties. *Nucleic Acids Research*. doi:10.1093/nar/gkx008

Hiscott, J., Lacoste, J., & Lin, R. (2006). Recruitment of an interferon molecular signaling complex to the mitochondrial membrane: Disruption by hepatitis C virus NS3-4A protease. *Biochemical Pharmacology*. doi:10.1016/j.bcp.2006.06.030

Hong, E., Halman, J. R., Shah, A. B., Khisamutdinov, E. F., Dobrovolskaia, M. A., & Afonin, K. A. (2018). Structure and composition define immunorecognition of nucleic acid nanoparticles. *Nano Letters*. doi:10.1021/acs.nanolett.8b01283

Hornung, V., Ellegast, J., Kim, S., Brzózka, K., Jung, A., Kato, H., ... Hartmann, G. (2006). 5′-Triphosphate RNA is the ligand for RIG-I. *Science*. doi:10.1126/science.1132505

Iurescia, S., Fioretti, D., & Rinaldi, M. (2018). Targeting cytosolic nucleic acid-sensing pathways for cancer immunotherapies. *Frontiers in Immunology*. doi:10.3389/fimmu.2018.00711

Jacobson, M. E., Wang-Bishop, L., Becker, K. W., & Wilson, J. T. (2019). Delivery of 5′-triphosphate RNA with endosomolytic nanoparticles potently activates RIG-I to improve cancer immunotherapy. *Biomaterials Science*. doi:10.1039/c8bm01064a

Johnson, M. B., Halman, J. R., Satterwhite, E., Zakharov, A. V., Bui, M. N., Benkato, K., ... Afonin, K. A. (2017). Programmable nucleic acid based polygons with controlled neuroimmunomodulatory properties for predictive QSAR modeling. *Small*, 1701255. doi:10.1002/smll.201701255

Jones, M., Cunningham, M. E., Wing, P., DeSilva, S., Challa, R., Sheri, A., ... Foster, G. R. (2017). SB 9200, a novel agonist of innate immunity, shows potent antiviral activity against resistant HCV variants. *Journal of Medical Virology*. doi:10.1002/jmv.24809

Kato, H., Takeuchi, O., Mikamo-Satoh, E., Hirai, R., Kawai, T., Matsushita, K., ... Akira, S. (2008). Length-dependent recognition of double-stranded ribonucleic acids by retinoic acid–inducible gene-I and melanoma differentiation–associated gene 5. *The Journal of Experimental Medicine*, 205(7), 1601–1610. doi:10.1084/jem.20080091

Kato, H., Takeuchi, O., Sato, S., Yoneyama, M., Yamamoto, M., Matsui, K., ... Akira, S. (2006). Differential roles of MDA5 and RIG-I helicases in the recognition of RNA viruses. *Nature*, 441(1), 101–105. doi:10.1038/nature04734

Ke, W., Hong, E., Saito, R. F., Rangel, M. C., Wang, J., Viard, M., ... Afonin, K. A. (2019). RNA-DNA fibers and polygons with controlled immunorecognition activate RNAi, FRET and transcriptional regulation of NF-κB in human cells. *Nucleic Acids Research*. doi:10.1093/nar/gky1215

Kim, K. H., Lee, E. S., Cha, S. H., Park, J. H., Park, J. S., Chang, Y. C., & Park, K. K. (2009). Transcriptional regulation of NF-κB by ring type decoy oligodeoxynucleotide in an animal model of nephropathy. *Experimental and Molecular Pathology*. doi:10.1016/j.yexmp.2008.11.011

Korolowicz, K. E., Iyer, R. P., Czerwinski, S., Suresh, M., Yang, J., Padmanabhan, S., ... Menne, S. (2016). Antiviral efficacy and host innate immunity associated with SB 9200 treatment in the woodchuck model of chronic hepatitis B. *PLoS ONE*. doi:10.1371/journal.pone.0161313

Kübler, K., Gehrke, N., Riemann, S., Böhnert, V., Zillinger, T., Hartmann, E., ... Barchet, W. (2010). Targeted activation of RNA helicase retinoic acid - Inducible gene-I induces proimmunogenic apoptosis of human ovarian cancer cells. *Cancer Research*. doi:10.1158/0008-5472.CAN-10-0825

Lee, J., Park, E. B., Min, J., Sung, S. E., Jang, Y., Shin, J. S., ... Choi, B. S. (2018). Systematic editing of synthetic RIG-I ligands to produce effective antiviral and anti-tumor RNA immunotherapies. *Nucleic Acids Research*. doi:10.1093/nar/gky039

Lee, Y., Urban, J. H., Xu, L., Sullenger, B. A., & Lee, J. (2016). 2′Fluoro modification differentially modulates the ability of RNAs to activate pattern recognition receptors. *Nucleic Acid Therapeutics*. doi:10.1089/nat.2015.0575

Loo, Y.-M., Fornek, J., Crochet, N., Bajwa, G., Perwitasari, O., Martinez-Sobrido, L., ... Gale, M. (2008). Distinct RIG-I and MDA5 Signaling by RNA Viruses in Innate Immunity. *Journal of Virology*, 82(1), 335–345. doi:10.1128/JVI.01080-07

Loo, Y.-M., Gale, M. (2011). Immune signaling by RIG-I-like receptors. *Immunity*, *34*(5), 680–692. doi:10.1016/j.immuni.2011.05.003

Öhman, T., Rintahaka, J., Kalkkinen, N., Matikainen, S., & Nyman, T. A. (2009). Actin and RIG-I/MAVS signaling components translocate to mitochondria upon influenza a virus infection of human primary macrophages. *The Journal of Immunology*. doi:10.4049/jimmunol.0803093

Olagnier, D., Scholte, F. E. M., Chiang, C., Albulescu, I. C., Nichols, C., He, Z., … Hiscott, J. (2014). Inhibition of dengue and chikungunya virus infections by RIG-I-Mediated type I interferon-independent stimulation of the innate antiviral response. *Journal of Virology*. doi:10.1128/jvi.03114-13

Poeck, H., Besch, R., Maihoefer, C., Renn, M., Tormo, D., Morskaya, S. S., … Hartmann, G. (2008). 5′-triphosphate-siRNA: Turning gene silencing and Rig-I activation against melanoma. *Nature Medicine*. doi:10.1038/nm.1887

Poeck, H., Bscheider, M., Gross, O., Finger, K., Roth, S., Rebsamen, M., … Ruland, J. (2010). Recognition of RNA virus by RIG-I results in activation of CARD9 and inflammasome signaling for interleukin 1B production. *Nature Immunology*, *11*(1), 63–69. doi:10.1038/ni.1824

Porciani, D., Tedeschi, L., Marchetti, L., Citti, L., Piazza, V., Beltram, F., & Signore, G. (2015). Aptamer-mediated codelivery of doxorubicin and NF-κB decoy enhances chemosensitivity of pancreatic tumor cells. *Molecular Therapy - Nucleic Acids*. doi:10.1038/mtna.2015.9

Ribas, A., & Wolchok, J. D. (2018). Cancer immunotherapy using checkpoint blockade. *Science*. doi:10.1126/science.aar4060

Robbins, M., Judge, A., Liang, L., McClintock, K., Yaworski, E., & MacLachlan, I. (2007). 2′-O-methyl-modified RNAs act as TLR7 antagonists. *Molecular Therapy*. doi:10.1038/sj.mt.6300240

Rosenfeld, M. R., Chamberlain, M. C., Grossman, S. A., Peereboom, D. M., Lesser, G. J., Batchelor, T. T., … Ye, X. (2010). A multi-institution phase II study of poly-ICLC and radiotherapy with concurrent and adjuvant temozolomide in adults with newly diagnosed glioblastoma. *Neuro-Oncology*. doi:10.1093/neuonc/noq071

Saito, T., Hirai, R., Loo, Y.-M., Owen, D., Johnson, C. L., Sinha, S. C., … Gale, M. (2007). Regulation of innate antiviral defenses through a shared repressor domain in RIG-I and LGP2. *Proceedings of the National Academy of Sciences*, *104*(2), 582–587. doi:10.1073/pnas.0606699104

Saito, T., Owen, D. M., Jiang, F., Marcotrigiano, J., & Gale, M. (2008). Innate immunity induced by composition-dependent RIG-I recognition of hepatitis C virus RNA. *Nature*, *454*(7203), 523–527. doi:10.1038/nature07106

Salazar, A. M., Levy, H. B., Ondra, S., Kende, M., Scherokman, B., Brown, D., … Ommaya, A. (1996). Long-term treatment of malignant gliomas with intramuscularly administered polyinosinic-polycytidylic acid stabilized with polylysine and carboxymethylcellulose: An open pilot study. *Neurosurgery*. doi:10.1097/00006123-199606000-00006

Sato, S., Li, K., Kameyama, T., Hayashi, T., Ishida, Y., Murakami, S., … Takaoka, A. (2015). The RNA sensor RIG-I dually functions as an innate sensor and direct antiviral factor for hepatitis B virus. *Immunity*. doi:10.1016/j.immuni.2014.12.016

Satoh, T., Kato, H., Kumagai, Y., Yoneyama, M., Sato, S., Matsushita, K., … Takeuchi, O. (2010). LGP2 is a positive regulator of RIG-I- and MDA5-mediated antiviral responses. *Proceedings of the National Academy of Sciences of the United States of America*. doi:10.1073/pnas.0912986107

Schmitt, F. C. F., Freund, I., Weigand, M. A., Helm, M., Dalpke, A. H., & Eigenbrod, T. (2017). Identification of an optimized 2′-O-methylated trinucleotide RNA motif inhibiting Toll-like receptors 7 and 8. *RNA*. doi:10.1261/rna.061952.117

Sioud, M. (2006). Single-stranded small interfering RNA are more immunostimulatory than their double-stranded counterparts: A central role for 2′-hyroxyl uridines in immune responses. *European Journal of Immunology*. doi:10.1002/eji.200535708

Smith, M., García-Martínez, E., Pitter, M. R., Fucikova, J., Spisek, R., Zitvogel, L., … Galluzzi, L. (2018). Trial Watch: Toll-like receptor agonists in cancer immunotherapy. *OncoImmunology*. doi:10.1080/2162402X.2018.1526250

Stewart, J. M., Viard, M., Subramanian, H. K. K., Roark, B. K., Afonin, K. A., & Franco, E. (2016). Programmable RNA microstructures for coordinated delivery of siRNAs. *Nanoscale*, *8*(40), 17542–17550. doi:10.1039/C6NR05085A

Strasfeld, L., & Chou, S. (2010). Antiviral drug resistance: Mechanisms and clinical implications. *Infectious Disease Clinics of North America*. doi:10.1016/j.idc.2010.07.001

Uzri, D., & Gehrke, L. (2009). Nucleotide sequences and modifications that determine RIG-I/RNA binding and signaling activities. *Journal of Virology*, *83*(9), 4174–4184. doi:10.1128/JVI.02449-08

Venkataraman, T., Valdes, M., Elsby, R., Kakuta, S., Caceres, G., Saijo, S., ... Barber, G. N. (2007). Loss of DExD/H Box RNA helicase LGP2 manifests disparate antiviral responses. *The Journal of Immunology*. doi:10.4049/jimmunol.178.10.6444

Wilkins, C., & Gale, M. (2010). Recognition of viruses by cytoplasmic sensors. *Current Opinion in Immunology*. doi:10.1016/j.coi.2009.12.003

Wu, B., Peisley, A., Tetrault, D., Li, Z., Egelman, E. H., Magor, K. E., ... Hur, S. (2014). Molecular imprinting as a signal-activation mechanism of the viral RNA sensor RIG-I. *Molecular Cell*. doi:10.1016/j.molcel.2014. 06.010

Wu, J., & Chen, Z. J. (2014). Innate immune sensing and signaling of cytosolic nucleic acids. *Annual Review of Immunology*, *32*(1), 461–488. doi:10.1146/annurev-immunol-032713-120156

Yoneyama, M., Kikuchi, M., Matsumoto, K., Imaizumi, T., Miyagishi, M., Taira, K., ... Fujita, T. (2005). Shared and unique functions of the DExD/H-Box helicases RIG-I, MDA5, and LGP2 in antiviral innate immunity. *The Journal of Immunology*, *175*(5), 2851–2858. doi:10.4049/jimmunol.175.5.2851

Yoneyama, M., Kikuchi, M., Natsukawa, T., Shinobu, N., Imaizumi, T., Miyagishi, M., ... Fujita, T. (2004). The RNA helicase RIG-I has an essential function in double-stranded RNA-induced innate antiviral responses. *Nature Immunology*. doi:10.1038/ni1087

Yong, H. Y., & Luo, D. (2018). RIG-I-like receptors as novel targets for pan-antivirals and vaccine adjuvants against emerging and re-emerging viral infections. *Frontiers in Immunology*. doi:10.3389/fimmu.2018.01379

29 Driving Dynamic Functions with Programmable RNA Nanostructures

Morgan Chandler, Dana Elasmar, Sam Eisen,
Luke Schmedake, and Kirill A. Afonin
University of North Carolina at Charlotte, Charlotte, USA

CONTENTS

Owing to their programmable and predictable base pairing, nucleic acids can bind one another in the cellular environment during the genetic flow of information but also fold on themselves forming diverse three-dimensional structures which are imbued with biological functions. Dynamic ribozymes, riboswitches, and aptamers are examples of functional subunits which nucleic acid nanotechnology can build from in the design of nucleic acid nanoparticles (NANPs) capable of carrying out similar roles. Understanding NANPs as more than simple static scaffolds can allow for such functions to be further implemented towards the design of smart therapeutic and diagnostic systems.

PROGRAMMABLE NANOMATERIALS

RIBOSWITCHES AND RIBOZYMES

In biological systems, nucleic acids have adapted to roles not just as carriers of the genetic code but also as its regulators. A prime example of this is the prevalence of riboswitches, which are sequences in RNA capable of binding to a target to change the production of proteins through a variety of mechanisms. Riboswitches have been found to be heavily utilized by bacteria, where over 40 different classes have been discovered and validated. They are not limited to bacteria, however, as they have also been characterized in archaea, fungi, algae, and even in some plants [1,2]. Riboswitches vary in structure, but many act in response to the same ligand and are thus grouped into families based on these criteria. An example of this is the SAM riboswitch family, which is composed of various RNA architectures responding to the metabolite *S*-adenosyl methionine (SAM). Once bound, the associated genes form a terminator loop, which halts further transcription of the gene products required for SAM biosynthesis [3].

    Along with regulating genetic processing, nucleic acids can also function as catalysts (enzymes) of reactions, termed ribozymes (for RNA-based structures) or DNAzymes (for DNA-based structures) [4]. These structures are of particular interest, especially in the field of theranostics, for their ability to mimic the action of enzymes while being relatively more stable and modifiable [5]. Riboswitches and ribozymes are typically found using bioinformatics approaches and describe roles that may exist in a single dynamic regulator. One of the most common examples of this can be found in the *glmS* gene present in some gram-positive bacteria. The *glmS* gene encodes the enzyme glutamine-fructose-6-phosphate amidotransferase, which is a catalyst in the production of glucosamine-6-phosphate (GlcN6P); however, the function of this gene is closely regulated by the presence of the *glmS* ribozyme located in its 5' untranslated region. If GlcN6P is present in high concentrations, it can bind to the ribozyme, inducing cleavage of the sequence so that production of GlcN6P is terminated [2]. In this case, *glmS* may be considered to be both a riboswitch capable of halting the production of GlcN6P, as well as a ribozyme capable of self-cleaving.

APTAMERS

A critical part of the riboswitch vital for its function is the domain where the small ligand can bind to the RNA structure. This recognition element is not just present in riboswitches but can also be found as short oligonucleotides which bind to any specific target molecule, and it is known as an aptamer. Aptamers are somehow the antibody equivalent for nucleic acids, providing selectivity that can be employed as a biological tool and engineered for specific use.

If not already preexisting, aptamers can be tailored to bind to a specific target molecule in a process called SELEX (systematic evolution of ligands by exponential enrichment) [6]. SELEX begins with a large pool of randomly generated sequences which are then introduced to the target molecule of interest and washed away. Any sequences which remain, having formed structures capable of binding to the target molecule, are isolated and sequenced to determine the best aptamer candidates. Oftentimes, the pool may be introduced to negative selection to ensure that no off-target binding can occur, or relative strengths of binding will be compared to select the best aptamer for the desired application [7]. In this way, a vast library of aptamers is selected with all aptamers to be specific to their target ligands and can be implemented as selective agents.

Aptamers have been established to carry out a variety of functions. For instance, fluorescence-activating aptamers have been shown to bind to inactive dyes, switching them on into fluorescent tags [8]. Malachite green was an early example of this and was implemented into various nanoparticles for their visualization [9,10]. Other fluorescence-activating aptamers, often named for their dye's fluorescent color upon activation, include Spinach [11,12], Broccoli [13–15], Corn [16], and Mango [17,18], with others in development. Aptamers have also been selected to bind exclusively to disease-specific targets for making therapeutics directed towards diseased cells. Macugen® is one such aptamer which has been FDA-approved for the treatment of age-related macular degeneration (AMD). It works by binding to and blocking the actions of an isoform of vascular endothelial growth factor (VEGF), a protein which, in the case of AMD, is responsible for the angiogenesis and permeability that results in vision loss [19]. A variety of other therapeutic aptamers are in the pipeline or are clinically available, including aptamers which promote coagulation [20] or inhibit tumor proliferation [19].

IMPLEMENTING FUNCTIONS INTO DYNAMIC NUCLEIC ACID NANOPARTICLES

NANPS AS SCAFFOLDS FOR FUNCTIONAL RNAS

Riboswitches, ribozymes, and aptamers have been implemented as the targeting moieties for nanoparticles or as the fluorescent tags for visualizing their uptake and biodistribution [21]. These functional subunits have been added to a variety of organic and inorganic nanoparticle platforms due to their biocompatibility and direct attachment, generally by electrostatic means [22]. However,

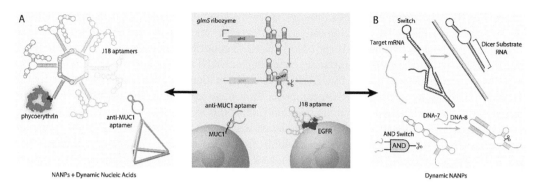

FIGURE 29.1 Dynamic nucleic acid nanomaterials can be implemented into NANPs by (a) using NANP scaffolds or (b) inspiring the design of dynamic NANPs. (a) Multifunctional NANP scaffolds can carry multiple targeting aptamers, such as J18 aptamers against EGFR or anti-MUC1 aptamers against MUC1. (b) Two-stranded NANP switches have been shown to release therapeutics upon interacting with a target mRNA. Engineered ribozyme AND logic gates have been shown to self-cleave only when two DNA inputs are present.

the use of functional RNAs is particularly advantageous when combined with the use of NANPs. The sequences assembled into a NANP scaffold may simply be extended with known functional sequences to implement them into NANP compositions [23]. So long as there are no more favorable secondary structures which may interfere with the functional architectures, their functions can be retained.

Once equipped with these functions, NANPs can carry therapeutic moieties in multiple combinations for their co-delivery into the same cellular environment [24–26]. For example, DNA tetrahedrons have been designed to carry aptamers targeting MUC1, a cell surface protein which is overexpressed in adenocarcinomas, for preferentially selecting breast cancer cells (Figure 29.1a) [27]. Additionally, hexameric RNA nanorings have been designed to carry up to five copies of the J18 RNA aptamer, which is specific to human epidermal growth factor receptor (EGFR). Using the sixth side of the nanoring to carry the fluorescent protein phycoerythrin, the binding of these NANPs can be visualized in human breast cancer cells (Figure 29.1a) [25].

Multiple combinations of these functional moieties can be added to NANP scaffolds. For example, on a 3WJ core with three handles for functionality, the combination of a fluorophore for tracking, an EGFR aptamer for targeting, and an anti-miR-21 for therapeutic activity was used to demonstrate the repression of tumor growth in triple negative breast cancer [28]. Other 3WJ-based NANPs have been used in conjunction with aptamers, including the annexin A2 aptamer for ovarian cancer cells and anti-gp120 aptamer for targeting HIV-1-infected cells based on surface recognition [29,30]. While the dynamic functions of these naturally occurring (or selected for) nucleic acids are promising as adornments on NANP scaffolds, they also offer a higher level of inspiration for designing NANPs which are directly capable of interacting dynamically in biological environments.

DYNAMIC NANPs

Much like riboswitches, the strand displacement between components of a NANP assembly can be used to switch between structural conformations in "inactive" and "active" states. Strand displacement occurs to maximize the numbers of bonds, thereby minimizing the free energy of the structure [31–34]. Introducing single-stranded toeholds into assemblies can be used to drive strand displacement in response to a programmed interaction with another, more complete sequence [35]. Many of these toehold-based systems have been used to construct NANPs which are inert until they can interact with a cognate NANP, thus kicking off completed duplexes which have therapeutic functions as inducers of RNA interference in cells [36–39]. By controlling the activation of the functionalities, this strategy is promising for reducing off-target effects and maximizing the activity of these constructs only in the area of interest. It can also be used to control the physicochemical and immune profile recognized by the cell. For instance, cube-shaped NANPs have been shown to interact with their reverse complement anti-cube NANPs to drive the formation of double-stranded duplexes [40]. By simply varying the DNA and RNA composition in these constructs, the melting temperature, stability, immune stimulation, and re-association rate of this system can be tailored. While inert, the cubes can be functionalized with split moieties which are only complete and active once the cubes re-associate with one another. As an example, fluorescence-activating aptamer sequences can be split across dynamic NANP subunits to achieve their conditional activation upon re-association [41–45]. Though first demonstrated with split Malachite Green, split Broccoli RNA aptamers have been shown to fluoresce only when the two halves of the aptamer are brought together through the interactions of cognate NANPs [9,40]. The split halves have also been used as functional outputs in the design of NANP logic gates. Only when the two halves are made available by the proper mechanisms is a fluorescent signal detectable [46].

Further approaches to dynamic NANPs aim to program structures which will only be active upon interaction with a disease-relevant sequence. This strategy benefits from the ubiquity of nucleic acids as disease biomarkers and the specificity with which single nucleotide polymorphisms can be detected, as demonstrated by biosensors such as molecular beacons [47–50]. Genetically encoded RNA-based molecular sensors (GERMS) represent systems which have been designed to incorporate both the recognition module and reporting system inspired by riboswitches into sensors capable of detecting various targets [51]. As a reporter system, fluorescence-activating aptamers are often employed. Synthetic ribozyme GERMS have been engineered to bind with target oligonucleotides and carry out functions which can be designated as Boolean logic gates (Figure 29.1b) [52]. Additionally, a two-stranded switch structure was shown which interacts with a trigger mRNA to induce a conformational change [53]. The change in structure results in the formation and release of a complete Dicer substrate RNA, which can be further diced into functional siRNA to undergo RNA interference (Figure 29.1b) [36].

FUTURE DIRECTIONS

The field of RNA nanotechnology has taken advantage of the natural dynamic diversity and programmability of nucleic acids to achieve the construction of NANPs which are capable of interacting specifically with biological environments. Riboswitches, ribozymes, and aptamers that can bind to specific molecules make excellent targeting platforms which nanomedicine can utilize; however, only NANPs composed of these same nucleic acid biomaterials can adapt to the same level of dynamicity for strand-dependent functionality. New advances in NANP design continue to evolve in a dynamic direction towards constructs which can be interactive rather than static, thereby increasing their specificity towards diagnostic and therapeutic applications.

ACKNOWLEDGMENTS

Research reported in this publication was supported by the National Institute of General Medical Sciences of the National Institutes of Health under Award Numbers R01GM120487 and R35GM139587 (to K.A.A.). The content is solely the responsibility of the authors and does not necessarily represent the official views of the National Institutes of Health.

REFERENCES

1. McCown, P. J.; Corbino, K. A.; Stav, S.; Sherlock, M. E.; Breaker, R. R., Riboswitch diversity and distribution. *RNA* 2017, *23* (7), 995–1011.

2. Serganov, A.; Nudler, E., A decade of riboswitches. *Cell* 2013, *152* (1–2), 17–24.
3. Breaker, R. R., Riboswitches and the RNA world. *Cold Spring Harbor Perspectives in Biology* 2012, *4* (2), a003566.
4. Breaker, R. R.; Joyce, G. F., A DNA enzyme that cleaves RNA. *Chemistry & Biology* 1994, *1* (4), 223–229.
5. Zhou, W.; Ding, J.; Liu, J., Theranostic DNAzymes. *Theranostics* 2017, *7* (4), 1010–1025.
6. Ellington, A. D.; Szostak, J. W., In vitro selection of RNA molecules that bind specific ligands. *Nature* 1990, *346* (6287), 818–822.
7. Gopinath, S. C. B., Methods developed for SELEX. *Analytical and Bioanalytical Chemistry* 2007, *387* (1), 171–182.
8. Ouellet, J., RNA Fluorescence with Light-Up Aptamers. *Frontiers in Chemistry* 2016, *4* (29).
9. Kolpashchikov, D. M., Binary malachite green aptamer for fluorescent detection of nucleic acids. *Journal of the American Chemical Society* 2005, *127* (36), 12442–12443.
10. Grate, D.; Wilson, C., Laser-mediated, site-specific inactivation of RNA transcripts. *Proceedings of the National Academy of Sciences* 1999, *96* (11), 6131.
11. Paige, J. S.; Nguyen-Duc, T.; Song, W.; Jaffrey, S. R., Fluorescence imaging of cellular metabolites with RNA. *Science* 2012, *335* (6073), 1194.
12. Paige, J. S.; Wu, K. Y.; Jaffrey, S. R., RNA mimics of green fluorescent protein. *Science* 2011, *333* (6042), 642–646.
13. Filonov Grigory, S.; Jaffrey Samie, R., RNA imaging with dimeric broccoli in live bacterial and mammalian cells. *Current Protocols in Chemical Biology* 2016, *8* (1), 1–28.
14. Filonov, G. S.; Kam, C. W.; Song, W.; Jaffrey, S. R., In-gel imaging of RNA processing using broccoli reveals optimal aptamer expression strategies. *Chemistry & Biology* 2015, *22* (5), 649–660.
15. Filonov, G. S.; Moon, J. D.; Svensen, N.; Jaffrey, S. R., Broccoli: Rapid selection of an RNA mimic of green fluorescent protein by fluorescence-based selection and directed evolution. *Journal of the American Chemical Society* 2014, *136* (46), 16299–16308.
16. Song, W.; Filonov, G. S.; Kim, H.; Hirsch, M.; Li, X.; Moon, J. D.; Jaffrey, S. R., Imaging RNA polymerase III transcription using a photostable RNA-fluorophore complex. *Nature Chemical Biology* 2017, *13* (11), 1187–1194.
17. Autour, A.; Jeng, S. C.; Cawte, A.D.; Abdolahzadeh, A.; Galli, A.; Panchapakesan, S. S. S.; Rueda, D.; Ryckelynck, M.; Unrau, P. J., Fluorogenic RNA mango aptamers for imaging small non-coding RNAs in mammalian cells. *Nature Communications* 2018, *9* (1), 656.
18. Dolgosheina, E. V.; Jeng, S. C. Y.; Panchapakesan, S. S. S.; Cojocaru, R.; Chen, P. S. K.; Wilson, P. D.; Hawkins, N.; Wiggins, P. A.; Unrau, P. J., RNA mango aptamer-fluorophore: A bright, high-affinity complex for RNA labeling and tracking. *ACS Chemical Biology* 2014, *9* (10), 2412–2420.
19. Zhou, J.; Rossi, J., Aptamers as targeted therapeutics: Current potential and challenges. *Nature Reviews Drug Discovery* 2017, *16* (3), 181–202.
20. Heckel, A.; Mayer, G., Light regulation of aptamer activity: An anti-thrombin aptamer with caged thymidine nucleobases. *Journal of the American Chemical Society* 2005, *127* (3), 822–823.
21. Shu, D.; Khisamutdinov, E. F.; Zhang, L.; Guo, P., Programmable folding of fusion RNA in vivo and in vitro driven by pRNA 3WJ motif of phi29 DNA packaging motor. *Nucleic Acids Research* 2013, *42* (2), e10–e10.
22. Alshaer, W.; Hillaireau, H.; Fattal, E., Aptamer-guided nanomedicines for anticancer drug delivery. *Advanced Drug Delivery Reviews* 2018, *134*, 122–137.
23. Panigaj, M.; Johnson, M. B.; Ke, W.; McMillan, J.; Goncharova, E. A.; Chandler, M.; Afonin, K. A., Aptamers as modular components of therapeutic nucleic acid nanotechnology. *ACS Nano* 2019.
24. Stewart, J. M.; Viard, M.; Subramanian, H. K. K.; Roark, B. K.; Afonin, K. A.; Franco, E., Programmable RNA microstructures for coordinated delivery of siRNAs. *Nanoscale* 2016, *8* (40), 17542–17550.
25. Afonin, K. A.; Viard, M.; Koyfman, A. Y.; Martins, A. N.; Kasprzak, W. K.; Panigaj, M.; Desai, R.; Santhanam, A.; Grabow, W. W.; Jaeger, L.; Heldman, E.; Reiser, J.; Chiu, W.; Freed, E. O.; Shapiro, B. A., Multifunctional RNA nanoparticles. *Nano Letters* 2014, *14* (10), 5662–5671.
26. Shu, D.; Shu, Y.; Haque, F.; Abdelmawla, S.; Guo, P., Thermodynamically stable RNA three-way junction for constructing multifunctional nanoparticles for delivery of therapeutics. *Nature Nanotechnology* 2011, *6* (10), 658–667.
27. Dai, B.; Hu, Y.; Duan, J.; Yang, X.-D., Aptamer-guided DNA tetrahedron as a novel targeted drug delivery system for MUC1-expressing breast cancer cells in vitro. *Oncotarget* 2016, *7* (25), 38257–38269.

28. Shu, D.; Li, H.; Shu, Y.; Xiong, G.; Carson, W. E.; Haque, F.; Xu, R.; Guo, P., Systemic delivery of anti-miRNA for suppression of triple negative breast cancer utilizing RNA nanotechnology. *ACS Nano* 2015, *9* (10), 9731–9740.

29. Pi, F.; Zhang, H.; Li, H.; Thiviyanathan, V.; Gorenstein, D. G.; Sood, A. K.; Guo, P., RNA nanoparticles harboring annexin A2 aptamer can target ovarian cancer for tumor-specific doxorubicin delivery. *Nanomedicine* 2017, *13* (3), 1183–1193.

30. Zhou, J.; Shu, Y.; Guo, P.; Smith, D. D.; Rossi, J. J., Dual functional RNA nanoparticles containing phi29 motor pRNA and anti-gp120 aptamer for cell-type specific delivery and HIV-1 inhibition. *Methods (San Diego, Calif.)* 2011, *54* (2), 284–294.

31. Srinivas, N.; Ouldridge, T. E.; Šulc, P.; Schaeffer, J. M.; Yurke, B.; Louis, A. A.; Doye, J. P.; Winfree, E., On the biophysics and kinetics of toehold-mediated DNA strand displacement. *Nucleic Acids Research* 2013, *41* (22), 10641–10658.

32. Qian, L.; Winfree, E., Scaling up digital circuit computation with DNA strand displacement cascades. *Science* 2011, *332* (6034), 1196–1201.

33. Qian, L.; Winfree, E.; Bruck, J., Neural network computation with DNA strand displacement cascades. *Nature* 2011, *475* (7356), 368–372.

34. Zhang, D. Y.; Winfree, E., Control of DNA strand displacement kinetics using toehold exchange. *Journal of the American Chemical Society* 2009, *131* (47), 17303–17314.

35. Yurke, B.; Turberfield, A. J.; Mills, A. P.; Simmel, F. C.; Neumann, J. L., A DNA-fuelled molecular machine made of DNA. *Nature* 2000, *406* (6796), 605–608.

36. Zakrevsky, P.; Parlea, L.; Viard, M.; Bindewald, E.; Afonin, K. A.; Shapiro, B. A., Preparation of a conditional RNA switch. In *Methods in molecular biology (Clifton, N.J.)*, 2017/07/22 ed.; 2017; Vol. 1632, pp. 303–324.

37. Afonin, K. A.; Viard, M.; Tedbury, P.; Bindewald, E.; Parlea, L.; Howington, M.; Valdman, M.; Johns-Boehme, A.; Brainerd, C.; Freed, E. O.; Shapiro, B. A., The use of minimal RNA toeholds to trigger the activation of multiple functionalities. *Nano Letters* 2016, *16* (3), 1746–1753.

38. Afonin, K. A.; Desai, R.; Viard, M.; Kireeva, M. L.; Bindewald, E.; Case, C. L.; Maciag, A. E.; Kasprzak, W. K.; Kim, T.; Sappe, A.; Stepler, M.; KewalRamani, V. N.; Kashlev, M.; Blumenthal, R.; Shapiro, B. A., Co-transcriptional production of RNA–DNA hybrids for simultaneous release of multiple split functionalities. *Nucleic Acids Research* 2014, *42* (3), 2085–2097.

39. Afonin, K. A.; Viard, M.; Martins, A. N.; Lockett, S. J.; Maciag, A. E.; Freed, E. O.; Heldman, E.; Jaeger, L.; Blumenthal, R.; Shapiro, B. A., Activation of different split functionalities upon re-association of RNA-DNA hybrids. *Nature Nanotechnology* 2013, *8* (4), 296–304.

40. Halman, J. R.; Satterwhite, E.; Roark, B.; Chandler, M.; Viard, M.; Ivanina, A.; Bindewald, E.; Kasprzak, W. K.; Panigaj, M.; Bui, M. N.; Lu, J. S.; Miller, J.; Khisamutdinov, E. F.; Shapiro, B. A.; Dobrovolskaia, M. A.; Afonin, K. A., Functionally-interdependent shape-switching nanoparticles with controllable properties. *Nucleic Acids Research* 2017, *45* (4), 2210–2220.

41. Sajja, S.; Chandler, M.; Striplin, C. D.; Afonin, K. A., Activation of split RNA aptamers: Experiments demonstrating the enzymatic synthesis of short RNAs and their assembly as observed by fluorescent response. *Journal of Chemical Education* 2018.

42. Chandler, M.; Lyalina, T.; Halman, J.; Rackley, L.; Lee, L.; Dang, D.; Ke, W.; Sajja, S.; Woods, S.; Acharya, S.; Baumgarten, E.; Christopher, J.; Elshalia, E.; Hrebien, G.; Kublank, K.; Saleh, S.; Stallings, B.; Tafere, M.; Striplin, C.; Afonin, A. K., Broccoli fluorets: Split aptamers as a user-friendly fluorescent toolkit for dynamic RNA nanotechnology. *Molecules* 2018, *23* (12).

43. Alam, K. K.; Tawiah, K. D.; Lichte, M. F.; Porciani, D.; Burke, D. H., A fluorescent split aptamer for visualizing RNA–RNA assembly in vivo. *ACS Synthetic Biology* 2017, *6* (9), 1710–1721.

44. Rogers, T. A.; Andrews, G. E.; Jaeger, L.; Grabow, W. W., Fluorescent monitoring of RNA assembly and processing using the split-spinach aptamer. *ACS Synthetic Biology* 2015, *4* (2), 162–166.

45. Debiais, M.; Lelievre, A.; Smietana, M.; Müller, S., Splitting aptamers and nucleic acid enzymes for the development of advanced biosensors. *Nucleic Acids Research* 2020, *48* (7), 3400–3422.

46. Goldsworthy, V.; LaForce, G.; Abels, S.; Khisamutdinov, E. F., Fluorogenic RNA aptamers: A nanoplatform for fabrication of simple and combinatorial logic gates. *Nanomaterials* 2018, *8* (12), 984.

47. Tan, W.; Wang, K.; Drake, T. J., Molecular beacons. *Current Opinion in Chemical Biology* 2004, *8* (5), 547–553.

48. Marras, S. A.; Kramer, F. R.; Tyagi, S., Multiplex detection of single-nucleotide variations using molecular beacons. *Genetic Analysis – Biomolecular Engineering* 1999, *14* (5–6), 151–156.

49. Tyagi, S.; Kramer, F. R., Molecular beacons: Probes that fluoresce upon hybridization. *Nature Biotechnology* 1996, *14* (3), 303.

50. Ludwig, J. A.; Weinstein, J. N., Biomarkers in cancer staging, prognosis and treatment selection. *Nature Reviews Cancer* 2005, *5* (11), 845.

51. Sun, Z.; Nguyen, T.; McAuliffe, K.; You, M., Intracellular imaging with genetically encoded RNA-based molecular sensors. *Nanomaterials* 2019, *9* (2), 233.

52. Penchovsky, R.; Breaker, R. R., Computational design and experimental validation of oligonucleotide-sensing allosteric ribozymes. *Nature Biotechnology* 2005, *23*, 1424.

53. Bindewald, E.; Afonin, K. A.; Viard, M.; Zakrevsky, P.; Kim, T.; Shapiro, B. A., Multistrand structure prediction of nucleic acid assemblies and design of RNA switches. *Nano Letters* 2016, *16* (3), 1726–1735.

30 Defining Immunological Properties of Nucleic Acid Nanoparticles Using Human Peripheral Blood Mononuclear Cells

Marina A. Dobrovolskaia
National Cancer Institute, Frederick, MD, USA

Kirill A. Afonin
University of North Carolina at Charlotte, Charlotte, NC, USA

CONTENTS

DOI: 10.1201/9781003001560-33

INTRODUCTION

NUCLEIC ACID-BASED TECHNOLOGIES

A variety of rapidly evolving nucleic acid-based technologies benefit from the ability of RNA and DNA to form both canonical and non-canonical base pairings [1]. Currently, existing libraries of RNA (and DNA) motifs [2–5] can be rationally combined to assemble various nucleic acid nanoparticles (NANPs) (Figure. 30.1a) that can be further decorated with therapeutic nucleic acids (TNAs) [6] and used as next generation TNA delivery platforms [7,8]. At its essence, the design strategy combines different motifs, similarly to Lego® blocks, and follows empirically rationalized rules to achieve a remarkable degree of structural control in bottom-up assemblies [9–13]. The one-pot assembly of NANPs designed with this approach requires two-step incubation, allowing for the correct folding of individual strands and further activation of magnesium-dependent long-range interactions (Figure 30.1b). Another common strategy (Figure 30.1c) is based solely on canonical Watson–Crick interactions with ssRNAs and/or ssDNAs computationally programmed to interact only with their partner strands while avoiding any intramolecular pairings [14]. Similar design principles are widely employed in structural DNA nanotechnology [15] and DNA origami [16] for the construction of a wide variety of functional DNA assemblies.

THERAPEUTIC NANPs

The increasing appreciation and use of biocompatible NANPs has led to the establishment of a new field where the innate biological functions of nucleic acids are reprogrammed to tackle specific biomedical challenges [8]. Additionally, the burgeoning NANP technology presents numerous advantages: i) NANPs can be designed to deliver cocktails of different TNAs to diseased cells [17–22] allowing for simultaneous targeting of several biological pathways with higher synergistic effects;

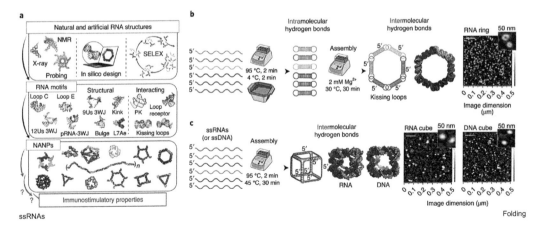

FIGURE 30.1 The main strategies for NANP design: (**a**) schematics showing the flow of NANP production with two main design strategies that rely on (**b**) intramolecular bonds within each monomer to promote the formation of magnesium-dependent interacting motifs required for NANPs' assembly and (**c**) individual monomers designed to only form intermolecular bonds with their cognate partner strands.

ii) the thermal and chemical stabilities of NANPs are tuned with chemical modifications [23,24] and NANPs' immunorecognition is controlled by their structure and composition [24–26]; iii) TNAs can be programmed into NANPs to induce responsive behavior [23,27]; iv) pyrogen-free NANPs can be produced with high batch-to-batch consistency [7] permitting their industrial-scale manufacture (endotoxin, a component of the cell wall of Gram-negative bacteria, is a common pyrogen that contaminates nanoformulations [28] precluding their clinical use); v) introduction of small molecules, aptamers, or antibodies into NANP structures makes them amenable for targeted delivery [29]; and vi) carrier-free NANPs are immunoquiescent and safe for systemic administration [25].

NANPs' Safety and Immune-mediated Efficacy Considerations

In order to further advance the translation of NANPs from bench to clinic, the field is in great need of reliable experimental protocols for the assessment of both safety and efficacy of these novel nano-materials. The assessment of immunological responses to NANPs has two-fold importance. First, it contributes to the understanding of NANPs' safety in that the systemic induction of inflammation is associated with the cytokine storm toxicity. Such assessment is especially crucial after several recent announcements of the US FDA halting clinical trials and biotech companies withdrawing TNAs formulated with lipid-based carriers due to adverse immune-mediated toxicities of those formulations (https://www.businesswire.com/news/home/20161110006612/en). Second, understanding immunological responses to NANPs will assist in exploring the alternative indications and routes of administrations to define conditions in which the local induction of inflammation is beneficial for the host as it improves the efficacy of vaccines and immunotherapies [30].

Development of the Protocol

Our labs initiated the very first systematic investigation of NANP recognition by immune cells [25]. In that work as well as in our other related studies [17,23,31–33], we used primary human peripheral blood mononuclear cells (PBMCs) collected from healthy human donors. We specifically developed and optimized the experimental settings for this particular model since PBMCs have been shown to produce the most predictive and reliable results for the potential cytokine storm toxicity when compared to common preclinical animal models such as rodents and primates. This conclusion comes from the recent experience of TeGenero Immuno Therapeutics' product TGN1412 that caused cytokine storm in patients after successfully passing all preclinical safety studies in rats and non-human primates. The cytokine storm toxicity of TGN1412, however, was predicted *in vitro* in PBMCs derived from healthy donors [34]. PBMCs were also proven to be predictive of the quality of vaccine adjuvants [35,36]. Consequently, we identified PBMCs as a sensitive and affordable model that can be employed for understanding how particular NANPs can trigger the immune system. We confirmed the reliable performance of this model for more than 60 different NANPs analyzed in PBMCs collected from more than 100 healthy human donors [17,23,25,26,31,33,37].

Advantages

The key advantage of this protocol is its modularity that allows various NANPs to be studied, together with a broad panel of induced cytokines assessed in clinically relevant settings. Other advantages include the ability of gaining mechanistic insights in NANPs' recognition by the human immune cells and standardization of the most critical procedure for generating supernatants and appropriate controls along with selecting critical biomarkers, while leaving the flexibility of choice of the cytokine detection platform to individual researchers based on resources available in their labs. Therefore, we anticipate that this protocol can be applied to a broader range of nucleic acid-containing materials [38–42].

Limitations

While this procedure is predictive of systemic effects of NANPs followed by injection into the blood, it does not assess the ability of NANPs to induce an inflammatory response in other cell types which may contribute to the local responses to NANPs upon distribution to certain tissues and, for non-systemically administered NANPs, at the site of injection. To analyze the response in other cell types (*e.g.*, fibroblasts and adipocytes), this procedure can be easily adapted by substituting PBMCs with another cell type of interest to the user. Likewise, to understand cytokine responses in PBMC of patients with certain diseases, the procedure can be adapted to PBMC derived from whole blood of patients. Institutional Review Board and Biosafety requirements for this adjustment are discussed in the Applications section.

Overview of the Procedure

Based on extensive experimental work over the course of the last six years [17,23,25,26,31,33,37], we introduce the comprehensive procedure required to accurately assess the critical biomarkers of immunostimulation as well as cells and molecular pathways responsible for the NANPs' immuno-recognition. We detail the protocols for pyrogen-free assembly and characterization of representative NANPs, isolation and handling of human PBMCs, transfections of NANPs, and assessment of biomarkers of NANPs' pro-inflammatory responses that include type I and III interferons (IFN-α, -β, -ω, and -λ). These cytokine responses were confirmed to be consistent for a large cohort (~100) of healthy human donors when the described procedures were performed by different technicians who had never worked with NANPs before [17,23,25,26,31,33,37].

Additionally, a set of pro-inflammatory cytokines (TNF-α, IL-1β, IL-6, and IL-8), also known as markers of pyrogenic responses, can be reproducibly measured in PBMCs using this protocol [23,25]. The regulated induction of pro-inflammatory cytokines and IFNs has become a powerful tool used in vaccines and immunotherapies [43,44]. However, excessive and uncontrolled production of cytokines, particularly TNF-α, may cause tissue necrosis at the site of injection [45]. Therefore, decreasing injection site reactions in patients is considered to be an essential safety goal [46–50].

The procedure, schematically summarized in Figure 30.2A, can be used to potentially detect any other cytokine in response to PBMC treatment with NANPs or other TNAs.

Applications

The empirically gained knowledge of relative immunostimulation contributes to establishing NANPs' safety profile and opens endless possibilities for NANPs' use not only as a drug delivery platform but also as adjuvants for vaccines and immunotherapies. Additionally, despite recent clinical successes (Onpattro® [51] and Givlaari™ [52]), the development of RNAi therapies has encountered several hurdles, not least of which was the severe inflammation and cytokine storm (https://www.businesswire.com/news/home/20161110006612/en). The current protocol offers a quick and affordable pre-clinical tool aiming to address the gaps in the understanding of immunological recognition of NANPs and TNAs, thus further accelerating their clinical translation. The described experimental settings can be easily adopted by the key stakeholders of NANP and TNA technologies, including the global community of basic researchers, regulatory scientists, international standard development organizations (*e.g.*, ISO and ASTM International), biotechnology companies, big pharma, and environmental and occupational health scientists to provide important information about NANPs safety, and, in relevant applications such as vaccines, efficacy to physicians, patients, and their families. If the user desires to investigate immunological responses to NANPs in the blood of patients with a particular disease (*e.g.*, cancer), this protocol can be adapted and would require collection of blood from relevant donors under approval by the Institutional Review Board protocol.

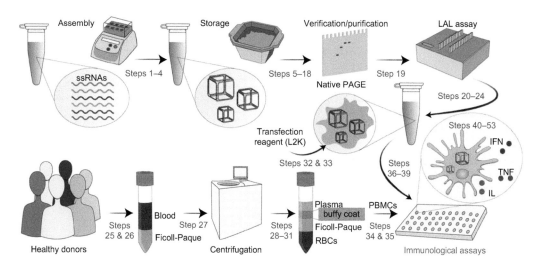

FIGURE 30.2 Schematic representation of the experimental design required to assess the immunological properties of NANPs and their interactions with PBMCs. Corresponding procedure steps are indicated in blue. Procedural steps (57–60) describing mechanistic studies are not shown.

While the minimum recommended number of donors in the described protocol is three, one may increase the number of donors to that desirable for the given experiment.

ALTERNATIVE METHODS

Human cell lines and engineered reporter cells (*e.g.*, HEK-293 cells overexpressing human TLR3, TLR7, TLR8, and TLR9) can be used in preliminary studies to reveal common trends in immunos- timulation by NANPs [24,53,54]. However, the cytokines' pleiotropy [55] and genetic diversity of human patients [26] in immune responses cannot be accurately addressed with these models, thus making their clinical relevance incomparable to studies with human PBMCs.

EXPERIMENTAL DESIGN

PBMCs

It is important to use freshly isolated PBMCs since cryopreservation procedures may alter biologi- cal responses of immune cells. For statistically meaningful results, we recommend using PMBCs isolated from at least three different donors; we also advise to use at least three technical replicates per sample per donor. The number of donors may be increased to address specific needs of the user's laboratory.

Controls

We recommend the use of RNA and DNA cubes (Figure 30.1C) as controls for all immunological studies since RNA cubes reliably demonstrate the highest immunostimulatory activity among all tested NANPs [25,26] while their DNA analogs only trigger the minimal activation of cytokine pro- duction. We also recommend using Lipofectamine 2000 as a standard carrier because it was inves- tigated most extensively, and the change in carrier may result in modified inflammatory response to NANPs. We also recommend using RNA cubes, DNA cubes, and RNA rings as controls since these NANPs represent alternative design strategies, chemical composition (RNA vs DNA), and dimen- sionality (globular cubes vs planar rings), and the immunological profiles of these NANPs are well studied with general trends summarized in Table 30.1

TABLE 30.1

Examples of IFN Responses to NANPs with Various Physicochemical Properties

	IFN-α (pg/ml)					
Supernatant	Donor 1		Donor 2		Donor 3	
Untreated	BLOQ	BLOQ	BLOQ	BLOQ	BLOQ	BLOQ
ODN2216	29287	28034	16963	15605	15179	13193
L2K	BLOQ	BLOQ	BLOQ	BLOQ	BLOQ	BLOQ
DNA cube/L2K	4889	4602	2375	2262	4198	5121
RNA cube/L2K	25530	19965	17564	17204	14915	15091
RNA ring/L2K	6937	5985	4552	5721	9398	7531
	IFN-β (pg/ml)					
Untreated	BLOQ	BLOQ	BLOQ	BLOQ	BLOQ	BLOQ
ODN2216	141	118	82	82	69	BLOQ
L2K	BLOQ	BLOQ	BLOQ	BLOQ	162	BLOQ
DNA cube/L2K	BLOQ	BLOQ	BLOQ	BLOQ	162	BLOQ
RNA cube/L2K	450	343	369	311	362	299
RNA ring/L2K	64	BLOQ	77	BLOQ	132	BLOQ
	IFN-ω (pg/ml)					
Untreated	BLOQ	BLOQ	BLOQ	BLOQ	BLOQ	12.4
ODN2216	1254	1209	738	774	633	607
L2K	BLOQ	BLOQ	BLOQ	BLOQ	BLOQ	BLOQ
DNA cube/L2K	201	202	118	105	214	231
RNA cube/L2K	875	730	623	587	599	686
RNA ring/L2K	194	223	132	178	290	291
	IFN-λ (pg/ml)					
Untreated	BLOQ	BLOQ	BLOQ	BLOQ	BLOQ	BLOQ
ODN2216	729	764	BLOQ	BLOQ	BLOQ	BLOQ
L2K	BLOQ	BLOQ	BLOQ	BLOQ	BLOQ	BLOQ
DNA cube/L2K	BLOQ	BLOQ	BLOQ	BLOQ	BLOQ	BLOQ
RNA cube/L2K	887	698	382	417	600	482
RNA ring/L2K	BLOQ	BLOQ	146	BLOQ	147	146

BLOQ = below lower limit of quantification. The data is reproduced from reference [25] with permission. Each value in the table represents an independent sample tested in duplicate on an ELISA plate and is the mean of a duplicate response (% CV < 25).

Preparation of Nucleic Acid Components

All DNAs, fluorescently labeled oligos, and RNAs can be purchased from Integrated DNA Technologies (https://www.idtdna.com). Alternatively, RNAs can be synthesized via *in vitro* run-off transcription with T7 RNA polymerase, as detailed elsewhere [7]. If purchased, we recommend including a purification step (denaturing PAGE) for each strand, as detailed elsewhere [7]. Also, all purchased RNAs, unless specified, have 5′-OH groups as opposed to 5′-triphosphates resulting from *in vitro* transcription. This fact needs to be taken into consideration during the immunological studies.

NANP Design

The design principles of six-stranded cubes and rings used as control NANPs are detailed in previously published Nature Protocols paper [7]. All sequences required for NANPs assemblies are provided in the supplementary information.

NANP Assembly

Importantly, since contamination with endotoxin (very common for nanotechnology-based formulations [28]) can induce false-positive responses, the assembly of endotoxin-free NANPs must be confirmed [7].

Cytokine Detection

There is no harmonized approach for what cytokine detection platform to use as well as to the choice between single- or multi-plex analysis. One should rely on their scientific judgment and the critical path of the project focusing on the certain types of nanoparticles to determine the types of the cytokines and method for analysis of supernatants. As an example, for this protocol, we show data generated using multi-plex chemiluminescent assays (Quansys Biosciences, Logan, UT). This protocol is used to detect type I and type III interferons (*e.g.*, IFN-α, -β, -ω, and -λ) when one needs to estimate NANPs' pro-inflammatory properties because NANPs stimulate these biomarkers according to their physicochemical properties and the responses are consistent between individual donors [25]. The detection of other pro-inflammatory cytokines (*e.g.*, TNF-α, IL-1β, IL-6, and IL-8) is recommended when one wants to screen NANP formulations for their pyrogenic properties. The analysis with and without a carrier and comparison to the benchmark carrier lipofectamine are recommended because immunological responses may change depending on NANPs' physicochemical properties and the routes of entry into immune cells. The supernatants prepared in steps 35–42 could also be used for the detection of any other biomarkers produced by PBMCs. For example, IL-2 and IFN-γ are the markers of T-cell activation and can be tested when the information regarding NANPs' ability to activate T-lymphocytes is wanted. When the analysis of other cytokines and secondary messengers produced by PBMCs in response to NANPs is desired, steps 1–34 are performed without modifications; in steps 35–42, one may need to add an additional positive control (*i.e.*, substances known to induce the biomarker of interest in PBMCs; some examples are provided in Table 30.2) and the

---

TABLE 30.2

Important Controls to Verify Functionality of the Test Model and Potential False-Negative or False-Positive Results

A. Positive controls shown in this table trigger different receptors and molecular pathways in peripheral blood cells, and, therefore, lead to the expression of different inflammatory markers. These controls provide important information about functionality of the cells used in the study. The procedure for setting up these controls is described in steps 38 and 39.

Description\Control	*E.coli* K12 LPS	ODN2216	PHA-M
Primary purpose	Positive control for Inflammatory Cytokines (TNF-α, IL1β, IL-6, IL-8, IL-10, IL-12)	Positive control for Type I (IFN-α, -β, -ω) and type III (IFN-λ)	Positive control for Type II interferon (IFN-γ)
Final concentration in assay	20 ng/mL	5 µg/mL	10 µg/mL

B. IEC and CFC help reveal the ability of NANPs present in the culture supernatants to alter cytokine detection and cause false-positive or false-negative results. VC helps to identify potential effects of the carrier. The procedure for setting up these controls is described in steps 38 and 39.

Description\Control	**IEC**	**CFC**	**VC**
Primary Purpose	Rule out false-negative results	Rule out false-positive results	Identify potential effects of the carrier
Content	a) Positive control supernatant collected at step 42 spiked with NANPs b) CFC collected at step 42 spiked with cytokine standards from ELISA kit	Complete cell culture medium and NANPs from step 35	Delivery carrier at the same concentration used for NANPs complexation in step 35

incubation time may also need to be adjusted from 24 hours to a shorter (*e.g.*, 6 or 8 h) or longer (*e.g.*, 48 or 72 h) duration depending on recommendations for that particular biomarker.

Any traditional single- or multi-plex ELISA assay can be used. In Section 30.3 (Biological Materials), we provide details about both commercial single- and multi-plex ELISA kits, and low-cost self-assembled single-plex ELISA assays that researchers can choose to use based on their resources and laboratory equipment available. Due to the differences in ranges and detection limits, one should follow the instructions from the cytokine detection kit to prepare a standard curve and sample dilution and follow the instructions for the reagent dilution and incubation time. Detailed protocols for self-assembled single-plex ELISA validated according to the "International Conference on Harmonization of Technical Requirements for Pharmaceuticals and Human Use" have been described by our labs earlier [56]; these protocols provide 10-fold reduction in cost per plate, therefore making these ELISAs more accessible for researchers with limited resources.

Understanding the Mechanism of NANP Recognition

To understand if endosomal TLRs are involved in the inflammatory response to NANPs, one can employ ODN2088 at final concentration of 5 µg/ml. This inhibitor is added to the RPMI in step 38. The results generated for NANP alone and NANP with ODN2088 are compared. To understand if the uptake of NANPs requires scavenger receptors, one may employ specific inhibitors such as fucoidan. This inhibitor is added to the cells in step 38. In addition to or instead of fucoidan, one may also use poly-I or dextran sulfate as inhibitor of scavenger receptors. Poly-C and chondroitin sulfate are used as non-specific controls for poly-I and dextran sulfate, respectively.

CONTROL NUCLEIC ACID NANOPARTICLES

Sequences of NANPs we propose for use as controls have been described earlier [25] and are also provided in the supplementary information. Each set of NANP sequences contains fluorescently labeled (with Alexa 488) strand for NANPs visualization.

BIOLOGICAL MATERIALS

Healthy human donor blood, anti-coagulated with Li-heparin, was obtained from at least 3 healthy donors. Whole blood should be collected from healthy donor volunteers who have not been on medication and are clear from infection for at least 2 weeks prior to blood donation. Li-heparin tubes should be used and the first 10 cc should be discarded. For the best results, whole blood should be used within 1 hour after collection. Prolonged storage (> 2h) of whole blood will lead to decrease in cell function.

CAUTION: The blood used in this protocol was obtained under the National Cancer Institute-at-Frederick protocol OH9-C-N046; equivalent approval from the user's home institution is required to work with human blood. Collection of blood requires institutional approval and certified personnel.

CRITICAL: Blood from donors with certain types of diseases (*e.g.*, cancer) can be used. CRITICAL: The use of cryopreserved blood or PBMCs is not recommended because cytokine response in these cells may be altered by the cryopreservation and storage procedures.

REAGENTS

- HyClone HyPure Water, Cell Culture Grade (GE LifeSciences, cat. no. SH30529.01)
- 10X Tris-Borate/EDTA (TBE) buffer (VWR, cat. no. 75800-952)
- Acrylamide-bisacrylamide (37.5:1), 40 % (VWR, cat. no. 97064-542) !CAUTION Toxic if ingested or absorbed through the skin.
- Glycerol (Life Technologies Inc., cat. no. 5514UA)
- Xylene Cyanol (Sigma, cat. no. X-4126) !CAUTION Avoid contact and inhalation

- Bromophenol Blue (Sigma, cat. no. B-8026) !CAUTION Avoid contact and inhalation
- Ammonium persulfate (APS) (Sigma, cat. no. A3678) !CAUTION Irritating to eyes
- TEMED (Sigma, cat. no. T7024) !CAUTION It may cause respiratory tract irritation if inhaled. It may also cause skin or eye irritation if absorbed through skin or if it comes into contact with eyes, and is harmful if swallowed
- Magnesium Chloride (Fisher, cat. no. M35-500)
- 10X Tris-Borate (TB) buffer (Sigma, cat. no. T1503-5KG, and VWR, cat. no. 97061-978)
- Triton X-100 (VWR, cat. no. 97062-208)
- Ultrapure LPS from K12 *E. coli* (Invivogen, cat. no. tlrl-peklps)
- ODN2216;a CpG DNA oligonucleotide with mixed backbone and the following sequence 5'-ggGGGACGATCGTCGggggG-3, where lowercase letters show phosphorothioate linkage and capital letters refer to phosphodiester linkage between nucleotides (Invivogen, cat. no. tlrl-2216)
- Phytohemagglutinin (PHA-M) (Sigma, cat.no. L8902)
- ODN2088; a CpG oligonucleotide that inhibits activity of endosomal TLRs (Miltenyi Biotec, cat.no. 130-105-815)
- Fucoidan, inhibitor of scavenger receptor (Sigma-Aldrich, cat.no. F8190-500MG)
- Dextran sulfate, inhibitor of scavenger receptor (Sigma-Aldrich, cat.no. D6001-1G)
- Chondroitin sulfate, control for dextran sulfate (Sigma-Aldrich, cat.no. C9819-5G)
- Poly-I, scavenger receptor inhibitor (Sigma-Aldrich, cat.no. P4154)
- Poly-C, control for Poly I (Sigma-Aldrich, cat.no.P4903)
- Phosphate-buffered saline (PBS) (GE Life Sciences, cat. no. SH30256.01)
- RPMI-1640 (Invitrogen, cat. no. 11835-055)
- Fetal bovine serum, GE Life Sciences (HyClone, cat. no. SH30070.03)
- Penicillin streptomycin solution (Invitrogen, cat. no.15140-148)
- L-glutamine (GE Life Sciences, cat.no. SH30034.01)
- Ficoll Paque Premium (GE Healthcare, cat. no.17-5442-02)
- Hank's balanced salt solution (HBSS) (Invitrogen, cat. no. 24020-117)
- OptiMEM (ThermoFisher, cat. no. 31985062)
- Lipofectamine 2000 (ThermoFisher, cat. no.11668019)
- Single- or multi-plex ELISA including the following analytes (TNF-α, IL-1β, IL-6, IL-8, IL-2, IL-12, IFN-α, -β, -ω, and -λ):
  - Quansys multi-plex cytokine and type II interferon (IFN-γ) panel, cat. no. 110433HU and custom 4-plex type I and III interferons are recommended to the labs that have chemiluminescence imaging multi-plex plate readers. CRITICAL: the procedure detailed in this protocol is for the multi-plex IFN kit from Quansys. The volumes, incubation time, incubation temperature, type and settings of a plate reader may vary when kits from other manufacturers are used.
  - MSD, multi-plex cytokines and type II interferon (IFN-γ) panel cat.no. K15049D and custom type I and type III panels are recommended to the labs that have fluorescence-based multi-plex readers.
  - R&D Systems ELISA kits, cat.no. DY9345-05, DIFNB0, D6050, DIA00D, and low cost self-assembled ELISAs are recommended to the labs that have regular single-plex readers capable of detecting absorbance at 450 nm).

REAGENT SETUP

5X NANP assembly buffer final concentrations: 5X Tris-borate buffer (pH 8.2), 10 mM MgCl$_2$, 250 mM KCl. Once made, this buffer can be stored at room temperature (20–22°C) for 2–3 months.

1X native-PAGE loading buffer final concentrations: 50 % glycerol, 1X Tris-borate buffer (pH 8.2), 0.01 % bromophenol blue and 0.01 % xylene cyanol tracking dyes. Once made, this buffer can be stored at room temperature for a year.

1X native-PAGE running buffer final concentrations: 1X Tris-borate buffer (pH 8.2), 2 mM MgCl$_2$. Once made, this buffer can be stored at 4 °C temperature for a year.

Complete RPMI-1640 medium: The complete RPMI medium should contain the following reagents: 10% FBS (heat inactivated), 2 mM L-glutamine, 100 U/mL penicillin, 100 µg/ml streptomycin. Store at 2–8 °C protected from light for no longer than 1 month. Before use, warm the media to 37 °C in a water bath.

Lipopolysaccharide for positive control in pro-inflammatory cytokines and chemokines analysis (LPS, 1 mg/mL stock): Add 1 mL of sterile water to 1 mg of LPS in the vial and vortex to mix. Aliquot 20 µL and store at a nominal temperature of –20°C. Avoid repeated freeze–thaw cycles. On the day of the experiment, thaw one aliquot and use such that its final concentration in PBMCs or whole blood (WB) culture is 20 ng/mL.

Phytohemagglutinin for positive control in type II interferon analysis (PHA-M, 1 mg/mL stock): Add 1 mL of sterile PBS or cell culture medium per 1 mg of PHA-M to the vial and gently rotate to mix. Store daily-use aliquots at a nominal temperature of –20°C. Avoid repeated freeze–thaw cycles. On the day of the experiment, dilute stock PHA-M solution in cell culture medium so that its final concentration in the positive control sample is 10 µg/mL.

ODN2216 for positive control in type I and type III interferon analysis (1 mg/mL stock): This mixed backbone oligonucleotide activates TLR9 and is supplied as lyophilized powder. Reconstitute in pyrogen-free, nuclease-free water to a final concentration of 1 mg/mL. Prepare single use 5 µL aliquots and store at −20 °C. On the day of the experiment, thaw an aliquot at room temperature and dilute in culture media so that its final concentration in the test sample is 5 µg/mL.

ODN2088 for mechanistic study to reveal potential involvement of endosomal TLRs (1 mg/mL stock): This oligonucleotide is a pan-TLR inhibitor; it is supplied as lyophilized powder. Reconstitute in pyrogen-free, nuclease-free water to a final concentration of 1 mg/mL. Prepare single-use 5 µL aliquots and store at -20 °C. On the day of the experiment, thaw an aliquot at room temperature and dilute in culture media so that its final concentration in the test sample is 5 µg/mL.

Fucoidan, dextran sulfate, poly-I, chondroitin sulfate, and poly-C preparation for mechanistic studies: Each of these materials is supplied as lyophilized powder. Dissolve the powder in PBS to prepare a stock with nominal concentration of 1 mg/mL. Prepare single-use 5 µL aliquots and store at -20 °C. On the day of the experiment, thaw an aliquot at room temperature and dilute in culture media so that their final concentrations in the test samples are 50 µg/mL. Fucoidan, poly-I, and dextran sulfate inhibit scavenger receptors, whereas chondroitin sulfate and poly-C do not. Chondroitin sulfate and poly-C are used as controls for dextran sulfate and poly-I, respectively.

Heat-inactivated fetal bovine serum: Thaw a 50 mL aliquot of fetal bovine serum and equilibrate to room temperature. Place the tube in a water bath set up to 56 °C and incubate with mixing for 35 min. The heat inactivation takes 30 min and the initial 5 min is used to bring the entire content of the vial to 56 °C. Chill the serum and use to prepare complete culture media.

EQUIPMENT

- Heating blocks with heated lids
- Vortex Mixer
- Ice machine
- Short wave UV-lamp
- Tube rotator
- Vertical electrophoresis system for native-PAGE
- NanoDrop 2000 UV spectrometer or similar
- BioRad ChemiDoc MP imaging system or similar
- Ethidium bromide for total gel staining (0.5 µg/mL)
- TLC plate
- Scalpel and cutting board

- Plastic wrap
- Refrigerated microcentrifuge
- Pipettes covering a range of 0.05 to 10 mL
- 96-well round bottom cell culture-grade plates
- Polypropylene tubes, 50 and 15 mL
- Microcentrifuge tubes
- Centrifuge
- Refrigerator, 2–8 °C
- Freezer, -20 °C
- Cell culture incubator with 5% CO_2 and 95% humidity.
- Biosafety cabinet approved for level II handling of biological materials
- Inverted microscope
- Hemocytometer
- Plate reader for cytokine detection (e.g., Quansys Biosciences ImagePro multi-plex reader and Molecular Devices SpectraMax M5 reader for single-plex ELISA)

BOX 1   HOW TO AVOID CONTAMINATION OF NANPS WITH PYROGENS

1. Always wear disposable gloves (nitrile preferred) and avoid reusing them.
2. Never touch your face or unclean surfaces and change gloves as needed during the process of NANP preparation.
3. Use pyrogen-free reagents and water.
4. Use autoclaved sterile pipette tips and avoid any cross contamination with other samples.
5. Do not use cellulose-based filters which can be a source of beta-glucan that will interfere with the LAL assay and further immunological studies.
6. Do not breathe, talk, cough, or sneeze around the open tubes during the NANPs' preparation.

PROCEDURE

Assembly of NANPs. TIMING: 1 hour

1. Measure the concentrations of solutions of individual RNA (or DNA) strands required for NANPs' assembly using a NanoDrop 2000 UV spectrometer or similar; the extinction coefficients of individual strands are calculated using the IDT tools (https://www.idtdna.com/calc/analyzer).
2. Specify the final concentration of NANPs and mix individual strands at equimolar concentrations in pyrogen-free water (*e.g.*, Lonza LAL grade water).
3. Place the tubes with samples on a heating block (95 °C) and incubate for two minutes. **CRITICAL STEP:** Do not incubate the samples at 95 °C longer than 2 minutes and avoid adding the assembly buffer to the hot samples since it can promote the degradation of RNAs. **CRITICAL STEP:** Heating blocks with heated lids are recommended to avoid condensation and changes in the final concentration of NANPs.
4. For *intramolecular* NANP assemblies (*e.g.*, RNA and DNA cubes), snap-cool the samples to 45 °C and incubate for two minutes, add 5X NANP assembly buffer to reach the appropriate concentration, and incubate at 45 °C for an additional 30 minutes; for *intra/intermolecular* NANP assemblies (*e.g.*, RNA rings), snap-cool the samples on ice for two minutes, add 5X NANP assembly buffer, and incubate for 30 minutes at 30 °C.

PAUSEPOINT: Once assembled, NANPs can be stored at 4 °C for several weeks. However, continuous execution of the protocol is recommended to avoid any potential degradation of NANPs

Verification of NANP assemblies by native-PAGE. TIMING: at least 2 hours

5. Pre-cast a mini gel for non-denaturing polyacrylamide gel electrophoresis, native-PAGE (8% acrylamide (37.5:1), 1X TB buffer (pH 8.2), 2 mM $MgCl_2$). We recommend using Mini-PROTEAN tetra vertical electrophoresis cells from Bio-Rad.
6. Pre-run the gel in a cold room or refrigerator (4 °C) in 1X native-PAGE running buffer for 3–5 minutes at 150 V.
7. While pre-running the gel, mix equal volumes of individual NANPs (final concentration of at least 1 μM) and 1X native-PAGE loading buffer, to up to 10 μL of final volume.
8. Load samples in individual lanes of a gel (5 μL per lane) and run it for 30 mins at 300 V at 4°C.
   CRITICAL STEP: Wash the loading wells several times with running buffer prior to loading.

TROUBLESHOOTING

9. Visualize the gel with a Bio-Rad ChemiDoc MP System (or similar) using total staining with ethidium bromide (or SYBR Green) or fluorescently labeled oligonucleotides entering NANPs' composition.
10. NANPs are expected to migrate as a single band on non-denaturing gels.
    CRITICAL STEP: The presence of any other bands should, in total, comprise no more than 20% of the entire band percentage as determined by following the product instructions for ChemiDoc analysis software.

Purification of NANPs by native-PAGE. TIMING: at least 24 hours

11. If after step 9, NANPs need to be further purified, prepare larger volumes and achieve at least 10 μM concentration following the steps 1–4 described above.
12. Pre-cast non-denaturing polyacrylamide gel electrophoresis, native-PAGE (8% acrylamide, 37.5:1). For purification, we recommend using a vertical gel of dimensions 16.5 cm X 22 cm with the spacer thickness of 1.5 mm for 10 wells.
13. Pre-run the gel in a cold room or refrigerator (4 °C) in 1X native-PAGE running buffer for 10 minutes at 150 V.
14. While pre-running the gel, mix equal volumes of NANPs with equal volumes of 1X native-PAGE loading buffer, to up to 100 μL of final volume.
15. Load samples in individual lanes of a gel (100 μL per lane) and run it for 90 mins at 300 V at 4 °C.
16. Disassemble gel and place it between the plastic wrap.
17. Place the TLC plate underneath the plastic-wrapped gel and visualize the NANP bands using a UV lamp on the short-wavelength setting.
18. Use a scalpel to cut out the NANP major band in each lane and place them in separate aliquots of 1X NANP assembly buffer; use as much buffer as needed to completely cover the gel pieces.
19. Allow elution overnight and assess the concentration of the purified NANPs using a NanoDrop 2000 UV spectrometer or similar; the extinction coefficients of NANPs are calculated as the summation of extinction coefficients of individual sequences comprising them.
20. Confirm the absence of endotoxins in assembled NANPs by LAL assay using protocols developed by our groups [7].
    PAUSEPOINT: Once purified, NANPs can be stored at 4 °C for several weeks. However, continuous execution of the protocol is recommended to avoid any potential degradation of NANPs.

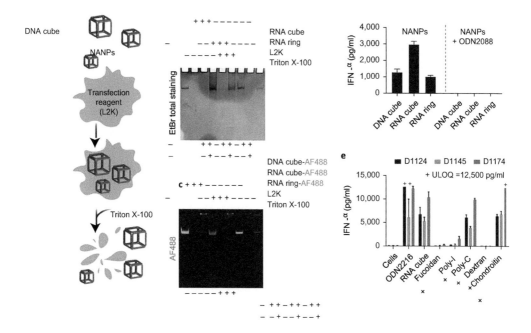

FIGURE 30.3 Verification of NANPs' retention of structural integrity upon complexation with Lipofectamine 2000 (L2K) and their cellular uptake. (**a**) Schematic representation of NANPs' association with L2K followed by their release upon detergent (Triton-X100) treatment. (**b**) Ethidium Bromide (EtBr) total staining native-PAGE indicating formation of NANPs, their complexation with L2K, and successful release upon treatments with Triton X-100. (**c**) NANPs labeled with Alexa-488 form complexes with L2K and retain their structural integrity upon detergent-mediated release. Note that in (**b**) and (**c**), NANPs' complexation with L2K prevents their entering the gel, while treatment with Triton-X100 restores NANPs' electrophoretic mobility. (**d**) Inhibition of NANPs' inflammatory response due to the TLR recognition. (**e**) Inhibition of the inflammatory response due to the NANPs uptake via scavenger receptor. Each bar shows a mean response of three independent samples and a standard deviation (N=3) for each of three donors. Each sample was tested in duplicate on ELISA plate (%CV < 25). The data used in sections (**d**) and (**e**) are reproduced from an earlier publication [25] with permission.

Assessment of structural integrity of NANPs upon complexation with delivery reagents (Figure 30.3A-B) [24,25,53]. TIMING: approximately 2 hours

21. Complex NANPs with a carrier. For the complexation with Lipofectamine 2000 (L2K) in a separate tube, combine 10 μL of NANP stock with concentration 1 μM and 2 μL of L2K; mix well by pipetting up and down repeatedly.
22. Incubate at room temperature for 30 minutes.
23. Aliquot 6 μL of NANP/L2K solution and add 2 μL of 10% TritonX-100. The remaining 6 μL of NANP/L2K solution will be used as a control in step 24.
24. Incubate at room temperature for an additional 30 minutes.
25. Analyze all samples (NANPs, NANPs/L2K, and NANPs/L2K + TritonX-100) on native-PAGE as described in steps 5–10. NANPs/L2K are expected to stay in the loading well, whereas Triton X-100 treated NANPs will migrate in the gel according to their molecular weight; structural integrity of NANPs, therefore, is confirmed by comparing the free NANPs with NANPs/L2K and NANPs/L2K+ TritonX-100 samples.

PAUSE POINT: Once mixed with L2K, NANPs can be stored at 4 °C for hours. However, continuous execution of the protocol is recommended to avoid any potential aggregation.

PBMC isolation. TIMING: approximately 4 hours

CRITICAL: Human blood may contain bloodborne pathogens. Therefore, *Biosafety in Microbiological and Biomedical Laboratories (BMBL)*, 5th edition, recommends following BSL-2 precautions during blood work. Laboratory staff should wear pants, close-toed shoes, disposable gloves, a laboratory coat, and eye protection. In addition, all procedures with human blood in which splashes or aerosols may be created should be conducted inside a biological safety cabinet. Bloodborne pathogen awareness training may be required and is strongly recommended. Any laboratory personnel working with human-derived blood should refer to the OSHA Bloodborne Pathogen Standard for specific required precautions. Always follow your local and institutional policies for working with human blood and, in case of any concerns or questions, contact your Biosafety Officer (BSO) or Institutional Biosafety Committee (IBC). Obtaining blood from patients with certain types of diseases also requires Institutional Review Board (IRB) approval.

26. Place freshly drawn blood into 15- or 50-mL conical centrifuge tubes separated per donor. Add an equal volume of room temperature PBS and mix well.
27. Use 3 mL of Ficoll-Paque solution per 4 mL of blood/PBS mixture. For example, 15 mL Ficoll-Paque per 20 mL of diluted blood in a 50 mL tube.
28. Slowly layer the Ficoll-Paque solution underneath the blood/PBS mixture by placing the tip of the pipet containing Ficoll-Paque at the bottom of the blood sample tube. Alternatively, the blood/PBS mixture may be slowly layered over the Ficoll-Paque solution.
    CRITICAL STEP: To maintain the Ficoll-blood interface, hold the tube at a 45° angle.
29. Centrifuge for 30 min at 900 x g, 18–20 °C, without brake.
    CRITICAL STEP: Turning brakes on the centrifuge off is needed to avoid altering the gradient which would result in a loss of the buffy coat. Depending on the type of centrifuge, one may also need to set acceleration speed to minimum.
30. Using a sterile pipet, remove the upper layer containing plasma and platelets and discard it. Using a fresh sterile pipet, transfer the mononuclear cell layer into a fresh 15 or 50 mL centrifuge tube separated per donor.
31. Wash cells by adding an excess (~3 times the volume of mononuclear layer) of HBSS and centrifuging for 10 min at 400 x g, 18-20 °C. CRITICAL STEP: The removal of excess Ficoll and platelet-rich plasma collected along with the PBMC fraction is critical to provide optimal cell viability and biological responses. Usually 4 mL of blood/PBS mixture results in ~2 mL fraction containing the cells of interest and requires at least 6 mL of HBSS for the wash step. We use 10 mL of HBSS per each 2 mL of cells.
32. Discard the supernatant by pipetting or using vacuum aspiration and repeat the wash step one more time.
33. Resuspend cells in 1 mL of complete RPMI-1640 medium. Dilute cells 1:5 or 1:10 with trypan blue, count cells, and determine viability using trypan blue exclusion [57].
34. If viability is at least 90%, continue to next steps.
    Exposing PBMCs to NANPs and collecting supernatants. TIMING: approximately 24 hours
35. Complex NANPs with a carrier. For complexation with L2K, in a separate tube, combine 20 µL of 1 µM NANP stock and 4 µL L2K (for a 96-well plate; adjust volume based on the manufacturer's recommendations if needed). Mix well by pipetting up and down repeatedly.
36. Incubate at room temperature for 5–30 minutes, then add 376 µL of OptiMEM.
37. While waiting on complexation in step 35, adjust PBMC concentration to 1.3 x 10^6 viable cells/mL using complete RPMI medium.
38. Dispense 20 µL of complete media (baseline), negative control (PBS), vehicle control (VC) (L2K in optimum at the same concentration as that used for particle complexation),

positive control (PC) and test samples (NANPs from step 36) into corresponding wells on a U-bottomed 96-well plate. Refer to Table 30.1 for the information regarding controls and Fig. S1 for an example of the plate map.

CRITICAL STEP: it is advised to set up 4–6 extra-replicates of positive control. Supernatants collected from these extra-PC samples will be used to prepare Inhibition Enhancement Controls (IECs) in ELISA assays (steps 43–56) to identify potential false-negative results. If one wants to understand whether NANPs may potentiate or inhibit the cellular response to the assay positive control (LPS, PHA-M, or ODN2216), the positive control should be co-cultured with NANPs in the presence of cells. In this case, each test well will receive 20 µL of the positive control, 20 µL of NANPs, 20 µL of complete RPMI and 40 µL of PBMC concentration to contain 2.6×10^6 viable cells/mL.

39. Dispense 80 µL of PBMC from step 37 per well into the 96-well plate containing 20 µL of NANPs, controls or medium in intended for Cell-Free Control (CFC) and prepared in step 38. Add 80 µL of complete RPMI instead of PBMC suspension to the CFC wells. The final number of cells per well is 100,000, and the total volume per well is 100 µL. Refer to the critical step after step 38, if the analysis of NANPs effects on the cellular response to PC is wanted.

40. Repeat steps 26–34 and 37 for cells obtained from each individual donor. CRITICAL STEP: There is no limit to the number of donors used in this test. It is advised to test each NANP formulation using blood derived from at least three healthy donors.

41. Incubate cells for 20 hours in a humidified 37 °C, 5% CO_2 incubator.

42. Spin the plate in a centrifuge at 400–700 × g for 5 minutes. Transfer supernatants to a fresh plate. Alternatively, the supernatants can be collected into Eppendorf tubes and cleared from cells by a brief centrifugation at 18,000 × g, then transferred to fresh tubes for storage.

    Pausepoint: After step 42, one may either proceed with ELISA analysis or aliquot and store supernatants at -20 or -80 °C up to 1 year.

    CRITICAL STEP: CFCs from step 39 are processed the same way as PBMC samples and serve as a control for false-positive results. To test for potential false-negative results, supernatant from the positive control (step 38) are spiked with nanoparticle at the final NANP concentration identical to that in the test sample. Alternatively, supernatants from CFC are spiked with relevant cytokine standards used in an ELISA or multi-plex. If NANPs inhibit detection of cytokine, a decrease in the cytokine level will be seen when compared to the level of cytokine in the positive control or quality control, respectively. If the user plans on using multiple ELISA assays, the volumes described in the protocol may be scaled up proportionally to generate higher volumes of supernatants. In this case, to avoid repeated freeze–thaw cycles, it is better to prepare multiple, single-use aliquots of each supernatant. The size of such aliquots depends on the volume and sample dilution that are specific to a given ELISA.

    Detection of biomarkers. TIMING: approximately 4 hours.

43. Prepare assay diluent, sample diluent, and wash buffer. Dilute the stock wash buffer provided with the kit 20-fold with distilled water. The assay and sample diluents are already at the ready-to-use concentrations. Store all buffers at room temperature for use on the same day that they are prepared.

44. Prepare calibration standards. First, add 200 µL of the stock calibrator to the first tube labeled Standard or Calibrator 1. Next, add 120 µL of the assay diluent to the additional 7 tubes labeled sequentially as Standard or Calibrator 2 through 7. After that perform serial, 3-fold dilution of the first standard by transferring 60 µL of the Standard 1 to the Standard 2, mixing by repeated up-and-down pipetting and transferring 60 µL into the next tube, and so on until Standard 7 is reached.

45. Thaw frozen culture supernatants at room temperature or use fresh supernatants collected at step 42.

46. Dilute culture supernatants 2-fold with sample diluent by mixing 60 µL of the culture supernatants with 60 µL of the sample diluent.

47. Add 50 μL of the assay diluent to each well on multi-plex ELISA plate.
48. Load 50 μL of standards from step 44 and culture supernatants from step 46 per well on the 96-well multi-plex plate loaded with assay diluent in step 47 and incubate at room temperature on a shaker set at approximately 500 rpm for 2 hours.
49. Wash the plate 3 times with the wash buffer prepared in step 43. Use 300 μL of the wash buffer per well. Tap the plate on a paper towel to remove excess buffer and immediately proceed to the next step.
50. Add 50 μL per well of the detection mix prepared by the reconstitution of lyophilized stock in 6 mL of distilled water.
51. Incubate the plate at room temperature on a shaker set to approximately 500 rpm for 1 hour.
52. Repeat step 49.
53. Add 50 μL per well of the ready-to-use Streptavidin-HRP conjugate provided with the kit.
54. Incubate the plate at room temperature on a shaker set to approximately 500 rpm for 15 minutes.
55. Repeat step 49 two times and immediately proceed to the next step.
56. Add 50 μL per well of the detection reagent prepared by mixing equal volumes of detection reagent A and detection reagent B and read the plate using a Quansys ImagePro reader equipped with licensed Q-View 3.112 (Quansys Biosciences Inc., 2017) or equivalent software. Process and analyze the data using Prism 8 (GraphPad Inc.,) or an equivalent software.

    CRITICAL STEP: Analyze supernatant from CFC samples prepared in step 39 to account for potential false-positive response. Analyze inhibition-enhancement controls prepared by spiking extra positive control supernatants from step 38 and further detailed in the critical step following step 42 with NANPs at the concentration equivalent to that used to prepare NANPs supernatants. This step will account for potential false-negative results.

    Mechanistic insight about particle uptake route and immune receptor recognition TIMING: at least 24 hours for the preparation of culture supernatants and 4 hours for the analysis of supernatants by ELISA.
57. Follow the same procedure described in steps 35–39 to set up culture supernatants.
58. When setting up treatments in step 38, add additional samples containing NANPs and inhibitors as follows. If the involvement of TLRs is of interest, add ODN2088 to the cells before adding NANPs. If the involvement of scavenger receptors is of interest, then apply fucoidan, dextran sulfate, chondroitin sulfate, poly-I, and poly-C to separate wells and then add NANPs.
59. Complete the procedure by following steps 39–42.
60. Follow steps 43–56 to analyze the culture supernatants prepared in steps 57–59.

Troubleshooting advice can be found in Table 30.3.

Pausepoint: after step 59 is complete, one may either proceed with ELISA analysis or aliquot and store supernatants up to one year at -20 or -80 °C.

ANTICIPATED RESULTS

Anticipated results for NANPs assembly and retention of NANPs' structural integrity upon interaction with transfection reagents. We recommend carrying out native-PAGE experiments (Figure 30.3) as a quick and low-cost way to verify the successful assembly of NANPs. The example shown in Figure 30.3 demonstrates the correct formation of three representative NANPs. As controls, previously extensively characterized NANPs (*e.g.*, RNA rings or RNA and DNA cubes) can be used. All bands can be quantified using ChemiDoc software; as an alternative, the publicly available software ImageJ can be used.

Anticipated results for endotoxin screening (step 20). The results should demonstrate that when the precautions described in Box 1 are followed, the endotoxin in NANP samples is undetectable (*i.e.*, the measured levels are below the assay lower limit of quantification of 0.001 EU/mL/200 nM).

TABLE 30.3

Troubleshooting

Step	Problem	Reason	Solution
17	NANP bands are not seen on total staining native-PAGE gel	Absence of divalent ions (Mg^{2+})	Verify that Mg^{2+} was added to NANP assembly buffer, native-PAGE gel, and native-PAGE running buffer
17	Additional bands besides NANP band are observed on native-PAGE gel	- Individual monomers (ssRNAs and ssDNAs) tend to stick to the test-tube walls - Monomer concentrations were not measured correctly	- Vortex all stock solutions of individual monomers for 5-10 seconds before NANP assembly - Measure the UV absorbance and calculate the monomer concentrations right before NANPs' assembly
17	NANPs stuck in the native-PAGE wells	Loading wells were not washed well enough	Wash the wells thoroughly prior to loading your samples
20	NANPs contain endotoxin	Pyrogen-free reagents and depyrogenated tools were not used	Use pyrogen-free water and sterile disposable consumables; depyrogenate non-disposable reagents and supplies by either baking at 230 °C for at least 30 minutes or cleaning them with Cavicide followed by rinsing with the excess of pyrogen-free water
25	NANPs do not get efficiently released from the carrier	Triton X-100 is the detergent that may not be an efficient releasing agent due to the non-lipid-like nature of the carrier	Use heparin (instead of Triton X-100) to outcompete electrostatic interactions between NANPs and other polymeric or inorganic carriers [74,75]
28–30	No buffy coat	Centrifuge acceleration and or deceleration speed was not adjusted to be minimal	Check centrifuge settings to set up the acceleration speed to minimal and turn deceleration speed off
30–34	Low PBMC yield	Incomplete collection of the buffy coat	Collect buffy coat using Eppendorf or equivalent pipette set for 1 mL volume. Avoid using serological pipettes, even when their volumes are 1–2 mL.
30–34	Low cell viability	- Blood was exposed to extreme temperatures during transportation or spent over 2 hours after the collection - Expired vacutainers - Diluted blood was held at room temperature over 2 hours - Purified cells were kept in HBSS over 1 hour	- During summer, it is better to transport the blood using cold packs equilibrated to room temperature to avoid overheating the blood. During winter months, it is better to transport the blood on warm packs equilibrated to 37 °C to avoid overcooling the blood - Check vacutainer expiration dates - Dilute blood with PBS immediately before loading onto Ficoll - After the last wash, replace wash buffer with complete culture medium. If the cells cannot be treated immediately after the isolation, place them into the incubator with 5% CO_2 and temperature of 37 °C until ready to treat with controls and NANPs
30–34	Altered cell morphology	Same as described above in the "low cell viability" section	Same as described above in the "low cell viability" section

(Continued)

TABLE 30.3 (CONTINUED)

Step	Problem	Reason	Solution
35–42	High level of cytokines in the negative control	- Blood was exposed to extreme heat during transportation - Cell culture medium or reagents used to prepare the medium are contaminated with endotoxin	- Avoid overheating the blood - Prepare the medium using fresh reagents
35–42	Low signal in the positive control	- Expired, degraded or inappropriately stored controls - Repeated freeze/thaw	- Prepare controls using fresh reagents. Avoid repeated subjecting stocks to freeze/thaw cycles. After the preparation, prepare single use aliquots of control stocks and store them at -20 °C. - To avoid repeated freeze–thaw of supernatants, prepare small aliquots at the time of supernatant collection and store them at -20 °C
35–42	No positive response in NANP samples	- Can be expected with small, planar and fibrous NANPs - When unexpected, can be due to the NANPs' degradation or failure to complex with a carrier. It can also be caused by repeated freeze–thaw cycles	- When unexpected, use fresh carrier (e.g., L2K) and store freshly prepared NANPs at 4 °C (for up to 1 week) or -20 °C (for longer time). Verify NANPs' stability by gel-electrophoresis - To avoid repeated freeze–thaw of supernatants, prepare small aliquots at the time of supernatant collection and store them at -20 °C
35–42	Unexpectedly high levels of IFNs and other cytokines in NANPs supernatants	Contamination of NANPs, L2K, or cell culture media with endotoxin and other innate immunity-modulating impurities	Test all components for the presence of endotoxin using LAL assay and replace contaminated materials with pyrogen-free ones.
43–56	High variability between replicates (% CV > 25)	Incubation was not performed on the shaker; loose pipet tips, especially when multichannel pipettes were used; the plate was allowed to dry after wash steps	Perform incubations on a shaker; verify that tips are tight on pipette; avoid long (> 5 min) intervals between plate wash and addition of reagents and keep the plate on a paper towel up-side-down until ready to add the reagents to the plate.
57–60	No decrease in cytokine level in the presence of an inhibitor	Nanoparticles' complexation with L2K did not work, a different carrier or no carrier was used, or the inhibitor quality was compromised during the storage or repeated freeze–thaw cycles	Verify that fresh L2K is used for the complexation. Prepare fresh inhibitors from commercial reagents as described in Section 30.1.1 (Experimental Design), and then prepare single use aliquots. Discard any unused inhibitor at the end of experiment; do not store to avoid repeated freeze–thaw cycles

Anticipated results for cytokine screening (steps 35–56). The results shown in Table 30.2 demonstrate that when NANPs' quality is achieved by appropriate performance of procedures described in steps 1–20, as well as when the complexation with a carrier and procedures described in steps 21–25 are followed, the induction of primary biomarkers of NANPs' immunostimulation (IFNα, IFNβ, IFNω, and IFNλ) detected in steps 43–56 is observed (Figure 4). When different carrier is used, the spectrum and the magnitude of cytokine response may be different and, therefore, will provide novel information about the influence of NANPs delivery carrier on their immunological recognition. The magnitude of the biomarker is determined by the NANP's physicochemical properties in that RNA-based NANPs, specifically RNA cubes, are the most potent immunostimulants, while DNA cubes and RNA rings are less immunostimulatory. Despite the difference in the magnitude of the interferon biomarker induction by NANPs with identical physicochemical properties, the type I interferons (IFN-α, -β, -ω) and type III interferons (IFN-λ) are consistently observed in all donors, as confirmed in our studies of over 100 healthy donors, provided the delivery was achieved by a control carrier lipofectamine. These interferons have beneficial therapeutic properties. For example, they were shown to induce dendritic cell (DC) maturation and support DC function [58–60]. Type I interferons are also used for therapy of viral infections (*e.g.*, Hepatitis C), cancer (*e.g.*, chronic myeloid leukemia), and immune-mediated disorders such as multiple sclerosis [61–65]. Type III interferons also have anti-cancer activity [66]. Clinical use of recombinant interferons is associated with fever-like reactions due to their systemic distribution and the ability to activate the immune cells in the blood [67] and is also complicated by an anti-drug antibody (ADA) response in some patients [68]. When the ADA is neutralizing, it affects both the drug efficacy and contributes to toxicity [69]. These toxicities occur despite protein engineering attempts and conjugation with hydrophilic poly(ethylene glycol) (PEG) [69]. Not only does PEGylation of recombinant proteins prevent immunogenicity of these products, but it may also create additional hurdles since the blood of healthy individuals contains pre-existing anti-PEG antibodies that may affect both the safety and efficacy of PEGylated drug products [70–73]. Due to their ability to induce type I and type III interferons, NANPs could potentially address these problems by stimulating the hosts' own type of interferon response.

If NANPs are tested without a carrier and a broad spectrum of pyrogenic cytokines (TNFα, IL-1β, IL-6, IL-8) is detected, it would be highly suggestive of endotoxin contamination, which was either left undetected or introduced during handling and storage of NANPs after the initial synthesis and analysis by LAL.

The protocol presented herein, therefore, allows for the screening of multiple NANPs and for selection of the one with a desirable magnitude of IFN response; understanding of potential contamination issues and investigating the role of various carriers and their impact on the magnitude and spectrum of NANPs-mediated cytokines in comparison to the benchmark carrier lipofectamine.

ANTICIPATED RESULTS FOR MECHANISTIC ANALYSIS (STEPS 57–60)

If the addition of ODN2088 resulted in a decrease in the cytokine response to the given NANPs, this response is due to the NANP's recognition by endosomal TLRs (TLR3, TLR7, TLR8, and TLR9) (Figure 30.3d). It is important to emphasize that this inhibitor cannot distinguish between individual endosomal TLRs.

If the addition of fucoidan resulted in a decrease in the cytokine response to the given NANPs, this response is due to the NANP's recognition by scavenger receptors. If the addition of either poly-I or dextran sulfate resulted in a decrease in the cytokine response to the given NANPs, whereas the addition of poly-C or chondroitin sulfate did not influence the result, this response is due to the NANP's recognition by scavenger receptors (Figure 30.3e). When an alternative to lipofectamine carrier is used to deliver NANPs to the immune cells, the resulting spectrum of cytokines and/or magnitude of the cellular response may be different, and, therefore, will provide novel information about mechanisms of NANPs recognition by the immune cells. Most importantly, other inhibitors specific to the pathway of interest for the user of this protocol can also be used. Therefore, this protocol provides a versatile research tool to NANPs researchers.

ACKNOWLEDGMENTS

The study was supported in part (to M.A.D.) by federal funds from the National Cancer Institute, National Institutes of Health, under contract HHSN261200800001E and 75N91019D00024. The content of this publication does not necessarily reflect the views or policies of the Department of Health and Human Services, nor does mention of trade names, commercial products, or organizations imply endorsement by the U.S. Government. Research reported in this publication was also supported by the National Institute of General Medical Sciences of the National Institutes of Health under Award Number R01GM120487 (to K.A.A.). The content is solely the responsibility of the authors and does not necessarily represent the official views of the National Institutes of Health. The authors would also like to thank Justin Halman and Morgan Chandler of the University of North Carolina at Charlotte and Enping Hong and Edward Cedrone of the Nanotechnology Characterization Lab for the excellent technical assistance.

REFERENCES

1. Leontis, N. B., Stombaugh, J. & Westhof, E. The non-Watson-Crick base pairs and their associated isostericity matrices. *Nucleic acids research* **30**, 3497–3531 (2002).
2. Bindewald, E., Hayes, R., Yingling, Y. G., Kasprzak, W. & Shapiro, B. A. RNAJunction: a database of RNA junctions and kissing loops for three-dimensional structural analysis and nanodesign. *Nucleic acids research* **36**, D392–397 (2008).
3. Parlea, L. et al. Ring Catalog: A resource for designing self-assembling RNA nanostructures. *Methods* **103**, 128–137, doi:10.1021/acscombsci.6b00073 (2016).
4. Parlea, L. G., Sweeney, B. A., Hosseini-Asanjan, M., Zirbel, C. L. & Leontis, N. B. The RNA 3D motif atlas: Computational methods for extraction, organization and evaluation of RNA motifs. *Methods* **103**, 99–119, doi:10.1016/j.ymeth.2016.04.025 (2016).
5. Geary, C., Chworos, A., Verzemnieks, E., Voss, N. R. & Jaeger, L. Composing RNA nanostructures from a syntax of RNA Structural Modules. *Nano letters* **17**, 7095–7101, doi:10.1021/acs.nanolett.7b03842 (2017).
6. Weng, Y. et al. Improved Nucleic Acid Therapy with Advanced Nanoscale Biotechnology. *Mol Ther Nucleic Acids* **19**, 581–601, doi:10.1016/j.omtn.2019.12.004 (2019).
7. Afonin, K. A. et al. Design and self-assembly of siRNA-functionalized RNA nanoparticles for use in automated nanomedicine. *Nat Protoc* **6**, 2022–2034, doi:10.1038/nprot.2011.418 (2011).
8. Jasinski, D., Haque, F., Binzel, D. W. & Guo, P. Advancement of the Emerging Field of RNA Nanotechnology. *ACS Nano* **11**, 1142–1164, doi:10.1021/acsnano.6b05737 (2017).
9. Jaeger, L., Westhof, E. & Leontis, N. B. TectoRNA: modular assembly units for the construction of RNA nano-objects. *Nucleic acids research* **29**, 455–463 (2001).
10. Jaeger, L. & Leontis, N. B. Tecto-RNA: One-Dimensional Self-Assembly through Tertiary Interactions This work was carried out in Strasbourg with the support of grants to N.B.L. from the NIH (1R15 GM55898) and the NIH Fogarty Institute (1-F06-TW02251-01) and the support of the CNRS to L.J. The authors wish to thank Eric Westhof for his support and encouragement of this work. *Angew Chem Int Ed Engl* **39**, 2521–2524 (2000).
11. Ohno, H. et al. Synthetic RNA-protein complex shaped like an equilateral triangle. *Nature nanotechnology* **6**, 116–120, doi:10.1038/nnano.2010.268 (2011).
12. Dibrov, S. M., McLean, J., Parsons, J. & Hermann, T. Self-assembling RNA square. *Proc Natl Acad Sci U S A* **108**, 6405–6408 (2011).
13. Boerneke, M. A., Dibrov, S. M. & Hermann, T. Crystal-Structure-Guided Design of Self-Assembling RNA Nanotriangles. *Angew Chem Int Ed Engl* **55**, 4097–4100, doi:10.1002/anie.201600233 (2016).
14. Afonin, K. A. et al. In vitro assembly of cubic RNA-based scaffolds designed in silico. *Nature nanotechnology* **5**, 676–682, doi:10.1038/nnano.2010.160 (2010).

15. Chidchob, P. & Sleiman, H. F. Recent advances in DNA nanotechnology. *Current opinion in chemical biology* **46**, 63–70, doi:10.1016/j.cbpa.2018.04.012 (2018).

16. Rothemund, P. W. K. Folding DNA to create nanoscale shapes and patterns. *Nature* **440**, 297–302, doi:10.1038/nature04586 (2006).

17. Afonin, K. A. et al. Triggering of RNA interference with RNA-RNA, RNA-DNA, and DNA-RNA nanoparticles. *ACS Nano* **9**, 251–259, doi:10.1016/bs.mie.2014.10.058 (2015).

18. Afonin, K. A. et al. Multifunctional RNA nanoparticles. *Nano Lett* **14**, 5662–5671, doi:10.1021/nn504508s (2014).

19. Lee, H. et al. Molecularly self-assembled nucleic acid nanoparticles for targeted in vivo siRNA delivery. *Nat Nanotechnol* **7**, 389–393, doi:10.1038/nnano.2012.73 (2012).

20. Li, H. et al. RNA as a stable polymer to build controllable and defined nanostructures for material and biomedical applications. *Nano Today* **10**, 631–655, doi:10.1016/j.nantod.2015.09.003 (2015).

21. Shu, D., Shu, Y., Haque, F., Abdelmawla, S. & Guo, P. Thermodynamically stable RNA three-way junction for constructing multifunctional nanoparticles for delivery of therapeutics. *Nat Nanotechnol* **6**, 658–667, doi:10.1038/nnano.2011.105 (2011).

22. Khisamutdinov, E. F. et al. Enhancing immunomodulation on innate immunity by shape transition among RNA triangle, square and pentagon nanovehicles. *Nucleic Acids Res* **42**, 9996–10004, doi:10.1093/nar/gku516 (2014).

23. Halman, J. R. et al. Functionally-interdependent shape-switching nanoparticles with controllable properties. *Nucleic Acids Res* **45**, 2210–2220, doi:10.1093/nar/gkx008 (2017).

24. Johnson, M. B. et al. Programmable Nucleic Acid Based Polygons with Controlled Neuroimmunomodulatory Properties for Predictive QSAR Modeling. *Small* **13**, doi:10.1002/smll.201701255 (2017).

25. Hong, E. et al. Structure and Composition Define Immunorecognition of Nucleic Acid Nanoparticles. *Nano Lett* **18**, 4309–4321, doi:10.1021/acs.nanolett.8b01283 (2018).

26. Hong, E. et al. Toll-Like Receptor-Mediated Recognition of Nucleic Acid Nanoparticles (NANPs) in Human Primary Blood Cells. *Molecules* **24**, doi:10.3390/nano9040611 (2019).

27. Bindewald, E. et al. Multistrand Structure Prediction of Nucleic Acid Assemblies and Design of RNA Switches. *Nano Lett* **16**, 1726–1735, doi:10.1016/j.ymeth.2016.04.016 (2016).

28. Dobrovolskaia, M. A. Pre-clinical immunotoxicity studies of nanotechnology-formulated drugs: Challenges, considerations and strategy. *J Control Release* **220**, 571–583, doi:10.1016/j.jconrel.2015.08.056 (2015).

29. Panigaj, M. et al. Aptamers as Modular Components of Therapeutic Nucleic Acid Nanotechnology. *ACS Nano* **13**, 12301–12321, doi:10.1021/acsnano.9b06522 (2019).

30. Chandler, M. & Afonin, K. A. Smart-Responsive Nucleic Acid Nanoparticles (NANPs) with the Potential to Modulate Immune Behavior. *Nanomaterials (Basel)* **9**, doi:10.1002/wrna.1452 (2019).

31. Sajja, S. et al. Dynamic Behavior of RNA Nanoparticles Analyzed by AFM on a Mica/Air Interface. *Langmuir*, doi:10.1021/acs.langmuir.8b00105 (2018).

32. Ke, W. et al. RNA-DNA fibers and polygons with controlled immunorecognition activate RNAi, FRET and transcriptional regulation of NF-kappaB in human cells. *Nucleic acids research*, doi:10.1093/nar/gky1215 (2018).

33. Rackley, L. et al. RNA Fibers as Optimized Nanoscaffolds for siRNA Coordination and Reduced Immunological Recognition. *Adv Funct Mater* **28**, doi:10.3390/nano9070951 (2018).

34. Vessillier, S. et al. Cytokine release assays for the prediction of therapeutic mAb safety in first-in man trials—Whole blood cytokine release assays are poorly predictive for TGN1412 cytokine storm. *J Immunol Methods* **424**, 43–52, doi:10.1016/j.jim.2015.04.020 (2015).

35. Gregg, K. A. et al. Rationally Designed TLR4 Ligands for Vaccine Adjuvant Discovery. *MBio* **8**, doi:10.1128/mBio.00492-17 (2017).

36. Oh, D. Y. et al. Adjuvant-induced Human Monocyte Secretome Profiles Reveal Adjuvant- and Age-specific Protein Signatures. *Mol Cell Proteomics* **15**, 1877–1894, doi:10.1074/mcp.M115.055541 (2016).

37. Ke, W. et al. RNA-DNA fibers and polygons with controlled immunorecognition activate RNAi, FRET and transcriptional regulation of NF-kappaB in human cells. *Nucleic Acids Res* **47**, 1350–1361, doi:10.1093/nar/gky1215 (2019).

38. Wei, M. et al. Polyvalent immunostimulatory nanoagents with self-assembled CpG oligonucleotide-conjugated gold nanoparticles. *Angew Chem Int Ed Engl* **51**, 1202–1206, doi:10.1002/anie.201105187 (2012).

39. Radovic-Moreno, A. F. et al. Immunomodulatory spherical nucleic acids. *Proc Natl Acad Sci U S A* **112**, 3892–3897, doi:10.1073/pnas.1502850112 (2015).

40. Li, J. et al. Self-assembled multivalent DNA nanostructures for noninvasive intracellular delivery of immunostimulatory CpG oligonucleotides. *ACS Nano* **5**, 8783–8789, doi:10.1021/nn202774x (2011).

41. Liu, X. et al. A DNA nanostructure platform for directed assembly of synthetic vaccines. *Nano letters* **12**, 4254–4259, doi:10.1021/nl301877k (2012).

42. Wang, S. et al. Rational vaccinology with spherical nucleic acids. *Proc Natl Acad Sci U S A* **116**, 10473–10481, doi:10.1073/pnas.1902805116 (2019).

43. Hong, E. & Dobrovolskaia, M. A. Addressing barriers to effective cancer immunotherapy with nanotechnology: achievements, challenges, and roadmap to the next generation of nanoimmunotherapeutics. *Adv Drug Deliv Rev*, doi:10.1016/j.addr.2018.01.005 (2018).

44. Dobrovolskaia, M. A. Nucleic Acid Nanoparticles at a Crossroads of Vaccines and Immunotherapies. *Molecules* **24**, doi:10.3390/molecules24244620 (2019).

45. Kondo, S. & Sauder, D. N. Tumor necrosis factor (TNF) receptor type 1 (p55) is a main mediator for TNF-alpha-induced skin inflammation. *European journal of immunology* **27**, 1713–1718, doi:10.1002/eji.1830270718 (1997).

46. Phillips, A., Patel, C., Pillsbury, A., Brotherton, J. & Macartney, K. Safety of Human Papillomavirus Vaccines: An Updated Review. *Drug Saf* **41**, 329–346, doi:10.1007/s40264-017-0625-z (2018).

47. Miller, E. R. et al. Post-licensure safety surveillance of zoster vaccine live (Zostavax(R)) in the United States, Vaccine Adverse Event Reporting System (VAERS), 2006-2015. *Hum Vaccin Immunother*, 1–23, doi:10.1080/21645515.2018.1456598 (2018).

48. Woo, E. J., Moro, P. L., Cano, M. & Jankosky, C. Postmarketing safety surveillance of trivalent recombinant influenza vaccine: Reports to the Vaccine Adverse Event Reporting System. *Vaccine* **35**, 5618–5621, doi:10.1016/j.vaccine.2017.08.047 (2017).

49. Gause, K. T. et al. Immunological Principles Guiding the Rational Design of Particles for Vaccine Delivery. *ACS Nano* **11**, 54–68, doi:10.1021/acsnano.6b07343 (2017).

50. Di Franco, S., Turdo, A., Todaro, M. & Stassi, G. Role of Type I and II Interferons in Colorectal Cancer and Melanoma. *Front Immunol* **8**, 878, doi:10.3389/fimmu.2017.00878 (2017).

51. Adams, D. et al. Patisiran, an RNAi Therapeutic, for Hereditary Transthyretin Amyloidosis. *The New England journal of medicine* **379**, 11–21, doi:10.1056/NEJMoa1716153 (2018).

52. Chan, A. et al. Preclinical Development of a Subcutaneous ALAS1 RNAi Therapeutic for Treatment of Hepatic Porphyrias Using Circulating RNA Quantification. *Mol Ther Nucleic Acids* **4**, e263, doi:10.1038/mtna.2015.36 (2015).

53. Halman, J. R. et al. A cationic amphiphilic co-polymer as a carrier of nucleic acid nanoparticles (Nanps) for controlled gene silencing, immunostimulation, and biodistribution. *Nanomedicine* **23**, 102094, doi:10.1016/j.nano.2019.102094 (2020).

54. Bui, M. N. et al. Versatile RNA tetra-U helix linking motif as a toolkit for nucleic acid nanotechnology. *Nanomedicine* **13**, 1137–1146, doi:10.1093/nar/gkx008 (2017).

55. Ozaki, K. & Leonard, W. J. Cytokine and cytokine receptor pleiotropy and redundancy. *J Biol Chem* **277**, 29355–29358, doi:10.1074/jbc.R200003200 (2002).

56. Potter, T. M., Neun, B. W., Rodriguez, J. C., Ilinskaya, A. N. & Dobrovolskaia, M. A. Analysis of Pro-inflammatory Cytokine and Type II Interferon Induction by Nanoparticles. *Methods Mol Biol* **1682**, 173–187, doi:10.1007/978-1-4939-7352-1_15 (2018).

57. Strober, W. Trypan blue exclusion test of cell viability. *Curr Protoc Immunol* **Appendix 3**, Appendix 3B, doi:10.1002/0471142735.ima03bs21 (2001).

58. Radvanyi, L. G., Banerjee, A., Weir, M. & Messner, H. Low levels of interferon-alpha induce CD86 (B7.2) expression and accelerates dendritic cell maturation from human peripheral blood mononuclear cells. *Scand J Immunol* **50**, 499–509, doi:10.1046/j.1365-3083.1999.00625.x (1999).

59. Tam, M. A. & Wick, M. J. MyD88 and interferon-alpha/beta are differentially required for dendritic cell maturation but dispensable for development of protective memory against Listeria. *Immunology* **128**, 429–438, doi:10.1111/j.1365-2567.2009.03128.x (2009).

60. Trepiakas, R., Pedersen, A. E., Met, O. & Svane, I. M. Addition of interferon-alpha to a standard maturation cocktail induces CD38 up-regulation and increases dendritic cell function. *Vaccine* **27**, 2213–2219, doi:10.1016/j.vaccine.2009.02.015 (2009).

61. Floros, T. & Tarhini, A. A. Anticancer Cytokines: Biology and Clinical Effects of Interferon-alpha2, Interleukin (IL)-2, IL-15, IL-21, and IL-12. *Semin Oncol* **42**, 539–548, doi:10.1053/j.seminoncol.2015.05.015 (2015).

62. Cheknev, S. B., Kobyakina, N. A., Mezentseva, M. V. & Skvortsova, V. I. Long-Term Study of Interferon System State in Patients with Multiple Sclerosis Received the Individual Immune Therapy with Human Recombinant IFN-alpha. *Russ J Immunol* **6**, 39–46 (2001).

63. Bongioanni, M. R. et al. Systemic high-dose recombinant-alpha-2a-interferon therapy modulates lymphokine production in multiple sclerosis. *J Neurol Sci* **143**, 91–99, doi:10.1016/s0022-510x(96)00176-1 (1996).

64. Kujawski, L. A. & Talpaz, M. The role of interferon-alpha in the treatment of chronic myeloid leukemia. *Cytokine Growth Factor Rev* **18**, 459–471, doi:10.1016/j.cytogfr.2007.06.015 (2007).

65. Rong, L. & Perelson, A. S. Modeling HIV persistence, the latent reservoir, and viral blips. *J Theor Biol.* **260**, 308–331 (2009).

66. Stiff, A. & Carson, W. Investigations of interferon-lambda for the treatment of cancer. *J Innate Immun* **7**, 243–250, doi:10.1159/000370113 (2015).

67. Filipi, M. L. et al. Nurses' perspective on approaches to limit flu-like symptoms during interferon therapy for multiple sclerosis. *Int J MS Care* **16**, 55–60 (2014).

68. Rosenberg, A. S. Immunogenicity of biological therapeutics: a hierarchy of concerns. *Dev Biol (Basel)* **112**, 15–21 (2003).

69. Baker, M. P., Reynolds, H. M., Lumicisi, B. & Bryson, C. J. Immunogenicity of protein therapeutics: The key causes, consequences and challenges. *Self Nonself*, 314–322 (2010).

70. Chang, C. J. et al. A genome-wide association study identifies a novel susceptibility locus for the immunogenicity of polyethylene glycol. *Nat Commun* **8**, 522, doi:10.1038/s41467-017-00622-4 (2017).

71. Chen, B. M. et al. Measurement of Pre-Existing IgG and IgM Antibodies against Polyethylene Glycol in Healthy Individuals. *Anal Chem* **88**, 10661–10666, doi:10.1021/acs.analchem.6b03109 (2016).

72. Chen, W. W., Zhang, X. & Huang, W. J. Role of neuroinflammation in neurodegenerative diseases (Review). *Mol Med Rep* **13**, 3391–3396, doi:10.3892/mmr.2016.4948 (2016).

73. Hsieh, Y. C. et al. Pre-existing anti-polyethylene glycol antibody reduces the therapeutic efficacy and pharmacokinetics of PEGylated liposomes. *Theranostics* **8**, 3164–3175, doi:10.7150/thno.22164 (2018).

74. Halman, J. R. et al. A cationic amphiphilic co-polymer as a carrier of nucleic acid nanoparticles (Nanps) for controlled gene silencing, immunostimulation, and biodistribution. *Nanomedicine*, 102094, doi:10.1016/j.nano.2019.102094 (2019).

75. Cruz-Acuna, M., Halman, J. R., Afonin, K. A., Dobson, J. & Rinaldi, C. Magnetic nanoparticles loaded with functional RNA nanoparticles. *Nanoscale* **10**, 17761–17770, doi:10.1039/c8nr04254c (2018).

Part IV

Delivery of Functional RNA Nanoparticles

31 The Emerging Field of RNA Nanotechnology

Peixuan Guo

University of Cincinnati, Cincinnati, Ohio, USA

CONTENTS

This chapter is adapted and reproduced from the full published article with permission from Guo P. The emerging field of RNA nanotechnology. *Nat Nanotechnol.* 2010; 5:833. Copyright © 2010, Nature Publishing Group, a division of Macmillan Publishers Limited. All Rights Reserved.

Macromolecules of DNA, RNA and proteins have intrinsically defined features on the nanoscale and may serve as powerful building blocks for the bottom-up fabrication of nanostructures and nanodevices. The field of DNA nanotechnology [1–3] is now well established, having its origins in work by Seeman some 30 years ago, and peptides and proteins have also been studied for applications in nanotechnology [4–7]. The concept of RNA nanotechnology [8–15] has been around for more than a decade, and the first evidence for the construction of RNA nanoparticles through the self-assembly of several re-engineered natural RNA molecules was reported in 1998 [8]. However, interest in RNA nanotechnology has increased in recent years as recognition of its potential for applications in nanomedicine – including the treatment of cancer, viral infection and genetic diseases – has grown (Figure 31.1).

RNA can be designed and manipulated with a level of simplicity that is characteristic of DNA, while displaying flexibility in structure and diversity in function (including enzymatic activities) that is similar to that of proteins. Although RNA nanotechnology is similar to that of DNA in a number of ways, there are important differences between the two disciplines (Table 31.1).

RNA is a polymer made up of four different nucleotides: adenine (A), cytosine (C), guanine (G) and uracil (U), whereas DNA contains thymine (T) rather than U. And as well as the Watson–Crick base pairing found in DNA (A with T, C with G), other forms of base pairing (referred to as non-canonical base pairing) are possible, such as G with A or U, which allow RNA to fold into rigid structural motifs that are distinct from those formed by single-stranded DNA [10, 12, 16–24] (Figure 31.2). At present, an RNA strand containing up to 80 nucleotides can be synthesized commercially, and an 80-nucleotide RNA strand can display up to 4^{80} (or 10^{48}) different structures. The availability of so many different structural building blocks is an advantage for many applications.

Moreover, RNA typically contains a large variety of single-stranded stem-loops for intra- and/or intermolecular interactions, and these can be used to make 'dovetail' joints between different building blocks, thus removing the need for an equivalent to dowels in RNA nanostructures and nanomachines. Loops and motifs also allow for the construction of a more complicated secondary structure.

DOI: 10.1201/9781003001560-35

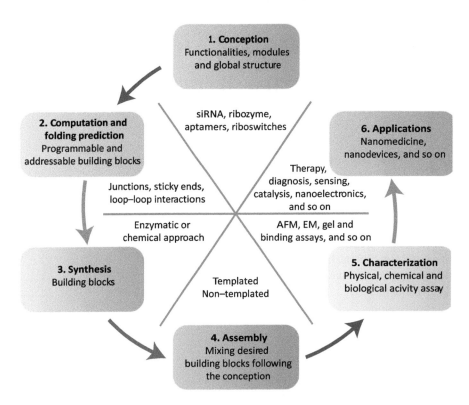

FIGURE 31.1 Approaches to RNA nanotechnology. The construction of RNA nanoparticles is a multistep process that starts with a conception step in which the desired properties of the nanoparticle are defined. A computational approach is then applied to predict the structure and folding of the building blocks and the consequences of inter-RNA interactions in the assembly of RNA nanoparticles. After the monomeric building blocks are synthesized (either by enzymatic or chemical approaches), the individual subunits assemble into quaternary architectures by either templated or non-templated methods. The assembled RNA nanostructures are characterized (by atomic force microscope (AFM), electron microscope (EM), gel electrophoresis and so on) to ensure proper folding with desired structural and functional capabilities. After thorough evaluation, the nanoparticles will be used for various applications.

Furthermore, RNA molecules such as aptamers, ribozymes and short interfering RNA (siRNA) can have special functionalities (see 'Applications of RNA nanotechnology' section).

Among the three helices (RNA/RNA, RNA/DNA and DNA/DNA), the RNA/RNA double helix is the most stable [25, 26]. RNA motifs and modules with special bends or stacks are particularly stable. The thermodynamic stability has been defined as the free energy, G, required for complex formation, or in some cases, to unwind the helix ($\Delta G^0 = -G^0_{helix} = G^0_{unwind}$); thus, the lower the free energy ($-G^0_{helix}$) the complex holds, the more stable it is. Because ΔG^0 is affected by neighbouring sequences, $-G^0_{helix}$ for RNA is calculated to be lower than DNA [25, 26] based on the nearest-neighbour model (Table 31.1). However, under physiological conditions, the RNA helix displays A-type configuration whereas the DNA helix is predominantly B-type. The 2'-OH in RNA ribose locks the ribose into a 3'-endo chair conformation that does not favour a B-helix. Base stacking is governed by van der Waals interaction, which contributes directly to the enthalpy. Though the difference in the stacking interaction is small between DNA and RNA, the sum over numerous base pairs can make a difference to the helix stability. Thus, RNA nanoparticles are more stable thermodynamically than their DNA counterparts. Like DNA tiles, stable RNA helices in solution can be produced using four to six nucleotides of RNA [12], but in certain cases as few as two nucleotides can promote complex formations in RNA [27–31].

TABLE 31.1
Differences between DNA and RNA

	DNA	RNA
Elements		
	%DVH A, C, G, T	%DVH A, C, G, U
	2' -deoxyribose	ribose
Base pairing	Canonical Watson–Crick (W–C)	Canonical and non-canonical W–C
Acidic effect	Depurination: apurine DNA sensitive to cleavage	Stable
Alkaline effect	Stable up to pH 12	Sensitive to alkaline hydrolysis
Configuration	Predominantly B-form:	A-form:
	– base pairs/turn of the helix: 10.5;	– base pairs/turn of the helix: 10.9;
	– pitch: 3.5 nm;	– pitch: 2.5 nm;
	– helix rise/base pairs: 0.314 nm;	– helix rise/base pairs: 0.275 nm;
	– humidity: nucleotide: H_2O = 1:1	– humidity: nucleotide: H_2O = 1:0.7
Chemical stability	Relatively stable but sensitive to DNase	Unstable, sensitive to RNase, but stable after chemical modification, for example, 2'-F or 2'-OMe modification
Thermal stability	G:C more stable than A:T	Thermally more stable than DNA, especially for RNA motifs and modules with particular bends or stacks
Free energy, ΔG^0	−1.4 KJmol^{-1} per base pair stack [25]	−3.6 to −8.5 KJmol^{-1} per base pair stack [25]
Helix formation	Needs a minimum of four nucleotides	Needs a minimum of two nucleotides [26, 27]
Intermolecular interactions	Cohesive ends, crossover motifs	Cohesive ends, crossover motifs, kissing loops, interlocking loops
In vivo replication		
Initiation	Origin of replication with primer	Promoter, exact nucleotide to start without primer Specific transcription terminators
Termination	No nature sequence for replication termination	
In vitro synthesis		
Enzymatic	DNA polymerase, polymerase chain reaction (PCR)	T7/SP6 transcription
Chemical	Up to 160 nucleotides; low cost	Up to 117 nucleotides; high cost and low yield

DISTINCT ATTRIBUTES OF RNA INSIDE THE BODY

Therapeutic particles are initially recognized by cell-surface receptor(s) before being internalized through the plasma membrane into vesicles (called endosomes) that sort the particles for either degradation or recycling. Escaping the endosome is an important consideration for *in vivo* delivery because most molecules cannot survive its acidic environment, with a pH ranging from 4.3 to 5.8 [32]. At this pH, protonation of DNA purine bases leads to their removal (a process known as depurination) and the resulting apurinic DNA is susceptible to cleavage [33]. The higher stability of RNA at low pHs is especially useful in therapy because it means that they will survive in the endosome and disperse throughout the cell after entry (Table 31.1).

Another intriguing property of RNA is the possibility of producing self-assembled RNA nanoparticles *in vivo*. In contrast to DNA, small RNA molecules are transcribed in the cell using DNA as a template. By using an inducible promoter [34] and appropriate terminators for transcription, small RNA molecules can be produced controllably. RNA can be processed into the desired length by including delta ribozymes at both the upstream and downstream terminals for *cis*-cleavage [35]. Natural RNA nanoparticles such as dimers [36–39] and hexamers [8, 9] have been discovered in

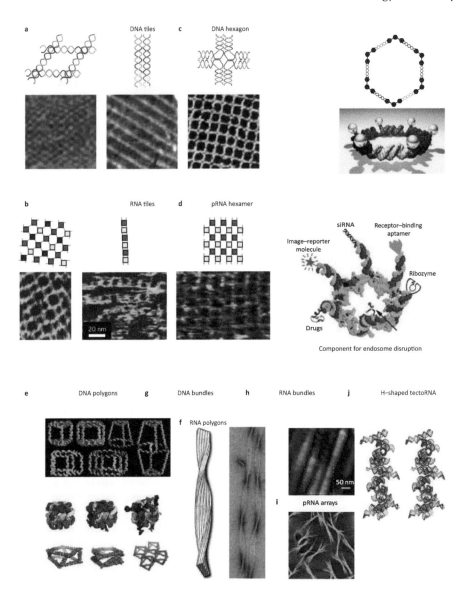

FIGURE 31.2 Comparison of self-assembled DNA and RNA nanoparticles. **(a,b)** Representative transmission electron microscope (TEM) and AFM images of DNA and RNA tiles are shown below the corresponding cartoon models. **a**, Left to right: TEM image of a parallelogram DNA tile formed by joining four Holliday junctions in parallel; TEM image of a double helix tile formed by the exchange of two DNA duplex strands and AFM image of a cross-shaped tile [1, 2] possessing four arms. **b**, Left to right: AFM images of tectosquare nanopatterns [13] (striped velvet, ladder and fishnet pattern). **(c)** Illustration of a hexagonal array of gold nanoparticles (yellow circles in lower image) on a DNA hexagon consisting of six non-identical molecules (triplets of grey hexagons in upper image) [1, 2], each linked with two single-stranded DNA molecules (coloured lines). **(d)** Illustration of a pRNA hexameric ring containing six positions that can carry different molecules [50, 87]. **(e,f)** Illustration of various 3D DNA polygons **(e)** [2] and RNA cubic scaffolds **(f)** [30, 77]. **(g)** TEM images of DNA bundles [70]. **(h,i)** AFM images of RNA bundles **(h)** [23] and pRNA arrays **(i)** [12]. **(j)** 3D model of H-shaped tectoRNA [21]. Figures reproduced with permission from: **a**, ref. 1, © 2009 ACS and ref. 2, © 2008 AAAS; **b**, ref. 13, © 2004 AAAS; **c**, ref. 1, © 2009 ACS and ref. 2, © 2008 AAAS; **d**, ref. 49, © 2005 Mary Ann Liebert; **e**, ref. 2, © 2008 AAAS; **f**, refs 30 and 77, © 2010 NPG; **g**, ref. 69, © 2009 AAAS; **h**, ref. 23, © 2009 ACS; **i**, ref. 12, © 2004 ACS; **j**, ref. 21 © 2006 Oxford University Press.

cells. Sequences such as packaging RNA (pRNA) [35] or transfer RNA (tRNA) [40, 41] for guiding the self-assembly of RNA nanoparticles with functionalities such as siRNA [42, 43], ribozymes [35] or aptamers [40] can be incorporated in the DNA template *in vivo* [35].

Small RNAs, such as riboswitch, with regulatory functions [44–47] in the cell may be viewed as Boolean networks based on logic operations [48, 49]. Input nodes can be seen as RNA nanostructures and the output (for example, the activation of a pathway) is based on logic functions of input RNA concentrations. Numerous small RNA regulators can be used to regulate the *in vivo* products and functional pathways, with controls by induction or repression through the *trans-* and *cis-*actions. Varieties of small RNA can work cooperatively, synergistically, or antagonistically – based on the design – to produce computational logic circuits as conjunctive or disjunctive normal forms, or other kinds of logic operation. By designing the logic network of AND/NOT/OR gates in the cell, an 'RNA computer' can theoretically be implemented and applied to bacterial, yeast and mammalian systems [48, 49].

TECHNIQUES FOR CONSTRUCTING RNA NANOPARTICLES

Construction of nanoparticles requires the use of programmable, addressable and predictable building blocks. Self-assembly of RNA building blocks in a predefined manner to form larger two-, three- and four-dimensional (2-, 3-, 4D) structures is a prominent bottom-up approach and represents an important means by which biological techniques and biomacromolecules can be successfully integrated into nanotechnology [12, 50, 51].

Within the realm of self-assembly there are two main subcategories: templated and non-templated assembly. Templated assembly involves the interaction of RNAs with one another under the influence of a specific external force, structure, or spatial constraint. RNA transcription, hybridization, replication, moulding and phi29 pRNA hexameric-ring formation are all in this category. Nontemplated assembly involves the formation of a larger structure by individual components without any external influence. Examples include ligation, chemical conjugation, covalent linkages, loop–loop interactions of RNA such as the HIV kissing loop and the phi29 pRNA dimer or trimer formation [10, 12, 50–52]. Various approaches available for RNA nanoparticle construction are discussed below (see Figure 31.3 for summary).

The first tactic uses the assembly mechanism of natural RNA nanoparticles that can form specific multimers *in vivo*. For example, the retrovirus kissing loops facilitate genomic RNA dimerization [36, 37]. The pRNA of the bacteriophage phi29 DNA-packaging motor assembles into dimers and hexamers through hand-in-hand interactions between two right and left interlocking loops [8, 12, 28, 39, 52, 53]. The *bcd* mRNA of *Drosophila* embryos forms dimers through handin-arm interactions [38]. *E. coli* non-coding RNA *dsrA* assembles into stripe patterns through their built-in palindrome sequence [23]. The assemblies of RNA nanoparticles *in vitro* that mimic their natural counterparts were reported twelve years ago [8]. The unusual HIV kissing-loop mechanism has also inspired the design of tectoRNA architectures [13, 29].

The second tactic is to import some of the well-developed principles from DNA nanotechnology. Although RNA is different from DNA, there are some common structural and chemical features that can be exploited for progressing RNA nanotechnology.

DNA nanotechnology uses the nature of DNA complementarity for the construction of nanomaterials by means of intermolecular interactions of DNA strands. A variety of elegant shapes have been created with precise control over their geometries, periodicities and topologies [1–3] (Figure 31.2). Various crossover motifs have been designed through reciprocal exchange of DNA backbones [3]. Branched DNA tiles have been constructed using sticky ends and crossover junction motifs, such as tensegrity triangles (rigid structures in periodic-array form) [54] and algorithmic self-assembled Sierpinski triangles (aperiodic arrays of fractal patterns) [55]. The DNA tiles can further self-assemble into nanotubes, helix bundles [56] and complex DNA motifs and arrays for positioning nanoparticles, proteins or dyes with precise control, such as polycatenated DNA ladders [57]. Elegant 3D DNA networks using a minimal set of DNA strands with topologies such as cubes,

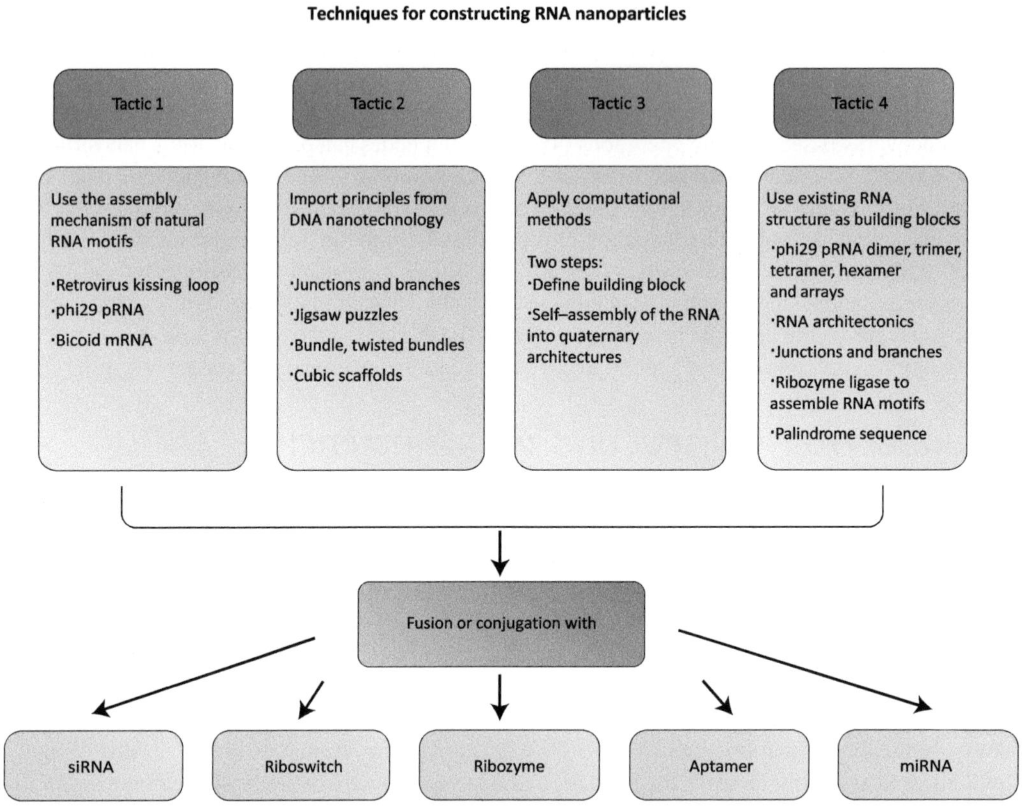

Techniques for constructing RNA nanoparticles

Tactic 1	Tactic 2	Tactic 3	Tactic 4
Use the assembly mechanism of natural RNA motifs	Import principles from DNA nanotechnology	Apply computational methods	Use existing RNA structure as building blocks
·Retrovirus kissing loop ·phi29 pRNA ·Bicoid mRNA	·Junctions and branches ·Jigsaw puzzles ·Bundle, twisted bundles ·Cubic scaffolds	Two steps: ·Define building block ·Self–assembly of the RNA into quaternary architectures	·phi29 pRNA dimer, trimer, tetramer, hexamer and arrays ·RNA architectonics ·Junctions and branches ·Ribozyme ligase to assemble RNA motifs ·Palindrome sequence

Fusion or conjugation with

siRNA	Riboswitch	Ribozyme	Aptamer	miRNA

FIGURE 31.3 Summary of different techniques for constructing RNA nanoparticles. See 'Techniques for constructing RNA nanoparticles' for complete description.

polyhedrons, prisms and buckyballs have also been fabricated based on junction flexibility and edge rigidity [3, 58]. Continuous growth of the tensegrity triangle in the periodic DNA module has resulted in the formation of DNA crystals diffracting to 4-Å resolution [59].

A striking illustration of the addressable and programmable properties of DNA is Rothemund's DNA origami [60], where a long single-stranded viral DNA is used as a scaffold for binding shorter strands to generate well-defined 2D and 3D configurations. DNA origami was subsequently applied to build 3D boxes that can be locked and unlocked [61], nano-arrays for label-free detection of substrates [62] and to elucidate the structure of organized proteins [63]. Rationally designed supramolecular DNA assemblies can be conjugated with organic and inorganic molecules, such as conjugation of porphyrins on parallel DNA helix bundles [64], nanomagnets [65] and elegant nanomachines [58, 66]. Replicable DNA architectures have been achieved to scale up the production of DNA nanostructures for practical applications by using enzymatic rolling-circle replication, bacterial cells infected with a viral vector [67] or chemical approaches for amplifying branched DNA arms [68].

Although the folding properties of RNA and DNA are not exactly the same, the fundamental principles in DNA nanotechnology are applicable to RNA nanotechnology. For example, the use of the three-way junction (3WJ) and four-way junction (4WJ) [18, 29] to build new and diverse RNA architectures is very similar to the branching approaches in DNA [1, 3] (Figure 31.2a,b,e,f). Both RNA and DNA can form jigsaw puzzles [13, 69] and be developed into bundles [12, 23, 30, 70] by combining elongation and expansion in the x–y direction (Figure 31.2a,b,g–j). The finding that insertion of bulges in the RNA helix leads to the formation of twisted bundles [12] (Figure 31.2i) was later demonstrated in DNA (Figure 31.2g), revealing that insertions and deletions of bases can form twisted DNA bundles with handedness [70], thereby illustrating the same basic principle. However, RNA is more rigid in bulge structure owing to non-canonical interactions, whereas in DNA the twisting requires the interaction of two DNA helices with four strands [70].

Recently, RNA cubic scaffolds [71] were constructed using several RNA sequences that do not fold on themselves but self-assemble with one another in a defined manner. This strategy is reminiscent of DNA nanotechnology, but in contrast to DNA strategies, RNA synthesis can be coupled to RNA self-assembly to generate fully assembled RNA cubes during *in vitro* transcription.

The third tactic is to apply computational methods in the construction of RNA nanoparticles. Computational approaches can be used to guide the design of new RNA assemblies and to optimize sequence requirements for the production of nanoscale fabrics with controlled direction and geometry [37, 72–75]. In contrast to traditional methods in which raw materials are selected rather than designed for a given application, the next generation of building blocks can be designed *a priori* for programmed assembly and synthesis. There are two steps involved in building RNA nanoparticles. The first is a computational approach (for example, using Kinefold [72]) using the spontaneous self-folding property of RNA into defined structures through base–base interactions dependent on their characteristic ΔG. [76] The second is the spontaneous assembly of the resulting RNA building blocks into larger assemblies based on the predicted architecture. This creates an effective computational pipeline for generating molecular models of RNA nanostructures. A recent example is the construction of cubic RNA-based scaffolds, whereby RNA sequence designs were optimized to avoid kinetic traps [77].

The fourth tactic is to use the existing RNA structure with known function as building blocks in RNA-nanoparticle construction. Varieties of mechanisms in RNA loop–loop interactions [8, 12, 31], tertiary architecture contacts [12, 15, 30] and formation of special motifs [12, 21, 29–31, 78–82] have been elucidated. Building blocks are first synthesized after computing intra- and intermolecular folding. Nanoparticles are built through spontaneous templated or non-templated self-assembly as planned. A rich resource of well-developed databases can be used to extract known RNA structural units for the construction of new RNA nanoparticles with desired properties [37, 83, 84].

Several methods have borrowed the properties of RNA in loop–loop interactions to construct RNA nanoparticles. The first method is based on the structural features of the pRNA of the bacteriophage phi29 DNA-packaging motor [8, 85], which uses a hexameric RNA ring to gear the machine [28, 86, 87]. The pRNA has been re-engineered to form dimers, trimers, tetramers, hexamers and arrays through hand-in-hand or foot-to-foot interactions between two interlocking loops [12, 52] (Figure 31.4). Dimers are formed using two building blocks with Ab′ (right and left hand, respectively) and Ba′ (Figure 31.4b). Trimers are formed using three building blocks with Ab′, Bc′ and Ca′ [12, 50, 51] (Figure 31.4c). Dimers of an extended configuration (twins) can also be efficiently self-assembled by introducing a palindrome sequence into the 3′-end of the pRNA [12]. These nanoparticles have been used successfully as polyvalent vehicles to deliver a variety of therapeutic molecules (see 'Applications of RNA nanotechnology' and 'Challenges and perspectives' sections) [12, 52]. The use of pRNA as building blocks for the construction of RNA arrays has also been achieved [12]. When three twins, Ab′, Bc′ and Ca′ are mixed, loop–loop interlocking makes the particles grow in three dimensions.

The second method is the RNA 'architectonics' [13], whereby structural modules specifying bends or stacks can be encoded in artificial RNA sequences for self-assembling higher-order specific shapes of RNA. Examples include RNA filaments [10, 21, 24] (Figure 31.2j), molecular jigsaw-puzzle units called tectosquares [13, 29] (Figure 31.2f) and tRNA antiprisms [88].

The third method is the application of 3- and 4WJs that are selected from known RNA structures or motifs [18, 19] to serve as the cornerstone in nanoparticle construction (Figure 31.2) [29, 76]. Some examples include RNA-structural motif (from ribosomal RNA; rRNA) to guide the tetramer assembly of L-shaped tectoRNAs; 3WJ motifs (from 23S rRNA) to construct a T-shaped arrangement of three helices; and tRNA motifs consisting of 4- and 5WJs to fold L-shaped tertiary structures [29, 37].

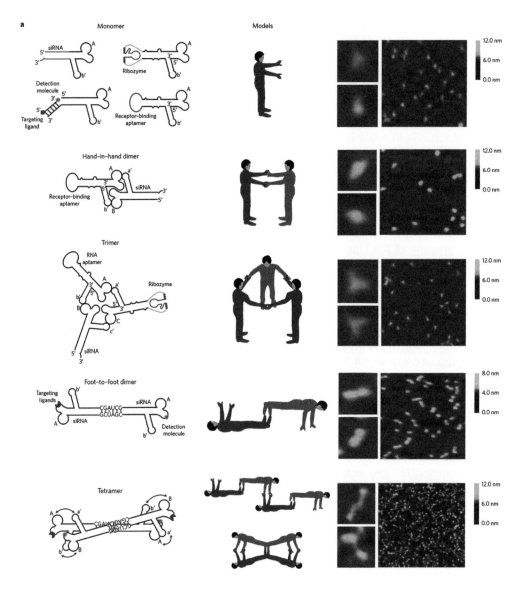

FIGURE 31.4 Applications of RNA nanotechnology. Left to right column: Schematic, models and AFM image showing the formation of different therapeutic nanoparticles containing siRNA, ribozymes, aptamers and other moieties using bacteriophage phi29 pRNA that possess left- and right-hand interlocking loops or a palindrome sequence [8, 12, 39]. (**a**) pRNA monomers bearing either a ribozyme, a receptor-binding aptamer or a targeting ligand and detection molecule. Uppercase and lowercase letters represent right and left hand, respectively. Same letter pair (e.g., Aa′) indicates complementarity [28]. (**b**) Monomer Ab′ that contains a receptor-binding aptamer and monomer Ba′, which contains an siRNA, assemble to form hand-in-hand dimers. (**c**) Trimers are formed between monomer Ab′ (which contains an RNA aptamer), Bc′ (contains an siRNA) and Ca′ (contains a ribozyme). (**d**) Foot-to-foot dimers form through the palindrome sequence at the end of two Ab′ monomers, with one bearing a targeting ligand and the other a detection molecule. (**e**) Tetramers assemble by the combination of interlocking loops and palindrome mechanism of two dimers (Ab′ and Ba′). The models illustrate how the various structures are held together. Frame size for AFM images: 200 × 200 nm (**a, b**); 300 × 300 nm (**c**); 250 × 250 nm (**d**); 500 × 500 nm (**e**). Figures reproduced with permission from: **a–e**, ref. 12, © 2004 ACS and ref. 51, © 2003 ASP.

The fourth method is to assemble non-natural functional RNAs with defined 3D structures using synthetic ribozyme ligase by employing the molecular design of RNA based on the *in vitro* selection technique [16, 17]. Conformational switch of RNA nanostructures can also be constructed using a peptide-binding RNA structural motif [20].

The fifth method is the use of a palindrome sequence that differs from the sticky end, at the 5'- or 3'-end of the RNA. The molecule will spontaneously assemble through self-annealing of the palindrome sequence immediately after *in vitro* transcription or chemical synthesis, before purification [12]. This method is useful for the creation of bundles, especially for designing 3D branches. As each of the 11 nucleotides of the A-RNA generates one helical turn of 360°, the angle or the direction of RNA-fibre extension is controllable by varying the number of nucleotides in the helix containing the palindrome sequence.

APPLICATIONS OF RNA NANOTECHNOLOGY

The versatility of RNA structure, low free energy in RNA annealing, amenability in sequence, options for structure control and the property of self-assembly make RNA an ideal material for nanotechnology applications. It is possible to adapt RNA to construct ordered, patterned or preprogrammed arrays or superstructures (Figure 31.2h,i). RNA sequences can mediate the growth of hexagonal palladium nanoparticles [89]; programmable self-assembling properties of RNA ladders can direct the arrangement of cationic gold nanoparticles; and periodically spaced RNA architectures can serve as a scaffold for nanocrowns [90]. Geometrically symmetrical shapes such as dimers, trimers or polygons can be constructed from RNA [12, 13, 52] (Figure 31.4). As symmetrical shapes facilitate the formation of crystals, RNA might serve as scaffolds for X-ray crystallography. Furthermore, self-assembly interaction between interlocking loops, self-linkages through a palindrome sequence, the continued growth into a hierarchical structure and ease in conjugation and biocompatibility make RNA a good candidate for the construction of scaffolds for tissue engineering [12, 21, 23]. Several laboratories have developed RNA aptamers as biosensors [91].

RNA's new role in nanomedicine applications includes cell recognition and binding for diagnosis [92]; targeted delivery through receptor-mediated endocytosis [93]; and intracellular control and computation through gene silencing and regulation [48, 49], nuclear membrane penetration and blood–brain barrier passing [94]. The most important therapeutic RNA moieties are discussed below.

An siRNA [42, 43] helix has 20–25 nucleotides and it interferes with gene expression through the cleavage of mRNA by a protein–RNA complex called RNA-induced silencing complex (RISC). The siRNA specifically suppresses the expression of a target protein whose mRNA includes a sequence identical to the sense strand of the siRNA. This discovery led to the award of the 2006 Nobel Prize to Andrew Fire and Craig Mello [42].

A ribozyme [95, 96] is an RNA molecule that has enzymatic activity. Ribozymes have significant therapeutic potentials capable of regulating gene function by intercepting and cleaving RNA substrates, such as mRNA, or the viral genome of RNA containing a sequence complementary to the catalytic centre of the ribozyme. This discovery also led to the award of the 1989 Nobel Prize to Thomas Cech and Sydney Altman.

RNA aptamers [97, 98] are a family of oligonucleotides with functions similar to that of antibodies in their ability to recognize specific ligands (organic compounds, nucleotides, or peptides) through the formation of binding pockets [92]. Systematic evolution of ligands by exponential enrichment (SELEX) [99] is the method used to screen for the aptamers from randomized RNA pools developed *in vitro* by Ellington and Szostak [97] and by Tuerk and Gold [98]. Using this technique, various aptamers have been selected for targeting markers relevant to diseases [92, 100, 101].

Riboswitches [102] are RNA components that bind small molecules and control gene expression in response to an organism's needs. As a biological control mechanism, riboswitches can recognize metabolites, induce premature termination of mRNA transcription, block ribosomes from translating mRNAs, cleave mRNAs and even trigger mRNA destruction. Therefore, RNA switches can be re-engineered to create a new generation of controllers regulated by drug-like molecules to tune the expression levels of targeted genes *in vivo*. Such RNA-based gene-control machines hold promise in future gene therapies by supplying nanoscale *cis*-acting modulation [103, 104].

Various RNA moieties including siRNAs, ribozymes, antisense RNAs, aptamers and ribo-switches, as well as other catalytic or editing RNAs can be easily fused or conjugated into RNA nanoparticles (Figure 31.4). The advantages of RNA nanomedicine include: (1) self-assembly (see 'Techniques for constructing RNA nanoparticles' section for self-assembly and self-processing *in vivo*); (2) multivalency; (3) targeted delivery; (4) protein-free; (5) nanoscale size; (6) controlled synthesis with defined structure and stoichiometry; and (7) combining therapy and detection of therapy effects into one particle.

Bottom-up assembly of RNA can lead to multivalency [51]. Each subunit may be separately functionalized to carry different therapeutic payloads, reporters and/or targeting ligands (Figures 31.2d and 31.4a). Cell-type-specific delivery allows a lower concentration of the drug to be administered, thus reducing the side effects. The multivalent approach is similar to that of cocktail therapy, in which a mixture of drugs is used to produce a synergistic effect. The multivalency offers a further advantage in that therapy and detection of therapeutic effects may be combined into one nanoparticle conducted under a single administration [12, 50, 51].

At present, a variety of other polyvalent nanoparticles have been developed; however, producing homologous particles and consistent reproduction of copy numbers within the population is challenging. Any uncertainty in structure and stoichiometry could cause unpredictable side effects or non-specific toxicity. Using RNA nanotechnology, the production of homogeneous nanoparticles can be 'manufactured' with high reproducibility, and defined structure and stoichiometry, thus facilitating quality and safety control.

The size of RNA particles on the nanometre scale is another advantage. For effective delivery to diseased tissues, many studies suggest that particles ranging from 10 to 50 nm are optimal for a non-viral vector because they are large enough to be retained by the body yet small enough to pass through the cell membrane by means of endocytosis, mediated by the cell-surface receptors [105].

Nanoparticle delivery has the potential to improve the pharmacokinetics, pharmacodynamics, biodistribution and safety of this newly emerging modality.

The protein-free nature will avoid the induction of antibodies, thus allowing repeated administration for the treatment of chronic diseases including cancers, viral infections and genetic ailments. Moreover, RNA nanoparticles are classified by the United States Food and Drug Administration (FDA) as chemical rather than biological entities, which will speed up the FDA approval.

The feasibility of RNA nanotechnology in disease therapy has been exemplified in the phi29 pRNA therapeutic system [14, 35, 50, 51, 106, 107]. Incubation of the synthetic polyvalent RNA nanoparticles containing receptor-binding aptamers or ligands resulted in cell binding and entry of the incorporated therapeutics, subsequently modulating apoptosis [50, 51]. The delivery efficiency and therapeutic effect were later confirmed in animal trials [50, 51]. The 3D design, circular permutation, folding energy alteration and nucleotide modification of RNA were applied to generate RNase-resistant RNA nanoparticles with low toxicity and to ensure processing of the chimeric RNA complexes into siRNA by Dicer after delivery.

CHALLENGES AND PERSPECTIVES

RNA nanoparticle construction involves conjugation of functionalities, crosslinking of modules, labelling of subunits and chemical modification of nucleotides. Methods of synthesizing RNA building blocks include both chemical and enzymatic approaches. Although great progress has been made, improvements are much needed.

Prediction of RNA structure or folding for particle assembly remains a challenge. Owing to the unusual folding properties such as non-canonical base pairing, the rules that elucidate RNA folding are yet to be sorted out. At present, using the RNA 2D prediction program by Zuker, typically only 70% of the 2D folding prediction is accurate, based on experimental data [74, 75]. Clearly, predicting the RNA 3D and 4D structures is even more elusive. Computer-aided programs in RNA-structure prediction and those for computing the intermolecular interactions of RNA subunits for quaternary nanostructure formation are still to be explored.

Natural RNA is sensitive to RNase and is especially unstable in serum or in the body. This instability has long hindered its application as a construction material. Improving the stability of RNA has progressed rapidly; chemical modification of the base (e.g., 5-Br-Ura and 5-I-Ura), phosphate linkage (e.g., phosphothioate, boranophosphate), and/or the C2' (e.g., 2'-fluorine, 2'-O-methyl or 2'-amine) [108] have all been explored. Other attempts include peptide nucleic acids, locked nucleic acids and their respective derivatives polycarbamate nucleic acids [109] or locked nucleic acids with a bridge at different positions (2'–4', 1'–3') [110]. The 3'-end capping also improved the base pairing selectivity in duplex formation [111]. For all these methods, the 2'-fluorine modification is the most appraisable because it has minimal detrimental effect on RNA folding and function [112].

Loop–loop interaction is one approach to assemble quaternary RNA nanoparticles; however, dissociation of loops can occur when the concentration is reduced. Crosslinking agents, such as psoralen, nitrogen mustard derivatives and transition metal compounds [113] can promote the formation of stable RNA complexes. Recent advancements include various bifunctional agents separated by linkers and phenolic derivatives [114] to increase the efficiency of crosslinking. Long-range (> 9 Å) and short-range (1.5 Å) photoaffinity crosslinking can be achieved using azidophenacyl derivatives and thionucleosides, such as 6-thioguanosine and 4-thiouridine, respectively.

For fluorescent labelling, single conjugation of fluorophores at the 5'- or 3'-end is preferable to prevent physical hindrance. End labelling is not difficult with chemical synthesis of small RNA; however, it is challenging for long RNA requiring enzymatic methods.

To meet this challenge, guanosine monophosphate (GMP) or adenosine monophosphate (AMP) derivatives that can only be used for transcription initiation, but not for chain elongation, have been used. Fluorescent RNA can also be easily synthesized *in vitro* with T7 RNA polymerase using a new agent tCTP [115].

The challenges of *in vivo* computation using RNA [48, 49] include scaling the logic operations with a large number of inputs, extending input signal types and eliminating nonspecific actions resulting in targeting unexpected or undesired pathways.

The results of modification related to RNA folding and *in vivo* toxicity of the nucleotide derivatives remain to be explored. Owing to metabolism and biocompatibility issues, the most stable RNA might not necessarily be the most desirable; retention of particles within an appropriate time period is more attractive.

The most challenging aspect of RNA therapeutics is the yield and cost of RNA production. Commercial RNA chemical synthesis can offer only 40 (conservative) to 80 nucleotides with low yield. Acetalester 2'-OH protecting groups, such as pivaloyloxymethyl, have been reported to enhance chemical synthesis of RNA [117]. RNase ligase II has been shown to be a good alternative over the traditional T4 DNA ligase to generate longer RNA by ligation of two shorter synthetic RNA fragments [116]. In enzymatic synthesis, heterogeneity of the 3'-end has been an issue [117]; this can be addressed by extending the transcribed sequence beyond the intended end and then cleaving the RNA at the desired site using ribozymes, DNAzymes, or RNase H [116–118]. Large-scale RNA complexes produced in bacteria escorted by a tRNA vector have also been reported [40, 41]. Based on the rapid reduction of cost over the history of DNA synthesis, it is expected that the cost of RNA synthesis will gradually decrease with the development of industrial-scale RNA production technologies.

In conclusion, natural or synthetic RNA molecules can fold into predefined structures that can spontaneously assemble into nanoparticles with numerous functionalities. The field of RNA nanotechnology is emerging but will play an increasingly important role in medicine, biotechnology, synthetic biology and nanotechnology.

ACKNOWLEDGEMENTS

This review is in part inspired by the 4th Annual Cancer Nanotechnology Think Tank: RNA Nanobiology (http://web.ncifcrf.gov/events/nanobiology/2009/) and it is an extension of the author's presentation at this think tank and his opening remark at the 2010 International Conference of RNA Nanotechnology and Therapeutics (http://www.eng.uc.edu/nanomedicine/RNA2010/). The author thanks John Rossi, Peter Stockley, Andrew Ellington, Shane Fimbel, Jason Lu, Farzin Haque, Anne Vonderheide, Randall Reif, Chaoping Chen, Mathieu Cinier and Feng Xiao for insightful comments; and Chad Schwartz, Yi Shu and Jia Geng for their assistance in preparation of this manuscript. The work in the author's laboratory is supported by National Institutes of Health (NIH) grants GM059944, EB003730 and NIH Nanomedicine Development Center entitled 'Phi29 DNA Packaging Motor for Nanomedicine' (PN2 EY018230) through the NIH Roadmap for Medical Research, as well as contract from Kylin Therapeutics, Inc., of which the author is a cofounder.

ADDITIONAL INFORMATION

The author declares competing financial interests: details accompany the paper at www.nature.com/naturenanotechnology.

REFERENCES

1. Lin, C., Liu, Y. & Yan, H. Designer DNA nanoarchitectures. *Biochemistry* **48**, 1663–1674 (2009).
2. Aldaye, F. A., Palmer, A. L. & Sleiman, H. F. Assembling materials with DNA as the guide. *Science* **321**, 1795–1799 (2008).
3. Seeman, N. C. Nanomaterials based on DNA. *Annu. Rev. Biochem.* **79**, 65–87 (2010).
4. Moll, D. et al. S-layer-streptavidin fusion proteins as template for nanopatterned molecular arrays. *Proc. Natl Acad. Sci. USA* **99**, 14646–14651 (2002).
5. Cui, H., Muraoka, T., Cheetham, A. G. & Stupp, S. I. Self-assembly of giant peptide nanobelts. *Nano. Lett.* **9**, 945–951 (2009).
6. Adler-Abramovich, L. et al. Self-assembled arrays of peptide nanotubes by vapour deposition. *Nature Nanotech.* **4**, 849–854 (2009).
7. Knowles, T. P. et al. Nanostructured films from hierarchical self-assembly of amyloidogenic proteins. *Nature Nanotech.* **5**, 204–207 (2010).
8. Guo, P. et al. Inter-RNA interaction of phage phi29 pRNA to form a hexameric complex for viral DNA transportation. *Mol. Cell* **2**, 149–155 (1998).
9. Zhang, F. et al. Function of hexameric RNA in packaging of bacteriophage phi29 DNA *in vitro*. *Mol. Cell* **2**, 141–147 (1998).
10. Jaeger, L. & Leontis, N. B. Tecto-RNA: one dimensional self-assembly through tertiary interactions. *Angew. Chem. Int. Ed.* **39**, 2521–2524 (2000).
11. Jaeger, L., Westhof, E. & Leontis, N. B. TectoRNA: modular assembly units for the construction of RNA nano-objects. *Nucleic Acids Res.* **29**, 455–463 (2001).
12. Shu, D. et al. Bottom-up assembly of RNA arrays and superstructures as potential parts in nanotechnology. *Nano Lett.* **4**, 1717–1723 (2004).
13. Chworos, A. et al. Building programmable jigsaw puzzles with RNA. *Science* **306**, 2068–2072 (2004).
14. Guo, P. RNA nanotechnology: engineering, assembly and applications in detection, gene delivery and therapy. *J. Nanosci. Nanotechnol.* **5**, 1964–1982 (2005).
15. Jaeger, L. & Chworos, A. The architectonics of programmable RNA and DNA nanostructures. *Curr. Opin. Struct. Biol.* **16**, 531–543 (2006).
16. Ikawa, Y., Tsuda, K., Matsumura, S. & Inoue, T. *De novo* synthesis and development of an RNA enzyme. *Proc. Natl Acad. Sci. USA* **101**, 13750–13755 (2004).
17. Matsumura, S. et al. Coordinated control of a designed trans-acting ligase ribozyme by a loop-receptor interaction. *FEBS Lett.* **583**, 2819–2826 (2009).
18. Leontis, N. B., Lescoute, A. & Westhof, E. The building blocks and motifs of RNA architecture. *Curr. Opin. Struct. Biol.* **16**, 279–287 (2006).

19. Schroeder, K. T., McPhee, S. A., Ouellet, J. & Lilley, D. M. A structural database for k-turn motifs in RNA. *RNA* **16**, 1463–1468 (2010).

20. Li, X., Horiya, S. & Harada, K. An efficient thermally induced RNA conformational switch as a framework for the functionalization of RNA nanostructures. *J. Am. Chem. Soc.* **128**, 4035–4040 (2006).

21. Nasalean, L., Baudrey, S., Leontis, N. B. & Jaeger, L. Controlling RNA selfassembly to form filaments. *Nucleic Acids Res* **34**, 1381–1392 (2006).

22. Liu, B., Baudrey, S., Jaeger, L. & Bazan, G. C. Characterization of tectoRNA assembly with cationic conjugated polymers. *J. Am. Chem. Soc.* **126**, 4076–4077 (2004).

23. Cayrol, B. et al. A nanostructure made of a bacterial noncoding RNA. *J. Am. Chem. Soc.* **131**, 17270–17276 (2009).

24. Geary, C., Chworos, A. & Jaeger, L. Promoting RNA helical stacking via A-minor junctions. *Nucleic Acids Res.* doi:10.1093/nar/gkq748 (2010).

25. Sugimoto, N. et al. Thermodynamic parameters to predict stability of RNA/DNA hybrid duplexes. *Biochemistry* **34**, 11211–11216 (1995).

26. Searle, M. S. & Williams, D. H. On the stability of nucleic acid structures in solution: enthalpy-entropy compensations, internal rotations and reversibility. *Nucleic Acids Res.* **21**, 2051–2056 (1993).

27. Kitamura, A. et al. Analysis of intermolecular base pair formation of prohead RNA of the phage phi29 DNA packaging motor using NMR spectroscopy. *Nucleic Acids Res.* **36**, 839–848 (2008).

28. Chen, C., Zhang, C. & Guo, P. Sequence requirement for hand-in-hand interaction in formation of pRNA dimers and hexamers to gear phi29 DNA translocation motor. *RNA* **5**, 805–818 (1999).

29. Severcan, I. et al. Square-shaped RNA particles from different RNA folds. *Nano Lett.* **9**, 1270–1277 (2009).

30. Severcan, I. et al. A polyhedron made of tRNAs. *Nature Chem.* **2**, 772–779 (2010).

31. Hansma, H. G., Oroudjev, E., Baudrey, S. & Jaeger, L. TectoRNA and 'kissingloop' RNA: atomic force microscopy of self-assembling RNA structures. *J. Microsc.* **212**, 273–279 (2003).

32. Lee, R. J., Wang, S. & Low, P. S. Measurement of endosome pH following folate receptor-mediated endocytosis. *Biochim. Biophys. Acta* **1312**, 237–242 (1996).

33. Pogocki, D. & Schoneich, C. Chemical stability of nucleic acid-derived drugs. *J. Pharm. Sci.* **89**, 443–456 (2000).

34. Laurenti, E. et al. Inducible gene and shRNA expression in resident hematopoietic stem cells *in vivo*. *Stem Cells* **28**, 1390–1398 (2010).

35. Hoeprich, S. et al. Bacterial virus phi29 pRNA as a hammerhead ribozyme escort to destroy hepatitis B virus. *Gene Ther.* **10**, 1258–1267 (2003).

36. Chang, K. Y. & Tinoco, I. Jr Characterization of a "kissing" hairpin complex derived from the human immunodeficiency virus genome. *Proc. Natl Acad. Sci. USA* **91**, 8705–8709 (1994).

37. Bindewald, E. et al. RNAJunction: a database of RNA junctions and kissing loops for three-dimensional structural analysis and nanodesign. *Nucleic Acids Res.* **36**, D392–D397 (2008).

38. Wagner, C., Ehresmann, C., Ehresmann, B. & Brunel, C. Mechanism of dimerization of bicoid mRNA: initiation and stabilization. *J. Biol. Chem.* **279**, 4560–4569 (2004).

39. Chen, C., Sheng, S., Shao, Z. & Guo, P. A dimer as a building block in assembling RNA: a hexamer that gears bacterial virus phi29 DNAtranslocating machinery. *J. Biol. Chem.* **275**, 17510–17516 (2000).

40. Ponchon, L., Beauvais, G., Nonin-Lecomte, S. & Dardel, F. A generic protocol for the expression and purification of recombinant RNA in Escherichia coli using a tRNA scaffold. *Nature Protoc.* **4**, 947–959 (2009).

41. Kuwabara, T. et al. Formation of a catalytically active dimer by tRNA-driven short ribozymes. *Nature Biotechnol.* **16**, 961–965 (1998).

42. Fire, A. et al. Potent and specific genetic interference by double-stranded RNA in Caenorhabditis elegans. *Nature* **391**, 806–811 (1998).

43. Li, H., Li, W. X. & Ding, S. W. Induction and suppression of RNA silencing by an animal virus. *Science* **296**, 1319–1321 (2002).

44. Breaker, R. R. Complex riboswitches. *Science* **319**, 1795–1797 (2008).

45. Fabian, M. R., Sonenberg, N. & Filipowicz, W. Regulation of mRNA translation and stability by microRNAs. *Annu. Rev. Biochem.* **79**, 351–379 (2010).

46. Zhang, C. Novel functions for small RNA molecules. *Curr. Opin. Mol. Ther.* **11**, 641–651 (2009).

47. Marvin, M. C. & Engelke, D. R. Broadening the mission of an RNA enzyme. *J. Cell Biochem.* **108**, 1244–1251 (2009).

48. Benenson, Y. RNA-based computation in live cells. *Curr. Opin. Biotechnol.* **20**, 471–478 (2009).
49. Shlyakhtenko, L. S. et al. Silatrane-based surface chemistry for immobilization of DNA, protein-DNA complexes and other biological materials. *Ultramicroscopy* **97**, 279–287 (2003).
50. Guo, S., Tschammer, N., Mohammed, S. & Guo, P. Specific delivery of therapeutic RNAs to cancer cells via the dimerization mechanism of phi29 motor pRNA. *Hum. Gene Ther.* **16**, 1097–1109 (2005).
51. Khaled, A., Guo, S., Li, F. & Guo, P. Controllable self-assembly of nanoparticles for specific delivery of multiple therapeutic molecules to cancer cells using RNA nanotechnology. *Nano Lett.* **5**, 1797–1808 (2005).
52. Shu, D., Huang, L., Hoeprich, S. & Guo, P. Construction of phi29 DNApackaging RNA (pRNA) monomers, dimers and trimers with variable sizes and shapes as potential parts for nano-devices. *J. Nanosci. Nanotechnol.* **3**, 295–302 (2003).
53. Turner, R. & Tijan, R. Leucine repeats and an adjacent DNA binding domain mediate the formation of functional c-Fos and c-Jun heterodimers. *Science* **243**, 1689–1694 (1989).
54. Liu, D. et al. Tensegrity: construction of rigid DNA triangles with flexible fourarm DNA junctions. *J. Am. Chem. Soc.* **126**, 2324–2325 (2004).
55. Rothemund, P. W., Papadakis, N. & Winfree, E. Algorithmic self-assembly of DNA Sierpinski triangles. *PLoS. Biol.* **2**, e424 (2004).
56. Park, S. H. et al. Three-helix bundle DNA tiles self-assemble into 2D lattice or 1D templates for silver nanowires. *Nano Lett.* **5**, 693–696 (2005).
57. Weizmann, Y. et al. A polycatenated DNA scaffold for the one-step assembly of hierarchical nanostructures. *Proc. Natl Acad. Sci. USA* **105**, 5289–5294 (2008).
58. Aldaye, F. A. & Sleiman, H. F. Modular access to structurally switchable 3D discrete DNA assemblies. *J. Am. Chem. Soc.* **129**, 13376–13377 (2007).
59. Zheng, J. et al. From molecular to macroscopic via the rational design of a selfassembled 3D DNA crystal. *Nature* **461**, 74–77 (2009).
60. Rothemund, P. W. K. Folding DNA to create nanoscale shapes and patterns. *Nature* **440**, 297–302 (2006).
61. Andersen, E. S. et al. Self-assembly of a nanoscale DNA box with a controllable lid. *Nature* **459**, 73–76 (2009).
62. Ke, Y. G. et al. Self-assembled water-soluble nucleic acid probe tiles for labelfree RNA hybridization assays. *Science* **319**, 180–183 (2008).
63. Douglas, S. M., Chou, J. J. & Shih, W. M. DNA-nanotube-induced alignment of membrane proteins for NMR structure determination. *Proc. Natl Acad. Sci. USA* **104**, 6644–6648 (2007).
64. Endo, M., Seeman, N. C. & Majima, T. DNA tube structures controlled by a four-way-branched DNA connector. *Angew. Chem. Int. Ed. Engl.* **44**, 6074–6077 (2005).
65. Tanaka, K. et al. A discrete self-assembled metal array in artificial DNA. *Science* **299**, 1212–1213 (2003).
66. Yurke, B. et al. A DNA-fuelled molecular machine made of DNA. *Nature* **406**, 605–608 (2000).
67. Lin, C. et al. *In vivo* cloning of artificial DNA nanostructures. *Proc. Natl Acad. Sci. USA* **105**, 17626–17631 (2008).
68. Eckardt, L. H. et al. DNA nanotechnology: chemical copying of connectivity. *Nature* **420**, 286 (2002).
69. Endo, M. et al. Programmed-assembly system using DNA jigsaw pieces. *Chemistry* **16**, 5362–5368 (2010).
70. Dietz, H., Douglas, S. M. & Shih, W. M. Folding DNA into twisted and curved nanoscale shapes. *Science* **325**, 725–730 (2009).
71. Cherny, D. I., Eperon, I. C. & Bagshaw, C. R. Probing complexes with single fluorophores: factors contributing to dispersion of FRET in DNA/RNA duplexes. *Eur. Biophys. J.* **38**, 395–405 (2009).
72. Shapiro, B. A. Computational design strategies for RNA nanostructures. *J. Biomol. Struct. Dyn.* **26**, 820 (2009).
73. Yingling, Y. G. & Shapiro, B. A. Computational design of an RNA hexagonal nanoring and an RNA nanotube. *Nano Lett.* **7**, 2328–2334 (2007).
74. Zuker, M. Mfold web server for nucleic acid folding and hybridization prediction. *Nucleic Acids Res.* **31**, 3406–3415 (2003).
75. Markham, N. R. & Zuker, M. UNAFold: software for nucleic acid folding and hybridization. *Methods Mol. Biol.* **453**, 3–31 (2008).
76. Bindewald, E. et al. Computational strategies for the automated design of RNA nanoscale structures from building blocks using NanoTiler. *J. Mol. Graph. Model.* **27**, 299–308 (2008).

77. Afonin, K. A. et al. *In vitro* assembly of cubic RNA-based scaffolds designed *in silico*. *Nature Nanotech.* **5**, 676–682 (2010).
78. Chakraborty, S., Modi, S. & Krishnan, Y. The RNA2-PNA2 hybrid i-motif-a novel RNA-based building block. *Chem. Commun.* 70–72 (2008).
79. Afonin, K. A., Cieply, D. J. & Leontis, N. B. Specific RNA self-assembly with minimal paranemic motifs. *J. Am. Chem. Soc.* **130**, 93–102 (2008).
80. Lescoute, A. & Westhof, E. Topology of three-way junctions in folded RNAs. *RNA* **12**, 83–93 (2006).
81. Ouellet, J. et al. Structure of the three-way helical junction of the hepatitis C virus IRES element. *RNA* **16**, 1597–1609 (2010).
82. de la, P. M., Dufour, D. & Gallego, J. Three-way RNA junctions with remote tertiary contacts: a recurrent and highly versatile fold. *RNA* **15**, 1949–1964 (2009).
83. Griffiths-Jones, S. et al. Rfam: annotating non-coding RNAs in complete genomes. *Nucleic Acids Res.* **33**, D121–D124 (2005).
84. Abraham, M., Dror, O., Nussinov, R. & Wolfson, H. J. Analysis and classification of RNA tertiary structures. *RNA* **14**, 2274–2289 (2008).
85. Guo, P., Erickson, S. & Anderson, D. A small viral RNA is required for *in vitro* packaging of bacteriophage phi29 DNA. *Science* **236**, 690–694 (1987).
86. Xiao, F., Demeler, B. & Guo, P. Assembly mechanism of the sixtysubunit nanoparticles via interaction of RNA with the reengineered protein connector of phi29 DNA-packaging motor. *ACS Nano* **4**, 3293–3301 (2010).
87. Shu, D., Zhang, H., Jin, J. & Guo, P. Counting of six pRNAs of phi29 DNApackaging motor with customized single molecule dual-view system. *EMBO J.* **26**, 527–537 (2007).
88. Woodson, S. A. Compact intermediates in RNA folding. *Annu. Rev. Biophys.* **39**, 61–77 (2010).
89. Gugliotti, L. A., Feldheim, D. L. & Eaton, B. E. RNA-mediated metal-metal bond formation in the synthesis of hexagonal palladium nanoparticles. *Science* **304**, 850–852 (2004).
90. Koyfman, A. Y. et al. Controlled spacing of cationic gold nanoparticles by nanocrown RNA. *J. Am. Chem. Soc.* **127**, 11886–11887 (2005).
91. Oguro, A., Ohtsu, T. & Nakamura, Y. An aptamer-based biosensor for mammalian initiation factor eukaryotic initiation factor 4A. *Anal. Biochem.* **388**, 102–107 (2009).
92. Mi, J. et al. *In vivo* selection of tumor-targeting RNA motifs. *Nature Chem. Biol.* **6**, 22–24 (2010).
93. Liu, Y. et al. Targeting hypoxia-inducible factor-1alpha with Tf-PEI-shRNA complex via transferring receptor-mediated endocytosis inhibits melanoma growth. *Mol. Ther.* **17**, 269–277 (2009).
94. Kumar, P. et al. Transvascular delivery of small interfering RNA to the central nervous system. *Nature* **448**, 39–43 (2007).
95. Kruger, K. et al. Self-splicing RNA: autoexcision and autocyclization of the ribosomal RNA intervening sequence of Tetrahymena. *Cell* **31**, 147–157 (1982).
96. Guerrier-Takada, C. et al. The RNA moiety of ribonuclease P is the catalytic subunit of the enzyme. *Cell* **35**, 849–857 (1983).
97. Ellington, A. D. & Szostak, J. W. *In vitro* selection of RNA molecules that bind specific ligands. *Nature* **346**, 818–822 (1990).
98. Tuerk, C. & Gold, L. Systematic evolution of ligands by exponential enrichment: RNA ligands to bacteriophage T4 DNA polymerase. *Science* **249**, 505–510 (1990).
99. Ellington, A. D. Back to the future of nucleic acid self-amplification. *Nature Chem. Biol.* **5**, 200–201 (2009).
100. Zhou, J., Li, H., Zaia, J. & Rossi, J. J. Novel dual inhibitory function aptamer-siRNA delivery system for HIV-1 therapy. *Mol. Ther.* **16**, 1481–1489 (2008).
101. Bunka, D. H. et al. Production and characterization of RNA aptamers specific for amyloid fibril epitopes. *J. Biol. Chem.* **282**, 34500–34509 (2007).
102. Sudarsan, N. et al. Riboswitches in eubacteria sense the second messenger cyclic di-GMP. *Science* **321**, 411–413 (2008).
103. Ogawa, A. & Maeda, M. An artificial aptazyme-based riboswitch and its cascading system in *E. coli*. *Chembiochem* **9**, 206–209 (2008).
104. Shahbabian, K., Jamalli, A., Zig, L. & Putzer, H. RNase Y, a novel endoribonuclease, initiates riboswitch turnover in *Bacillus subtilis*. *EMBO J.* **28**, 3523–3533 (2009).
105. Prabha, S., Zhou, W. Z., Panyam, J. & Labhasetwar, V. Size-dependency of nanoparticle-mediated gene transfection: studies with fractionated nanoparticles. *Int. J. Pharm.* **244**, 105–115 (2002).

106. Guo, S., Huang, F. & Guo, P. Construction of folate-conjugated pRNA of bacteriophage phi29 DNA packaging motor for delivery of chimeric siRNA to nasopharyngeal carcinoma cells. *Gene Ther.* **13**, 814–820 (2006).

107. Zhang, H. M. et al. Target delivery of anti-coxsackievirus siRNAs using ligand-conjugated packaging RNAs. *Antivir. Res.* **83**, 307–316 (2009).

108. Watts, J. K., Deleavey, G. F. & Damha, M. J. Chemically modified siRNA: tools and applications. *Drug Discov. Today* **13**, 842–855 (2008).

109. Madhuri, V. & Kumar, V. A. Design, synthesis and DNA/RNA binding studies of nucleic acids comprising stereoregular and acyclic polycarbamate backbone: polycarbamate nucleic acids (PCNA). *Org. Biomol. Chem.* **8**, 3734–3741 (2010).

110. Mathe, C. & Perigaud, C. Recent approaches in the synthesis of conformationally restricted nucleoside analogues. *Eur. J. Org. Chem.* 1489–1505 (2008).

111. Patra, A. & Richert, C. High fidelity base pairing at the 3′-terminus. *J. Am. Chem. Soc.* **131**, 12671–12681 (2009).

112. Liu, J. et al. Fabrication of stable and RNase-resistant RNA nanoparticles active in gearing the nanomotors for viral DNA packaging. *ACS Nano* (in press).

113. Efimov, V. A., Fediunin, S. V. & Chakhmakhcheva, O. G. Cross-linked nucleic acids: formation, structure, and biological function. *Bioorg. Khim.* **36**, 56–80 (2010).

114. Song, Z. et al. Synthesis and oxidation-induced DNA cross-linking capabilities of bis(catechol) quaternary ammonium derivatives. *Chemistry* **14**, 5751–5754 (2008).

115. Stengel, G., Urban, M., Purse, B. W. & Kuchta, R. D. Incorporation of the fluorescent ribonucleotide analogue tCTP by T7 RNA polymerase. *Anal. Chem.* **82**, 1082–1089 (2010).

116. Solomatin, S. & Herschlag, D. Methods of site-specific labeling of RNA with fluorescent dyes. *Methods Enzymol.* **469**, 47–68 (2009).

117. Lavergne, T., Bertrand, J. R., Vasseur, J. J. & Debart, F. A base-labile group for 2′-OH protection of ribonucleosides: a major challenge for RNA synthesis. *Chemistry* **14**, 9135–9138 (2008).

118. Hoeprich, S. & Guo, P. Computer modeling of three-dimensional structure of DNA-packaging RNA(pRNA) monomer, dimer, and hexamer of phi29 DNA packaging motor. *J. Biol. Chem.* **277**, 20794–20803 (2002).

32 Thermodynamically Stable RNA Three-Way Junction for Constructing Multifunctional Nanoparticles for Delivery of Therapeutics[1]

Dan Shu, Yi Shu, and Farzin Haque
The Ohio State University, Columbus, Ohio, USA
Nanobio Delivery Pharmaceutical Co. Ltd., Columbus, Ohio, USA

Sherine Abdelmawla
Kylin Therapeutics, West Lafayette, Indiana, USA

Peixuan Guo
The Ohio State University, Columbus, Ohio, USA
Nanobio Delivery Pharmaceutical Co. Ltd., Columbus, Ohio, USA

CONTENTS

DOI: 10.1201/9781003001560-36

Living organisms produce a variety of highly ordered structures made of DNA, RNA and proteins to perform diverse functions. DNA has been widely used as a biomaterial (Seeman, 2010). Even though RNA has many of the attributes of DNA that make it useful as a biomaterial, such as ease of manipulation, it has received less attention (Guo *et al.*, 1998; Guo, 2010; Shukla *et al.*, 2011). RNA also permits non-canonical base pairing and offers catalytic functions similar to some proteins (Guo, 2010). Typically, RNA molecules contain a large variety of single-stranded stem-loops for inter- or intramolecular interactions (Cruz and Westhof, 2009). These loops serve as mounting dovetails, which eliminates the need for external linking dowels during fabrication and assembly (Guo *et al.*, 1998; Jaeger *et al.*, 2009). Since the discovery of siRNA (Fire *et al.*, 1998), nanoparticles of siRNA (Guo *et al.*, 2005; Khaled *et al.*, 2005; Guo *et al.*, 2006), ribozymes (Sarver *et al.*, 1990; Simpson *et al.*, 2001; Liu *et al.*, 2007), riboswitches (Winkler *et al.*, 2004; Mulhbacher *et al.*, 2010) and microRNAs (Chen *et al.*, 2010; Pegtel *et al.*, 2010; Ye *et al.*, 2011) have been explored for the treatment of cancers and viral infections.

One of the problems in the field of RNA nanotechnology is that RNA nanoparticles are relatively unstable; the lack of covalent binding or crosslinking in the particles causes dissociation at ultra-low concentrations in animal and human circulation systems after systemic injection. This has hindered the efficiency of delivery and therapeutic applications of RNA nanoparticles (Guo, 2010). Although not absolutely necessary for RNA helix formation, tens of millimoles of magnesium are required for optimum folding of nanoparticles such as phi29 pRNA (Chen and Guo, 1997; Chen *et al.*, 2000). Because the concentration of magnesium under physiological conditions is generally less than 1 mM, misfolding and dissociation of nanostructures that use RNA as a scaffold can occur at these low concentrations.

The DNA packaging motor of bacteriophage phi29 is geared by a pRNA ring (Guo *et al.*, 1987), which contains two functional domains (Reid *et al.*, 1994; Zhang *et al.*, 1994). The central domain of a pRNA subunit contains two interlocking loops, denoted as right- and left-handed loops, which can be engineered to form dimers, trimers or hexamers (Guo *et al.*, 1998; Chen *et al.*, 2000; Xiao *et al.*, 2005; Shu *et al.*, 2007). Since the two domains fold separately, replacing the helical domain with an siRNA does not affect the structure, folding or intermolecular interactions of the pRNA (Zhang *et al.*, 1995; Shu *et al.*, 2004; Khaled *et al.*, 2005). Such a pRNA/siRNA chimera has been shown to be useful for gene therapy (Hoeprich *et al.*, 2003; Guo *et al.*, 2005; Khaled *et al.*, 2005; Guo *et al.*, 2006). The two domains are connected by a three-way junction (3WJ) region (Figure 32.1c,d), and this unique structure has motivated its use in RNA nanotechnology. Here we show that the 3WJ region of pRNA can be assembled from three pieces of small RNA oligomers with high affinity. The resulting complex is stable and resistant to denaturation in the presence of 8 M urea. Incubation of three RNA oligomers, each carrying an siRNA, receptor-binding aptamer or ribozyme, resulted in trivalent RNA nanoparticles that are suitable as therapeutic agents. Of the 25 3WJ motifs obtained from different biological systems, we found the 3WJ-pRNA to be most stable.

PROPERTIES OF 3WJ-pRNA

The 3WJ domain of phi29 pRNA was constructed using three pieces of RNA oligos denoted as a_{3WJ}, b_{3WJ} and c_{3WJ} (Figure 31.1d). Two of the oligos, a_{3WJ} and c_{3WJ}, were resistant to staining by ethidium bromide (Figure 32.2a) and weakly stained by SYBR Green II; c_{3WJ} remained unstainable (Figure 32.2a). Ethidium bromide is an intercalating agent that stains double-stranded (ds) RNA and

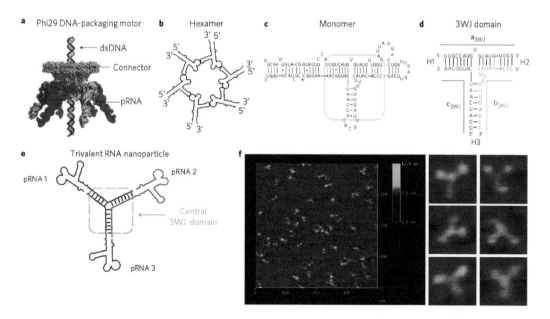

FIGURE 32.1 Sequence and secondary structure of phi29 DNA-packaging RNA. (**a**) Illustration of the phi29 packaging motor geared by six pRNAs (cyan, purple, green, pink, blue and orange structures). (**b**) Schematic showing a pRNA hexamer assembled through hand-in-hand interactions of six pRNA monomers. (**c**) Sequence of pRNA monomer Ab′ (Guo *et al.*, 1998). Green box: central 3WJ domain. In pRNA Ab′, A and b′ represent right- and left-hand loops, respectively. (**d**) 3WJ domain composed of three RNA oligomers in black, red and blue. Helical segments are represented as H1, H2, H3. (**e,f**) A trivalent RNA nanoparticle consisting of three pRNA molecules bound at the 3WJ-pRNA core sequence (black, red and blue) (**e**) and its accompanying AFM images (**f**). Ab′ indicates non-complementary loops (Chen *et al.*, 1999).

dsDNA or short-stranded (ss) RNA containing secondary structures or base stacking. SYBR Green II stains most ss- and ds-RNA or DNA. The absence of, or weak, staining indicates novel structural properties.

The mixing of the three oligos, a_{3WJ}, b_{3WJ} and c_{3WJ}, at a 1:1:1 molar ratio at room temperature in distilled water resulted in efficient formation of the 3WJ domain. Melting experiments suggest that the three components of the 3WJ-pRNA core (T_m of 58°C) had a much higher affinity to interact favorably in comparison with any of the two components (Figure 32.2b). The 3WJ domain remained stable in distilled water without dissociating at room temperature for weeks. If one of the oligos was omitted (Figure 32.2a, lanes 4–6), dimers were observed, as seen by the faster migration rates compared with the 3WJ domain (Figure 32.2a, lane 7). Generally, dsDNA and dsRNA are denatured and dissociate in the presence of 5 M (Carlson *et al.*, 1975) or 7 M urea (Pagratis, 1996). In the presence of 8 M urea, the 3WJ domain remained stable without dissociation (Figure 32.2d), thereby demonstrating its robust nature.

The lengths of helices H1, H2 and H3 were 8, 9 and 8 base pairs, respectively. RNA complexes with the deletion of two base pairs in H1 and H3 (Figures 32.1d and 32.2d) seem to have no effect on complex formation (Figure 32.2d, lanes 8, 9). However, deletion of two base pairs at H2 (Figures 32.1d and 32.2d) did not affect complex formation but made the 3WJ domain unstable in the presence of 8 M urea (Figure 32.2d, lanes 7, 10). These results demonstrate that although six base pairs are sufficient in two of the stem regions, eight bases are necessary for H2 to keep the junction domain stable under strongly denaturing conditions.

To further evaluate the chemical and thermodynamic properties of 3WJ-pRNA, the same sequences were used to construct a DNA 3WJ domain. In native gel, when the three DNA oligos

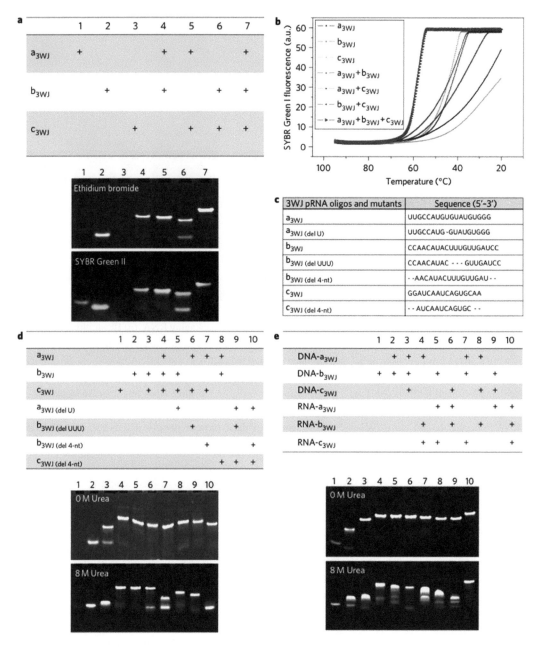

FIGURE 32.2 Assembly and stability studies of 3WJ-pRNA. In the tables, '+' indicates the presence of the strand in samples of the corresponding lanes. (**a**) 15% native PAGE showing the assembly of the 3WJ core, stained by ethidium bromide (upper) and SYBR Green II (lower). (**b**) Tm melting curves for the assembly of the 3WJ core. Melting curves for the individual strands (brown, green, silver), the two-strand combinations (blue, cyan, pink) and the three-strand combination (red) are shown. (**c**) Oligo sequences of 3WJ-pRNA cores and mutants. 'del U', deletion of U bulge; 'del UUU', deletion of UUU bulge; 'del 4-nt', deletion of two nucleotides at the 3' and 5' ends, respectively. (**d**) Length requirements for the assembly of 3WJ cores and stability assays by urea denaturation. (**e**) Comparison of DNA and RNA 3WJ core in native and urea gel.

are mixed in a 1:1:1 molar ratio, the 3WJ-DNA assembled (Figure 32.2e). However, the DNA 3WJ complex dissociated in the presence of 8 M urea (Figure 32.2e, bottom). DNA–RNA hybrid 3WJ domains exhibited increasing stability as more RNA strands were incorporated. In essence, by controlling the ratio of DNA to RNA in the 3WJ domain region, the stability can be tuned accordingly.

To assess the stability of 3WJ-pRNA, we conducted competition experiments in the presence of urea and at different temperatures as a function of time. For a candidate therapeutic RNA nanoparticle, it is necessary to evaluate whether it would dissociate at a physiological temperature of 37°C. A fixed concentration of the Cy3-labeled 3WJ-pRNA core was incubated with unlabeled b_{3WJ} at 25, 37 and 55°C. At 25°C, there is no exchange of labeled and unlabeled b_{3WJ} (Figure 32.3a). At a physiological temperature of 37°C, only a very small amount of exchange is observed in the presence of a 1,000-fold higher concentration of labeled b_{3WJ} (Figure 32.3a). At 55°C (close to the T_m of 3WJ-pRNA), there is approximately half-and-half exchange at a 10-fold excess concentration and near-complete exchange at a 1,000-fold higher concentration of labeled b_{3WJ} (Figure 32.3a). These results are consistent with the T_m measurements.

A fixed concentration of the Cy3-labelled 3WJ-pRNA core was incubated with unlabeled b_{3WJ} at room temperature in the presence of 0–6 M urea. At equimolar concentrations (Cy3-$[ab^*c]_{3WJ}$:unlabelled b_{3WJ} = 1:1), there was little or no exchange under all the urea conditions investigated (Figure 32.3b). At a fivefold higher concentration (Cy3-$[ab^*c]_{3WJ}$:unlabelled b_{3WJ} = 1:5),

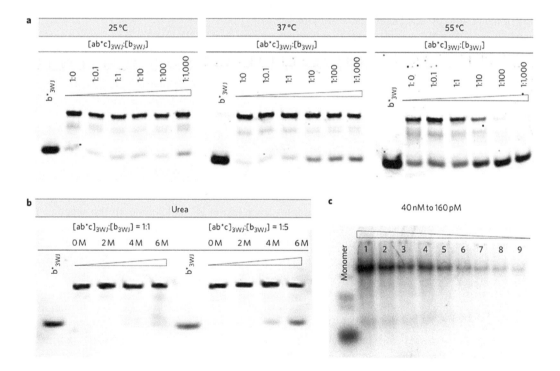

FIGURE 32.3 Competition and dissociation assays of 3WJ-pRNA. (**a**) Temperature effects on the stability of the 3WJ-pRNA core, denoted as $[ab^*c]_{3WJ}$, evaluated by 16% native gel. A fixed concentration of Cy3-labeled $[ab^*c]_{3WJ}$ was incubated with varying concentrations of unlabeled b_{3WJ} at 25, 37 and 55°C. (**b**) Urea denaturing effects on the stability of $[ab^*c]_{3WJ}$ evaluated by 16% native gel. A fixed concentration of labeled $[ab^*c]_{3WJ}$ was incubated with unlabeled b_{3WJ} at ratios of 1:1 and 1:5 in the presence of 0–6 M urea at 25°C. (**c**) Dissociation assay for the $[^{32}P]$-3WJ-pRNA complex harboring three monomeric pRNAs by twofold serial dilution (lanes 1–9). The monomer unit is shown on the left.

there was little or no exchange under 2 and 4 M urea conditions, and ~20% exchange at 6 M urea (Figure 32.3b). Hence, 6 M urea 'destabilizes' the 3WJ-pRNA complex to only an insignificant extent.

PROPERTIES OF 3WJ-pRNA WITH THERAPEUTIC MODULES

It has previously been demonstrated that the extension of phi29 pRNA at the 3'-end does not affect the folding of the pRNA global structure (Zhang *et al.*, 1995; Shu *et al.*, 2004). Sequences of each of the three RNA oligos, a_{3WJ}, b_{3WJ} and c_{3WJ}, were placed at the 3'-end of the pRNA monomer Ab'. Mixing the three resulting pRNA chimeras containing a_{3WJ}, b_{3WJ} and c_{3WJ} sequences, respectively, at equimolar concentrations led to the assembly of 3WJ branched nanoparticles harboring one pRNA at each branch. Atomic force microscopy (AFM) images strongly confirmed the formation of larger RNA complexes with three branches (Figure 32.1e,f), which were consistent with gel shift assays. This nanoparticle can also be co-transcribed and assembled in one step during transcription with high yield (data not shown).

When RNA nanoparticles are delivered systemically to the body, these particles are subjected to dilution at a very low concentration due to the circulating blood. Only those RNA particles that are intact at low concentrations can be considered as therapeutic agents for systemic delivery. To determine whether the larger structure with three branches harboring multi-module functionalities is dissociated at low concentration, this [^{32}P]-labeled complex was serially diluted to extremely low concentrations: the concentration for dissociation was below the detection limit of [^{32}P]-labeling technology. Even at 160 pM in TMS buffer, which was the lowest concentration tested, the dissociation of nanoparticles was undetectable (Figure 32.3c).

Multi-module RNA nanoparticles were constructed using this 3WJ-pRNA domain as a scaffold (Figure 32.4a). Each branch of the 3WJ carried one RNA module with defined functionality, such as a cell-receptor-binding ligand, aptamer, siRNA or ribozyme. The presence of modules or therapeutic moieties did not interfere with the formation of the 3WJ domain, as demonstrated by AFM imaging (Figure 32.4c). Furthermore, the chemically modified (2'-F U/C) 3WJ-pRNA therapeutic complex was resistant to degradation in cell culture medium with 10% serum even after 36 h of incubation, whereas the unmodified RNA degraded within 10 min (Figure 32.5).

IN VITRO AND *IN VIVO* ASSESSMENTS OF MULTI-MODULE 3WJ-pRNA

Making fusion complexes of DNA or RNA is not hard to achieve but ensuring the appropriate folding of individual modules within the complex after fusion is challenging. To test whether the incorporated RNA moieties retain their original folding and functionality after being fused and incorporated, hepatitis B virus (HBV)-cleaving ribozyme (Hoeprich *et al.*, 2003) and MG (malachite green dye, triphenylmethane)-binding aptamer (Baugh *et al.*, 2000) were used as model systems for structure and function verification. Free MG is not fluorescent by itself but emits fluorescent light after binding to the aptamer.

HBV ribozyme was able to cleave its RNA substrate after being incorporated into the nanoparticles (Figure 32.4d), and fused MG-binding aptamer retained its capacity to bind MG, as demonstrated by its fluorescence emission (Figure 32.4f). The activity results are comparable to optimized positive controls, thus confirming that individual RNA modules fused into the nanoparticles retaining their original folding after incorporation into the RNA nanoparticles.

Several cancer cell lines, especially of epithelial origin, overexpress the folate receptor on the surface by a factor of 1,000. Folate has been used extensively as a cancer cell delivery agent through folate-receptor-mediated endocytosis (Lu and Low, 2002). The 2'-F U/C-modified fluorescent 3WJ-pRNA nanoparticles with folate conjugated into one of their branches were tested for cell-binding efficiency. One fragment of the 3WJ-pRNA core was labeled with folic acid for targeted delivery (Guo *et al.*, 2006); the second fragment was labeled with Cy3; and the third fragment was fused to siRNA that could silence the gene of the anti-apoptotic factor, Survivin (Ambrosini *et al.*, 1997).

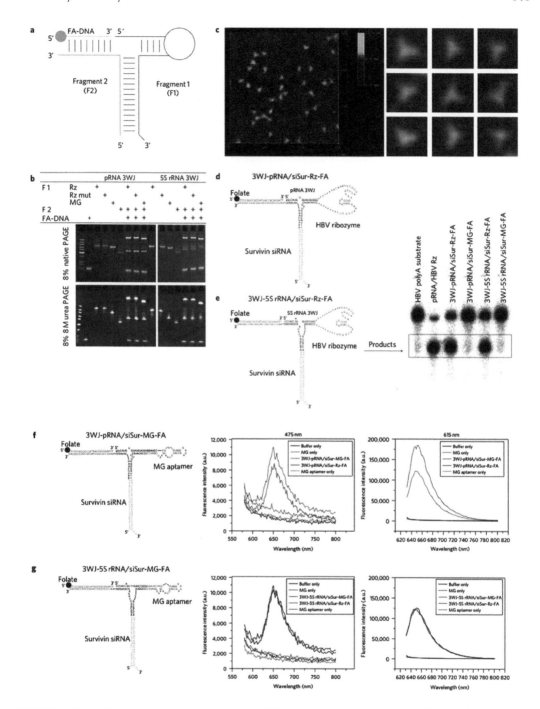

FIGURE 32.4 Construction of multi-module RNA nanoparticles harboring siRNA, ribozyme and aptamer. (**a–c**) Assembly of RNA nanoparticles with functionalities using 3WJ-pRNA and 3WJ-5S rRNA as scaffolds. (**a–c**) Illustration (**a**), 8% native (upper) and denaturing (lower) PAGE gel (**b**) and AFM images (**c**) of 3WJ-pRNA-siSur-Rz-FA nanoparticles. (**d,e**) Assessing the catalytic activity of the HBV ribozyme incorporated into the 3WJ-pRNA (**d**) and 3WJ-5S rRNA (**e**) cores, evaluated in 10% 8 M urea PAGE. The cleaved RNA product is boxed. Positive control: pRNA/HBV-Rz; negative control: 3WJ-RNA/siSur-MG-FA. (**f,g**) Functional assay of the MG aptamer incorporated in RNA nanoparticles using the 3WJ-pRNA (**f**) and 3WJ-5S rRNA (**g**) cores. MG fluorescence was measured using excitation wavelengths of 475 and 615 nm.

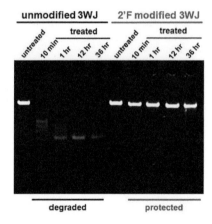

FIGURE 32.5 RNase and serum stability assay of trivalent RNA nanoparticles 3WJ-3pRNA (unmodified and 2′-F modified). RNA nanoparticles were incubated in RPMI-1640 medium containing 10% fetal bovine serum (Sigma). 200 ng of RNA samples were taken at 10 min, 1 h, 12 h and 36 h time points after incubation at 37°C, followed by analysis using 8% native PAGE gel.

Negative controls included RNA nanoparticles that contained folate but a scrambled siRNA sequence, and a 3WJ-pRNA core with active siRNA but without folate. Flow cytometry data showed that the folate-3WJ-pRNA nanoparticles bound to the cell with almost 100% binding efficiency (Figures 32.6a and 32.7). Confocal imaging indicated strong binding of the RNA nanoparticles and efficient entry into targeted cells, as demonstrated by the excellent co-localization and overlap of fluorescent 3WJ-pRNA nanoparticles (red) and the cytoplasm (green) (Figure 32.6b).

Two 3WJ-RNA nanoparticles were constructed for assaying the gene-silencing effect. Particle [3WJ-pRNA-siSur-Rz-FA] harbors folate and Survivin siRNA, and particle [3WJ-pRNA-siScram-Rz-FA] harbors folate and Survivin siRNA scramble as control. After 48 h post-transfection, both quantitative reverse transcription-polymerase chain reaction (qRT–PCR) and Western Blot assays confirmed a reduced Survivin gene expression level for 3WJ-pRNA-siSur-Rz-FA compared with the scramble control on both messenger RNA and protein levels. The silencing potency is comparable to the positive Survivin siRNA-only control, although the reduction of both the RNA complexes was modest (Figure 32.6c).

Two key factors that may affect the pharmacokinetic profile are metabolic stability and renal filtration. It has been reported that regular siRNA molecules have extremely poor pharmacokinetic properties because they have a short half-life ($T_{1/2}$) and fast kidney clearance as a result of metabolic instability and small size (<10 nm) (Soutschek et al., 2004). The pharmacokinetic profile of Alex Fluor 647-2′-F-pRNA nanoparticles that use the 3WJ domain as a scaffold was studied in mice by systemic administration via a single tail vein intravenous injection, followed by blood collection (Carey and Uhlenbeck, 1983). The concentration of the fluorescent nanoparticle in serum was determined. The $T_{1/2}$ of the pRNA nanoparticles was determined to be 6.5–12.6 h, compared with control 2′-F-modified siRNA, which could not be detected beyond 5 min post-injection, which is close to the $T_{1/2}$ of 35 min reported in the literature (Behlke, 2006).

To confirm that RNA nanoparticles were not dissociated into individual subunits in vivo, these nanoparticles were constructed by a bipartite approach (Abdelmawla et al., 2011; Shu et al., 2011) with one subunit carrying the folate to serve as a ligand for binding to the cancer cells, and the other subunit carrying a fluorescent dye. The nanoparticles were systemically injected into mice through the tail vein (Abdelmawla et al., 2011). Whole-body imaging showed that fluorescence was located specifically at the xenographic cancer expressing the folate receptor and was not detected in other organs of the body (Figure 32.6e), indicating that the particles did not dissociate in vivo after systemic delivery.

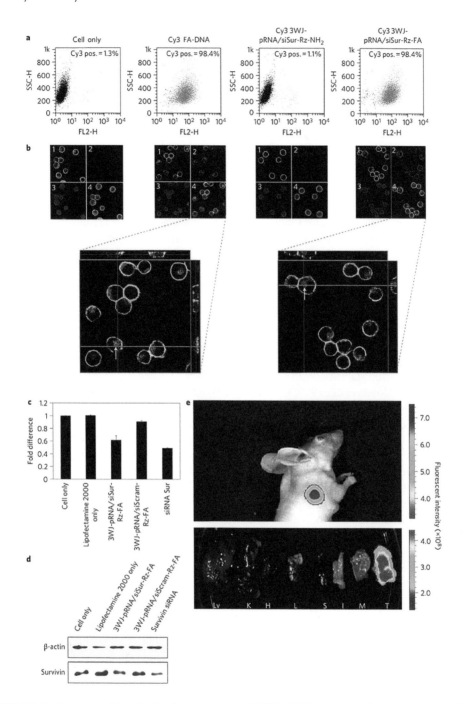

FIGURE 32.6 *In vitro* and *in vivo* binding and entry of 3WJ-pRNA nanoparticles into targeted cells. (a) Flow cytometry revealed the binding and specific entry of fluorescent-[3WJ-pRNA-siSur-Rz-FA] nanoparticles into folate-receptor-positive (FA+) cells. Positive and negative controls were Cy3-FA-DNA and Cy3-[3WJ-pRNA-siSur-Rz-NH2] (without FA), respectively. (b)Confocal images showed targeting of FA+-KB cells by co-localization (overlap, 4) of cytoplasm (green, 1) and RNA nanoparticles (red, 2) (magnified, bottom panel). Blue–nuclei, 3. (c,d) Target gene knock-down effects shown by (c) qRT–PCR with GADPH as endogenous control and by (d) Western Blot assay with b-actin as endogenous control. (e) 3WJ-pRNA nanoparticles target FA+ tumor xenografts on systemic administration in nude mice. Upper panel: whole body; lower panel: organ imaging (Lv, liver; K, kidney; H, heart; L, lung; S, spleen; I, intestine; M, muscle; T, tumor).

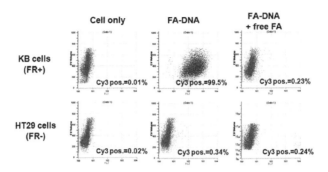

FIGURE 32.7   Folate-labeled particle specifically target folate receptor positive cell lines. The KB cell line (FR+) is folate receptor positive, and HT29 cell line (FR-) is the folate receptor negative control. Free folate served as the competitor in this assay. Red: Cy3- negative cells; Purple: Cy3-positive cells.

COMPARING 3WJ-pRNA WITH OTHER BIOLOGICAL 3WJ MOTIFS

There are many 3WJ motifs in biological RNA, some of which are stabilized by extensive tertiary interactions and non-canonical base pairings and base stacking (Chen *et al.*, 1999; Lilley, 2000; Leontis and Westhof, 2003; Lescoute and Westhof, 2006; de la Pena *et al.*, 2009; Afonin *et al.*, 2010). To assess whether the properties of the 3WJ-pRNA core are unique, we thoroughly investigated the assembly and stability of 25 3WJ motifs (Tables 32.1 and 32.2) reported in the literature (Chen *et al.*, 1999; Honda *et al.*, 1999; Lescoute and Westhof, 2006; Wakeman *et al.*, 2009; Kulshina *et al.*, 2010). Of the 25 motifs, 14 were impractical to study using core sequences; for example, some were too short (less than 10 nt for one of their fragments) for chemical synthesis. Using synthesized RNA fragments with the exact sequences as reported, with appropriate controls, the other 11 motifs were thoroughly investigated. Only 6 of the 11 structures were able to assemble into a 3WJ complex, based on gel shift assays (Table 32.1 and Figure 32.8). However, in the presence of 8 M urea, only the 3WJ-pRNA core and the 3WJ-5S ribosomal RNA core were stable. The Alu SRP appears to have assembled; however, with appropriate controls (Figure 32.9), it was found that the band was from the strong folding of one individual RNA fragment (a_{3WJ}) by itself, rather than the assembly of a 3WJ.

Moreover, 25 different RNA nanoparticles were constructed using each of the central 3WJ motifs as the scaffold to test their potential for constructing RNA nanoparticles harboring three functionalities with extended sequences (Figure 32.10). Here, we used individual phi29 pRNA subunits as modules (Zhang *et al.*, 1994; Guo *et al.*, 1998; Shu *et al.*, 2004, 2007; Xiao *et al.*, 2005). The sequences for each of the three oligos comprising individual 3WJ were placed at the 3′-end of the 117-nt pRNA, thereby serving as sticky ends. On co-transcription (of three pRNA strands harboring the sticky end sequences of 3WJ, respectively), 10 of the 25 constructs were able to assemble into a 3WJ complex through the sticky ends representing the three fragments of 3WJ, as demonstrated by gel shift assays (Figure 32.10). However, only two of the constructs (3WJ-pRNA and 3WJ-5S rRNA) were resistant to 8 M urea denaturation, which is consistent with RNA oligo assembly data (Figure 32.8). These results suggest that only 3WJ-5S rRNA is comparable to 3WJ-pRNA, and therefore 3WJ-5S rRNA can also serve as a platform to organize RNA modules bearing different functionalities.

To test whether the functionalities incorporated in the nanoparticles with the 3WJ core display catalytic or binding function, HBV ribozyme and MG aptamer were incorporated into RNA nanocomplexes. Both HBV ribozyme (Figure 32.4d,e) and MG aptamer were functional after being incorporated into 3WJ-pRNA or 3WJ-5S rRNA, respectively (Figure 32.4f,g), suggesting that

TABLE 32.1
Comparison of Biophysical Properties of Various 3WJ Cores

Family	Name	Sequence 5'→3'	Assembly of 3WJ-RNA Core		Assembly of 3WJ-pRNA with three pRNA monomers		T_m(°C)
			Native Gel	8M Urea Denaturing Gel	Native Gel	8M Urea Denaturing Gel	TMS Buffer
A	16s H34-H35-H38	a) GGG GAC GAC GUC b) CGA GCG CAA CCC CC c) GUC GUC AGC UCG	Weak	No	Yes	No	45.3 ± 6.7
	23s H75-H76-H79	a) GAG GAC ACC GA b) GGC UCU CAC UC c) UCG CUG AGC C	No	No	No	No	33.3 ± 0.6
B	23s H83-H84-H85	a) AGC AAA AGA U b) CCC GGC GAA GAG UG c) AUC UCA GCC GGG	No	No	No	No	53.7 ± 0.6
C	5S rRNA	a) CCC GGU UCG CCG CCA b) CCC ACC AGC GUU CCG GG c) AGG CGG CCA UAG CGG UGG G	Very Strong	Very Strong	Yes	Yes	54.3 ± 3.1

(Continued)

TABLE 32.1 (CONTINUED)

Family	Name	Sequence 5′→3′	Assembly of 3WJ-RNA Core		Assembly of 3WJ-pRNA with three pRNA monomers		T_m(°C)
			Native Gel	8M Urea Denaturing Gel	Native Gel	8M Urea Denaturing Gel	TMS Buffer
	G-Riboswitch (Type I)	a) GGA CAU AUA AUC GCG UG b) AUG UCC GAC UAU GUC C c) CAC GCA AGU UUC UAC CGG GCA	Medium	No	Yes	No	46.0 ± 3.5
	TPP Riboswitch (Type II)	a) GCG ACU CGG GGU GCC CUU C b) GAA GGC UGA GAA AUA CCC GUA UCA CCU GAU CUG G c) CCA GCG UAG GGA AGU CGC	Strong	No	Yes	No	52.0 ± 4.4
	M-box Riboswitch (Type II)	a) GAC GCC AAU GGG UCA ACA GAA AUC AUC G b) AGG UGA UUU UUA AUG CAG CU c) ACG CUG CUG CCC AAA AAU GUC	Strong	No	Yes	No	45.3 ± 5.5
	Hammerhead ribozyme	a) CUG UCA CCG GAU b) GGA CGA AAC AG c) UUC CGG UCU GAU GAG UCC	No	No	No	No	49.7 ± 1.5
	Alu SRP	a) GGG CCG GGC GCG GU b) UCG GGA GGC UC c) GGC GCG CGC CUG UAG UCC CAG C	No	No	No	No	45.3 ± 4.6
Unknown	HCV	a) UCA UGG UGU UCC GGA AAG CGC b) GUG AUG AGC CGA UCG UCA GA c) UCU GGU GAU ACC GAG A	No	No	No	No	49.7 ± 1.5
	pRNA	a) UUG CCA UGU GUA UGU GGG b) CCC ACA UAC UUU GUU GAU CC c) GGA UCA AUC AUG GCA A	Very Strong	Very Strong	Yes	Yes	58.0 ± 0.5

Note: The sequences of the 3WJ cores were obtained from References (Chen *et al.*, 1999; Honda *et al.*, 1999; Lescoute and Westhof, 2006; Wakeman *et al.*, 2009; Kulshina *et al.*, 2010). Families A, B and C are based on Lescoute and Westhof classification (Lescoute and Westhof, 2006).

TABLE 32.2

Comparison of Assembly and Stability of the Other 14 3WJ Cores That Was Not Practical for Thorough Investigations

Family	Name	Sequence 5'→3'	Assembly of 3WJ-pRNA harboring three pRNA monomers	
			Native Gel	8M Urea Denaturing Gel
A	16s H22-H23-H23a	a) GGA ACG CCG AUG GCG b) GAG AGG GUG GUG GAA U c) GGC AGC CAC CUG GU	No	No
	16s H25-H25-H26a	a) CGC GUU AAG CGC b) GGG CCG AAG CU c) GCG CUA GGU CU	No	No
	23s H48-Hx-H60	a) GCC UAA UGG AU b) AAG CCA AGG C c) GUC CAU GGC GG	No	No
	23s H99-H100-H101	a) GAC GCG GUC GAU AGA CU b) UCC CGC GUA C c) AGC ACU AAC AGA	No	No
B	16s H33-H33a-H33b	a) AGG GAA CCC GG GU b) GGG AGC CCU c) GCC UGG GGU GCC C	Yes	No
	23s H33-H34-H35	a) UGU GUA GGG GUG b) GAC GAU CUA CGC A c) GGC CCA UCG AGU C	No	No
	16s H28-H29-H43	a) AUC GCU AGU AAU b) GUG AAU ACG UUC CCG GG c) CCC GCA CAA GCG GU	No	No
	16S H32-H33-H34	a) UCA GCA UGG CCC UUA CGG CCU GGG C b) CAC AGG UGC UGC AUG G c) UUA CCA GGC CUU GAC AUG	No	No
C	RNase P B-type	a) AGG GCA GGA b) GGU AAA CCC CU c) UCC UUG AAA GUG CC	No	No
	L11-rRNA	a) GCC AGG AUG UAG GCU b) AGC UCA CUG GU c) GCA GCC AUC AUU UAA AGA AAG CG	No	No
Unknown	M2/NF	a) UAG UAU GGC ACA UGA UUG GG b) CCC ACA UGU CAC GGG G c) CCC UCU UAC UA	Yes	No
	SF5	a) UAA UGU AUG UGU GUC GG b) CCG ACA GCA GGG GAG c) CUC UUG CAU UA	Yes	No

(Continued)

TABLE 32.2 (CONTINUED)

Comparison of Assembly and Stability of the Other 14 3WJ Cores That Was Not Practical for Thorough Investigations

B103	a) UAGUAUGGUGCGUGAUUGGG		Yes	No
	b) CCCACACGCCACGGGG			
	c) CCCUCUUACUA			
GA1	a) AUA UAU GGC UGU GCA ACG G		No	No
	b) CCG UUG ACA GGU UGU UGC			
	c) GCA AUA CUA UAU AU			

Note: The sequences of the 3WJ cores were obtained from References (Chen *et al.*, 1999; Honda *et al.*, 1999; Lescoute and Westhof, 2006; Wakeman *et al.*, 2009; Kulshina *et al.*, 2010). Families A, B and C are based on Lescoute and Westhof classification (Lescoute and Westhof, 2006).

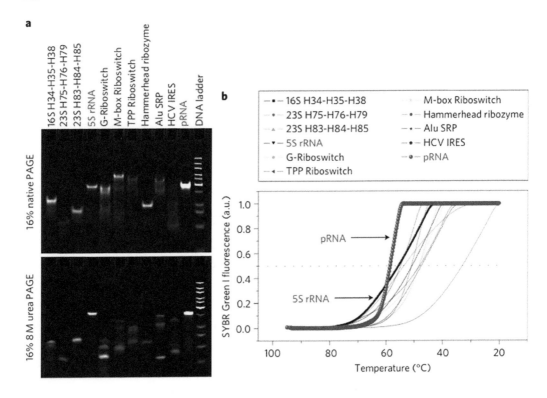

FIGURE 32.8 Comparison of different 3WJ-RNA cores. (**a**) Assembly and stability of 11 3WJ-RNA core motifs assayed in 16% native (upper) and 16% 8 M urea (lower) PAGE gel. (**b**) Melting curves for each of the 11 RNA 3WJ core motifs assembled from three oligos for each 3WJ motif under physiological buffer TMS. Refer to Table 32.1 for the respective Tm values.

3WJ-5S rRNA is comparable to 3WJ-pRNA for constructing complexes harboring different RNA functionalities for cell delivery.

One of the most important parameters for evaluating therapeutic RNA nanoparticles is their thermodynamic stability under physiological conditions *in vivo*. T_m studies on three oligos for each of the 11 3WJ motifs were conducted in physiological buffer containing 5 mM MgCl$_2$ and 100 mM NaCl at pH 7.6 (Figure 32.8 and Table 32.1). Among the assembled 3WJ structures, pRNA showed the highest T_m (58°C). The T_m closest to 3WJ-pRNA was that of 3WJ-5S rRNA (54.3°C).

The affinity and efficiency of assembly were further investigated by both gel retardation assay and melting experiments (data shown only for 3WJ-pRNA (Figure 32.2a,b), 3WJ-5S rRNA and 3WJ-Alu SRP cores) (Figure 32.9). 3WJ-pRNA displayed a very smooth high-slope temperature-dependent melting curve and clean bands in the gel, clearly indicating the assembly of monomer, dimer and 3WJ with little or no residual RNA fragments (Figure 32.2a,b). The results suggest that the three components of 3WJ-pRNA have a much higher affinity to interact favorably in comparison with any of the two components. Furthermore, the sharp melting transition indicates cooperative simultaneous folding of the three helical stems. In contrast, 3WJ-5S rRNA and all the other 3WJ motifs display temperature-dependent T_m curve with lower slopes (Figure 32.8b). Titration of the three oligos of the 3WJ-5S rRNA system showed that mixing of only two of the three RNA fragments resulted in the formation of a urea-sensitive band with a migration rate even slower than that of the entire 3WJ complex (Figure 32.9). This suggests that the individual fragment or the two-fragment combination of 3WJ-5S rRNA might have undesired binding affinities that interfere with the final 3WJ assembly. Nevertheless, the affinity of the three-component interaction is sufficiently higher than that of the two-component interaction in the 3WJ-5S rRNA system to drive the assembly of the final 3WJ structure. For Alu SRP, the folding of individual strands significantly interferes with the formation of 3WJ, and hence the complex does not assemble (Figure 32.9).

Although systematic comparison of the assembly and stability of different 3WJs has not been reported, there are several limited studies on T_m measurements of individual 3WJ motifs (Rettberg et al., 1999; Klostermeier and Millar, 2000; Diamond et al., 2001; Mathews and Turner, 2002; Liu et al., 2011). Some studies explain the thermodynamic factors that govern the folding of 3WJ RNA motifs, such as hairpin ribozyme (Klostermeier and Millar, 2000) and intact stem–loop messenger RNA (Rettberg et al., 1999). Thermodynamic parameters of a variety of constructs (mutations and insertions) based on the structure and sequence of the 3WJ core of 5S-rRNA have been reported (Diamond et al., 2001; Mathews and Turner, 2002; Liu et al., 2011), but they have only used a two-strand system instead of the three-fragment approach, and hence the results are not directly comparable. Nevertheless, the results are consistent with our findings.

In conclusion, these results suggest that the phi29 3WJ domain has the potential to serve as a platform for the construction of RNA nanoparticles containing multiple functionalities for the delivery of therapeutics to specific cells for the treatment of cancer, viral infection and genetic diseases. We thoroughly evaluated 25 3WJ motifs in biological RNA and identified 3WJ-5S rRNA as the only 3WJ motif comparable to 3WJ-pRNA for constructing complexes harboring different functionalities. Nevertheless, we found that 3WJ-pRNA is the most stable nanoparticle with the sharpest slope in the T_m curve (Figure 32.8b).

METHODS

SYNTHESIS AND PURIFICATION OF PRNA

The pRNAs were synthesized by enzymatic methods as described previously (Zhang et al., 1994). RNA oligos were synthesized chemically by IDT. 2′-deoxy-2′-Fluoro (2′-F)-modified RNAs were synthesized by in vitro transcription with the mutant Y639F T7 RNA polymerase (Sousa and Padilla, 1995) using the 2-F modified dCTP and dUTP (Liu et al., 2010) (TriLink).

CONSTRUCTION AND PURIFICATION OF PRNA COMPLEXES

Dimers were constructed by mixing equal molar ratios of pRNA Ab′ and Ba′ in presence of 5 mM magnesium. Similarly, trimers were assembled by mixing pRNA Ab′, Bc′ and Ca′. The 3WJ domain was constructed using three pieces of RNA oligos, denoted as $a3_{WJ}$, $b3_{WJ}$ and $c3_{WJ}$ that were mixed at 1:1:1 molar ratio in DEPC treated water or TMS buffer (89 mM Tris, 5 mM $MgCl_2$, pH 7.6) (Figure 32.1d).

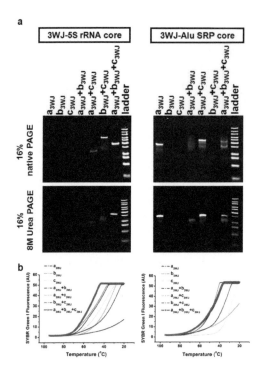

FIGURE 32.9 Comparison of 3WJ-5S rRNA and 3WJ-Alu SRP cores. (**a**) Assembly and urea denaturation assay of the 3WJ-5S rRNA and 3WJ-Alu SRP cores using three pieces of small RNA oligos (denoted a_{3WJ}, b_{3WJ} and c_{3WJ}) assayed in 16% native (upper) and 16% 8 M urea (lower) PAGE gel. (**b**) Melting curves showing the assembly of the 3WJ-5S rRNA and 3WJ-Alu SRP cores using the three pieces of RNA oligos under the physiological buffer TMS.

To purify the complex, the band corresponding to the dimer, trimer and 3WJ domain was excised from 8 to 15% native PAGE gel, running in TBM (89 mM Tris, 200 mM Boric Acid, 5 mM $MgCl_2$, pH 7.6) buffer, eluted in 0.5 M NH_4OAc, 0.1 mM EDTA, 0.1% SDS and 5 mM $MgCl_2$ for ~4 h at 37°C, followed by ethanol precipitation overnight. The dried pellet was then rehydrated in DEPC treated water or TMS buffer.

The formation of the complexes was then analyzed by 8–15% native PAGE or 8M urea PAGE gel in TBM running buffer, as specified in the Results Section. After running at 4°C for 3 h, the RNA was visualized by ethidium bromide or SBYR Green II staining.

CONSTRUCTION OF MULTI-MODULE RNA NANOPARTICLES

Sequences for each of the RNA strands, a_{3WJ}, b_{3WJ} and c_{3WJ}, were added to the 3'-end of each 117-nt pRNA-Ab' (Figure 32.1e). pRNA-a_{3WJ}, pRNA-b_{3WJ} and pRNA-c_{3WJ} were then synthesized *in vitro* by transcription of the corresponding DNA template by T7 RNA polymerase. The 3WJ-pRNA harboring three monomeric pRNAs was then self-assembled by mixing the three subunits in equal molar concentrations. Alternatively, the three individual templates can be co-transcribed and assembled in one step followed by purification in 8% native polyacrylamide gel electrophoresis (PAGE).

Sequences for siRNA, HBV ribozyme, MG-binding aptamer and folate-labeled RNA were rationally designed with sequences of the strands a_{3WJ}, b_{3WJ} and c_{3WJ}, respectively (Figure 32.4 and Table 32.3). Multi-module 3WJ-pRNA-HBV ribozyme-Survivin siRNA-folate (3WJ-pRNA-siSur-Rz-FA) or 3WJ-pRNA-MG aptamer-Survivin siRNA-folate (3WJ-pRNA-siSur-MG-FA) was assembled from four individual fragments, including a 26-nt folate-labeled RNA (TriLink) or

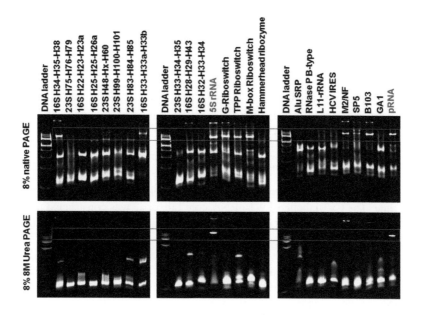

FIGURE 32.10 Assembly and stability of 25 3WJ-3pRNA constructs. Three individual phi29 pRNAs subunits with three protruding ends were used as modules. The sequences of each of the three RNA oligos of the respective 3WJs were placed at the 3′-end of the pRNA monomer Ba′, and the complex was assembled into a three-way branched nanoparticles harboring one pRNA at each arm, as demonstrated by 8% native (upper) and 8 M urea (lower) PAGE gel. The bands corresponding to the assembled 3WJ structures are boxed in red.

TABLE 32.3
Sequences Used for Constructing the RNA Nanoparticles with Functionalities

Fragment Name		Sequence (5′→3′)
Folate-DNA oligo		folate-CTCCCGGCCGCCATGGCCGCGGGATT
siRNA antisense strand		siRNA: UGACAGAUAAGGAACCUGCUU
		Scramble: AUAGUGGGACCAAUCAAGCUU
3WJ-pRNA	Fragment-1 (F1)	GGCCAUGUGUAUGUGGGAAAAAAAAACAAAUUCUUUACUGAU
siSur-FA-Rz-FA		GAGUCCGUGAGGACGAAACGGGUCAAAAAAAACCCAC
		AUACUUUGUUGAUCCAAUGACAGAUAAGGAACCUGCUUU
	Fragment-2 (F2)	GGCAGGUUCCUUAUCUGUCAAAGGAUCAAUCAUG
		GCCAAUCCCGCGGCCAUGGCGGCCGGGAG
3WJ-pRNA	Fragment-1 (F1)	GGCCAUGUGUAUGUGGGGGGAUCCCGACUGGCGAG
siSur-FA-MG-FA		AGCCAGGUAACGAAUGGAUCCCCCACAUACUUUGU
		UGAUCCAAUGACAGAUAAGGAACCUGCUUU
	Fragment-2 (F2)	GGCAGGUUCCUUAUCUGUCAAAGGAUCAAUCAUG
		GCCAAUCCCGCGGCCAUGGCGGCCGGGAG
3WJ-5S rRNA	Fragment-1 (F1)	ggCACCAGCGUUCCGGGAAAAAAAAACAAAUUCUUUACUG(A)
siSur-FA-Rz-FA		AUGAGUCCGUGAGGACGAAACGGGUCAAAAAAAAACCCG
		GUUCGCCGCCAAAUGACAGAUAAGGAACCUGCUUU
	Fragment-2 (F2)	GGCAGGUUCCUUAUCUGUCAAAAGGCGGCCAUAG
		CGGUGccAAUCCCGCGGCCAUGGCGGCCGGGAG
3WJ-5S rRNA	Fragment-1 (F1)	ggCACCAGCGUUCCGGGGGAUCCCGACUGGCGAGA
siSur-FA-Rz-FA		GCCAGGUAACGAAUGGAUCCCCCGGUUCGCCGCC
		AAAUGACAGAUAAGGAACCUGCUUU
	Fragment-2 (F2)	GGCAGGUUCCUUAUCUGUCAAAAGGCGGCCAUAG
		CGGUGccAAUCCCGCGGCCAUGGCGGCCGGGAG

folate-DNA strand (synthesized in-house), and a chemically synthesized 21-nt siRNA or scramble siRNA anti-sense strand (IDT). The 106-nt strand harboring HBV ribozyme sequence, the 96-nt strand harboring MG-binding aptamer and the 41-nt strand harboring siRNA sense strand were transcribed from DNA template amplified by PCR (Table 32.3). Fluorescent dyes were labeled on the 106-nt RNA strand by using the Label IT siRNA Tracker Intracellular Localization Kit, Cy3™ (Mirus Bio LLC). The four RNA strands were mixed after purification in TMS buffer at equal molar ratios, and then heated up to 80°C for 5 min followed by slow cooling to 4°C. The assembled nanoparticles were then purified from 8% native PAGE gel.

COMPETITION ASSAYS AND RADIOLABEL CHASING

Competition experiments were carried out in the presence of urea and at different temperatures as a function of time. The Cy3-labelled 3WJ-pRNA core $[ab*c]_{3WJ}$ was constructed using three RNA oligos, a_{3WJ}, Cy3-b_{3WJ} and c_{3WJ}, mixed in a 1:1:1 molar ratio in diethylpyrocarbonate (DEPC)-treated water or TMS buffer.

Presence of urea: the concentration of labeled $[ab*c]_{3WJ}$ was fixed; unlabeled b_{3WJ} was incubated with labeled $[ab*c]_{3WJ}$ for 30 min at room temperature in the presence of variable concentrations of urea (0–6 M). The samples were then loaded onto 16% native gel. Two concentration ratios were evaluated: $[ab*c]_{3WJ}$: unlabeled b_{3WJ} = 1:1 and 1:5.

Different temperatures: the concentration of labeled $[ab*c]_{3WJ}$ was fixed, and varying concentration ratios of unlabeled b_{3WJ} (1:0–1:1,000) were incubated with labeled $[ab*c]_{3WJ}$ for 30 min at 25, 37 and 55°C and then loaded onto 16% native gel.

Dilution assay to test dissociation at extremely low concentrations: the stability of the 3WJ-pRNA complex harboring three monomeric pRNAs was evaluated by radiolabel assays. The purified $[^{32}P]$-complexes were serially diluted from 40 to 160 pM in TMS buffer and then loaded onto 8% native PAGE gel.

MELTING EXPERIMENTS FOR T_M

Melting experiments were conducted by monitoring the fluorescence of the 3WJ RNAs using the LightCycler® 480 Real-Time PCR System (Roche). 1× SYBR Green I dye (Invitrogen) (emission 465–510 nm), which binds double-stranded nucleic acids but not single-stranded ones, was used for all the experiments. The respective RNA oligonucleotides (IDT) were mixed at room temperature in physiological TMS buffer. The 3WJ RNA samples were slowly cooled from 95 to 20°C at a ramping rate of 0.11°C s^{-1}. Data were analyzed by LightCycler® 480 Software using the first derivative of the melting profile. The T_m value represents the mean and standard deviation of three independent experiments.

STABILITY ASSAY IN SERUM

RNA nanoparticles were synthesized in the presence 2'-F dCTP and dUTP (Liu et al., 2010) and incubated in RPMI-1640 medium containing 10% fetal bovine serum (Sigma). 200 ng of RNA were taken at 10 min, 1 h, 12 h and 36 h time points after incubation at 37°C, followed by analysis using 8% native PAGE gel.

HBV RIBOZYME ACTIVITY ASSAY

HBV ribozyme is an RNA enzyme that cleaves the genomic RNA of HBV genome (Hoeprich et al., 2003). HBV RNA substrate was radiolabeled by $[\alpha\text{-}^{32}P]$ UTP (PerkinElmer) and incubated with the 3WJ-pRNA or 3WJ-5S rRNA core harboring HBV ribozyme at 37°C for 60 min in a buffer containing 20 mM MgCl$_2$, 20 mM NaCl and 50 mM Tris-HCl (pH 7.5). The pRNA/HBV ribozyme served as a positive control (Hoeprich *et al.*, 2003), and 3WJ RNA harboring MG aptamer was used as a negative control (Figure 32.4). The samples were then loaded on 8 M urea/10% PAGE gel for autoradiography.

MG Aptamer Fluorescence Assay

3WJ-pRNA or 3WJ-5S rRNA trivalent RNA nanoparticles harboring MG-binding aptamer (Baugh *et al.*, 2000) (100 nM) were mixed with MG (2 mM) in binding buffer containing 100 mM KCl, 5 mM $MgCl_2$ and 10 mM HEPES (pH 7.4) and incubated at room temperature for 30 min (Figure 32.4f,g). Fluorescence was measured using a fluorospectrometer (Horiba Jobin Yvon), excited at 475 nm (540–800 nm scanning for emission) and 615 nm (625–800 nm scanning for emission).

Methods for synthesis and purification of pRNA; construction and purification of pRNA complexes; serum stability assays; flow cytometry analysis of folate-mediated cell binding (Shu *et al.*, 2011); confocal microscopy imaging (Shu *et al.*, 2011); assays for the silencing of genes in a cancer cell model; stability and systemic pharmacokinetic analysis in animals (Abdelmawla *et al.*, 2011); targeting of tumor xenograft by systemic injection in animals (Abdelmawla *et al.*, 2011); and AFM imaging (Lyubchenko and Shlyakhtenko, 2009) can be found in the Supplementary Information.

Flow Cytometry Analysis of Folate-Mediated Cell Binding

Human nasopharyngeal carcinoma KB cells [American Type Culture Collection (ATCC)] were maintained in folate-free RPMI-1640 medium (Gibco), then trypsinized and rinsed with PBS (137 mM NaCl, 2.7 mM KCl, 100 mM Na_2HPO_4, 2 mM KH_2PO_4, pH 7.4). 200 nM Cy3 labeled 3WJ-pRNA/siSur-Rz-FA and the folate-free control 3WJ-pRNA/siSur-Rz-NH2 were each incubated with 2×10^5 KB cells at 37°C for 1 h. After washing with PBS, the cells were resuspended in PBS buffer. Flow cytometry (Beckman Coulter) was used to observe the cell-binding efficacy of the Cy3 3WJ-RNA nanoparticles.

Confocal Microscopy

KB cells were grown on glass coverslips in folate-free medium overnight. Cy3-labeled 3WJ-pRNA/siSur-Rz-FA and the folate-free control 3WJ-pRNA/siSur-Rz-NH2 were each incubated with the cells at 37°C for 2 h. After washing with PBS, the cells were fixed by 4% paraformaldehyde and stained by Alexa Fluor® 488 phalloidin (Invitrogen) for cytoskeleton and TO-PRO®-3 iodide (642/661) (Invitrogen) for nucleus. The cells were then assayed for binding and cell entry by Zeiss LSM 510 laser scanning confocal microscope.

Assay for the Silencing of Genes in Cancer Cell Model

Two 3WJ-RNA constructs were assayed for the subsequent gene silencing effects: one harboring folate and Survivin siRNA; and, the other harboring folate and Survivin siRNA scramble control.

KB cells were transfected with 25 nM of the individual 3WJ-RNAs and a positive Survivin siRNA control (Ambion, Inc.) using Lipofectamine 2000 (Invitrogen). After 48 h treatment, cells were collected and target gene silencing effects were assessed by both qRT-PCR and Western Blot assays.

Cells were processed for total RNA using Illustra RNAspin Mini kits (GE healthcare). The first cDNA strand was synthesized on mRNA (500 ng) from KB cells with the various 3WJ-RNAs treatment using SuperScript™ III First-Strand Synthesis System (Invitrogen). Real-time PCR was performed using Roche Universal Probe Library Assay. All reactions were carried out in a final volume of 10 μL and assayed in triplicate.

Primers for human GAPDH and Survivin are:

GAPDH left: 5′-AGCCACATCGCTCAGACAC-3′;
GAPDH right: 5′-GCCCAATACGACCAAATCC-3′;
Survivin left: 5′-CACCGCATCTCTACATTCAAGA-3′;
Survivin right: 5′-CAAGTCTGGCTCGTTCTCAGT-3′.

PCR was performed on LightCycler® 480 for 45 cycles. The data were analyzed by the comparative C_T Method ($\Delta\Delta C_T$ Method). For Western Blot assays, cells were lysed by RIPA lysis buffer (Sigma) and the cell total protein was extracted for the assay. Equal amounts of proteins were then loaded onto 15% SDS-PAGE and electrophoretically transferred to Immun-Blot PVDF membranes (Bio-Rad). The membrane was probed with Survivin antibody (R&D) (1:4000 diluted) and β-actin antibody (Sigma) (1:5000 diluted) overnight, followed by 1:10000 anti-rabbit secondary antibody conjugated with HRP (Millipore) for 1 h. Membranes were blotted by ECL kits (Millipore) and exposed to film for autoradiography.

AFM Imaging

RNA was imaged using specially modified mica surfaces (APS mica) (Lyubchenko and Shlyakhtenko, 2009) with MultiMode AFM NanoScope IV system (Veeco/Digital Instruments, Santa Barbara, CA), operating in tapping mode.

Stability and Systemic Pharmacokinetic Analysis in Animals

Each mouse was injected via the tail vein with 150 μg (6 mg/kg) of 2′-F U/C modified Alexa Fluor® 647-labeled pRNA nanoparticles that contain the 3WJ domain as a scaffold. DY647-labeled siRNA with equal molar concentration (40 μg/mice or 1.6 mg/kg) was used as a control. Up to 20 μL of blood samples were collected via the lateral saphenous vein at 1, 4, 8, 12, 16 and 24 h post injection. Blood samples were collected in BD Vacutainer® SSTTM Serum Separation Tubes, and samples were mixed by inverting the tubes five times and kept for 30 min at room temperature. After clotting, serum was collected by centrifugation at 1000–1300 × g for 10 min. Each 2 μL of serum was mixed with 1 μL of proteinase K and 37 μL of water and incubated at 37°C for 30 min before loading for capillary gel electrophoresis (CGE-Beckman Coulter, dsDNA 1000 Kit) for measuring the fluorescence intensity using a P/ACE MDQ capillary electrophoresis system, equipped with a 635 nm laser (Beckman Coulter, Inc., Fullerton, California). Samples were loaded into a 32 cm × 100 μm capillary, by voltage injection for 40 sec at 4 kV and ran for 25 min at 7.8 kV. The results were integrated using the software "32 Karat" version 7 (Beckman Coulter). The pRNA concentration was calculated from a standard curve. The serum concentration profiles were fitted with an IV bolus non-compartmental model using the Kinetica program (Fisher Scientific, Inc.) to deduce the key secondary pharmacokinetic parameters, including $T_{1/2}$, AUC (area under the curve), V_d (volume of distribution), Cl (clearance) and MRT (mean residence time).

Targeting Tumor Xenograft by Systemic Injection in Animals

For biodistribution, stability assay and specific tumor targeting studies on pRNA nanoparticles with the 3WJ domain as a scaffold, 6-week old male nude mice (*nu/nu*) (NCI/Frederick) were fed on a folate-free diet for 2 weeks before the experiment. The mice were injected with cancer cells (KB cells ~3 × 10⁶ cells per mouse) in a 40% matrigel in a folate-free RPMI-1640 medium. When the tumors grew up to about 500 mm³, the mice were injected intravenously through the tail vein with a single dose of 600 μg of 2′-F U/C -modified FA-Alexa Fluor® 647-labeled pRNA nanoparticle (about 15 nmol in PBS buffer, equal to 24 mg/kg). After 24 h post injection, the mouse was euthanized by CO_2 asphyxiation. Whole-body imaging was carried out with an IVIS® Lumina station with the body of the mouse lying sideways in the imaging chamber. After whole-body imaging, the tumors, liver, spleen, heart, lung, intestine, kidney and skeleton muscle of the mice were dissected and individually imaged.

ACKNOWLEDGMENTS

This research was mainly supported by the National Institutes of Health (NIH; grants EB003730, GM059944 and CA151648 to P.G.). P.G. is also a co-founder of Kylin Therapeutics Inc. The authors thank L. Shlyakhtenko and Y. Lyubchenko for AFM images via the Nanoimaging Core Facility supported by the NIH SIG Program and the UNMC Program of ENRI, as well as N. Abdeltawab and Z. Zhu from M. Kotb's laboratory at the University of Cincinnati for help with qRT-PCR assays.

AUTHOR CONTRIBUTIONS

P.G. conceived, designed and led the project. D.S., Y.S. and F.H. designed and conducted the *in vitro* experiments. S.A. performed animal imaging experiments. P.G., D.S., Y.S. and F.H. analyzed the data and co-wrote the manuscript. C.X. made modifications to the manuscript for publication on the 2nd edition of RNA Nanotechnology and Therapeutics book.

ADDITIONAL INFORMATION

The authors declare competing financial interests. Details accompany the full-text HTML version of the chapter at http://www.nature.com/naturenanotechnology. Supplementary information accompanies this chapter at http://www.nature.com/naturenanotechnology. Reprints and permission information is available online at http://www.nature.com/reprints. Correspondence and requests for materials should be addressed to P.G.

NOTE

1 This chapter is adapted from the full published article with permission from Nature Publishing Group, a division of Macmillan Publishers Limited. © 2011.
Original citation: Shu D, Shu Y, Haque F, Abdelmawla S, Guo P (2011) Thermodynamically stable RNA three-way junctions as platform for constructing multifunctional nanoparticles for delivery of therapeutics. *Nature Nanotechnology* 6: 658–667.

REFERENCES

Abdelmawla S, Guo S, Zhang L, Pulukuri S, Patankar P, Conley P, Trebley J, Guo P, Li QX (2011) Pharmacological characterization of chemically synthesized monomeric pRNA nanoparticles for systemic delivery. *Mol Ther* **19:** 1312–1322.

Afonin KA, Bindewald E, Yaghoubian AJ, Voss N, Jacovetty E, Shapiro BA, Jaeger L (2010) In vitro assembly of cubic RNA-based scaffolds designed in silico. *Nat Nanotechnol* **5:** 676–682.

Ambrosini G, Adida C, Altieri DC (1997) A novel anti-apoptosis gene, survivin, expressed in cancer and lymphoma. *Nat Med* **3:** 917–921.

Baugh C, Grate D, Wilson C (2000) 2.8 A crystal structure of the malachite green aptamer. *J Mol Biol* **301:** 117–128.

Behlke MA (2006) Progress towards in vivo use of siRNAs. *Mol Ther* **13:** 644–670.

Carey J, Uhlenbeck OC (1983) Kinetic and thermodynamic characterization of the R17 coat protein-ribonucleic acid interaction. *Biochemistry* **22:** 2610–2615.

Carlson RD, Olins AL, Olins DE (1975) Urea denaturation of chromatin periodic structure. *Biochemistry* **14:** 3122–3125.

Chen C, Guo P (1997) Magnesium-induced conformational change of packaging RNA for procapsid recognition and binding during phage phi29 DNA encapsidation. *J Virol* **71:** 495–500.

Chen C, Sheng S, Shao Z, and Guo P (2000) A dimer as a building block in assembling RNA: A hexamer that gears bacterial virus phi29 DNA-translocating machinery. *J Biol Chem* **275:** 17510–17516.

Chen C, Zhang C, Guo P (1999) Sequence requirement for hand-in-hand interaction in formation of pRNA dimers and hexamers to gear phi29 DNA translocation motor. *RNA* **5:** 805–818.

Chen Y, Zhu X, Zhang X, Liu B, Huang L (2010) Nanoparticles modified with tumor-targeting scFv deliver siRNA and miRNA for cancer therapy. *Mol Ther* **18:** 1650–1656.

Cruz JA, Westhof E (2009) The dynamic landscapes of RNA architecture. *Cell* **136:** 604–609.

de la Pena M, Dufour D, Gallego J (2009) Three-way RNA junctions with remote tertiary contacts: a recurrent and highly versatile fold. *RNA* **15:** 1949–1964.

Diamond JM, Turner DH, Mathews DH (2001) Thermodynamics of three-way multibranch loops in RNA. *Biochemistry* **40:** 6971–6981.

Fire A, Xu S, Montgomery MK, Kostas SA, Driver SE, Mello CC (1998) Potent and specific genetic interference by double-stranded RNA in *Caenorhabditis elegans*. *Nature* **391:** 806–811.

Guo P (2010) The emerging field of RNA nanotechnology. *Nat Nanotechnol* **5:** 833–842.

Guo P, Erickson S, Anderson D (1987) A small viral RNA is required for *in vitro* packaging of bacteriophage phi29 DNA. *Science* **236:** 690–694.

Guo P, Zhang C, Chen C, Trottier M, Garver K (1998) Inter-RNA interaction of phage phi29 pRNA to form a hexameric complex for viral DNA transportation. *Mol Cell* **2:** 149–155.

Guo S, Tschammer N, Mohammed S, Guo P (2005) Specific delivery of therapeutic RNAs to cancer cells via the dimerization mechanism of phi29 motor pRNA. *Hum Gene Ther* **16:** 1097–1109.

Guo S, Huang F, Guo P (2006) Construction of folate-conjugated pRNA of bacteriophage phi29 DNA packaging motor for delivery of chimeric siRNA to nasopharyngeal carcinoma cells. *Gene Ther* **13:** 814–820.

Hoeprich S, Zhou Q, Guo S, Qi G, Wang Y, Guo P (2003) Bacterial virus phi29 pRNA as a hammerhead ribozyme escort to destroy hepatitis B virus. *Gene Ther* **10:** 1258–1267.

Honda M, Beard MR, Ping LH, Lemon SM (1999) A phylogenetically conserved stem-loop structure at the 5′ border of the internal ribosome entry site of hepatitis C virus is required for cap-independent viral translation. *J Virol* **73:** 1165–1174.

Jaeger L, Verzemnieks EJ, Geary C (2009) The UA_handle: a versatile submotif in stable RNA architectures. *Nucleic Acids Res* **37:** 215–230.

Khaled A, Guo S, Li F, Guo P (2005) Controllable Self-Assembly of Nanoparticles for Specific Delivery of Multiple Therapeutic Molecules to Cancer Cells Using RNA Nanotechnology. *Nano Lett* **5:** 1797–1808.

Klostermeier D Millar DP (2000) Helical junctions as determinants for RNA folding: origin of tertiary structure stability of the hairpin ribozyme. *Biochemistry* **39:** 12970–12978.

Kulshina N, Edwards TE, Ferre-D'Amare AR (2010) Thermodynamic analysis of ligand binding and ligand binding-induced tertiary structure formation by the thiamine pyrophosphate riboswitch. *RNA* **16:** 186–196.

Leontis NB, Westhof E (2003) Analysis of RNA motifs. *Curr Opin Struct Biol* **13:** 300–308.

Lescoute A, Westhof E (2006) Topology of three-way junctions in folded RNAs. *RNA* **12:** 83–93.

Lilley DM (2000) Structures of helical junctions in nucleic acids. *Q Rev Biophys* **33:** 109–159.

Liu B, Diamond JM, Mathews DH, Turner DH (2011) Fluorescence competition and optical melting measurements of RNA three-way multibranch loops provide a revised model for thermodynamic parameters. *Biochemistry* **50:** 640–653.

Liu H, Guo S, Roll R, Li J, Diao Z, Shao N, Riley MR, Cole AM, Robinson JP, Snead NM, Shen G, Guo P (2007) Phi29 pRNA vector for efficient escort of hammerhead ribozyme targeting Survivin in multiple cancer cells. *Cancer Biol Ther* **6:** 697–704.

Liu J, Guo S, Cinier M, Shlyakhtenko L, Shu Y, Chen C, Shen G, Guo P (2010) Fabrication of stable and RNase-resistant RNA nanoparticles active in gearing the nanomotors for viral DNA packaging. *ACS Nano* **5:** 237–246.

Lu Y, Low PS (2002) Folate-mediated delivery of macromolecular anticancer therapeutic agents. *Adv Drug Deliv Rev* **54:** 675–693.

Lyubchenko YL, Shlyakhtenko LS (2009) AFM for analysis of structure and dynamics of DNA and protein-DNA complexes. *Methods* **47:** 206–213.

Mathews DH, Turner DH (2002) Experimentally derived nearest-neighbor parameters for the stability of RNA three- and four-way multibranch loops. *Biochemistry* **41:** 869–880.

Mulhbacher J, St-Pierre P, Lafontaine DA (2010) Therapeutic applications of ribozymes and riboswitches. *Curr Opin Pharmacol* **10:** 551–556.

Pagratis NC (1996) Rapid preparation of single stranded DNA from PCR products by streptavidin induced electrophoretic mobility shift. *Nucleic Acids Res* **24:** 3645–3646.

Pegtel DM, Cosmopoulos K, Thorley-Lawson DA, van Eijndhoven MA, Hopmans ES, Lindenberg JL, de Gruijl TD, Wurdinger T, Middeldorp JM (2010) Functional delivery of viral miRNAs via exosomes. *Proc Natl Acad Sci U S A* **107:** 6328–6333.

Reid RJD, Bodley JW, Anderson D (1994) Characterization of the prohead-pRNA interaction of bacteriophage phi29. *J Biol Chem* **269:** 5157–5162.

Rettberg CC, Prere MF, Gesteland RF, Atkins JF, Fayet O (1999) A three-way junction and constituent stem-loops as the stimulator for programmed -1 frameshifting in bacterial insertion sequence IS911. *J Mol Biol* **286:** 1365–1378.

Sarver NA, Cantin EM, Chang PS, Zaia JA, Ladne PA, Stephens DA, Rossi JJ (1990) Ribozymes as potential anti-HIV-1 therapeutic agents. *Science* **247:** 1222–1225.

Seeman NC (2010) Nanomaterials based on DNA. *Annu Rev Biochem* **79:** 65–87.

Shu D, Moll WD, Deng Z, Mao C, Guo P (2004) Bottom-up assembly of RNA arrays and superstructures as potential parts in nanotechnology. *Nano Lett* **4:** 1717–1723.

Shu D, Zhang H, Jin J, Guo P (2007) Counting of six pRNAs of phi29 DNA-packaging motor with customized single molecule dual-view system. *EMBO J* **26:** 527–537.

Shu Y, Cinier M, Fox SR, Ben-Johnathan N, Guo P (2011) Assembly of therapeutic pRNA-siRNA nanoparticles using bipartite approach. *Mol Ther* **19:** 1304–1311.

Shukla GC, Haque F, Tor Y, Wilhelmsson LM, Toulme JJ, Isambert H, Guo P, Rossi JJ, Tenenbaum SA, Shapiro BA (2011) A boost for the emerging field of RNA nanotechnology. *ACS Nano* **5:** 3405–3418.

Simpson AA, Leiman PG, Tao Y, He Y, Badasso MO, Jardine PJ, Anderson DL, Rossman MG (2001) Structure determination of the head-tail connector of bacteriophage phi29. *Acta Cryst* **D57:** 1260–1269.

Sousa R Padilla R (1995) A mutant T7 RNA polymerase as a DNA polymerase. *EMBO J* **14:** 4609–4621.

Soutschek J, Akinc A, Bramlage B, Charisse K, Constien R, Donoghue M, Elbashir S, Geick A, Hadwiger P, Harborth J, John M, Kesavan V, Lavine G, Pandey RK, Racie T, Rajeev KG, Rohl I, Toudjarska I, Wang G, Wuschko S, Bumcrot D, Koteliansky V, Limmer S, Manoharan M, Vornlocher HP (2004) Therapeutic silencing of an endogenous gene by systemic administration of modified siRNAs. *Nature* **432:** 173–178.

Wakeman CA, Ramesh A, Winkler WC (2009) Multiple metal-binding cores are required for metalloregulation by M-box riboswitch RNAs. *J Mol Biol* **392:** 723–735.

Winkler WC, Nahvi A, Roth A, Collins JA, Breaker RR (2004) Control of gene expression by a natural metabolite-responsive ribozyme. *Nature* **428:** 281–286.

Xiao F, Moll D, Guo S, Guo P (2005) Binding of pRNA to the N-terminal 14 amino acids of connector protein of bacterial phage phi29. *Nucleic Acids Res* **33:** 2640–2649.

Ye X, Liu Z, Hemida M, and Yang D (2011) Mutation tolerance and targeted delivery of anti-coxsackievirus artificial microRNAs using folate conjugated bacteriophage phi29 pRNA. *PLoS ONE,* **6:** e21215.

Zhang CL, Lee C-S, Guo P (1994) The proximate 5′ and 3′ ends of the 120-base viral RNA (pRNA) are crucial for the packaging of bacteriophage φ29 DNA. *Virology* **201:** 77–85.

Zhang CL, Trottier M, Guo PX (1995) Circularly permuted viral pRNA active and specific in the packaging of bacteriophage Phi29 DNA. *Virology* **207:** 442–451.

33 RNAi Nanotherapeutics for Localized Cancer Therapy

Anu Puri and Bruce A. Shapiro
RNA Structure and Design Section, RNA Biology Laboratory,
NCI-Frederick, Frederick, MD, USA

CONTENTS

Cancer nanomedicine is a promising field for improved delivery of drugs and bioactive agents, including nucleic acids. Research areas for clinical translation of formulated nanomedicine have been recognized and multidisciplinary research activities are underway. The concept of exploiting RNA interference (RNAi) as a potential therapeutic stems from the seminal discovery by Fire and Mellow describing modulation of gene expression by RNAi in mammalian cells [1]. Subsequently, RNAi-based nanotherapeutics have begun to display a positive impact towards disease treatments. In 2018, the first lipid-based RNAi therapeutic, ONPATTRO™ was approved by the FDA for polyneuropathy caused by hereditary transthyretin-mediated amyloidosis (hATTR amyloidosis) in adults [2].

Naked unmodified siRNA and RNA nanoparticles (RNA-NPs) are met with barriers for their *in vivo* utility due to their plasma instability, limited biodistribution, off-target effects, renal clearance, immunological responses, and limited endosomal escape. Therefore, intense research areas for the clinical suitability of RNAi-nanomedicine include (but are not limited to): (i) desirable biodistribution characteristics, (ii) improved plasma stability *in vivo*, and (iii) viable technologies for on-demand spatial and temporal cargo release. Some popular strategies to this end are (i) utilization of siRNA (including RNA-NPs) containing modified bases coupled with targeting ligands (such as aptamers); (ii) exploration of molecules that can serve as suitable delivery agents (carriers), designed to carry the siRNA unharmed to the sites of its action [3–7].

On-site, intracellular delivery of formulated siRNA can be achieved by the development of tunable RNA-NPs. Heat and/or light-sensitive tunable RNA-NPs are considered promising platforms [8, 9] among other approaches. Design of these RNA-NPs relies on strategic selection of the tunable components responsive to a given stimuli. For example, light-sensitive RNA-NPs for site-specific cargo release can be developed by utilizing suitable light sources compatible with tissue-penetrating wavelengths (Figure 33.1a). Recent studies show that photo-switchable siRNA assemblies (composed of sulfonated polymers as delivery agents) can facilitate endosomal escape upon light activation for efficient and localized RNAi therapy [9] (Figure 33.1b). Here, a photosensitive molecule covalently linked to a polymer (Figure 33.1c) is used for photoactivation. The main features of this technology include:

- A positively charged polymer, polyethylene imine (PEI) covalently conjugated with a photodynamic therapy (PDT) molecule, Pyropheophorbide-a (Pyro).
- Modification of PEI by the conversion of a selected number of amines to sulfonate groups allowing for overall reduction of positive charges.
- This tunable modification enables the blockage of endosomal escape until photoactivation occurs.

DOI: 10.1201/9781003001560-37

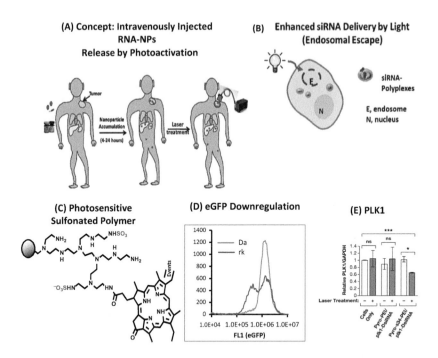

FIGURE 33.1 Concept and Utility of Photoactivable RNA-NPs. (a) Intravenously injected nanoparticles accumulate in tumors and other organs. Selective light treatments in the tumor area result in enhanced siRNA delivery to diseased cells. (b) Photo-switchable nanoparticles are taken up by endocytosis and can only be released into the cytoplasm upon photoactivation resulting in disruption of the endosomes. (c) Photosensitive Sulfonated Polymer (sulfonate groups are shown in red). (d) Pyro-conjugated PEI are complexed with anti-eGFP siRNA, preincubated with MDA-MB-eGFP cells, subsequently exposed to 661 nm for 5 minutes and gene silencing was monitored at 48 hours by FACS. A shift in the fluorescent peak upon laser-treatment of the samples (red) as compared to untreated samples (blue) indicates an enhancement of eGFP silencing upon photoactivation (reprinted with permission from reference [9]). (e) PLK1 downregulation in MDA-MB-231 cells by anti-PLK1 DsiRNA-NPs. MDA-MB-231 cell suspensions were incubated with anti-PLK1 DsiRNA-NPs (at a low dose concentration corresponding to 1 pmol RNA per 10^5 cells) treated with the laser for 5 minutes. Cells were lysed post 48-hour incubations and PLK1 expression was determined by Western Blot analysis. GAPDH was used as a housekeeping gene control. The statistical significance of differences in PLK1 expression was determined from multiple experimental replicates by two-tailed Student's t-test. P-values are indicated as follows: n.s. (not significant) indicates $p > 0.05$; * indicates $p < 0.05$, *** indicates $p < 0.001$. Error bars represent ± S.D. (Reprinted with permission from reference [9]).

This dual-function single polymer system delivers the siRNA and confers specific light triggered activity. The pyro-sulfo-PEI/siRNA nanoplexes are shown to only exhibit gene silencing upon photoactivation. Selective downregulation of a reporter gene eGFP was reported (Figure 33.1d). Similarly, photoactivation resulted in enhanced downregulation of a cancer-causing gene, PLK1 (Figure 33.1e), overexpressed in many types of cancers.

This platform has a built-in engineered triggering mechanism for localized endosomal disruption. The expected outcome from these efforts will be the minimization of off target effects while increasing the efficacy of RNA-NPs and siRNA activity *in vivo*. The development of photoactivatable nano delivery systems to accomplish enhanced intracellular siRNA release within defined time and space *in vivo* is likely to significantly enhance the intended applications of RNAi-based nanomedicine and will have direct impact on improving treatment of patients suffering from cancer.

Targeted RNAi therapies mediated through this technology are likely to improve cancer treatment due to the selective action of RNAi at the tumor site.

ACKNOWLEDGMENTS

This research was supported [in part] by the Intramural Research Program of the NIH, Center for Cancer Research, NCI-Frederick. The content of this publication does not necessarily reflect the views or policies of the Department of Health and Human Services, nor does mention of trade names, commercial products, or organizations imply endorsement by the U.S. Government. This research was supported by the National Institutes of Health (R01EB017270 and DP5OD017898), the National Science Foundation (1555220), and the University at Buffalo Clinical and Translational Science Institute.

REFERENCES

1. Fire, A., et al., Potent and specific genetic interference by double-stranded RNA in Caenorhabditis elegans. *Nature*, 391(6669): 806–811, 1998.
2. https://www.onpattro.com/
3. Shukla GC, Haque F, Tor Y, Wilhelmsson LM, Toulmé JJ, Isambert H, Guo P, Rossi JJ, Tenenbaum SA, Shapiro BA. A boost for the emerging field of RNA nanotechnology. *ACS Nano*, 5(5): 3405–3418, 2011.
4. Khaled A, Guo S, Li F, Guo P. Controllable self-assembly of nanoparticles for specific delivery of multiple therapeutic molecules to cancer cells using RNA nanotechnology. *Nano Letters*, 5(9): 1797–1808, 2005.
5. Parlea L, Puri A, Kasprzak W, Bindewald E, Zakrevsky P, Satterwhite E, Joseph K, Afonin KA, Shapiro BA. Cellular Delivery of RNA Nanoparticles. *ACS Combinatorial Science*, 18 (9): 527–547, 2016.
6. Afonin KA, Viard M, Koyfman AY, Martins AN, Kasprzak WK, Panigaj M, Desai R, Santhanam A, Grabow WW, Jaeger L, Heldman E, Reiser J, Chiu W, Freed EO, and Shapiro BA. Multifunctional RNA nanoparticles. *Nano Letters*, 14: 5662–5671, 2014.
7. Puri A, Viard M, Zakrevsky P, Parlea L, Singh KP, Shapiro BA. Alliance of Lipids with siRNA: Opportunities and Challenges for RNAi Therapy in "Molecular Medicines for Cancer: Concepts and Applications of Nanotechnology" CRC Press/Taylor & Francis Group. Editors R.I. Mahato, D. Chitkara & A. Mittal, chapter 14, pp. 419–442, 2018.
8. Puri A. Phototriggerable Liposomes: Current Research and Future Perspectives (Invited Review). *Pharmaceutics*, 6: 1–15, 2014; Photoactivatable lipid-based nanoparticles as vehicles for dual agent delivery: U.S. Patent 10117942,2018.
9. Puri A, Viard M, Zakrevsky P, Zampino S, Chen A, Isemann C, Alvi S, Clogston J, Chitgupi U, Lovell JF, Shapiro BA Photoactivation of Sulfonated Polyplexes Enables Localized Gene Silencing by DsiRNA in Breast Cancer Cells. *Nanomedicine: Nanotechnology, Biology and Medicine*, 26: pp. 102176 (1–21), 2020.

34 Delivery of RNA Nanoparticles

Weina Ke, Seraphim S. Kozlov, and Kirill A. Afonin
The University of North Carolina at Charlotte, Charlotte, NC, USA

CONTENTS

Nucleic acids are an important class of macromolecules that are essential to all forms of life. Both deoxyribonucleic acid (DNA) and ribonucleic acid (RNA) are simple linear polymers that consist of four major monomeric subunits and yet these molecules carry out assorted functions that determine the inherited characteristics of every living organism. In nature, various forms of nucleic acids exist and, while eukaryotic cells contain both RNA and DNA, some biological entities carry only one type of nucleic acid in their composition. For example, the genome of SARS-coV-2 virus, responsible for the recent outbreak of COVID-19, is represented by a single-stranded RNA [1] while the nucleocapsid of hepatitis B virus only encloses the viral DNA [2]. Alongside this diversity and the essential role of various nucleic acids as information storage for life, a class of therapeutic nucleic acids (TNAs) has made significant progress in the treatment of myriad diseases over the past few decades. Antisense oligonucleotides, DNA and RNA aptamers, mRNAs, RNA interference (RNAi) inducers, and ribozymes have been studied extensively in clinical development [3]. Some TNAs are already in clinical use with the latest successful RNAi-based therapeutic agent, givosiran, (GIVLAARI, Alnylam Pharmaceuticals, Inc.) approved by the Food and Drug Administration for adults with acute hepatic porphyria (AHP) in 2019 [4]. With the increasing interests and advanced development of nanotechnology, a novel class of nan-TNA – nucleic acid nanoparticles (NANPs) that are formulated by self-assembly of nanoscale-size oligonucleotides and designed to fold into unique predetermined three-dimensional structures is under particular examination [5–8]. RNA NANPs are those in which the scaffold, ligand and therapeutic agent are all made of RNA [9]. They can be programmed to carry multiple therapeutic modules among which are short interfering RNAs (siRNAs) required for RNAi-mediated targeting of specific mRNAs [5,10,11], ribozymes that exhibit catalytic sequence-specific cleavage of the target [12], and fluorogenic aptamers utilized as imaging agents [13], just to name a few [14]. Besides these intrinsic benefits, NANPs functionalization with RNA aptamers can also promote targeted delivery to diseased cells [15,16]. Despite these advantages, negatively charged RNA NANPs are inefficient in crossing biological membranes, relatively unstable against nuclease degradation in blood serum and in a biological environment, and are prone to detrimental immune responses [17]. Therefore, cellular delivery of RNA NANPs has become a major hurdle in the transition from benchtop to bedside.

Numerous studies have been carried out in search of stable, efficient, and safe carriers for delivery of RNA NANPs. Most of these carriers are lipid-based and possess a positive charge to enhance electrostatic interactions with negatively charged NANPs. Also, lipid-based carriers are prone to interaction with the cell membrane, thus facilitating the cellular uptake of NANPs. Additionally, biocompatible and biodegradable features of carriers can be advantageous for efficient delivery of NANPs [18]. Some of the most common lipid-based carriers that are routinely used for delivery of exogenous RNA NANPs in research settings are liposomes, lipid nanoparticles (LNPs), and commercially available lipofectamine [11,19–22]. However, high net positive charge associated with the headgroup induces cytotoxicity and innate immune activation becoming a major barrier for *in vivo* use.

DOI: 10.1201/9781003001560-38

There has been great effort focused on the rational design of lipids as well as the development of using amphiphilic carriers to reduce toxicity for application in nucleic acid delivery [23]. For example, bola-amphiphiles are amphiphilic molecules that have two hydrophilic headgroups at both ends of a hydro-phobic hydrocarbon chain [24,25]. Molecular dynamic simulations show that the efficient complexation between bolaamphiphiles and RNAs is stabilized by the electrostatic and hydrophobic interactions as well as by hydrogen bonding [17]. Furthermore, bolaamphiphiles confer very little toxicity, high thermal and chemical stability, and the ability to cross the blood– brain barrier [26,27]. Therefore, bolaamphiphiles have become a potent candidate for delivering siRNAs and functional RNA NANPs *in vivo* [5,28,29].

Another example of amphiphilic polymer carrier is poly-(lactide-co-glycolide)-graft-polyethyl-enimine (referred to as PgP), which has been characterized and demonstrated as an efficient carrier for delivering functional RNA NANPs of various sizes, shapes, and composition. The stability, intracellular uptake, hemocompatibility, and biodistribution of novel formulations have been evalu-ated [30]. Results indicate that PgP was able to efficiently deliver various RNA NANPs without being affected by their distinct size or shape. PgP also showed a decreased immunorecognition compared to other carriers used to deliver the same RNA NANPs.

Magnetofection, an advanced technique, improves the transfection efficiency both *in vitro* and *in vivo*. The complexes formed through the electrostatic interaction between positively charged magnetic nanoparticles (MNPs) and NANPs are subsequently added to cell culture while applying a magnetic field gradient which, in turn, results in an enhancement of endocytosis [31,32]. This tech-nique has shown an ability to deliver RNA NANPs functionalized with Dicer substrate RNAs using iron oxide MNPs in human breast cancer cells with high transfection efficiency, nuclease degrada-tion protection, and selective targeting of intracellular protein expression [33].

Besides the aforementioned examples, other synthetic carriers have shown promising results in delivery of RNA nanoparticles, including micelles [34], dendrimer [35], gold nanoparticles [36], carbon nanotube [37], mesoporous silica-based nanoparticles (MSN) [38], superparamagnetic iron oxide nanoparticles (SPION) [39], and quantum-dots (QDs) [40].

A promising avenue of delivery vehicles already present in the body naturally – exosomes – have garnered significant attention. These little extracellular vesicles facilitate the transfer of genetic material which could be exploited for efficient delivery of RNA NANPs. The first published study indicated exosomes were capable of efficient delivery of exogenous siRNA cargos to the brain of mice and successfully knockdown mRNA and protein of BACE1 with very little immune response [41]. Subsequently, myriad examples of using exosomes for successful delivery of RNA NANPs and knockdown targeted genes in different diseases have been reported [42–45]. Exosomes' surfaces can also be decorated with image agents or targeting ligands for diagnostic and therapeutic purposes. Exosomes displaying folate conjugated to the three-way junction (3WJ) arrowtail RNA nanopar-ticles were proven to efficiently deliver siRNA against survivin, resulting in enhanced cancer sup-pression without endosome trapping [46] (Figure 34.1).

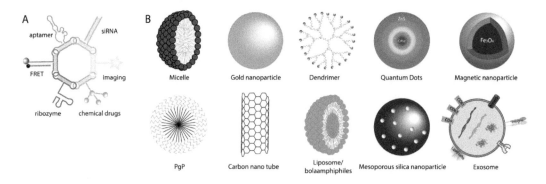

FIGURE 34.1   (a) Schematic representation of multifunctional RNA NANP designed for simultaneous delivery of various functionalities. (b) Common carries used in RNA nanoparticle delivery.

Regardless of whether carriers are synthetic or natural, the delivery vehicles for NANPs necessitate safety and efficacy, bioavailability, stability, membrane permeation capacity including the blood–brain barrier, low toxicity, low immunogenicity, and low off-target effect. To achieve these requirements, further research need to be conducted.

ACKNOWLEDGMENTS

The study was supported by the National Institute of General Medical Sciences of the National Institutes of Health under Award Numbers R01GM120487 and R35GM139587 (to K.A.A.). The content of this publication does not necessarily reflect the views or policies of the Department of Health and Human Services, nor does mention of trade names, commercial products, or organizations imply endorsement by the U.S. Government.

REFERENCES

1. Forster, P., et al., Phylogenetic network analysis of SARS-CoV-2 genomes. *Proceedings of the National Academy of Sciences*, 2020: p. 202004999.
2. Valenzuela, P., Hepatitis A, B, C, D and E viruses: Structure of their genomes and general properties. *Gastroenterologia Japonica*, 1990. **25**(S2): pp. 62–71.
3. Sridharan, K. and N.J. Gogtay, Therapeutic nucleic acids: Current clinical status. *British Journal of Clinical Pharmacology*, 2016. **82**(3): pp. 659–672.
4. https://www.fda.gov/drugs/resources-information-approved-drugs/fda-approves-givosiran-acute-hepatic-porphyria.
5. Afonin, K.A., et al., *Multifunctional RNA Nanoparticles. Nano Letters*, 2014. **14**(10): pp. 5662–5671.
6. Afonin Kirill, A., B. Lindsay, and A. Shapiro Bruce, *Engineered RNA Nanodesigns for Applications in RNA Nanotechnology*, in *DNA and RNA Nanotechnology*. 2013.
7. Grabow, W.W. and L. Jaeger, RNA Self-Assembly and RNA Nanotechnology. *Accounts of Chemical Research*, 2014. **47**(6): pp. 1871–1880.
8. Lee, H., et al., Molecularly self-assembled nucleic acid nanoparticles for targeted in vivo siRNA delivery. *Nature Nanotechnology*, 2012. **7**(6): pp. 389–393.
9. Shu, Y., et al., Stable RNA nanoparticles as potential new generation drugs for cancer therapy. *Advanced Drug Delivery Reviews*, 2014. **66**: pp. 74–89.
10. Ke, W., et al., RNA-DNA fibers and polygons with controlled immunorecognition activate RNAi, FRET and transcriptional regulation of NF-kappaB in human cells. *Nucleic Acids Research*, 2019. **47**(3): pp. 1350–1361.
11. Rackley, L., et al., RNA Fibers as Optimized Nanoscaffolds for siRNA Coordination and Reduced Immunological Recognition. *Advanced Functional Materials*, 2018. **28**(48): p. 1805959.
12. Hoeprich, S., et al., Bacterial virus phi29 pRNA as a hammerhead ribozyme escort to destroy hepatitis B virus. *Gene Therapy*, 2003. **10**(15): pp. 1258–1267.
13. Autour, A., et al., Fluorogenic RNA Mango aptamers for imaging small non-coding RNAs in mammalian cells. *Nature Communications*, 2018. **9**(1): p. 656.
14. Germer, K., M. Leonard, and X. Zhang, RNA aptamers and their therapeutic and diagnostic applications. *International Journal of Biochemistry and Molecular Biology*, 2013. **4**(1): pp. 27–40.
15. Bagalkot, V., et al., Quantum Dot–Aptamer Conjugates for Synchronous Cancer Imaging, Therapy, and Sensing of Drug Delivery Based on Bi-Fluorescence Resonance Energy Transfer. *Nano Letters*, 2007. **7**(10): pp. 3065–3070.
16. Ray, P. and R.R. White, Aptamers for Targeted Drug Delivery. *Pharmaceuticals* (Basel, Switzerland), 2010. **3**(6): pp. 1761–1778.
17. Afonin, K.A., et al., In silico design and enzymatic synthesis of functional RNA nanoparticles. *Accounts of Chemical Research*, 2014. **47**(6): pp. 1731–1741.
18. Xue, H., et al., Lipid-based nanocarriers for RNA delivery. *Current Pharmaceutical Design*, 2015. **21**(22): pp. 3140–3147.
19. Hong, E., et al., Structure and composition define immunorecognition of nucleic acid nanoparticles. *Nano Letters*, 2018. **18**(7): pp. 4309–4321.
20. Moss, K.H., et al., Lipid nanoparticles for delivery of therapeutic RNA oligonucleotides. *Molecular Pharmaceutics*, 2019. **16**(6): pp. 2265–2277.

21. Xue, H.Y., et al., Lipid-based nanocarriers for RNA delivery. *Current Pharmaceutical Design*, 2015. **21**(22): pp. 3140–3147.

22. Dash, T.K. and V.B. Konkimalla, Nanoformulations for delivery of biomolecules: Focus on liposomal variants for siRNA delivery. *Critical Reviews in Therapeutic Drug Carrier Systems*, 2013. **30**(6): pp. 469–493.

23. Ramishetti, S., et al., A combinatorial library of lipid nanoparticles for RNA delivery to leukocytes. *Advanced Materials*, **32**(12): pp. e1906128–e1906128.

24. Fuhrhop, J.-H. and T. Wang, Bolaamphiphiles. *Chemical Reviews*, 2004. **104**(6): pp. 2901–2938.

25. Qiu, F., C.K. Tang, and Y.Z. Chen, Amyloid-like aggregation of designer bolaamphiphilic peptides: Effect of hydrophobic section and hydrophilic heads. *Journal of Peptide Science*, 2018. **24**(2).

26. Popov, M., et al., Cationic vesicles from novel bolaamphiphilic compounds. *Journal of Liposome Research*, 2010. **20**(2): pp. 147–159.

27. Kim, T., et al., Characterization of Cationic Bolaamphiphile Vesicles for siRNA Delivery into Tumors and Brain. *Molecular Therapy – Nucleic Acids*, 2020. **20**: pp. 359–372.

28. Kim, T., et al., In silico, in vitro, and in vivo studies indicate the potential use of bolaamphiphiles for therapeutic siRNAs delivery. *Molecular Therapy – Nucleic Acids*, 2013. **2**(3): p. e80.

29. Gupta, K., et al., Bolaamphiphiles as carriers for siRNA delivery: From chemical syntheses to practical applications. *Journal of Controlled Release*, 2015. **213**: pp. 142–151.

30. Halman, J.R., et al., A cationic amphiphilic co-polymer as a carrier of nucleic acid nanoparticles (Nanps) for controlled gene silencing, immunostimulation, and biodistribution. *Nanomedicine-Nanotechnology Biology and Medicine*, 2020. **23**.

31. Schillinger, U., et al., Advances in magnetofection—magnetically guided nucleic acid delivery. *Journal of Magnetism and Magnetic Materials*, 2005. **293**(1): pp. 501–508.

32. Scherer, F., et al., Magnetofection: enhancing and targeting gene delivery by magnetic force in vitro and in vivo. *Gene Therapy*, 2002. **9**(2): pp. 102–109.

33. Cruz-Acuña, M., et al., Magnetic nanoparticles loaded with functional RNA nanoparticles. *Nanoscale*, 2018. **10**(37): pp. 17761–17770.

34. Falamarzian, A., et al., Polymeric micelles for siRNA delivery. *Journal of Drug Delivery Science and Technology*, 2012. **22**(1): pp. 43–54.

35. Dzmitruk, V., et al., Dendrimers Show Promise for siRNA and microRNA Therapeutics. *Pharmaceutics*, 2018. **10**(3): p. 126.

36. Paul, A.M., et al., Delivery of antiviral small interfering RNA with gold nanoparticles inhibits dengue virus infection in vitro. *Journal of General Virology*, 2014. **95**: pp. 1712–1722.

37. Wang, X., J. Ren, and X. Qu, Targeted RNA interference of cyclin A(2) mediated by functionalized single-walled carbon nanotubes induces proliferation arrest and apoptosis in chronic myelogenous leukemia K562 cells. *Chemmedchem*, 2008. **3**(6): pp. 940–945.

38. Hom, C., et al., Mesoporous silica nanoparticles facilitate delivery of siRNA to shutdown signaling pathways in mammalian cells. *Small (Weinheim an der Bergstrasse, Germany)*, 2010. **6**(11): pp. 1185–1190.

39. Dey, S. and T.K. Maiti, Superparamagnetic nanoparticles and RNAi-mediated gene silencing: Evolving class of cancer diagnostics and therapeutics. 2012. **2012**: pp. 1–15.

40. Lin, G., et al., Quantum dots-siRNA nanoplexes for gene silencing in central nervous system tumor cells. *Frontiers in Pharmacology*, 2017. **8**.

41. Alvarez-Erviti, L., et al., Delivery of siRNA to the mouse brain by systemic injection of targeted exosomes. *Nature Biotechnology*, 2011. **29**(4): pp. 341–U179.

42. Aqil, F., et al., Milk exosomes – Natural nanoparticles for siRNA delivery. *Cancer Letters*, 2019. **449**: pp. 186–195.

43. Banizs, A.B., et al., In vitro evaluation of endothelial exosomes as carriers for small interfering ribonucleic acid delivery. *International Journal of Nanomedicine*, 2014. **9**: pp. 4223–4230.

44. Shtam, T.A., et al., Exosomes are natural carriers of exogenous siRNA to human cells in vitro. *Cell Communication and Signaling*, 2013. **11**: p. 10.

45. Yang, R., et al., Targeted exosome-mediated delivery of Survivin Sirna for the treatment of bladder cancer. *Journal of Urology*, 2017. **197**(4): pp. E1180–E1180.

46. Zheng, Z., et al., Folate-displaying exosome mediated cytosolic delivery of siRNA avoiding endosome trapping. *Journal of Controlled Release*, 2019. **311–312**: pp. 43–49.

35 Recommendations for Planning In Vivo Studies for RNAi Therapeutics

Lorena Parlea and Bruce A. Shapiro

RNA Structure and Design Section, RNA Biology Laboratory,
NCI-Frederick, Frederick, Maryland, USA

CONTENTS

Remarkable progress has been made during the past decades in drug discovery and development for treating various cancers and diseases. With an increased life expectancy, there is a need for new drugs and novel approaches to treatment. Yet from the concept to market, or "bench to the bedside", a drug's translational process is not a linear path, and it involves a close collaboration between basic researchers, clinicians and the pharmaceutical industry. A given drug's concept and design may be devised in silico; however, it has to undergo rigorous experimental research in vitro and in vivo. Initial validation of the drug in vitro is conducted in biological and/or chemical laboratories, and reproducible experiments and multiple validations have to be performed before in vivo testing. During the in vivo studies, there are several aspects that have to be characterized, and failure to meet the required criteria does not allow drugs to pass beyond preclinical testing. (Some steps in the preclinical validation process are presented in Figure 35.1a.) When the in vitro or in vivo testing fails or is only partially successful, there is a need to go back to the bench, or even to the "drawing board" to revise the drug's concept and design. Some considerations for planning in vivo studies, especially for a particular class of therapeutics acting through the RNA interference (RNAi) pathway[1] are discussed here.

There are certain characteristics desirable in a drug: safety – or low toxicity, great potency – or high efficacy at low doses, minimum side effects, ease of administration, and availability – or scalable production.[2] Natural and naturally derived drug components may be preferable choices due to their biocompatibility, low(er) toxicity, programmable functions, and low(er) production costs. An example of a naturally derived material used for drug design are nucleic acids, and in particular RNA. There are several advantages in using RNA for nanomedicine: it is a biodegradable polymer, it can encompass structure and function within the same molecule, it can be designed for distinct purposes, and it can modulate several cellular processes simultaneously.[3] It is important to note that there are several nucleic acid- or RNA-based therapeutics: vaccines[4] and vaccine adjuvants,[5] mRNA or DNA plasmids,[4] DNA and RNA ribozymes,[6] nucleic acid aptamers,[6] antisense oligonucleotides,[7] CRISPR-CAS9 platforms for gene modulation,[8] and RNAi-based drugs[9] that are at different stages of clinical development approved in clinical trials or in preclinical evaluation. Numerous laboratories, commercial and academic, are working with nucleic acid-based drugs that are deliverable in vivo, but there are many hurdles that still have to be overcome to reach the bedside.

RNAi-based therapeutics is a field still in its infancy.[10] RNAi-based drugs come in a variety of flavors: microRNAs (miRNAs) and miRNA mimics, miRNA antagomirs, silencing RNAs (siRNAs),

DOI: 10.1201/9781003001560-39

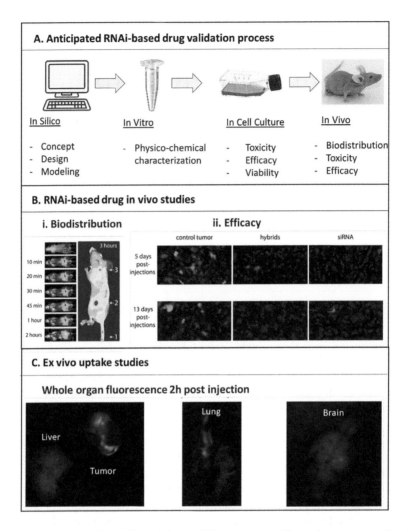

FIGURE 35.1 (a) Anticipated RNAi-based drug validation process. From the concept to in vivo studies, RNAi-based drugs go through rigorous in vitro and in cell culture characterizations; (b) RNAi-based drug in vivo studies. i. Biodistribution of a fluorescently labeled formulation is monitored through whole body imaging. Here, IRDye700 labeled DNA/RNA hybrids, a potential platform for RNAi drugs, targeting Green Fluorescent Protein (eGFP) expressed in a subcutaneously injected flank xenograft of human breast cancer (MDA-MB-231/ eGFP) tumors, were scanned in mice at various timepoints. One can see significant fluorescence in 1 – the tail vein point of injection; 2 – the tumor and 3 – the point of blood extraction. ii. Efficacy of the hybrids was monitored through downregulation of the eGFP, expressed in the tumor cells. (*reproduced with permission from Afonin & al. Nat Nanotechnol. 2013, 8(4), 296–304.*) (c) Ex vivo uptake studies. A formulation of Bola amphiphile GLH-19/GHL-20 vesicles complexed with siRNA labeled with IRDye 700 was injected into a mouse tail vein. The organs collected 2 hours post injection display fluorescence in the tumor and in the brain (green color over background), but not in the liver and lungs, a desirable attribute in these experiments. (*reproduced with permission from Kim et al. Nucleic acids. 2020, 20, 359–372*).

Dicer substrate RNAs (DsiRNAs) and more complex nucleic acid structures – or nanoparticles – that incorporate therapeutic RNAs. The latter can be assembled from structural and functional elements that can address simultaneously sensing, targeting, delivery, imaging, and therapeutic capabilities.[11–18] Oncogenes, aberrant or mutated genes, viruses, and genes overexpressed in a particular cancer and diseases can be targeted for downregulation through RNAi with interfering RNAs (miRNA, siRNAs, DsiRNAs, etc.).

Careful consideration has to be given to the appropriate in vivo organism model as well as the disease type.[19] First in vivo proof-of-concept is usually done in athymic nude mice, with a disease-specific xenograft or orthotopic mouse model. These studies are followed by patient-derived xenografts and/or immune-competent mice. Genetically engineered mice can be used to investigate mechanistic aspects of the therapeutic or to study a particular disease type (e.g. arthritis, muscular dystrophy, etc.). A timeline for development of an RNAi drug from the concept to in vivo is illustrated in Figure 35.1a. The design of RNAi therapeutics can be computationally modeled and validated against the desired target(s) and verified for no off-target effects with various software platforms. Initial in vitro validation has to include rigorous physico-chemical characterization and has to prove that RNAi-based therapeutics have high efficacy with no toxicity and no off-target effects in cell cultures. In vitro validation of a drug does not necessarily correlate with its efficacy in vivo, as in vivo experiments bring another level of complexity. In vivo studies can determine the biodistribution, cytotoxicity, and efficacy of the drug.

Additionally, the possibility of using a delivery agent needs to be carefully considered as well, as it can greatly affect the properties and characteristics of the therapeutic in vivo. The carrier's charge plays an important role as it can significantly influence its ability to complex with the RNAi therapeutic, the formulation's biodistribution, immunogenicity, toxicity, and metabolic availability.[17,20] The route of administration is usually chosen according to the targeted organ, location of the tumor, and the goals of the study. Various common routes have been used: nasal, ocular, inhaler, subcutaneous, intratumoral and intravenous. Local administration routes do not necessarily require preceding pharmacokinetic studies, but intravenous treatments do.[19] An example of a biodistribution study is illustrated in Figure 35.1b. DNA/RNA hybrids were investigated for their potential as RNAi therapeutic carriers.[21,22] Functional double-stranded RNAs can be split into two strands and associated with their complementary DNA with a toehold. Hybrids are inactive in this form, and they release the functional RNA solely upon reassociation. The DNA/RNA hybrids were fluorescently labeled with IRDye 700 and complexed with Bolaamphiphile GLH-19, a lipid-based carrier. This formulation was injected into the tail vein and the mice were imaged at various timepoints. Such dynamic scan studies can reveal organ uptake and the persistence of the formulation in the body. Here, relatively high amounts of hybrid-Bolaamphiphile complexes appear to accumulate mainly in the tumor. Additionally, the organs can be imaged ex vivo, as seen in Figure 35.1c.[23] Mice injected with Bolaamphiphile vesicles complexed with siRNA were imaged ex vivo 2 hours post injection. As the images indicate, at this time point the tumor and the brain display the highest amount of fluorescence. This formulation passes the blood–brain barrier, and the hybrids can be tailored towards treating diseases of the brain. Generally, the observed fluorescence correlates with the amount of drug or formulation taken up in the respective organs or tumor. For a metastatic cancer type, a general biodistribution and uptake of the drug in all organs might be preferable. For localized diseases and solid tumors, preferential uptake into the diseased cells might be desired.

A thorough efficacy study includes control groups, such as an untreated group, a negative control group, a positive control group, and groups treated with the active drug. The negative controls can be either a formulation with a "mock" drug or a formulation with everything included but the active drug. In vivo studies should be performed in a relevant sample size per group, and power calculations aid in revealing the minimum number of animals needed for a statistically relevant predicted effect.[24] Statistically relevant differences between the control group(s) and the treated groups have to be demonstrated to prove the efficacy of a drug.

The next steps in the validation process should evaluate the RNAi drug's toxicity, and/or the toxicity of the formulation. There are various ways to define in vivo toxicity, but generally toxicity is regarded as an undesired effect at the cellular, tissue, or organism level. If a reasonable uptake into the targeted organ or xenograft tumors is observed in biodistribution studies, efficacy studies can be planned at doses where the therapeutic effect is expected to greatly surpass any toxic side effect. Planning the in vivo study must be done according to the end goals and the questions needed to be addressed. For example, to investigate the drug's effectiveness in vivo, a range of doses can determine the drug's potency and efficacy. Formulations of RNAi therapeutics studied in vivo include doses as low as 0.05 mg/kg, and as high as 24 mg/kg, and commonly range between 1 and 5 mg/kg dose.[25] Since RNAi-based drugs target genes for downregulation, their therapeutic effect can be assessed by measuring the level of target knockdown and/or the overall effect of the drug on the cancer or disease. Efficient targeting of complementary transcripts is measured ex vivo by collecting the tissue of interest, or the tumor, through histopathology studies, antibody staining, DNA paint, RNA Scope, or FISH. In Figure 35.1b.i, the biodistribution and efficacy of the siRNA and hybrid DNA/RNA complexed with Bolaamphiphiles were assessed ex vivo with fluorescence microscopy. For functionality proof-of-concept, the siRNA and the hybrids were designed to downregulate eGFP (Green Fluorescence Protein) expression. The subcutaneous breast cancer tumors (cell line MDA-MB-231) had eGFP stably expressed. The mice were injected once with the siRNA- or hybrid-Bolaamphiphile formulation, and the tumors collected 5 days and 13 days post-injection. As seen from the fluorescent microscopy images, both siRNA and hybrids downregulated eGFP, and the hybrids appear to have had a greater effect than the siRNA (Figure 35.1b.ii).

The RNAi therapeutics field achieved a great success with the release of the first interfering RNA drug (https://www.fda.gov/news-events/press-announcements/fda-approves-first-its-kind-targeted-rna-based-therapy-treat-rare-disease[26]), but it is a long way from large-scale applications. Some of the roadblocks encountered along the way have been addressed; others are yet to be conquered. Specific tissue or tumor targeting can be achieved by incorporating antibodies, aptamers, or other targeting agents into the drug design or formulation. A drug is considered to have high potency if it has maximum efficacy at low doses and minimum side effects, desirably no toxicity, no immunogenicity, and no off-target effects. There are currently several RNAi-based drugs in development, which hold immense potential in treating various diseases and cancer types.[27]

ACKNOWLEDGMENTS

This research was supported [in part] by the Intramural Research Program of the NIH, Center for Cancer Research, NCI-Frederick. This work has been funded in whole or in part with Federal funds from the Frederick National Laboratory for Cancer Research, National Institutes of Health, under contract no. HHSN261200800001E. The content of this publication does not necessarily reflect the views or policies of the Department of Health and Human Services, nor does mention of trade names, commercial products, or organizations imply endorsement by the U.S. Government.

REFERENCES

1. Fire, A.; Xu, S.; Montgomery, M. K.; Kostas, S. A.; Driver, S. E.; Mello, C. C. Potent and specific genetic interference by double-stranded RNA in Caenorhabditis elegans. *Nature*. 1998, *391(6669)*, 806–811.

2. Strovel, J.; Sittampalam, S.; Coussens, N. P.; Hughes, M.; Inglese, J.; Kurtz, A.; Andalibi, A.; Patton, L.; Austin, C.; Baltezor, M.; Beckloff, M.; Weingarten, M.; Weir, S., Early drug discovery and development guidelines: For academic researchers, collaborators, and start-up companies, In *Assay Guidance Manual*, Sittampalam, G. S., et al., Editors. 2004: Bethesda, MD.

3. Guo, P.; Haque, F.; Hallahan, B.; Reif, R.; Li, H. Uniqueness, advantages, challenges, solutions, and perspectives in therapeutics applying RNA nanotechnology. *Nucleic Acid Ther.* 2012, *22(4)*, 226–245.

4. Zhang, C.; Maruggi, G.; Shan, H.; Li, J. Advances in mRNA vaccines for infectious diseases. *Front Immunol.* 2019, *10*, 594.

5. Wang, Z.-B.; Xu, J. Better adjuvants for better vaccines: Progress in adjuvant delivery systems, modifications, and adjuvant-antigen codelivery. *Vaccines*. 2020, *8(1)*, 128.

6. Weng, Y.; Huang, Q.; Li, C.; Yang, Y.; Wang, X.; Yu, J.; Huang, Y.; Liang, X.-J. Improved nucleic acid therapy with advanced nanoscale biotechnology. *Mol. Ther. Nucleic Acids*. 2020, *19*, 581–601.

7. Bennett, C. F. Therapeutic antisense oligonucleotides are coming of age. *Ann. Rev. Med.* 2019, *70(1)*, 307–321.

8. Li, H.; Yang, Y.; Hong, W.; Huang, M.; Wu, M.; Zhao, X. Applications of genome editing technology in the targeted therapy of human diseases: mechanisms, advances and prospects. *Signal Transduct. Target. Ther.* 2020, *5(1)*, 1.

9. Bajan, S.; Hutvagner, G. RNA-Based Therapeutics: From Antisense Oligonucleotides to miRNAs. *Cells.* 2020, *9(1)*, 137.

10. Setten, R. L.; Rossi, J. J.; Han, S.-P. The current state and future directions of RNAi-based therapeutics. *Nat. Rev. Drug Discov.* 2019, *18(6)*, 421–446.

11. Afonin, K. A.; Bindewald, E.; Yaghoubian, A. J.; Voss, N.; Jacovetty, E.; Shapiro, B. A.; Jaeger, L. In vitro assembly of cubic RNA-based scaffolds designed in silico. *Nat. Nanotechnol.* 2010, *5(9)*, 676–682.

12. Afonin, K. A.; Grabow, W. W.; Walker, F. M.; Bindewald, E.; Dobrovolskaia, M. A.; Shapiro, B. A.; Jaeger, L. Design and self-assembly of siRNA-functionalized RNA nanoparticles for use in automated nanomedicine. *Nat. Protoc.* 2011, *6(12)*, 2022–2034.

13. Afonin, K. A.; Kasprzak, W.; Bindewald, E.; Puppala, P. S.; Diehl, A. R.; Hall, K. T.; Kim, T. J.; Zimmermann, M. T.; Jernigan, R. L.; Jaeger, L.; Shapiro, B. A. Computational and experimental characterization of RNA cubic nanoscaffolds. *Methods.* 2014, *67(2)*, 256–265.

14. Afonin, K. A.; Viard, M.; Koyfman, A. Y.; Martins, A. N.; Kasprzak, W. K.; Panigaj, M.; Desai, R.; Santhanam, A.; Grabow, W. W.; Jaeger, L.; Heldman, E.; Reiser, J.; Chiu, W.; Freed, E. O.; Shapiro, B. A. Multifunctional RNA nanoparticles. *Nano Lett.* 2014, *14(10)*, 5662–5671.

15. Jasinski, D.; Haque, F.; Binzel, D. W.; Guo, P. Advancement of the Emerging Field of RNA Nanotechnology. *ACS Nano.* 2017, *11(2)*, 1142–1164.

16. Parlea, L.; Bindewald, E.; Sharan, R.; Bartlett, N.; Moriarty, D.; Oliver, J.; Afonin, K. A.; Shapiro, B. A. Ring Catalog: A resource for designing self-assembling RNA nanostructures. *Methods.* 2016, *103*, 128–137.

17. Parlea, L.; Puri, A.; Kasprzak, W.; Bindewald, E.; Zakrevsky, P.; Satterwhite, E.; Joseph, K.; Afonin, K. A.; Shapiro, B. A. Cellular delivery of RNA nanoparticles. *ACS Comb Sci.* 2016, *18(9)*, 527–547.

18. Zakrevsky, P.; Kasprzak, W. K.; Heinz, W. F.; Wu, W.; Khant, H.; Bindewald, E.; Dorjsuren, N.; Fields, E. A.; de Val, N.; Jaeger, L.; Shapiro, B. A. Truncated tetrahedral RNA nanostructures exhibit enhanced features for delivery of RNAi substrates. *Nanoscale.* 2020, *12(4)*, 2555–2568.

19. Ireson, C. R.; Alavijeh, M. S.; Palmer, A. M.; Fowler, E. R.; Jones, H. J. The role of mouse tumour models in the discovery and development of anticancer drugs. *Br. J. Cancer.* 2019, *121(2)*, 101–108.

20. Lamberti, G.; Barba, A. A. Drug delivery of siRNA therapeutics. *Pharmaceutics.* 2020, *12(2)*, 178.

21. Afonin, K. A.; Viard, M.; Martins, A. N.; Lockett, S. J.; Maciag, A. E.; Freed, E. O.; Heldman, E.; Jaeger, L.; Blumenthal, R.; Shapiro, B. A. Activation of different split functionalities on re-association of RNA-DNA hybrids. *Nat. Nanotechnol.* 2013, *8(4)*, 296–304.

22. Afonin, K. A.; Desai, R.; Viard, M.; Kireeva, M. L.; Bindewald, E.; Case, C. L.; Maciag, A. E.; Kasprzak, W. K.; Kim, T.; Sappe, A.; Stepler, M.; Kewalramani, V. N.; Kashlev, M.; Blumenthal, R.; Shapiro, B. A. Co-transcriptional production of RNA-DNA hybrids for simultaneous release of multiple split functionalities. *Nucleic Acids Res.* 2014, *42(3)*, 2085–2097.

23. Kim, T.; Viard, M.; Afonin, K. A.; Gupta, K.; Popov, M.; Salotti, J.; Johnson, P. F.; Linder, C.; Heldman, E.; Shapiro, B. A. Characterization of cationic Bolaamphiphile vesicles for siRNA delivery into tumors and brain. *Mol. Ther. Nucleic Acids.* 2020, *20*, 359–372.

24. Aberson, C. L. *Applied power analysis for the behavioral sciences.* 2010, New York, NY: Routledge/ Taylor & Francis Group. xiv, 257.

25. Chernikov, I. V.; Vlassov, V. V.; Chernolovskaya, E. L. Current development of siRNA bioconjugates: From research to the clinic. *Front Pharmacol.* 2019, *10*, 444.

26. Second RNAi drug approved. *Nat. Biotechnol.* 2020, *38(4)*, 385–385.

27. Zhou, L. Y.; Qin, Z.; Zhu, Y. H.; He, Z. Y.; Xu, T. Current RNA-based therapeutics in clinical trials. *Curr. Gene Ther.* 2019, *19(3)*, 172–196.

36 Mesoporous Silica Nanoparticles for Efficient siRNA Delivery

Mubin Tarannum and Juan L. Vivero-Escoto
The University of North Carolina at Charlotte, Charlotte, NC, USA

CONTENTS

Nucleic acids (NA), including DNA, small interfering RNA (siRNA), microRNA (miRNA), and antisense oligonucleotides, exhibit huge therapeutic potential, especially against undruggable targets for treating a wide range of diseases such as cancer, metabolic disorders, infections, inflammation, and degenerative conditions (1–4). These NA therapeutics can selectively knock down genes, decrease the expression of gene products, alter mRNA splicing, and thereby regulate cellular processes dependent on the specific genes or its products. In particular, RNA interference (RNAi) process is a transient gene silencing mechanism where siRNAs in the cytoplasm can recognize endogenous mRNA through the complementary sequence, followed by its degradation accomplished via the RISC complex, thus, preventing the translation of specific mRNA into proteins (5,6). The success of RNAi is faced by obstacles in siRNA delivery due to its lack of systemic stability, degradation by nucleases, tissue penetration, off-target effects, and ineffective cellular internalization (7,8). The development of various siRNA delivery systems has received extensive attention to confronting these limitations (9–11). Also, the FDA approval of the first siRNA therapeutic, a liposome-based siRNA delivery system, validates the use of delivery platforms (12).

Mesoporous silica nanoparticles (MSNs) are inorganic materials with a highly ordered network of pores within a silicon oxide framework. The unique properties of MSNs make this delivery system attractive for its application in siRNA delivery. MSNs exhibit excellent biocompatibility, highly ordered porous structure, tunable particle and pore size, large surface area and pore volume, chemically modifiable surfaces, and facile functionalization (13–16). This chapter focuses on describing the multiple advantages of using MSNs for the delivery of siRNA. The rational design of an efficient siRNA delivery system is built on the following criteria: high loading of siRNA, protection against nucleases, targeting toward specific cells, and the effective release of siRNA into the cytoplasm.

ENGINEERING MSNs TO ADDRESS MAJOR CHALLENGES IN siRNA DELIVERY

The MSN material can be rationally designed to address the above-mentioned limitations and develop an effective platform for the delivery of siRNA. To achieve efficient loading of siRNA, the negative charge of MSNs associated with the silanol groups is modified with either amine functional

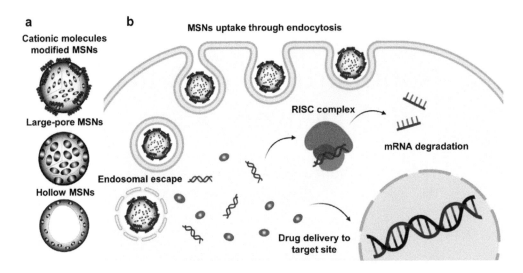

FIGURE 36.1 (a) Structural variations of the MSN platform used for siRNA delivery. MSNs can be modified with cationic molecules or polymers to electrostatically bind siRNA; large-pore MSNs have been fabricated to load siRNA inside the pores, and hollow MSNs (HMSNs) were developed to carry siRNA in the interior core or on the exterior surface of the nanoparticles. (b) Schematic representation of the cellular internalization of siRNA/drug-loaded MSNs. The scheme depicts the nanoparticle uptake through the endo-lysosomal pathway, endosomal escape by different mechanisms, and the efficient release of siRNA and/or drugs in the cytoplasm.

groups or cationic polymers (Figure 36.1a). The cationic modifications change the surface charge of the MSNs to complex anionic siRNA through electrostatic interactions. Cationic polyethyleneimine (PEI) polymer is the most frequently used modification to enhance the siRNA loading (17–20). Other strategies involve the use of PAMAM dendrimers (21), poly-L-lysine (PLL) (22), and other cationic biomaterials (23). Nevertheless, due to the potential toxicity of cationic polymers, other strategies are being investigated. For example, the use of calcium ion-mediated interconnection between the phosphates of RNA and silicates of MSNs is reported as a promising alternative for siRNA loading (24). To expand the application of the MSN platform for siRNA delivery, the structural properties of the material have been modified. Large pore MSNs (25,26) and hollow MSNs (27,28) have been developed to increase the pore capacity with the purpose of increasing siRNA loading and efficient delivery (Figure 36.1a).

The protection of siRNA against nucleases in the systemic circulation is a key consideration for a successful siRNA delivery system. Complexing of siRNA on the surface of MSNs has shown to protect the siRNA payload from enzymatic degradation (29). In order to offer improved protection, siRNA loading into the large pores was shown to increase the level of protection against enzymatic attack compared to surface complexed siRNA (30). Also, polymers or lipid bilayer coated MSNs have demonstrated to further increase the protection of siRNA encapsulated into the porous structure (31,32).

Targeting of siRNA-loaded MSNs toward specific cells is achieved by functionalizing their surface with moieties like antibodies or peptides, which are specific to the targeted cells (31,33,34).

To accomplish the release of siRNA into the cytoplasm, MSNs are designed to incorporate various strategies for endosomal escape (Figure 36.1b). These include the use of cationic polymers, which assist the delivery of siRNA through proton sponge effect. The principle behind this effect is based on the self-buffering capacity of cationic polymers, which ruptures the endosomes releasing siRNA in the cytoplasm (20). Moreover, MSNs can be loaded with near-infrared dyes such as indocyanine green (ICG). Upon light irradiation, the ICG loaded in MSNs locally generates reactive oxygen species (ROS) that break the endosomal vesicles. As a result, endosomal escape of siRNA into cytoplasm is produced. This phenomenon is termed as photochemical internalization (PCI) (35). Besides, MSNs can be modified with redox-responsive moieties to enhance the delivery of siRNA in reducing environments. For example, Sun et al. reported the design of MSNs consisting of organo-silica framework that is hybridized with disulfide bonds. This material is degraded under the reducing environment of cancer cells to achieve controlled release of siRNA (36).

THERAPEUTIC APPLICATIONS OF siRNA-LOADED MSN PLATFORM IN HUMAN DISEASES

MSNs have been investigated for siRNA delivery for various disorders such as cancer, fibrosis, and osteoporosis (8). MSNs carrying siRNA against HER2 oncogene, which is amplified in human cancers leading to tumor progression, have been investigated. The MSNs' surface was further functionalized with trastuzumab, a specific antibody to target HER2 positive cancer cells. The siHER2-MSN system was evaluated in an orthotopic model of breast cancer. The siHER2-MSNs showed efficient HER2 knockdown and tumor growth inhibition (33). Similarly, Lio et al. reported the topical delivery of siRNA targeting TGFβR-1 for skin squamous cell carcinoma (SCC) therapy. The MSNs were loaded with TGFβR-1 siRNA and further coated with PLL polymer to complex siRNA and improve cellular internalization. The transdermal delivery of topically applied MSNs was investigated in the SCC xenograft model, exhibiting minimal growth of the tumors with attenuated TGFβR-1 in treated tumors (22). Other groups have successfully applied MSNs against TWIST (37) and K-ras (29) gene silencing for treating ovarian and pancreatic cancer, respectively. A recent breakthrough for RNA therapeutics against cancer is the development of programmable nucleic acid nanoparticles (NANPs) (38). These NANPs can be designed with different morphologies such as globular, planar, or fibrous. Our group engineered MSNs for the efficient delivery of fibrous RNA NANPs. The internalization of RNA fibers using MSNs and subsequent GFP silencing was observed in a triple-negative breast cancer (TNBC) cell line (17).

Despite RNA therapeutics are mainly investigated for cancer treatment, their applications go beyond cancer. Our group has recently reported on the use of MSNs for the delivery of siRNA against Tenascin C (TnC). This glycoprotein plays a major role in fibrogenesis. Inhibition of TnC has been studied as a promising target to treat hepatic cirrhosis. The siTnC-MSNs showed efficient knockdown of TnC expression in hepatic stellate cells (HSC) leading to their decreased invasion as well as diminished downstream inflammatory cytokines (18). Similarly, PEI-coated MSNs were used for the delivery of siRNA against heat shock protein 47 (HSP47) to reduce fibrosis. The intradermal administration of siHSP47-MSNs in the scleroderma mouse model exhibited an efficient knockdown of HSP47 proteins. This resulted in alleviated skin fibrosis evident from the decreased expression of fibrotic markers and skin thickness in mice. Besides, the intrinsic antioxidant ability of the MSN core can act synergistically with gene therapy in such inflammatory diseases (39).

MULTIFUNCTIONAL APPLICATIONS OF siRNA-LOADED MSNs FOR CANCER TREATMENT

The outstanding structural features associated with the MSN platform bestow them with superior performance for combination therapies. In particular, by having two distinctive surfaces, interior pores and an exterior surface, MSNs permit the loading of siRNA on the external surface and the simultaneous encapsulation of anticancer drugs within the pores. This strategy makes MSNs highly suitable to develop multifunctional systems to combine chemo and siRNA therapies (40). Meng et al. used PEG-PEI copolymer coated MSNs for the co-delivery of doxorubicin (DOX) and siRNA to overcome drug resistance by silencing the P-glycoprotein (P-gp) drug exporter. The intravenous administration of DOX/siP-gp-loaded MSNs allowed P-gp knockdown and synergistic inhibition of tumor growth (19). Han et al. reported MSNs loaded with DOX and siRNA against vascular endothelial growth factor (VEGF). VEGF is an important cytokine required for angiogenesis in tumors. MSNs were further functionalized with galactose for increased targeting toward hepatocellular carcinoma (HCC) cells. The DOX/siVEGF-loaded MSNs showed potent tumor efficacy along with VEGF gene silencing in the HCC tumors (41). Similarly, other groups have successfully applied MSNs-based platform for combining DOX or sorafenib with siVEGF (42,43).

Large-pore MSNs and hollow MSNs have been extensively used to deliver chemotherapeutic drugs combined with siRNA that silence genes associated with major challenges for the treatment of cancer, for instance, the genes responsible for multidrug resistance proteins, including P-gp (19,28,44,45), MDR1 (46), and T-type Ca^{2+} channels (47). Also, antiapoptotic proteins, including Bcl2 (21,23,27,48), and survivin, (49) have been used as targets. Apart from chemotherapy drugs, MSNs have also been used to combine siRNA with molecular inhibitors and miRNA agents for cancer therapy (34,50). In particular, Wang et al. reported MSNs for the co-delivery of siRNA against Polo-like kinase1 (siPlk1) and miR-200c. Plk1 is a cell cycle regulator overexpressed in tumors, whereas miR-200c plays a vital role in the epithelial mesenchymal transition (EMT) essential for cancer metastasis. The Plk1 plus miR-200c-loaded MSNs showed superior regression of the primary tumor as well as exhibited antimetastatic ability in an aggressive metastatic TNBC model (35).

THERANOSTIC APPLICATIONS OF siRNA-LOADED MSNs FOR CANCER IMAGING AND THERAPY

Theranostic platforms integrate simultaneous bioimaging and drug delivery for the treatment of a disease using a single nanocarrier (51,52). MSNs are ideal candidates for theranostic applications owing to the facile functionalization of their internal/external surfaces. Chen et al. reported the use of MSNs modified with magnetic nanoparticles in the MSN core for the delivery of siRNA against VEGF as an anti-angiogenesis therapy to treat lung cancer. The magnetic core of these MSNs rendered T2-weighted contrast magnetic resonance imaging (MRI) properties to the platform. MSNs were further functionalized with KALA fusogenic peptide as targeting ligands. The intravenous administration of the nanocomposite successfully achieved VEGF knockdown and antitumor effect in ectopic and orthotopic mice models. Moreover, the MSN-based nanocomposite was successfully utilized in the detection of the metastatic tumor nodules (53). The same group applied this MSN system for the effective treatment and imaging of ovarian cancer (54). Yang et al. also reported the use of MSNs for co-delivery of DOX/siP-gp. In this work, MSNs contain Fe_3O_4-Au nanocrystals as the core for simultaneous MRI and a photosensitizer (chlorin-e6, Ce6) for photodynamic therapy. *In vivo* therapeutic efficacy of the multimodal nanocomposite was evaluated in a breast cancer mice model. The multimodal approach, which combined chemo, RNA, and photodynamic therapy, showed a significant decrease in tumor growth. In addition, the MRI capability of the platform allowed the monitoring of tumor growth inhibition.

CONCLUSIONS

RNA therapeutics is a burgeoning field in medicine that has the potential to target a wide variety of diseases. However, the successful clinical application of RNA therapeutics is hindered by their inefficient delivery. MSN platform has emerged as a promising nanocarrier for siRNA delivery. MSNs have shown to protect siRNA against enzymatic degradation, efficiently transport and deliver siRNA. In addition, due to their outstanding structural and chemical features, the use of MSNs for siRNA therapy has been expanded to develop multimodal and theranostic platforms. MSNs are used as a delivery system *in vitro* and *in vivo* to demonstrate the effective therapeutic capacity of siRNA against various diseases, including fibrosis, liver dysfunction, osteoporosis, neuron degeneration, and different types of cancer. We envision that the promising results obtained with MSNs as siRNA delivery system and their multimodal/theranostic applications will encourage future studies that pave the way for the translation of this platform for clinical use.

REFERENCES

1. Alvarez-Salas LM. Nucleic acids as therapeutic agents. *Current Topics in Medicinal Chemistry*. 2008;8(15):1379–404. Epub 2008/11/11. doi: 10.2174/156802608786141133. PubMed PMID: 18991725.
2. Teo PY, Cheng W, Hedrick JL, Yang YY. Co-delivery of drugs and plasmid DNA for cancer therapy. *Advanced Drug Delivery Reviews*. 2016;98:41–63. Epub 2015/11/04. doi: 10.1016/j.addr.2015.10.014. PubMed PMID: 26529199.
3. Li F, Mahato RI. miRNAs as targets for cancer treatment: therapeutics design and delivery. Preface. *Advanced Drug Delivery Reviews*. 2015;81:v–vi. Epub 2014/12/17. doi: 10.1016/j.addr.2014.11.005. PubMed PMID: 25500272; PMCID: PMC4778390.
4. Du L, Gatti RA. Progress toward therapy with antisense-mediated splicing modulation. *Current Opinion in Molecular Therapeutics*. 2009;11(2):116–23. Epub 2009/03/31. PubMed PMID: 19330717; PMCID: PMC2753608.
5. Reynolds A, Leake D, Boese Q, Scaringe S, Marshall WS, Khvorova A. Rational siRNA design for RNA interference. *Nature Biotechnology*. 2004;22(3):326–30. Epub 2004/02/06. doi: 10.1038/nbt936. PubMed PMID: 14758366.
6. Preall JB, Sontheimer EJ. RNAi: RISC gets loaded. *Cell*. 2005;123(4):543–5. Epub 2005/11/16. doi: 10.1016/j.cell.2005.11.006. PubMed PMID: 16286001.
7. Haussecker D. Current issues of RNAi therapeutics delivery and development. *Journal of Controlled Release: Official Journal of the Controlled Release Society*. 2014;195:49–54. Epub 2014/08/12. doi: 10.1016/j.jconrel.2014.07.056. PubMed PMID: 25111131.
8. Kim B, Park J-H, Sailor MJ. Rekindling RNAi Therapy: Materials Design Requirements for In Vivo siRNA Delivery. *Advanced Materials*. 2019;31(49):1903637. doi: 10.1002/adma.201903637.
9. Giacca M, Zacchigna S. Virus-mediated gene delivery for human gene therapy. *Journal of Controlled Release: Official Journal of the Controlled Release Society*. 2012;161(2):377–88. Epub 2012/04/21. doi: 10.1016/j.jconrel.2012.04.008. PubMed PMID: 22516095.
10. de Jesus MB, Zuhorn IS. Solid lipid nanoparticles as nucleic acid delivery system: properties and molecular mechanisms. *Journal of Controlled Release: Official Journal of the Controlled Release Society*. 2015;201:1–13. Epub 2015/01/13. doi: 10.1016/j.jconrel.2015.01.010. PubMed PMID: 25578828.
11. Lehto T, Ezzat K, Wood MJA, El Andaloussi S. Peptides for nucleic acid delivery. *Advanced Drug Delivery Reviews*. 2016;106(Pt A):172–82. Epub 2016/10/31. doi: 10.1016/j.addr.2016.06.008. PubMed PMID: 27349594.
12. Adams D, Gonzalez-Duarte A, O'Riordan WD, Yang CC, Ueda M, Kristen AV, Tournev I, Schmidt HH, Coelho T, Berk JL, Lin KP, Vita G, Attarian S, Plante-Bordeneuve V, Mezei MM, Campistol JM, Buades J, Brannagan TH, 3rd, Kim BJ, Oh J, Parman Y, Sekijima Y, Hawkins PN, Solomon SD, Polydefkis M, Dyck PJ, Gandhi PJ, Goyal S, Chen J, Strahs AL, Nochur SV, Sweetser MT, Garg PP, Vaishnaw AK, Gollob JA, Suhr OB. Patisiran, an RNAi Therapeutic, for Hereditary Transthyretin Amyloidosis. *The New England Journal of Medicine*. 2018;379(1):11–21. Epub 2018/07/05. doi: 10.1056/NEJMoa1716153. PubMed PMID: 29972753.
13. Slowing II, Vivero-Escoto JL, Wu C-W, Lin VSY. Mesoporous silica nanoparticles as controlled release drug delivery and gene transfection carriers. *Advanced Drug Delivery Reviews*. 2008;60(11):1278–88. doi: 10.1016/j.addr.2008.03.012. PubMed PMID: 18514969.
14. Keasberry NA, Yapp CW, Idris A. Mesoporous silica nanoparticles as a carrier platform for intracellular delivery of nucleic acids. *Biochemistry* (Moscow). 2017;82(6):655–62. doi: 10.1134/S0006297917060025.
15. Vivero-Escoto JL, Slowing II, Trewyn BG, Lin VS-Y. Mesoporous Silica Nanoparticles for Intracellular Controlled Drug Delivery. *Small* (Weinheim an der Bergstrasse, Germany). 2010;6(18):1952–67. doi: 10.1002/smll.200901789.
16. Cha W, Fan R, Miao Y, Zhou Y, Qin C, Shan X, Wan X, Li J. Mesoporous Silica Nanoparticles as Carriers for Intracellular Delivery of Nucleic Acids and Subsequent Therapeutic Applications. *Molecules* (Basel, Switzerland). 2017;22(5). Epub 2017/05/12. doi: 10.3390/molecules22050782. PubMed PMID: 28492505; PMCID: PMC6154527.
17. Rackley L, Stewart JM, Salotti J, Krokhotin A, Shah A, Halman JR, Juneja R, Smollett J, Lee L, Roark K, Viard M, Tarannum M, Vivero-Escoto J, Johnson PF, Dobrovolskaia MA, Dokholyan NV, Franco E, Afonin KA. RNA Fibers as Optimized Nanoscaffolds for siRNA Coordination and Reduced Immunological Recognition. *Advanced Functional Materials*. 2018;28(48):1805959. Epub 10/09. doi: 10.1002/adfm.201805959. PubMed PMID: 31258458.

18. Vivero-Escoto JL, Vadarevu H, Juneja R, Schrum LW, Benbow JH. Nanoparticle mediated silencing of tenascin C in hepatic stellate cells: effect on inflammatory gene expression and cell migration. *Journal of Materials Chemistry B*. 2019;7(46):7396–405. doi: 10.1039/C9TB01845J.

19. Meng H, Mai WX, Zhang H, Xue M, Xia T, Lin S, Wang X, Zhao Y, Ji Z, Zink JI, Nel AE. Codelivery of an optimal drug/siRNA combination using mesoporous silica nanoparticles to overcome drug resistance in breast cancer in vitro and in vivo. *ACS Nano*. 2013;7(2):994–1005. Epub 2013/01/08. doi: 10.1021/nn3044066. PubMed PMID: 23289892; PMCID: PMC3620006.

20. Xia T, Kovochich M, Liong M, Meng H, Kabehie S, George S, Zink JI, Nel AE. Polyethyleneimine Coating Enhances the Cellular Uptake of Mesoporous Silica Nanoparticles and Allows Safe Delivery of siRNA and DNA Constructs. *ACS Nano*. 2009;3(10):3273–86. doi: 10.1021/nn900918w.

21. Chen AM, Zhang M, Wei D, Stueber D, Taratula O, Minko T, He H. Co-delivery of doxorubicin and Bcl-2 siRNA by mesoporous silica nanoparticles enhances the efficacy of chemotherapy in multidrug-resistant cancer cells. *Small* (Weinheim an der Bergstrasse, Germany). 2009;5(23):2673–7. Epub 2009/09/26. doi: 10.1002/smll.200900621. PubMed PMID: 19780069; PMCID: PMC2833276.

22. Lio DCS, Liu C, Oo MMS, Wiraja C, Teo MHY, Zheng M, Chew SWT, Wang X, Xu C. Transdermal delivery of small interfering RNAs with topically applied mesoporous silica nanoparticles for facile skin cancer treatment. *Nanoscale*. 2019;11(36):17041–51. doi: 10.1039/C9NR06303J.

23. Pan Q-S, Chen T-T, Nie C-P, Yi J-T, Liu C, Hu Y-L, Chu X. In Situ Synthesis of Ultrathin ZIF-8 Film-Coated MSNs for Codelivering Bcl 2 siRNA and Doxorubicin to Enhance Chemotherapeutic Efficacy in Drug-Resistant Cancer Cells. *ACS Applied Materials & Interfaces*. 2018;10(39):33070–7. doi: 10.1021/acsami.8b13393.

24. Choi E, Lee J, Kwon IC, Lim DK, Kim S. Cumulative directional calcium gluing between phosphate and silicate: A facile, robust and biocompatible strategy for siRNA delivery by amine-free non-positive vector. *Biomaterials*. 2019;209:126–37. Epub 2019/04/30. doi: 10.1016/j.biomaterials.2019.04.006. PubMed PMID: 31034981.

25. Hartono SB, Gu W, Kleitz F, Liu J, He L, Middelberg APJ, Yu C, Lu GQ, Qiao SZ. Poly-l-lysine Functionalized Large Pore Cubic Mesostructured Silica Nanoparticles as Biocompatible Carriers for Gene Delivery. *ACS Nano*. 2012;6(3):2104–17. doi: 10.1021/nn2039643.

26. Meka AK, Niu Y, Karmakar S, Hartono SB, Zhang J, Lin CXC, Zhang H, Whittaker A, Jack K, Yu M, Yu C. Facile Synthesis of Large-Pore Bicontinuous Cubic Mesoporous Silica Nanoparticles for Intracellular Gene Delivery. *ChemNanoMat*. 2016;2(3):220–5. doi: 10.1002/cnma.201600021.

27. Ma X, Zhao Y, Ng KW, Zhao Y. Integrated Hollow Mesoporous Silica Nanoparticles for Target Drug/siRNA Co-Delivery. *Chemistry – A European Journal*. 2013;19(46):15593–603. doi: 10.1002/chem.201302736.

28. Wu M, Meng Q, Chen Y, Zhang L, Li M, Cai X, Li Y, Yu P, Zhang L, Shi J. Large pore-sized hollow mesoporous organosilica for redox-responsive gene delivery and synergistic cancer chemotherapy. *Advanced Materials* (Deerfield Beach, Fla). 2016;28(10):1963–9. Epub 2016/01/09. doi: 10.1002/adma.201505524. PubMed PMID: 26743228.

29. Hom C, Lu J, Liong M, Luo H, Li Z, Zink JI, Tamanoi F. Mesoporous silica nanoparticles facilitate delivery of siRNA to shutdown signaling pathways in mammalian cells. *Small* (Weinheim an der Bergstrasse, Germany). 2010;6(11):1185–90. Epub 2010/05/13. doi: 10.1002/smll.200901966. PubMed PMID: 20461725; PMCID: PMC2953950.

30. Na H-K, Kim M-H, Park K, Ryoo S-R, Lee KE, Jeon H, Ryoo R, Hyeon C, Min D-H. Efficient functional delivery of siRNA using mesoporous silica nanoparticles with ultralarge pores. *Small* 2012;8(11):1752–61. doi: 10.1002/smll.201200028.

31. Ashley CE, Carnes EC, Epler KE, Padilla DP, Phillips GK, Castillo RE, Wilkinson DC, Wilkinson BS, Burgard CA, Kalinich RM, Townson JL, Chackerian B, Willman CL, Peabody DS, Wharton W, Brinker CJ. Delivery of small interfering RNA by peptide-targeted mesoporous silica nanoparticle-supported lipid bilayers. *ACS Nano*. 2012;6(3):2174–88. doi: 10.1021/nn204102q.

32. Gao F, Botella P, Corma A, Blesa J, Dong L. Monodispersed Mesoporous Silica Nanoparticles with Very Large Pores for Enhanced Adsorption and Release of DNA. *The Journal of Physical Chemistry B*. 2009;113(6):1796–804. doi: 10.1021/jp807956r.

33. Ngamcherdtrakul W, Morry J, Gu S, Castro DJ, Goodyear SM, Sangvanich T, Reda MM, Lee R, Mihelic SA, Beckman BL, Hu Z, Gray JW, Yantasee W. Cationic polymer modified mesoporous silica nanoparticles for targeted SiRNA delivery to HER2+ breast cancer. *Advanced Functional Materials*. 2015;25(18):2646–59. doi: 10.1002/adfm.201404629. PubMed PMID: 26097445.

34. Zheng G, Zhao R, Xu A, Shen Z, Chen X, Shao J. Co-delivery of sorafenib and siVEGF based on mesoporous silica nanoparticles for ASGPR mediated targeted HCC therapy. *European Journal of Pharmaceutical Sciences*. 2018;111:492–502. doi: 10.1016/j.ejps.2017.10.036.

35. Wang Y, Xie Y, Kilchrist KV, Li J, Duvall CL, Oupický D. Endosomolytic and Tumor-Penetrating Mesoporous Silica Nanoparticles for siRNA/miRNA Combination Cancer Therapy. *ACS Applied Materials & Interfaces*. 2020;12(4):4308–22. doi: 10.1021/acsami.9b21214.

36. Sun L, Wang D, Chen Y, Wang L, Huang P, Li Y, Liu Z, Yao H, Shi J. Core-shell hierarchical meso-structured silica nanoparticles for gene/chemo-synergetic stepwise therapy of multidrug-resistant cancer. *Biomaterials*. 2017;133:219–28. Epub 2017/04/26. doi: 10.1016/j.biomaterials.2017.04.028. PubMed PMID: 28441616.

37. Shahin SA, Wang R, Simargi SI, Contreras A, Parra Echavarria L, Qu L, Wen W, Dellinger T, Unternaehrer J, Tamanoi F, Zink JI, Glackin CA. Hyaluronic acid conjugated nanoparticle delivery of siRNA against TWIST reduces tumor burden and enhances sensitivity to cisplatin in ovarian cancer. *Nanomedicine: Nanotechnology, Biology, and Medicine*. 2018;14(4):1381–94. Epub 2018/04/18. doi: 10.1016/j.nano.2018.04.008. PubMed PMID: 29665439; PMCID: PMC6186509.

38. Afonin KA, Viard M, Koyfman AY, Martins AN, Kasprzak WK, Panigaj M, Desai R, Santhanam A, Grabow WW, Jaeger L, Heldman E, Reiser J, Chiu W, Freed EO, Shapiro BA. Multifunctional RNA Nanoparticles. *Nano Letters*. 2014;14(10):5662–71. doi: 10.1021/nl502385k.

39. Morry J, Ngamcherdtrakul W, Gu S, Goodyear SM, Castro DJ, Reda MM, Sangvanich T, Yantasee W. Dermal delivery of HSP47 siRNA with NOX4-modulating mesoporous silica-based nanoparticles for treating fibrosis. *Biomaterials*. 2015;66:41–52. doi: 10.1016/j.biomaterials.2015.07.005.

40. Zhou Y, Quan G, Wu Q, Zhang X, Niu B, Wu B, Huang Y, Pan X, Wu C. Mesoporous silica nanoparticles for drug and gene delivery. *Acta Pharmaceutica Sinica B*. 2018;8(2):165–77. doi: 10.1016/j.apsb.2018.01.007.

41. Han L, Tang C, Yin C. Dual-targeting and pH/redox-responsive multi-layered nanocomplexes for smart co-delivery of doxorubicin and siRNA. *Biomaterials*. 2015;60:42–52. doi: 10.1016/j.biomaterials.2015.05.001.

42. Li T, Shen X, Geng Y, Chen Z, Li L, Li S, Yang H, Wu C, Zeng H, Liu Y. Folate-Functionalized Magnetic-Mesoporous Silica Nanoparticles for Drug/Gene Codelivery To Potentiate the Antitumor Efficacy. *ACS Applied Materials & Interfaces*. 2016;8(22):13748–58. doi: 10.1021/acsami.6b02963.

43. Zheng G, Zhao R, Xu A, Shen Z, Chen X, Shao J. Co-delivery of sorafenib and siVEGF based on mesoporous silica nanoparticles for ASGPR mediated targeted HCC therapy. *European Journal of Pharmaceutical Sciences: Official Journal of the European Federation for Pharmaceutical Sciences*. 2018;111:492–502. Epub 2017/11/07. doi: 10.1016/j.ejps.2017.10.036. PubMed PMID: 29107835.

44. Meng H, Liong M, Xia T, Li Z, Ji Z, Zink JI, Nel AE. Engineered design of mesoporous silica nanoparticles to deliver doxorubicin and P-glycoprotein siRNA to overcome drug resistance in a cancer cell line. *ACS Nano*. 2010;4(8):4539–50. doi: 10.1021/nn100690m. PubMed PMID: 20731437.

45. Cheng W, Liang C, Wang X, Tsai HI, Liu G, Peng Y, Nie J, Huang L, Mei L, Zeng X. A drug-self-gated and tumor microenvironment-responsive mesoporous silica vehicle: "four-in-one" versatile nanomedicine for targeted multidrug-resistant cancer therapy. *Nanoscale*. 2017;9(43):17063–73. Epub 2017/11/01. doi: 10.1039/c7nr05450e. PubMed PMID: 29085938.

46. Wang D, Xu X, Zhang K, Sun B, Wang L, Meng L, Liu Q, Zheng C, Yang B, Sun H. Codelivery of doxo-rubicin and MDR1-siRNA by mesoporous silica nanoparticles-polymerpolyethylenimine to improve oral squamous carcinoma treatment. *International Journal of Nanomedicine*. 2017;13:187–98. doi: 10.2147/IJN.S150610. PubMed PMID: 29343957.

47. Wang S, Liu X, Chen S, Liu Z, Zhang X, Liang XJ, Li L. Regulation of Ca(2+) Signaling for Drug-Resistant Breast Cancer Therapy with Mesoporous Silica Nanocapsule Encapsulated Doxorubicin/siRNA Cocktail. *ACS Nano*. 2019;13(1):274–83. Epub 2018/12/20. doi: 10.1021/acsnano.8b05639. PubMed PMID: 30566319.

48. Zhao S, Xu M, Cao C, Yu Q, Zhou Y, Liu J. A redox-responsive strategy using mesoporous silica nanoparticles for co-delivery of siRNA and doxorubicin. *Journal of Materials Chemistry B*. 2017;5(33):6908–19. doi: 10.1039/C7TB00613F.

49. Li Z, Zhang L, Tang C, Yin C. Co-Delivery of Doxorubicin and Survivin shRNA-Expressing Plasmid Via Microenvironment-Responsive Dendritic Mesoporous Silica Nanoparticles for Synergistic Cancer Therapy. *Pharmaceutical Research*. 2017;34(12):2829–41. Epub 2017/09/28. doi: 10.1007/s11095-017-2264-6. PubMed PMID: 28948461.

50. Shi XL, Li Y, Zhao LM, Su LW, Ding G. Delivery of MTH1 inhibitor (TH287) and MDR1 siRNA via hyaluronic acid-based mesoporous silica nanoparticles for oral cancers treatment. *Colloids and Surfaces B, Biointerfaces*. 2019;173:599–606. Epub 2018/10/24. doi: 10.1016/j.colsurfb.2018.09.076. PubMed PMID: 30352381.

51. Jafari S, Derakhshankhah H, Alaei L, Fattahi A, Varnamkhasti BS, Saboury AA. Mesoporous silica nanoparticles for therapeutic/diagnostic applications. *Biomedicine & Pharmacotherapy*. 2019;109:1100–11. doi: 10.1016/j.biopha.2018.10.167.

52. Ravindran Girija A, Balasubramanian S. Theragnostic potentials of core/shell mesoporous silica nanostructures. *Nanotheranostics*. 2019;3(1):1–40. doi: 10.7150/ntno.27877. PubMed PMID: 30662821.

53. Chen Y, Gu H, Zhang DS-Z, Li F, Liu T, Xia W. Highly effective inhibition of lung cancer growth and metastasis by systemic delivery of siRNA via multimodal mesoporous silica-based nanocarrier. *Biomaterials*. 2014;35(38):10058–69. doi: 10.1016/j.biomaterials.2014.09.003.

54. Chen Y, Wang X, Liu T, Zhang DS, Wang Y, Gu H, Di W. Highly effective antiangiogenesis via magnetic mesoporous silica-based siRNA vehicle targeting the VEGF gene for orthotopic ovarian cancer therapy. *International Journal of Nanomedicine*. 2015;10:2579–94. Epub 2015/04/08. doi: 10.2147/ijn.s78774. PubMed PMID: 25848273; PMCID: PMC4386807.

37 Method of Large-Scale Exosome Purification and Its Use for Pharmaceutical Applications

Fengmei Pi
ExonanoRNA LLC, Columbus, USA

Yangwu Fang and Dongsheng Li
ExonanoRNA Biomedicine LTD, Foshan, China

Peixuan Guo
The Ohio State University, Columbus, USA

CONTENTS

INTRODUCTION

Exosomes are a class of extracellular vesicles (EVs) that originated from inward budding of endosome membranes into the multivesicular bodies. They contain endosomal trafficking markers such as tumor susceptibility gene 101 (TSG101), apoptosis linked gene-2 interacting protein (Alix), tetraspanin (CD63) and flotillins (Raab-Traub and Dittmer, 2017). Originally, exosomes were thought to act as cellular garbage disposals. Recent studies have found exosomes are natural carriers for mRNA, lncRNA, miRNA, siRNA, protein, DNA and peptide for long- distance intercellular communication. EVs also involve in regulating the cell differentiation, angiogenesis, metabolic reprogramming, tumor progression, immune modulation and pathogen challenges (Willms et al., 2018). Substantial studies have demonstrated that EVs are a potential source of cancer biomarkers. There has been a great progress from discovery of these biomarkers to validation and clinical application as cancer liquid biopsy (Zhao et al., 2019). EVs have also been studied as a natural delivery vehicle for small RNA therapeutics. Exosomes isolated from dendritic cells reengineered with RVG peptide can systemically deliver GAPDH siRNA to the brain after systemic injection into mice (Alvarez-Erviti et al., 2011). EVs can be also post genesis decorated with artificial RNA nanoparticle-based ligand to enhance its targeting selectively to cancer cells, and it has been used for delivery of siRNA for cancer regression in prostate cancer, breast cancer and colorectal cancer mice model (Pi et al., 2018).

Isolation of exosomes and extracellular RNA is of great importance for biomarkers for early disease diagnosis (Nazarenko, 2020) and is being used for therapeutics development. Significant obstacles to purify exosomes in large quantity and high purity still exist. The gold standard for exosome isolation has been differential ultracentrifugation method developed by Thery (Thery et al., 2006). Due to the multistep process and requirement of ultracentrifugation, other methods based on the principle of size exclusion or affinity capture have also been developed. Here, we review the recent updates on techniques for exosome purification and isolation and discuss about large-scale exosome purification and quality control for future pharmaceutical development.

CURRENT METHOD FOR EXOSOME PURIFICATION

The three major groups of EVs described by its mechanism of generation are microvesicles, apoptotic bodies and exosomes. The three groups are overlapping in size. When considering getting exosomes as ingredient for pharmaceutical development, the primary impurities to remove in exosome production will be cell debris, primarily proteins and host cell DNAs, and the other types of EVs including microvesicles and apoptotic bodies. Usually, the removal of virus infection and endotoxins are required.

DIFFERENTIAL ULTRACENTRIFUGATION

The traditional differential ultracentrifugation method (Thery et al., 2006) is still the gold standard for exosome purification and is used by many labs, which involves several successive steps of centrifugation and ultracentrifugation. As summarized in Figure 37.1, the step of centrifuge at 300g was to remove cell, at 2,000g was to remove dead cells, at 10,000g was to remove cell debris and an extra washing step with 100,000g was to remove contaminating proteins which are associated with exosomes or formed aggregates during the ultracentrifugation step. This method has been the most reliable method and been tested by many different labs. But the concern with it is the disruption of exosome structure integrity after two times pelleting under high ultracentrifugation force and also low yield.

Modified versions of this method have been reported in the past, instead of doing the stepwise centrifuge at 300g, 2000g and 10,000g to remove cells, dead cells and cell debris; you can use sterile filtration process by passing through the conditioned cell culture medium through 0.22μm filter under vacuum (Thery et al., 2006).

CUSHION-MODIFIED ULTRACENTRIFUGATION

30% sucrose in heavy water D_2O (Thery et al., 2006) and Optiprep™ 60% iodixanol in water (Pi et al., 2018) have been reported to be used as cushion to add to the bottom for exosome ultracentrifugation step. Since these cushion materials have higher density than exosomes, the isolated exosomes will form a layer on the top of the cushion, while the contaminating proteins with an average density around 1.35 g/ml are usually precipitated down to the bottom of the tube; on the other hand, the cushion layer can help to protect the exosomes' intact structure under the ultracentrifugation force to prevent them from deformation; while there is no significant difference in particle size distribution and zeta potential (Figure 37.2) (Pi et al., 2018). To get exosomes with high purity for pharmaceutical application, removal of the sucrose or Optiprep from the exosome sample will be a concern since

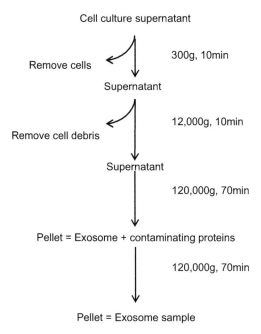

Cell culture supernatant

Remove cells ← 300g, 10min

Supernatant

Remove cell debris ← 12,000g, 10min

Supernatant

120,000g, 70min

Pellet = Exosome + contaminating proteins

120,000g, 70min

Pellet = Exosome sample

FIGURE 37.1 Schematics for the steps used in differential ultracentrifugation method for exosome purification {Thery, 2006 #239}.

they are having either high viscosity or high molecular weight. ExoJuice™ is a cushion material introduced by ExonanoRNA Company recently. Its osmotic pressure is 431 mOsm/kg, which is basically harmless to cells in the concentration below 3% (Figure 37.3, unpublished data). Its low molecular weight and low viscosity allow easy removal of the cushion material from the final puri-fied exosome samples by simply dialysis. The density of ExoJuice™ is closer to exosomes, which allows the exosomes to go into the cushion layer after ultracentrifugation, and we can get exosomes with less contamination with microvesicle contamination using ExoJuice as cushion for ultracentri-fugation (Figure 37.4a), comparing to using OptiPrep as cushion (Figure 37.4b) which usually show as the fraction of bigger particle size (unpublished data).

DENSITY GRADIENT ULTRACENTRIFUGATION

A relatively pure exosome isolation can be achieved by the combination of techniques using differ-ent separation principles, orthogonal method. The density of most human cell-derived exosomes is between 1.12 and 1.17g/ml (Onodi et al., 2018). To get enough EVs from blood plasma for in vivo experiment, the EVs should be separated from its major impurities such as lipoprotein particles and other abundant plasma protein albumin and fibrinogen. Onodi et al. loaded plasma supernatant onto a 10-30-50% iodixanol gradient layers, performed ultracentrifugation for 24 hrs and found that exo-some can be greatly separated from most of the contaminant proteins such as ApoA1, ApoB100, APoB48 which are lighter than EVs. Other protein with similar density such as fibrinogen must be removed by Size Exclusion Chromatography (SEC) method.

Density gradient was also used with cushion-modified ultracentrifugation pelleting method to purify exosomes. Li et al. (Li et al., 2018) isolated exosomes from macrophage cell culture by running ultracentrifuge at 100,000g with 60% iodixanol at the bottom as cushion, after collecting the cushion layers with exosomes, the crude exosomes were further purified with 5-10-20-40% iodixanol with ultracentrifugation at 100,000g for 18hrs to reach density equilibrium and found that exosomes with TSG101 and Alix expression are mainly at the fractions with density of 1.08 to

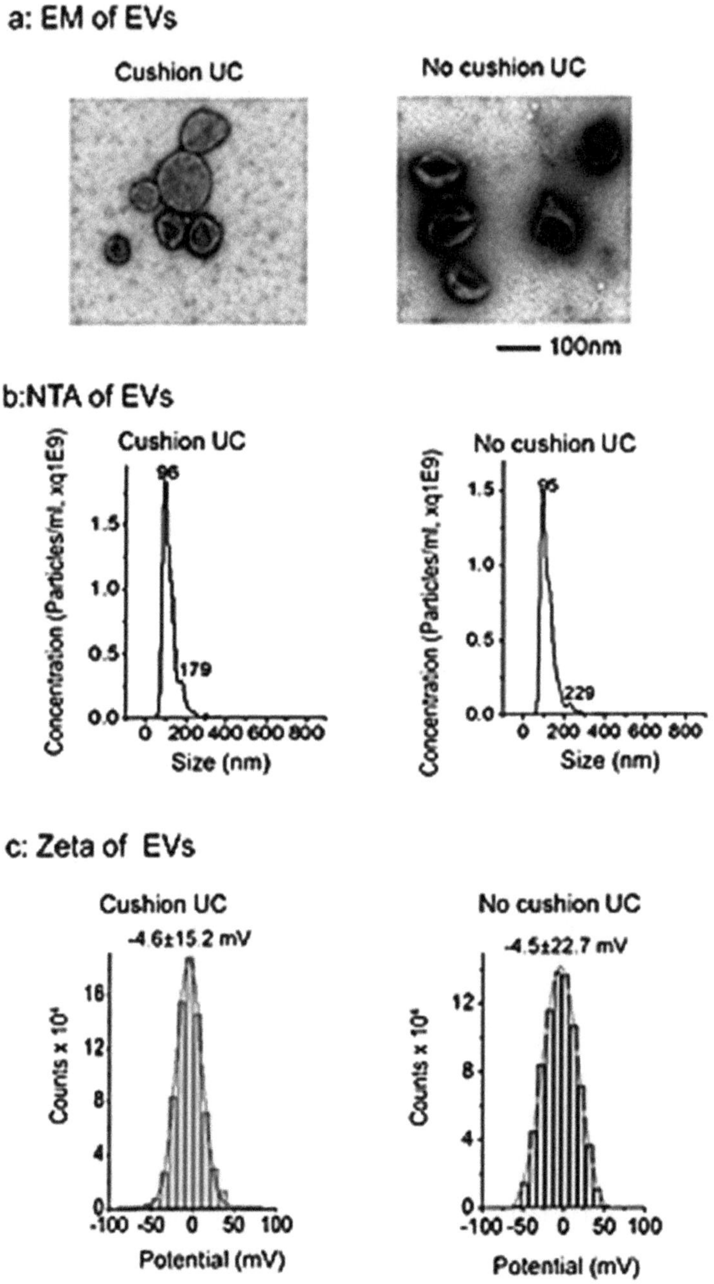

FIGURE 37.2   Cushion-modified ultracentrifugation (UC) method can prevent exosomes from physical disruption due to shear force, with keeping a more intact structure of isolated EVs under electron microscopy (a); there is no significant difference in the size distribution (b) and zeta potential (c) for EVs purified by differential UC or cushion-modified UC. Figure is produced with permission from Ref {Pi, 2018 #238} © Nature Publishing Group.

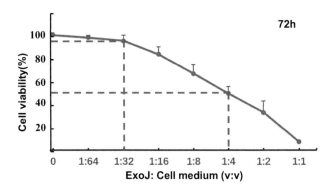

FIGURE 37.3   Cell viability of HEK293T cells incubated with different concentrations of ExoJuice for 72h (unpublished data).

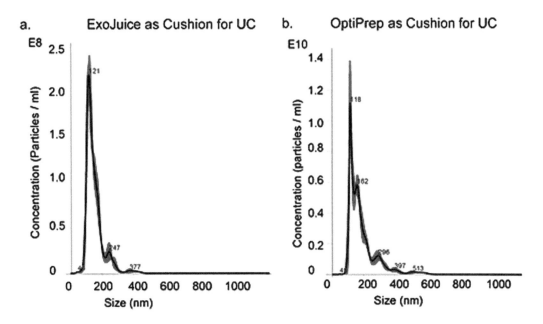

FIGURE 37.4   NTA of HEK293T cell-derived exosomes purified by using ExoJuice as cushion with ultracentrifugation method (a) or using OptiPrep as cushion with ultracentrifugation method (b) (unpublished data).

1.11mg/ml. The contaminating proteins are mainly in the later fractions with higher density. Similar approach was also used in purifying exosome-like particles from ginger roots (Li et al., 2018).

The advantage of above-mentioned orthogonal purification method is to get exosomes with higher purity, and the concern will be long time process with ultracentrifuge. While using ExoJuice as cushion at the bottom of tube and centrifuge at 100,000g for only 70min, it can form density gradient at the bottom fractions (Table 37.1). Therefore, the use of ExoJuice in the ultracentrifugation process gave the advantage of purifying exosomes by density difference and also by sedimentation property; it can greatly simplify the process for exosome purification and improve its purity.

TABLE 37.1

Using ExoJuice for Cushion After 100,000G Centrifuge for 70Min with Sw55 Rotor Formed Density Gradient. The Fractions After Ultracentrifugation was Taken Out and Measured for Their Density. Exosomes are Located in the Fractions of Density Near 1.1g/ml.

Fractions(150ul)	Density(g/ml)
When add 500ul for cushion	
4	1.045
3	1.1
2	1.129
1	1.146
When add 1000ul for cushion	
7	**1.045**
6	1.089
5	1.11
4	1.119
3	1.132
2	1.133
1	1.156

Size Exclusion Chromatography

The total EVs can be isolated through by SEC since most EVs are large than 700KDa, although EVs with the three populations of exosomes, apoptotic bodies and microvesicles cannot be separated. The core bead chromatography has been utilized successfully to purify EVs using SEC mechanism. BE-SEC column (HiScreen Capto Core 700 column, GE Lifescience) has been used with AKTA Prime plus system to purify EVs. The EVs larger than 700kDa bypassed the beads core in the column come out as early fractions while most contaminating proteins came in later fractions as a sharp peak. The result showed that about 80% EVs were collected using this method while keeping a better integrity comparing to traditional ultracentrifugation method. (Corso et al., 2017)

LARGE-SCALE EXOSOMES PRODUCTION AND PURIFICATION

Production of Exosomes

Large-scale exosome production relies on scaling up the cell culture. Most recent studies have focused on increasing the cell culture surface area. The cost-efficient and easy to operate mode is to culture cells in flasks or dishes, but this method is very labor intensive and has big quality variations among different plates. For scalable expansion of cells for exosome isolation, many bioreactors are already available, the multilayered vesicles and stirred suspension bioreactors have been used most in preparing exosomes for clinical trials (Panchalingam et al., 2015).

Multilayer cell culture factories (i.e., Nunc Cell Factory, Thermo Fisher Scientific, Waltham, MA, USA; Corning Cell-STACK; Corning, Corning, NY, USA), are typically stacking 4–40 stacks of flask layer into one unit, thus reducing the labor for space requirement for operation. It has been reported that culturing hMSCs with multilayer cell factories is similar to that of T-225 flasks, 10 of four-layered cell factories were used to obtain more than 200×10^6 of hMSCs for a clinical dose (Panchalingam et al., 2015).

Stirred suspension bioreactors are usually simple vesicles with a centrally located impeller to agitate the medium for uniform condition. Stirred suspension bioreactors at different volumes are available now, such as the DASGIP Parallel Bioreactor system and Celligen (Eppendorf, Hauppauge, NY, USA), PADReactor (Pall Life Sciences). With stirred suspension bioreactors, a large number of cells can be expanded in one vessel uniformly, thus, to avoid vessel-to-vessel variability and minimizing costs related to labor and consumables.

PURIFICATION OF EXOSOMES

The traditional differential ultracentrifugation method had been utilized for exosomes' isolation at small scale. It can also be utilized to prepare exosome at large scale at a GMP level though the process {Mendt, 2018 #248}{Dai, 2008 #247}. Polymer-based precipitation method is widely used for both small-scale and large-scale exosome purification. The principle is to use a reagent that binds to water molecules, thereby forcing the less soluble parts such as EVs to precipitate out of the solution. This method allows quick isolation of exosomes and eliminates the requirement for ultracentrifuge equipment. It was reported that precipitating exosomes from cell culture medium with adding PEG8000 could preserve the best activity of EVs comparing to other methods including ultracentrifugation, ultrafiltration (Niu et al., 2017). For large-scale production especially for pharmaceutical development, the removal and residue of polymer will be a major concern.

For large-scale exosome purification, Tangential Flow Filtration (TFF) has shown to be a promising approach for exosome industrialization. About 1liter of conditioned medium is collected from large-scale cell culture system; then it is filtered through 0.2 μm polyether sulfone (PES) membrane to remove the cell debris and macrovesicles, and then the medium is subjected to concentrating with ultrafiltration in a TFF equipment. A cartridge with 500 kDa cutoff was used to concentrate exosome from mesenchymal stem cells (Haraszti et al., 2018). The exosomes were concentrated 9 fold and then buffer exchanged with 6 volumes of PBS before final storage at −80°C. During the TFF process, the transmembrane pressure should be maintained very low (around 2 PSI) to minimize the loss of small exosomes. It was proposed to add a third step filtration by passing through the concentrated sample from TFF with a sterile 100nm track-etch polycarbonate membrane to remove the larger populations to get exosomes with higher purity (Heinemann et al., 2014). Comparing the quality of exosomes purified from ultracentrifugation method or ultrafiltration method, the ultrafiltration method showed to preserve more biological function of exosomes {Bari, 2019 #246}. It might be so because high shear force during ultracentrifugation caused physical damage to the exosomes. Also ultrafiltration showed to be able to recover higher yield of exosomes compared to ultracentrifugation method.

PROGRESS ON EXOSOME-BASED PHARMACEUTICS DEVELOPMENT

Although the field of exosome research is still in its infancy stage, there is still more to explore to understand the biological function of exosomes, as well as for impurities. The development of exosome for pharmaceutical application has shown great progress so far. There are already more than 10 formulations with exosome-gained FDA's approval for phase 1 or phase 2 clinical trial stages so far (Table 37.2). The formulations containing exosomes can be lyophilized to powder format for longer term storage {Bari, 2019 #246}. MSC-derived exosomes possess the similar immune stimulatory property to parent cells; it will be more advantageous to use cell-free formulation to achieve the similar clinical efficacy. Phase 1 clinical trial of ascites-derived exosomes showed that subcutaneous injection of exosomes purified from density gradient ultracentrifugation method in combination with the granulocyte-macrophage colony-stimulating factor is safe and well tolerated and can induce beneficial tumor-specific antitumor cytotoxic T lymphocyte response {Dai, 2008 #247}. Exosomes produced at GMP level by ultracentrifugation method were used as carriers to load siRNA against KRAS and have been shown to be able to regulate KRAS expression and extend the survival of PDX pancreatic cancer mice model {Mendt, 2018 #248}. Zhang et al. recently reported a breakthrough finding that aging rates are essentially controlled by exosomes in hypothalamic NSCs (htNSCs)

TABLE 37.2

Recent Clinical Trials for Developing Exosome-Based Pharmaceutics

Study Title	Conditions	Organization
Phase 1 Clinical trial investigating the Ability of Plant Exosomes to Deliver Curcumin to Normal and Maligant Colon Tissue	Colon Cancer	University of Louisville
Preliminary Clinical Trial Investigating the Ability of Plant Exosomes to Abrogate Oral Mucositis Induced by Combined Chemotherapy and Radiation in Head and Neck Cancer Patients	Head and Neck Cancer	University of Louisville
Effect of Plasma Derived Exosomes on Intractable Cutaneous Wound Healing Prospective Trial	Ulcer	Kumamoto University
A prelinilary Clinical Trial Investing the Ability of Plant Exosomes to Mitigate Insulin Resistance and Chronic Inflammation in Patients Diagnosed With Polycystic Ovary Syndrome (PCOS)	Polycystic Ovary Syndrome	University of Louisville
Allogenic Mesenchymal Stem Cell Derived Exosomes in Patients With Acute Ishemic Stroke	Cerebrovascular Disorders	Isfahan University of Medical Sciences
Phase 2 trial of a Vaccination with Tumor Antigen-loaded Dendritic Cell-Derived Exosomes on Patients With Unrespectable Non-Small Cell Lung Cancer Responding to Induction Chemotherapy	Non Small Cell Lung Cancer	Gustave Roussy, Cancer Campus, Grand Paris
Phase 1 Study of The Effect of Cell-Free Cord Blood Derived Microvescicles on ß-cell Massive type 1 Diabetes Mellitus (T1DM) Patients	Diabetes Mellitus Type 1	General Committee of Teaching Hospitals and Institute, Egypt
Mesenchymal Stem Cells Derived Exosomes Promote Healing of Large and Refractory Macular Holes	Macular Holes	Tianjin Medical University
Phase 1 Study of Mesenchymal Stromal Cells-Derived Exosomes with KrasG12D siRNA for Metastatic Pancreas Cancer Patients Harboring KrasG12D Mutation	Pancreatic Cancer	M.D. Anderson Cancer Center

[Zhang et al., 2017]. Mechanistically, htNSCs produce exosomal miRNAs present in the cerebro-spinal fluid of which more than 20 miRNAs are reduced during aging. Importantly, treatment with exosomes secreted by healthy htNSCs can delay aging, suggesting that these exosomes miRNAs are functionally associated with aging. Therefore, this work clearly emphasizes the role of exosomes in mediating anti-aging.

Genetically engineered T cells expressing the chimeric antigen receptor (CAR) are rapidly emerging and are expected to be used in new treatments for hematological and non-hematological malignancies. CAR-T cells release EVs, most of which are exosomes that carry CAR on their surface. CAR-containing exosomes express high levels of cytotoxic molecules and inhibit tumor growth. Compared to CAR-T cells, CAR exosomes do not express programmed cell death protein 1 (PD1), and recombinant PD-L1 treatment does not impair its anti-tumor effect. In a preclinical in vivo model of cytokine release syndrome, administration of CAR exosomes is relatively safe compared to CAR-T therapy (Wenyan Fu et al., 2019).

The progress in clinical trial on exosome-based pharmaceutics shows promising helping human beings to battle against cancer.

CONCLUSION

Exosome purification method development is undergoing continuous improvement along with the knowledge of understanding exosome contents, function. Here, we reviewed the recent progress in exosome purification method, especially for scalable processes and for pharmaceutical development. Exosomes isolated from stem cells can be used as regenerative medicine for immune therapy development; exosomes are also advanced drug delivery vehicles especially for intracellular delivery of siRNA since it can escape from endosome {Zheng, 2019 #249}. The recent progress in 3D cell culture system and bioreactors provides method for GMP production of cell-generated exosomes; a variety of methods for purifying exosomes at large scale are also available. Although large-scale exosome purification is still challenging, with the increasing use of new technologies and methods, it is becoming easier and more convenient to study exosomes.

ACKNOWLEDGMENTS

We thank members of P. Guo laboratories for valuable comments and suggestions. **This project was also supported by the National Institutes of Health (NIH; grants EB003730, GM059944 and CA151648 to P.G.). P.G. is also a co-funder of ExonanoRNA**.

REFERENCES

Alvarez-Erviti, L., Seow, Y., Yin, H., Betts, C., Lakhal, S., and Wood, M.J. (2011). Delivery of siRNA to the mouse brain by systemic injection of targeted exosomes. Nature Biotechnology 29, 341–345.

Corso, G., Mager, I., Lee, Y., Gorgens, A., Bultema, J., Giebel, B., Wood, M.J.A., Nordin, J.Z., and Andaloussi, S.E. (2017). Reproducible and scalable purification of extracellular vesicles using combined bind-elute and size exclusion chromatography. Scientific Reports 7, 11561.

Fu, W., Lei, C., Liu, S., Cui, Y., Wang, C., Qian, K., Li, T., Shen, Y., Fan, X., Lin, F., Ding, M., Pan, M., Ye, X., Yang, Y., and Hu, S. (2019). CAR exosomes derived from effector CAR-T cells have potent antitumour effects and low toxicity. Nature Communications 10, 1–2.

Haraszti, R.A., Miller, R., Stoppato, M., Sere, Y.Y., Coles, A., Didiot, M.C., Wollacott, R., Sapp, E., Dubuke, M.L., Li, X., et al. (2018). Exosomes produced from 3D cultures of MSCs by tangential flow filtration show higher yield and improved activity. Molecular Therapy: The Journal of the American Society of Gene Therapy 26, 2838–2847.

Heinemann, M.L., Ilmer, M., Silva, L.P., Hawke, D.H., Recio, A., Vorontsova, M.A., Alt, E., and Vykoukal, J. (2014). Benchtop isolation and characterization of functional exosomes by sequential filtration. Journal of Chromatography A 1371, 125–135.

Li, Z., Wang, H., Yin, H., Bennett, C., Zhang, H.G., and Guo, P. (2018). Arrowtail RNA for ligand display on ginger exosome-like nanovesicles to systemic deliver siRNA for cancer suppression. Scientific Reports 8, 14644.

Li, K., Wong, D.K., Hong, K.Y., and Raffai, R.L. (2018). Cushioned-density gradient ultracentrifugation (C-DGUC): A refined and high performance method for the isolation, characterization, and use of exosomes. Methods in Molecular Biology 1740, 69–83.

Nazarenko, I. (2020). Extracellular vesicles: Recent developments in technology and perspectives for cancer liquid biopsy. Recent results in cancer research Fortschritte der Krebsforschung Progres dans les recherches sur le cancer. Tumor Liquid Biopsies 215, 319–344.

Niu, Z., Pang, R.T.K., Liu, W., Li, Q., Cheng, R., and Yeung, W.S.B. (2017). Polymer-based precipitation preserves biological activities of extracellular vesicles from an endometrial cell line. PLoS One 12, e0186534.

Onodi, Z., Pelyhe, C., Terezia Nagy, C., Brenner, G.B., Almasi, L., Kittel, A., Mancek-Keber, M., Ferdinandy, P., Buzas, E.I., and Giricz, Z. (2018). Isolation of high-purity extracellular vesicles by the combination of iodixanol density gradient ultracentrifugation and bind-elute chromatography from blood plasma. Frontiers in Physiology 9, 1479.

Panchalingam, K.M., Jung, S., Rosenberg, L., and Behie, L.A. (2015). Bioprocessing strategies for the large-scale production of human mesenchymal stem cells: A review. Stem Cell Research & Therapy 6, 225.

Pi, F., Binzel, D.W., Lee, T.J., Li, Z., Sun, M., Rychahou, P., Li, H., Haque, F., Wang, S., Croce, C.M., et al. (2018). Nanoparticle orientation to control RNA loading and ligand display on extracellular vesicles for cancer regression. Nature Nanotechnology 13, 82–89.

Raab-Traub, N., and Dittmer, D.P. (2017). Viral effects on the content and function of extracellular vesicles. Nature Reviews. Microbiology 15, 559–572.

Thery, C., Amigorena, S., Raposo, G., and Clayton, A. (2006). Isolation and characterization of exosomes from cell culture supernatants and biological fluids. Current Protocols in Cell Biology 30, 3–22.

Willms, E., Cabanas, C., Mager, I., Wood, M.J.A., and Vader, P. (2018). Extracellular vesicle heterogeneity: Subpopulations, isolation techniques, and diverse functions in cancer progression. Frontiers in Immunology 9, 738.

Zhang, Y., Kim, M.S., Jia, B., Yan, J., Zuniga-Hertz, J.P., Han, C., Cai, D., (2017) Hypothalamic stem cells control ageing speed partly through exosomal miRNAs, Nature 548(7665) 52–57.

Zhao, Z., Fan, J., Hsu, Y.S., Lyon, C.J., Ning, B., and Hu, T.Y. (2019). Extracellular vesicles as cancer liquid biopsies: From discovery, validation, to clinical application. Lab on a Chip 19, 1114–1140.

38 Engineered Extracellular Vesicle-Based Therapeutics for Liver Cancer

Kaori Ishiguro and Tushar Patel
Mayo Clinic, Jacksonville, USA

CONTENTS

INTRODUCTION

RNA nanotechnology provides an opportunity to contribute to the development of new therapeutic approaches for human diseases. Using RNA nanotechnology, biological nanoparticles can be engineered to optimize or improve their targeted delivery to cells or tissues of interest. The versatility provided by the use of RNA can support ligand display for targeting as well as scaffold properties for ligand display. In this chapter, we describe the clinical therapeutic potential of using RNA nanotechnology for enhancing targeting specificity in the development of a therapeutic delivery approach that involves the generation of targeted biological nanoparticles. Extracellular vesicles (EVs) are used as the basis of a delivery platform for RNA therapeutics. This application exploits the high stability and biocompatibility of EVs and their innate capacity to communicate with target cells and pre-existing biological mechanisms of gene transfer to these cells. These properties enhance their capability for specific and efficient delivery of cargo into the cytosol of target tumor cells in expectation of an enhanced permeability and retention effect. These therapeutic EVs use RNA nanotechnology to incorporate RNA aptamers as ligands for cell surface receptors on target cells. The attachment of ligand-displaying RNA aptamers to these EVs enables their targeting to appropriate target cells. This approach is illustrated with reference to its use for hepatocellular carcinoma (HCC), which is among the most common types of cancers worldwide. The application further describes the utility of this approach to develop therapeutics that can target cancer stem cells as

DOI: 10.1201/9781003001560-42

well as intracellular targets such as beta-catenin that have previously been challenging to effectively target for therapeutic purposes.

ENGINEERED BIOLOGICAL NANOPARTICLES AS CARRIERS OF RNA-BASED THERAPEUTICS

The use of RNA-based approaches to modulate immune responses or gene expression offers the prospect of novel therapeutics. Preventive approaches can involve the use of RNA as neo-antigens for vaccine development. Therapeutic strategies can involve the use of small interference RNA (siRNA) or microRNA to modulate gene expression through target gene silencing. The use of the latter as RNA therapeutics has been hampered by the instability of RNA molecules. Delivery approaches using viral vector-based approaches are hampered by size limitations, immunogenicity and genomic insertion risks. Approaches using synthetic lipid or protein-based nanoparticles have also been used successfully. While these can offer enhanced stability, specificity and reduced toxicity, their use is hampered by immunogenicity of nanoparticles, innate immune responses, phagocytosis and sequestration within tissues, such as the liver, spleen, lung and bone marrow. The clinical application of their use is further limited by difficulty in reproducibility and toxicity issues (Bamrungsap et al. 2012).

The use of engineered biological nanoparticles such as EVs offers an attractive alternative for drug delivery. EVs can be loaded with RNA or protein therapeutics. This exploits the innate cell-uptake ability of EVs and protects their RNA cargoes from degradative nucleases as well. These biological nanoparticles offer several advantages over the use of viral vectors or synthetic nanoparticles due to their low cytotoxicity, reduced immunogenicity and high biocompatibility. EVs loaded with siRNA had better biocompatibility and higher efficacy compared to synthetic siRNA nanoparticles in vivo (Valadi et al. 2007; Skog et al. 2008). The therapeutic potential of EVs was shown by the use of self-derived dendritic EVs targeted with Lamp2b protein to mediate siRNA delivery to the brain after intravenous injection (Alvarez-Erviti et al. 2011). Other studies have shown the potential for use as an anticancer agent. Delivery of GE11-targeted EVs containing let-7a miRNA to epidermal growth factor (EGFR)-overexpressing breast cancer cells enhanced their anti-tumor effect *in vitro* and *in vivo* (Ohno et al. 2013). EV loaded with anticancer agents or miRNA cargoes also had anti-tumor effects in liver cancers *in vitro* and *in vivo* (George, Yan, and Patel 2018). These examples indicated that EVs are an effective method for therapeutic gene delivery. However, the therapeutic efficacy of siRNA delivered via EV *in vivo* study is short-lived (Matsuda and Patel 2018). However, the duration of effects and therapeutic efficacy can be enhanced by improving the specificity of delivery into target cells and tissues. The surface membrane of EVs can be modified to enhance desired properties such as targeting and uptake (Luan et al. 2017). In this chapter, we will highlight the use of RNA nanotechnology to optimally engineer EVs for therapeutic applications by providing opportunities for targeting EVs to specific tissues to enhance their therapeutic efficacy and reduce off-target effects.

EXTRACELLULAR VESICLES AS BIOLOGICAL NANOPARTICLES

EVs are nano-sized membrane vesicles that are released by many cell types and may contain a variety of different types of molecular cargo (Valadi et al. 2007; Skog et al. 2008; Duijvesz et al. 2011). EVs typically refer to a heterogeneous group of secreted vesicles that include exosomes, microvesicles and other secreted vesicles of varying sizes and biogenesis (Shah, Patel, and Freedman 2018). Physiologically, EV could play a significant role in intercellular communication by providing a mechanism for the transfer of RNAs, lipids and proteins from one cell to another (Raposo and Stoorvogel 2013; van Dommelen et al. 2012). By exploiting these biological roles, EVs are promising nanocarriers of therapeutic agents. Their attractiveness for use as drug delivery vehicles is enhanced by intrinsic advantages related to their small size and capability to penetrate into deep tissues, stability in the circulation, cell targeting properties and ability to overcome immune system barriers compared to the use of synthetic nanoparticles.

A scalable, safe and cost-effective source of EVs is necessary for the development of new therapeutic applications and to serve as a nanocarrier of cancer therapeutic agents. However, isolating bulk quantities of EVs from small amount of biological fluids or cultured medium for clinical application have been a major challenge. Mammalian EVs can be, and are, regularly isolated from cell culture medium as well as from biological fluids, such as blood (Caby et al. 2005), saliva (Palanisamy et al. 2010), semen (Petersen et al. 2014), urine (Pisitkun, Shen, and Knepper 2004), breast milk (Admyre et al. 2007), cerebrospinal fluid (Street et al. 2012) or bronchoalveolar fluid (Prado et al. 2008). EVs can also be isolated from agricultural products such as fruits and plants. Plant-derived-nanoparticles have been isolated from grapefruit juice (Wang et al. 2013, 2016), ginger (Li et al. 2018), sunflower seeds (Regente et al. 2009) and leaves (Rutter and Innes 2017). These sources have generated interest as alternative options for clinical use because of their scalable source, economical advantage and safety profiles.

Bovine milk consumption is safe and provides nutritional benefits (Haug, Hostmark, and Harstad 2007). EV can be isolated from bovine milk. Indeed, bovine milk provides an economical and scalable source of EVs for the production of EVs on a large scale. Milk-derived nanovesicles (MNV) have been demonstrated to be highly efficient for delivering a variety of therapeutic agents, including RNA therapeutics such as antisense oligonucleotides or small interfering RNA, and chemotherapeutic drugs such as doxorubicin (Munagala et al. 2016; George, Yan, and Patel 2018; Matsuda et al. 2019). We have developed a standardized scalable production approach for MNV that uses membrane filtration and differential ultracentrifugation and results in yields in the range of 5.0×10^{12} – 5.0×10^{13} particles from 200ml of fat free milk. The MNV isolated using this approach are homogenous with >90% within a size range of 100–200nm when quantitated using Nanoparticle Tracking Analysis (George, Yan, and Patel 2018). This provides a magnitude scale up of overall yield compared to isolation of EV from cells in culture. Thus, the use of MNV as a source of biological nanovesicles can meet an essential requirement for scalability at economical cost of large-scale production. Milk-derived EV are safe *in vitro* and are also biocompatible with no observable systemic toxicity *in vivo* following systemic administration into mice (Maji et al., 2017; Somiya, Yoshioka, and Ochiya 2018). Not only do they provide a scalable and cost-effective source of EVs but also a favorable short-term safety profile after intravenous administration.

PHYSICOCHEMICAL CHARACTERIZATION OF EXTRACELLULAR VESICLES

The therapeutic use of nanovesicles as drug delivery vehicles is dictated by several considerations and the need to have favorable physicochemical properties that can promote enhanced permeability and retention within tissues. These include both size and zeta-potential, an electro-kinetic potential in colloidal systems (Carvalho et al. 2018). Interactions with the reticuloendothelial system within the circulation are important because the utility of conventional nanoparticles is impacted by their rapid recognition and clearance by macrophages. Negatively charged nanocarriers can favorably lead to binding and internalization for cellular uptake at positively charged regions through electrostatic interaction (Win and Feng 2005; Carvalho et al. 2018). In contrast, positively charged nanocarriers have challenging concerns such as cytotoxicity and can lead to lysosomal degradation (Nel et al. 2009).

The use of biological nanoparticles such as EVs as therapeutic delivery agents and nanocarriers provide a distinctive advantage as they may escape from degradation or clearance by the immune system and thus can have a longer circulation (Johnstone 2006). MNV derived from commercially obtained fat-free milk are highly negatively charged (−33.1 mV) and retained a negative charge, albeit smaller, even after loading with siRNA, anti-sense oligonucleotides or doxorubicin (Matsuda and Patel 2018) (George, Yan, and Patel 2018). Likewise, the zeta potential of ligand displaying

EVs, isolated from engineered HEK293T cells were also shown to have a negative zeta potential (−15.6 mV) which enables better colloidal stability and avoids aggregation (Pi et al. 2018). Similarly, the size of MNV after loading with drugs or RNA was not significantly different from that of unloaded MNV.

ENGINEERING EVs THROUGH SURFACE MODIFICATIONS

EVs are nanovesicles with a lipid bilayer containing embedded proteins (Yang et al. 2018) and enclosing a hydrophilic content (Frydrychowicz et al. 2015). Cell- or tissue-specific targeting could be achieved through the modification of the EV surface. Cells can be engineered to produce modified EVs. Alternatively, EVs could be engineered through surface modification with membrane-binding moieties. For the latter, both covalent (Smyth et al. 2014) and non-covalent modifications have been described (Nakase and Futaki 2015; Qi et al. 2016). An example of a non-covalent modification of EV surfaces involves hydrophobic interactions with lipophilic species such as the use of a cholesteryl-TEG anchor (Pfeiffer and Hook 2004). Cholesterol-TEG-based anchor molecules can be readily inserted into lipid membranes without significant condensation of the membrane lipids or disturbance of their structure and dynamics. We have conjugated cholesteryl-TEG to RNA molecules and have observed spontaneous integration of the lipophilic cholesteryl-TEG oligonucleotide moiety onto the EV surface. This interaction can be achieved using a simple co-incubation at 37°C. By tethering phospholipid bilayers to EV surfaces, cholesterol-TEG-based anchors can provide robust attachment and a stable platform. Thus, the use of cholesterol-TEG-based biomolecules could provide a versatile and flexible method for incorporating RNA oligonucleotides such as aptamers for targeting specificity.

Assemblies between EV membranes and RNA oligonucleotides can be designed to provide a versatile platform through the incorporation of various types of structural skeletons, scaffolds or direct target binding ligands onto the surface of EVs. For example, structural stability can be provided by the use of RNA nanoparticles based on the three-way junction (3WJ) motif derived from bacteriophage phi29 packaging RNA (pRNA) (Shu et al. 2011; Haque et al. 2012). These pRNA-3WJ nanoparticles are composed of three fragments with high affinity for each other. Thus, they can be constructed via bottom-up self-assembly. When each fragment is mixed in equal molar ratio, the complex assembles by self-annealing with high efficiency. These 3WJ nanoparticles are thermodynamically stable and can remain intact at ultralow concentrations and moreover are non-immunogenic *in vivo* (Haque et al. 2012; Binzel, Khisamutdinov, and Guo 2014). In order to enhance their chemical stability, 2′-F modified U and C nucleotides are used on each RNA strand backbone. This structure provides resistance against RNase degradation and furthermore avoids unwanted RNA folding thereby maintaining the desired functionalities of RNA modules (Binzel, Khisamutdinov, and Guo 2014; Shu et al. 2013). 3WJ can be conjugated with a variety of molecules such as with fluorophores for labeling, cholesterol-TEG for membrane insertion and RNA aptamers for specific target binding. In proof of concept studies, a modular design of a planar arrangement with three angles of 60°, 120°, 180° for original 3WJ structure (Pi et al. 2018) in an arrow-tail configuration with cholesterol-TEG conjugation was anchored onto EVs and not internalized into the EVs (Li et al. 2018; Pi et al. 2018). Additionally, ligand displaying EVs were constructed in a so-called "3D- flower shape" to enhance the delivery and therapeutic effect to specifically targeted cancer cells. These physical properties and their anionic nature support the use of engineered EV with RNA scaffolds for use in *in vivo* delivery applications.

USE OF RNA APTAMERS FOR TARGETING SPECIFICITY

An advantage of using RNA scaffolds attached to the EV surface is that it can provide the basis for incorporation of targeting or detection oligonucleotides. Aptamers are single-stranded RNA or DNA oligonucleotides with small molecular weight that can bind to and recognize specific target molecules. The formation of 3D structures enables them to bind strongly to targets on cancer cells without the involvement of covalent bonds and thus makes them suitable for use for cell surface molecule targeted therapy (Zhou and Rossi 2014; Lundin, Gissberg, and Smith 2015). Aptamers can be generated using Systematic Evolution of Ligands by Exponential Enrichment (SELEX) approaches, which were first described in the early 1990s and which enabled chemically synthesized oligonucleotides with minimal batch variation (Tuerk and Gold 1990; Ellington and Szostak 1990). The use of RNA aptamers for clinical applications has been challenged by their susceptibility to nuclease degradation in biological systems. Aptamer stabilization can be provided by using 2′-F modifications and thereby provide stability and enhance function when used in clinical settings (Binzel, Khisamutdinov, and Guo 2014; Piao et al. 2018). With these design considerations, RNA aptamers to desired targets could be incorporated onto the membranes of EVs to generate a targeted delivery system.

AN ILLUSTRATIVE APPLICATION – DEVELOPING TARGETED THERAPIES FOR HCC

Putting this all together, we will highlight a test case to illustrate the use of RNA nanotechnology for developing targeted EV-based therapies. HCC is the fifth most common type of cancer worldwide. There is a lack of effective therapies for the patients with advanced HCC, which is further complicated by therapeutic resistance, in part related to the complex and heterogeneous underlying etiologies and genetic and epigenetic alterations involved in HCC (Mittal and El-Serag 2013; Wallace et al. 2015). Subsequently, patients with HCC have a poor prognosis (Zender et al. 2006). The use of engineered EV offers a novel approach to the treatment of these cancers by enabling therapies that are targeted to specific genetic causes.

CHOICE OF CELLULAR TARGETS

HCC is remarkably heterogeneous and can be related to alterations in several signaling pathways (Breuhahn, Longerich, and Schirmacher 2006). Several genetic alterations have been described in HCC. Among the most commonly encountered mutations in HCC are those associated with the β-catenin gene and the activation of Wnt/β-catenin pathway. This pathway plays an important role in tumor proliferation, maintenance of tumor initiating cells and metastasis (Thompson and Monga 2007; Anastas and Moon 2013). Aberrant activation of Wnt signaling has been associated with carcinogenesis. Mutations in genes encoding for β-catenin or *CTNNB1* (Gao et al. 2018; de La Coste et al. 1998) occur in up to 40% of HCCs. The presence of *CTNNB1* mutations is related to aberrant β-catenin signaling and tumor aggressiveness (Cieply et al. 2009). Specific therapy targeting β-catenin could therefore be beneficial for a subset of HCC or other types of cancers that are characterized by mutations or aberrant activations in this pathway. However, there are no current therapeutic agents that can directly target this pathway. We have developed and reported the use of a therapeutic approach based on delivery of β-catenin siRNA to tumor cells using MNV as a biological nanoparticle delivery carrier (Matsuda and Patel 2018). The therapeutic MNV (tMNV) were generated by

TABLE 38.1
Potential Targets for HCC

Cell types	Hepatocellular cancer cells
	Stromal cells
	Liver cancer stem cells
Surface targets	AFP
	EpCAM, CD133, CD90, CD44, Osteopontin
Intercellular targets	β-catenin, c-myc

loading MNV with β-catenin siRNA using lipid-based transfection. The use of tMNV provides an efficient and effective means of hepatic delivery of RNA-based therapeutics (Table 38.1).

CHOICE OF TARGETED CELL TYPES

Several types of cells can be targeted for the therapy of HCC. The tumor cell itself represents a major target. Additional opportunities for therapeutic targeting can involve targeting either stromal cells that maintain favorable micro-environmental conditions for tumor growth or liver cancer stem cells (CSC) that maintain tumor cell renewal and proliferation. Tumor heterogeneity can reflect the hierarchical organization of the tumor mediated by a small population of CSC (Jordan, Guzman, and Noble 2006). These CSC are a subset of cells within tumors, which possess self-renewal capability and unlimited proliferation ability and have been related to therapy resistance. CSCs express hepatic stem cell markers such as EpCAM, CD133, CD90 and CD44 (Chiba et al. 2007). These markers can be used to develop CSC specific targeting approaches. Moreover, targeting CSC's could be performed concomitantly with other therapies to improve treatment responses.

CHOICE OF TARGETING APTAMER

The use of aptamers offers an opportunity for targeting. However, selection of aptamers for specific targeting to HCC is challenging because of the heterogeneity of disease and presence of distinctive molecular profiles. Although most aptamer applications that target HCC have been based on DNA, there are some reports of RNA aptamers (Ladju et al. 2018). An RNA aptamer targeting osteopontin significantly downregulated the epithelial-mesenchymal transition and tumor growth of HCC in an animal model (Bhattacharya et al. 2012). Similarly, an RNA aptamer targeting alpha-fetoprotein (AFP) specifically inhibited the proliferation of AFP-related HCC cells (Lee and Lee 2012). An RNA aptamer targeting epithelial cell adhesion molecule (EpCAM) enhanced the antitumor effect with high binding ability to EpCAM positive HCC cells *in vitro* and *in vivo*.

The use of EpCAM is attractive for targeting CSC populations. EpCAM (CD326) is a 40 kD glycoprotein that is expressed at a low level on normal epithelial cells and predominantly expressed on CSC. It is associated with tumorigenesis (Trzpis et al. 2007) and poor prognosis (Stoecklein et al. 2006). As a mitogenic transducer, it can trigger several intracellular signaling pathways and can regulate cell cycle progression and cell proliferation (Maaser and Borlak 2008; Maetzel et al. 2009; Munz, Baeuerle, and Gires 2009; Manhas et al. 2016). EpCAM expression is associated with CSC and cancer-initiating cells-enriched populations in many different solid cancers, including HCC (Yamashita et al. 2009; Kimura et al. 2010; Li et al. 2016). An RNA-aptamer targeting EpCAM was isolated by SELEX (Shigdar et al. 2011). A 40-base RNA aptamer was first isolated by SELEX and then further truncated to 19 bases (binding affinity:55nM) (Shigdar et al. 2011). Various types of EpCAM targeting aptamer-nanoparticles mediated targeting cancer gene or drug delivery system have been developed (Subramanian et al. 2012; Mohammadi et al. 2015). EpCAM aptamer conjugated with recombinant adenovirus carrying the tumor suppressor gene PTEN using polyethelene glycol (PEG) as a linker was shown to have anti-tumor effects in HCC (Xiao et al. 2017; Liu et al. 2018). Therefore, it is feasible to consider the use of aptamers targeting EpCAM as a targeting strategy for specifically targeting EPCAM expressing CSC in the liver.

GENERATION OF TARGETED BIOLOGICAL NANOPARTICLES

The expression of EpCAM is associated with the Wnt/β-catenin pathway which can control the proliferation of hepatic stem cells (Yamashita et al. 2007; Li et al. 2016; Katoh 2017). β-catenin

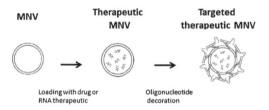

FIGURE 38.1 Generation of targeted therapeutic EV for delivery of RNA therapeutics from milk-derived nanovesicles (MNVs).

stabilization has been associated with tumor-initiating cells or CSC renewal. EpCAM positive cell populations isolated from AFP positive HCC have hepatic CSC-like properties with the capacity of self-renewal and differentiation and feature activation of Wnt/β-catenin signaling (Yamashita et al. 2009). Furthermore, silencing of β-catenin by siRNA could inhibit proliferation and self-renewal characteristics of liver CSC derived from CD133(+) sphere forming cells *in vitro* and *in vivo* (Quan et al. 2013).

Using RNA nanotechnology, therapeutic MNVs containing siRNA to β-catenin were engineered to display EPCAM targeting ligand for the delivery of specific siRNA to liver CSC (Figure 38.1). EpCAM targeting therapeutic MNVs can increase the targeting specificity for the delivery of siRNA to HCC cells. Importantly, the RNA aptamer can be internalized after binding to cell surface EpCAM, thus providing an additional desirable property.

The therapeutic engineered EVs can have broader utility. Further applications for the use of this approach for other cancer types will depend on their basal level of EpCAM expression and the contributions of β-catenin activation on tumor growth and progression. In addition to HCC, EpCAM is expressed in many other cancers. Among these, adenocarcinoma of colon, pancreas and prostate are especially promising for EpCAM targeting therapy (Went et al. 2004). Other than HCC, mutations of β-catenin are commonly observed in many other types of cancers such as breast, lung, colorectal cancer. Thus, the use of EpCAM targeted β-catenin siRNA-based therapeutics may have broader possibility of clinical application beyond HCC therapy.

CONCLUSIONS AND FUTURE PERSPECTIVES

The rapidly evolving interest in the use of EVs as therapeutics offers several promising treatment opportunities for human diseases. In particular, previously poorly accessible and difficult to target tissues such as brain, cell types such as CSC or cell targets such as β-catenin may be amenable to EV-based therapeutics. While the further development of these approaches is challenging and will require a broader foundational and technical knowledge, these opportunities provide a compelling justification for the continued development of engineered EV. Clearly there is a need for a more complete understanding of basic biological processes such as the mechanisms by which EV are taken up by target cells, their cell and tissue specificity, stability in the circulation, effective longevity following systemic administration and immunomodulatory capabilities. Effective therapeutic manipulations will require the development and refinement of targetable technologies. The use of therapeutic EVs will require a comprehensive assessment of optimal dosage and administration strategies. As this field expands in scope, the application of RNA nanotechnology will not only enable but will also support the ability to effectively engineer EVs, and thereby not only broaden our understanding of nano-biological interactions, but importantly lead to new approaches for preventing or treating human disease.

ACKNOWLEDGMENTS

This work was supported by funding provided by the Mayo Clinic Center for Regenerative Medicine and by grants from the National Institutes of Health (CA217833).

REFERENCES

Admyre, C., S. M. Johansson, K. R. Qazi, J. J. Filen, R. Lahesmaa, M. Norman, E. P. Neve, A. Scheynius, and S. Gabrielsson. 2007. 'Exosomes with immune modulatory features are present in human breast milk', *J Immunol*, 179: 1969–78.

Alvarez-Erviti, L., Y. Seow, H. Yin, C. Betts, S. Lakhal, and M. J. Wood. 2011. 'Delivery of siRNA to the mouse brain by systemic injection of targeted exosomes', *Nat Biotechnol*, 29: 341–5.

Anastas, J. N., and R. T. Moon. 2013. 'WNT signalling pathways as therapeutic targets in cancer', *Nat Rev Cancer*, 13: 11–26.

Bamrungsap, S., Z. Zhao, T. Chen, L. Wang, C. Li, T. Fu, and W. Tan. 2012. 'Nanotechnology in therapeutics: A focus on nanoparticles as a drug delivery system', *Nanomedicine (London)*, 7: 1253–71.

Bhattacharya, S. D., Z. Mi, V. M. Kim, H. Guo, L. J. Talbot, and P. C. Kuo. 2012. 'Osteopontin regulates epithelial mesenchymal transition-associated growth of hepatocellular cancer in a mouse xenograft model', *Ann Surg*, 255: 319–25.

Binzel, D. W., E. F. Khisamutdinov, and P. Guo. 2014. 'Entropy-driven one-step formation of Phi29 pRNA 3WJ from three RNA fragments', *Biochemistry*, 53: 2221–31.

Breuhahn, K., T. Longerich, and P. Schirmacher. 2006. 'Dysregulation of growth factor signaling in human hepatocellular carcinoma', *Oncogene*, 25: 3787–800.

Caby, M. P., D. Lankar, C. Vincendeau-Scherrer, G. Raposo, and C. Bonnerot. 2005. 'Exosomal-like vesicles are present in human blood plasma', *Int Immunol*, 17: 879–87.

Carvalho, P. M., M. R. Felicio, N. C. Santos, S. Goncalves, and M. M. Domingues. 2018. 'Application of light scattering techniques to nanoparticle characterization and development', *Front Chem*, 6: 237.

Chiba, T., Y. W. Zheng, K. Kita, O. Yokosuka, H. Saisho, M. Onodera, H. Miyoshi, M. Nakano, Y. Zen, Y. Nakanuma, H. Nakauchi, A. Iwama, and H. Taniguchi. 2007. 'Enhanced self-renewal capability in hepatic stem/progenitor cells drives cancer initiation', *Gastroenterology*, 133: 937–50.

Cieply, B., G. Zeng, T. Proverbs-Singh, D. A. Geller, and S. P. Monga. 2009. 'Unique phenotype of hepatocellular cancers with exon-3 mutations in beta-catenin gene', *Hepatology*, 49: 821–31.

de La Coste, A., B. Romagnolo, P. Billuart, C. A. Renard, M. A. Buendia, O. Soubrane, M. Fabre, J. Chelly, C. Beldjord, A. Kahn, and C. Perret. 1998. 'Somatic mutations of the beta-catenin gene are frequent in mouse and human hepatocellular carcinomas', *Proc Natl Acad Sci U S A*, 95: 8847–51.

Duijvesz, D., T. Luider, C. H. Bangma, and G. Jenster. 2011. 'Exosomes as biomarker treasure chests for prostate cancer', *Eur Urol*, 59: 823–31.

Ellington, A. D., and J. W. Szostak. 1990. 'In vitro selection of RNA molecules that bind specific ligands', *Nature*, 346: 818–22.

Frydrychowicz, M., A. Kolecka-Bednarczyk, M. Madejczyk, S. Yasar, and G. Dworacki. 2015. 'Exosomes - structure, biogenesis and biological role in non-small-cell lung cancer', *Scand J Immunol*, 81: 2–10.

Gao, C., Y. Wang, R. Broaddus, L. Sun, F. Xue, and W. Zhang. 2018. 'Exon 3 mutations of CTNNB1 drive tumorigenesis: A review', *Oncotarget*, 9: 5492–508.

George, J., I. K. Yan, and T. Patel. 2018. 'Nanovesicle-mediated delivery of anticancer agents effectively induced cell death and regressed intrahepatic tumors in athymic mice', *Lab Investig*, 98: 895–910.

Haque, F., D. Shu, Y. Shu, L. S. Shlyakhtenko, P. G. Rychahou, B. M. Evers, and P. Guo. 2012. 'Ultrastable synergistic tetravalent RNA nanoparticles for targeting to cancers', *Nano Today*, 7: 245–57.

Haug, A., A. T. Hostmark, and O. M. Harstad. 2007. 'Bovine milk in human nutrition – a review', *Lipids Health Dis*, 6: 25.

Johnstone, R. M. 2006. 'Exosomes biological significance: A concise review', *Blood Cells Mol Dis*, 36: 315–21.

Jordan, C. T., M. L. Guzman, and M. Noble. 2006. 'Cancer stem cells', *N Engl J Med*, 355: 1253–61.

Katoh, M. 2017. 'Canonical and non-canonical WNT signaling in cancer stem cells and their niches: Cellular heterogeneity, omics reprogramming, targeted therapy and tumor plasticity (Review)', *Int J Oncol*, 51: 1357–69.

Kimura, O., T. Takahashi, N. Ishii, Y. Inoue, Y. Ueno, T. Kogure, K. Fukushima, M. Shiina, Y. Yamagiwa, Y. Kondo, J. Inoue, E. Kakazu, T. Iwasaki, N. Kawagishi, T. Shimosegawa, and K. Sugamura. 2010.

'Characterization of the epithelial cell adhesion molecule (EpCAM)+ cell population in hepatocellular carcinoma cell lines', *Cancer Sci*, 101: 2145–55.

Ladju, R. B., D. Pascut, M. N. Massi, C. Tiribelli, and C. H. C. Sukowati. 2018. 'Aptamer: A potential oligo-nucleotide nanomedicine in the diagnosis and treatment of hepatocellular carcinoma', *Oncotarget*, 9: 2951–61.

Lee, Y. J., and S. W. Lee. 2012. 'Regression of hepatocarcinoma cells using RNA aptamer specific to alpha-fetoprotein', *Biochem Biophys Res Commun*, 417: 521–7.

Li, Y., R. W. Farmer, Y. Yang, and R. C. Martin. 2016. 'Epithelial cell adhesion molecule in human hepatocel-lular carcinoma cell lines: A target of chemoresistence', *BMC Cancer*, 16: 228.

Li, Z., H. Wang, H. Yin, C. Bennett, H. G. Zhang, and P. Guo. 2018. 'Arrowtail RNA for ligand display on ginger exosome-like nanovesicles to systemic deliver siRNA for cancer suppression', *Sci Rep*, 8: 14644.

Liu, Z., X. Sun, X. Xiao, Y. Lin, C. Li, N. Hao, M. Zhou, R. Deng, S. Ke, and Z. Zhong. 2018. 'Characterization of aptamer-mediated gene delivery system for liver cancer therapy', *Oncotarget*, 9: 6830–40.

Luan, X., K. Sansanaphongpricha, I. Myers, H. Chen, H. Yuan, and D. Sun. 2017. 'Engineering exosomes as refined biological nanoplatforms for drug delivery', *Acta Pharmacol Sin*, 38: 754–63.

Lundin, K. E., O. Gissberg, and C. I. Smith. 2015. 'Oligonucleotide therapies: The past and the present', *Hum Gene Ther*, 26: 475–85.

Maaser, K., and J. Borlak. 2008. 'A genome-wide expression analysis identifies a network of EpCAM-induced cell cycle regulators', *Br J Cancer*, 99: 1635–43.

Maetzel, D., S. Denzel, B. Mack, M. Canis, P. Went, M. Benk, C. Kieu, P. Papior, P. A. Baeuerle, M. Munz, and O. Gires. 2009. 'Nuclear signalling by tumour-associated antigen EpCAM', *Nat Cell Biol*, 11: 162–71.

Maji S, Yan IK, Parasramka M, Mohankumar S, Matsuda A, Patel T. 2017. In vitro toxicology studies of extra-cellular vesicles. *J Appl Toxicol* 37(3):310–318.

Manhas, J., A. Bhattacharya, S. K. Agrawal, B. Gupta, P. Das, S. V. Deo, S. Pal, and S. Sen. 2016. 'Characterization of cancer stem cells from different grades of human colorectal cancer', *Tumour Biol*, 37: 14069–81.

Matsuda, A., K. Ishiguro, I. K. Yan, and T. Patel. 2019. 'Extracellular vesicle-based therapeutic targeting of beta-catenin to modulate anticancer immune responses in hepatocellular cancer', *Hepatol Commun*, 3: 525–41.

Matsuda, A., and T. Patel. 2018. 'Milk-derived extracellular vesicles for therapeutic delivery of small interfer-ing RNAs', *Methods Mol Biol*, 1740: 187–97.

Mittal, S., and H. B. El-Serag. 2013. 'Epidemiology of hepatocellular carcinoma: Consider the population', *J Clin Gastroenterol*, 47 Suppl: S2–6.

Mohammadi, M., Z. Salmasi, M. Hashemi, F. Mosaffa, K. Abnous, and M. Ramezani. 2015. 'Single-walled carbon nanotubes functionalized with aptamer and piperazine-polyethylenimine derivative for targeted siRNA delivery into breast cancer cells', *Int J Pharm*, 485: 50–60.

Munagala, R., F. Aqil, J. Jeyabalan, and R. C. Gupta. 2016. 'Bovine milk-derived exosomes for drug delivery', *Cancer Lett*, 371: 48–61.

Munz, M., P. A. Baeuerle, and O. Gires. 2009. 'The emerging role of EpCAM in cancer and stem cell signal-ing', *Cancer Res*, 69: 5627–9.

Nakase, I., and S. Futaki. 2015. 'Combined treatment with a pH-sensitive fusogenic peptide and cationic lipids achieves enhanced cytosolic delivery of exosomes', *Sci Rep*, 5: 10112.

Nel, A. E., L. Madler, D. Velegol, T. Xia, E. M. Hoek, P. Somasundaran, F. Klaessig, V. Castranova, and M. Thompson. 2009. 'Understanding biophysicochemical interactions at the nano-bio interface', *Nat Mater*, 8: 543–57.

Ohno, S., M. Takanashi, K. Sudo, S. Ueda, A. Ishikawa, N. Matsuyama, K. Fujita, T. Mizutani, T. Ohgi, T. Ochiya, N. Gotoh, and M. Kuroda. 2013. 'Systemically injected exosomes targeted to EGFR deliver antitumor microRNA to breast cancer cells', *Mol Ther*, 21: 185–91.

Palanisamy, V., S. Sharma, A. Deshpande, H. Zhou, J. Gimzewski, and D. T. Wong. 2010. 'Nanostructural and transcriptomic analyses of human saliva derived exosomes', *PLoS One*, 5: e8577.

Petersen, K. E., E. Manangon, J. L. Hood, S. A. Wickline, D. P. Fernandez, W. P. Johnson, and B. K. Gale. 2014. 'A review of exosome separation techniques and characterization of B16-F10 mouse melanoma exosomes with AF4-UV-MALS-DLS-TEM', *Anal Bioanal Chem*, 406: 7855–66.

Pfeiffer, I., and F. Hook. 2004. 'Bivalent cholesterol-based coupling of oligonucletides to lipid membrane assemblies', *J Am Chem Soc*, 126: 10224–5.

Pi, F., D. W. Binzel, T. J. Lee, Z. Li, M. Sun, P. Rychahou, H. Li, F. Haque, S. Wang, C. M. Croce, B. Guo, B. M. Evers, and P. Guo. 2018. 'Nanoparticle orientation to control RNA loading and ligand display on extracellular vesicles for cancer regression', *Nat Nanotechnol*, 13: 82–89.

Piao, X., H. Wang, D. W. Binzel, and P. Guo. 2018. 'Assessment and comparison of thermal stability of phos-phorothioate-DNA, DNA, RNA, 2'-F RNA, and LNA in the context of Phi29 pRNA 3WJ', *RNA*, 24: 67–76.

Pisitkun, T., R. F. Shen, and M. A. Knepper. 2004. 'Identification and proteomic profiling of exosomes in human urine', *Proc Natl Acad Sci U S A*, 101: 13368–73.

Prado, N., E. G. Marazuela, E. Segura, H. Fernandez-Garcia, M. Villalba, C. Thery, R. Rodriguez, and E. Batanero. 2008. 'Exosomes from bronchoalveolar fluid of tolerized mice prevent allergic reaction', *J Immunol*, 181: 1519–25.

Qi, H., C. Liu, L. Long, Y. Ren, S. Zhang, X. Chang, X. Qian, H. Jia, J. Zhao, J. Sun, X. Hou, X. Yuan, and C. Kang. 2016. 'Blood exosomes endowed with magnetic and targeting properties for cancer therapy', *ACS Nano*, 10: 3323–33.

Quan, M. F., L. H. Xiao, Z. H. Liu, H. Guo, K. Q. Ren, F. Liu, J. G. Cao, and X. Y. Deng. 2013. '8-bromo-7-me-thoxychrysin inhibits properties of liver cancer stem cells via downregulation of beta-catenin', *World J Gastroenterol*, 19: 7680–95.

Raposo, G., and W. Stoorvogel. 2013. 'Extracellular vesicles: Exosomes, microvesicles, and friends', *J Cell Biol*, 200: 373–83.

Regente, M., G. Corti-Monzon, A. M. Maldonado, M. Pinedo, J. Jorrin, and L. de la Canal. 2009. 'Vesicular fractions of sunflower apoplastic fluids are associated with potential exosome marker proteins', *FEBS Lett*, 583: 3363–6.

Rutter, B. D., and R. W. Innes. 2017. 'Extracellular vesicles isolated from the leaf apoplast carry stress-response proteins', *Plant Physiol*, 173: 728–41.

Shah, R., T. Patel, and J. E. Freedman. 2018. 'Circulating extracellular vesicles in human disease', *N Engl J Med*, 379: 958–66.

Shigdar, S., J. Lin, Y. Yu, M. Pastuovic, M. Wei, and W. Duan. 2011. 'RNA aptamer against a cancer stem cell marker epithelial cell adhesion molecule', *Cancer Sci*, 102: 991–8.

Shu, D., Y. Shu, F. Haque, S. Abdelmawla, and P. Guo. 2011. 'Thermodynamically stable RNA three-way junction for constructing multifunctional nanoparticles for delivery of therapeutics', *Nat Nanotechnol*, 6: 658–67.

Shu, Y., F. Haque, D. Shu, W. Li, Z. Zhu, M. Kotb, Y. Lyubchenko, and P. Guo. 2013. 'Fabrication of 14 differ-ent RNA nanoparticles for specific tumor targeting without accumulation in normal organs', *RNA*, 19: 767–77.

Skog, J., T. Wurdinger, S. van Rijn, D. H. Meijer, L. Gainche, M. Sena-Esteves, W. T. Curry, Jr., B. S. Carter, A. M. Krichevsky, and X. O. Breakefield. 2008. 'Glioblastoma microvesicles transport RNA and proteins that promote tumour growth and provide diagnostic biomarkers', *Nat Cell Biol*, 10: 1470–6.

Smyth, T., K. Petrova, N. M. Payton, I. Persaud, J. S. Redzic, M. W. Graner, P. Smith-Jones, and T. J. Anchordoquy. 2014. 'Surface functionalization of exosomes using click chemistry', *Bioconjug Chem*, 25: 1777–84.

Somiya, M., Y. Yoshioka, and T. Ochiya. 2018. 'Biocompatibility of highly purified bovine milk-derived extra-cellular vesicles', *J Extracell Vesicles*, 7: 1440132.

Stoecklein, N. H., A. Siegmund, P. Scheunemann, A. M. Luebke, A. Erbersdobler, P. E. Verde, C. F. Eisenberger, M. Peiper, A. Rehders, J. S. Esch, W. T. Knoefel, and S. B. Hosch. 2006. 'Ep-CAM expression in squa-mous cell carcinoma of the esophagus: A potential therapeutic target and prognostic marker', *BMC Cancer*, 6: 165.

Street, J. M., P. E. Barran, C. L. Mackay, S. Weidt, C. Balmforth, T. S. Walsh, R. T. Chalmers, D. J. Webb, and J. W. Dear. 2012. 'Identification and proteomic profiling of exosomes in human cerebrospinal fluid', *J Transl Med*, 10: 5.

Subramanian, N., V. Raghunathan, J. R. Kanwar, R. K. Kanwar, S. V. Elchuri, V. Khetan, and S. Krishnakumar. 2012. 'Target-specific delivery of doxorubicin to retinoblastoma using epithelial cell adhesion molecule aptamer', *Mol Vis*, 18: 2783–95.

Thompson, M. D., and S. P. Monga. 2007. 'WNT/beta-catenin signaling in liver health and disease', *Hepatology*, 45: 1298–305.

Trzpis, M., P. M. McLaughlin, L. M. de Leij, and M. C. Harmsen. 2007. 'Epithelial cell adhesion molecule: More than a carcinoma marker and adhesion molecule', *Am J Pathol*, 171: 386–95.

Tuerk, C., and L. Gold. 1990. 'Systematic evolution of ligands by exponential enrichment: RNA ligands to bacteriophage T4 DNA polymerase', *Science*, 249: 505–10.

Valadi, H., K. Ekstrom, A. Bossios, M. Sjostrand, J. J. Lee, and J. O. Lotvall. 2007. 'Exosome-mediated trans-fer of mRNAs and microRNAs is a novel mechanism of genetic exchange between cells', *Nat Cell Biol*, 9: 654–9.

van Dommelen, S. M., P. Vader, S. Lakhal, S. A. Kooijmans, W. W. van Solinge, M. J. Wood, and R. M. Schiffelers. 2012. 'Microvesicles and exosomes: Opportunities for cell-derived membrane vesicles in drug delivery', *J Control Release*, 161: 635–44.

Wallace, M. C., D. Preen, G. P. Jeffrey, and L. A. Adams. 2015. 'The evolving epidemiology of hepatocellular carcinoma: A global perspective', *Expert Rev Gastroenterol Hepatol*, 9: 765–79.

Wang, Q., X. Zhuang, J. Mu, Z. B. Deng, H. Jiang, L. Zhang, X. Xiang, B. Wang, J. Yan, D. Miller, and H. G. Zhang. 2013. 'Delivery of therapeutic agents by nanoparticles made of grapefruit-derived lipids', *Nat Commun*, 4: 1867.

Wang, Q., X. Zhuang, J. Mu, Z. B. Deng, H. Jiang, L. Zhang, X. Xiang, B. Wang, J. Yan, D. Miller, and H. G. Zhang. 2016. 'Corrigendum: Delivery of therapeutic agents by nanoparticles made of grapefruit-derived lipids', *Nat Commun*, 7: 11347.

Went, P. T., A. Lugli, S. Meier, M. Bundi, M. Mirlacher, G. Sauter, and S. Dirnhofer. 2004. 'Frequent EpCam protein expression in human carcinomas', *Hum Pathol*, 35: 122–8.

Win, K. Y., and S. S. Feng. 2005. 'Effects of particle size and surface coating on cellular uptake of polymeric nanoparticles for oral delivery of anticancer drugs', *Biomaterials*, 26: 2713–22.

Xiao, S., Z. Liu, R. Deng, C. Li, S. Fu, G. Chen, X. Zhang, F. Ke, S. Ke, X. Yu, S. Wang, and Z. Zhong. 2017. 'Aptamer-mediated gene therapy enhanced antitumor activity against human hepatocellular carcinoma in vitro and in vivo', *J Control Release*, 258: 130–45.

Yamashita, T., A. Budhu, M. Forgues, and X. W. Wang. 2007. 'Activation of hepatic stem cell marker EpCAM by Wnt-beta-catenin signaling in hepatocellular carcinoma', *Cancer Res*, 67: 10831–9.

Yamashita, T., J. Ji, A. Budhu, M. Forgues, W. Yang, H. Y. Wang, H. Jia, Q. Ye, L. X. Qin, E. Wauthier, L. M. Reid, H. Minato, M. Honda, S. Kaneko, Z. Y. Tang, and X. W. Wang. 2009. 'EpCAM-positive hepato-cellular carcinoma cells are tumor-initiating cells with stem/progenitor cell features', *Gastroenterology*, 136: 1012–24.

Yang, Y., Y. Hong, E. Cho, G. B. Kim, and I. S. Kim. 2018. 'Extracellular vesicles as a platform for membrane-associated therapeutic protein delivery', *J Extracell Vesicles*, 7: 1440131.

Zender, L., M. S. Spector, W. Xue, P. Flemming, C. Cordon-Cardo, J. Silke, S. T. Fan, J. M. Luk, M. Wigler, G. J. Hannon, D. Mu, R. Lucito, S. Powers, and S. W. Lowe. 2006. 'Identification and validation of onco-genes in liver cancer using an integrative oncogenomic approach', *Cell*, 125: 1253–67.

Zhou, J., and J. J. Rossi. 2014. 'Cell-type-specific, Aptamer-functionalized agents for targeted disease therapy', *Mol Ther Nucleic Acids*, 3: e169.

39 Extracellular Vesicles (EVs)
An Innovative Approach to Engineering Nucleic Acid Delivery

Senny Nordmeier, Frank Hsiung, and Victoria Portnoy
System Biosciences, Palo Alto, USA

CONTENTS

WHAT ARE EXTRACELLULAR VESICLES?

Most cell types secrete multiple distinct classes of membranous particles, collectively referred to as extracellular vesicles (EVs), as a conduit of intercellular communication (El Andaloussi, Mäger, Breakefield, & Wood, 2013; Maas, Breakefield, & Weaver, 2017). EVs are biological nanoparticles comprising exosomes, microvesicles (MVs), and apoptotic bodies (Raposo & Stoorvogel, 2013). MVs represent one of the major classes of EVs, are generated at the plasma membrane, from which they are shed off, and typically range in the sizes from 200 nm to 1 µm in diameter. MVs have also been named shedding vesicles, microparticles, ectosomes, and, when shown to contain oncogenic cargo, oncosomes. The second major class of EVs, exosomes, are formed as intraluminal vesicles within endosomal multivesicular bodies (MVBs) and are released from cells upon MVBs fusion with the cell membrane (Stoorvogel, Kleijmeer, Geuze, & Raposo, 2002). Exosomes typically range from 30 to 150 nm in diameter. The third class of EVs, apoptotic bodies, is released as blebs from cells undergoing apoptosis and ranges in size from 0.8 µm to 5 µm (Battistelli & Falcieri, 2020). All EVs have been shown to contain metabolites, proteins, and nucleic acids, such as RNA and/or DNA (Yáñez-Mó et al., 2015). This bioactive cargo can be transferred to other cells and thereby trigger a broad range of cellular activities and biological responses (Ståhl, Johansson, Mossberg, Kahn, & Karpman, 2019). EV biogenesis has been reported to be dysregulated in many different pathologies, particularly in cancer (Han, Lam, & Sun, 2019; Whiteside, 2016), where it has been suggested to promote tumor cell growth (Tai et al., 2019), therapy resistance (O'Neill, Gilligan, & Dwyer, 2019), angiogenesis (Todorova, Simoncini, Lacroix, Sabatier, & Dignat-George, 2017), and metastasis (Chin & Wang, 2016).

Unfortunately, there is a good deal of confusion, and in some cases, misinformation regarding which class of EVs is responsible for specific biological functions (Pegtel & Gould, 2019). To be specific, the term exosome has often been used as a benchmark of all EV-mediated processes, and

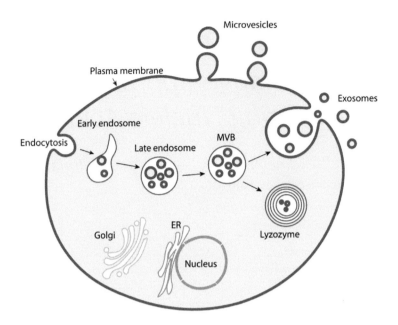

FIGURE 39.1 Extracellular vesicles biogenesis.
Schematic representation of EVs biogenesis: formation of MVs and exosomes. MVs are shedding directly from the plasma membrane. Exosomes are formed through maturation of early endosome as a result of endocytosis. Late endosomes undergo inward budding leading to formation of MVBs with accumulating intraluminal vesicles inside the lumen. The MVBs then will merge with lysosome for degradation; otherwise they will merge with the plasma membrane, releasing intraluminal vesicles into extracellular spaces, which are exosomes.

many researchers have assumed that only exosomes have significant biological functions and importance (Witwer & Théry, 2019). However, there is an increase in the generation of both exosomes and MVs from diseased, damaged, and cancerous cells (Abels & Breakefield, 2016; Patton, Zubair, Khan, Singh, & Singh, 2020; Whiteside, 2016). Since both exosomes and MVs share similar physical and biological properties, it remains challenging to completely segregate these EV populations with current isolation methods (Pegtel & Gould, 2019). Thus for the remainder of this chapter and as per the recommendations laid out by the International Society for Extracellular Vesicles (Lötvall et al., 2014), we will be referring to exosomes and MVs as EVs (Figure 39.1).

THERAPEUTIC POTENTIAL OF EVs AS A DRUG DELIVERY VEHICLE

EVs play significant roles in intercellular communication via cell–cell transfer of metabolites, proteins, and in particular nucleic acids, such as microRNAs (miRNAs), long noncoding RNAs (lncRNAs), and mRNAs (Valadi et al., 2007). The notion of EVs as natural nucleic acid carriers has warranted investigation in their utilization for short interfering RNA (siRNA) delivery. In one important study shown by Wood *et al.*, it was demonstrated that exosomes could target the brain to deliver therapeutic siRNA (Alvarez-Erviti et al., 2011). Additional work has shown that EVs could be utilized for targeted delivery of miRNAs and small molecule drugs to battle cancer (Bunggulawa et al., 2018; Vader, Mol, Pasterkamp, & Schiffelers, 2016), thus paving the way for EVs to be applied as drug carriers for treatment of numerous diseases, as well as in a variety of applications in tissue repair and regeneration (Taverna, Pucci, & Alessandro, 2017; Teixeira, Silva, Almeida, Barbosa, & Santos, 2016).

Ultimately, the issue on how to effectively load EVs with therapeutic nucleic acids has not been settled, and further explorations of the potential methods are needed.

EV LOADING METHODS

ELECTROPORATION

Electroporation is a well-known technology used to introduce DNA into cells (Shigekawa & Dower, 1988). This transfection method utilizes electrical pulses to transiently open holes on the cell membrane, which allow for macromolecules, such as drugs or nucleic acids, to pass into the cells. Similarly, EV membranes can be subjected to electroporation to load drugs or nucleic acids (Kim et al., 2016; Lamichhane & Jay, 2018). In a study done by Wood *et al.*, EVs were loaded with siRNA targeting GAPDH via electroporation (Alvarez-Erviti et al., 2011). The major drawback of this technology is that the electrical pulses can stimulate EV aggregation (Johnsen et al., 2016; Pomatto et al., 2019) or precipitate siRNA (Kooijmans et al., 2013), which can greatly limit trans-fection efficiency into EVs. Some studies have reported an inability to efficiently load EVs with siRNA (Kooijmans et al., 2013) or miRNA (Ohno et al., 2013) using this approach, potentially due to electric field-induced aggregation of these short RNA molecules. In addition, parameters such as electric wave form (i.e., square vs. exponential wave), voltage, buffer conductivity, duration, and number of electric pulses delivered are all factors that may need to be optimized per experiment in order to achieve the desired outcome. Nonetheless, electroporation has been a classically established methodology for the transfer of functional nucleic acids, and there are solutions to reduce nucleic acid aggregation, for instance lipid complexation and trehalose disaccharide (Bagai, Sun, & Tang, 2014; Johnsen et al., 2016).

SONOPORATION/SONICATION

Sonoporation or sonication has been implemented since the 1980s to transfect plasmid DNA into cells (Fechheimer et al., 1987). This technique uses sonic waves to modify the permeability of the cell membrane (Zolochevska & Figueiredo, 2012). Isolated EVs are mixed with drugs or RNA and subsequently sonicated with a probe. The physical shear force from the sonication will weaken the membrane integrity and allow the cargo to diffuse into the EV during this membrane deformation. However, in the case from Kim *et al.* paclitaxel was encapsulated not only inside the EVs but also-outside the membrane (Kim et al., 2016). The EV membrane microviscosity dramatically decreased after sonication; however, membrane integrity was able to recover within an hour after incubation at 37°C. Another limitation of sonication is that it can cause damage to the nucleic acid cargo (Lamichhane et al., 2016; Sukreet, Silva, Adamec, Cui, & Zempleni, 2019).

PRODUCER CELL ENGINEERING

An alternative method explored by researchers is engineering of a producer cell line by transfection (Akao et al., 2011; Ohno et al., 2013). These transient or stably transfected cells would then produce EVs carrying the designed cargo. The benefits of the transfection-based approaches are efficiency in loading and molecular stability compared to electroporation. However, this approach can be limited due to toxicity and safety concerns associated with the transfection reagents to possibly alter the producer cell gene expression and thereby negatively affect downstream changes in EV cargo and their intrinsic/nascent bioactivity (Kooijmans et al., 2013). In addition, there is variability in cell transfection efficiency, which could alter EV production or lead to inconsistent levels of a desired therapeutic molecule.

SMALL RNA/DNA LOADING KIT- EXO-FECT siRNA/miRNA TRANSFECTION KIT

The only available commercial kit for loading of EVs with small RNAs, therapeutic oligonucle-otides, and nucleic acid nanoparticles (manuscript in preparation) is the Exo-Fect siRNA/miRNA

TABLE 39.1

Current Methods for EV Loading with Functional Cargo

EV Loading Method	Advantages	Disadvantages
Electroporation	Low cost	Requires optimizations, causes aggregation of small RNA and EVs,
Sonication	Low cost	Might cause damage of the cargo
Producer cells	Efficient loading, molecular stability of the cargo and EVs	Toxicity and safety of the transfection affects gene expression and therefore EVs bioactivity
Commercial kit (CPP-based)	Efficient loading, molecular stability of the cargo and EVs, no change in EVs bioactivity, no optimizations required	Moderate cost

EV loading kit available from System Biosciences. The Exo-Fect transfection reagent utilizes a proprietary cell-penetrating peptide to package and directly load small RNAs/DNAs into EVs. A subsequent clean up column helps to remove excess reagent and nucleic acids, resulting in a preparation with only loaded EVs. The advantage of this method is that there is no need for specialized equipment or optimizing instrument settings (Table 39.1).

CELLULAR UPTAKE

In many cases, the functionality of the EV contents depends on where the cargo is dispatched, whether is it in the cytoplasm or potentially even into the nucleus (Rappa et al., 2017). Direct entry into the cytoplasm can be achieved by fusion of EVs to the plasma membrane of the recipient cells, but some form of endocytosis seems to be the most common mode of entry (Joshi, de Beer, Giepmans, & Zuhorn, 2020). So far, it has been proposed that the cells internalize EVs either by fusion with the plasma membrane or via endocytosis (Mulcahy, Pink, & Carter, 2014). Uptake via endocytosis can be categorized into different types of processes, including clathrin-mediated endocytosis, caveolin-mediated endocytosis, lipid raft-mediated endocytosis, macropinocytosis, and phagocytosis (Costa Verdera, Gitz-Francois, Schiffelers, & Vader, 2017; Joshi et al., 2020). The uptake of EVs may be dependent on the cell type and its physiologic condition (Gonda, Moyron, Kabagwira, Vallejos, & Wall, 2020). Ligands expressed on the surface of the EV will also recognize specific receptors on the cell surface or vice versa, which illustrates EVs' ability of cell targeted delivery (Gonda, Kabagwira, Senthil, & Wall, 2019). In short, various mechanisms are involved in the internalization of EVs for either the transfer of functional cargo(s) or disposal of the cellular waste content. These directional trafficking events appear to be cell-specific and are mediated by ligand/receptor interaction to dictate where the EV packages are addressed.

CONCLUSIONS

EVs are natural nanoscopic particles produced by most cells that hold immense promise for utilization as drug carriers in personalized medicine. They can be isolated from a patient's own cells or biofluid, packaged with therapeutic reagents, and reintroduced into the body, thus designing a personalized drug delivery system with minimized potential immunogenicity. While the capacity to employ EVs as delivery vehicles is in high demand, one of the limitations is the ability to efficiently load EVs with therapeutic cargo. Future designs in drug delivery will have to focus on improved or new methods to productively generate designer EVs. To conclude, efforts are being made to advance the drug delivery area of EVs, thus adding validity to a significant biological agent that was once thought of as an artifact.

REFERENCES

Abels, E. R., & Breakefield, X. O. (2016, April 1). Introduction to extracellular vesicles: Biogenesis, RNA cargo selection, content, release, and uptake. *Cellular and molecular neurobiology.* Springer New York LLC. doi:10.1007/s10571-016-0366-z

Akao, Y., Iio, A., Itoh, T., Noguchi, S., Itoh, Y., Ohtsuki, Y., & Naoe, T. (2011). Microvesicle-mediated RNA molecule delivery system using monocytes/macrophages. *Molecular Therapy, 19*(2), 395–399. doi:10.1038/mt.2010.254

Alvarez-Erviti, L., Seow, Y., Yin, H., Betts, C., Lakhal, S., & Wood, M. J. A. (2011). Delivery of siRNA to the mouse brain by systemic injection of targeted exosomes. *Nature Biotechnology, 29*(4), 341–345. doi:10.1038/nbt.1807

Bagai, S., Sun, C., & Tang, T. (2014). Lipid-modified polyethylenimine-mediated DNA attraction evaluated by molecular dynamics simulations. *Journal of Physical Chemistry B, 118*(25), 7070–7076. doi:10.1021/jp503381r

Battistelli, M., & Falcieri, E. (2020). Apoptotic bodies: Particular extracellular vesicles involved in intercellular communication. *Biology, 9*(1). doi:10.3390/biology9010021

Bunggulawa, E. J., Wang, W., Yin, T., Wang, N., Durkan, C., Wang, Y., & Wang, G. (2018). Recent advancements in the use of exosomes as drug delivery systems. *Journal of Nanobiotechnology, 16*(1), 81. doi:10.1186/s12951-018-0403-9

Chin, A. R., & Wang, S. E. (2016). Cancer-derived extracellular vesicles: The 'soil conditioner' in breast cancer metastasis? *Cancer and Metastasis Reviews, 35*(4), 669–676. doi:10.1007/s10555-016-9639-8

Costa Verdera, H., Gitz-Francois, J. J., Schiffelers, R. M., & Vader, P. (2017). Cellular uptake of extracellular vesicles is mediated by clathrin-independent endocytosis and macropinocytosis. *Journal of Controlled Release, 266*, 100–108. doi:10.1016/j.jconrel.2017.09.019

El Andaloussi, S., Mäger, I., Breakefield, X. O., & Wood, M. J. A. (2013, May 15). Extracellular vesicles: Biology and emerging therapeutic opportunities. *Nature Reviews Drug Discovery.* Nature Publishing Group. doi:10.1038/nrd3978

Fechheimer, M., Boylan, J. F., Parker, S., Sisken, J. E., Patel, G. L., & Zimmer, S. G. (1987). Transfection of mammalian cells with plasmid DNA by scrape loading and sonication loading. *Proceedings of the National Academy of Sciences of the United States of America, 84*(23), 8463–8467. doi:10.1073/pnas.84.23.8463

Gonda, A., Kabagwira, J., Senthil, G. N., & Wall, N. R. (2019, February 1). Internalization of exosomes through receptor-mediated endocytosis. *Molecular Cancer Research.* American Association for Cancer Research Inc. doi:10.1158/1541-7786.MCR-18-0891

Gonda, A., Moyron, R., Kabagwira, J., Vallejos, P. A., & Wall, N. R. (2020). Cellular-defined microenvironmental internalization of exosomes. In *Extracellular vesicles and their importance in human health.* IntechOpen. doi:10.5772/intechopen.86020

Han, L., Lam, E. W. F., & Sun, Y. (2019, March 30). Extracellular vesicles in the tumor microenvironment: Old stories, but new tales. *Molecular Cancer.* BioMed Central Ltd. doi:10.1186/s12943-019-0980-8

Johnsen, K. B., Gudbergsson, J. M., Skov, M. N., Christiansen, G., Gurevich, L., Moos, T., & Duroux, M. (2016). Evaluation of electroporation-induced adverse effects on adipose-derived stem cell exosomes. *Cytotechnology, 68*(5), 2125–2138. doi:10.1007/s10616-016-9952-7

Joshi, B. S., de Beer, M. A., Giepmans, B. N. G., & Zuhorn, I. S. (2020). Endocytosis of extracellular vesicles and release of their cargo from endosomes. *ACS Nano* doi:10.1021/acsnano.9b10033

Kim, M. S., Haney, M. J., Zhao, Y., Mahajan, V., Deygen, I., Klyachko, N. L., … Batrakova, E. V. (2016). Development of exosome-encapsulated paclitaxel to overcome MDR in cancer cells. *Nanomedicine: Nanotechnology, Biology, and Medicine, 12*(3), 655–664. doi:10.1016/j.nano.2015.10.012

Kooijmans, S. A. A., Stremersch, S., Braeckmans, K., De Smedt, S. C., Hendrix, A., Wood, M. J. A., … Vader, P. (2013). Electroporation-induced siRNA precipitation obscures the efficiency of siRNA loading into extracellular vesicles. *Journal of Controlled Release, 172*(1), 229–238. doi:10.1016/j.jconrel.2013.08.014

Lamichhane, T. N., & Jay, S. M. (2018). Production of extracellular vesicles loaded with therapeutic cargo. *Methods in Molecular Biology (Clifton, N.J.), 1831*, 37–47. doi:10.1007/978-1-4939-8661-3_4

Lamichhane, T. N., Jeyaram, A., Patel, D. B., Parajuli, B., Livingston, N. K., Arumugasaamy, N., … Jay, S. M. (2016). Oncogene knockdown via active loading of small RNAs into extracellular vesicles by sonication. *Cellular and Molecular Bioengineering, 9*(3), 315–324. doi:10.1007/s12195-016-0457-4

Lötvall, J., Hill, A. F., Hochberg, F., Buzás, E. I., Vizio, D. Di, Gardiner, C., … Théry, C. (2014). Minimal experimental requirements for definition of extracellular vesicles and their functions: A position statement from the international society for extracellular vesicles. *Journal of Extracellular Vesicles.* Co-Action Publishing. doi:10.3402/jev.v3.26913

Maas, S. L. N., Breakefield, X. O., & Weaver, A. M. (2017, March 1). Extracellular vesicles: Unique intercellular delivery vehicles. *Trends in Cell Biology*. Elsevier Ltd. doi:10.1016/j.tcb.2016.11.003

Mulcahy, L. A., Pink, R. C., & Carter, D. R. F. (2014). Routes and mechanisms of extracellular vesicle uptake. *Journal of Extracellular Vesicles*. Co-Action Publishing. doi:10.3402/jev.v3.24641

O'Neill, C. P., Gilligan, K. E., & Dwyer, R. M. (2019). Role of extracellular vesicles (EVs) in cell stress response and resistance to cancer therapy. *Cancers*, *11*(2). doi:10.3390/cancers11020136

Ohno, S. I., Takanashi, M., Sudo, K., Ueda, S., Ishikawa, A., Matsuyama, N., … Kuroda, M. (2013). Systemically injected exosomes targeted to EGFR deliver antitumor microRNA to breast cancer cells. *Molecular Therapy*, *21*(1), 185–191. doi:10.1038/mt.2012.180

Patton, M. C., Zubair, H., Khan, M. A., Singh, S., & Singh, A. P. (2020). Hypoxia alters the release and size distribution of extracellular vesicles in pancreatic cancer cells to support their adaptive survival. *Journal of Cellular Biochemistry*, *121*(1), 828–839. doi:10.1002/jcb.29328

Pegtel, D. M., & Gould, S. J. (2019). Exosomes. *Annual Review of Biochemistry*, *88*(1), 487–514. doi:10.1146/annurev-biochem-013118-111902

Pomatto, M. A. C., Bussolati, B., D'Antico, S., Ghiotto, S., Tetta, C., Brizzi, M. F., & Camussi, G. (2019). Improved loading of plasma-derived extracellular vesicles to encapsulate antitumor miRNAs. *Molecular Therapy - Methods and Clinical Development*, *13*, 133–144. doi:10.1016/j.omtm.2019.01.001

Raposo, G., & Stoorvogel, W. (2013, February). Extracellular vesicles: Exosomes, microvesicles, and friends. *Journal of Cell Biology*. The Rockefeller University Press. doi:10.1083/jcb.201211138

Rappa, G., Santos, M. F., Green, T. M., Karbanová, J., Hassler, J., Bai, Y., … Lorico, A. (2017). Nuclear transport of cancer extracellular vesicle-derived biomaterials through nuclear envelope invagination-associated late endosomes. *Oncotarget*, *8*(9), 14443–14461. doi:10.18632/oncotarget.14804

Shigekawa, K., & Dower, W. J. (1988). Electroporation of eukaryotes and prokaryotes: A general approach to the introduction of macromolecules into cells. *BioTechniques*, *6*(8), 742–751.

Ståhl, A. L., Johansson, K., Mossberg, M., Kahn, R., & Karpman, D. (2019). Exosomes and microvesicles in normal physiology, pathophysiology, and renal diseases. *Pediatric Nephrology*, *34*(1), 11–30. doi:10.1007/s00467-017-3816-z

Stoorvogel, W., Kleijmeer, M. J., Geuze, H. J., & Raposo, G. (2002). The biogenesis and functions of exosomes. *Traffic*, *3*(5), 321–330. doi:10.1034/j.1600-0854.2002.30502.x

Sukreet, S., Silva, B. V. R. E., Adamec, J., Cui, J., & Zempleni, J. (2019). Sonication and short-term incubation alter the content of bovine milk exosome cargos and exosome bioavailability (OR26-08-19). *Current Developments in Nutrition*, *3*(Suppl 1). doi:10.1093/CDN/NZZ033.OR26-08-19

Tai, Y. L., Chu, P. Y., Lee, B. H., Chen, K. C., Yang, C. Y., Kuo, W. H., & Shen, T. L. (2019). Basics and applications of tumor-derived extracellular vesicles. *Journal of Biomedical Science*, *26*(1), 1–17. doi:10.1186/s12929-019-0533-x

Taverna, S., Pucci, M., & Alessandro, R. (2017). Extracellular vesicles: Small bricks for tissue repair/regeneration. *Annals of Translational Medicine*, *5*(4). doi:10.21037/atm.2017.01.53

Teixeira, J. H., Silva, A. M., Almeida, M. I., Barbosa, M. A., & Santos, S. G. (2016). Circulating extracellular vesicles: Their role in tissue repair and regeneration. *Transfusion and Apheresis Science*, *55*(1), 53–61. doi:10.1016/j.transci.2016.07.015

Todorova, D., Simoncini, S., Lacroix, R., Sabatier, F., & Dignat-George, F. (2017, May 12). Extracellular vesicles in angiogenesis. *Circulation Research*. Lippincott Williams and Wilkins. doi:10.1161/CIRCRESAHA.117.309681

Vader, P., Mol, E. A., Pasterkamp, G., & Schiffelers, R. M. (2016, November 15). Extracellular vesicles for drug delivery. *Advanced Drug Delivery Reviews*. Elsevier B.V. doi:10.1016/j.addr.2016.02.006

Valadi, H., Ekström, K., Bossios, A., Sjöstrand, M., Lee, J. J., & Lötvall, J. O. (2007). Exosome-mediated transfer of mRNAs and microRNAs is a novel mechanism of genetic exchange between cells. *Nature Cell Biology*, *9*(6), 654–659. doi:10.1038/ncb1596

Whiteside, T. L. (2016). Tumor-derived exosomes and their role in cancer progression. *Advances in Clinical Chemistry*, *74*, 103–141. doi:10.1016/bs.acc.2015.12.005

Witwer, K. W., & Théry, C. (2019). Extracellular vesicles or exosomes? On primacy, precision, and popularity influencing a choice of nomenclature. *Journal of Extracellular Vesicles*, *8*(1), 1648167. doi:10.1080/20013078.2019.1648167

Yáñez-Mó, M., Siljander, P. R.-M., Andreu, Z., Bedina Zavec, A., Borràs, F. E., Buzas, E. I., … De Wever, O. (2015). Biological properties of extracellular vesicles and their physiological functions. *Journal of Extracellular Vesicles*, *4*(1), 27066. doi:10.3402/jev.v4.27066

Zolochevska, O., & Figueiredo, M. L. (2012). Advances in sonoporation strategies for cancer. *Frontiers in Bioscience - Scholar*, *4 S*(3), 988–1006. doi:10.2741/s313

40 Extracellular Vesicles (EVs)
Naturally Occurring Vehicles for RNA Nanotherapeutics

Alice Braga and Giulia Manferrari
University of Cambridge, Cambridge, UK

Jayden A. Smith
Cambridge Innovation Technologies Consulting (CITC) Ltd, Cambridge, UK

Stefano Pluchino
University of Cambridge, Cambridge, UK

CONTENTS

DOI: 10.1201/9781003001560-44

INTRODUCTION

Our understanding of the biological role of RNA has evolved substantially in recent decades, from that of a mere intermediate in the central dogma of molecular biology up to a multifunctional platform that is active in the storage, transmission, translation, expression, and even destruction of genetic information. Indeed, there is a compelling argument that life itself emerged from RNA-based precursors. As we have come to appreciate the myriad roles played by RNA *in vivo*, so too have we characterized a diversity and complexity of the RNA variants that facilitate these actions.

The versatility intrinsic to this biopolymer has fostered its adoption as a promising platform for biotechnological applications in contexts such as materials science, molecular computation, and, perhaps most prominently, therapeutics. This latter application has typically had its basis in the modulation of endogenous RNA levels, with concomitant effects on the expression of (patho) physiological proteins. Such an approach has been substantially inspired by the discovery of RNA interference in the 1990s and the advent of nanotechnology, providing a means through which to influence biological activity through the introduction of exogenous modulating RNAs.

Nevertheless, there have been hurdles toward developing RNA-based therapeutics, largely arising from concerns regarding the deliverability of RNA given its reputation for instability. Many different delivery vehicles have been explored in the search for safe, efficacious, and targeted means to deliver RNA therapeutics to their site/tissue of action: inorganics, dendrimers, polymers, polyplexes, even RNA-based vehicles (e.g. 3WJs). Liposomal carriers, synthetic membrane vesicles encapsulating an aqueous core, have received particular attention in this regard. Indeed, the recently approved first RNA interference (RNAi)-based drug *patisiran*, intended for the treatment of transthyretin-mediated amyloidosis, is delivered in a liposomal envelope. This approach will certainly see further use in new RNAi-based therapies and, ultimately, clustered regularly interspaced short palindromic repeat (CRISPR)-based gene editing.

But liposomes have a natural analog in extracellular vesicles (EVs), a heterogenous group of cell-secreted, membranous particles which are intrinsic carriers of biomolecules, including RNA, and often reflect the nature of their parent cell and/or the pathophysiological state under which they were obtained. Dismissed early as a waste disposal mechanism, understanding has evolved to encompass their role in intercellular communication and thus their potential as diagnostic/prognostic biomarkers or even routes of therapeutic intervention. Being inherently biocompatible, there have been substantial efforts to exploit EVs as drug delivery vehicles, including modification of their cargoes and selectivity. Thus, EVs are commonly repurposed or re-engineered as vehicles of exogenous RNAs.

In this chapter, we briefly describe the biogenesis of classification of different categories of EV, exploring their composition with a particular focus on RNA content. We then provide an overview of their reported roles in homeostasis and disease and detail their applications as therapeutic tools, giving attention to RNA delivery, with an exploration of notable preclinical and clinical studies.

THE BIOGENESIS OF EVs

A heterogeneous variety of carrier vesicles are released by virtually all cell types, and their role in the transport of biomolecules between intracellular compartments is widely established. However, only relatively recently has the concept emerged that EVs can also actively mediate communications and induce functional responses in adjacent and distant target cells. Being a relatively nascent area of research complicated by many technical challenges regarding the isolation, characterization, and study of EVs, there are still great uncertainties regarding their biogenesis, release, and intercellular signaling mechanisms.

EVs are typically characterized by a lipid bilayer, embedded with transmembrane proteins and enclosing a lumen of hydrophilic soluble biomolecules. The inner and structural composition of EVs and their cargos (and thus their functional outcome on acceptor cells) is largely dictated by their biogenesis and by their parental cell lineage. The secretion of EVs can be either constitutive or inducible in response to particular stimuli, such as receptor activation or response to cellular stress, and it is contingent to cell and EV types (Théry, Ostrowski, and Segura 2009).

A broad definition of EVs is usually adopted to describe the various membrane structures that are secreted into the extracellular space by donor cells as a form of cellular signaling (or, perhaps, waste disposal). However, this *"catch-all"* definition comprises a diversity of vesicle types that reflects not only a physical heterogeneity of EVs but also a confounding and confusing nomenclature stemming from a non-standardized plethora of isolation and classification methods (Gould and Raposo 2013).

Various types of vesicles described in the literature include microvesicles (Cocucci, Racchetti, and Meldolesi 2009), ectosomes (Cocucci and Meldolesi 2015), exosomes (Vlassov et al. 2012), microparticles (Curtis et al. 2013), gesicles (Mangeot et al. 2011), tolerosomes (Karlsson et al. 2001), oncosomes (Al-Nedawi et al. 2008) (as distinct from *large* oncosomes (Minciacchi, You et al. 2015b)), prostasomes (Aalberts, Stout, and Stoorvogel 2014), migrasomes (Ma et al. 2014), dexosomes (Le Pecq 2005), and human endogenous retroviral particles (Balaj et al. 2011), among others (Gyorgy et al. 2011).

The classification of EVs has often been grounded on the basis of their morphology, contents, and surface chemistry (lipid and protein composition), or otherwise reflects the EV's lineage. The labelling of EVs according to their parent cell has resulted in substantial redundancy, giving rise to a variety of names describing what are essentially the same class of vesicle (Cocucci, Racchetti, and Meldolesi 2009). Alternatively, the use of low-resolution isolation techniques has led to heterogeneous mixtures of vesicles being labelled as a distinct type, thus potentially masking any subtle diversity of function.

More recently, a systematic classification focused on general mechanisms of EV biogenesis (van Niel, D'Angelo, and Raposo 2018; Raposo and Stoorvogel 2013) and release has been adopted, providing greater consistency.

According to this system, three broad categories of EV have been characterized (as depicted in Figure 40.1): (i) *exosomes*, which originate from multivesicular bodies (MVBs) associated with the endosomal pathway; (ii) *microvesicles* (MVs; also known as shedding vesicles or ectosomes), which arise from direct budding and shedding of the cellular plasma membrane; and (iii) *apoptotic bodies*, plasma membrane-derived vesicles arising from apoptosis-induced blebbing. Each category likely encompasses a variety of different EV types; the identification of type-specific biochemical markers reflecting their pathway of biogenesis and further elucidation of specific EV subtypes is an active and contentious area of research (Figure 40.1).

It is important to consider that cells secrete all these vesicular subtypes simultaneously as part of their *"secretome"*. Most of the studies identify the isolated vesicles as exosomes on the basis of their protein contents. However, populations of circulating vesicles are likely composed of both exosomes and MVs, and regularly employed purification methods do not allow for full discrimination between exosomes and MVs (let alone the diversity within each class). As such, a major ongoing challenge is to establish new high-resolution purification methods able to distinguish between types of EV (Witwer et al. 2013). Differences in properties such as size, morphology, buoyant density, and protein composition, often considered independently, seem insufficient for a clear distinction (Bobrie et al. 2011), (Raposo and Stoorvogel 2013). Therefore, well-defined EV classification has proven elusive due to the heterogeneity of EV species and the assortment of non-specific isolation techniques.

Several different techniques are routinely used to isolate exosomes from biological fluids, including differential centrifugation, density-gradient centrifugation, and ultrafiltration. Each technique presents its own advantages and disadvantages, and the resultant preparations crucially differ in purity and heterogeneity. Thus, it is critical to appreciate that any classification based on the composition, cargo, and functionality of these EV populations is likely to reflect the presence of diverse mix of vesicular and non-vesicular components, rather than a specific class of vesicles. The heterogeneity of EV preparations is at the basis of several controversies surrounding the properties of specific vesicle types, including, notably, whether exosomes can deliver functional nucleic acid

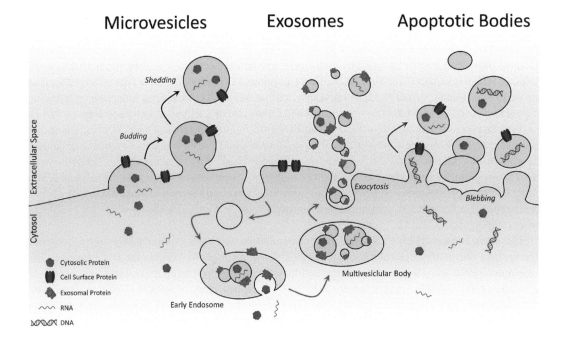

FIGURE 40.1 Mechanisms of biogenesis of extracellular vesicles: *microvesicles* are generated from budding and shedding of the cell membrane; *exosomes* are generated via an endosomal pathway in which a multivesicular body merges with the cell membrane, releasing the vesicles in to the extracellular space; and apoptotic bodies are generated through the blebbing and disintegration of apoptotic cells.

cargoes. It should therefore be noted that many of the studies reported in this chapter have not clearly defined the specific class of EVs under investigation. Indeed, the term *"exosome"* is often used in the literature to describe any small EV.

There is a clear need for more discriminating and/or standardized purification methods for the isolation and characterization of EVs. Indeed, a better understanding of EV heterogeneity, and a better understanding of their biogenesis and secretion mechanisms, is of fundamental importance to appreciate their biological function and thus their potential use in therapeutic or diagnostic applications (Murphy et al. 2019; Shah, Patel, and Freedman 2018). Furthermore, actual physiological effects *in vivo* will be as a result of exposure to a mixture of EV types from a diversity of parent cells, and so a clear understanding of the biological relevance of any one subtype versus bulk heterogeneity is needed to exploit EVs in a therapeutic context.

Here, we provide a brief overview of each of the three broad classes of EV, as delineated by their pathway of biogenesis:

Exosomes

Exosomes are a specific subtype of EVs first identified in mammalian reticulocytes (Johnstone et al. 1991). Although a definitive physical characterization of what classifies as an exosome has not yet been demarcated, they are currently identified as a homogenous collection of EVs (appearing cup-shaped in transmission electron micrographs) with diameters ranging from ca. 40 to 150 nm. Depending on the source, they are typically attributed a density between 1.05 and 1.19 g/cm^3 and collected with a sedimentation force of 100,000–200,000 x g. Although commonly (but not uniquely) distinguished by their size range, exosomes differ from other EVs primarily by their route of biogenesis (and, as a result, their composition). Even though their biogenesis and intracellular trafficking is still the subject of investigation, exosomes are believed to originate from a multive-sicular endosomal (MVE) pathway. Their formation starts with the invagination of clathrin-coated microdomains on the plasma membrane, which enter the cell to be matured by the Endosomal Sorting Complex Required for Transport (ESCRT) protein complex into early endosomes. Subsequently, the budding of intraluminal vesicles (ILVs) into the endosomes themselves results in the maturation of the early endosomes complex into large MVBs. Ultimately, these MVBs can be either trafficked to lysosomes for degradation (*degradative* MVBs) or fuse with the plasma membrane of the cell (*exo-cytic* MVBs), releasing their ILVs, at this stage termed exosomes, into extracellular space. While the ESCRT machinery plays a pivotal role, particularly the accumulation of vesicles destined for a degradative fate, ESCRT-independent exosome biogenesis is also reported (Juan and Furthauer 2018). In the latter case, the sorting of ILVs destined for exocytic release appears to be instead gov-erned by the distribution of the sphingolipid ceramide within raft-based microdomains on the MVB (Trajkovic et al. 2008). The peculiar enrichment of ceramide, occurring at this stage, together with other specific lipids and proteins derived from the MVB membrane, is considered a distinctive exo-somal feature. The abundance of endosome-associated proteins, such as ALG-2-interacting protein X (ALIX) and tumor susceptibility gene 101 (TSG101), has commonly been seen as a signature of endosomal origin; however, how specific cytosolic proteins are sorted into exosomes lumen remains to be elucidated. Compared to other types of EV, exosomes typically exhibit lower levels of mem-brane phosphatidylserine (PS) as they do not arise directly from the plasma membrane; however, they are abundant in surface proteins of the tetraspanin family, especially cluster of differentiation 9 (CD9), CD63, and CD81, which are considered canonical exosome markers (Lee et al. 2011). They are also reported to be enriched in membrane trafficking proteins of the Ras-associated protein (RAB) GTPase family (Bobrie et al. 2011), believed to be involved in exosome secretion.

Nevertheless, a compelling re-examination of EV heterogeneity recently reported by Jeppesen et al. (2019) has cast doubt on the specificity of some well-accepted exosomal markers. The authors have employed a combination of high-resolution density gradient fractionation and Direct Immunoaffinity Capture (DIC) to purify *bona fide* exosomes, thus proposing a surprising reassess-ment of exosome composition and establishing a differential distribution of protein, RNA, and DNA between small extracellular vesicles (sEVs), a broad category inclusive of exosomes, and higher-density non-vesicular (NV) extracellular matter. EVs isolated by DIC of canonical tetraspanin exo-some markers (CD63, CD81, and/or CD9) were found to be absent in many of the previously widely accepted exosomal markers and cargo constituents (van Niel, D'Angelo, and Raposo 2018; Kowal et al. 2016). While ALIX was abundant in DIC-isolated exosomes, TSG101 was absent and instead attributed to another class of sEVs called ARMMs (see the next section on MVs). Furthermore, the RAB GTPases, while implicated in the trafficking of MVBs to the plasma membrane for exo-some release, were found to be a poor marker of exosomes. High-abundance cytosolic proteins such as glyceraldehyde 3-phosphate dehydrogenase (GAPDH), pyruvate kinase M1/2 (PKM), enolase 1 (ENO1), and 14-3-3 are widely acknowledged as typical components of the exosome lumen. However, Jeppesen et al. instead found these proteins enriched in non-exosomal sEV or even NV fractions. This suggests that the uptake of cytosolic constituents during classical exosome biogen-esis is not the result of a random sampling of the cytosol but rather a highly controlled process. Moreover, this study distinguished sEVs (including exosomes) by their lack of nuclear or cyto-solic microRNA (miRNA) protein machinery (such as Drosha, DiGeorge syndrome critical region 8 (DGCR8), Dicer, TAR RNA binding protein (TRBP), and glycine-tryptophan protein of 182 kDa (GW182)). These new findings refute the idea that exosomes hold the ability to carry out cell-independent miRNA biogenesis as previously suggested (Melo et al. 2014). Indeed, Jeppesen et al. found that many of the most abundant miRNAs showed substantial 3′-trimming and were associated with the NV fraction, rather than being present in exosomes (or any other vesicles). Additionally, neither exosomes nor any other type of sEVs were found associated with double-stranded DNA in the extracellular environment, suggesting that sEVs might not be directly involved in chromatin release (Jeppesen et al. 2019). In summary, classical tetraspanin-enriched exosomes might contain only a limited repertoire of the diversity of biomolecules present in the extracellular milieu and thus a better elucidation of the heterogeneity of sEVs, and indeed NV fractions, is essential (Pluchino and Smith 2019).

Exosome biogenesis is intricately interconnected to the endomembrane system (EMS), a complex series of interconnected membranous organelles playing important roles in responding to stress and maintaining cell homeostasis during health and disease. Another important pathway associated with EMS is autophagy, a highly regulated intracellular lysosome-mediated mechanism of self-digestion and recycling. Exosome biogenesis and autophagy are therefore strictly linked by the endolysosomal pathway and autophagy-related proteins have been found involved in exosome biogenesis. Indeed, several proteins of the autophagy-related (ATG) class have been shown to contribute to the formation and trafficking, and ultimate extracellular release, of endosomal-origin vesicles (Guo et al. 2017; Murrow, Malhotra, and Debnath 2015). Another recent report has demonstrated autophagic clearance of aberrant endocytic vacuoles caused by CD63 knockout, where inhibition of autophagy partially rescued exosome biogenesis in the CD63-null cells (Hurwitz et al. 2018). However, EVs bearing phosphatidylethanolamine-conjugated light chain 3, a marker of the autophagosome, have been determined to be distinct from DIC-purified *bona fide* exosomes expressing the canonical tetraspanins (Jeppesen et al. 2019). Nevertheless, these observations further support an overlap between autophagy-mediated secretion and exosome release; both autophagy and exosome release display functional redundancy in eliminating unwanted proteins whereby each pathway may compensate for a deficiency in the other. In fact, defective MVBs and their contents may be disposed via autophagic degradation, and autophagy inhibition may rescue exosome release from MVBs that would otherwise be degraded (Villarroya-Beltri et al. 2013). Alternatively, autophagy and exosome release may operate in concert to counter cellular stress (Kumar et al. 2015), with both processes adapting in response to external (patho)physiological conditions.

MICROVESICLES

MVs (also known as shedding vesicles or ectosomes) are a heterogeneous class of EVs, highly diverse in shape and size, although typically ranging from 100 to 1,000 nm in diameter. While morphologically diverse, MVs arise via a broadly similar phenomenon in that they are directly shed into the extracellular space from outward budding and subsequent fission of the cellular plasma membrane. The specific mechanisms underlying their biogenesis still require elucidation; however, the general process involves an increase in intracellular Ca^{2+} concentration leading to changes in plasma membrane phospholipid distribution and subsequent enzymatic cytoskeletal remodelling (Taylor and Bebawy 2019). The resultant shedding vesicles are believed to originate from cholesterol-rich lipid rafts in the plasma membrane (Del Conde et al. 2005). Accordingly, MVs commonly exhibit high levels of PS exposure and are enriched in lipid raft-associated proteins such as tissue factor (TF) and flotillin-1, along with various selectins and integrins, CD40 ligand, complement receptor-1, and the matrix metalloproteinases MMP-2 and MMP-9 (Théry, Ostrowski, and Segura 2009; Lee et al. 2011). The membrane-associated protein Annexin A1 has often been identified in exosomes. However, the recent re-appraisal by Jeppesen et al. instead found Annexin A1 (and potentially Annexin A2) to be a novel and specific marker of classical microvesicles (Jeppesen et al. 2019). ADP-ribosylation factor 6 (ARF6), a small GTPase, is understood to play a key role in the shedding of many types of MVs (Muralidharan-Chari et al. 2009; Tricarico, Clancy, and D'Souza-Schorey 2017), including *large* oncosomes ca. 1–10 μm in size (Minciacchi, Freeman, and Di Vizio 2015a).

MVs are constitutively yet slowly released by healthy cells, but, like exosomes, their release can notably increase upon appropriate stimulation of the parent cell. For instance, a particular category of MVs known as gesicles are known to shed only from cells with induced overexpression the vesicular stomatitis virus spike glycoprotein (VSVG) (Mangeot et al. 2011). Still not fully characterized, gesicles are regarded as a heterogeneous population averaging approximately 100 nm in diameter.

The biosynthetic pathway of these peculiar vesicles is still undetermined, although they do not present any markers correlated with endosomal origin, a typical feature of exosomes.

The substantial size range encompassed by the MV designation includes a number of types of sEV that may overlap with exosomes in conventional isolation procedures. One such example described as *membrane particles* originates from the shedding of subdomains on the plasma membrane enriched in the stem cell marker prominin-1 (CD133). These particles range 50–80 nm in diameter with a density of 1.04–1.07 g/cm^3. Like other plasma membrane-derived EVs, they do not exhibit surface expression of the tetraspanin exosomal markers (i.e. CD63), but their lipid composition is still the object of ongoing study (Théry, Ostrowski, and Segura 2009; Marzesco et al. 2005). Another variety of small EV arising from plasma membrane budding is the Arrestin-domain-containing protein 1 (ARRDC1)-mediated microvesicle (ARMM) (Nabhan et al. 2012). These MVs are ca. 40 to 100 nm in size and originate from an interaction between TSG101 from the endosomal machinery and ARRDC1 at the plasma membrane. This leads to the direct budding of sEVs expressing an endosomal protein (TSG101), despite originating via an MV-like shedding process (Jeppesen et al. 2019).

Apoptotic Bodies

Apoptotic bodies (ABs) are a heterogeneous class of EVs which are shed only during cell fragmentation in late apoptosis, distinguishing them from constitutively secreted MVs and exosomes. ABs are typically very large EVs (1–5 μm in diameter), although a smaller population of 'apoptotic vesicles' (ca. 100–1000 nm) have also been described (Hauser, Wang, and Didenko 2017). Much like MVs, ABs are marked by high levels of PS on their outer surface along with associated Annexin V; the biogenesis of ABs involves blebbing of the plasma membrane of an apoptotic cell, yielding vesicular parcels encapsulating remnants of the cell, including cytosolic contents, organelles, or even fragmented nuclear material (Xu, Lai, and Hua 2019). As such, the contents of ABs can be highly variable dependent on what happens to be in the cytoplasmic protrusion from which they arise. The presence-or-absence of dense chromatin within ABs contributes to their physical diversity, with a higher average density (1.16–1.28 g/cm^3) than other EV types. Nevertheless, the presence of histones (Théry, Ostrowski, and Segura 2009; Mathivanan, Ji, and Simpson 2010) and/or genomic DNA (Holmgren 2010) appears unique to ABs among EVs and suggests a role as potential vectors of horizontal gene transfer in cell-to-cell communication. Indeed, intercellular signaling is a key function of ABs as they convey an "*eat me*" signal to attract phagocytes, enabling apoptotic cell clearance (Caruso and Poon 2018).

Other EVs

The EV categories described above represent a diversity of different vesicle types, but there remain some that currently defy or transcend such categories. One emerging area of interest involves mitochondrial vesicles, containing mitochondrial DNA (mtDNA), proteins, or even entire mitochondrial particles. Mitochondrial material has been observed in a diversity of EVs, but the mechanisms by which these originate remains to be elucidated. Larger EVs can contain either mtDNA or entire mitochondrial particles, as observed in mesenchymal stem cells and astrocytes (Guescini et al. 2010; Hayakawa et al. 2016; Phinney et al. 2015), possibly as a route of disposal for dysfunctional mitochondria. Platelet-derived MVs have likewise been reported to contain mitochondria or mitochondrial components (Boudreau et al. 2014), while lipopolysaccharide (LPS)-stimulated monocytes have been found to release MVs containing mitochondrial proteins (Bernimoulin et al. 2009). As extracellular mitochondria and mitochondria-derived damage associated molecular patterns (DAMPs) are renowned inducers of inflammation (Krysko et al. 2011), one hypothesis is that EV-associated mitochondria shed by activated immune cells may function as a signaling hub able to actively mediate the induction of proinflammatory responses in target cells. Indeed, a mouse model of hepatic inflammation was found to correlate with increased levels of circulating MVs containing mtDNA (Garcia-Martinez et al. 2016). Moreover, a recent study by Puhm et al. showed that LPS-stimulated monocytes release MVs containing mitochondrial components as well as free mitochondria (translocase of the outer membrane 22-positive vesicles, representing a subset of EVs) (Puhm et al. 2019). These stress-induced extracellular signaling vehicles were found to be strong inducers of tumor necrosis factor (TNF) and Type I interferon inflammatory responses in recipient endothelial cells, an effect largely mediated by vesicular TNFα and oxidized mitochondrial 16S rRNA. The involvement of EVs in the horizontal transfer of mitochondria has also recently been explored in the context of tissue regeneration (Spees et al. 2006; Islam et al. 2012). One of the hypotheses regarding the transfer of whole mitochondrial particles and their subsequent incorporation in the mitochondrial network of the receiving cell is that this is a means of rescuing the mitochondrial function of dysfunctional recipient cells. While this phenomenon, and the role played by EVs in it, has yet to be verified, it does suggest a compelling mechanism of action for stem cell transplants in regenerative immunology.

While EVs containing mitochondrial material generally appear to fall under the MV class of vesicles, mitochondria themselves give rise to so-called mitochondria-derived vesicles (MDVs) that enter the endosomal pathway. MDVs are 70–150 nm in size and constitute a fourth mechanism of mitochondrial quality control alongside the actions of mitochondrial proteases, ubiquitin-mediated proteasomal degradation, and mitophagy. Generated through selective incorporation of protein cargoes into a membranous structure, MDVs are trafficked intracellularly to either the late endosome/MVB for degradation in the lysosome (Soubannier et al. 2012) or to peroxisomes (Neuspiel et al. 2008) in an outer membrane marker-dependent manner. Since MVBs can also be routed to the cell surface, this former scenario offers one potential explanation for the reported observation of extracellular mitochondrial material in exosomes. Further studies are needed to test this hypothesis and to elucidate the details governing the generation, packaging, and fate of MDVs.

Finally, the study of EVs and their prospective functions is further complicated by the presence of non-vesicular carriers of functional biomolecules (notably RNAs) in EV preparations. These include the non-membranous nanoparticles called exomeres (Zhang et al. 2019a), Argonaute complexes (Arroyo et al. 2011), ribonucleoproteins (Wei et al. 2017), and lipoproteins (Vickers et al. 2011). The co-purification of these complexes with *bona fide* EVs further illustrates the need for standardized isolation and characterization methods when ascribing function to extracellular messengers (Mateescu et al. 2017).

THE RNA CARGO OF EVs

RNA is a highly dynamic and versatile biomolecular tool capable of influencing numerous cellular activities. Over the last century, elucidation of the structural and functional diversity of RNA has shed light on its multifunctionality, being able to both store genetic information and catalyze biological reactions, linking genotype and phenotype. While the majority of nucleic acids are located within cells, pioneering works demonstrating the presence of circulating DNA within human blood samples (Mandel and Metais 1948; Tan et al. 1966) paved the way for the investigations into the potential biological role of RNAs as extracellular signaling molecules within organisms (Dinger, Mercer, and Mattick 2008).

Several studies have demonstrated the capability of RNA molecules to be transferred intercellularly within an organism (Fire et al. 1998) or even between different organisms (Knip, Constantin, and Thordal-Christensen 2014), establishing this phenomenon as a powerful and universal mechanism for cellular communication. While free RNA is quickly degraded by ribonucleases, groundbreaking work from Valadi et al. demonstrated the presence of functional mRNA and miRNA in exosomes derived from both murine and human mast cells, asserting a role for such EVs in complex cell-to-cell communication and suggesting potential applications in gene therapy (Valadi et al. 2007). Importantly, this study showed that exosome-trafficked mouse mRNAs were found to effect *de novo* synthesis of mouse proteins when delivered to human cells. Moreover, the presence of specific circulating EV-resident RNAs associated with different pathologies has suggested their use as potential biomarkers (Lasser 2012). Advances in high-throughput next-generation RNA sequencing allowed for the identification of a plethora of non-coding RNAs within EV structures, including vault RNA, Y-RNA, small nuclear RNA (snRNA), small nucleolar RNA (snoRNA), signal recognition particle RNA (SRP-RNA), transfer RNA (tRNA), and long non-coding RNA (lncRNA) (Turchinovich, Drapkina, and Tonevitsky 2019). The relative abundance of these different RNA classes differs between EVs and donor cells, which suggests precise RNA-sorting mechanisms from parental cells into EVs that remain to be fully elucidated.

Considerable attention has been paid to investigations of the RNA cargo of different EV populations since understanding the mechanisms that regulate their sorting, targeting, and biological effects will improve future utilization of EVs as therapeutic tools for the delivery of RNA (or other agents). One of the major caveats to consider in such studies is the method(s) used to isolate specific types of vesicles, with the heterogeneity of EVs, and co-purification with non-vesicular species, making it difficult to obtain pure samples for characterization. Notably, a recent study employed two different methods to specifically isolate and characterize *bona fide* exosomes from other EVs and non-vesicular contaminants: high-resolution density gradient fractionation and DIC demonstrated significant differences in both the proteomic and nucleic acid content of EVs as compared to the previously established and commonly accepted makeup of such vesicles (Jeppesen et al. 2019). The authors demonstrated that EVs – exosomes in particular – are largely devoid of nuclear and cytosolic miRNA machinery, such as Argonautes 1-4, Drosha, Dicer, DiGeorge syndrome chromosomal region 8 (DGCR8), TAR RNA-binding protein (TRBP), and glycine-tryptophan protein of 182 kDa (GW182), as well as other RNA-binding proteins like ribosomal proteins S3 or S8. Moreover, this study suggested that extracellular chromatin is not associated with EVs, but rather DNA and histones are released into the extracellular space by a vesicle-independent endosomal pathway.

This evidence contradicts previous findings in the literature, some controversial and others dogmatic, but it is indicative of a more limited or regulated EV payload than previously thought and sheds light on the importance of improved methods of EV isolation and characterization.

mRNA

Several studies have reported the presence of mRNA in EVs. mRNA, protein-coding RNA, is synthetized in the nucleus as pre-mRNA and subsequently goes through splicing, end-modification, and cytosolic translocation for eventual protein translation. In a work published in 2006 it was demonstrated that MVs obtained from various human carcinoma cell lines carried mRNA for growth factors, along with commonly reported EV cargo proteins, and were capable of influencing the immunophenotype and biological activity of target monocytes (Baj-Krzyworzeka et al. 2006). The implication that EV-trafficked mRNA could be translated in recipient cells, potentially affecting their function, was further evidenced in another report from co-authors of the first (Ratajczak et al. 2006). MVs derived from murine embryonic stem cells (ES) were found to be enriched with mRNA corresponding to early pluripotent transcription factors. This ES-MV mRNA was demonstrated not only to be transferred to target/recipient cells, but more importantly to be efficiently translated into the corresponding protein product, underlying the possibility of a horizontal transfer from parental cells to target ones. The functionality of EV-mRNAs was elegantly confirmed in a later study from Skog et al. (2008). Parental cells were transfected with a lentivirus encoding for luciferase and luciferase mRNA levels were measured in isolated MVs prior to transfer to recipient cells. The expression of functional luciferase protein was subsequently demonstrated in these recipient cells. In the same study the authors also showed the presence of mRNAs that were exclusively present in MVs compared to parental glioblastoma cells, indicative of selective enrichment. The most abundant MV mRNAs, more likely to influence recipient cells, were related through gene ontology analysis to metabolic processes, cell proliferation, immune response, and histone modification. While a clear mechanism for mRNA sorting into EVs has not yet been characterized, some specific motifs have been identified as promoting a specific enrichment of mRNAs in EVs over parental cells. Moreover, the mRNA content of EVs has been shown to reflect the state of the parental cell, as demonstrated by work from Cossetti et al. that found EVs from pro-inflammatory Th1-activated NSCs were enriched in mRNAs coding for the interferon gamma (IFNγ) signaling pathway (Cossetti et al. 2014). Another study on human glioblastoma-derived MVs revealed that mRNAs enriched in MVs were character-ized by zipcode-like 25-nt sequence (Bolukbasi et al. 2012). The authors showed the binding of miRNA-1289 to the zipcode sequence as an orchestrator of mRNA sorting into the MVs, suggesting that this communal sequence might be employed to increase mRNA levels in MVs. While the litera-ture is replete with examples of exosomal mRNAs from a variety of cell types, the recent stringent re-evaluation of exosome cargo by Jeppesen et al. suggests that the majority of extracellular (vesicu-lar and non-vesicular) are intronic in nature (Jeppesen et al. 2019). An interesting and potentially complementary phenomenon was described in a study by Batagov et al. wherein exosomes secreted from human glioblastoma cells were characterized by the presence of fragmented mRNA, enriched in 3′-ends of transcripts (Batagov and Kurochkin 2013). Since such 3′-untranslated regions (UTRs) are able to regulate translational efficiency and localization of mRNAs and are rich in miRNA bind-ing sites, the RNA fragments delivered in exosomes may influence protein translation in recipient cells through competition with endogenous mRNA rather than solely through transfer of genetic information.

MicroRNA

MicroRNAs are highly conserved small non-coding RNAs known to play a role in many physiologi-cal and pathophysiological process by altering the stability or translation of mRNA with which they share partial complementarity at the 3′-UTRs (Bushati and Cohen 2007). Circulating miRNA in bio-fluids can be found in different forms, such as high-density lipoprotein particles (LPPs), Argonaute 2 (Ago2) protein complexes, or loaded in EVs (Turchinovich et al. 2013). The sorting and loading of miRNA into EVs appears to be a selective process, dependent on cell type and condition, rather than a random sampling of the parental cell miRNA population; this implies a likely role in modulating specific functions in recipient cells. Conversely, the promiscuity of a given miRNA makes it capable of simultaneously regulating multiple target messengers, and thus potentially influencing a wide range of (patho)physiological processes (Hu, Drescher, and Chen 2012).

Additionally the identification of specific enrichment of miRNAs in EVs has positioned them as possible biomarkers for specific pathophysiological states, with the sorting and packing of EV cargo, including miRNA, differing between healthy and diseased states (Iguchi, Kosaka, and Ochiya 2010; Katsuda et al. 2014). Indeed, not only have miRNAs isolated from EVs been used to discrimi-nate and classify specific tumors (Kinoshita et al. 2017), but they have also enabled identification of tumor severity and progression (Camacho, Guerrero, and Marchetti 2013), placing them in a unique position as prognostic tools. Their potential as biomarkers has been supported by several lines of evidence: for example, EVs isolated from the cerebrospinal fluid (CSF) of Parkinson's disease patients showed a distinct extracellular miRNA profile that differed from healthy controls, thus leading to the identification of six miRNAs as biomarkers with high specificity and sensitivity for the pathology (Gui et al. 2015). Indeed, the identification of miRNAs in EVs and their key roles in various biological processes has been a key driver of efforts to repurpose EVs as vehicles for the delivery of RNA-based therapeutics.

While the mechanisms of vesicular miRNA packaging have yet to be fully elucidated, several studies have provided compelling clues as to the biomolecular agents and processes involved. One such study, using human peripheral blood mononuclear cells, showed that the specific binding of heterogeneous nuclear ribonucleoprotein A2/B1 (hnRNPA2B1) to miR-198 controlled the sorting of the miRNA to exosomes (Villarroya-Beltri et al. 2013). Specifically, sumoylated hnRNPA2B1, where sumoylation is believed to be important for determining protein function, stability, and localization, was found to drive the loading of miRNAs presenting the GGAG "*EXOmotif*". Similar mechanisms have also been described in a hepatocyte *in vitro* model, where the synaptotagmin-binding cytoplasmic RNA-interacting protein (SYNCRIP) has been identified as one of the components regulating miRNA sorting into exosomes (Santangelo et al. 2016). SYNCRIP was shown to directly bind miRNA enriched with a specific sequence, GGCU (called a "*hEXO motif*"), supporting the potential of specific proteins to discriminate specific miRNAs as exosomal (or as EV cargo more generally). Interestingly, SYNCRIP was also previously demonstrated to be present in astrocytoma-derived GW bodies, cellular components involved in RNA trafficking (Mayer et al. 2007). While their biological and functional importance remains unclear, the identification of such motifs and corresponding proteins might be exploited for the engineering of exosomes for therapeutic applications.

It is poorly understood how EV-trafficked miRNAs might engage with the RNA-silencing machinery of a recipient cell. miRNAs are known to be functional and thus able to accomplish gene silencing, upon forming the RNA-induced silencing complex (RISC)-loading complex with the proteins Dicer, TAR RNA- binding protein (TRBP), and Ago2. However, uncomplexed, single-stranded miRNAs have poor silencing efficiency and are likely to be outcompeted by abundant endogenous miRNAs in the recipient cell. Melo et al. therefore postulated that miRNAs in EVs, specifically in exosomes, should be associated with such proteins in order to function in target cells (Melo et al. 2014). The authors showed, by comparing metastatic to non-metastatic cancer-derived exosomes that exosomes from the former cells were enriched with functional Dicer protein, along with Ago2 and TRBP; in particular, Dicer-containing exosomes were able to process pre-miRNA into mature miRNA. This work, together with similar evidence from different cancer cell lines, showing the presence of RISC-associated proteins (McKenzie et al. 2016), suggested a possible mechanism for miRNA delivery through exosomes. Nevertheless, the presence RNA-binding proteins, especially those associated with miRNA biogenesis and function in exosomes (or EVs more generally) remains controversial (Van Deun et al. 2014; Jeppesen et al. 2019). Indeed, while great advances have been made in unraveling the mechanisms involved in miRNA delivery via EVs and advancing their use as screening or therapeutic tools, it is increasingly clear that standardized isolation and purification methods are necessary to properly distinguish and characterize different types of EVs (Jeppesen et al. 2019; Konoshenko et al. 2018; Pluchino and Smith 2019; Jakymiw et al. 2005) (as well as their RNA cargo (Van Deun et al. 2014)). Similarly, the functionality of vesicular RNAs has been challenged by recent work from Wei et al. that describes a diverse landscape of extracellular RNAs and suggests that massive uptakes of EVs would be required for EV-trafficked RNAs to have a biological impact in recipient cells (Wei et al. 2017). Beyond sorting by established RNA-binding proteins, other mechanisms have been proposed in the loading of miRNAs into EVs. For example, the lipidic composition of exosomes, consisting largely of cholesterol, sphingomyelin, glycosphin-golipids, and phosphatidylcholine, might play a role in the selection of miRNA cargo, with specific nucleotide sequences showing high affinity to phospholipid bilayers (Janas et al. 2015). Notably, neutral sphingomyelinase 2 (nSMase2), an enzyme involved in the generation of ceramide, has been demonstrated to regulate the loading of miR-120 into exosomes from metastatic cancer cells (Kosaka et al. 2013). Finally, by using a combination of *in vitro* experiments and computational analysis, Squadrito et al. characterized a passive mechanism for miRNA sorting into exosomes in which overly abundant endogenous miRNAs are routed to vesicles for extracellular release in an "*miRNA relocation effect*" (Squadrito et al. 2014).

The RISC-silencing machinery relies on the endonuclease activity of Argonaute proteins that can cleave the target transcript. Argonaute proteins are divided into two different families: the *bona fide* Argonaute and P-element-induced wimpy testis (PIWI) families, where the former binds to miRNA and siRNA and PIWIs bind to a small population of small RNAs known as PIWI-interacting RNA (piRNA) (Peters and Meister 2007). piRNAs arise from clusters of intergenic repetitive elements and they regulate the activity of transposons (Iwasaki, Siomi, and Siomi 2015). Evidence from exosomes derived from hypothalamic neurons *in vitro* through the use of two different isolation methods (ultracentrifugation and Optiprep velocity gradient centrifugation) revealed the presence of several RNAs, including piRNAs (Quek et al. 2017). Exosomes from human colon cancer and glioblastoma cell lines have also been shown to be abundant in various small RNAs, with piRNA prominent among them (Jeppesen et al. 2019).

OTHER NON-CODING RNAS

Another class of small non-coding RNA that has been found to be enriched in EVs are the Y-RNAs (Driedonks and Nolte-'t Hoen 2018). Y-RNAs were first characterized in serum from lupus patients but they have been demonstrated to be highly conserved among vertebrate species, and a homologous structure is also present in nematodes. In humans, four different Y-RNAs have been described, all containing a long stem but differing in their primary and secondary structures (Kowalski and Krude 2015). Y-RNAs are implicated in several crucial cellular functions, such as initiation of chromosomal DNA replication (Gardiner et al. 2009), as well as playing a role in RNA stability and quality control (Stein et al. 2005). EV-trafficked Y-RNAs were first identified in EVs derived from cultured dendritic cells (DCs). Interestingly, host Y-RNAs are encapsulated in the capsids of viruses along with the viral genome, likely enhancing viral assembly, stability, or infectivity (Nolte-'t Hoen et al. 2012). Based on this evidence and on the high amount of Y-RNA observed in EVs, the release of Y-RNA as an evolutionarily conserved mechanism, by which cellular RNAs are stabilized in vesicles allowing functional RNA transfer into target cells, was proposed. Similar enrichment has been observed in EVs derived from different cultured cell types, including breast cancer lines (Tosar et al. 2015) or glioblastoma cells (Wei et al. 2017). Nevertheless, these studies identified only fragments of Y-RNA arms, likely due to technical limitations; Y-RNAs are characterized by complex RNA structures that hamper efficient reverse transcriptases and lead to overestimation of fragments compared to full-length Y-RNAs (Driedonks and Nolte-'t Hoen 2018). A variety of proteins have also been described as interacting with Y-RNAs, with their functions inferred as a consequence. Among these proteins, Y-box binding protein 1 (YBX1) has been found to play a crucial role in the sorting of miR-223 into exosomes (Shurtleff et al. 2016): while miR-223 was found enriched in exosomes derived from HEK293T cells, its packaging was compromised when YBX1 was knocked down, suggesting a role as chaperone-mediating sorter. The density gradient purified vesicle fraction from this work showed the absence of Ago2 protein, confirming and supporting more recent findings from Jeppesen et al. (2019) in redefining the nature of exosomes. While packaged miR-223 was found to be decreased upon deletion of YBX1, YBX1 is also known to interact with other RNA types, suggesting a complexity of RNA trafficking from cells to EVs and the need for a further and deeper characterization of such mechanisms.

Circular RNAs (circRNAs) have attracted more attention in recent years as a result of the development of new high-throughput sequencing techniques. Their name derives from their peculiar conformation, where the 3' and 5' ends are jointed in a covalently closed continuous loop. The absence of 5' and 3' ends confers upon circRNAs resistance to major degrading enzymes, resulting in a higher stability and longer half-lives compared to linear RNAs. Proposed functions for circRNAs include RNA-binding protein sponges, miRNA sponges, and, more generally, regulation of transcription (Dong et al. 2017). Specific circRNAs have been identified in EVs isolated from both human glioblastoma cells and from radioresistant cells from the same cell line used to investigate the potential role of EVs in drug resistance (Zhao et al. 2019). RNA-seq and pathway analysis revealed a potential candidate, circATP8B4, and predicted its interaction with miR-766-5p as a potential promoter for cell radioresistance. Several other emerging studies reported an enrichment of circRNAs in EVs (Li et al. 2015), but whether they can propagate their function in target cells through EV-horizontal transfer or whether EVs just represent a possible mechanism for their elimination from parental cells is still under debate (Lasda and Parker 2016).

Vault RNAs (vtRNAs), short polymerase III transcripts 80 to 150 nucleotides (nt) in length, are also found in association with EVs. Typically, vtRNAs comprise some 5% of the mass of vault particles, large, highly conserved ribonucleoprotein complexes that are found in the cytoplasm of many eukaryotic cells. While implicated in intracellular transport, a thorough understanding of the function of vault particles, and indeed vtRNAs, remains elusive; experimental degradation of vtRNAs has identified them as a functional component rather than merely a structural one (van Zon et al. 2003). Notably, the most abundant and enriched small RNAs in MVs isolated from *in vitro* cultured human glioblastoma cells was found to be a full-length vtRNA, VTRNA1-1, and small RNA reads corresponding to processed regulatory vtRNAs recently discovered to be generated through a DICER-dependent and DROSHA-independent mechanism (Li et al. 2013). The exposure of human brain microvascular endothelial cells to glioblastoma cell-derived MVs did not result in any expected regulatory influence on the predicted targets of vtRNA fragments which, together with the apparent absence of proteins characterizing vault particles, suggested alternative functions of vtRNAs besides their role in vault complexes. Similar enrichment of vtRNAs in EVs, in the absence of associated vault proteins, was also observed by Jeppesen et al. (2019). The specificity of EV trafficking of vtRNAs was evidenced in vesicles derived from DCs where an EV enrichment in only one of the stem loops of the vtRNA structure was found compared to cellular counterparts, suggesting a likely preferential export of a specific cleaved part of the vtRNA into the extracellular space (Nolte-'t Hoen et al. 2012).

A number of long non-coding RNAs (lncRNAs) have also been identified in EVs. lncRNAs are >200 nt non-coding transcripts that are divided into different categories such as sense/antisense, divergent/convergent or intronic/intergenic, based on their position relative to a protein-coding gene. The growing innovation in sequencing techniques has highlighted their important role in several biological mechanisms, such as regulation of post-transcriptional processes, chromatin remodeling, gene transcription, and mRNA turnover (Bhat et al. 2016). lncRNAs have been characterized in EVs derived from several cell lines associated with various pathological states. Among early evidence, exosomes derived from cultured HeLa and breast cancer cell lines, upon exposure to a DNA damaging agent, revealed an enrichment of specific lncRNAs compared to parental cells, reflecting a damage response (Gezer et al. 2014). This specific enrichment implied a possible function of shuttle lncRNAs: for example, highly differential expression of *lincRNA-p21* in exosomes from both cell lines suggested a possible regulatory role in target cells as a repressor of both p53-dependent transcriptional responses and target mRNA. Similarly, exosomal HOX transcript antisense RNA (*HOTAIR*) enrichment was hypothesized to result in modulation of gene expression in recipient cells through interaction with chromatin remodeling complexes (Gezer et al. 2014). Other studies on cancer cell lines have described similar specific lncRNA enrichments in EVs (Fan et al. 2018). Exosomes from human glioma cell lines were shown to be highly enriched in *lincRNA-CCAT2*, previously described to be involved in glioma progression (Lang et al. 2017). Knockdown of CCAT2 combined with the transfer of exosome-containing medium from glioma cells to human umbilical vein endothelial cells was employed to investigate the effect of contact-independent cell-to-cell communication and the role of the identified lncRNA. Results revealed the capability of transferred exosomes to promote angiogenesis in target cells through *lincRNA-CCAT2*, highlighting the role of both exosomes and the target lncRNAs as possible future therapeutic agents. On a related note, Balaj et al. have reported high levels of retrotransposon RNA in tumor EVs, suggesting a route through which these mobile genetic elements can interact with remote, healthy cells, potentially transmitting carcinogenic factors (Balaj et al. 2011).

THE (PATHO)PHYSIOLOGICAL ROLE OF EVs

Intercellular signaling is a finely regulated process fundamental to sustaining homeostasis and proper functioning of multicellular organisms. Cell-to-cell communication occurs through a diversity of chemical messengers, including small molecules, proteins, glycoproteins, lipids, growth factors, and even nucleic acids. These signaling molecules can act on the source cell via autocrine signaling, or adjacent cells via juxtracrine signaling, targeting neighbors direct contact via the formation of cell junctions (e.g. gap junctions). Soluble messengers like hormones, cytokines, and chemokines can also be released to act over longer distances through paracrine and endocrine signaling. In recent times, relatively long-distance (> 100 μm) cell-to-cell connections via membrane nanotubes have been described as another putative means of intercellular signaling for the exchange of genetic information or transfer of pathogens between distal cells (Davis and Sowinski 2008). Interestingly, these nanotubules seem to impede small molecules trafficking, facilitating instead the transfer of vesicle-packaged material (Rustom et al. 2004). Signaling vesicle secretion has been described in a multitude of different cell types, yet only a few specialized cases of short-range cell-to-cell communication mediated by EVs (e.g. synaptic exchanges between neurons (Sudhof 2004)) were formally recognized as canonical.

Despite being first described more than half a century ago (Wolf 1967), with the first supporting evidence implicating their role in coagulation (Bastida et al. 1984) and tumor growth (Taylor and Black 1986) emerging in the following decades, EVs have long been dismissed as passive debris arising from damaged or necrotic cells. Indeed, EVs and their contents are often regarded as bio-markers of pathology and, to date, a great part of the research surrounding EVs has focused upon their role in tumor antigen presentation to T cells (Schorey and Bhatnagar 2008). However, it is now becoming clear that EVs are not just a random sampling of biomolecules from their parent cell, but that, in fact, their biogenesis is the result of highly controlled sorting and trafficking mechanisms (Hu, Drescher, and Chen 2012; Pant, Hilton, and Burczynski 2012). Moreover, upon secretion from the parent cell, EVs have been shown to interact with acceptor cells in a specific, targeted manner (Cocucci, Racchetti, and Meldolesi 2009), enlightening their crucial role in intercellular signaling (Camussi et al. 2010; Simons and Raposo 2009; Mathivanan, Ji, and Simpson 2010; Ratajczak et al. 2006; Raposo and Stahl 2019; Mause and Weber 2010). For example, platelet-derived EVs (also known as *microparticles*) seemingly only target monocytes but not neutrophils (Losche et al. 2004); conversely, neutrophil-derived EVs (*ectosomes*) specifically target macrophages, dendritic cells (DCs), and platelets (Gasser et al. 2003; Eken et al. 2008; Pluskota et al. 2008).

The importance of EVs in a multitude of physiological and pathological processes is becoming increasingly clear, with their functions contingent upon their cargo, which in turn reflects the unique characteristics of the parent cell (Al-Nedawi, Meehan, and Rak 2009). EVs are able to exert their action at large distances from their origin cell by circulating widely in biological fluids such as blood plasma (Caby et al. 2005), serum (Taylor, Akyol, and Gercel-Taylor 2006), saliva (Palanisamy et al. 2010), urine (Pisitkun, Shen, and Knepper 2004), milk (Admyre et al. 2007), semen (Poliakov et al. 2009), and cerebrospinal fluid (CSF) (Street et al. 2012). When EVs encounter a recipient cell, their interaction occurs via adhesion to the cell surface (mediated by lipid or ligand–receptor interactions) (Segura et al. 2007; Nolte-'t Hoen et al. 2009; Koppler et al. 2006), internalization through (potentially receptor-mediated) endocytic uptake (Miyanishi et al. 2007; Gasser and Schifferli 2004; Morelli, Larregina, and Shufesky 2004), or by direct fusion of the vesicle and cell membranes (Mangeot et al. 2011; Parolini et al. 2009). EV–cell interactions result in the modulation of the target cell's physiology through several different pathways, ultimately affecting the biology of the tissue/organism as a whole (Stahl and Raposo 2019).

IMMUNE MODULATION

Changes in the numbers and cellular origin of circulating EVs have been described in different pathologies wherein they have been proposed to promote inflammation (Buzas et al. 2014). The cargo (e.g. nucleic acids and proteins) of EVs may reflect the state of activation of the cells from which they originate, and different mechanisms by which EVs activate or repress the immune status of target cells have been described. Indeed, changes in the microenvironment are the main drivers of the EVs pro- or anti-inflammatory commitment (Valadi et al. 2007). For instance, Zhang et al. demonstrated an immunosuppressive role for exosomes isolated from embryonic stem cell (ESC)-derived mesenchymal stromal/stem cell (MSC) cultures, which may be envisaged as leading to inflammation resolution (Zhang, Yin, et al. 2014a). The same EVs increased levels of the anti-inflammatory cytokine interleukin-10 (*IL10*) transcript and decreased expression levels of pro-inflammatory cytokines *IL1B* and *IL12P40* in human monocytes.

Mokarizadeh et al. provided further evidence supporting the role of MSC-EVs in immune-modulation (Mokarizadeh et al. 2012). In their work, they demonstrated that MSC-EVs exert an immunoregulatory effect on splenic mononuclear cells derived from experimental autoimmune encephalomyelitis (EAE) affected mice. Here, MSC-EVs elicited an induction of tolerogenic signals in two ways: (i) by triggering apoptosis in activated T cells to inhibit the auto-reactive lymphocyte proliferation and (ii) by promoting the secretion of anti-inflammatory cytokines such as IL-10 and tumor growth factor beta (TGF-β) (Mokarizadeh et al. 2012). MSCs also promoted immunosuppression by shedding EVs enriched in miRNA targeting the transcript of zinc finger protein 36 (*Zfp36*), which encodes for a protein that regulates TNF-α production (Nargesi, Lerman, and Eirin 2017).

In some instances, EVs appear to display enhanced immunomodulatory effects compared to their cellular counterparts. For instance, in a series of murine models of experimental asthma, systemically injected human and mouse MSC-derived EVs were shown to be more effective than their parent cells in ameliorating Th2/Th17-mediated allergic airway inflammation and lung inflammation by modulating T cell activation (Monsel et al. 2015). Differences in the immune system modulation between cells and adipose MSC-derived EVs has also been reported in another recent study (de Castro et al. 2017), thus affirming the beneficial potential of EVs in asthma treatment. Finally, EVs have been shown to also modulate compartmentalized innate immunity. Transplanted neural stem cells (NSCs) exert their therapeutic function largely by attenuating brain inflammation and promoting neuroprotection through the *in situ* release of immune modulatory molecules and neurotrophic factors (Pluchino et al. 2005), with EVs likely to be a key contributor to this phenomenon (Rong et al. 2019; Vogel, Upadhya, and Shetty 2018; Webb et al. 2018). Indeed, NSC-derived EVs have been shown to hold the potential to convey inflammatory cues to target cells via an IFNγ/IFNγ receptor 1-mediated pathway (Cossetti et al. 2014).

EXTRACELLULAR SIGNALING

EV-mediated communication can be initiated through either direct ligand–receptor stimulation of the target cell or through transfer of surface receptors from the EV (potentially via membrane fusion) (Cocucci, Racchetti, and Meldolesi 2009; Lee et al. 2011). Extensive literature covers such interactions in the context of immune responses (Théry, Ostrowski, and Segura 2009; Bobrie et al. 2011; Mitchell et al. 2009; Chaput and Thery 2011; Pilzer et al. 2005) and in the propagation of pathogenic signals (Lee et al. 2011; Schorey and Bhatnagar 2008; Al-Nedawi, Meehan, and Rak 2009). Evidence of EVs carrying antigenic material, or even peptide-major histocompatibility complexes (MHC), supports their putative role as mediators for antigen presenting cells (APCs) in immune response signaling. For example, B lymphocyte-derived EVs are highly enriched in proteins involved in antigen presentation and can mediate T cell stimulation *in vitro* (Raposo et al. 1996; Escola et al. 1998; Muntasell, Berger, and Roche 2007). Furthermore, tumor-derived exosomes can induce antigen-specific T cell activation in the presence of DCs which had not been exposed to the tumor antigen (Wolfers et al. 2001; Andre et al. 2002), and B cells-derived EVs have been found to present MHC Class II to follicular DCs which otherwise don't express MHC Class II (Denzer et al. 2000). EV-mediated leukocyte-to-endothelial cell signaling appears to play a major role in vascular endothelium activation, which is a substantial component of inflammatory responses. MVs released from THP-1 monocytic cells and peripheral blood mononuclear cells upon LPS stimulation have been shown to activate a pro-inflammatory response in endothelial cells (Wang et al. 2011). It has also been shown that EVs derived from human endothelial progenitor cells actively participate in the activation of an angiogenic program in endothelial cells via interaction with α4- and β1-integrins. Deregibus et al. found in the endothelial progenitor cell-derived EVs a subset of cellular mRNAs specifically associated with angiogenic pathways (e.g. phosphatidylinositol 3-kinase/Akt and endothelial nitric oxide synthetase signaling pathways), thus implying a targeted mRNA-mediated triggering of the angiogenic program (Deregibus et al. 2007).

EV-mediated signaling via transfer of membrane components has also been widely described in the cellular proliferation of disease. Examples of this phenomenon include an increased susceptibility to HIV and enhanced resistance to apoptosis in macrophages upon uptake of EV-shuttled C-C chemokine receptor type 5 (Mack et al. 2000) or an increased tendency to enter apoptosis among T lymphocytes receiving the "*cell death ligand*" FasL via EV-mediated transfer (Kim et al. 2005). Tumor-derived EVs have been ascribed myriad functions in the progression of cancer (Willms et al. 2018), such as via immune modulation of T and natural killer cells (Whiteside Theresa 2013). Additionally, it has been shown that the EV-delivery of the oncogenic epidermal growth factor receptor (EGFR) variant III to unaffected cells stimulates the propagation of glioma growth (Al-Nedawi et al. 2008). Moreover, EVs may also act as supramolecular interaction platforms for the coordination of more complex multi-signaling processes (Gasser and Schifferli 2004; Cocucci, Racchetti, and Meldolesi 2009), as exemplified by the contribution of TF-bearing EVs in the induction of coagulation (Müller et al. 2003; Del Conde et al. 2005). Another cellular process highly influenced by EV-mediated signaling is cell maturation; for example, neutrophil-derived ectosomes have been shown to retard the maturation of DCs upon binding to their surface (Eken et al. 2008), whereas EVs from activated DCs are capable of triggering maturation in resting DCs by increasing their antigen presentation capabilities (Obregon et al. 2006).

CARGO TRAFFICKING

Another way by which EVs can influence a recipient cell's biology is via the delivery of their bioactive cargo into the target's cytosol following vesicle uptake/internalization. This phenomenon has been mostly studied with regard to EV-associated protein cargos, believed to affect a variety of cellular functions upon trafficking to recipient cells (Lee et al. 2011; Camussi et al. 2010; Mause and Weber 2010; Ratajczak et al. 2006; Simpson, Jensen, and Lim 2008). Affected phenomena include modulation of development and differentiation (Greco, Hannus, and Eaton 2001; Quesenberry et al. 2015; Bidarimath et al. 2017; Entchev and Gonzalez-Gaitan 2002), gene expression regulation via delivery of transcription factors (Ray et al. 2008; Zhou et al. 2018; Lee, EL Andaloussi, and Wood 2012), transport of myokines during physical activity (Trovato, Di Felice, and Barone 2019), influence over circadian rhythm (Tao and Guo 2018), and conveyance of drug resistance proteins (Namee and O'Driscoll 2018; Conde-Vancells et al. 2010).

A recent study has also highlighted the ability of EVs to exert their function in the extracellular space, with NSC-derived EVs shown to act as independent metabolic units, consuming and producing metabolites. Functional and metabolic studies revealed the presence of L-asparaginase activity, catalyzed by the enzyme asparaginase-like protein 1 (Asrgl1) in both human and mouse NSC-derived EVs (Iraci et al. 2017). In addition, EV-associated Asrgl1 displays high selectivity for asparagine metabolism while lacking glutaminase activity, and EV-trafficked Asrgl1 showed greater activity than the NSC-resident enzyme. These findings pave the way to investigations of such enzymatically active EVs as a putative therapy in asparagine-avid cancers such as ALL, for which the gold-standard bacterial L-asparaginase treatment is burdened with toxic and immunogenic side-effects. This important observation not only suggests that EVs secreted by NSCs are metabolically active but also reinforces the existence of a highly selective enrichment for specific enzymatic cargos. In the same work, the neurotransmitters glutamate and gamma-aminobutyric acid were also found altered by NSC-secreted EVs, as well as potent signaling metabolites such as lactate, involved in cancer proliferation and inflammation (Iraci et al. 2017).

NUCLEIC ACID TRAFFICKING

The repurposing of exosomal cellular processes is a well-established strategy by which viruses can fuel the propagation of their own genetic material (Meckes et al. 2010; Pegtel et al. 2010; Meckes and Raab-Traub 2011; Gourzones et al. 2010). Interestingly, a relatively recent finding has suggested that endogenous EVs can also shape intercellular communications and support homeostasis through the transfer of nucleic acids. As described, both DNA and RNA have been detected in secreted EVs, with studies suggesting the involvement of an active and energy-dependent sorting and packaging mechanism that accounts for at least some of this nucleic acid content (Gibbings et al. 2009; Kosaka et al. 2010; Zhang et al. 2010). The trafficking of both coding and non-coding nucleic acids has been reported, indicating both horizontal transfer of genetic material and modulation of gene expression, respectively (Skog et al. 2008; Belting and Wittrup 2008; Valadi et al. 2007; Bergsmedh et al. 2001). In particular, RNA seems to cover a key role in extracellular signaling (Mittelbrunn and Sanchez-Madrid 2012; Dinger, Mercer, and Mattick 2008), particularly controlling gene expression via EV-mediated intercellular transfer of mRNAs and miRNAs (Hu, Drescher, and Chen 2012; Turchinovich, Drapkina, and Tonevitsky 2019).

While EV-mediated trafficking of *functional* miRNA remains controversial, many studies have reported changes in recipient cells attributable to the transfer of miRNA. EVs secreted by ESCs have been shown to deliver a subset of miRNAs to mouse embryonic fibroblasts *in vitro* (Yuan et al. 2009). Epstein Barr virus-infected lymphoblasts secrete exosomes containing viral miRNAs which are functional in uninfected recipient primary immature monocyte-derived DCs (Pegtel et al. 2010). Another example is reported whereby T cell-derived exosomes transfer functional miRNAs to APCs in an antigen-dependent and unidirectional fashion to modulate immune synapsis (Mittelbrunn et al. 2011). Moreover, EV-delivered miRNAs have been also shown to influence the metabolic state of the recipient cell. As demonstrated by Müller and colleagues, large adipocytes in rats can communicate with small recipient adipocytes via the exchange of EV-packed miRNAs (and mRNAs), promoting the upregulation of lipogenesis and cell growth (Mueller et al. 2011).

CANCER PROGRESSION

The exchange of EV-shuttled miRNAs has been also largely studied in the context of tumor diffusion. As described above, Skog et al. found that glioblastoma cells transfer mRNAs, miRNAs, and proteins through vesicular transport, responsible for inducing tubule formation in normal brain microvascular endothelial cells. Hence, EVs derived from glioblastoma cells drive self-promotion proliferation of the glioblastoma cell line (Skog et al. 2008). Since this discovery, a diverse variety of non-coding RNAs have been identified in brain tumor-derived EVs (Spinelli et al. 2018), and in the context of cancer more generally (Bhome et al. 2018; Kogure, Kosaka, and Ochiya 2019). Interestingly, Skog et al. also showed that glioma-specific subset of mRNAs and miRNAs could be detected in serum EV of patients and may be used as a diagnostic marker (Skog et al. 2008); EV miRNAs have been identified in various other biological fluids (Simpson, Jensen, and Lim 2008), making extracellular miRNA profiles a popular and potentially powerful biomarker of many pathological conditions (Ciesla et al. 2011; De Smaele, Ferretti, and Gulino 2010; Taylor and Gercel-Taylor 2008; Shah, Patel, and Freedman 2018; Yoshioka, Katsuda, and Ochiya 2018). Indeed, while EV-enclosed RNA content largely reflects that of the parental cell, very specific miRNAs subsets have been found enriched or depleted in vesicles in response to environmental and/or (patho)physiological conditions, hinting that miRNAs secretion via EVs is the result of an active sorting process (Kosaka et al. 2010; Zhang et al. 2010; Gibbings et al. 2009). Nonetheless, preliminary and emerging observations covering EV-mediated RNA signaling are still intense investigation and the field is still in its infancy.

The role of EVs in hematological malignancies has also seen considerable attention (Ohyashiki, Umezu, and Ohyashiki 2018). EVs derived from leukemia cells have been found to effect various biological and phenotypic changes on recipient stromal cells, including enhanced proliferation and the promotion of vascularization, leading to a more favorable niche for the leukemia cells (Farahani et al. 2015; Corrado et al. 2016; Fei et al. 2015). Furthermore, chronic lymphocytic leukemia (CLL)-derived EVs were found to be enriched in miR-202-3p compared to their parent cells, suggesting an active means of disposal of this anti-tumorigenic miRNA (Farahani et al. 2015). Conversely,

MSC-derived EVs were found to rescue recipient CLL cells from spontaneous or drug-induced apoptosis and promote migration (Crompot et al. 2017). Nevertheless, reports of the effect of MSC-derived EVs on tumors have been contradictory, with pro- or anti-tumorigenic results reported depending on variability in experimental conditions and MSC source (Borgovan et al. 2019; Bruno et al. 2014). Furthermore, EVs have been found to convey multidrug resistance among acute myeloid leukemia (AML) cells by promoting multidrug resistance protein 1 expression in recipient cells, possibly through mechanisms involving the transfer of miR-19b and miR-20a (Bouvy et al. 2017).

BIOMARKERS OF DISEASE

EV-mediated signals are being extensively investigated as putative indicators of disease and or injury, with the biomolecular makeup of specific populations of EVs thought to reflect underlying pathologies (Shah, Patel, and Freedman 2018). For instance, specific subpopulations of circulating endothelial microparticles (EMPs) associate with specific cardiometabolic risk factors (Amabile et al. 2014), with EV abundance found to be increased in a number of cardiovascular/cardiometabolic disorders (Dickhout and Koenen 2018). Salivary EVs obtained from traumatic brain injury patients are reportedly enriched in the mRNA of several genes (e.g. cathepsin D and casein kinase) also upregulated in Alzheimer's disease patients (Cheng et al. 2019). In a case-controlled study from Fiandaca et al., exosomes were found to contain levels of phosphorylated tau or amyloid beta peptide that were predictive of future clinical Alzheimer's development (Fiandaca et al. 2015). Furthermore, CSF from Alzheimer's and Parkinson's disease patients has been found to contain exosomes with altered miRNA profiles (Gui et al. 2015). miR-155 levels in EVs isolated from body fluids has been identified as a potential biomarker for hematologic malignancies such as CLL and AML (Caivano et al. 2017). Alternatively, changes in EV composition may reflect responsiveness to therapeutic intervention, as is reportedly the case for the TGF-β1 content of EVs isolated from the plasma of AML patients (Hong et al. 2014). This is also evident in infectious disease: the abundance and size of plasma EVs in HIV-1-infected patients correlate to viral activity and disease progression, while successful antiretroviral therapy results in decreased miR-155 and miR-223 expression in EVs (Hubert et al. 2015).

DEVELOPMENT AND REGENERATION

Although EV signaling has been most extensively studied in the context of circulating biomarkers of disease, numerous studies have now also highlighted that their role in (patho)physiological processes can be exploited in therapeutic use. Of particular interest is the role played by EVs in mediating paracrine exchanges of soluble factors, a process repurposed through the use of stem cells in regenerative medicine and tissue repair (Zhang, Li, et al. 2014b; Riazifar et al. 2017; Nawaz et al. 2016). Pluripotent stem cells, including both ESCs and induced pluripotent stem cells (iPSCs), having substantial proliferation and differentiation potential hold a central role in the developing field of regenerative medicine. Both iPSCs and iPSC-derived cells have been observed to secrete EVs in the same way as other cells do (Zhou et al. 2016); however, only recently has the regenerative potential of non-cellular ESC/iPSC components such as EVs been appreciated.

On a related note, the role of EVs in development is a growing area of interest, with strong evidence supporting a key role for EV-mediated signaling in cell programming and fate. The migration of ESCs and the induction of differentiation in developing embryoid bodies are highly regulated by intercellular signaling within ESC colonies. Both human (Giacomini et al. 2017) and murine (Cruz et al. 2018) ESCs have been reported to exert these pleiotropic effects by the secretion of EVs. A study by Ratajczak et al. was one of the first studies on RNA horizontal transfer between cells that showed the presence of RNAs in EVs secreted by ESCs (Ratajczak et al. 2006). The authors identified high levels of transcripts of early pluripotent transcription factors, as well as the protein wingless-type MMTV integration site family, member in ESC-derived EVs, capable of driving cell reprogramming and expansion of hematopoietic progenitor cells. In the same study it was also observed that ESC-derived EVs can also act as signaling messengers in driving self-renewal and adult stem cells expansion. Treatment with ESC-EVs resulted in enhanced survival and improved expansion of murine hematopoietic stem cells *in vitro*. This was accompanied by an upregulation of early pluripotent and early hematopoietic stem cells markers and induced phosphorylation of mitogen-activated protein kinases p42/44 and the serine-threonine kinase Akt (Ratajczak et al. 2006).

Another early characterization of ESC-EV content and trafficking is described in the work of Yuan et al. (2009). By using engineered ESC-EVs, they were able to demonstrate the ability of these EVs to dock, fuse, and transfer contents to other ESCs and embryonic fibroblasts *in vitro*. ESC-specific miRNAs were isolated from ESC-EVs in addition to other abundant ubiquitously expressed mRNAs. Given the pivotal role of miRNA in developmental regulation, these findings open up the attractive possibility that stem cells can alter the gene expression and cell fate of neighboring cells through the EV-mediated transfer of miRNA.

Moreover, intriguing evidence underscore the key role of EVs in one of the earliest and most important steps in early embryogenesis, the communication between the maternal uterine environment and the blastocyst. In fact, a study by Desrochers et al. uncovered a unique mechanism by which ESCs derived from the inner cell mass shed MVs able to stimulate trophoblast migration into the uterus during embryo implantation (Desrochers et al. 2016). In this study, proteins from the extracellular matrix, like laminin and fibronectin, have been detected in MVs released by inner cell matrix ESCs *in vitro*. Their interaction with integrins expressed on the trophoblast surface promotes the activation of two signaling kinases, c-Jun N-terminal kinases and focal adhesion kinase, involved in cell cycle control. *In vivo* injection of ESC-MVs in blastocysts revealed enhanced trophoblast migration and an overall increase of implantation efficiency.

MSC-derived EVs have also seen considerable attention with respect to their trafficking and the extent to which they influence cell differentiation and proliferation. Ostensibly capable of differentiation into several mesenchymal lineages (i.e. adipocytes, osteoblasts, chondrocytes, and endothelial cells), MSCs are reportedly capable of generating non-mesenchymal lineage cells, such as neurons and myocytes, under the appropriate conditions. EVs derived from MSCs are therefore expected to mirror and convey the differentiation-influencing properties of the parental cells. Accordingly, transcriptome characterization on human bone marrow MSC-derived EVs has revealed the content of a wide range of mRNAs involved in neural, bone, endo/epithelial, and hematopoietic cell differentiation (Bruno et al. 2009). More generally, recent findings suggest that MSC-EVs may induce both differentiation and de-differentiation of terminally differentiated cells, thus triggering a proliferative program that may contribute to tissue formation and/or repair of tissue injury (Tan et al. 2014). For example, it has been demonstrated that EVs from human bone marrow are able to act on target cells to accelerate kidney repair in a mouse model of acute kidney injury. This reportedly occurs through EV-based delivery of specific subset of cellular mRNAs correlated with the MSC phenotype, which may trigger a proliferative program in surviving tubular cells post injury (Bruno et al. 2009). Similarly, EV-mediated transfer of a specific array of mRNA derived from human liver pluripotent stem cells has been shown to induce proliferation and apoptosis resistance in cultured human hepatocytes and to favor liver regeneration in hepatectomized rats (Herrera et al. 2010). These studies strongly imply the existence of a bidirectional genetic transfer between stem and injured cells, mediated by the exchange of highly regulatory RNA molecules, promoting the activation of tissue repair pathways.

CENTRAL NERVOUS SYSTEM FUNCTION AND DISEASE

EV trafficking has also been observed in the CNS, notably from NSCs, self-renewing, multipotent cells which are able to be expanded in culture indefinitely or to generate terminally differentiated neurons and/or glia cells (*in vitro* or during CNS development). While this process diminishes with age, some NSCs persist in the adult vertebrate brain and continue to produce neurons throughout life (Gage 2019). NSCs support neurogenesis and gliogenesis via the release of soluble mediators (e.g. cytokines, chemokines, metalloproteases, adhesion molecules, etc.). Likewise, NSC-derived EVs, as part of the NSC "*secretome*" (Drago et al. 2013), might act in parallel as modulators of the CNS stem cell niche (Batiz et al. 2015). Still, the potential role of NSC-EVs in the neurogenic niches remains essentially unexplored.

In the broader context of the CNS, there have been a substantial number of studies supporting the hypothesis that EVs play a significant role in nervous system signaling, transforming the field of neuroscience over the last decade (Smalheiser 2007, 2009). EVs having been identified from nearly all neural cell types (Von Bartheld and Altick 2011; Lai and Breakefield 2012), including neurons (Faure et al. 2006; Schiera et al. 2007; Putz et al. 2008), astrocytes (Taylor et al. 2007; Guescini et al. 2010), oligodendrocytes (Trajkovic et al. 2008; Hsu et al. 2010; Fitzner, Schnaars, and van Rossum 2011), microglia (Bianco et al. 2009; Bianco et al. 2005; Potolicchio et al. 2005; Tamboli et al. 2010), and, importantly, NSCs (Marzesco et al. 2005; Huttner et al. 2008; Cossetti et al. 2012; Cossetti et al. 2014; Iraci et al. 2017).

Remarkably, high levels of EVs enriched in the NSC marker prominin-1 (CD133) were found in the CSF of patients affected by glioblastoma or partial epilepsy, two pathological conditions associated with impairment of adult neurogenesis (Ming and Song 2011). Accordingly, neural-derived EVs are receiving a great deal of attention both as putative indicators of neural progenitor cell fitness and markers of disease (Huttner et al. 2008; Huttner et al. 2012). Indeed, as previously mentioned, most of the studies regarding neural EVs have been carried out in various glioma models (Guescini et al. 2010; Balaj et al. 2011; Al-Nedawi, Meehan, and Rak 2009; Trams et al. 1981; van der Vos et al. 2011; Graner et al. 2009; Svensson et al. 2011), with a particular focus on tumor-derived EVs and their potential roles as passive diagnostic markers (Skog et al. 2008; Pelloski et al. 2007).

An important function in the maintenance of brain homeostasis might be provided by endothelium-derived EVs (EMPs). EMPs are normally circulating in healthy subjects; however, their levels increase in response to pathological states, such as during infection or thrombotic disease, implying a probable regulatory function in inflammatory or coagulatory processes (Rabelink, de Boer, and van Zonneveld 2010; Morel et al. 2011). The role of EMPs is particularly relevant in the context of cerebrovascular disease, as circulating EMP levels arising from endothelial injury might correlate to important factors reflecting damage severity, such as brain lesion volume or predicting stroke outcome (Simak et al. 2006; Jung et al. 2009). Additionally, EMPs seem to play a role in eliciting inflammatory responses in multiple sclerosis (MS) pathogenesis by binding and activating monocytes (Jy et al. 2004), or conversely promoting proliferation, motility, and survival of oligodendrocyte precursor cells in the healthy CNS (Kurachi, Mikuni, and Ishizaki 2016).

An increasing quantity of evidence is now implicating EVs in the diffusion and formation of pathogenic proteins and peptides involved in neurodegenerative disorders. This includes prions (Fevrier et al. 2004; Vella et al. 2007; Alais et al. 2008; Hartmann et al. 2017), responsible for Kuru and variant Creutzfeldt-Jakob disease, and the circulation of misfolded proteins associated with Alzheimer's disease, Parkinson's disease, dementia, and amyotrophic lateral sclerosis (ALS) (Goedert, Clavaguera, and Tolnay 2010; Saman et al. 2012; Bulloj et al. 2010; Emmanouilidou et al. 2010; Rajendran et al. 2006; Sharples et al. 2008; Guest et al. 2011; Pérez, Avila, and Hernández 2019).

Despite being widely characterized in several neuropathological states, EVs in the brain also appear to play a decisive role in a variety of fundamental physiological events, such as the sequestration of redundant metal cation-transporting proteins during cellular stress (Putz et al. 2008), biogenesis regulation of myelin (Bakhti, Winter, and Simons 2011), downregulation of toxic levels of ATP following ischemia injury (Ceruti et al. 2011), and regional transfer of proteins and mRNAs in neuronal synaptic activity (Twiss and Fainzilber 2009). Furthermore, as also indicated in the activity of glioma-derived EVs, neural EVs also appear to be intimately involved in the regulation of physiological immune responses in the CNS (Cossetti et al. 2012). Finally, there is emerging evidence that intercellular communication via EVs may play an important role in mental health (Saeedi et al. 2019).

EVs AS THERAPEUTIC TOOLS

Nanomedicine is an emerging field incorporating nanotechnology with pharmacology/medicine, with the aim of improving the efficacy and specificity of drug delivery. Of particular interest is the therapeutic use of nucleic acids, but *in vivo* delivery is often hampered by limitations such as their charged nature, sensitivity to degradation, and the need for target-specific ligands (Dykxhoorn and Lieberman 2006). Thus, research over the last couple of decades has focused on the development of scalable, resistant, and bio-compatible carriers for the delivery of therapeutic cargoes such as small-molecule drugs or siRNAs; while few such nanomedicines have received regulatory approval to date, the field is growing in popularity (Farjadian et al. 2019). While a thorough review is beyond the scope of this chapter, substantial progress has been made in the development of *synthetic* EV analogs such as liposomes (Antimisiaris, Mourtas, and Marazioti 2018). Liposomes are spherical lipid bilayers encapsulating and protecting a bioactive payload, with currently marketed nanoliposomal drugs, including Doxil (containing doxorubicin), Marqibo (vincristine), AmBisome (amphotericin B), and Visudyne (benzoporphyrin derivative mono acid ring A) (Farjadian et al. 2019). However, as synthetic agents' liposomes can elicit immunogenic reactions with subsequent targeting and elimination by the immune system (Sercombe et al. 2015). EVs, as naturally occurring liposomes, embody an ideal delivery platform by exploiting an existing biological delivery mechanism employed for cell-to-cell communication. Thus, nanomedicine is converging on employing EVs, be they wild type or bioengineered, as vehicles for drug delivery. Indeed, the therapeutic use of EVs has several advantages over other synthetic nanomedicinal vehicles, including low immunogenicity (Zhu et al. 2017), the capability to cross tissue and cell barriers accompanied with an intrinsic tropism (crucial features for efficient drug delivery) (Antes et al. 2018; Kooijmans, Schiffelers, et al. 2016a), and an extended shelf life, allowing for long storage (Jeyaram and Jay 2017). EV-engineering includes various approaches aimed at improving EV functionality, such as cargo modification, improvement of kinetics, biodistribution, cell specificity, or internalization (Melling et al. 2019).

EV BIODISTRIBUTION: ENGINEERING FOR ENHANCED UPTAKE

EV biodistribution is poorly understood; the increasing use of EVs has underlined the need for better characterization of their distribution upon *in vivo* delivery in order to minimize off-target accumulation. To achieve such an aim, various EV labeling methods have been employed (Gangadaran, Hong, and Ahn 2018). For example, in a study by Wiklander et al. EVs from three different cell types were labeled with a near-infrared lipophilic dye, 1,1-dioctadecyl-3,3,3,3-tetramethylindotricarbocyanine iodide (DiR), which is fluorescent when present in a lipid membrane (Wiklander et al. 2015). The biodistribution of EVs isolated from bone marrow-derived DCs and mesenchymal stem cells, as well as HEK293T (kidney), C2C12 (muscle) and B16F10 (melanoma) cell lines was tracked following *in vivo* administration. A comparison of different routes of injection revealed interesting differences in tissue distribution: HEK293T-derived DiR-EVs were found to preferentially accumulate in the liver and spleen 24 h after intravenous (i.v.) injection. Conversely, intraperitoneal and subcutaneous injections resulted in a decreased accumulation in the aforementioned organs, but an increased presence in the gastro-intestinal tract and pancreas. Intriguingly, the source cell type of the transplanted EVs was also found to influence biodistribution (e.g. an increased distribution of DC-EVs in the spleen), suggesting a possible tropism and highlighting the importance of the cell source when anticipating therapeutic EV approaches. Finally, as a proof of concept for targeted therapies, the authors transfected DCs with LAMP2b-RVG (Wiklander et al. 2015), a fusion protein that includes the rabies virus glycoprotein (RVG) moiety known to bind to acetylcholine receptors. Following i.v. administration, RVG DiR-EVs were found to have a significantly increased accumulation in the brain, muscle, and heart (tissues with high acetylcholine receptor expression) of mice when compared to non-functionalized control EVs; distribution in other organs did not reveal any significant differences between targeted and non-targeted EVs. This skewed distribution convincingly demonstrates the potential of improving EV specificity and/or uptake through functionalization.

A variety of different approaches have been explored in an effort to impart upon EVs tissue specificity or the capability to cross biological barriers (Lu et al. 2018a; Luan et al. 2017; Das et al. 2019; Murphy et al. 2019). One such approach involves decorating EVs with antibodies and nanobodies. For example, the functionalization of HEK293-EVs with an anti-human epidermal growth factor receptor 2 (HER2) antibody has been used to deliver chemotherapeutic EVs to HER2+ breast cancer tumor xenografts in mice, effectively reducing tumor growth (Wang et al. 2018). The authors sought to enhance the efficacy of chemotherapy by taking advantage of cell-specific gene-delivered enzyme activation of an inert prodrug compound. Specifically, 6-chloro-9-nitro-5-oxo-5H-benzo-(a)-phenoxazine (CNOB) was used in combination with the HChrR6 bacterial enzyme, delivered to tumor cells as a functional mRNA via EVs. To implement HER2+ cell-specificity, a chimeric targeting protein was created consisting of a leader sequence directing its migration to the EV surface, a high-affinity anti-HER2 single-chain variable fragment (scFv) antibody domain, lactadherin C1-C2 domains for membrane interaction, and a His-tag for purification. Two different methods for EV generation were then tested: direct transfection of HEK293 cells and subsequent EV collection, or incubation of EVs obtained from non-transfected cells with the pure chimeric protein, with the latter resulting in a significantly higher yield. To achieve efficient loading of HChrR6-encoding mRNA, 293FT cells were transfected with a plasmid containing the *HChrR6* sequence followed by a zip code, previously shown to orchestrate miRNA loading into EVs (Bolukbasi et al. 2012). Isolated EVs were then functionalized with the HER2 antibody. The efficacy of these EVs was tested on BT474 tumors implanted in athymic mice, with the combination of CNOB and targeted EVs resulting in a significant reduction of tumor volume (Wang et al. 2018).

Similar to antibodies, nanobodies, small single-variable domains obtained from heavy-chain antibodies from Camelidae have been used to endow EVs with targeting capabilities. For instance, such structures have been used to target the epidermal growth factor receptor (EGFR) (Kooijmans, Aleza, et al. 2016b), overexpressed in many tumors. The EGa1 nanobody was used due to its high affinity for EGFR and its ability to competitively inhibit natural ligand binding, preventing receptor activation. To facilitate membrane binding of EGa1, fusion to a C-terminal glycosylphosphatidylinositol (GPI) signal peptide from human decay-accelerating factor (DAF) was used. EVs are enriched in lipid raft-associated lipids that include GPI-anchored proteins, with GPI-anchored DAF previously described to be secreted in EVs; thus, human DAF-derived GPI-anchor signal peptide was fused to the nanobodies, and its expression was shown to otherwise not affect EV characteristics. To investigate the efficacy of such-functionalized EVs, derived from Neuro2A mouse neuroblastoma cells, a far-red fluorescent dye was used to track their uptake by recipient Neuro2A, HeLa, or A431 cells. These lines vary in EGFR expression, with levels being undetectable in Neuro2A cells but greatly overexpressed in A431 cells. Concomitantly, incubation with EVs expressing DAF and EGa1 resulted in a significant greater *binding* to A431 cells, but the *uptake* of the EVs did not significantly differ from non-functionalized control EVs or among cell types. This unexpected result highlights the need for further exploration of the uptake mechanisms underlying EV-based cargo delivery.

Another approach that has been explored with regard to enhancing EV uptake relies on alteration of the tropism of the vesicles through pseudotyping, a method commonly used in virology. This technique, based on the packaging of the genetic component of a particular virus into the envelope proteins derived from different viruses, results in an increased infectivity of the resultant recombinant virus (Cronin, Zhang, and Reiser 2005). Adapting this approach, Meyer et al. demonstrated the possibility of augmenting exosome uptake in different cell types (Meyer et al. 2017). The authors expressed VSVG in HEK293 cells, based on the knowledge that VSVG is often used for viral pseudotyping due to its broad tropism and high efficacy in transduction. VSVG expression was then validated by fluorescent co-localization of red signal associated with VSVG and green with the vesicular tetraspanin CD63, revealing similar intracellular and temporal pattern and thus suggesting the incorporation of VSVG into exosomes through the endocytic pathway. Finally, after demonstrating the efficient release of VSVG exosomes from parental cells by using VSVG-luciferase fusion, increased uptake of VSVG-expressing exosomes was measured in various cell types (HEPG2, a human liver cancer cell line; U87, a human glioblastoma cell line; HEK293; and commercial human iPSCs).

Aptamers, nucleic acid-based specific-recognition ligands, are becoming an increasingly popular means to increase the specificity of EVs. One particularly novel method, based on the versatility of the three-way junction (3WJ) derived from packaging RNA (pRNA) of the phi29 bacteriophage, has recently been developed to control both RNA loading and the ligand properties of EVs from HEK293T cells (Pi et al. 2018). First, siRNA against *survivin*, an inhibitor of cell apoptosis, was loaded into EVs in order to decrease tumor growth upon delivery. To promote integration of the 3WJs into EVs, a single cholesterol-triethylene glycol molecule was conjugated to the arrow-tail of 3WJ structure, exploiting the spontaneous capability of cholesterol to integrate into EV membranes through its hydrophobicity. Modified 2′F-3WJs were then conjugated with three different targeting moieties to achieve cell-specific delivery in three different cancer lines: folate (FA) for targeting KB cells (a HeLa subline), and RNA aptamers targeting prostate-specific membrane antigen (PSMA) for LNCaP-LN3 cells (human prostate carcinoma) and EGFR for MDA-MB-468 cells (a human breast cancer line). For imaging, 3WJs were also labeled with a fluorescent moiety. Analysis of loaded EVs compared to non-loaded controls showed no differences in size distribution or zeta potential, suggesting the absence of alteration of EV characteristics upon functionalization with 3WJs. Initial *in vitro* screening of PSMA-siRNA-EVs in LNCaP-LN3 cells showed efficiency in targeting, delivery, and silencing; even more interestingly, *in vivo* delivery of the three different targeted EVs (FA-siRNA-EVs, PSMA-siRNA and EGFR-siRNA-EVs) led to significant reductions in their respective tumor-xenograft sizes in conjunction with significant reductions in *survivin* mRNA levels. A follow-up study using ginger-derived, *survivin* siRNA-loaded exosomes functionalized with the FA-3WJs also successfully inhibited xenografted tumor growth upon systemic delivery (Li et al. 2018). In another example of aptamer-mediated EV specificity, Wang et al. recently employed a DNA aptamer, AS1411, targeting nucleolin, which has been described to be overexpressed in breast cancer cells (Wang et al. 2017). EVs isolated form murine DCs were loaded (through electroporation) with the miRNA let-7, shown to play a crucial role in cell proliferation, differentiation, and apoptosis; these EVs were then functionalized with cholesterol-modified AS1411 via incubation. Both *in vitro* and *in vivo* assays using MDA-MB-231 cells showed efficient uptake of AS1411-EVs into target cells, reduction of tumor size, and, notably, no off-target tissue damage (toxicity) in other organs such as heart, liver, spleen, and kidney. Methods for the rapid functionalization EVs with AS1411 via extrusion have been described (Wan et al. 2018) and are likely to be broadly applicable to other EV and aptamer species. The potential for cancer-specific aptamer-functionalized EVs was further demonstrated in a study by Zou et al. (2019). Again exploiting lipophilic modified aptamers for membrane integration, the authors decorated doxorubicin-loaded DC-derived EVs with sgc8, an aptamer recognizing the T-cell acute lymphoblastic leukemia (ALL) marker protein tyrosine kinase 7 (PTK7). As compared to non-functionalized EVs, the sgc8-EVs were selective in their recognition of and uptake by PTK7-overexpressing lymphoblastic leukemia CEM cells over low-PTK7 Ramos cells, resulting in an enhanced cytotoxicity (Zou et al. 2019).

The exact molecular mechanism underpinning the fate of material transferred through EVs still remains elusive, particularly the possibility of shuttling material into the nucleus. In an attempt to elucidate such a process, Rappa et al. described the presence of nuclear envelope invagination-associated late endosomes (N-ALE), sub-nuclear compartments arising from interactions between Rab7+ late endosomes and nuclear envelope invaginations, which allow for the nuclear delivery of EV material (Rappa et al. 2017). Interestingly, the authors observed a differential uptake of EV-derived material between quiescent and proliferating cancer cell lines, with the former lacking in EV-material compared to the latter. These findings, as a possible result of the presence of modifications in nuclear pore complexes between the two cell types, suggest that EVs in cancer might contribute to the spread and growth of tumorigenicity through this nuclear transfer. While a clear elucidation of the puzzling pathways involved in the uptake of EVs and internalization of their cargo is still lacking, these data shed light on a novel mechanism for modulating the nuclear transfer of EV material.

EV Cargo-Loading Methods

With their biocompatibility and potential for tissue specificity, EVs are compelling candidates for nanomedicinal drug-delivery vehicles. Efforts to engineer therapeutic EVs have focused not only on the introduction of targeting ligands but also on the incorporation of biologically active payloads. Methodologies for loading cargos into EVs are broadly divided into indirect or direct approaches. Indirect methods involve manipulation of the parental cells from which EVs are isolated, mainly through lentiviral vectors for specific (over)expression of a biological target of interest. As described earlier, this method allows for EV membrane engineering in order to enhance their delivery efficiency, as in the example of transfection with LAMP2b-RVG to enhance delivery to cells expressing the acetylcholine receptor (Wiklander et al. 2015). On the other hand, the luminal cargo of the EV can be modified by taking advantage of the sequence that has been identified to be trafficked to EVs, as previously mentioned (Bolukbasi et al. 2012), thus providing an enrichment of an endogenous molecule. Indeed, advances in synthetic biology are enabling the production of designer exosomes (Kojima et al. 2018). However, indirect methods require genetic manipulation and often result in poor packaging yields, and therefore several alternative strategies have been developed for direct manipulation of EV payloads. Direct methods can be further divided into active and passive encapsulation, where active methods involve the brief disruption of the EV membrane to facilitate cargo entry, whereas passive methods are based on spontaneous membrane interactions and uptake by the EV. Among the active methods, one well-established technique involves electroporation: the application of an electrical field to EVs resuspended in a medium results in the provisional opening of pores in the bilayer membranes, allowing for the entrance of molecular cargo. Electroporation has been used to efficiently load glyceraldehyde 3-phosphate dehydrogenase (*Gapdh*) and beta secretase 1 (*BACE1*) siRNA into exosomes isolated from murine immature DCs, without altering the properties of the exosomes (Alvarez-Erviti et al. 2011). Nevertheless, attention should be paid to the settings and materials employed in such technique, as cautioned by Kooijmans et al. (2013). Their work showed that electroporation might result in siRNA precipitation, misinterpreted as efficient loading. Similarly, sonication of EVs induces the temporary formation of membrane pores which has been exploited in the loading of siRNA into EVs. Water bath sonication of EVs derived from HEK293T with siRNA against *Gapdh* and *Her2* led to high efficiency loading into EVs with a significant reduction of siRNA aggregation compared to electroporation (Lamichhane et al. 2016). Haney et al. also compared different methods for loading catalase into murine macrophages-derived exosomes (Haney et al. 2015). The authors compared saponin, freeze-thaw cycles, sonication, and extrusion methods, finding that sonication and extrusion resulted in the most efficient incorporation. As an example of the passive loading strategy, Sun et al. enhanced the efficiency of curcumin delivery through its inclusion within EVs: upon incubation with EL-4 (a mouse lymphoma cell line)-derived exosomes, hydrophobic curcumin was able to interact with and assemble into the lipid bilayer of the EVs (Sun et al. 2010). Integration of curcumin into exosomes resulted in a greater bioavailability *in vivo*. In a similar manner, Aqil et al. obtained celatrol-loaded bovine milk-derived exosomes through incubation with a low percentage of ethanol, hypothesizing a possible interaction between the drug and EV-surface lipids or proteins (Aqil et al. 2016). Lipid-conjugation has also been exploited as a strategy to enhance siRNA-loading into EVs (O'Loughlin et al. 2017). Thus, while the mechanisms dictating the natural cargo loading of EVs are poorly understood, there exist multiple routes through which to generate a vesicular payload of interest for therapeutic applications.

Preclinical Studies of EV Therapeutic Applications

EVs have been demonstrated to mediate cell-to-cell communications under both physiological and pathological conditions, and there is a significant interest in repurposing them as therapeutic tools (Katsuda et al. 2014). Indeed, their growing use as a putative nanomedicine has been driven by the advantageous properties of EVs as drug delivery vehicles, including biocompatibility, low immunogenicity, and the possibility to introduce cell or tissue specificity. The majority of studies to date have explored the use of EVs in oncology, exploiting vesicles as delivery vehicles for chemotherapeutic agents. However, there is growing interest in applications to the treatment of central nervous system (CNS) disorders, capitalizing on the ability of EVs to cross biological barriers (i.e. the blood–brain barrier, BBB) that are often a hurdle to conventional pharmaceuticals. An *in vivo* study from Yang et al. exemplifies both of these applications, with brain endothelial cell-derived EVs used to traffic the anticancer drugs paclitaxel and doxorubicin across the BBB and suppress tumor growth in a zebrafish model (Yang et al. 2015).

The diversity of payloads successfully delivered via EVs covers siRNA, miRNA, proteins, and small molecule drugs. As an example, preclinical data from Gehrmann et al. showed the feasibility and efficacy of loading alpha-galactosylceramide (αGC) and ovalbumin into EVs derived from murine bone marrow cells, which were subsequently used to induce activation of invariant NKT cells (Gehrmann et al. 2013). This, in turn, further boosted the immune system response as an antitumor therapy. In a similar manner, it has been demonstrated that EVs derived from different stem cell types are characterized by a specific cargo profile that is able to influence and modulate the surrounding microenvironment. This phenomenon is typically in the context of immune-modulation. In one supporting study, Cossetti et al. described the molecular signature of EVs derived from NSCs exposed to pro- or anti-inflammatory cues (Cossetti et al. 2014). In fact, by exposing parental cells to cytokine cocktails mimicking either pro-inflammatory (Th1-like) or anti-inflammatory (Th2-like) milieu, the authors observed a specific enrichment in proteins and RNA in the EVs that mirrored the changes in their parent cells. Interestingly, when target cells were exposed to such EVs, significant changes in gene and protein expression were induced in response to their pro- or anti-inflammatory nature, suggesting the capability of cells to respond to the external environment and to signal accordingly through secreted EVs. Many different EV sources have been explored for potential therapeutic applications, including plant-derived exosomes. As a proof of concept, Ju et al. showed the beneficial effects of grape-derived EVs in protecting mice from dextran sulfate sodium-induced colitis through stimulation of intestinal stem cell proliferation (Ju et al. 2013).

In addition to bioactive cargos, the composition of the EVs themselves has been exploited to augment their therapeutic effects. Lentiviral transduction of parental cells to overexpress TNF-related apoptosis-inducing ligand (TRAIL), led to the functional transport of TRAIL through exosomes to target cells (Rivoltini et al. 2016). Systemic *in vivo* delivery of such exosomes into three different murine tumor models led to a significant inhibition of the tumor growth owing to the selective sensitivity of cancer cells to TRAIL-mediated apoptosis. These results, together with other findings, such as the presence of MHC proteins transported on EVs, ostensibly confer upon them the capacity to interfere with both innate and adaptive immunity (Buschow et al. 2009). Thus, research has also focused on exploring the immunotherapeutic potential of EVs. Several studies have investigated the use of DC-derived EVs, and preclinical data has demonstrated their efficacy in treating models of melanoma (Naslund et al. 2013), mastocytoma, and adenocarcinoma (Zitvogel et al. 1998). In addition to DC-derived EVs, other immune cell sources have been considered, including macrophages. For example, in order to increase the efficacy of paclitaxel, a commonly used chemotherapeutic agent, exosomes derived from murine macrophages were loaded with this agent (Kim et al. 2016). *In vivo* delivery in a mouse model of pulmonary metastases resulted in a significant reduction in tumor growth.

EVs have also been extensively studied in the context of tissue repair. It has been demonstrated that EVs from human bone marrow are able to act on target cells to accelerate kidney repair in a mouse model of acute kidney injury. This reportedly occurs through EV-based delivery of specific subset of cellular mRNAs correlated with the MSC phenotype, which may trigger a proliferative program in surviving tubular cells post injury (Bruno et al. 2009). Similarly, EV-mediated transfer of a specific array of mRNA derived from human liver pluripotent stem cells has been shown to induce proliferation and apoptosis resistance in cultured human hepatocytes and to favor liver regeneration in hepatectomized rats (Herrera et al. 2010). These studies strongly imply the existence of a bidirectional genetic transfer between stem and injured cells, mediated by the exchange of highly regulatory RNA molecules, promoting the activation of tissue repair pathways.

As described by Khan et al. (2015), delivery of mouse ESC-exosomes into the heart after myocardial infarction stimulates and expands cardiac progenitor cell (CPC) and cardiomyocyte proliferation. Observed beneficial effects of ESC-derived exosomes also include enhanced neovascularization, cardiomyocyte survival, and reduced fibrosis post-infarction, along with augmented CPC cardiac commitment (Barile and Vassalli 2017). Another interesting study from Zhang et al. (2015) showed that in a rat model of wound healing exosomes isolated from human iPSC-derived MSCs have the ability to accelerate wound re-epithelialization, to reduce scar width, promoting collagen maturation and accelerating the genesis and maturation of newly formed vessels. *In vitro*, they show that human iPSC-MSC exosomes can stimulate in a dose-dependent way migration, proliferation, and tube formation of human umbilical vein endothelial cells. They also induced the migration, proliferation, and secretion of collagen and elastin by human fibroblasts. Exosomes from the same cells were tested *in vitro* and *in vivo* on a rat model of bone damage (Qi et al. 2016). Human iPSC-MSC-derived exosomes were shown to enhance proliferation and mRNA and protein expression of osteoblast-related genes in MSCs isolated from the rat model. When transplanted into critical-sized calvarial defects, the exosomes stimulated bone regeneration and angiogenesis in a dose-dependent manner (Qi et al. 2016). Furthermore, the transplantation of human iPSC-MSC-derived EVs in a rat model of acute kidney injury demonstrated that they have an anti-necroptotic effect that is mediated by the delivery of specificity protein 1 to target renal cells, with subsequent activation of sphingosine kinase 1 and generation of sphinganine-1-phosphate (Yuan et al. 2017). MSCs-EVs also promote hepatocytes differentiation and replication by modulating their cell cycle progression *in vivo* and increase cell survival in an *in vitro* model of liver injury (Tan et al. 2014).

iPSCs are particularly resilient upon transplantation in an ischemic environment. One supporting study investigated the possibility that iPSC-derived exosomes may improve cardiac function during myocardial infarction (Wang et al. 2015). The authors demonstrated that iPSC exosomes protect H9C2 cardiac cell line cells against oxidative stress-induced apoptosis *in vitro* and that they protect ischemic cardiomyocytes in a rat model of myocardial infarction. Also, human iPSC-derived EVs transfer RNA and proteins to cardiac MSCs that promote proliferation and increase survival of target cells when under stress (Bobis-Wozowicz et al. 2015). Similar results were obtained in a recent study where iPSC-derived EVs were found to contain proangiogenic and cryoprotective proteins and to increase migration and angiogenic properties of murine cardiac endothelial cells *in vitro*, reducing their apoptosis. In the same study iPSC-EVs also improved cardiac function in a murine model of myocardial infarction, reducing ventricle mass and increasing perfusion and cardiomyocytes apoptosis (Adamiak et al. 2018).

The potential of EVs as therapeutic application has been exploited also in regenerative medicine for CNS diseases (Zhang, Buller, and Chopp 2019b; Galieva et al. 2019; Vogel, Upadhya, and Shetty 2018). In one study, the acute injection of exosomes derived from MSCs in a rat model of stroke was found to invoke functional improvements reminiscent of those obtained during MSC transplantation, thus suggesting exosomes as a novel cell-free therapy (Xin et al. 2013). Similar positive effects have been observed in models of multiple sclerosis (MS). For instance, exosomes derived from IFNγ-stimulated DCs were found to induce promotion of remyelination in demyelinated organotypic slices, as well as in rat brains *in vivo*, possibly through the delivery of miR-219 (Pusic et al. 2014). Furthermore, in a recent example, MSC-derived exosomes were functionalized with the myelin-binding aptamer LJM-3064; the LJM-exosomes were found to promote proliferation of oligodendrocytes *in vitro* and produce a prophylactic effect in a mouse model of MS, suppressing inflammation and demyelination (Hosseini Shamili et al. 2019). Moreover, the delivery of α-synuclein siRNA-loaded RVG-expressing exosomes derived from DCs has been found to substantially reduce the expression of α-synuclein, commonly associated with Parkinson's disease progression, in a murine model of the disease (Cooper et al. 2014). Thus, there is compelling preclinical evidence to support the potential of EVs as valid therapeutic approaches in treating a variety of human diseases and disorders.

EVs FOR RNA THERAPEUTICS

With the discovery of RNAi came the possibility to develop new therapeutic approaches by targeting any mRNA of interest and modulating protein expression; nevertheless, delivery of naked RNA molecules is hindered by factors such as rapid degradation and difficulties to reach cell cytoplasm due to their hydrophilicity and polyanionic nature (Dykxhoorn and Lieberman 2006). Being natural nucleic acid carriers, EVs have been extensively explored as a vehicle for RNA-based interventions (Lu et al. 2018b). In particular, their cargo has commonly been augmented with specific RNAi agents, siRNAs, and miRNAs, enhancing their delivery to recipient cells where they can produce an altered phenotype (Jiang, Vader, and Schiffelers 2017). This approach has been successfully applied in a number of preclinical animal models of disease, but again anti-cancer applications are most common. Ohno et al. isolated exosomes from transfected HEK293 cells overexpressing GE11, a peptide shown to specifically bind EGFR, and let-7, a putative tumor suppressor miRNA (Ohno et al. 2013). Systemic delivery of the derivative exosomes to a mouse model of carcinoma yielded a significant reduction in tumor growth. Similarly, MVs isolated from cells transfected to express transforming growth factor-β1 siRNA showed an efficient reduction in tumor growth and metastasis following intravenous delivery to mice (Zhang, Li, et al. 2014b). As previously described, efficacious CNS delivery of siRNA has been demonstrated by *in vivo* intravenous administration of DC-derived exosomes loaded with siRNA for *BACE1* (Alvarez-Erviti et al. 2011) or α-synuclein (Cooper et al. 2014). Also within the context of CNS disorders, Didiot et al. employed glioblastoma-derived exosomes loaded with siRNA targeting the *huntingtin* gene (*htt*), producing a significant reduction of *htt* upon exosome delivery into the cortex of treated mice, paving the way for a new therapeutic approach to treating Hungtington's disease (Didiot et al. 2016). EV-based RNAi therapies have also produced promising outcomes in other *in vivo* disease models, including lung inflammation (Zhang et al. 2018) and renal fibrosis (Wang et al. 2016). Several notable examples of EV-mediated RNA delivery *in vivo* are described in Table 40.1.

Proof-of-concept of EV-mediated RNA therapeutic delivery has expanded beyond siRNA/miRNA. One notable recent example from Usman et al. beautifully demonstrated the use of EVs in delivering a diversity of putative RNA drugs (Usman et al. 2018). Employing red blood cell (RBC)-derived EVs, specifically those obtained from universal donors, the authors demonstrated the inhibition of oncogenic miR-125b in cancer cell lines using RBC-EV delivered antisense oligonucleotides and confirmed its therapeutic potential by suppressing cancer growth in leukemia and breast cancer models *in vivo*. Furthermore, RBC-EVs were successfully used to conduct CRISPR/*CRISPR-associated protein 9* (CRISPR-Cas9) gene editing *in vitro*, efficiently delivering *Cas9* mRNA and anti-miR-125b guide RNA to a leukemia cell line, and knocking out the oncogenic miRNA (Usman et al. 2018). In summary, preclinical studies confirm EVs as a promising vector for RNAi and gene therapies, offering biocompatibility, safety, targeted delivery and the possibility to engineer their composition for the development of personalized medicine.

EVs IN CLINICAL TRIALS

While an appreciation of the physiological importance of EVs has been growing for some decades, only relatively recently has their therapeutic potential become recognized, with an increasing

TABLE 40.1

Preclinical (in vivo) Examples of EV-Mediated RNA Delivery

EV Source (EV Type)	Targeting Moiety	RNA Cargo	Animal Model (Target Tissue)	Route of Delivery	References
HEK-293 kidney cells (exosomes)	GE11 peptide (EGFR ligand)	let-7a miRNA	RAG2 KO mice (subcutaneous HCC70 breast cancer xenograft)	Intravenous injection	Ohno et al. (2013)
L929 mouse fibroblasts (MVs)		*TGF-β1* siRNA	Wild-type BALB/c mice (subcutaneous S180 sarcoma xenograft)	Intravenous injection	Zhang, Li, et al. (2014b)
Mouse bone marrow (exosomes)	Lamp2b-RVG	*BACE1* siRNA	Wild-type C57BL/6 mice (cortical brain tissue)	Intravenous injection	Alvarez-Erviti et al. (2011)
Mouse bone marrow dendritic cells (exosomes)	Lamp2b-RVG	*α-Syn* siRNA	Transgenic mouse model expressing human S129D *α-Syn* cDNA under the PrP promoter (brain)	Intravenous injection	Cooper et al. (2014)
U87 glioblastoma cells (exosomes)		*Huntingtin* siRNA (hydrophobically modified, Cy3-fluorescent)	Wild-type FVB/NJ mice (brain striatum)	Stereotactic injection into the striata	Didiot et al. (2016)
Mouse serum (exosomes)		*Myd88* siRNA or miR-15a	LPS-induced lung inflammation in C57BL/6 mice (alveolar macrophages)	Intratracheal instillation	Zhang et al. (2018)
Human bone marrow MSCs (exosomes)		let-7c miRNA	Male C57BL/6J mice with unilateral ureteral obstruction (kidneys)	Intravenous injection	Wang et al. (2016)
Human red blood cells of group O patients (EVs)		Anti-miR-125b antisense oligonucleotides (or *Cas9* mRNA and gRNA*)	Acute myeloblastic leukemia model busulfan-conditioned NSG mice (MOLM13 tail-vein xenograft) or breast cancer model female nude mice (MCF10CA1a flank xenograft)	Intratumoral injection (leukemia model), intraperitoneal injection (breast cancer model)	Usman et al. (2018)
Mouse and human primary astrocytes, and A172 human astrocytes (exosomes)		*lincRNA-cox2* siRNA	Morphine- or LPS administrated wild-type (brain tissue)	Intranasal administration	Hu et al. (2018), Liao et al. (2019)
HEK-293T kidney cells (MVs)		*CD-UPRT* mRNA (and protein)	Athymic mice (HEI-193FC schwannoma sciatic nerve xenograft)	Intratumoral injection	Mizrak et al. (2013)

* Delivery of *Cas9* mRNA and gRNA through exosomes was evaluated *in vitro*.

Abbreviations: α-Syn, α-synuclein; BACE1, beta-secretase 1; CD, cytosine deaminase; EGFP, enhanced green fluorescent protein; EGFR, epidermal growth factor receptor; Lamp2b, lysosome-associated membrane glycoprotein 2b; lincRNA, long intergenic non-coding RNA; LPS, lipopolysaccharide; MSC, mesenchymal stromal cells; MV, microvesicles; Myd88, myeloid differentiation primary response 88; NSG, NOD-SCID gamma; PrP, prion protein; RAG2, recombination activating gene 2; RVG, rabies virus glycoprotein; TGF-β1, transforming growth factor β1; UPRT, uracil phophoribosyltransferase.

number of research studies aimed at employing EVs as treatments for a variety of diseases. Thus, while preclinical data has hinted at promising results, full and clear characterization of the myriad EV species, as well as gold standards for regulating EV production, isolation, and delivery for clinical trials are still missing, limiting their clinical translation (Lener et al. 2015). At the time of writing a search of the ClinicalTrials.gov database using the term "*extracellular vesicles*" reveals 29 registered clinical trials involving EVs, the majority of which exploit EVs as biomarkers. Only one study, NCT03857841 (clinicaltrials.gov identifier) by United Therapeutics, aims at employing EVs (methods of isolation not specified) from human bone marrow-derived MSCs for treating broncho-pulmonary dysplasia in infants. This Phase I study is currently recruiting.

With respect to the keyword "*exosomes*", some 60 clinical trials are registered, but only 9 are based on the use of exosomes as a therapy (again, the remaining studies employ exosomes as diagnostic measurements). Among these exosome trials, the Phase II NCT01159288 (Gustave Roussy, France) study employed exosomes from lipopolysaccharide- or IFNγ-matured DCs in patients bearing non-operable non-small cell lung cancer (Besse et al. 2016); a prior Phase I trial using these exosomes showed the feasibility of large-scale production of the therapy and confirmed the safety of the administration in patients (Escudier et al. 2005). The primary endpoint of the Phase II study was to observe at least 50% of patients with progression-free survival at 4 months after cessation of chemotherapy, while secondary aims were to assess the safety of the therapy and identify biomarkers for efficacy, such as natural killer (NK) cell activation, restoration of NKG2D transmembrane protein expression and peptide vaccine-specific T cell response. However, the main endpoint of the study was not reached, despite results showing that the immunotherapy was well tolerated. No cancer-specific T cell immune response was observed, nor was there a reduction in NKG2D positive NK cells compared to age-matched healthy volunteers.

Among the active and enrolling trials, NCT02565264 (Kumamoto University, Japan), an early Phase I trial, aims at evaluating the effects of autologous plasma-derived exosomes for the treatment of intractable cutaneous ulcers, such as in rheumatic disease, peripheral arterial disease, chronic venous insufficiency, decubitus, or burns. Exosomes are planned to be applied to the ulcers and a 28-day time course of observations used to evaluate wound healing. Phase I study NCT01294072 (University of Louisville, USA) aims to investigate the effect of dietary, plant-derived exosomes loaded with curcumin on normal colon tissue and colon tumors (Wu et al. 2017). Primary outcomes of this trial will be the comparison of tissue curcumin concentrations when administered alone or conjugated with exosomes, 7 days after ingestion; secondary outcomes include safety and tolerability, immune system response by measurement of serum cytokines levels, effects on colon cells, and *ex vivo* evaluation of curcumin and exosomes on colon cancer cells. Similarly, Phase I NCT01668849 (University of Louisville, USA) will evaluate the capability of grape-derived exosomes, given as a grape powder, to reduce the incidence of oral mucositis during treatment of head/neck tumors with radiation and chemotherapy. Grape extract will be self-administered for 35 days and pain caused by oral mucositis will be evaluated weekly as a primary outcome, while immune biomarkers in blood and mucosal tissue will be measured as secondary outcomes. A further plant exosome study from the University of Louisville (USA), NCT03493984 will evaluate changes in glucose tolerance as primary outcomes for the treatment of polycystic ovary syndrome with ginger or aloe plant-derived exosomes. Changes in serum testosterone, sex hormone binding globulin, gut microbiota, and inflammatory markers such as CD4, CD8, and forkhead box P3 will be measured as secondary outcomes.

MSCs are the most common cell source for therapeutic transplant therapies in clinical trials, across a great diversity of diseases and conditions. Such transplants typically do not promote recovery via cell replacement, but rather through modulation of the microenvironment through secreted factors (Hsuan et al. 2016). Accordingly, there is a growing interest in the potential of MSC-derived EVs. The early Phase I study NCT03437759 (Tianjin Medical University, China) aims at assessing the safety and efficacy of human umbilical cord MSCs or MSC-derived exosome injections in large and refractory macular holes. 50 μg or 20 μg of MSC-exosomes are to be dripped into the vitreous cavity around the macular hole, and patients are then followed for at least 6 months. A minimum linear diameter of the hole represents the primary outcome, accompanied by best-corrected visual activity as secondary measure. Cell-free therapies such as EVs represent an appealing therapeutic approach as they can circumvent many of the uncertainties associated with cellular transplantation. Based on the promising results obtained in preclinical studies evidencing the beneficial effects of exosomes in murine models of brain injury (Xin et al. 2013), and the capability of miR-124 loaded exosomes to promote neurogenesis upon brain injury (Yang et al. 2017), the Phase I and II study NCT03384433 (Isfahan University of Medical Sciences, Iran) aim at investigating the effect of MSC-derived exosomes loaded with miR-124 on patients with acute ischemic stroke. Patients will receive 200 μg total protein of MSC-exosomes via stereotaxis, 1 month after injury. Treatment will be primarily evaluated based on stroke recurrences, brain oedema, seizures and hemorrhagic transformation. Measurements of the degree of disabilities in patients will be also recorded. Umbilical cord-blood derived MSC MVs are to be employed for another registered Phase II/III clinical trial, NCT02138331 (General Committee of Teaching Hospitals and Institutes, Egypt), intended to assess the capability of such MVs to reduce inflammation in type 1 diabetes mellitus. Two intravenous injections of MVs are planned: first an injection of exosomes (40–180 nm), followed 7 days later by MVs (180–1000 nm). Required total daily insulin dose is to be used as a primary outcome measure, while pancreatic β-cell mass, before and 3 months after the treatment, will serve as a secondary measure. While this trial represents an early test of the clinical potential of EV-delivered RNA, its status is currently unknown. At the time of writing, the synergistic potential of EVs and RNA therapeutics is being examined in a Phase I study NCT03608631 from the M.D. Anderson Cancer Center (USA): MSC-exosomes loaded with siRNA targeting the *KRAS* G12D mutation (iExosomes) are to be used in the treatments of patients with pancreatic cancer to test for the best dose and assess any side effects. Intravenous injections are to be delivered at days 1, 4, and 10, and treatment will be repeated every 14 days, up to 3 times in the absence of disease progression or toxicity. Patients will then be followed up for 30 days, and subsequently every 3 months for up to 1 year. Secondary measurements will evaluate the pharmacokinetics of the treatment, the disease control rate, and the median progression-free survival.

CONCLUSIONS

There is a compelling and growing body of evidence for the biological relevance of EVs under both physiological and pathological conditions, opening up highly promising opportunities for their use in diagnostic, prognostic, or therapeutic applications. However, the heterogeneity of the EV menagerie, a lack of a harmonious nomenclature in the literature, and the ongoing challenge of developing a gold standard for isolation and purification has led to ambiguity and controversy regarding the composition and function of specific EV types. Moreover, a thorough understanding of the phenomena underlying EV biogenesis and their mechanisms of intercellular communication remains elusive. Full exploitation of EVs will require a comprehensive understanding of specific aspects of their activity, particularly relating to cargo packaging and delivery.

Nevertheless, the appeal of a naturally occurring, biocompatible drug delivery vehicle is self-evident and is providing a promising new avenue through which to effect safer and more efficacious therapies. Advances in biotechnological techniques are allowing for the re-purposing and re-engineering of "*natural*" vesicles, manipulating their makeup and cargo, or the synthesis of synthetic analogs, for tailored and targeted therapeutic use.

EV-trafficked RNA appears to play an integral role in the intricate network of intercellular communications, with transferred messengers or non-coding RNAs modulating a diversity of regulatory processes in recipient cells. Interceding in such processes, or emulating them through the introduction of exogenous RNA, is a compelling therapeutic approach that promises the generation of novel nanobiotechnologies capable of intervening in heretofore undruggable conditions. Ultimately, a better comprehension of the origin, composition, and fate of EVs will further our understanding of related diseases while simultaneously enhancing the utility of a potent nanotherapeutic platform.

REFERENCES

Aalberts M, Stout TAE, Stoorvogel W (2014). Prostasomes: extracellular vesicles from the prostate. *Reproduction* **147(1):**R1–14.

Adamiak M, Cheng G, Bobis-Wozowicz S, Zhao L, Kedracka-Krok S, Samanta A, Karnas E, Xuan YT, Skupien-Rabian B, Chen X, Jankowska U, Girgis M, Sekula M, Davani A, Lasota S, Vincent RJ, Sarna M, Newell KL, Wang OL, Dudley N, Madeja Z, Dawn B, Zuba-Surma EK (2018). Induced Pluripotent Stem Cell (iPSC)-Derived Extracellular Vesicles Are Safer and More Effective for Cardiac Repair Than iPSCs. *Circ Res* **122(2):**296–309.

Admyre C, Johansson SM, Qazi KR, Filen JJ, Lahesmaa R, Norman M, Neve EP, Scheynius A, Gabrielsson S (2007). Exosomes with immune modulatory features are present in human breast milk. *J Immunol* **179(3):**1969–1978.

Al-Nedawi K, Meehan B, Micallef J, Lhotak V, May L, Guha A, Rak J (2008). Intercellular transfer of the oncogenic receptor EGFRvIII by microvesicles derived from tumour cells. *Nat Cell Biol* **10(5):**619–624.

Al-Nedawi K, Meehan B, Rak J (2009). Microvesicles: messengers and mediators of tumor progression. *Cell Cycle* **8(13):**2014–2018.

Alais S, Simoes S, Baas D, Lehmann S, Raposo G, Darlix JL, Leblanc P (2008). Mouse neuroblastoma cells release prion infectivity associated with exosomal vesicles. *Biol Cell* **100(10):**603–615.

Alvarez-Erviti L, Seow Y, Yin H, Betts C, Lakhal S, Wood MJ (2011). Delivery of siRNA to the mouse brain by systemic injection of targeted exosomes. *Nat Biotechnol* **29(4):**341–345.

Amabile N, Cheng S, Renard JM, Larson MG, Ghorbani A, McCabe E, Griffin G, Guerin C, Ho JE, Shaw SY, Cohen KS, Vasan RS, Tedgui A, Boulanger CM, Wang TJ (2014). Association of circulating endothelial microparticles with cardiometabolic risk factors in the Framingham Heart Study. *Eur Heart J* **35(42):**2972–2979.

Andre F, Schartz NE, Movassagh M, Flament C, Pautier P, Morice P, Pomel C, Lhomme C, Escudier B, Le Chevalier T, Tursz T, Amigorena S, Raposo G, Angevin E, Zitvogel L (2002). Malignant effusions and immunogenic tumour-derived exosomes. *Lancet* **360(9329):**295–305.

Antes TJ, Middleton RC, Luther KM, Ijichi T, Peck KA, Liu WJ, Valle J, Echavez AK, Marban E (2018). Targeting extracellular vesicles to injured tissue using membrane cloaking and surface display. *J Nanobiotechnology* **16(1):**61.

Antimisiaris SG, Mourtas S, Marazioti A (2018). Exosomes and Exosome-Inspired Vesicles for Targeted Drug Delivery. *Pharmaceutics* **10(4):**218.

Aqil F, Kausar H, Agrawal AK, Jeyabalan J, Kyakulaga AH, Munagala R, Gupta R (2016). Exosomal formulation enhances therapeutic response of celastrol against lung cancer. *Exp Mol Pathol* **101(1):**12–21.

Arroyo JD, Chevillet JR, Kroh EM, Ruf IK, Pritchard CC, Gibson DF, Mitchell PS, Bennett CF, Pogosova-Agadjanyan EL, Stirewalt DL, Tait JF, Tewari M (2011). Argonaute2 complexes carry a population of circulating microRNAs independent of vesicles in human plasma. *Proc Natl Acad Sci USA* **108(12):**5003–5008.

Baj-Krzyworzeka M, Szatanek R, Weglarczyk K, Baran J, Urbanowicz B, Branski P, Ratajczak MZ, Zembala M (2006). Tumour-derived microvesicles carry several surface determinants and mRNA of tumour cells and transfer some of these determinants to monocytes. *Cancer Immunol Immunother* **55(7):**808–818.

Bakhti M, Winter C, Simons M (2011). Inhibition of myelin membrane sheath formation by oligodendrocyte-derived exosome-like vesicles. *J Biol Chem* **286(1):**787–796.

Balaj L, Lessard R, Dai L, Cho Y-J, Pomeroy SL, Breakefield XO, Skog J (2011). Tumour microvesicles contain retrotransposon elements and amplified oncogene sequences. *Nat Commun* **2:**180.

Barile L, Vassalli G (2017). Exosomes: Therapy delivery tools and biomarkers of diseases. *Pharmacol Ther* **174:**63–78.

Bastida E, Ordinas A, Escolar G, Jamieson GA (1984). Tissue factor in microvesicles shed from U87MG human glioblastoma cells induces coagulation, platelet aggregation, and thrombogenesis. *Blood* **64(1):**177–184.

Batagov AO, Kurochkin IV (2013). Exosomes secreted by human cells transport largely mRNA fragments that are enriched in the 3′-untranslated regions. *Biol Direct* **8:**12.

Batiz LF, Castro MA, Burgos PV, Velasquez ZD, Munoz RI, Lafourcade CA, Troncoso-Escudero P, Wyneken U (2015). Exosomes as novel regulators of adult neurogenic niches. *Front Cell Neurosci* **9:**501.

Belting M, Wittrup A (2008). Nanotubes, exosomes, and nucleic acid–binding peptides provide novel mechanisms of intercellular communication in eukaryotic cells: implications in health and disease. *J Cell Biol* **183(7):**1187–1191.

Bergsmedh A, Szeles A, Henriksson M, Bratt A, Folkman MJ, Spetz AL, Holmgren L (2001). Horizontal transfer of oncogenes by uptake of apoptotic bodies. *Proc Natl Acad Sci USA* **98(11):**6407–6411.

Bernimoulin M, Waters EK, Foy M, Steele BM, Sullivan M, Falet H, Walsh MT, Barteneva N, Geng JG, Hartwig JH, Maguire PB, Wagner DD (2009). Differential stimulation of monocytic cells results in distinct populations of microparticles. *J Thromb Haemost* **7(6):**1019–1028.

Besse B, Charrier M, Lapierre V, Dansin E, Lantz O, Planchard D, Le Chevalier T, Livartoski A, Barlesi F, Laplanche A, Ploix S, Vimond N, Peguillet I, Thery C, Lacroix L, Zoernig I, Dhodapkar K, Dhodapkar M, Viaud S, Soria JC, Reiners KS, Pogge von Strandmann E, Vely F, Rusakiewicz S, Eggermont A, Pitt JM, Zitvogel L, Chaput N (2016). Dendritic cell-derived exosomes as maintenance immunotherapy after first line chemotherapy in NSCLC. *Oncoimmunology* **5(4):**e1071008.

Bhat SA, Ahmad SM, Mumtaz PT, Malik AA, Dar MA, Urwat U, Shah RA, Ganai NA (2016). Long noncoding RNAs: Mechanism of action and functional utility. *Noncoding RNA Res* **1(1):**43–50.

Bhome R, Del Vecchio F, Lee G-H, Bullock MD, Primrose JN, Sayan AE, Mirnezami AH (2018). Exosomal microRNAs (exomiRs): Small molecules with a big role in cancer. *Cancer Lett* **420:**228–235.

Bianco F, Perrotta C, Novellino L, Francolini M, Riganti L, Menna E, Saglietti L, Schuchman EH, Furlan R, Clementi E, Matteoli M, Verderio C (2009). Acid sphingomyelinase activity triggers microparticle release from glial cells. *EMBO J* **28(8):**1043–1054.

Bianco F, Pravettoni E, Colombo A, Schenk U, Moller T, Matteoli M, Verderio C (2005). Astrocyte-derived ATP induces vesicle shedding and IL-1 beta release from microglia. *J Immunol* **174(11):**7268–7277.

Bidarimath M, Khalaj K, Kridli RT, Kan FW, Koti M, Tayade C (2017). Extracellular vesicle mediated intercellular communication at the porcine maternal-fetal interface: A new paradigm for conceptus-endometrial cross-talk. *Sci Rep* **7:**40476.

Bobis-Wozowicz S, Kmiotek K, Sekula M, Kedracka-Krok S, Kamycka E, Adamiak M, Jankowska U, Madetko-Talowska A, Sarna M, Bik-Multanowski M, Kolcz J, Boruczkowski D, Madeja Z, Dawn B, Zuba-Surma EK (2015). Human Induced Pluripotent Stem Cell-Derived Microvesicles Transmit RNAs and Proteins to Recipient Mature Heart Cells Modulating Cell Fate and Behavior. *Stem Cells* **33(9):**2748–2761.

Bobrie A, Colombo M, Raposo G, Théry C (2011). Exosome Secretion: Molecular Mechanisms and Roles in Immune Responses. *Traffic* **12(12):**1659–1668.

Bolukbasi MF, Mizrak A, Ozdener GB, Madlener S, Strobel T, Erkan EP, Fan JB, Breakefield XO, Saydam O (2012). miR-1289 and "Zipcode"-like Sequence Enrich mRNAs in Microvesicles. *Mol Ther Nucleic Acids* **1:**e10.

Borgovan T, Crawford L, Nwizu C, Quesenberry P (2019). Stem cells and extracellular vesicles: biological regulators of physiology and disease. *Am J Physiol Cell Physiol* **317(2):**C155–C166.

Boudreau LH, Duchez AC, Cloutier N, Soulet D, Martin N, Bollinger J, Pare A, Rousseau M, Naika GS, Levesque T, Laflamme C, Marcoux G, Lambeau G, Farndale RW, Pouliot M, Hamzeh-Cognasse H, Cognasse F, Garraud O, Nigrovic PA, Guderley H, Lacroix S, Thibault L, Semple JW, Gelb MH, Boilard E (2014). Platelets release mitochondria serving as substrate for bactericidal group IIA-secreted phospholipase A2 to promote inflammation. *Blood* **124(14):**2173–2183.

Bouvy C, Wannez A, Laloy J, Chatelain C, Dogne JM (2017). Transfer of multidrug resistance among acute myeloid leukemia cells via extracellular vesicles and their microRNA cargo. *Leuk Res* **62:**70–76.

Bruno S, Collino F, Iavello A, Camussi G (2014). Effects of mesenchymal stromal cell-derived extracellular vesicles on tumor growth. *Front Immunol* **5:**382.

Bruno S, Grange C, Deregibus MC, Calogero RA, Saviozzi S, Collino F, Morando L, Busca A, Falda M, Bussolati B, Tetta C, Camussi G (2009). Mesenchymal stem cell-derived microvesicles protect against acute tubular injury. *J Am Soc Nephrol* **20(5):**1053–1067.

Bulloj A, Leal MC, Xu H, Castano EM, Morelli L (2010). Insulin-degrading enzyme sorting in exosomes: a secretory pathway for a key brain amyloid-beta degrading protease. *J Alzheimers Dis* **19(1):**79–95.

Buschow SI, Nolte-'t Hoen EN, van Niel G, Pols MS, ten Broeke T, Lauwen M, Ossendorp F, Melief CJ, Raposo G, Wubbolts R, Wauben MH, Stoorvogel W (2009). MHC II in dendritic cells is targeted to lysosomes or T cell-induced exosomes via distinct multivesicular body pathways. *Traffic* **10(10):**1528–1542.

Bushati N, Cohen SM (2007). microRNA functions. *Annu Rev Cell Dev Biol* **23:**175–205.

Buzas EI, Gyorgy B, Nagy G, Falus A, Gay S (2014). Emerging role of extracellular vesicles in inflammatory diseases. *Nat Rev Rheumatol* **10(6):**356–364.

Caby MP, Lankar D, Vincendeau-Scherrer C, Raposo G, Bonnerot C (2005). Exosomal-like vesicles are present in human blood plasma. *Int Immunol* **17(7):**879–887.

Caivano A, La Rocca F, Simeon V, Girasole M, Dinarelli S, Laurenzana I, De Stradis A, De Luca L, Trino S, Traficante A, D'Arena G, Mansueto G, Villani O, Pietrantuono G, Laurenti L, Del Vecchio L, Musto P (2017). MicroRNA-155 in serum-derived extracellular vesicles as a potential biomarker for hematologic malignancies - a short report. *Cell Oncol* **40(1):**97–103.

Camacho L, Guerrero P, Marchetti D (2013). MicroRNA and protein profiling of brain metastasis competent cell-derived exosomes. *PLoS One* **8(9):**e73790.

Camussi G, Deregibus MC, Bruno S, Cantaluppi V, Biancone L (2010). Exosomes/microvesicles as a mechanism of cell-to-cell communication. *Kidney Int* **78(9):**838–848.

Caruso S, Poon IKH (2018). Apoptotic cell-derived extracellular vesicles: more than just debris. *Front Immunol* **9:**1486.

Ceruti S, Colombo L, Magni G, Vigano F, Boccazzi M, Deli MA, Sperlagh B, Abbracchio MP, Kittel A (2011). Oxygen-glucose deprivation increases the enzymatic activity and the microvesicle-mediated release of ectonucleotidases in the cells composing the blood-brain barrier. *Neurochem Int* **59(2):**259–271.

Chaput N, Thery C (2011). Exosomes: immune properties and potential clinical implementations. *Semin Immunopathol* **33(5):**419–440.

Cheng Y, Pereira M, Raukar N, Reagan JL, Queseneberry M, Goldberg L, Borgovan T, LaFrance WC, Jr., Dooner M, Deregibus M, Camussi G, Ramratnam B, Quesenberry P (2019). Potential biomarkers to detect traumatic brain injury by the profiling of salivary extracellular vesicles. *J Cell Physiol* **234(8):**14377–14388.

Ciesla M, Skrzypek K, Kozakowska M, Loboda A, Jozkowicz A, Dulak J (2011). MicroRNAs as biomarkers of disease onset. *Anal Bioanal Chem* **401(7):**2051–2061.

Cocucci E, Meldolesi J (2015). Ectosomes and exosomes: shedding the confusion between extracellular vesicles. *Trends Cell Biol* **25(6):**364–372.

Cocucci E, Racchetti G, Meldolesi J (2009). Shedding microvesicles: artefacts no more. *Trends Cell Biol* **19(2):**43–51.

Conde-Vancells J, Gonzalez E, Lu SC, Mato JM, Falcon-Perez JM (2010). Overview of extracellular microvesicles in drug metabolism. *Expert Opin Drug Metab Toxicol* **6(5):**543–554.

Cooper JM, Wiklander PB, Nordin JZ, Al-Shawi R, Wood MJ, Vithlani M, Schapira AH, Simons JP, El-Andaloussi S, Alvarez-Erviti L (2014). Systemic exosomal siRNA delivery reduced alpha-synuclein aggregates in brains of transgenic mice. *Mov Disord* **29(12):**1476–1485.

Corrado C, Saieva L, Raimondo S, Santoro A, De Leo G, Alessandro R (2016). Chronic myelogenous leukaemia exosomes modulate bone marrow microenvironment through activation of epidermal growth factor receptor. *J Cell Mol Med* **20(10):**1829–1839.

Cossetti C, Iraci N, Mercer TR, Leonardi T, Alpi E, Drago D, Alfaro-Cervello C, Saini HK, Davis MP, Schaeffer J, Vega B, Stefanini M, Zhao C, Muller W, Garcia-Verdugo JM, Mathivanan S, Bachi A, Enright AJ, Mattick JS, Pluchino S (2014). Extracellular vesicles from neural stem cells transfer IFN-gamma via Ifngr1 to activate Stat1 signaling in target cells. *Mol Cell* **56(2):**193–204.

Cossetti C, Smith JA, Iraci N, Leonardi T, Alfaro-Cervello C, Pluchino S (2012). Extracellular membrane vesicles and immune regulation in the brain. *Front Physiol* **3:**117.

Crompot E, Van Damme M, Pieters K, Vermeersch M, Perez-Morga D, Mineur P, Maerevoet M, Meuleman N, Bron D, Lagneaux L, Stamatopoulos B (2017). Extracellular vesicles of bone marrow stromal cells rescue chronic lymphocytic leukemia B cells from apoptosis, enhance their migration and induce gene expression modifications. *Haematologica* **102(9):**1594–1604.

Cronin J, Zhang XY, Reiser J (2005). Altering the tropism of lentiviral vectors through pseudotyping. *Curr Gene Ther* **5(4):**387–398.

Cruz L, Arevalo Romero JA, Brandao Prado M, Santos TG, Hohmuth Lopes M (2018). Evidence of Extracellular Vesicles Biogenesis and Release in Mouse Embryonic Stem Cells. *Stem Cell Rev Rep* **14(2):**262–276.

Curtis AM, Edelberg J, Jonas R, Rogers WT, Moore JS, Syed W, Mohler ER, 3rd (2013). Endothelial microparticles: sophisticated vesicles modulating vascular function. *Vasc Med* **18(4):**204–214.

Das CK, Jena BC, Banerjee I, Das S, Parekh A, Bhutia SK, Mandal M (2019). Exosome as a Novel Shuttle for Delivery of Therapeutics across Biological Barriers. *Mol Pharm* **16(1):**24–40.

Davis DM, Sowinski S (2008). Membrane nanotubes: dynamic long-distance connections between animal cells. *Nat Rev Mol Cell Biol* **9(6):**431–436.

de Castro LL, Xisto DG, Kitoko JZ, Cruz FF, Olsen PC, Redondo PAG, Ferreira TPT, Weiss DJ, Martins MA, Morales MM, Rocco PRM (2017). Human adipose tissue mesenchymal stromal cells and their extracellular vesicles act differentially on lung mechanics and inflammation in experimental allergic asthma. *Stem Cell Res Ther* **8(1):**151.

De Smaele E, Ferretti E, Gulino A (2010). MicroRNAs as biomarkers for CNS cancer and other disorders. *Brain Res* **1338:**100–111.

Del Conde I, Shrimpton CN, Thiagarajan P, Lopez JA (2005). Tissue-factor-bearing microvesicles arise from lipid rafts and fuse with activated platelets to initiate coagulation. *Blood* **106(5):**1604–1611.

Denzer K, van Eijk M, Kleijmeer MJ, Jakobson E, de Groot C, Geuze HJ (2000). Follicular dendritic cells carry MHC class II-expressing microvesicles at their surface. *J Immunol* **165(3):**1259–1265.

Deregibus MC, Cantaluppi V, Calogero R, Lo Iacono M, Tetta C, Biancone L, Bruno S, Bussolati B, Camussi G (2007). Endothelial progenitor cell derived microvesicles activate an angiogenic program in endothelial cells by a horizontal transfer of mRNA. *Blood* **110(7):**2440–2448.

Desrochers LM, Bordeleau F, Reinhart-King CA, Cerione RA, Antonyak MA (2016). Microvesicles provide a mechanism for intercellular communication by embryonic stem cells during embryo implantation. *Nat Commun* **7:**11958.

Dickhout A, Koenen RR (2018). Extracellular Vesicles as Biomarkers in Cardiovascular Disease; Chances and Risks. *Front Cardiovasc Med* **5:**113.

Didiot MC, Hall LM, Coles AH, Haraszti RA, Godinho BM, Chase K, Sapp E, Ly S, Alterman JF, Hassler MR, Echeverria D, Raj L, Morrissey DV, DiFiglia M, Aronin N, Khvorova A (2016). Exosome-mediated Delivery of Hydrophobically Modified siRNA for Huntingtin mRNA Silencing. *Mol Ther* **24(10):**1836–1847.

Dinger ME, Mercer TR, Mattick JS (2008). RNAs as extracellular signaling molecules. *J Mol Endocrinol* **40(4):**151–159.

Dong Y, He D, Peng Z, Peng W, Shi W, Wang J, Li B, Zhang C, Duan C (2017). Circular RNAs in cancer: an emerging key player. *J Hematol Oncol* **10(1):**2.

Drago D, Cossetti C, Iraci N, Gaude E, Musco G, Bachi A, Pluchino S (2013). The stem cell secretome and its role in brain repair. *Biochimie* **95(12):**2271–2285.

Driedonks TAP, Nolte-'t Hoen ENM (2018). Circulating Y-RNAs in Extracellular Vesicles and Ribonucleoprotein Complexes; Implications for the Immune System. *Front Immunol* **9:**3164.

Dykxhoorn DM, Lieberman J (2006). Knocking down disease with siRNAs. *Cell* **126(2):**231–235.

Eken C, Gasser O, Zenhaeusern G, Oehri I, Hess C, Schifferli JA (2008). Polymorphonuclear neutrophil-derived ectosomes interfere with the maturation of monocyte-derived dendritic cells. *J Immunol* **180(2):** 817–824.

Emmanouilidou E, Melachroinou K, Roumeliotis T, Garbis SD, Ntzouni M, Margaritis LH, Stefanis L, Vekrellis K (2010). Cell-produced alpha-synuclein is secreted in a calcium-dependent manner by exosomes and impacts neuronal survival. *J Neurosci* **30(20):**6838–6851.

Entchev EV, Gonzalez-Gaitan MA (2002). Morphogen gradient formation and vesicular trafficking. *Traffic* **3(2):**98–109.

Escola JM, Kleijmeer MJ, Stoorvogel W, Griffith JM, Yoshie O, Geuze HJ (1998). Selective enrichment of tetraspan proteins on the internal vesicles of multivesicular endosomes and on exosomes secreted by human B-lymphocytes. *J Biol Chem* **273(32):**20121–20127.

Escudier B, Dorval T, Chaput N, Andre F, Caby MP, Novault S, Flament C, Leboulaire C, Borg C, Amigorena S, Boccaccio C, Bonnerot C, Dhellin O, Movassagh M, Piperno S, Robert C, Serra V, Valente N, Le Pecq JB, Spatz A, Lantz O, Tursz T, Angevin E, Zitvogel L (2005). Vaccination of metastatic melanoma patients with autologous dendritic cell (DC) derived-exosomes: results of the first phase I clinical trial. *J Transl Med* **3:**10.

Fan Q, Yang L, Zhang X, Peng X, Wei S, Su D, Zhai Z, Hua X, Li H (2018). The emerging role of exosome-derived non-coding RNAs in cancer biology. *Cancer Lett* **414:**107–115.

Farahani M, Rubbi C, Liu L, Slupsky JR, Kalakonda N (2015). CLL Exosomes Modulate the Transcriptome and Behaviour of Recipient Stromal Cells and Are Selectively Enriched in miR-202-3p. *PLoS One* **10(10):**e0141429.

Farjadian F, Ghasemi A, Gohari O, Roointan A, Karimi M, Hamblin MR (2019). Nanopharmaceuticals and nanomedicines currently on the market: challenges and opportunities. *Nanomedicine* **14(1):**93–126.

Faure J, Lachenal G, Court M, Hirrlinger J, Chatellard-Causse C, Blot B, Grange J, Schoehn G, Goldberg Y, Boyer V, Kirchhoff F, Raposo G, Garin J, Sadoul R (2006). Exosomes are released by cultured cortical neurones. *Mol Cell Neurosci* **31(4):**642–648.

Fei F, Joo EJ, Tarighat SS, Schiffer I, Paz H, Fabbri M, Abdel-Azim H, Groffen J, Heisterkamp N (2015). B-cell precursor acute lymphoblastic leukemia and stromal cells communicate through Galectin-3. *Oncotarget* **6(13):**11378–11394.

Fevrier B, Vilette D, Archer F, Loew D, Faigle W, Vidal M, Laude H, Raposo G (2004). Cells release prions in association with exosomes. *Proc Natl Acad Sci USA* **101(26):**9683–9688.

Fiandaca MS, Kapogiannis D, Mapstone M, Boxer A, Eitan E, Schwartz JB, Abner EL, Petersen RC, Federoff HJ, Miller BL, Goetzl EJ (2015). Identification of preclinical Alzheimer's disease by a profile of pathogenic proteins in neurally derived blood exosomes: A case-control study. *Alzheimers Dement* **11(6):**600–607.

Fire A, Xu S, Montgomery MK, Kostas SA, Driver SE, Mello CC (1998). Potent and specific genetic interference by double-stranded RNA in Caenorhabditis elegans. *Nature* **391(6669):**806–811.

Fitzner D, Schnaars M, van Rossum D (2011). Selective transfer of exosomes from oligodendrocytes to microglia by macropinocytosis. *J Cell Sci* **124(3):**447–458.

Gage FH (2019). Adult neurogenesis in mammals. *Science* **364(6443):**827–828.

Galieva LR, James V, Mukhamedshina YO, Rizvanov AA (2019). Therapeutic Potential of Extracellular Vesicles for the Treatment of Nerve Disorders. *Front Neurosci* **13:**163.

Gangadaran P, Hong CM, Ahn BC (2018). An update on *in vivo* imaging of extracellular vesicles as drug delivery vehicles. *Front Pharmacol* **9:**169.

Garcia-Martinez I, Santoro N, Chen Y, Hoque R, Ouyang X, Caprio S, Shlomchik MJ, Coffman RL, Candia A, Mehal WZ (2016). Hepatocyte mitochondrial DNA drives nonalcoholic steatohepatitis by activation of TLR9. *J Clin Invest* **126(3):**859–864.

Gardiner TJ, Christov CP, Langley AR, Krude T (2009). A conserved motif of vertebrate Y RNAs essential for chromosomal DNA replication. *RNA* **15(7):**1375–1385.

Gasser O, Hess C, Miot S, Deon C, Sanchez JC, Schifferli JA (2003). Characterisation and properties of ectosomes released by human polymorphonuclear neutrophils. *Exp Cell Res* **285(2):**243–257.

Gasser O, Schifferli JA (2004). Activated polymorphonuclear neutrophils disseminate anti-inflammatory microparticles by ectocytosis. *Blood* **104(8):**2543–2548.

Gehrmann U, Hiltbrunner S, Georgoudaki AM, Karlsson MC, Naslund TI, Gabrielsson S (2013). Synergistic induction of adaptive antitumor immunity by codelivery of antigen with alpha-galactosylceramide on exosomes. *Cancer Res* **73(13):**3865–3876.

Gezer U, Ozgur E, Cetinkaya M, Isin M, Dalay N (2014). Long non-coding RNAs with low expression levels in cells are enriched in secreted exosomes. *Cell Biol Int* **38(9):**1076–1079.

Giacomini E, Vago R, Sanchez AM, Podini P, Zarovni N, Murdica V, Rizzo R, Bortolotti D, Candiani M, Viganò P (2017). Secretome of in vitro cultured human embryos contains extracellular vesicles that are uptaken by the maternal side. *Sci Rep* **7(1):**5210.

Gibbings DJ, Ciaudo C, Erhardt M, Voinnet O (2009). Multivesicular bodies associate with components of miRNA effector complexes and modulate miRNA activity. *Nat Cell Biol* **11(9):**1143–1149.

Goedert M, Clavaguera F, Tolnay M (2010). The propagation of prion-like protein inclusions in neurodegenerative diseases. *Trends Neurosci* **33(7):**317–325.

Gould SJ, Raposo G (2013). As we wait: coping with an imperfect nomenclature for extracellular vesicles. *J Extracell Vesicles* **2:**20389.

Gourzones C, Gelin A, Bombik I, Klibi J, Verillaud B, Guigay J, Lang P, Temam S, Schneider V, Amiel C, Baconnais S, Jimenez AS, Busson P (2010). Extra-cellular release and blood diffusion of BART viral micro-RNAs produced by EBV-infected nasopharyngeal carcinoma cells. *Virol J* **7:**271.

Graner MW, Alzate O, Dechkovskaia AM, Keene JD, Sampson JH, Mitchell DA, Bigner DD (2009). Proteomic and immunologic analyses of brain tumor exosomes. *FASEB J* **23(5):**1541–1557.

Greco V, Hannus M, Eaton S (2001). Argosomes: a potential vehicle for the spread of morphogens through epithelia. *Cell* **106(5):**633–645.

Guescini M, Genedani S, Stocchi V, Agnati LF (2010). Astrocytes and Glioblastoma cells release exosomes carrying mtDNA. *J Neural Trans* **117(1):**1–4.

Guest WC, Silverman JM, Pokrishevsky E, O'Neill MA, Grad LI, Cashman NR (2011). Generalization of the prion hypothesis to other neurodegenerative diseases: an imperfect fit. *J Toxicol Environ Health A* **74(22–24):**1433–1459.

Gui Y, Liu H, Zhang L, Lv W, Hu X (2015). Altered microRNA profiles in cerebrospinal fluid exosome in Parkinson disease and Alzheimer disease. *Oncotarget* **6(35):**37043–37053.

Guo H, Chitiprolu M, Roncevic L, Javalet C, Hemming FJ, Trung MT, Meng L, Latreille E, Tanese de Souza C, McCulloch D, Baldwin RM, Auer R, Cote J, Russell RC, Sadoul R, Gibbings D (2017). Atg5 Disassociates the V1V0-ATPase to Promote Exosome Production and Tumor Metastasis Independent of Canonical Macroautophagy. *Dev Cell* **43(6):**716–730.

Gyorgy B, Szabo TG, Pasztoi M, Pal Z, Misjak P, Aradi B, Laszlo V, Pallinger E, Pap E, Kittel A, Nagy G, Falus A, Buzas EI (2011). Membrane vesicles, current state-of-the-art: emerging role of extracellular vesicles. *Cell Mol Life Sci* **68(16):**2667–2688.

Haney MJ, Klyachko NL, Zhao Y, Gupta R, Plotnikova EG, He Z, Patel T, Piroyan A, Sokolsky M, Kabanov AV, Batrakova EV (2015). Exosomes as drug delivery vehicles for Parkinson's disease therapy. *J Control Release* **207:**18–30.

Hartmann A, Muth C, Dabrowski O, Krasemann S, Glatzel M (2017). Exosomes and the prion protein: more than one truth. *Frontiers in Neuroscience* **11:**194.

Hauser P, Wang S, Didenko VV (2017). Apoptotic bodies: selective detection in extracellular vesicles. *Methods Mol Biol* **1554:**193–200.

Hayakawa K, Esposito E, Wang X, Terasaki Y, Liu Y, Xing C, Ji X, Lo EH (2016). Transfer of mitochondria from astrocytes to neurons after stroke. *Nature* **535(7613):**551–555.

Herrera MB, Fonsato V, Gatti S, Deregibus MC, Sordi A, Cantarella D, Calogero R, Bussolati B, Tetta C, Camussi G (2010). Human liver stem cell-derived microvesicles accelerate hepatic regeneration in hepatectomized rats. *J Cell Mol Med* **14(6b):**1605–1618.

Holmgren L (2010). Horizontal gene transfer: you are what you eat. *Biochem Biophys Res Commun* **396(1):**147–151.

Hong CS, Muller L, Whiteside TL, Boyiadzis M (2014). Plasma exosomes as markers of therapeutic response in patients with acute myeloid leukemia. *Front Immunol* **5:**160.

Hosseini Shamili F, Alibolandi M, Rafatpanah H, Abnous K, Mahmoudi M, Kalantari M, Taghdisi SM, Ramezani M (2019). Immunomodulatory properties of MSC-derived exosomes armed with high affinity aptamer toward mylein as a platform for reducing multiple sclerosis clinical score. *J Control Release* **299:**149–164.

Hsu C, Morohashi Y, Yoshimura S, Manrique-Hoyos N, Jung S, Lauterbach MA, Bakhti M, Gronborg M, Mobius W, Rhee J, Barr FA, Simons M (2010). Regulation of exosome secretion by Rab35 and its GTPase-activating proteins TBC1D10A-C. *J Cell Biol* **189(2):**223–232.

Hsuan YC, Lin CH, Chang CP, Lin MT (2016). Mesenchymal stem cell-based treatments for stroke, neural trauma, and heat stroke. *Brain Behav* **6(10):**e00526.

Hu G, Drescher KM, Chen X-M (2012). Exosomal miRNAs: biological properties and therapeutic potential. *Front Genet* **3:**56.

Hu G, Liao K, Niu F, Yang L, Dallon BW, Callen S, Tian C, Shu J, Cui J, Sun Z, Lyubchenko YL, Ka M, Chen XM, Buch S (2018). Astrocyte EV-Induced lincRNA-Cox2 Regulates Microglial Phagocytosis: Implications for Morphine-Mediated Neurodegeneration. *Mol Ther Nucleic Acids* **13:**450–463.

Hubert A, Subra C, Jenabian MA, Tremblay Labrecque PF, Tremblay C, Laffont B, Provost P, Routy JP, Gilbert C (2015). Elevated Abundance, Size, and MicroRNA Content of Plasma Extracellular Vesicles in Viremic HIV-1+ Patients: Correlations With Known Markers of Disease Progression. *J Acquir Immune Defic Syndr* **70(3):**219–227.

Hurwitz SN, Cheerathodi MR, Nkosi D, York SB, Meckes DG, Jr. (2018). Tetraspanin CD63 Bridges Autophagic and Endosomal Processes To Regulate Exosomal Secretion and Intracellular Signaling of Epstein-Barr Virus LMP1. *J Virol* **92(5):**e01969–17.

Huttner HB, Corbeil D, Thirmeyer C, Coras R, Kohrmann M, Mauer C, Kuramatsu JB, Kloska SP, Doerfler A, Weigel D, Klucken J, Winkler J, Pauli E, Schwab S, Hamer HM, Kasper BS (2012). Increased membrane shedding – indicated by an elevation of CD133-enriched membrane particles – into the CSF in partial epilepsy. *Epilepsy Res* **99(1–2):**101–106.

Huttner HB, Janich P, Kohrmann M, Jaszai J, Siebzehnrubl F, Blumcke I, Suttorp M, Gahr M, Kuhnt D, Nimsky C, Krex D, Schackert G, Lowenbruck K, Reichmann H, Juttler E, Hacke W, Schellinger PD, Schwab S, Wilsch-Brauninger M, Marzesco AM, Corbeil D (2008). The stem cell marker prominin-1/CD133 on membrane particles in human cerebrospinal fluid offers novel approaches for studying central nervous system disease. *Stem Cells* **26(3):**698–705.

Iguchi H, Kosaka N, Ochiya T (2010). Secretory microRNAs as a versatile communication tool. *Commun Integr Biol* **3(5):**478–481.

Iraci N, Gaude E, Leonardi T, Costa ASH, Cossetti C, Peruzzotti-Jametti L, Bernstock JD, Saini HK, Gelati M, Vescovi AL, Bastos C, Faria N, Occhipinti LG, Enright AJ, Frezza C, Pluchino S (2017). Extracellular vesicles are independent metabolic units with asparaginase activity. *Nat Chem Biol* **13(9):**951–955.

Islam MN, Das SR, Emin MT, Wei M, Sun L, Westphalen K, Rowlands DJ, Quadri SK, Bhattacharya S, Bhattacharya J (2012). Mitochondrial transfer from bone-marrow-derived stromal cells to pulmonary alveoli protects against acute lung injury. *Nat Med* **18(5):**759–765.

Iwasaki YW, Siomi MC, Siomi H (2015). PIWI-Interacting RNA: Its Biogenesis and Functions. *Annu Rev Biochem* **84:**405–433.

Jakymiw A, Lian S, Eystathioy T, Li S, Satoh M, Hamel JC, Fritzler MJ, Chan EK (2005). Disruption of GW bodies impairs mammalian RNA interference. *Nat Cell Biol* **7(12):**1267–1274.

Janas T, Janas MM, Sapon K, Janas T (2015). Mechanisms of RNA loading into exosomes. *FEBS Lett* **589(13):**1391–1398.

Jeppesen DK, Fenix AM, Franklin JL, Higginbotham JN, Zhang Q, Zimmerman LJ, Liebler DC, Ping J, Liu Q, Evans R, Fissell WH, Patton JG, Rome LH, Burnette DT, Robert J. Coffey M (2019). Reassessment of Exosome Composition. *Cell* **177(2):**428–445.

Jeyaram A, Jay SM (2017). Preservation and Storage Stability of Extracellular Vesicles for Therapeutic Applications. *AAPS J* **20(1):**1.

Jiang L, Vader P, Schiffelers RM (2017). Extracellular vesicles for nucleic acid delivery: progress and prospects for safe RNA-based gene therapy. *Gene Ther* **24(3)**:157–166.

Johnstone RM, Mathew A, Mason AB, Teng K (1991). Exosome formation during maturation of mammalian and avian reticulocytes: evidence that exosome release is a major route for externalization of obsolete membrane proteins. *J Cell Physiol* **147(1)**:27–36.

Ju S, Mu J, Dokland T, Zhuang X, Wang Q, Jiang H, Xiang X, Deng ZB, Wang B, Zhang L, Roth M, Welti R, Mobley J, Jun Y, Miller D, Zhang HG (2013). Grape exosome-like nanoparticles induce intestinal stem cells and protect mice from DSS-induced colitis. *Mol Ther* **21(7)**:1345–1457.

Juan T, Furthauer M (2018). Biogenesis and function of ESCRT-dependent extracellular vesicles. *Semin Cell Dev Biol* **74**:66–77.

Jung KH, Chu K, Lee ST, Park HK, Bahn JJ, Kim DH, Kim JH, Kim M, Kun Lee S, Roh JK (2009). Circulating endothelial microparticles as a marker of cerebrovascular disease. *Annals of Neurology* **66(2)**:191–199.

Jy W, Minagar A, Jimenez JJ, Sheremata WA, Mauro LM, Horstman LL, Bidot C, Ahn YS (2004). Endothelial microparticles (EMP) bind and activate monocytes: elevated EMP-monocyte conjugates in multiple sclerosis. *Front Biosci* **9**:3137–3144.

Karlsson M, Lundin S, Dahlgren U, Kahu H, Pettersson I, Telemo E (2001). "Tolerosomes" are produced by intestinal epithelial cells. *Eur J Immunol* **31(10)**:2892–2900.

Katsuda T, Ikeda S, Yoshioka Y, Kosaka N, Kawamata M, Ochiya T (2014). Physiological and pathological relevance of secretory microRNAs and a perspective on their clinical application. *Biol Chem* **395(4)**:365–373.

Khan M, Nickoloff E, Abramova T, Johnson J, Verma SK, Krishnamurthy P, Mackie AR, Vaughan E, Garikipati VN, Benedict C, Ramirez V, Lambers E, Ito A, Gao E, Misener S, Luongo T, Elrod J, Qin G, Houser SR, Koch WJ, Kishore R (2015). Embryonic stem cell-derived exosomes promote endogenous repair mechanisms and enhance cardiac function following myocardial infarction. *Circ Res* **117(1)**:52–64.

Kim MS, Haney MJ, Zhao Y, Mahajan V, Deygen I, Klyachko NL, Inskoe E, Piroyan A, Sokolsky M, Okolie O, Hingtgen SD, Kabanov AV, Batrakova EV (2016). Development of exosome-encapsulated paclitaxel to overcome MDR in cancer cells. *Nanomedicine* **12(3)**:655–664.

Kim SH, Lechman ER, Bianco N, Menon R, Keravala A, Nash J, Mi Z, Watkins SC, Gambotto A, Robbins PD (2005). Exosomes derived from IL-10-treated dendritic cells can suppress inflammation and collagen-induced arthritis. *J Immunol* **174(10)**:6440–6448.

Kinoshita T, Yip KW, Spence T, Liu FF (2017). MicroRNAs in extracellular vesicles: potential cancer biomarkers. *J Hum Genet* **62(1)**:67–74.

Knip M, Constantin ME, Thordal-Christensen H (2014). Trans-kingdom cross-talk: small RNAs on the move. *PLoS Genet* **10(9)**:e1004602.

Kogure A, Kosaka N, Ochiya T (2019). Cross-talk between cancer cells and their neighbors via miRNA in extracellular vesicles: an emerging player in cancer metastasis. *J Biomed Sci* **26(1)**:7.

Kojima R, Bojar D, Rizzi G, Hamri GCE, El-Baba MD, Saxena P, Ausländer S, Tan KR, Fussenegger M (2018). Designer exosomes produced by implanted cells intracerebrally deliver therapeutic cargo for Parkinson's disease treatment. *Nat Commun* **9(1)**:1305.

Konoshenko MY, Lekchnov EA, Vlassov AV, Laktionov PP (2018). Isolation of Extracellular Vesicles: General Methodologies and Latest Trends. *Biomed Res Int* **2018**:8545347.

Kooijmans SA, Aleza CG, Roffler SR, van Solinge WW, Vader P, Schiffelers RM (2016b). Display of GPI-anchored anti-EGFR nanobodies on extracellular vesicles promotes tumour cell targeting. *J Extracell Vesicles* **5**:31053.

Kooijmans SAA, Schiffelers RM, Zarovni N, Vago R (2016a). Modulation of tissue tropism and biological activity of exosomes and other extracellular vesicles: New nanotools for cancer treatment. *Pharmacol Res* **111**:487–500.

Kooijmans SAA, Stremersch S, Braeckmans K, de Smedt SC, Hendrix A, Wood MJA, Schiffelers RM, Raemdonck K, Vader P (2013). Electroporation-induced siRNA precipitation obscures the efficiency of siRNA loading into extracellular vesicles. *J Control Release* **172(1)**:229–238.

Koppler B, Cohen C, Schlondorff D, Mack M (2006). Differential mechanisms of microparticle transfer to B cells and monocytes: anti-inflammatory propertiesof microparticles. *Eur J Immunol* **36(3)**:648–660.

Kosaka N, Iguchi H, Hagiwara K, Yoshioka Y, Takeshita F, Ochiya T (2013). Neutral sphingomyelinase 2 (nSMase2)-dependent exosomal transfer of angiogenic microRNAs regulate cancer cell metastasis. *J Biol Chem* **288(15)**:10849–10859.

Kosaka N, Iguchi H, Yoshioka Y, Takeshita F, Matsuki Y, Ochiya T (2010). Secretory Mechanisms and Intercellular Transfer of MicroRNAs in Living Cells. *J Biol Chem* **285(23)**:17442–17452.

Kowal J, Arras G, Colombo M, Jouve M, Morath JP, Primdal-Bengtson B, Dingli F, Loew D, Tkach M, Thery C (2016). Proteomic comparison defines novel markers to characterize heterogeneous populations of extracellular vesicle subtypes. *Proc Natl Acad Sci USA* **113(8)**:E968–E977.

Kowalski MP, Krude T (2015). Functional roles of non-coding Y RNAs. *The International Journal of Biochemistry & Cell Biology* **66:**20–29.

Krysko DV, Agostinis P, Krysko O, Garg AD, Bachert C, Lambrecht BN, Vandenabeele P (2011). Emerging role of damage-associated molecular patterns derived from mitochondria in inflammation. *Trends Immunol* **32(4):**157–164.

Kumar D, Gupta D, Shankar S, Srivastava RK (2015). Biomolecular characterization of exosomes released from cancer stem cells: Possible implications for biomarker and treatment of cancer. *Oncotarget* **6(5):**3280–3291.

Kurachi M, Mikuni M, Ishizaki Y (2016). Extracellular Vesicles from Vascular Endothelial Cells Promote Survival, Proliferation and Motility of Oligodendrocyte Precursor Cells. *PLoS One* **11(7):**e0159158.

Lai CP, Breakefield XO (2012). Role of exosomes/microvesicles in the nervous system and use in emerging therapies. *Front Physiol* **3:**228.

Lamichhane TN, Jeyaram A, Patel DB, Parajuli B, Livingston NK, Arumugasaamy N, Schardt JS, Jay SM (2016). Oncogene Knockdown via Active Loading of Small RNAs into Extracellular Vesicles by Sonication. *Cell Mol Bioeng* **9(3):**315–324.

Lang HL, Hu GW, Zhang B, Kuang W, Chen Y, Wu L, Xu GH (2017). Glioma cells enhance angiogenesis and inhibit endothelial cell apoptosis through the release of exosomes that contain long non-coding RNA CCAT2. *Oncol Rep* **38(2):**785–798.

Lasda E, Parker R (2016). Circular RNAs Co-Precipitate with Extracellular Vesicles: A Possible Mechanism for circRNA Clearance. *PLoS One* **11(2):**e0148407.

Lasser C (2012). Exosomal RNA as biomarkers and the therapeutic potential of exosome vectors. *Expert Opin Biol Ther* **12(Suppl 1):**S189–S197.

Le Pecq JB (2005). Dexosomes as a therapeutic cancer vaccine: from bench to bedside. *Blood Cells Mol Dis* **35(2):**129–135.

Lee TH, D'Asti E, Magnus N, Al-Nedawi K, Meehan B, Rak J (2011). Microvesicles as mediators of intercellular communication in cancer—the emerging science of cellular 'debris'. *Springer Semin Immunopathol* **33(5):**455–467.

Lee Y, EL Andaloussi S, Wood MJA (2012). Exosomes and microvesicles: extracellular vesicles for genetic information transfer and gene therapy. *Human Mol Genet* **21(R1):**R125–R134.

Lener T, Gimona M, Aigner L, Borger V, Buzas E, Camussi G, Chaput N, Chatterjee D, Court FA, Del Portillo HA, O'Driscoll L, Fais S, Falcon-Perez JM, Felderhoff-Mueser U, Fraile L, Gho YS, Gorgens A, Gupta RC, Hendrix A, Hermann DM, Hill AF, Hochberg F, Horn PA, de Kleijn D, Kordelas L, Kramer BW, Kramer-Albers EM, Laner-Plamberger S, Laitinen S, Leonardi T, Lorenowicz MJ, Lim SK, Lotvall J, Maguire CA, Marcilla A, Nazarenko I, Ochiya T, Patel T, Pedersen S, Pocsfalvi G, Pluchino S, Quesenberry P, Reischl IG, Rivera FJ, Sanzenbacher R, Schallmoser K, Slaper-Cortenbach I, Strunk D, Tonn T, Vader P, van Balkom BW, Wauben M, Andaloussi SE, Thery C, Rohde E, Giebel B (2015). Applying extracellular vesicles based therapeutics in clinical trials - an ISEV position paper. *J Extracell Vesicles* **4:**30087.

Li CC, Eaton SA, Young PE, Lee M, Shuttleworth R, Humphreys DT, Grau GE, Combes V, Bebawy M, Gong J, Brammah S, Buckland ME, Suter CM (2013). Glioma microvesicles carry selectively packaged coding and non-coding RNAs which alter gene expression in recipient cells. *RNA Biol* **10(8):**1333–1344.

Li Y, Zheng Q, Bao C, Li S, Guo W, Zhao J, Chen D, Gu J, He X, Huang S (2015). Circular RNA is enriched and stable in exosomes: a promising biomarker for cancer diagnosis. *Cell Res* **25:**981.

Li Z, Wang H, Yin H, Bennett C, Zhang H-g, Guo P (2018). Arrowtail RNA for Ligand Display on Ginger Exosome-like Nanovesicles to Systemic Deliver siRNA for Cancer Suppression. *Sci Rep* **8(1):**14644.

Liao K, Niu F, Dagur RS, He M, Tian C, Hu G (2019). Intranasal Delivery of lincRNA-Cox2 siRNA Loaded Extracellular Vesicles Decreases Lipopolysaccharide-Induced Microglial Proliferation in Mice. *J Neuroimmune Pharmacol* in press.

Losche W, Scholz T, Temmler U, Oberle V, Claus RA (2004). Platelet-derived microvesicles transfer tissue factor to monocytes but not to neutrophils. *Platelets* **15(2):**109–115.

Lu M, Xing H, Xun Z, Yang T, Ding P, Cai C, Wang D, Zhao X (2018a). Exosome-based small RNA delivery: Progress and prospects. *Asian J Pharm Sci* **13(1):**1–11.

Lu M, Xing H, Xun Z, Yang T, Zhao X, Cai C, Wang D, Ding P (2018b). Functionalized extracellular vesicles as advanced therapeutic nanodelivery systems. *Eur J Pharm Sci* **121:**34–46.

Luan X, Sansanaphongpricha K, Myers I, Chen H, Yuan H, Sun D (2017). Engineering exosomes as refined biological nanoplatforms for drug delivery. *Acta Pharmacol Sin* **38(6):**754–763.

Ma L, Li Y, Peng J, Wu D, Zhao X, Cui Y, Chen L, Yan X, Du Y, Yu L (2014). Discovery of the migrasome, an organelle mediating release of cytoplasmic contents during cell migration. *Cell Res* **25:**24–38.

Mack M, Kleinschmidt A, Bruhl H, Klier C, Nelson PJ, Cihak J, Plachy J, Stangassinger M, Erfle V, Schlondorff D (2000). Transfer of the chemokine receptor CCR5 between cells by membrane-derived microparticles: a mechanism for cellular human immunodeficiency virus 1 infection. *Nat Med* **6(7):**769–775.

Mandel P, Metais P (1948). Les acides nucléiques du plasma sanguin chez l'homme. *C R Seances Soc Biol Fil* **142(3–4):**241–243.

Mangeot PE, Dollet S, Girard M, Ciancia C, Joly S, Peschanski M, Lotteau V (2011). Protein transfer into human cells by VSV-G-induced nanovesicles. *Mol Ther* **19(9):**1656–1666.

Marzesco AM, Janich P, Wilsch-Brauninger M, Dubreuil V, Langenfeld K, Corbeil D, Huttner WB (2005). Release of extracellular membrane particles carrying the stem cell marker prominin-1 (CD133) from neural progenitors and other epithelial cells. *J Cell Sci* **118(13):**2849–2858.

Mateescu B, Kowal EJ, van Balkom BW, Bartel S, Bhattacharyya SN, Buzas EI, Buck AH, de Candia P, Chow FW, Das S, Driedonks TA, Fernandez-Messina L, Haderk F, Hill AF, Jones JC, Van Keuren-Jensen KR, Lai CP, Lasser C, Liegro ID, Lunavat TR, Lorenowicz MJ, Maas SL, Mager I, Mittelbrunn M, Momma S, Mukherjee K, Nawaz M, Pegtel DM, Pfaffl MW, Schiffelers RM, Tahara H, Thery C, Tosar JP, Wauben MH, Witwer KW, Nolte-'t Hoen EN (2017). Obstacles and opportunities in the functional analysis of extracellular vesicle RNA - an ISEV position paper. *J Extracell Vesicles* **6(1):**1286095.

Mathivanan S, Ji H, Simpson RJ (2010). Exosomes: extracellular organelles important in intercellular communication. *J Proteomics* **73(10):**1907–1920.

Mause SF, Weber C (2010). Microparticles: protagonists of a novel communication network for intercellular information exchange. *Circ Res* **107(9):**1047–1057.

Mayer WJ, Irschick UM, Moser P, Wurm M, Huemer HP, Romani N, Irschick EU (2007). Characterization of antigen-presenting cells in fresh and cultured human corneas using novel dendritic cell markers. *Invest Ophthalmol Vis Sci* **48(10):**4459–4467.

McKenzie AJ, Hoshino D, Hong NH, Cha DJ, Franklin JL, Coffey RJ, Patton JG, Weaver AM (2016). KRAS-MEK Signaling Controls Ago2 Sorting into Exosomes. *Cell Rep* **15(5):**978–987.

Meckes DG, Jr., Shair KHY, Marquitz AR, Kung C-P, Edwards RH, Raab-Traub N (2010). Human tumor virus utilizes exosomes for intercellular communication. *Proc Natl Acad Sci USA* **107(47):**20370–20375.

Meckes DG, Raab-Traub N (2011). Microvesicles and viral infection. *J Virol* **85(24):**12844–12854.

Melling GE, Carollo E, Conlon R, Simpson JC, Raul Francisco Carter D (2019). The Challenges and Possibilities of Extracellular Vesicles as Therapeutic Vehicles. *Eur J Pharma Biopharma in press*.

Melo SA, Sugimoto H, O'Connell JT, Kato N, Villanueva A, Vidal A, Qiu L, Vitkin E, Perelman LT, Melo CA, Lucci A, Ivan C, Calin GA, Kalluri R (2014). Cancer exosomes perform cell-independent microRNA biogenesis and promote tumorigenesis. *Cancer Cell* **26(5):**707–721.

Meyer C, Losacco J, Stickney Z, Li L, Marriott G, Lu B (2017). Pseudotyping exosomes for enhanced protein delivery in mammalian cells. *Int J Nanomedicine* **12:**3153–3170.

Minciacchi VR, Freeman MR, Di Vizio D (2015a). Extracellular vesicles in cancer: exosomes, microvesicles and the emerging role of large oncosomes. *Semin Cell Dev Biol* **40:**41–51.

Minciacchi VR, You S, Spinelli C, Morley S, Zandian M, Aspuria PJ, Cavallini L, Ciardiello C, Reis Sobreiro M, Morello M, Kharmate G, Jang SC, Kim DK, Hosseini-Beheshti E, Tomlinson Guns E, Gleave M, Gho YS, Mathivanan S, Yang W, Freeman MR, Di Vizio D (2015b). Large oncosomes contain distinct protein cargo and represent a separate functional class of tumor-derived extracellular vesicles. *Oncotarget* **6(13):**11327–11341.

Ming GL, Song H (2011). Adult neurogenesis in the mammalian brain: significant answers and significant questions. *Neuron* **70(4):**687–702.

Mitchell PJ, Welton J, Staffurth J, Court J, Mason MD, Tabi Z, Clayton A (2009). Can urinary exosomes act as treatment response markers in prostate cancer? *J Transl Med* **7:**4.

Mittelbrunn M, Gutiérrez-Vázquez C, Villarroya-Beltri C, González S, Sánchez-Cabo F, González MÁ, Bernad A, Sánchez-Madrid F (2011). Unidirectional transfer of microRNA-loaded exosomes from T cells to antigen-presenting cells. *Nat Commun* **2:**282.

Mittelbrunn M, Sanchez-Madrid F (2012). Intercellular communication: diverse structures for exchange of genetic information. *Nat Rev Mol Cell Biol* **13(5):**328–335.

Miyanishi M, Tada K, Koike M, Uchiyama Y, Kitamura T, Nagata S (2007). Identification of Tim4 as a phosphatidylserine receptor. *Nature* **450(7168):**435–439.

Mizrak A, Bolukbasi MF, Ozdener GB, Brenner GJ, Madlener S, Erkan EP, Strobel T, Breakefield XO, Saydam O (2013). Genetically engineered microvesicles carrying suicide mRNA/protein inhibit schwannoma tumor growth. *Mol Ther* **21(1):**101–108.

Mokarizadeh A, Delirezh N, Morshedi A, Mosayebi G, Farshid AA, Mardani K (2012). Microvesicles derived from mesenchymal stem cells: potent organelles for induction of tolerogenic signaling. *Immunol Lett* **147(1–2):**47–54.

Monsel A, Zhu YG, Gennai S, Hao Q, Hu S, Rouby JJ, Rosenzwajg M, Matthay MA, Lee JW (2015). Therapeutic Effects of Human Mesenchymal Stem Cell-derived Microvesicles in Severe Pneumonia in Mice. *Am J Respir Crit Care Med* **192(3):**324–336.

Morel O, Morel N, Jesel L, Freyssinet JM, Toti F (2011). Microparticles: a critical component in the nexus between inflammation, immunity, and thrombosis. *Springer Semin Immunopathol* **33(5):**469–486.

Morelli AE, Larregina AT, Shufesky WJ (2004). Endocytosis, intracellular sorting, and processing of exosomes by dendritic cells. *Blood* **104(10):**3257–3266.

Mueller G, Schneider M, Biemer-Daub G, Wied S (2011). Microvesicles released from rat adipocytes and harboring glycosylphosphatidylinositol-anchored proteins transfer RNA stimulating lipid synthesis. *Cell Signal* **23(7):**1207–1223.

Müller I, Klocke A, Alex M, Kotzsch M, Luther T, Morgenstern E, Zieseniss S, Zahler S, Preissner K, Engelmann B (2003). Intravascular tissue factor initiates coagulation via circulating microvesicles and platelets. *FASEB J* **17(3):**476–478.

Muntasell A, Berger AC, Roche PA (2007). T cell-induced secretion of MHC class II-peptide complexes on B cell exosomes. *EMBO J* **26(19):**4263–4272.

Muralidharan-Chari V, Clancy J, Plou C, Romao M, Chavrier P, Raposo G, D'Souza-Schorey C (2009). ARF6-regulated shedding of tumor cell-derived plasma membrane microvesicles. *Curr Biol* **19(22):**1875–1885.

Murphy DE, de Jong OG, Brouwer M, Wood MJ, Lavieu G, Schiffelers RM, Vader P (2019). Extracellular vesicle-based therapeutics: natural versus engineered targeting and trafficking. *Exp Mol Med* **51(3):**32.

Murrow L, Malhotra R, Debnath J (2015). ATG12-ATG3 interacts with Alix to promote basal autophagic flux and late endosome function. *Nat Cell Biol* **17(3):**300–310.

Nabhan JF, Hu R, Oh RS, Cohen SN, Lu Q (2012). Formation and release of arrestin domain-containing protein 1-mediated microvesicles (ARMMs) at plasma membrane by recruitment of TSG101 protein. *Proc Natl Acad Sci USA* **109(11):**4146–4151.

Namee NM, O'Driscoll L (2018). Extracellular vesicles and anti-cancer drug resistance. *Biochim Biophys Acta Rev Cancer* **1870(2):**123–136.

Nargesi AA, Lerman LO, Eirin A (2017). Mesenchymal Stem Cell-derived Extracellular Vesicles for Renal Repair. *Curr Gene Ther* **17(1):**29–42.

Naslund TI, Gehrmann U, Qazi KR, Karlsson MC, Gabrielsson S (2013). Dendritic cell-derived exosomes need to activate both T and B cells to induce antitumor immunity. *J Immunol* **190(6):**2712–2719.

Nawaz M, Fatima F, Vallabhaneni KC, Penfornis P, Valadi H, Ekstrom K, Kholia S, Whitt JD, Fernandes JD, Pochampally R, Squire JA, Camussi G (2016). Extracellular Vesicles: Evolving Factors in Stem Cell Biology. *Stem Cells Int* **2016:**1073140.

Neuspiel M, Schauss AC, Braschi E, Zunino R, Rippstein P, Rachubinski RA, Andrade-Navarro MA, McBride HM (2008). Cargo-selected transport from the mitochondria to peroxisomes is mediated by vesicular carriers. *Curr Biol* **18(2):**102–108.

Nolte-'t Hoen EN, Buermans HP, Waasdorp M, Stoorvogel W, Wauben MH, t Hoen PA (2012). Deep sequencing of RNA from immune cell-derived vesicles uncovers the selective incorporation of small non-coding RNA biotypes with potential regulatory functions. *Nucleic Acids Res* **40(18):**9272–9285.

Nolte-'t Hoen ENM, Buschow SI, Anderton SM, Stoorvogel W, Wauben MHM (2009). Activated T cells recruit exosomes secreted by dendritic cells via LFA-1. *Blood* **113(9):**1977–1981.

O'Loughlin AJ, Mager I, de Jong OG, Varela MA, Schiffelers RM, El Andaloussi S, Wood MJA, Vader P (2017). Functional Delivery of Lipid-Conjugated siRNA by Extracellular Vesicles. *Mol Ther* **25(7):**1580–1587.

Obregon C, Rothen-Rutishauser B, Gitahi SK, Gehr P, Nicod LP (2006). Exovesicles from human activated dendritic cells fuse with resting dendritic cells, allowing them to present alloantigens. *Am J Pathol* **169(6):**2127–2136.

Ohno S, Takanashi M, Sudo K, Ueda S, Ishikawa A, Matsuyama N, Fujita K, Mizutani T, Ohgi T, Ochiya T, Gotoh N, Kuroda M (2013). Systemically injected exosomes targeted to EGFR deliver antitumor microRNA to breast cancer cells. *Mol Ther* **21(1):**185–191.

Ohyashiki JH, Umezu T, Ohyashiki K (2018). Extracellular vesicle-mediated cell-cell communication in haematological neoplasms. *Philos Trans R Soc Lond B Biol Sci* **373(1737):**20160484.

Palanisamy V, Sharma S, Deshpande A, Zhou H, Gimzewski J, Wong DT (2010). Nanostructural and transcriptomic analyses of human saliva derived exosomes. *PLoS One* **5(1):**e8577.

Pant S, Hilton H, Burczynski ME (2012). The multifaceted exosome: biogenesis, role in normal and aberrant cellular function, and frontiers for pharmacological and biomarker opportunities. *Biochem Pharmacol* **83(11):**1484–1494.

Parolini I, Federici C, Raggi C, Lugini L, Palleschi S, De Milito A, Coscia C, Iessi E, Logozzi M, Molinari A, Colone M, Tatti M, Sargiacomo M, Fais S (2009). Microenvironmental pH is a key factor for exosome traffic in tumor cells. *J Biol Chem* **284(49):**34211–34222.

Pegtel DM, Cosmopoulos K, Thorley-Lawson DA, van Eijndhoven MAJ, Hopmans ES, Lindenberg JL, de Gruijl TD, Wurdinger T, Middeldorp JM (2010). Functional delivery of viral miRNAs via exosomes. *Proc Natl Acad Sci USA* **107(14):**6328–6333.

Pelloski CE, Ballman KV, Furth AF, Zhang L, Lin E, Sulman EP, Bhat K, McDonald JM, Yung WK, Colman H, Woo SY, Heimberger AB, Suki D, Prados MD, Chang SM, Barker FG, 2nd, Buckner JC, James CD, Aldape K (2007). Epidermal growth factor receptor variant III status defines clinically distinct subtypes of glioblastoma. *J Clin Oncol* **25(16):**2288–2294.

Pérez M, Avila J, Hernández F (2019). Propagation of Tau via Extracellular Vesicles. *Front Neurosci* **13:**698.

Peters L, Meister G (2007). Argonaute proteins: mediators of RNA silencing. *Mol Cell* **26(5):**611–623.

Phinney DG, Di Giuseppe M, Njah J, Sala E, Shiva S, St Croix CM, Stolz DB, Watkins SC, Di YP, Leikauf GD, Kolls J, Riches DW, Deiuliis G, Kaminski N, Boregowda SV, McKenna DH, Ortiz LA (2015). Mesenchymal stem cells use extracellular vesicles to outsource mitophagy and shuttle microRNAs. *Nat Commun* **6:**8472.

Pi F, Binzel DW, Lee TJ, Li Z, Sun M, Rychahou P, Li H, Haque F, Wang S, Croce CM, Guo B, Evers BM, Guo P (2018). Nanoparticle orientation to control RNA loading and ligand display on extracellular vesicles for cancer regression. *Nat Nanotechnol* **13(1):**82–89.

Pilzer D, Gasser O, Moskovich O, Schifferli JA, Fishelson Z (2005). Emission of membrane vesicles: roles in complement resistance, immunity and cancer. *Springer Semin Immunopathol* **27(3):**375–387.

Pisitkun T, Shen RF, Knepper MA (2004). Identification and proteomic profiling of exosomes in human urine. *Proc Natl Acad Sci USA* **101(36):**13368–13373.

Pluchino S, Smith JA (2019). Explicating Exosomes: Reclassifying the Rising Stars of Intercellular Communication. *Cell* **177(2):**225–227.

Pluchino S, Zanotti L, Rossi B, Brambilla E, Ottoboni L, Salani G, Martinello M, Cattalini A, Bergami A, Furlan R, Comi G, Constantin G, Martino G (2005). Neurosphere-derived multipotent precursors promote neuroprotection by an immunomodulatory mechanism. *Nature* **436(7048):**266–271.

Pluskota E, Woody NM, Szpak D, Ballantyne CM, Soloviev DA, Simon DI, Plow EF (2008). Expression, activation, and function of integrin alphaMbeta2 (Mac-1) on neutrophil-derived microparticles. *Blood* **112(6):**2327–2335.

Poliakov A, Spilman M, Dokland T, Amling CL, Mobley JA (2009). Structural heterogeneity and protein composition of exosome-like vesicles (prostasomes) in human semen. *Prostate* **69(2):**159–167.

Potolicchio I, Carven GJ, Xu X, Stipp C, Riese RJ, Stern LJ, Santambrogio L (2005). Proteomic analysis of microglia-derived exosomes: metabolic role of the aminopeptidase CD13 in neuropeptide catabolism. *Journal of Immunology* **175(4):**2237–2243.

Puhm F, Afonyushkin T, Resch U, Obermayer G, Rohde M, Penz T, Schuster M, Wagner G, Rendeiro AF, Melki I, Kaun C, Wojta J, Bock C, Jilma B, Mackman N, Boilard E, Binder CJ (2019). Mitochondria Are a Subset of Extracellular Vesicles Released by Activated Monocytes and Induce Type I IFN and TNF Responses in Endothelial Cells. *Circ Res* **125(1):**43–52.

Pusic AD, Pusic KM, Clayton BL, Kraig RP (2014). IFNgamma-stimulated dendritic cell exosomes as a potential therapeutic for remyelination. *J Neuroimmunol* **266(1–2):**12–23.

Putz U, Howitt J, Lackovic J, Foot N, Kumar S, Silke J, Tan SS (2008). Nedd4 family-interacting protein 1 (Ndfip1) is required for the exosomal secretion of Nedd4 family proteins. *J Biol Chem* **283(47):**32621–32627.

Qi X, Zhang J, Yuan H, Xu Z, Li Q, Niu X, Hu B, Wang Y, Li X (2016). Exosomes Secreted by Human-Induced Pluripotent Stem Cell-Derived Mesenchymal Stem Cells Repair Critical-Sized Bone Defects through Enhanced Angiogenesis and Osteogenesis in Osteoporotic Rats. *Int J Biol Sci* **12(7):**836–849.

Quek C, Bellingham SA, Jung CH, Scicluna BJ, Shambrook MC, Sharples RA, Cheng L, Hill AF (2017). Defining the purity of exosomes required for diagnostic profiling of small RNA suitable for biomarker discovery. *RNA Biol* **14(2):**245–258.

Quesenberry PJ, Aliotta J, Deregibus MC, Camussi G (2015). Role of extracellular RNA-carrying vesicles in cell differentiation and reprogramming. *Stem Cell Res Ther* **6(1):**153.

Rabelink TJ, de Boer HC, van Zonneveld AJ (2010). Endothelial activation and circulating markers of endothelial activation in kidney disease. *Nat Rev Nephrol* **6(7):**404–414.

Rajendran L, Honsho M, Zahn TR, Keller P, Geiger KD, Verkade P, Simons K (2006). Alzheimer's disease β-amyloid peptides are released in association with exosomes. *Proc Natl Acad Sci USA* **103(30):** 11172–11177.

Raposo G, Nijman HW, Stoorvogel W, Liejendekker R, Harding CV, Melief CJ, Geuze HJ (1996). B lymphocytes secrete antigen-presenting vesicles. *J Exp Med* **183(3):**1161–1172.

Raposo G, Stahl PD (2019). Extracellular vesicles: a new communication paradigm? *Nat Rev Mol Cel Biol*
20(9):509–510

Raposo G, Stoorvogel W (2013). Extracellular vesicles: exosomes, microvesicles, and friends. *J Cell Biol*
200(4):373–383.

Rappa G, Santos MF, Green TM, Karbanova J, Hassler J, Bai Y, Barsky SH, Corbeil D, Lorico A (2017).
Nuclear transport of cancer extracellular vesicle-derived biomaterials through nuclear envelope invagination-associated late endosomes. *Oncotarget* **8(9):**14443–14461.

Ratajczak J, Miekus K, Kucia M, Zhang J, Reca R, Dvorak P, Ratajczak MZ (2006). Embryonic stem cell-
derived microvesicles reprogram hematopoietic progenitors: evidence for horizontal transfer of mRNA
and protein delivery. *Leukemia* **20(5):**847–856.

Ray DM, Spinelli SL, Pollock SJ, Murant TI, O'Brien JJ, Blumberg N, Francis CW, Taubman MB, Phipps RP
(2008). Peroxisome proliferator-activated receptor gamma and retinoid X receptor transcription factors
are released from activated human platelets and shed in microparticles. *Thromb Haemost* **99(1):**86–95.

Riazifar M, Pone EJ, Lötvall J, Zhao W (2017). Stem Cell Extracellular Vesicles: Extended Messages of
Regeneration. *Annu Rev Pharmacol Tox* **57(1):**125–154.

Rivoltini L, Chiodoni C, Squarcina P, Tortoreto M, Villa A, Vergani B, Burdek M, Botti L, Arioli I, Cova A,
Mauri G, Vergani E, Bianchi B, Della Mina P, Cantone L, Bollati V, Zaffaroni N, Gianni AM, Colombo
MP, Huber V (2016). TNF-Related Apoptosis-Inducing Ligand (TRAIL)-Armed Exosomes Deliver
Proapoptotic Signals to Tumor Site. *Clin Cancer Res* **22(14):**3499–3512.

Rong Y, Liu W, Wang J, Fan J, Luo Y, Li L, Kong F, Chen J, Tang P, Cai W (2019). Neural stem cell-derived
small extracellular vesicles attenuate apoptosis and neuroinflammation after traumatic spinal cord injury
by activating autophagy. *Cell Death Dis* **10(5):**340.

Rustom A, Saffrich R, Markovic I, Walther P, Gerdes HH (2004). Nanotubular highways for intercellular organelle transport. *Science* **303(5660):**1007–1010.

Saeedi S, Israel S, Nagy C, Turecki G (2019). The emerging role of exosomes in mental disorders. *Transl
Psychiatry* **9(1):**122.

Saman S, Kim W, Raya M, Visnick Y, Miro S, Saman S, Jackson B, McKee AC, Alvarez VE, Lee NC, Hall
GF (2012). Exosome-associated tau is secreted in tauopathy models and is selectively phosphorylated in
cerebrospinal fluid in early Alzheimer disease. *J Biol Chem* **287(6):**3842–3849.

Santangelo L, Giurato G, Cicchini C, Montaldo C, Mancone C, Tarallo R, Battistelli C, Alonzi T, Weisz A,
Tripodi M (2016). The RNA-Binding Protein SYNCRIP Is a Component of the Hepatocyte Exosomal
Machinery Controlling MicroRNA Sorting. *Cell Rep* **17(3):**799–808.

Schiera G, Proia P, Alberti C, Mineo M, Savettieri G, Di Liegro I (2007). Neurons produce FGF2 and VEGF
and secrete them at least in part by shedding extracellular vesicles. *J Cell Mol Med* **11(6):**1384–1394.

Schorey JS, Bhatnagar S (2008). Exosome function: from tumor immunology to pathogen biology. *Traffic*
9(6):871–881.

Segura E, Guerin C, Hogg N, Amigorena S, Thery C (2007). CD8+ dendritic cells use LFA-1 to capture MHC-
peptide complexes from exosomes in vivo. *J Immunol* **179(3):**1489–1496.

Sercombe L, Veerati T, Moheimani F, Wu SY, Sood AK, Hua S (2015). Advances and Challenges of Liposome
Assisted Drug Delivery. *Front Pharmacol* **6:**286.

Shah R, Patel T, Freedman JE (2018). Circulating Extracellular Vesicles in Human Disease. *N Engl J Med*
379(10):958–966.

Sharples RA, Vella LJ, Nisbet RM, Naylor R, Perez K, Barnham KJ, Masters CL, Hill AF (2008). Inhibition
of gamma-secretase causes increased secretion of amyloid precursor protein C-terminal fragments in
association with exosomes. *FASEB J* **22(5):**1469–1478.

Shurtleff MJ, Temoche-Diaz MM, Karfilis KV, Ri S, Schekman R (2016). Y-box protein 1 is required to sort
microRNAs into exosomes in cells and in a cell-free reaction. *eLife* **5:**e19276.

Simak J, Gelderman MP, Yu H, Wright V, Baird AE (2006). Circulating endothelial microparticles in acute
ischemic stroke: a link to severity, lesion volume and outcome. *J Thromb Haemost* **4(6):**1296–1302.

Simons M, Raposo G (2009). Exosomes--vesicular carriers for intercellular communication. *Curr Opin Cell
Biol* **21(4):**575–581.

Simpson RJ, Jensen SS, Lim JW (2008). Proteomic profiling of exosomes: current perspectives. *Proteomics*
8(19):4083–4099.

Skog J, Würdinger T, van Rijn S, Meijer D, Gainche L, Sena-Esteves M, Curry W, Carter B, Krichevsky
A, Breakefield X (2008). Glioblastoma microvesicles transport RNA and proteins that promote tumour
growth and provide diagnostic biomarkers. *Nat Cell Biol* **10(12):**1470–1476.

Smalheiser NR (2007). Exosomal transfer of proteins and RNAs at synapses in the nervous system. *Biol Direct* **2:**35.

Smalheiser NR (2009). Do Neural Cells Communicate with Endothelial Cells via Secretory Exosomes and Microvesicles? *Cardiovasc Psychiatry Neurol* **2009:**383086.

Soubannier V, McLelland GL, Zunino R, Braschi E, Rippstein P, Fon EA, McBride HM (2012). A vesicular transport pathway shuttles cargo from mitochondria to lysosomes. *Curr Biol* **22(2):**135–141.

Spees JL, Olson SD, Whitney MJ, Prockop DJ (2006). Mitochondrial transfer between cells can rescue aerobic respiration. *Proc Natl Acad Sci USA* **103(5):**1283–1288.

Spinelli C, Adnani L, Choi D, Rak J (2018). Extracellular Vesicles as Conduits of Non-Coding RNA Emission and Intercellular Transfer in Brain Tumors. *Non-Coding RNA* **5:**1.

Squadrito ML, Baer C, Burdet F, Maderna C, Gilfillan GD, Lyle R, Ibberson M, De Palma M (2014). Endogenous RNAs modulate microRNA sorting to exosomes and transfer to acceptor cells. *Cell Rep* **8(5):**1432–1446.

Stahl PD, Raposo G (2019). Extracellular Vesicles: Exosomes and Microvesicles, Integrators of Homeostasis. *Physiology* **34(3):**169–177.

Stein AJ, Fuchs G, Fu C, Wolin SL, Reinisch KM (2005). Structural insights into RNA quality control: the Ro autoantigen binds misfolded RNAs via its central cavity. *Cell* **121(4):**529–539.

Street JM, Barran PE, Mackay CL, Weidt S, Balmforth C, Walsh TS, Chalmers RT, Webb DJ, Dear JW (2012). Identification and proteomic profiling of exosomes in human cerebrospinal fluid. *J Transl Med* **10:**5.

Sudhof TC (2004). The synaptic vesicle cycle. *Annu Rev Neurosci* **27:**509–547.

Sun D, Zhuang X, Xiang X, Liu Y, Zhang S, Liu C, Barnes S, Grizzle W, Miller D, Zhang HG (2010). A novel nanoparticle drug delivery system: the anti-inflammatory activity of curcumin is enhanced when encapsulated in exosomes. *Mol Ther* **18(9):**1606–1614.

Svensson KJ, Kucharzewska P, Christianson HC, Skold S, Lofstedt T, Johansson MC, Morgelin M, Bengzon J, Ruf W, Belting M (2011). Hypoxia triggers a proangiogenic pathway involving cancer cell microvesicles and PAR-2-mediated heparin-binding EGF signaling in endothelial cells. *Proc Natl Acad Sci USA* **108(32):**13147–13152.

Tamboli IY, Barth E, Christian L, Siepmann M, Kumar S, Singh S, Tolksdorf K, Heneka MT, Lutjohann D, Wunderlich P, Walter J (2010). Statins promote the degradation of extracellular amyloid {beta}-peptide by microglia via stimulation of exosome-associated insulin-degrading enzyme (IDE) secretion. *J Biol Chem* **285(48):**37405–37414.

Tan CY, Lai RC, Wong W, Dan YY, Lim SK, Ho HK (2014). Mesenchymal stem cell-derived exosomes promote hepatic regeneration in drug-induced liver injury models. *Stem Cell Res Ther* **5(3):**76.

Tan EM, Schur PH, Carr RI, Kunkel HG (1966). Deoxybonucleic acid (DNA) and antibodies to DNA in the serum of patients with systemic lupus erythematosus. *J Clin Invest* **45(11):**1732–1740.

Tao SC, Guo SC (2018). Extracellular Vesicles: Potential Participants in Circadian Rhythm Synchronization. *Int J Biol Sci* **14(12):**1610–1620.

Taylor AR, Robinson MB, Gifondorwa DJ, Tytell M, Milligan CE (2007). Regulation of heat shock protein 70 release in astrocytes: role of signaling kinases. *Dev Neurobiol* **67(13):**1815–1829.

Taylor DD, Akyol S, Gercel-Taylor C (2006). Pregnancy-associated exosomes and their modulation of T cell signaling. *J Immunol* **176(3):**1534–1542.

Taylor DD, Black PH (1986). Shedding of plasma membrane fragments Neoplastic and developmental importance. *Dev Biol* **3:**33–57.

Taylor DD, Gercel-Taylor C (2008). MicroRNA signatures of tumor-derived exosomes as diagnostic biomarkers of ovarian cancer. *Gynecol Oncol* **110(1):**13–21.

Taylor J, Bebawy M (2019). Proteins Regulating Microvesicle Biogenesis and Multidrug Resistance in Cancer. *Proteomics* **19(1–2):**1800165.

Théry C, Ostrowski M, Segura E (2009). Membrane vesicles as conveyors of immune responses. *Nat Rev Immunol* **9(8):**581–593.

Tosar JP, Gambaro F, Sanguinetti J, Bonilla B, Witwer KW, Cayota A (2015). Assessment of small RNA sorting into different extracellular fractions revealed by high-throughput sequencing of breast cell lines. *Nucleic Acids Res* **43(11):**5601–5616.

Trajkovic K, Hsu C, Chiantia S, Rajendran L, Wenzel D, Wieland F, Schwille P, Brugger B, Simons M (2008). Ceramide triggers budding of exosome vesicles into multivesicular endosomes. *Science* **319(5867):**1244–1247.

Trams EG, Lauter CJ, Salem N, Jr., Heine U (1981). Exfoliation of membrane ecto-enzymes in the form of micro-vesicles. *Biochim Biophys Acta* **645(1):**63–70.

Tricarico C, Clancy J, D'Souza-Schorey C (2017). Biology and biogenesis of shed microvesicles. *Small GTPases* **8(4):**220–232.

Trovato E, Di Felice V, Barone R (2019). Extracellular Vesicles: Delivery Vehicles of Myokines. *Front Physiol* **10:**522.

Turchinovich A, Drapkina O, Tonevitsky A (2019). Transcriptome of Extracellular Vesicles: State-of-the-Art. *Front Immunol* **10:**202.

Turchinovich A, Samatov TR, Tonevitsky AG, Burwinkel B (2013). Circulating miRNAs: cell-cell communication function? *Front Genet* **4:**119.

Twiss JL, Fainzilber M (2009). Ribosomes in axons--scrounging from the neighbors? *Trends Cell Biol* **19(5):**236–243.

Usman WM, Pham TC, Kwok YY, Vu LT, Ma V, Peng B, Chan YS, Wei L, Chin SM, Azad A, He AB-L, Leung AYH, Yang M, Shyh-Chang N, Cho WC, Shi J, Le MTN (2018). Efficient RNA drug delivery using red blood cell extracellular vesicles. *Nat Commun* **9(1):**2359.

Valadi H, Ekström K, Bossios A, Sjöstrand M, Lee JJ, Lötvall JO (2007). Exosome-mediated transfer of mRNAs and microRNAs is a novel mechanism of genetic exchange between cells. *Nat Cell Biol* **9(6):**654–659.

van der Vos KE, Balaj L, Skog J, Breakefield XO (2011). Brain tumor microvesicles: insights into intercellular communication in the nervous system. *Cell Mol Neurobiol* **31(6):**949–959.

Van Deun J, Mestdagh P, Sormunen R, Cocquyt V, Vermaelen K, Vandesompele J, Bracke M, De Wever O, Hendrix A (2014). The impact of disparate isolation methods for extracellular vesicles on downstream RNA profiling. *Journal Extracell Vesicles* **3(1):**24858.

van Niel G, D'Angelo G, Raposo G (2018). Shedding light on the cell biology of extracellular vesicles. *Nat Rev Mol Cell Biol* **19(4):**213–228.

van Zon A, Mossink MH, Scheper RJ, Sonneveld P, Wiemer EA (2003). The vault complex. *Cell Mol Life Sci* **60(9):**1828–1837.

Vella LJ, Sharples RA, Lawson VA, Masters CL, Cappai R, Hill AF (2007). Packaging of prions into exosomes is associated with a novel pathway of PrP processing. *J Pathol* **211(5):**582–590.

Vickers KC, Palmisano BT, Shoucri BM, Shamburek RD, Remaley AT (2011). MicroRNAs are transported in plasma and delivered to recipient cells by high-density lipoproteins. *Nat Cell Biol* **13(4):**423–433.

Villarroya-Beltri C, Gutierrez-Vazquez C, Sanchez-Cabo F, Perez-Hernandez D, Vazquez J, Martin-Cofreces N, Martinez-Herrera DJ, Pascual-Montano A, Mittelbrunn M, Sanchez-Madrid F (2013). Sumoylated hnRNPA2B1 controls the sorting of miRNAs into exosomes through binding to specific motifs. *Nat Commun* **4:**2980.

Vlassov A, Magdaleno S, Setterquist R, Conrad R (2012). Exosomes: current knowledge of their composition, biological functions, and diagnostic and therapeutic potentials. *Biochim Biophys Acta* **1820(7):**940–948.

Vogel A, Upadhya R, Shetty AK (2018). Neural stem cell derived extracellular vesicles: Attributes and prospects for treating neurodegenerative disorders. *EBioMedicine* **38:**273–282.

Von Bartheld CS, Altick AL (2011). Multivesicular bodies in neurons: distribution, protein content, and trafficking functions. *Prog Neurobiol* **93(3):**313–340.

Wan Y, Wang L, Zhu C, Zheng Q, Wang G, Tong J, Fang Y, Xia Y, Cheng G, He X, Zheng SY (2018). Aptamer-Conjugated Extracellular Nanovesicles for Targeted Drug Delivery. *Cancer Res* **78(3):**798–808.

Wang B, Yao K, Huuskes BM, Shen HH, Zhuang J, Godson C, Brennan EP, Wilkinson-Berka JL, Wise AF, Ricardo SD (2016). Mesenchymal Stem Cells Deliver Exogenous MicroRNA-let7c via Exosomes to Attenuate Renal Fibrosis. *Mol Ther* **24(7):**1290–1301.

Wang JG, Williams JC, Davis BK, Jacobson K, Doerschuk CM, Ting JP, Mackman N (2011). Monocytic microparticles activate endothelial cells in an IL-1beta-dependent manner. *Blood* **118(8):**2366–2374.

Wang JH, Forterre AV, Zhao J, Frimannsson DO, Delcayre A, Antes TJ, Efron B, Jeffrey SS, Pegram MD, Matin AC (2018). Anti-HER2 scFv-Directed Extracellular Vesicle-Mediated mRNA-Based Gene Delivery Inhibits Growth of HER2-Positive Human Breast Tumor Xenografts by Prodrug Activation. *Mol Cancer Ther* **17(5):**1133–1142.

Wang Y, Chen X, Tian B, Liu J, Yang L, Zeng L, Chen T, Hong A, Wang X (2017). Nucleolin-targeted Extracellular Vesicles as a Versatile Platform for Biologics Delivery to Breast Cancer. *Theranostics* **7(5):**1360–1372.

Wang Y, Zhang L, Li Y, Chen L, Wang X, Guo W, Zhang X, Qin G, He SH, Zimmerman A, Liu Y, Kim IM, Weintraub NL, Tang Y (2015). Exosomes/microvesicles from induced pluripotent stem cells deliver cardioprotective miRNAs and prevent cardiomyocyte apoptosis in the ischemic myocardium. *Int J Cardiol* **192:**61–69.

Webb RL, Kaiser EE, Scoville SL, Thompson TA, Fatima S, Pandya C, Sriram K, Swetenburg RL, Vaibhav K, Arbab AS, Baban B, Dhandapani KM, Hess DC, Hoda MN, Stice SL (2018). Human Neural Stem Cell Extracellular Vesicles Improve Tissue and Functional Recovery in the Murine Thromboembolic Stroke Model. *Transl Stroke Res* **9(5):**530–539.

Wei Z, Batagov AO, Schinelli S, Wang J, Wang Y, El Fatimy R, Rabinovsky R, Balaj L, Chen CC, Hochberg F, Carter B, Breakefield XO, Krichevsky AM (2017). Coding and noncoding landscape of extracellular RNA released by human glioma stem cells. *Nat Commun* **8(1):**1145.

Whiteside Theresa L (2013). Immune modulation of T-cell and NK (natural killer) cell activities by TEXs (tumour-derived exosomes). *Biochem Soc Trans* **41(1):**245–251.

Wiklander OP, Nordin JZ, O'Loughlin A, Gustafsson Y, Corso G, Mager I, Vader P, Lee Y, Sork H, Seow Y, Heldring N, Alvarez-Erviti L, Smith CI, Le Blanc K, Macchiarini P, Jungebluth P, Wood MJ, Andaloussi SE (2015). Extracellular vesicle in vivo biodistribution is determined by cell source, route of administration and targeting. *J Extracell Vesicles* **4:**26316.

Willms E, Cabañas C, Mäger I, Wood MJA, Vader P (2018). Extracellular Vesicle Heterogeneity: Subpopulations, Isolation Techniques, and Diverse Functions in Cancer Progression. *Front Immunol* **9:**738.

Witwer KW, Buzas EI, Bemis LT, Bora A, Lasser C, Lotvall J, Nolte-'t Hoen EN, Piper MG, Sivaraman S, Skog J, Thery C, Wauben MH, Hochberg F (2013). Standardization of sample collection, isolation and analysis methods in extracellular vesicle research. *J Extracell Vesicles* **2:**20360.

Wolf P (1967). The nature and significance of platelet products in human plasma. *Br J Haematol* **13(3):**269–288.

Wolfers J, Lozier A, Raposo G, Regnault A, Thery C, Masurier C, Flament C, Pouzieux S, Faure F, Tursz T, Angevin E, Amigorena S, Zitvogel L (2001). Tumor-derived exosomes are a source of shared tumor rejection antigens for CTL cross-priming. *Nat Med* **7(3):**297–303.

Wu K, Xing F, Wu SY, Watabe K (2017). Extracellular vesicles as emerging targets in cancer: Recent development from bench to bedside. *Biochim Biophys Acta Rev Cancer* **1868(2):**538–563.

Xin H, Li Y, Cui Y, Yang JJ, Zhang ZG, Chopp M (2013). Systemic administration of exosomes released from mesenchymal stromal cells promote functional recovery and neurovascular plasticity after stroke in rats. *J Cereb Blood Flow Metab* **33(11):**1711–1715.

Xu X, Lai Y, Hua Z-C (2019). Apoptosis and apoptotic body: disease message and therapeutic target potentials. *Bioscience Reports* **39(1):**BSR20180992.

Yang J, Zhang X, Chen X, Wang L, Yang G (2017). Exosome Mediated Delivery of miR-124 Promotes Neurogenesis after Ischemia. *Mol Ther Nucleic Acids* **7:**278–287.

Yang T, Martin P, Fogarty B, Brown A, Schurman K, Phipps R, Yin VP, Lockman P, Bai S (2015). Exosome delivered anticancer drugs across the blood-brain barrier for brain cancer therapy in Danio rerio. *Pharm Res* **32(6):**2003–2014.

Yoshioka Y, Katsuda T, Ochiya T (2018). Extracellular vesicles and encapusulated miRNAs as emerging cancer biomarkers for novel liquid biopsy. *Jpn J Clin Oncol* **48(10):**869–876.

Yuan A, Farber EL, Rapoport AL, Tejada D, Deniskin R, Akhmedov NB, Farber DB (2009). Transfer of microRNAs by embryonic stem cell microvesicles. *PLoS One* **4(3):**e4722.

Yuan X, Li D, Chen X, Han C, Xu L, Huang T, Dong Z, Zhang M (2017). Extracellular vesicles from human-induced pluripotent stem cell-derived mesenchymal stromal cells (hiPSC-MSCs) protect against renal ischemia/reperfusion injury via delivering specificity protein (SP1) and transcriptional activating of sphingosine kinase 1 and inhibiting necroptosis. *Cell Death Dis* **8(12):**3200.

Zhang B, Yin Y, Lai RC, Tan SS, Choo AB, Lim SK (2014a). Mesenchymal stem cells secrete immunologically active exosomes. *Stem Cells Dev* **23(11):**1233–1244.

Zhang D, Lee H, Wang X, Rai A, Groot M, Jin Y (2018). Exosome-Mediated Small RNA Delivery: A Novel Therapeutic Approach for Inflammatory Lung Responses. *Mol Ther* **26(9):**2119–2130.

Zhang J, Guan J, Niu X, Hu G, Guo S, Li Q, Xie Z, Zhang C, Wang Y (2015). Exosomes released from human induced pluripotent stem cells-derived MSCs facilitate cutaneous wound healing by promoting collagen synthesis and angiogenesis. *J Transl Med* **13(1):**49.

Zhang Q, Higginbotham JN, Jeppesen DK, Yang Y-P, Li W, McKinley ET, Graves-Deal R, Ping J, Britain CM, Dorsett KA, Hartman CL, Ford DA, Allen RM, Vickers KC, Liu Q, Franklin JL, Bellis SL, Coffey RJ (2019a). Transfer of Functional Cargo in Exomeres. *Cell Rep* **27(3):**940–954.

Zhang Y, Li L, Yu J, Zhu D, Zhang Y, Li X, Gu H, Zhang CY, Zen K (2014b). Microvesicle-mediated delivery of transforming growth factor beta1 siRNA for the suppression of tumor growth in mice. *Biomaterials* **35(14):**4390–4400.

Zhang Y, Liu D, Chen X, Li J, Li L, Bian Z, Sun F, Lu J, Yin Y, Cai X, Sun Q, Wang K, Ba Y, Wang Q, Wang D, Yang J, Liu P, Xu T, Yan Q, Zhang J, Zen K, Zhang C-Y (2010). Secreted Monocytic miR-150 Enhances Targeted Endothelial Cell Migration. *Mol Cell* **39(1):**133–144.

Zhang ZG, Buller B, Chopp M (2019b). Exosomes — beyond stem cells for restorative therapy in stroke and neurological injury. *Nat Rev Neurol* **15(4):**193–203.

Zhao M, Xu J, Zhong S, Liu Y, Xiao H, Geng L, Liu H (2019). Expression profiles and potential functions of circular RNAs in extracellular vesicles isolated from radioresistant glioma cells. *Oncol Rep* **41(3):**1893–1900.

Zhou J, Benito-Martin A, Mighty J, Chang L, Ghoroghi S, Wu H, Wong M, Guariglia S, Baranov P, Young M, Gharbaran R, Emerson M, Mark MT, Molina H, Canto-Soler MV, Selgas HP, Redenti S (2018). Retinal progenitor cells release extracellular vesicles containing developmental transcription factors, microRNA and membrane proteins. *Sci Rep* **8(1):**2823.

Zhou J, Ghoroghi S, Benito-Martin A, Wu H, Unachukwu UJ, Einbond LS, Guariglia S, Peinado H, Redenti S (2016). Characterization of Induced Pluripotent Stem Cell Microvesicle Genesis, Morphology and Pluripotent Content. *Sci Rep* **6:**19743.

Zhu X, Badawi M, Pomeroy S, Sutaria DS, Xie Z, Baek A, Jiang J, Elgamal OA, Mo X, Perle K, Chalmers J, Schmittgen TD, Phelps MA (2017). Comprehensive toxicity and immunogenicity studies reveal minimal effects in mice following sustained dosing of extracellular vesicles derived from HEK293T cells. *J Extracell Vesicles* **6(1):**1324730.

Zitvogel L, Regnault A, Lozier A, Wolfers J, Flament C, Tenza D, Ricciardi-Castagnoli P, Raposo G, Amigorena S (1998). Eradication of established murine tumors using a novel cell-free vaccine: dendritic cell-derived exosomes. *Nat Med* **4(5):**594–600.

Zou J, Shi M, Liu X, Jin C, Xing X, Qiu L, Tan W (2019). Aptamer-Functionalized Exosomes: Elucidating the Cellular Uptake Mechanism and the Potential for Cancer-Targeted Chemotherapy. *Anal Chem* **91(3):**2425–2430.

41 Harnessing Exosomes and Bioinspired Exosome-Like Nanoparticles for siRNA Delivery

Mei Lu, Haonan Xing, and Yuanyu Huang
Beijing Institute of Technology, Beijing, P. R. China
Beijing Institute of Pharmacology and Toxicology, Beijing, P. R. China

CONTENTS

INTRODUCTION

The potential of RNA interference (RNAi) to treat or prevent a disease at its genetic root has long been fascinating for both the scientific community and general public (Fire et al. 1998). A critical mediator of RNAi is the small interfering RNAs (siRNAs) that suppress the expression of target genes by binding to the RNA-inducing silencing complex (RISC) and directly mediating cleavage of the complementary mRNAs in the cytoplasm (Castanotto and Rossi 2009; Hu et al. 2019). For years, RNAi-based therapy has been considered as one of the most strategically promising biotechnologies. To date, four siRNA therapeutics have been approved by the Food and Drug Administration (FDA), namely Onpattro, Givlaari, Leqvio and Oxlumo developed by Alnylam Pharmaceuticals, Inc. Compared with therapeutics based on small chemical molecules and proteins, there are several advantages of using siRNAs for gene therapy, including ease of synthesis, high specificity of the association between siRNAs and the complementary mRNA targets, and the capability of silencing any genes of interest, even those coding for undruggable protein products (Castanotto and Rossi 2009; Setten, Rossi, and Han 2019; Weng et al. 2019). Notwithstanding these favorable features, translation of siRNA molecules into clinical therapeutics still remains a challenging task, which is mainly attributed to the lack of efficient and safe carriers for delivery of siRNAs to target sites of action.

Multiple obstacles need to be addressed to achieve effective siRNA delivery (Setten, Rossi, and Han 2019; Weng et al. 2019), such as evading degradation by nuclease and phagocytosis by the reticuloendothelial system, accumulation to target tissues, permeation through biological membranes, and escaping from endosomal/lysosomal compartments to the cytosol. However, naked siRNA molecules are susceptible to enzymatic degradation. This, together with several structural characteristics

DOI: 10.1201/9781003001560-45

of siRNAs, such as small particle size, hydrophilicity, and high molecular weight, leads to their fast clearance with a half-life less than 20 min in the bloodstream (Antimisiaris, Mourtas, and Papadia 2017; Huang et al. 2011; Huang et al. 2016). In addition, siRNA molecules possess anionic charge and have low membrane permeability, which may substantially limit their ability to achieve efficient cellular internalization. After uptake by target cells *via* endocytosis, siRNA molecules are entrapped in endosomal/lysosomal lumens and thus may be rapidly degraded by lysosomal proteases and the acidic pH (Gilleron et al. 2013). In addition to delivery obstacles, siRNAs with specific sequences or chemical structures could trigger off-target immune responses (Naito and Ui-Tei 2013). To this end, additional assistance is necessary to deliver siRNA molecules to target sites of action for efficient and safe therapeutic effects.

Extracellular vesicles (EVs) are heterogeneous membrane-enclosed vesicles that are secreted by (almost) all cell types and play important roles in intercellular communication. Defined by size and sub-cellular origin, EVs are generally classified into three subtypes, including exosomes, microvesicles, and apoptotic bodies (Lu, Xing, Xun et al. 2018a; Raposo and Stoorvogel 2013). Among these vesicles, current research interest primarily focuses on exosomes and microvesicles, as the large size of apoptotic bodies (50–5000 nm) rules out their potential for therapeutic application. Exosomes with size ranging from 50 to 150 nm are the smallest vesicles originated from intraluminal budding of multivesicular bodies (MVBs), while microvesicles (50–1000 nm) are secreted by directly budding and fission of the plasma membrane. Despite the apparent differences, it still remains difficult to distinguish different subsets of EVs for current isolation techniques (Piffoux et al. 2019; Kowal et al. 2016). Therefore, in this chapter, we used the term exosomes that is mainly referred to exosomes and microvesicles.

The past decade has witnessed the rapid development of exosome-based siRNA delivery vehicles. Actually, they are characterized by several advantages over conventional nonviral vectors. Exosomes are endogenous RNA carriers and display excellent biocompatibility (Lu, Xing, Xun et al. 2018a). It has been demonstrated that exosomes are enriched in "self-marker" proteins, such as CD47, which may act as "invisibility cloak" for loaded drugs and protect them from enzymatic digestion and clearance by the reticuloendothelial system (Kamerkar et al. 2017). In addition, exosomes derived from specific cell types may possess intrinsic targeting properties (Vader et al. 2016). Moreover, exosomes are composed of specific lipids and membrane proteins that may guide the loaded cargoes across natural membranous barriers (Montecalvo et al. 2012; Parolini et al. 2009). For example, it has been reported that dendritic cell-derived exosomes could directly fuse with target cells through their CD9 tetraspanin interaction with glycoproteins on target cells, which could largely bypass the endocytic pathway and contribute to a direct siRNA delivery to the cytosol of target cells (van den Boorn et al. 2011). Despite these appealing advantages, major challenges concerning low production yield, difficulty in drug loading, and considerable complexity are ahead before reaching the maximum potential of exosomes in the clinic. To this end, bioinspired exosome-like nanoparticles that could recapitulate the distinct biological properties of exosomes and beneficial advantages of synthetic nanomaterials have sparked considerable interests of scientists in this field. In this chapter, we attempt to give an updated overview of exosome-based siRNA delivery platforms with focus on state-of-the-art strategies to harness natural exosomes, functionalized exosomes, as well as bioinspired exosome-like nanoparticles for siRNA delivery.

EXOSOME COMPOSITION

Given that the biological properties of exosome-based nanoplatforms are closely associated with the unique composition of exosomes, a greater understanding of exosome composition may facilitate the development of more excellent exosome-based siRNA delivery vehicles. In this section, the composition of exosomes will be briefly described (Figure 41.1). Exosomes are typically composed of luminal cargoes, including proteins, RNAs and DNAs, surrounded by a lipid bilayer membrane with a rich repertoire of membrane proteins (Colombo, Raposo, and Thery 2014). Although there is

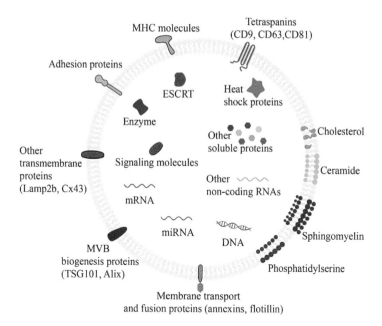

FIGURE 41.1   Schematic illustrating the composition of exosomes. Exosomes carry a wide array of luminal cargoes, including proteins, RNAs and DNAs. The lipid bilayer of exosomes is commonly enriched in cholesterol, sphingomyelin, phosphatidylserine, and ceramide and is integrated with a repertoire of membrane proteins such as tetraspanins, MHC molecules, membrane transport and fusion proteins, adhesion proteins, and proteins involved in MVB biogenesis.

a wide variation across exosomes derived from different cell types, exosomes are considered to share a common set of proteins, lipids, and RNAs, irrespective of the cell type of origin (Vlassov et al. 2012; Arenaccio et al. 2019; Lu, Zhao, Xing et al. 2018d; Thery et al. 2018). For example, exosomes are generally enriched in membrane transport and fusion proteins (i.e., Annexins, GTPases, and flotillin), tetraspanins (i.e., CD9, CD63, CD81, CD82), heat shock proteins (HSC70 and HSC90), components of endosomal sorting complex required for transport (ESCRTs), proteins involved in the biogenesis of MVBs (Alix, TSG101), and lipid-related proteins (Thery, Ostrowski, and Segura 2009; Conde-Vancells et al. 2008; Wubbolts et al. 2003; Vlassov et al. 2012). Therefore, proteins such as CD9, CD63, CD81, and TSG101 are routinely used as positive markers for exosome identification. Exosomal proteins involve in a multitude of functions, ranging from basic cellular processes such as cell adhesion and signal transduction to functional cargo delivery. For example, exosomal proteins such as CD47, CD55, and CD59 could mediate "don't eat me" signal, thus endowing exosomes with antiphagocytic property (Wubbolts et al. 2003; Kamerkar et al. 2017). CD9 involves in the fusion of exosomes with recipient cells, thus facilitating a direct cytosolic cargo delivery (van den Boorn et al. 2011). Recently, it has been reported that connexin 43 (Cx43) was enriched in exosomes, and it was a potential cytosolic delivery helper that favors intracellular delivery of the carried plasmid (Soares et al. 2015). Besides, exosomal Cx43 has been shown to be capable of recruiting RNA molecules through RNA-binding motifs, implying RNA transport by exosomes through gap junction channels (Varela-Eirin et al. 2017).

In general, exosomes share a common feature in lipid type with their parental cells but are enriched in particular lipids. Exosomes appear to be enriched in raft-associated lipids, such as cholesterol, phosphatidylserine, sphingomyelin, and ceramide (Llorente et al. 2013; Subra et al. 2007; Colombo, Raposo, and Thery 2014). As compared to donor cells, exosomes contain significantly lower amount of phosphatidylcholine, while the level of phosphatidylethanolamine remains relatively unchanged

(Laulagnier et al. 2004). Of note, sphingomyelin and cholesterol typically form tight packing in the lipid bilayer, which may contribute to the excellent rigidity and stability of exosomes (Kooijmans et al. 2012). In addition, the unique lipid composition of exosomes could increase their cellular internalization than conventional liposomes (Smyth et al. 2015). For instance, similar to envelope viruses, high level of raft-associated lipids may facilitate exosomes to fuse with target cells, which is beneficial for cytosolic delivery of their cargoes (Montecalvo et al. 2012).

Additionally, it has been reported that exosomes contain a considerable repertoire of RNAs, including miRNA, mRNA, and other non-coding RNAs (Valadi et al. 2007). In a pioneering work, Valadi et al. reported that mRNA transcripts could be functionally shuttled by exosomes between cells, leading to the alteration of recipient cells' phenotype (Valadi et al. 2007). Later on, it was found that DNA carried by exosomes could be functionally transferred between cells, revealing a novel mechanism for intercellular transfer of genetic materials (Cai et al. 2013). Exosomal miRNAs are reliable biomarkers for disease diagnosis due to the stability of miRNAs in exosomes and the abundance of exosomes in diverse body fluids (Salehi and Sharifi 2018). Together, considering exosomes are endogenous carriers of RNAs and possess a unique membrane composition, they could serve as promising delivery vehicles for therapeutic siRNA molecules.

LOADING EXOSOMES WITH siRNA

Despite the beneficial composition and properties of exosomes, loading these vesicles with siRNA in a robust and reproducible manner remains a challenging task in this field, primarily due to the relatively tight and ordered lipid bilayer of exosomes and the hydrophilic, macromolecular, and anionic characteristic of siRNA (Lu et al. 2017). Currently, the most commonly used approach to load siRNA into exosomes is electroporation, which mainly depends on creating transient pores on exosome membrane to promote the encapsulation of siRNA into the lumen of exosomes (Alvarez-Erviti et al. 2011). To date, several groups have reported proof-of-concept results for utilizing this approach to achieve gene silencing *in vivo* (Alvarez-Erviti et al. 2011; El-Andaloussi et al. 2012). Typically, the loading efficiency is not dependent on the sequence of siRNA, but the electroporation condition such as voltage and pulse time is required to be investigated to obtain the optimized loading efficiency. For example, El-Andaloussi et al. showed approximately 25% of the electroporated siRNA could be loaded into exosomes under the optimal electroporation condition (El-Andaloussi et al. 2012). However, publications have pointed out that there were some difficulties in the application of this approach, among which was the high degree of variability (Kooijmans et al. 2013; Wahlgren et al. 2012). Given that the electroporation equipment, buffers, and settings vary considerably among different laboratories, divergent results may be obtained. Besides, electroporation is often accompanied by extensive siRNA aggregate formation, which may result in an overestimation of siRNA loading efficiency (Wahlgren et al. 2012; Kooijmans et al. 2013). But, such effect can be ameliorated to some extent by altering the electroporation media, using cuvettes with polymer electrodes, or adding trehalose (Kooijmans et al. 2013). In addition to electroporation, transfection-based methods have also been used for loading exosomes with siRNA. In this approach, conventional transfection reagents are generally introduced to transfect the donor cells or the purified exosomes. However, a major challenge of this technique is the difficulty to completely remove contaminating transfection reagent from the loaded exosomes (Wahlgren et al. 2012). Additionally, some groups have attempted to hydrophobically modify siRNA before the loading process, where lipid (i.e., cholesterol) moieties were conjugated to the 3' end of the passenger strand of siRNA to allow the display of siRNA on exosome surface *via* insertion into the outer vesicular membrane (Didiot et al. 2016). In this approach, the interaction between lipid-conjugated siRNA and the lipid bilayer of exosomes is of great importance, which could be enhanced by optimizing experimental conditions such as ratio of lipid-conjugated siRNA to exosomes, and incubation temperature and time. For example, O'Loughlin et al. reported that incubation of 15 molecules of siRNA per exosome at 37°C for one hour could contribute to the highest retention of siRNA on exosome surface, which further promotes a concentration-dependent downregulation of human antigen R (*HuR*), a therapeutic target in cancer (O'Loughlin et al. 2017). In a study conducted by Stremersch and coworkers, cholesterol-conjugated siRNA was shown to bind efficiently on the surface of exosomes; however, exosomes containing cholesterol-conjugated siRNA were found to be less effective for functional siRNA delivery than anionic fusogenic liposomes (Stremersch et al. 2016). Therefore, a more in-depth understanding of the loading mechanism and details such as siRNA conjugation and incubation conditions is required before this approach could be regarded as a reliable loading approach. Together, these results highlight the necessity for establishing more robust and reliable protocols to prepare siRNA-loaded exosomes.

HARNESSING NATURAL EXOSOMES FOR siRNA DELIVERY

The innate delivery properties of exosomes have been widely exploited to enhance the therapeutic efficacy of siRNA. Overall, natural exosomes could be considered as a viable nanocarrier for siRNA, especially for the treatment of tumors and neurological diseases. A summary of siRNA delivery systems based on natural exosomes is shown in Table 41.1. Zhang et al., for example, reported the use of mouse fibroblast L929 cell-derived exosomes to deliver siRNA against transforming growth factor β1 (TGF-β1), a therapeutic target for tumor suppression. Exosomes were loaded with TGF-β1 siRNA by using transfection-based approaches. As a consequence, a successful inhibition of the growth and metastasis of murine sarcomas cells was observed. In accordance with *in vitro* results, exosomes containing TGF-β1 siRNA significantly attenuated the expression level of TGF-β1 in tumor tissue, resulting in effective inhibition of tumor growth and lung metastasis in mice (Zhang et al. 2014). In another study, Shtam et al. explored the capability of exosomes as carriers of siRNA specific to RAD51 and RAD52. After incubating with siRNA-loaded exosomes, the expression of RAD51 but not RAD52 proteins were significantly suppressed. Furthermore, the inhibition of RAD51 recombinase reduced the recruitment of RAD51 as a critical player of homologous

TABLE 41.1
Efforts to Harness Natural Exosomes for siRNA Delivery

siRNA Therapeutics	Exosome Origins	Disease Models	Therapeutic Outcomes	References
siRNA against TGF-β1	Mouse fibroblast L929 cells	Murine sarcomas tumor	Effective inhibition of tumor growth and lung metastasis in mice	Zhang et al. (2014)
siRNA against RAD51 and RAD52	Hela and HT1080 human fibosarcoma cells	Hela tumor cells	Significant knockdown of RAD51 expression and induction of massive cancer cell death *in vitro*	Shtam et al. (2013)
siRNA against luciferase	Primary endothelial cells	Luciferase-expressing endothelial cells	*In vitro* inhibition of luciferase expression in target cells	Banizs et al. (2014)
siRNA against PLK-1	HEK 293 cells and MSCs	Bladder cancer cells	Significant suppression of PLK-1 expression and cell proliferation *in vitro*	Greco et al. (2016)
siRNA against VEGF	Brain endothelial cells	Zebrafish neuronal glioblastoma-astrocytoma tumor	Carrying siRNA across the BBB, and significant knockdown of tumor growth in zebrafish brain	Yang et al. (2017)
siRNA against PTEN	MSCs	Mouse spinal cord injury	Significant inhibition of PTEN expression in injured spinal cord region; Promoting axonal growth and neovascularization	Guo et al. (2019)

Notes: TGF-β1, transforming growth factor β1; PLK-1, polo-like kinase-1; VEGF, vascular endothelial growth factor; PTEN, phosphatase and tensin homolog recombination at double-stranded breaks, leading to S/G2 cell cycle arrest and massive cell death (Shtam et al. 2013). Additional evidences were provided by Banizs et al., who suggested that endothelial exosomes containing siRNA against luciferase were capable of silencing the expression of luciferase with an efficiency higher than 40%, when compared with the scramble siRNA group (Banizs et al. 2014). Likewise, exosomes derived from human embryonic kidney (HEK) 293 and mesenchymal stem cells (MSCs) were explored for delivery of siRNA against polo-like kinase-1 (PLK-1) to bladder cancer cells. As a result, significant suppression of PLK-1 expression and inhibition of cell proliferation were observed (Greco et al. 2016). Taken together, siRNA delivered by exosomes can be considered as a promising strategy to suppress various aspects of tumor progression.

Apart from tumors, siRNA therapeutics delivered by natural exosomes have also been employed for therapy of neurological diseases. For instance, Yang et al. recently harnessed the intrinsic capacity of exosomes to surmount the blood–brain barrier (BBB) for siRNA delivery. The authors found that siRNA targeting vascular endothelial growth factor (VEGF) was efficiently delivered by exosomes derived from brain endothelial cells to cross the BBB and inhibited the growth of xenograft neuronal glioblastoma-astrocytoma tumor in zebrafish model. The study implied that brain endothelial exosomes were viable carriers of siRNA to overcome the BBB and treat neurological disorders (Yang et al. 2017). More recently, Guo et al. confirmed that exosomes derived from MSCs were capable of crossing the BBB and migrating to the injured spinal cord area in rats after intranasal administration. Moreover, exosomes containing siRNA specific to phosphatase and tensin homolog (PTEN) could significantly suppress the expression of PTEN in the injured spinal cord region, leading to enhancement of axonal growth and neovascularization while reducing microgliosis and astrogliosis. Overall, these results indicated that intranasal administration of MSC-derived exosomes containing anti-PTEN siRNA (ExoPTEN) could elicit functional recovery in rats with complete spinal cord injury (SCI), suggesting the great potential of ExoPTEN for therapy of SCI in the clinic (Guo et al. 2019). Collectively, exosomes hold great potential to serve as siRNA carriers for downregulation of specific genes and treatment of neurological diseases, owing to their distinct traits such as high efficiency to deliver siRNA crossing biological membranes (i.e., the BBB).

MODIFIED EXOSOME FOR TARGETED siRNA DELIVERY

Targeting delivery has long been sought by researchers to enhance the therapeutic efficacy and reduce systemic toxicity of drugs. It has been reported that exosomes secreted by natural cells possess intrinsic targeting ability to some extent (Vader et al. 2016). To be exemplified, exosomes expressing Tspan8 could recognize CD11b- and CD54-positive cells through receptor–ligand interactions (Rana et al. 2012). However, the intrinsic targeting ability of native exosomes is generally thought to be insufficient for therapeutic application. Therefore, functionalization of exosomes with targeting moieties to endow them with better specific cell recognition has attracted substantial attention of researchers (Wang, Zheng, and Zhao 2016). Typically, exosome-modification techniques can be categorized into cell engineering and exosome engineering, as illustrated in Figure 41.2 (Lu, Xing, Xun, et al. 2018a). Cell engineering approach takes advantage of the biosynthesis mechanism of exosomes to obtain modified exosomes. Given that molecules on cell membrane are naturally incorporated in exosome membrane through the MVB pathway, modified exosomes may be designed by engineering the donor cells to express candidate proteins or peptides fused with exosome-enriched membrane proteins, such as lysosome-associated membrane glycoprotein 2b (Lamp2b), platelet-derived growth factor receptor (PDGFR), and tetraspanins (i.e., CD9, CD63) (Stickney et al. 2016). Genetic engineering of producer cells helps to preserve the structural and functional integrity of proteins or peptides. However, some concerns have been raised about this approach, such as the stability of Lamp2b-fused ligands, as proteolytic degradation of the fused proteins or peptides may occur in the endosomal environment before being sorted in exosomes (Hung and Leonard 2015). Other issues include the relatively complex decoration process, variable performance, and inapplicability to all proteins/peptides (Armstrong, Holme, and Stevens 2017; Zhao et al. 2016).

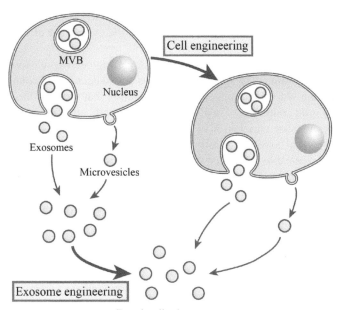

FIGURE 41.2. Schematic illustration of exosome-modification strategies. Functionalized exosomes can be generated by using "cell engineering", in which parental cells are genetically engineered to express functional ligands and allowed to secrete modified exosomes. Alternatively, exosomes can be directly functionalized through "exosome engineering". Adapted with permission from (Lu, Xing, Xun, et al. 2018c). Copyright Elsevier, 2018.

In a proof-of-concept study by Alvarez-Erviti et al., dendritic cells were engineered to express rabies virus glycoprotein (RVG) and Lamp2b fusion protein and allowed to secrete modified exosomes with RVG peptide on their surface. These functionalized exosomes containing siRNA could specifically recognize acetylcholine receptors on neuronal cells, causing an effective knockdown of BACE1 gene (a therapeutic target of Alzheimer's disease) in neurons, microglia, and oligodendrocytes with efficiency of about 60%. In addition, exposure to RVG-modified exosomes did not lead to non-specific uptake in other tissues (Alvarez-Erviti et al. 2011). Later on, the same RVG display technique was utilized in an array of studies and similar results were obtained. In one of the studies, Liu et al. utilized exosomes with RVG functionalization to deliver siRNA against opioid receptor for morphine addiction treatment. Following intravenous injection, the engineered exosomes dramatically promoted the delivery of siRNA to the central nervous system of treated mice and specifically downregulated the expression level of opioid receptor (Liu et al. 2015). In another study, Cooper and colleagues suggested RVG-modified exosomes could induce significant reduction of alpha-synuclein (α-Syn) aggregates, a characteristic pathological feature of Parkinson's disease, in the brain of treated mouse, including dopaminergic neurons of the substantia nigra (Cooper et al. 2014). Additionally, Izco et al. designed shRNA minicircles against α-Syn to prolong the effectiveness of siRNA and used RVG-functionalized exosomes as the vector for targeted delivery of siRNA to the brain. Similar results were obtained, including long-term downregulation of α-Syn aggregation in the brain, attenuated loss of dopaminergic neurons in the substantia nigra pars compacta, and improved pathological symptoms (Izco et al. 2019). Therefore, these results collectively suggested RVG-modified exosomes could serve as a powerful siRNA delivery vehicle for neurological disease treatment. In addition to RVG peptide, recently, Limoni et al. engineered HEK 293T cells to express designed ankyrin repeat protein (DARPin G3) which had high affinity to human epidermal growth factor receptor (HER-2), allowing the engineered cells to produce exosomes with DARPin

G3 peptide on their surface. As a result, exosomes with DARPin G3 could specifically bind to HER2 overexpressing breast cancer cells and effectively silence the expression of target gene by specific delivery of siRNA into these cells (Limoni et al. 2019).

Alternatively, surface functionalization could be by direct engineering of exosomes, which is generally achieved through covalent conjugation or noncovalent association. In conjugation approaches, targeting ligands are introduced through conjugation reaction that directly occurs on exosome surface under relatively mild conditions (Smyth et al. 2014). Specifically, functional ligands are generally linked to amines that are enriched in exosome membrane by standard amide chemistry (Aqil et al. 2019). Provided that exosomes are nonliving entities, some chemical reagents or reaction conditions that are inapplicable to cells could be used in exosome engineering. In the case of noncovalent exosome engineering strategies, targeting ligands are previously conjugated with lipids to form lipid-conjugated moieties which could spontaneously insert into exosome membrane through hydrophobic interaction. Compared with covalent conjugation, noncovalent strategies involve more mild reaction conditions and are applicable to more types of functional ligands (Kooijmans et al. 2013). Together, direct exosome engineering technique appears to be more efficient and controllable and has general applicability to a wide array of proteins and peptides, as compared to cell engineering approach (Armstrong, Holme, and Stevens 2017).

Folic acid (FA) is a widely used targeting moiety to functionalize nanocarriers with the capacity of specifically recognizing folate receptor that is overexpressed on the surface of many cancer cells (Parker et al. 2005). Recently, Aqil et al. covalently attached FA to milk exosomes through standard stable amide chemistry. FA-functionalized exosomes carrying siRNA against KRAS gene showed efficient anti-proliferative effects on A549 cells and significantly inhibited the growth of lung tumor xenograft in nude mouse model after being administered intravenously (Aqil et al. 2019). In a recent study, Pi et al. reported the application of RNA nanotechnology to display FA on exosome surface. As pointed by the authors, the orientation of arrow-shaped RNA could be modulated to control ligand display on exosome surface for specific targeting or to alter the intracellular trafficking of small RNAs. Specifically, placing cholesterol at the tail of the arrow allowed for displaying of FA on the outer surface of exosomes. Building off these observations, the authors conjugated cholesterol to the arrowtail of the three-way junction of the bacteriophage phi29 motor packaging RNA (pRNA-3WJ) for anchoring 3WJ on exosome surface. Targeting ligands (i.e., FA) were further conjugated to the 3WJ to display them on exosome surface. The generated FA-displaying exosomes were used to deliver anti-survivin siRNA and showed efficient targeting capacity, siRNA delivery and cancer inhibition in patient-derived colorectal cancer xenograft model (Pi et al. 2018). The mechanism underlying the high potency of FA-displaying exosomes constructed by RNA nanotechnology was elucidated by Zheng et al. in a following study. By using fluorescence microscopic technique, the authors found that FA-displaying exosomes could fuse with cell membrane following binding to the specific folate receptors, thus contributing to a direct cytosolic release of siRNA without endosome trapping (Zheng et al. 2019). These results suggest that RNA nanotechnology holds great promise to display targeting moieties on exosome surface for targeted siRNA delivery. Efforts to modify exosomes for targeted delivery of siRNA are summarized in Table 41.2.

DESIGNING BIOINSPIRED EXOSOME-LIKE NANOPARTICLES FOR siRNA DELIVERY

Although exosomes represent promising candidate for siRNA delivery, their potential for clinical applications is tempered by some challenges. One of the major hurdles is the lack of standard, scalable, and cost-effective approach to isolate sufficient quantities of exosomes with desirable purity for clinical application (Lener et al. 2015). Besides, efficient encapsulation of exogenous siRNA into exosomes remains another impediment that substantially constrains the development of exosome-based therapeutics (Lu et al. 2017). Additionally, it is worth noting that exosomes contain a complex structure and composition, which is difficult for thorough characterization to ensure

TABLE 41.2

A Summary of Exosome Modifications for Targeted siRNA Delivery

siRNA Therapeutics	Exosome Origin	Targeting Ligands	Modification Approaches	Therapeutic Outcomes	References
siRNA against BACE1	Murine dendritic cells	RVG peptide	Cell engineering	Strong knockdown of BACE1 expression in mouse Alzheimer's disease model	Alvarez-Erviti et al. (2011)
siRNA against opioid receptor	HEK 293T cells	RVG peptide	Cell engineering	Efficient delivery of siRNA to mouse brain; Downregulation of MOR expression and inhibition of morphine relapse	Liu et al. (2015)
siRNA against α-Syn	Murine dendritic cells	RVG peptide	Cell engineering	Significantly reducing the expression of α-Syn aggregates in mouse Parkinson's disease model	Cooper et al. (2014)
shRNA minicircles against α-Syn	Murine dendritic cells	RVG peptide	Cell engineering	Long-term downregulation of α-Syn aggregation in the brain; Reduced loss of dopaminergic neurons and improved pathological symptoms	Izco et al. (2019)
siRNA against TPD52	HEK293 cells	DARPin G3 peptide	Cell engineering	Efficient downregulation of target gene expression with efficiency of 70% *in vitro*	Limoni et al. (2019)
siRNA against KRAS	Bovine milk	Folic acid	Covalent conjugation	Significant inhibition of lung tumor xenograft in mouse model	Aqil et al. (2019)
siRNA targeting survivin	HEK293T cells	Folic acid	RNA nanotechnology	Efficient cancer inhibition in patient-derived colorectal cancer xenograft model	Pi et al. (2018)
siRNA targeting survivin	HEK293T cells	Folic acid	RNA nanotechnology	Able to fuse with target cells and avoid endosomal trapping; Potent silencing efficacy	Zheng et al. (2019)

Notes: RVG, rabies virus glycoprotein; α-Syn, alpha-synuclein; DARPin, designed ankyrin repeat protein

reproducibility and safety from a pharmaceutical point of view (Kooijmans et al. 2012). In contrast, synthetic siRNA delivery vehicles (i.e., liposomes and polymers) are currently more developed. Actually, synthetic nanocarriers possess some advantages such as scalability, efficient drug loading, well-characterized composition, controllable manipulation, and facile surface modification (Johnsen et al. 2018). In this circumstance, a positive feedback could be facilitated between the two fields to construct bioinspired exosome-like delivery systems, where superiority from exosomes regarding unique composition beneficial for efficient delivery, circulation, and targeting could be combined with the delivery capability of synthetic nanocarriers. In recent years, construction of exosome-inspired nanovectors for siRNA delivery has attracted extensive interest of researchers. According to two trends in biomimetic strategies, bioinspired exosome-like nanoparticles can be primarily categorized into four types, including fully artificial exosome-like nanoparticles fabricated by bottom-up biomimicry methodologies, physical origin nanovesicles (also termed as cell-derived nanovesicles) generated through top-down biomimicry methodologies, and hybrid exosome-like nanovesicles as well as exosome membrane-camouflaged nanoparticles based on combined bottom-up and top-down biomimicry methodologies (Figure 41.3) (Lu and Huang 2020).

Provided that not all components in exosomes are essential for their proper functionality, fully artificial exosome-like nanoparticles could be fabricated by assembling only crucial components in natural exosomes. Therefore, these nanovectors have lower complexity, which helps to increase

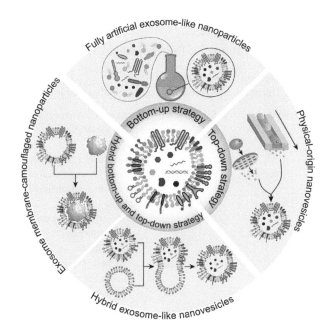

FIGURE 41.3. Schematic illustrating some typical types of bioinspired exosome-like nanoparticles, including fully artificial exosome-like nanoparticles, physical-origin nanovesicles, hybrid exosome-like nanovesicles, and exosome membrane camouflaged-nanoparticles. Exosome-like nanoparticles can be constructed by bottom-up, top-down, and hybrid bottom-up and top-down biomimicry strategies. Reproduced with permission from (Lu and Huang 2020). Copyright Elsevier, 2020.

the pharmaceutical acceptability of exosomes (Kooijmans et al. 2012). By using thin-film hydration approach commonly used for liposome preparation, our group assembled exosome-mimetic liposomes to mimic the lipid composition of natural exosomes. The ability of exosome-mimetic liposomes and conventional liposomes to deliver siRNA against VEGF was assessed in two cell lines, namely adenocarcinomic human alveolar basal epithelial A549 cells and endothelial HUVEC cells. The results showed that the silencing efficiency of exosome-mimetic liposomes was improved to a significant degree (> three-fold) by mimicking the unique lipid composition of exosomes, but it was still far from adequate for efficient siRNA delivery when compared with cationic liposomes (Lu, Zhao, Xing et al. 2018d). To further enhance the delivery efficiency of siRNA, we constructed connexin-43-embedded, exosome-mimetic nanoparticles in a following study. Chitosan nanoparticles (CS NPs) were introduced to enhance the loading ability of siRNA, which were then coated by exosome-mimicking lipid bilayers to obtain L/CS NPs. Afterwards, a cytosolic delivery helper protein Cx43 enriched in natural exosomes was expressed *in vitro* and integrated in the lipid bilayer of L/CS NPs to form Cx43/L/CS NPs by using cell-free protein synthesis technique (Lu, Zhao, Xing et al. 2018d). As a result, the *de novo* synthesized Cx43 was able to facilitate the cytosolic delivery of anti-VEGF siRNA by forming gap junction with Cx43-expressing recipient cells, thus contributing to an enhancement of siRNA silencing effect with efficiency comparable to that of chemically loaded exosomes. However, the delivery efficiency of exosome-like Cx43/L/CS NPs was still significantly lower than that of commercial Lipofectamine 2000, which was possibly due to the inadequacy of bottom-up biomimicry approach to reproduce the complex lipid and protein makeup of natural exosomes (Lu et al. 2019).

As an effort to overcome the low production of exosomes, physical-origin nanovesicles are generated by crushing cell membranes into small membrane fragments which then self-assemble into nanovesicles. Therefore, these exosome-like vesicles are also called cell-derived nanovesicles. One of the most fascinating means to obtain bioinspired exosome-like nanovesicles is membrane extrusion, where cultured cells are consecutively extruded through a series of polycarbonate membrane filters with diminishing pore sizes followed by density gradient ultracentrifugation (Jang et al. 2013). The resulted nanovesicles may share some similarities with natural exosomes, such as physicochemical properties, surface markers, *in vitro* and *in vivo* behaviors, and therapeutic efficacy (Goh et al. 2017; Nasiri Kenari et al. 2019; Jang et al. 2013). However, the production yield of nanovesicles could be more than 100-fold greater than that of exosomes secreted from the same number of cells (Jo et al. 2014). To be exemplified, in a study by Yang et al., exosome-mimetic nanovesicles were generated by serial extrusion of epithelial MCF-10A cells through membrane filters. The production yield is around 150 folds of spontaneous exosomes. Moreover, these nanovesicles were harnessed for delivery of siRNA specific to cyclin-dependent kinase 4 (CDK4), a crucial factor for cell cycle progression in many tumors. The authors suggested that regardless of whether secreted by cells naturally or formed by extrusion artificially, these nanovesicles possessed similar surface protein markers and were efficient and safe for siRNA delivery to tumor cells both *in vitro* and *in vivo*. As a consequence, a significant downregulation of CDK4 expression and inhibition of tumor growth with similar efficiency was observed for siRNA-loaded naovesicles and exosomes. All these results implied that physical-origin exosome-like nanovesicles could serve as a promising nanovector for siRNA delivery (Yang et al. 2016).

Both hybrid exosome-like nanovesicles and exosome membrane-camouflaged nanoparticles are developed through hybrid bottom-up and top-down biomimicry strategies in order to capitalize on the intrinsic functionality of exosomes and the siRNA carrying capability of synthetic nanocarriers. Composing a lipid bilayer integrated with exosomal proteins, hybrid exosome-like nanovesicles are commonly constructed by membrane fusion between liposomes and exosomes driven by techniques such as extrusion. For example, Jhan et al. recently produced hybrid exosome-like nanovesicles by fusing exosome membrane with a library of cationic, zwitterionic, and anionic liposomes. Hybridization between cationic liposomes and exosomes increased the knockdown efficiency of about 49%, which was comparable to commercial Lipofectamine RNAiMax. Besides, fusion exosomes with zwitterionic or anionic liposomes could endow them with downregulation effect, as simple siRNA complexes with these liposomes were unable to induce gene silencing. Noteworthy, the number of vesicles increased by 6 to 43 folds after membrane fusion *via* extrusion. These results showed that hybrid exosome-like nanovesicles could be a potential siRNA carrier with scalability to a certain degree while combining the beneficial features of exosomes and liposomes (Jhan et al. 2020). Exosome membrane-camouflaged nanoparticles are generally formulated by extrusion or sonication approaches. Attachment of exosome membrane onto the surface of inner cores allows the transfer of exosomes' beneficial delivery properties to the synthetic inner cores (Cheng et al. 2018). As an example, Zhao et al. recently used exosome membrane from autologous breast cancer cells to encapsulate cationic bovine serum albumin (CBSA) containing siRNA against S100A4, an important protein contributing to tumor metastasis by using extrusion approach. The generated biomimetic CBSA/siS100A4@exosome nanoparticles could protect siRNA from degradation, exhibited good biocompatibility, higher affinity to lung cancer, and thus efficient gene silencing and pulmonary metastasis suppression in mouse model (Zhao et al. 2020). In another study, Zhupanyn et al. prepared exosome membrane-modified polyethylenimine (PEI)/siRNA complexes by sonication. They found that exosome membrane modification enhanced the physicochemical properties (i.e., storage stability) and biological functionalities (i.e., circulation stability, targeting capacity, gene knockdown, and therapeutic efficacy) of PEI/siRNA nanocomplexes. Of note, the authors suggested there was a marked difference between the source of exosome membrane regarding the capacity to enhance silencing efficacy, and the difference was independent of target cells (Zhupanyn et al. 2020). Taken together, both hybrid exosome-like nanovesicles and exosome membrane-camouflaged nanoplatforms are appealing delivery vehicles for siRNA therapeutics. Efforts to design bioinspired exosome-like nanoparticles for siRNA are summarized in Table 41.3.

TABLE 41.3

Efforts to Construct Bioinspired Exosome-Like Nanovectors for siRNA Delivery

siRNA Therapeutics	Exosome-like Nanoparticles	Exosome/Nanovesicle Origins	Therapeutic Outcomes	References
siRNA against VEGF	Exosome-like liposomes	HEK293T cells	Significant enhanced uptake and silencing efficiency	Lu, Zhao, Xing et al. (2018d)
siRNA against VEGF	Cx43-integrated, lipid bilayer-coated nanoparticles	HEK293T cells	Delivery efficiency comparable to chemically loaded exosomes	Lu et al. (2019)
siRNA against CDK4	Physical-origin nanovesicles	Epithelial MCF-10A cells	Similar downregulation of CDK4 expression and inhibition of tumor growth for exosome-mimetic naovesicles and native exosomes	Yang et al. (2016)
siRNA against green fluorescent protein	Hybrid exosome-like nanovesicles fused with liposomes	Mouse fibroblast NIH 3T3 and human lung carcinoma A549 cells	Enhanced knockdown efficiency and vesicle yield to a greater degree	Jhan et al. (2020)
siRNA against S100A4	Exosome membrane-camouflaged CBSA nanoparticles	Autologous breast cancer cells	Possessing enhanced biocompatibility and targeting ability; Efficient inhibition of tumor growth and metastasis in mice	Zhao et al. (2020)
siRNA against survivin	Exosome membrane-modified PEI/siRNA nanocomplexes	Ovarian carcinoma SKOV3, prostate carcinoma PC3, prostate carcinoma HCT-116, and osteosarcoma Saos-2 cells	Enhancing the physicochemical properties and biological activities of PEI/siRNA nanocomplexes	Zhupanyn et al. (2020)

Notes: VEGF, vascular endothelial growth factor; CDK4, cyclin-dependent kinase 4; CBSA, cationic bovine serum albumin

CONCLUSION

Exosomes have been extensively tailored as promising siRNA delivery vehicles since the discovery that they have the intrinsic ability to traverse biological barriers and functionally transport RNAs between cells. In addition, to harness exosomes for targeted siRNA delivery, the intrinsic targeting properties of exosomes could be further tuned with targeting moieties through cell engineering or direct exosome engineering. Displaying RVG, FA or other homing ligands on exosome surface has been proven able to endow exosomes with greater accumulation at target sites and enhanced therapeutic efficacy. Furthermore, by borrowing a leaf from nature's book, biomimetic exosome-like nanoparticles have been designed to exploit the beneficial features of exosomes, while circumventing some of their drawbacks. Fabrication of biomimetic nanoparticles harboring only crucial components of native exosomes leads to the formation of fully artificial exosome-like nanoparticles with defined composition, facile characterization, and better pharmaceutical acceptability. Generation of physical-origin nanovesicles is an effective strategy to enhance the production yield of exosomes. Moreover, construction of hybrid exosome-like nanovesicles and exosome membrane-camouflaged nanoparticles can improve the loading capacity, delivery efficiency, and therapeutic efficacy of exosomes. However, for better harnessing exosome-based nanocarriers for siRNA delivery, a lot of issues still remain open and need to be explored more in detail: (I) complete understanding of exosome biogenesis, composition, biological functionality, and targeting, (II) thoroughly exploring the immunogenicity, *in vivo* behaviors as well as potential safety concerns of exosomes and exosome-like nanoparticles, (III) strengthening regulatory requirements in production and quality control processes. Advance in these aspects could contribute to further improvements in the development of exosomes and exosome-inspired nanovehicles for siRNA delivery.

ACKNOWLEDGMENTS

This work was supported by the Beijing Natural Science Foundation (7214283, 7214302), the National Natural Science Foundation of China (32101157, 82104105, 31871003, 32171394, 32001008), China Postdoctoral Science Foundation (2020M670169), Beijing-Tianjin-Hebei Basic Research Cooperation Project (19JCZDJC64100), the National Key R & D Program of China (2021YFE0106900, 2021YFA1201000, 2021YFC2302400), and the Natural Science Foundation of Guangdong Province (2019A1515010776). M. L. designed the project and wrote the chapter. H. X. involved in information gathering and figure preparation. Y. H. led the project and gave important advice.

REFERENCES

Alvarez-Erviti, L., Y. Seow, H. Yin, C. Betts, S. Lakhal, and M. J. Wood. 2011. Delivery of siRNA to the mouse brain by systemic injection of targeted exosomes. *Nat Biotechnol* 29 (4):341–345.

Antimisiaris, S., S. Mourtas, and K. Papadia. 2017. Targeted si-RNA with liposomes and exosomes (extracellular vesicles): How to unlock the potential. *Int J Pharm* 525 (2):293–312.

Aqil, F., R. Munagala, J. Jeyabalan, et al. 2019. Milk exosomes – natural nanoparticles for siRNA delivery. *Cancer Lett* 449:186–195.

Arenaccio, C., C. Chiozzini, F. Ferrantelli, P. Leone, E. Olivetta, and M. Federico. 2019. Exosomes in therapy: engineering, pharmacokinetics and future applications. *Curr Drug Targets* 20 (1):87–95.

Armstrong, J. P., M. N. Holme, and M. M. Stevens. 2017. Re-engineering extracellular vesicles as smart nanoscale therapeutics. *ACS Nano* 11 (1):69–83.

Banizs, A. B., T. Huang, K. Dryden, et al. 2014. In vitro evaluation of endothelial exosomes as carriers for small interfering ribonucleic acid delivery. *Int J Nanomedicine* 9:4223–4230.

van den Boorn, J. G., M. Schlee, C. Coch, and G. Hartmann. 2011. SiRNA delivery with exosome nanoparticles. *Nat Biotechnol* 29 (4):325–326.

Cai, J., Y. Han, H. Ren, et al. 2013. Extracellular vesicle-mediated transfer of donor genomic DNA to recipient cells is a novel mechanism for genetic influence between cells. *J Mol Cell Biol* 5 (4):227–238.

Castanotto, D., and J. J. Rossi. 2009. The promises and pitfalls of RNA-interference-based therapeutics. *Nature* 457 (7228):426–433.

Cheng, G., W. Li, L. Ha, et al. 2018. Self-Assembly of Extracellular Vesicle-like Metal-Organic Framework Nanoparticles for Protection and Intracellular Delivery of Biofunctional Proteins. *J Am Chem Soc* 140 (23):7282–7291.

Colombo, M., G. Raposo, and C. Thery. 2014. Biogenesis, secretion, and intercellular interactions of exosomes and other extracellular vesicles. *Annu Rev Cell Dev Biol* 30:255–289.

Conde-Vancells, J., E. Rodriguez-Suarez, N. Embade, et al. 2008. Characterization and comprehensive proteome profiling of exosomes secreted by hepatocytes. *J Proteome Res* 7 (12):5157–5166.

Cooper, J. M., P. B. Wiklander, J. Z. Nordin, et al. 2014. Systemic exosomal siRNA delivery reduced alpha-synuclein aggregates in brains of transgenic mice. *Mov Disord* 29 (12):1476–1485.

Didiot, M. C., L. M. Hall, A. H. Coles, et al. 2016. Exosome-mediated delivery of hydrophobically modified siRNA for Huntingtin mRNA silencing. *Mol Ther* 24 (10):1836–1847.

El-Andaloussi, S., Y. Lee, S. Lakhal-Littleton, et al. 2012. Exosome-mediated delivery of siRNA in vitro and in vivo. *Nat Protoc* 7 (12):2112–2126.

Fire, A., S. Xu, M. K. Montgomery, S. A. Kostas, S. E. Driver, and C. C. Mello. 1998. Potent and specific genetic interference by double-stranded RNA in Caenorhabditis elegans. *Nature* 391 (6669):806–811.

Gilleron, J., W. Querbes, A. Zeigerer, et al. 2013. Image-based analysis of lipid nanoparticle-mediated siRNA delivery, intracellular trafficking and endosomal escape. *Nat Biotechnol* 31 (7):638–646.

Goh, W. J., S. Zou, W. Y. Ong, et al. 2017. Bioinspired cell-derived nanovesicles versus exosomes as drug delivery systems: A cost-effective alternative. *Sci Rep* 7 (1):14322.

Greco, K. A., C. A. Franzen, K. E. Foreman, R. C. Flanigan, P. C. Kuo, and G. N. Gupta. 2016. PLK-1 silencing in bladder cancer by siRNA delivered with exosomes. *Urology* 91:241.e1–241.e7.

Guo, S., N. Perets, O. Betzer, et al. 2019. Intranasal delivery of mesenchymal stem cell derived exosomes loaded with phosphatase and tensin homolog siRNA repairs complete spinal cord injury. *ACS Nano* 13 (9):10015–10028.

Hu, B., Y. Weng, X. H. Xia, X. J. Liang, and Y. Huang. 2019. Clinical advances of siRNA therapeutics. *J Gene Med* 21 (7):e3097.

Huang, Y., Q. Cheng, J. L. Ji, et al. 2016. Pharmacokinetic Behaviors of Intravenously Administered siRNA in Glandular Tissues. *Theranostics* 6 (10):1528–1541.

Huang, Y., J. Hong, S. Zheng, et al. 2011. Elimination pathways of systemically delivered siRNA. *Mol Ther* 19 (2):381–385.

Hung, M. E., and J. N. Leonard. 2015. Stabilization of exosome-targeting peptides via engineered glycosylation. *J Biol Chem* 290 (13):8166–8172.

Izco, M., J. Blesa, M. Schleef, et al. 2019. Systemic exosomal delivery of shRNA minicircles prevents Parkinsonian pathology. *Mol Ther* 27 (12):2111–2122.

Jang, S. C., O. Y. Kim, C. M. Yoon, et al. 2013. Bioinspired exosome-mimetic nanovesicles for targeted delivery of chemotherapeutics to malignant tumors. *ACS Nano* 7 (9):7698–7710.

Jhan, Y. Y., D. Prasca-Chamorro, G. Palou Zuniga, et al. 2020. Engineered extracellular vesicles with synthetic lipids via membrane fusion to establish efficient gene delivery. *Int J Pharm* 573:118802.

Jo, W., J. Kim, J. Yoon, et al. 2014. Large-scale generation of cell-derived nanovesicles. *Nanoscale* 6 (20):12056–12064.

Johnsen, K. B., J. M. Gudbergsson, M. Duroux, T. Moos, T. L. Andresen, and J. B. Simonsen. 2018. On the use of liposome controls in studies investigating the clinical potential of extracellular vesicle-based drug delivery systems - A commentary. *J Control Release* 269:10–14.

Kamerkar, S., V. S. LeBleu, H. Sugimoto, et al. 2017. Exosomes facilitate therapeutic targeting of oncogenic KRAS in pancreatic cancer. *Nature* 546 (7659):498–503.

Kooijmans, S. A. A., S. Stremersch, K. Braeckmans, et al. 2013. Electroporation-induced siRNA precipitation obscures the efficiency of siRNA loading into extracellular vesicles. *J Control Release* 172 (1):229–238.

Kooijmans, S. A., P. Vader, S. M. van Dommelen, W. W. van Solinge, and R. M. Schiffelers. 2012. Exosome mimetics: A novel class of drug delivery systems. *Int J Nanomedicine* 7:1525–1541.

Kowal, Joanna, Guillaume Arras, Marina Colombo, et al. 2016. Proteomic comparison defines novel markers to characterize heterogeneous populations of extracellular vesicle subtypes. *Proc Natl Acad Sci USA* 113:E968–E977.

Laulagnier, K., C. Motta, S. Hamdi, et al. 2004. Mast cell- and dendritic cell-derived exosomes display a specific lipid composition and an unusual membrane organization. *Biochem J* 380 (Pt 1):161–171.

Lener, T., M. Gimona, L. Aigner, et al. 2015. Applying extracellular vesicles based therapeutics in clinical trials - an ISEV position paper. *J Extracell Vesicles* 4:30087.

Limoni, S. K., M. F. Moghadam, S. M. Moazzeni, H. Gomari, and F. Salimi. 2019. Engineered exosomes for targeted transfer of siRNA to HER2 positive breast cancer cells. *Appl Biochem Biotechnol* 187 (1):352–364.

Liu, Y., D. Li, Z. Liu, et al. 2015. Targeted exosome-mediated delivery of opioid receptor Mu siRNA for the treatment of morphine relapse. *Sci Rep* 5:17543.

Llorente, A., T. Skotland, T. Sylvanne, et al. 2013. Molecular lipidomics of exosomes released by PC-3 prostate cancer cells. *Biochim Biophys Acta* 1831 (7):1302–1309.

Lu, M., and Y. Huang. 2020. Bioinspired exosome-like therapeutics and delivery nanoplatforms. *Biomaterials* 242:119925.

Lu, M., H. Xing, Z. Xun, et al. 2018a. Functionalized extracellular vesicles as advanced therapeutic nanodelivery systems. *Eur J Pharm Sci* 121:34–46.

Lu, Mei, Haonan Xing, Zhe Xun, et al. 2018c. Exosome-based small RNA delivery: Progress and prospects. *Asian J Pharm Sci* 13 (1):1–11.

Lu, Mei, Haonan Xing, Zhen Yang, et al. 2017. Recent advances on extracellular vesicles in therapeutic delivery: Challenges, solutions, and opportunities. *Eur J Pharm Biopharm* 119:381–395.

Lu, M., X. Zhao, H. Xing, et al. 2018b. Liposome-chaperoned cell-free synthesis for the design of proteoliposomes: Implications for therapeutic delivery. *Acta Biomater* 76:1–20.

Lu, M., X. Zhao, H. Xing, et al. 2018d. Comparison of exosome-mimicking liposomes with conventional liposomes for intracellular delivery of siRNA. *Int J Pharm* 550 (1–2):100–113.

Lu, M., X. Zhao, H. Xing, et al. 2019. Cell-free synthesis of connexin 43-integrated exosome-mimetic nanoparticles for siRNA delivery. *Acta Biomater* 96:517–536.

Montecalvo, A., A. T. Larregina, W. J. Shufesky, et al. 2012. Mechanism of transfer of functional microRNAs between mouse dendritic cells via exosomes. *Blood* 119 (3):756–766.

Naito, Y., and K. Ui-Tei. 2013. Designing functional siRNA with reduced off-target effects. *Methods Mol Biol* 942:57–68.

Nasiri Kenari, A., K. Kastaniegaard, D. W. Greening, et al. 2019. Proteomic and post-translational modification profiling of exosome-mimetic nanovesicles compared to exosomes. *Proteomics* 19 (8):e1800161.

O'Loughlin, A. J., I. Mager, O. G. de Jong, et al. 2017. Functional Delivery of Lipid-Conjugated siRNA by Extracellular Vesicles. *Mol Ther* 25 (7):1580–1587.

Parker, N., M. J. Turk, E. Westrick, J. D. Lewis, P. S. Low, and C. P. Leamon. 2005. Folate receptor expression in carcinomas and normal tissues determined by a quantitative radioligand binding assay. *Anal Biochem* 338 (2):284–293.

Parolini, I., C. Federici, C. Raggi, et al. 2009. Microenvironmental pH is a key factor for exosome traffic in tumor cells. *J Biol Chem* 284 (49):34211–34222.

Pi, F., D. W. Binzel, T. J. Lee, et al. 2018. Nanoparticle orientation to control RNA loading and ligand display on extracellular vesicles for cancer regression. *Nat Nanotechnol* 13 (1):82–89.

Piffoux, Max, Alba Nicolás-Boluda, Vladmir Mulens-Arias, et al. 2019. Extracellular vesicles for personalized medicine: The input of physically triggered production, loading and theranostic properties. *Adv Drug Deliv Rev* 138:247–258.

Rana, S., S. Yue, D. Stadel, and M. Zoller. 2012. Toward tailored exosomes: the exosomal tetraspanin web contributes to target cell selection. *Int J Biochem Cell Biol* 44 (9):1574–1584.

Raposo, G., and W. Stoorvogel. 2013. Extracellular vesicles: exosomes, microvesicles, and friends. *J Cell Biol* 200 (4):373–383.

Salehi, Mahsa, and Mohammadreza Sharifi. 2018. Exosomal miRNAs as novel cancer biomarkers: Challenges and opportunities. *J Cell Physiol* 233 (9):6370–6380.

Setten, R. L., J. J. Rossi, and S. P. Han. 2019. The current state and future directions of RNAi-based therapeutics. *Nat Rev Drug Discov* 18 (6):421–446.

Shtam, T. A., R. A. Kovalev, E. Y. Varfolomeeva, E. M. Makarov, Y. V. Kil, and M. V. Filatov. 2013. Exosomes are natural carriers of exogenous siRNA to human cells in vitro. *Cell Commun Signal* 11:88.

Smyth, T., M. Kullberg, N. Malik, P. Smith-Jones, M. W. Graner, and T. J. Anchordoquy. 2015. Biodistribution and delivery efficiency of unmodified tumor-derived exosomes. *J Control Release* 199:145–155.

Smyth, T., K. Petrova, N. M. Payton, et al. 2014. Surface functionalization of exosomes using click chemistry. *Bioconjug Chem* 25 (10):1777–1784.

Soares, A. R., T. Martins-Marques, T. Ribeiro-Rodrigues, et al. 2015. Gap junctional protein Cx43 is involved in the communication between extracellular vesicles and mammalian cells. *Sci Rep* 5:13243.

Stickney, Z., J. Losacco, S. McDevitt, Z. Zhang, and B. Lu. 2016. Development of exosome surface display technology in living human cells. *Biochem Biophys Res Commun* 472 (1):53–59.

Stremersch, S., R. E. Vandenbroucke, E. Van Wonterghem, A. Hendrix, S. C. De Smedt, and K. Raemdonck. 2016. Comparing exosome-like vesicles with liposomes for the functional cellular delivery of small RNAs. *J Control Release* 232:51–61.

Subra, C., K. Laulagnier, B. Perret, and M. Record. 2007. Exosome lipidomics unravels lipid sorting at the level of multivesicular bodies. *Biochimie* 89 (2):205–212.

Thery, C., M. Ostrowski, and E. Segura. 2009. Membrane vesicles as conveyors of immune responses. *Nat Rev Immunol* 9 (8):581–593.

Thery, C., K. W. Witwer, E. Aikawa, et al. 2018. Minimal information for studies of extracellular vesicles 2018 (MISEV2018): a position statement of the International Society for Extracellular Vesicles and update of the MISEV2014 guidelines. *J Extracell Vesicles* 7 (1):1535750.

Vader, Pieter, Emma A. Mol, Gerard Pasterkamp, and Raymond M. Schiffelers. 2016. Extracellular vesicles for drug delivery. *Adv Drug Deliver Rev* 106:148–156.

Valadi, H., K. Ekström, A. Bossios, M. Sjöstrand, J. J. Lee, and J. O. Lötvall. 2007. Exosome-mediated transfer of mRNAs and microRNAs is a novel mechanism of genetic exchange between cells. *Nat Cell Biol* 9 (6):654–659.

Varela-Eirin, M., A. Varela-Vazquez, M. Rodriguez-Candela Mateos, et al. 2017. Recruitment of RNA molecules by connexin RNA-binding motifs: Implication in RNA and DNA transport through microvesicles and exosomes. *Biochim Biophys Acta Mol Cell Res* 1864 (4):728–736.

Vlassov, A. V., S. Magdaleno, R. Setterquist, and R. Conrad. 2012. Exosomes: current knowledge of their composition, biological functions, and diagnostic and therapeutic potentials. *Biochim Biophys Acta* 1820 (7):940–948.

Wahlgren, J., L. Karlson T. De, M. Brisslert, et al. 2012. Plasma exosomes can deliver exogenous short interfering RNA to monocytes and lymphocytes. *Nucleic Acids Res* 40 (17):e130.

Wang, J., Y. Zheng, and M. Zhao. 2016. Exosome-based cancer therapy: implication for targeting cancer stem cells. *Front Pharmacol* 7:533.

Weng, Y., H. Xiao, J. Zhang, X. J. Liang, and Y. Huang. 2019. RNAi therapeutic and its innovative biotechnological evolution. *Biotechnol Adv* 37 (5):801–825.

Wubbolts, R., R. S. Leckie, P. T. Veenhuizen, et al. 2003. Proteomic and biochemical analyses of human B cell-derived exosomes. potential implications for their function and multivesicular body formation. *J Biol Chem* 278 (13):10963–10972.

Yang, T., B. Fogarty, B. LaForge, et al. 2017. Delivery of small interfering RNA to inhibit vascular endothelial growth factor in zebrafish using natural brain endothelia cell-secreted exosome nanovesicles for the treatment of brain cancer. *AAPS J* 19 (2):475–486.

Yang, Z., J. Xie, J. Zhu, et al. 2016. Functional exosome-mimic for delivery of siRNA to cancer: in vitro and in vivo evaluation. *J Control Release* 243:160–171.

Zhang, Y., L. Li, J. Yu, et al. 2014. Microvesicle-mediated delivery of transforming growth factor beta1 siRNA for the suppression of tumor growth in mice. *Biomaterials* 35 (14):4390–4400.

Zhao, C., D. J. Busch, C. P. Vershel, and J. C. Stachowiak. 2016. Multifunctional transmembrane protein ligands for cell-specific targeting of plasma membrane-derived vesicles. *Small* 12 (28):3837–3848.

Zhao, L., C. Gu, Y. Gan, L. Shao, H. Chen, and H. Zhu. 2020. Exosome-mediated siRNA delivery to suppress postoperative breast cancer metastasis. *J Control Release* 318:1–15.

Zheng, Z., Z. Li, C. Xu, B. Guo, and P. Guo. 2019. Folate-displaying exosome mediated cytosolic delivery of siRNA avoiding endosome trapping. *J Control Release* 311–312:43–49.

Zhupanyn, P., A. Ewe, T. Buch, et al. 2020. Extracellular vesicle (ECV)-modified polyethylenimine (PEI) complexes for enhanced siRNA delivery in vitro and in vivo. *J Control Release* 319:63–76.

Part V

Application and Exploitation in RNA Nanotechnology

42 RNA Structural Modeling for Therapeutic Applications

Yi Cheng, Dong Zhang, Travis Hurst, and Xiaoqin Zou
University of Missouri, Columbia, Missouri, USA

Paloma H Giangrande
University of Iowa, Iowa City, IA, USA

Wave Life Sciences Inc, Cambridge, MA, USA

Shi-Jie Chen
University of Missouri, Columbia, Missouri, USA

CONTENTS

INTRODUCTION

RNA molecules play critical roles in cellular functions at the level of transcription, translation, and gene regulation (Guo 2010, Jasinski et al. 2017). In particular, non-coding RNAs attract ever-increasing attention for many promising applications in RNA therapeutics because of their regulatory and enzymatic functions in gene expression (Janssen et al. 2013, Deng et al. 2014, Valdmanis and Kay 2017). For example, microRNA (miRNA), small interfering RNA (siRNA), and small hairpin RNA (shRNA) can silence targeted gene expression (Wilson and Doudna 2013, Fellmann and Lowe 2014). However, despite remarkable success, RNA interference therapeutics has been hindered by off-target effects and a lack of efficient delivery methods due to nonspecific cell type targeting and the short half-life of these molecules in vivo (McManus et al. 2002, McManus and Sharp 2002, Huesken et al. 2005). Rapidly advancing in recent years, RNA nanotechnology offers a highly promising approach to utilize regulatory RNA for disease treatment by addressing the above issues (McCaffrey et al. 2002, Lienert et al. 2014, Black, Perez-Pinera, and Gersbach 2017). The great promise of RNA nanotechnology stems from the designability of RNA three-dimensional structures (Guo 2005, Afonin, Lindsay, and Shapiro 2013, Yu et al. 2019). RNA molecules of different sequences can be designed to optimize binding to targets and to self-assemble into various specific structures. For instance, we can design siRNA and shRNA sequences to minimize off-target effects or design RNA nanoparticles to form specific structures for the efficient delivery of siRNA (Afonin et al. 2011, Afonin et al. 2014, Grabow and Jaeger 2014, Sharma et al. 2016).

DOI: 10.1201/9781003001560-47

To improve delivery efficiency by targeting specific cell types and molecules, RNA aptamers are designed as short, single-stranded oligonucleotide ligands that are synthesized to bind to target molecules with high affinity and specificity, and they represent an emerging class of biologics that can be easily adapted for targeted cancer diagnostics and therapy (Dassie et al. 2009, Keefe, Pai, and Ellington 2010). At the bench, RNA aptamers have been successfully used as inhibitors of their targets (Thiel and Giangrande 2009) and serve to deliver chemotherapeutic agents (Bagalkot et al. 2006, Dhar et al. 2008, Gu et al. 2008, Cao et al. 2009), nanoparticles (Farokhzad et al. 2004), radionuclides (Hicke et al. 2006), and siRNA (McNamara Ii et al. 2006, Zhou et al. 2008, Dassie et al. 2009, Zhou et al. 2009, Pastor et al. 2010) to specific cell types in culture and *in vivo*. Compared to antibodies, RNA aptamers have several advantages as targeted therapeutic agents, including smaller size, better tissue penetration, ease of chemical synthesis/modification, and lack of immune system stimulation. However, the clinical potential of the majority of RNA aptamers has yet to be realized. A significant hurdle to the clinical adoption of these aptamers is the limited information on their secondary (2D) and tertiary (3D) structures. Structural studies of these RNA aptamers and their complexes with targets can greatly facilitate and expedite the post-selection optimization steps required for translation, including truncation to increase manufacturing efficiency, chemical modification to enhance stability and improve safety, and chemical conjugation to improve drug properties for combinatorial therapy.

The laborious and costly process of experimentally testing for efficiency and potency of each designed RNA—accounting for off-target effects—has increased demand for more efficient methods. Computational modeling and design are highly desirable approaches to improve the efficiency of RNA nanotechnology. For example, we can utilize computational models that consider RNA/RNA interactions to identify targets on mRNA and then find the most potent siRNA/shRNA for specific gene knockdown (Lai and Meyer 2016). Efficient computational methods such as GUUGle (Gerlach and Giegerich 2006), RNAhybrid (Rehmsmeier et al. 2004, Kruger and Rehmsmeier 2006), and RNAduplex (Hofacker et al. 1994) were developed to predict miRNA targets by considering the intermolecular interactions of complementary base pairing or free energy of stacking. Then, RNAup (Muckstein et al. 2006), IntaRNA (Busch, Richter, and Backofen 2008, Richter et al. 2010, Smith et al. 2010, Mann, Wright, and Backofen 2017), and RNAplex (Amman et al. 2011) models were developed to improve predictions by accounting for intramolecular interactions. By allowing for intramolecular interactions, these models include the accessibility of the corresponding mRNA site of miRNA binding in their predictions. In addition to computing miRNA targets, these algorithms can also consider siRNA, shRNA, and other small RNA (sRNA) targets. Based on these algorithms, RNApredator (Eggenhofer et al. 2011) and CopraRNA (Wright et al. 2013, Wright et al. 2014) models further improve prediction by accounting for small bacterial sRNA targets.

While miRNA are generally endogenous, siRNA are usually synthesized. Alone, the half-life of siRNA is short due to instability in the serum, and siRNA cannot efficiently cross the cellular membrane. To improve siRNA delivery efficiency, a plethora of methods have been devised (Taxman et al. 2010, Fellmann et al. 2013, Ling, Fabbri, and Calin 2013, Pelossof et al. 2017). Recombinant Adeno-Associated Virus (rAAV) gene therapy uses an engineered AAV vector to generate knock-ins that facilitate production of siRNA in cells (Snyder et al. 1997, Valdmanis and Kay 2017). Aptamer-siRNA chimeras also constitute a realistic and effective way to deliver siRNA to targeted cells, which increases precision of siRNA delivery and reduces off-target effects (McNamara Ii et al. 2006). However, the small size of aptamer-siRNA chimeras makes them more susceptible to renal clearance than larger nanoparticles. To treat this problem, properly sized, thermodynamically stable RNA 3-way junctions (3WJs), nanorings, and dendrimers are designed to provide a core for attachment of siRNA, aptamers, and other drugs to improve delivery specificity and potency while increasing the half-life of these molecules (Grabow et al. 2011, Shu et al. 2011, Afonin, Lindsay, and Shapiro 2013, Sharma et al. 2016, Afonin et al. 2014). The mechanism and design of these special, multi-stranded, RNA nanostructures has been the subject of considerable research (Jaeger and Chworos 2006, Yingling and Shapiro 2007, Bindewald et al. 2008, Severcan et al. 2009, Dibrov et al. 2011). At least two factors increase the half-life of these nanostructures: the increased size reduces renal clearance, and enhanced thermostability coupled with ribonuclease resistance reduces degradation. Additionally, combining cell-specific aptamers with nanostructures facilitates crossing of targeted cellular membranes, increasing potency and reducing off-target effects. Design of Aptamer-siRNA chimeras and RNA nanoparticles to carry the siRNA to target cells requires models that consider RNA/RNA interactions. With the development of computational models, such as RNAcofold (Bernhart et al. 2006), VfoldCPX (Cao and Chen 2006, 2011b, Cao, Xu, and Chen 2014), PairFold, MultiFold (Andronescu, Zhang, and Condon 2005), NUPACK (Dirks et al. 2007, Zadeh et al. 2011), NanoFolder (Bindewald et al. 2011), and others, we are approaching the goals of siRNA application and improving understanding of RNA regulation mechanisms.

COMPUTATIONAL MODELS FOR RNA NANOTECHNOLOGY AND THERAPEUTICS

Computational models have rapidly grown with RNA nanotechnology. Single-stranded RNA structures—including pseudoknots—can now be predicted by both 2D and 3D structure prediction models. Accurate prediction of structure from the RNA sequence normally follows a hierarchical paradigm, where 2D structures predicted from sequences are used to supply constraints for prediction of 3D structures. First, predictive RNA 2D structure models calculate the minimum free energy 2D structure for a specific sequence or use sequence alignment to determine the structure from conserved motifs within the sequence. Then, given the predicted 2D structure, 3D structure models predict RNA structures with motifs extracted from experimental data or using trained (coarse grained) force fields. Extension of RNA 2D structure algorithms has facilitated prediction of structures containing RNA/RNA interactions, which are widely found in RNA nanotechnology. Here we focus on common models dealing with RNA/RNA interactions. An overview of these models is shown in Table 42.1.

Predicting miRNA, siRNA, and shRNA targets usually involves a short RNA and a long mRNA target. In the first type of approach, the structure and accessibility of the long mRNA is neglected. In order to calculate target sites in long sequences, the first type of model focuses on intermolecular base pairing by locating complementary sequences. GUUGle predicts perfectly hybridized targets by considering A-U, C-G, and G-U base pairing, and it is a fast way to find potential targets in genome-wide sequences. However, full sequence complementarity is not required for miRNA, siRNA, or shRNA to bind to targets. To account for this, RNAduplex, RNAhybrid, UNAFold (Markham and Zuker 2008), and RIsearch (Wenzel, Akbaşli, and Gorodkin 2012) compute both optimal and suboptimal 2D structures for hybridization. In particular, RIsearch is designed to rapidly screen genome-wide, RNA/RNA interactions and act as a pre-filter for more computationally demanding methods.

By including the accessibility and structure of the target and the small RNA, the second type of model leads to better prediction of target sites. The earliest form of this type of model was OligoWalk (Mathews et al. 1999), which predicts targets by considering the affinity of the RNA to target sites and comparing the stability of the oligonucleotide-target helix to the predicted 2D structure of both the target and the oligonucleotide. However, many simplified models do not account for the fact that hybridization between the target and small RNA requires multiple steps to unfold the separate structures, hybridize the two strands, and relax the complex. To incorporate this concept, RNAup, RNAplex, RNApredator, and IntaRNA include intramolecular interactions to predict target sites for miRNA, siRNA, and shRNA, while accounting for the more complex and realistic hybridization process. RNAup performs this task by extending computation of the partition function to assess the ability for sequence intervals to remain unpaired. RNAplex utilizes a simplified energy algorithm

TABLE 42.1

The RNA/RNA Interaction Prediction Models and Their Features. Models Labeled from 1 to 15 are RNAduplex (Hofacker et al. 1994), GUUGle (Gerlach and Giegerich 2006), OligoWalk (Mathews et al. 1999), RIsearch (Wenzel, Akbaşli, and Gorodkin 2012), RNAup (Muckstein et al. 2006), RNAplex (Amman et al. 2011), RNAhybrid (Kruger and Rehmsmeier 2006), IntaRNA (Busch, Richter, and Backofen 2008), CopraRNA (Wright et al. 2013), RNApredator (Eggenhofer et al. 2011), PairFold (Andronescu, Zhang, and Condon 2005), RNAcofold (Bernhart et al. 2006), NUPACK (Zadeh et al. 2011), VfoldMCPX (Cao and Chen 2006, 2011b, Cao, Xu, and Chen 2014), and NanoFolder (Bindewald et al. 2011), respectively. Models that Consider More Complex Interactions Cannot Easily Handle RNA Above ~1000 Nucleotides (Labeled "No" in the "Long RNA" Row). Some Simplified Models only Consider Intermolecular Duplex Formation Without Accounting for any Interactions of the Molecules with Themselves (Labeled "No" in the "Intramolecular" Row). More Complete Models Consider the Global Conformational Ensemble, Rather than a Simplified Description (Labeled "Yes" in the "Global" Row), and Some of these also Account for Pseudoknot Formation (Labeled "Yes" in the "Pseudoknot" Row).

Models	1	2	3	4	5	6	7	8	9	10	11	12	13	14	15
Long RNA	Yes										No				
Intramolecular	No				Yes										
Global	No										Yes				
Pseudoknot	No														Yes
Application	Simplified small RNA target prediction; focus on intermolecular duplex formation				Consider some intramolecular interactions for small RNA target prediction; more computationally demanding than simplified models						Two-stranded RNA complex structure prediction		Multi-stranded RNA structure prediction		

to rapidly calculate folding conformations and implicitly includes accessibility effects by reading in results from RNAup or RNAplfold (Tafer et al. 2011). RNApredator uses RNAplex machinery for fast run times and offers a graphical overview of sRNA binding accessibility, along with Gene Ontology enrichment analysis. IntaRNA explicitly accounts for accessibility effects while also including a seed region of user-defined length to account for the fact that oftentimes perfect Watson-Crick pairing of seven or eight consecutive bases at the 5' end of animal miRNA produces sufficient mRNA regulation. CopraRNA builds on the IntaRNA platform to construct combined target predictions for groups of organisms with homologous sRNA. Compared with the first type of model, these models consume more computational resources.

Though calculations of intramolecular interactions are included, the previous models still quickly predict RNA/RNA interaction by focusing on local interactions. For the purpose of predicting interactions between short RNA more precisely, a third type of model, implemented in PairFold and RNAcofold, calculates the partition function of the structural ensemble to determine the structure with minimum free energy. MultiFold, NUPACK, Nanofolder, and VfoldMCPX further extend these models to predict multi-stranded RNA structures. MultiFold builds on PairFold to predict multi-stranded structures. While MultiFold and NUPACK can deal with non-pseudoknotted structures for multi-stranded RNA, Nanofolder and VfoldMCPX can also account for pseudoknotted structures, which are important in the design of nanostructures. Although Nanofolder can account for more complexly pseudoknotted structures, it mainly considers helix-free energies in its estimations. In contrast, VfoldMCPX also accounts for both loop and helix contributions to the overall free energy but is limited to predictions of simpler pseudoknots. Based on these models, the stability of a designed nanoparticle and the minimum free-energy structure of multi-stranded RNA can be calculated.

miRNA, siRNA, shRNA, AND sRNA BINDING SITE PREDICTION

Regulatory RNA, such as endogenous miRNA, exogenous siRNA, and shRNA can trigger gene-silencing. An siRNA normally consists of 21–23 base pairs, and shRNA are hairpin-like extended siRNA. They bind to mRNA sites with complementary or near-complementary sequences and are incorporated into the RNA-induced silencing complex (RISC). While siRNA or shRNA with completely complementary sequences will cause mRNA degradation, others, with near-complementary sequences, will block mRNA translation, like miRNA. Due to the long sequence length of mRNA, siRNA and shRNA may bind to multiple target sites and cause unnecessary gene-silencing. To minimize off-target effects, we need to account for potential off-target sites when designing siRNA and shRNA.

GUUGle and BLAST provide fast ways to screen siRNA/shRNA since base pairs between siRNA/shRNA and their targets are mostly complementary, and the complex structures are probably the helical with some bulge or internal loops. Though taking both intermolecular and intramolecular interactions into account requires more computational resources, models that include these interactions improve the accuracy of predictions. For example, CopraRNA precisely predicts both bacterial small RNA targets on mRNA and interaction domains on sRNA by using comparative genomics. The pipeline of CopraRNA is shown in Figure 42.1a (Wright et al. 2013). Based on IntaRNA, genome-wide target predictions given by CopraRNA result from summarizing orthologous genes to find conserved interactions across multiple distinct organisms.

The true positive rate of CopraRNA prediction for the selected set and benchmark set is shown in Figure 42.1b with solid and dashed lines, respectively. Results from statistical analysis of predictions for GcvB, a major regulator of amino acid metabolism, are shown in Figure 42.1c–d, indicating the interaction domain in GcvB and the tendency to target the region near the start codon of mRNA.

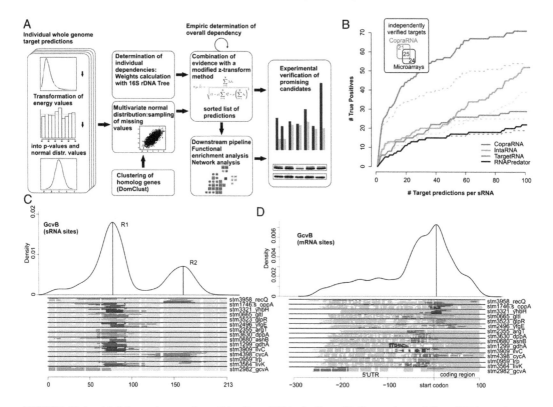

FIGURE 42.1 (a) Schematic overview of the CopraRNA pipeline. (b) Comparison of CopraRNA predictions with microarray results and other target prediction methods. (c) Visualization of the predicted interaction domains in GcvB. (d) The predicted mRNA targets of GcvB. (Source: Wright et al. 2013)

RNA STRUCTURE PREDICTION FOR DOUBLE-STRANDED
AND MULTI-STRANDED RNA

Aptamer-siRNA chimeras are designed to guide siRNA bound to the aptamer to a specific cell surface receptor, which facilitates target-specificity and crossing of the cell membrane. Usually, the siRNA is connected to the 3' end of the aptamer. To retain functionality, we need to ensure the stability of both the aptamer and the siRNA structures. To predict the structure of these chimeras, we can use PairFold or RNAcofold, which calculates the partition function of the whole ensemble of complex structures between two RNA. Given the free energy of the predicted structures, we can determine the stability of the designed aptamer-siRNA chimeras.

Due to their stability in serum and reduced renal clearance due to their larger size, branched RNA nanostructures are designed as direct gene-interfering RNA or siRNA carriers. NUPACK, NanoFolder, and VfoldMCPX models were established to calculate free energies for these multi-stranded RNA nanostructures. NUPACK calculates the pseudoknot-free multi-stranded RNA structures based on thermodynamic parameters and provides special treatment for breaks in the RNA strands. However, pseudoknot structures are often important components of nanostructure design. Nanofolder accounts for pseudoknots but mainly focuses on helical structures that contribute most of the free energy to the complex structure. To more precisely predict multi-stranded RNA structure, free energy contributions of helices and loops are considered in VfoldMCPX. By computing the free energy of optimal and suboptimal multi-stranded RNA structures, the probability and stability of each conformation can be more accurately estimated.

As shown in Figure 42.2, multi-stranded RNA structure prediction relies on connecting RNA strands into one strand. Then, the multi-stranded RNA is transformed into a one-RNA system with breakpoints that cannot be paired. The free energies of loops with breakpoints are set to zero. H-type pseudoknotted structures can be predicted by VfoldMCPX by including the free energy for pseudoknot loops. By increasing the number of free energy parameters calculated from the Vfold model, more motifs, such as hairpin–hairpin kissing, can be properly accounted for.

Multi-stranded RNA structure prediction models can currently assess many kinds of nanoparticles. NanoFolder can accurately predict triangular RNA structures (Bindewald et al. 2011) and RNA squares (Dibrov et al. 2011). NUPACK and VfoldMCPX can be utilized to predict branched nanostructures such as 3WJs, "Module-2" (Sharma et al. 2016), trimer RNA, tetramer RNA (Nakashima et al. 2011), and others. Module-2 illustrates the modular ability of designed 3WJ motifs and is formed from the self-assembly of a thermodynamically stable, core 3WJ motif that was taken from a packaging RNA of the phi29 bacteriophage DNA packaging motor. The Module-2 nanoparticle contains additional 3WJ motifs attached to the three ends of the core motif. Extension of the modular 3WJ scaffold can be used to form a stable delivery dendrimer for many functional moieties, such as aptamers, miRNA, siRNA, drugs, and other biologics. Figure 42.3a–d shows these RNA nanostructures, as predicted by VfoldMCPX. Based on the predicted 2D structures, 3D structures of Module-2 were also predicted. The top-ranked 3D structures are shown in Figure 42.3f, which agree with experimental observations (Sharma et al. 2016).

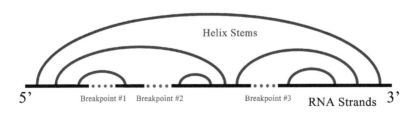

FIGURE 42.2   A schematic representation of RNA multi-stranded structure.

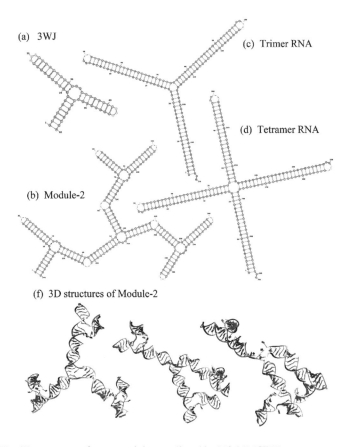

(a) 3WJ

(c) Trimer RNA

(d) Tetramer RNA

(b) Module-2

(f) 3D structures of Module-2

FIGURE 42.3   The 2D structure of nanoparticles predicted by VfoldMCPX.

RNA NANOSTRUCTURE DESIGN

When directly delivered, siRNA are easily degraded by the serum and cleared by the kidneys. In order to increase the efficiency of siRNA delivery, one method is to connect the siRNA to RNA nanoparticles (Afonin et al. 2014, Grabow and Jaeger 2014), which increases their size and prevents renal clearance. For example, 3WJs, 4WJs, nanorings (Grabow et al. 2011), and RNA dendrimers can form a core that siRNA, aptamers, and drugs can be attached to for targeted delivery to particu-lar cell types. These nanostructures are mostly self-assembled and designed by extracting motifs from experimental structures (Jasinski et al. 2017, Dibrov et al. 2011, Grabow et al. 2011). As detailed in the previous section, computational methods have been used to model the structure for multi-stranded, modular extensions of the 3WJ to form dendrimers. Computational optimization and screening of RNA dendrimers could help reduce the time and costs associated with producing these nanoparticles for many therapeutic applications.

To computationally facilitate design of self-assembling RNA ring structures, Parlea et al. devel-oped the Ring Catalog to tabulate RNA ring structures that are predicted to form stable closed rings, which can guide generation of nanostructures. One recently designed RNA ring nanostructure (Parlea et al. 2016) is shown in Figure 42.4. Shapiro et al. use nanoTiler (Bindewald et al. 2008) to perform a combinatorial search of chosen motifs and NanoFolder to predict multi-stranded RNA structure. Figure 42.4a,b shows the 3D structures colored by the motifs and strands, respectively. The 3WJ shown in Figure 42.4c and 42.4d was extracted from a crystal structure of the large ribo-somal subunit (PDB identifier: 2OGM). The 2D structure of the square predicted with NanoFolder, colored by strands, is shown in Figure 42.4e.

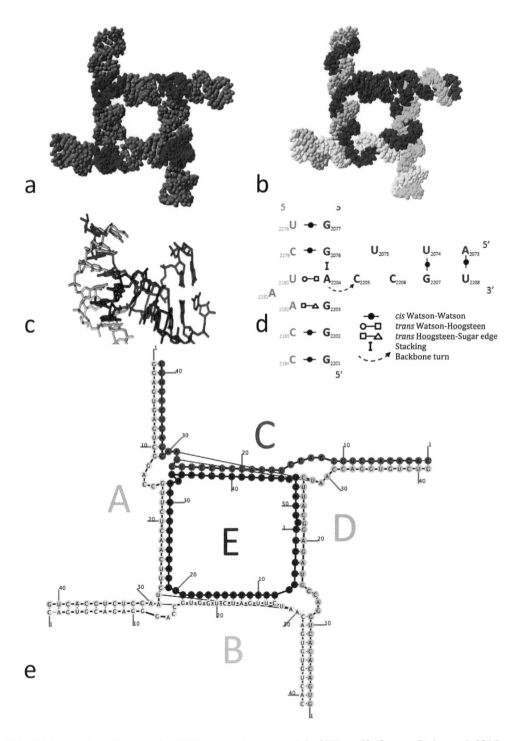

FIGURE 42.4 The self-assembling RNA square structure and the 3WJ motif. (Source: Parlea et al. 2016).

By combining experimental methods with computational structure prediction, design efficiency of nanostructures with aptamers, siRNA, and drugs can be radically improved. Where possible, trial-and-error experimental methods should be replaced with computationally guided experiments to avoid unnecessary costs. As seen in the next section, computational modeling already aids in the optimal design of RNA aptamers, where trial-and-error approaches previously failed.

COMPUTATIONALLY MODELING THE STRUCTURE OF RNA APTAMERS

To lower the high cost of manufacturing long RNA molecules through chemical synthesis for clinical adoption of RNA aptamers, we desire to design truncated aptamers that retain high functionality in binding to target molecules. Enzymatic activity of prostate-specific membrane antigen (PSMA) has been linked to metastatic cancer, and two unique, nuclease-resistant RNA aptamers (A9 and A10) have been shown to bind to PSMA, inhibiting its carcinogenic properties. However, these full-length aptamers are too long for the large-scale synthesis required for therapeutic application. Previously, a trial-and-error approach was applied to successfully truncate the A10 RNA aptamer from 71 to 56 nucleotides while retaining functionality (Lupold et al. 2002). However, applying a similar truncation approach to the A9 aptamer—a better inhibitor of PSMA enzymatic activity— rendered the aptamer inactive.

Demonstrating the utility of computational approaches, a rational truncation approach guided by computational RNA structural modeling and RNA/protein docking algorithms (Rockey et al. 2011) was developed to guide the truncation of the A9 PSMA RNA aptamer, a 70 nucleotide aptamer with great potential for targeted therapeutics. Seven initial truncations of the A9 aptamer with different lengths (A9a through A9g) were designed, in which terminal nucleotides were removed so that 2D structures of the remaining oligonucleotides predicted by RNAStructure (Mathews 2006, Bellaousov et al. 2013), Vfold2D (Cao and Chen 2005, Xu, Zhao, and Chen 2014), or Mfold (Zuker and Sankoff 1984, Zuker 2003) were as similar as possible to that of the full-length A9 molecule. The NAALADase assay (Xiao et al. 2000) was utilized to assess inhibition of PSMA enzymatic activity by those truncations. The results suggest that even the most severely truncated aptamer—A9g, with 43 nucleotides—preserves key sequence/structure features important for inhibition of PSMA enzymatic activity, and it binds to PSMA with high affinity and specificity. The essential sequence and structural elements within A9g that facilitate PSMA binding were further checked through introducing a series of base changes that either preserve or disrupt various structural elements involved in A9g binding. Additionally, a 3D structure of A9g was built by Vfold3D model (Cao and Chen 2011a, Xu et al. 2016) and then was computationally docked (Huang and Zou 2010) to an experimental structure of PSMA (Davis et al. 2005). Those structural studies uncovered key interactions between the A9g RNA aptamer and PSMA and suggested that sequence conservation of uridine at position 39 of the A9g RNA aptamer may be more important than the overall conformation of the L1 loop for conferring the inhibitory function of the aptamer. This analysis resulted in another truncated derivative of the A9 aptamer (A9L) that (i) retains PSMA binding activity/specificity and functionality, (ii) can be effectively internalized into cells expressing PSMA, and (iii) is amenable to large-scale chemical synthesis due to its reduced length of 41 nucleotides.

To expedite the optimization of RNA aptamers for clinical applications, we recently summarized a general approach (Xu et al. 2016), in which a structural computational modeling methodology is coupled to a standard functional assay to determine key sequence/structural motifs of an RNA aptamer. This general approach was successfully applied to computationally guide the truncation of a PSMA RNA aptamer that previous, trial-and-error approaches failed to functionally truncate. Based on the predicted 2D structure (by Vfold2D, RNAStructure, Mfold, etc.) for the full-length A9 aptamer sequence, terminal nucleotides were gradually removed to shorten the sequence while retaining the functional 2D folding structure. This method may be readily transferred to optimize other RNA aptamers with therapeutic potential. From experimental functional inhibition data, such as NAALADase activity, the minimal functional structure that retains activity similar to the original sequence can be identified, analogously to the A9g truncation. By combining experimental results from systematic base deletions, insertions, or mutations to the minimal function structure with computational 2D structure studies, the structure–activity relationship for those altered sequences can be used to identify critical nucleotides and essential structural features that contribute to inhibitory activity of the RNA aptamer. Finally, to gain extra insights into the interactions between an RNA aptamer and its target, 3D all-atom structures of the proposed minimal functional structure can be modeled by the Vfold3D algorithm and bound to its target through a protein-RNA docking program, such as MDockPP (Huang and Zou 2010).

Here, based on the aforementioned general approach, we provide a systematic overview of key sequence and structural elements of the A9g aptamer that lead to inhibition of PSMA activity. Firstly, the A9g aptamer was further truncated to either retain sequence and structural loop elements (A9h and A9i in Figure 42.5a) or retain sequence and structural stem elements (A9j and A9k in Figure 42.5a).

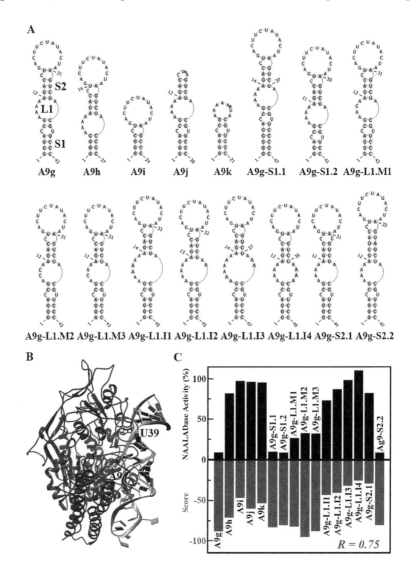

FIGURE 42.5 Structural characterization of A9g aptamer binding to PSMA. (a) Systematic base deletions, insertions, and/or mutations (highlighted by red) within A9g to probe essential sequence/structural features that contribute to the inhibitory activity of RNA aptamer. The 2D structures for those altered sequences are predicted by RNAStructure. (b) Modeled 3D structure of A9g (orange) docked to an experimental structure of PSMA (blue). The bases A9 and U39 (highlighted by red) within A9g are predicted to form direct interactions with PSMA structure. (c) Effects of altered sequences on PSMA NAALADase activity (top panel) and the binding scores for the corresponding modeled 3D structures docking to experimental PSMA structure by ITScorePR (Huang and Zou 2014) (bottom panel). Correlation between experimental activities and predicted scores is $R = 0.75$.

The fact that none of these four additional truncations (A9h-A9k) inhibit activity under the given assay condition (200 nM RNA concentration) indicates the A9g aptamer should be treated as a whole for PSMA inhibition (Figure 42.5c). Secondly, the length requirement of the stem S1 for PSMA inhibition was tested through elongating (A9g-S1.1) and shortening (A9g-S1.2) the helix length (see Figure 42.5a). The NAALADase activities for those two altered aptamers are very close to that for A9g (see Figure 42.5c), so the length requirement of the stem S1 for inhibition of PSMA activity is somewhat flexible. Thirdly, the sequence specificity of the L1 loop (A9 and A10) for PSMA activity was checked by introducing some designed mutations (A9g-L1.M1, A9g-L1.M2, and A9g-L1.M3), and those mutations resulted in notable but decreased PSMA inhibitions. Then, insertion of several adenines into the internal loop L1 (A8-A11, A36) was used to check the effect of loop L1-bending on inhibition of PSMA activity, and the NAALADase activities for these altered sequences all largely increased (see A9g-L1.I1-A9g-L1.I4 in Figure 42.5a). These results clearly demonstrate the structural/sequence importance of L1 loop for inhibition of PSMA activity. Finally, the importance of stem S2 rigidity was tested by inserting nucleotides into the stem, and the effects of A9g-S2.1 and A9g-S2.2 aptamers on PSMA NAALADase activity show that the C16 bulge may also play a key role in inhibiting PSMA enzymatic activity. Similar to the structural study for A9g aptamer (see Figure 42.5b), the 3D structures for all the modified sequences given in Figure 42.5a were built with the Vfold3D model and were computationally docked to the experimentally determined PSMA structure.

CONCLUSION

With the aid of computational models, RNA nanotechnology and therapeutics are rapidly developing. The power of RNA nanotechnology resides in its ability to overcome problems of delivery and transience through efficient design of resilient nanostructures capable of targeting and crossing cellular boundaries to improve potency, reduce off-target effects, and resist degradation and excretion. Computational models have been successfully applied in siRNA target site prediction, aptamer design, multi-stranded RNA structure prediction, and RNA nanoparticle design. In comparison to arduous, traditional trial-and-error experiments, computational guidance of nanostructure design markedly improves optimization efficiency and should be utilized where possible to reduce costs associated with experimentation and production of RNA nanotherapeutics. However, application of these models to solve problems in RNA nanotechnology and therapeutics is still in its nascent stage. Current algorithms are insufficient to treat genome-wide sequences with high precision. In the future, we should improve our models to calculate long sequences with high precision and establish prediction models to deal with more structure types. Then, we should extend the predictive models to aid in the design of siRNA.

ACKNOWLEDGMENT

This work was supported by grants to SJC from the National Institutes of Health through grants R01-GM117059 (SJC) and R01-GM063732 (SJC), to PHG from Department of Defense Congressionally Directed Medical Research Programs (PR150627P1), University of Iowa Carver College of Medicine (CCOM) (Carver Collaborative Pilot Grant Award 2015), and University of Iowa Award from The Office of the Vice President for Research and Economic Development (OVPRED 2015), and to XZ from the National Institutes of Health through grants R01-GM109980 (XZ), R01-HL126774 and R01-HL142301 (Cui).

REFERENCES

Afonin, Kirill A., Wade W. Grabow, Faye M. Walker, Eckart Bindewald, Marina A Dobrovolskaia, Bruce A. Shapiro, and Luc Jaeger. 2011. "Design and self-assembly of siRNA-functionalized RNA nanoparticles for use in automated nanomedicine." *Nature protocols* 6 (12):2022.

Afonin, Kirill A., Mathias Viard, Alexey Y. Koyfman, Angelica N. Martins, Wojciech K. Kasprzak, Martin Panigaj, Ravi Desai, Arti Santhanam, Wade W. Grabow, and Luc Jaeger. 2014. "Multifunctional RNA nanoparticles." *Nano letters* 14 (10):5662–5671.

Afonin, Kirill A., Brian Lindsay, and Bruce A. Shapiro. 2013. "Engineered RNA Nanodesigns for Applications in RNA Nanotechnology." *DNA and RNA Nanotechnology* 1 (1). doi:10.2478/rnan-2013-0001.

Amman, Fabian, Florian Eggenhofer, Hakim Tafer, Ivo L. Hofacker, and Peter F. Stadler. 2011. "Fast accessibility-based prediction of RNA–RNA interactions." *Bioinformatics* 27 (14):1934–1940. doi:10.1093/bioinformatics/btr281.

Andronescu, Mirela, Zhi C. Zhang, and Anne Condon. 2005. "Secondary structure prediction of interacting RNA molecules." *Journal of Molecular Biology* 345 (5):987–1001.

Bagalkot, Vaishali, Omid C. Farokhzad, Robert Langer, and Sangyong Jon. 2006. "An Aptamer–Doxorubicin Physical Conjugate as a Novel Targeted Drug-Delivery Platform." *Angewandte Chemie International Edition* 45 (48):8149–8152. doi:10.1002/anie.200602251.

Bellaousov, Stanislav, Jessica S. Reuter, Matthew G. Seetin, and David H. Mathews. 2013. "RNAstructure: web servers for RNA secondary structure prediction and analysis." *Nucleic Acids Research* 41 (W1):W471–W474. doi:10.1093/nar/gkt290.

Bernhart, Stephan H., Hakim Tafer, Ulrike Mückstein, Christoph Flamm, Peter F. Stadler, and Ivo L. Hofacker. 2006. "Partition function and base pairing probabilities of RNA heterodimers." *Algorithms for Molecular Biology* 1 (1):3.

Bindewald, Eckart, Kirill Afonin, Luc Jaeger, and Bruce A. Shapiro. 2011. "Multistrand RNA secondary structure prediction and nanostructure design including pseudoknots." *ACS Nano* 5 (12):9542–9551.

Bindewald, Eckart, Calvin Grunewald, Brett Boyle, Mary O'Connor, and Bruce A. Shapiro. 2008. "Computational strategies for the automated design of RNA nanoscale structures from building blocks using NanoTiler." *Journal of Molecular Graphics and Modelling* 27 (3):299–308.

Black, Joshua B., Pablo Perez-Pinera, and Charles A. Gersbach. 2017. "Mammalian Synthetic Biology: Engineering Biological Systems." *Annual Review of Biomedical Engineering* 19 (1):249–277. doi:10.1146/annurev-bioeng-071516-044649.

Busch, A., A. S. Richter, and R. Backofen. 2008. "IntaRNA: efficient prediction of bacterial sRNA targets incorporating target site accessibility and seed regions." *Bioinformatics* 24 (24):2849–2856. doi:10.1093/bioinformatics/btn544.

Cao, Song, and Shi-Jie Chen. 2005. "Predicting RNA folding thermodynamics with a reduced chain representation model." *RNA* 11 (12):1884–1897.

Cao, Song, and Shi-Jie Chen. 2006. "Free energy landscapes of RNA/RNA complexes: with applications to snRNA complexes in spliceosomes." *Journal of Molecular Biology* 357 (1):292–312. doi:10.1016/j.jmb.2005.12.014.

Cao, Song, and Shi-Jie Chen. 2011a. "Physics-Based De Novo Prediction of RNA 3D Structures." *The Journal of Physical Chemistry B* 115 (14):4216–4226. doi:10.1021/jp112059y.

Cao, Song, and Shi-Jie Chen. 2011b. "Structure and stability of RNA/RNA kissing complex: with application to HIV dimerization initiation signal." *RNA* 17 (12):2130–2143. doi:10.1261/rna.026658.111.

Cao, Song, Xiaojun Xu, and Shi-Jie Chen. 2014. "Predicting structure and stability for RNA complexes with intermolecular loop-loop base-pairing." *RNA* 20 (6):835–845. doi:10.1261/rna.043976.113.

Cao, Zehui, Rong Tong, Abhijit Mishra, Weichen Xu, Gerard C. L. Wong, Jianjun Cheng, and Yi Lu. 2009. "Reversible Cell-Specific Drug Delivery with Aptamer-Functionalized Liposomes." *Angewandte Chemie International Edition* 48 (35):6494–6498. doi:10.1002/anie.200901452.

Dassie, Justin P., Xiu-ying Liu, Gregory S. Thomas, Ryan M. Whitaker, Kristina W. Thiel, Katie R. Stockdale, David K. Meyerholz, Anton P. McCaffrey, James O. McNamara Ii, and Paloma H. Giangrande. 2009. "Systemic administration of optimized aptamer-siRNA chimeras promotes regression of PSMA-expressing tumors." *Nature Biotechnology* 27:839. doi:10.1038/nbt.1560

Davis, Mindy I., Melanie J. Bennett, Leonard M. Thomas, and Pamela J. Bjorkman. 2005. "Crystal structure of prostate-specific membrane antigen, a tumor marker and peptidase." *Proceedings of the National Academy of Sciences* 102 (17):5981–5986. doi:10.1073/pnas.0502101102.

Deng, Yan, Chi Chiu Wang, Kwong Wai Choy, Quan Du, Jiao Chen, Qin Wang, Lu Li, Tony Kwok Hung Chung, and Tao Tang. 2014. "Therapeutic potentials of gene silencing by RNA interference: principles, challenges, and new strategies." *Gene* 538 (2):217–227.

Dhar, Shanta, Frank X. Gu, Robert Langer, Omid C. Farokhzad, and Stephen J. Lippard. 2008. "Targeted delivery of cisplatin to prostate cancer cells by aptamer functionalized Pt(IV) prodrug-PLGA–PEG nanoparticles." *Proceedings of the National Academy of Sciences* 105 (45):17356–17361. doi:10.1073/pnas.0809154105.

Dibrov, Sergey M, Jaime McLean, Jerod Parsons, and Thomas Hermann. 2011. "Self-assembling RNA square." *Proceedings of the National Academy of Sciences* 108 (16):6405–6408.

Dirks, Robert M., Justin S. Bois, Joseph M. Schaeffer, Erik Winfree, and Niles A. Pierce. 2007. "Thermodynamic Analysis of Interacting Nucleic Acid Strands." *SIAM Review* 49 (1):65–88. doi:10.1137/060651100.

Eggenhofer, F., H. Tafer, P. F. Stadler, and I. L. Hofacker. 2011. "RNApredator: fast accessibility-based prediction of sRNA targets." *Nucleic Acids Res* 39 (Web Server issue):W149–W154. doi:10.1093/nar/gkr467.

Farokhzad, Omid C., Sangyong Jon, Ali Khademhosseini, Thanh-Nga T. Tran, David A. LaVan, and Robert Langer. 2004. "Nanoparticle-Aptamer Bioconjugates." *A New Approach for Targeting Prostate Cancer Cells* 64 (21):7668–7672. doi:10.1158/0008-5472.can-04-2550.

Fellmann, Christof, Thomas Hoffmann, Vaishali Sridhar, Barbara Hopfgartner, Matthias Muhar, Mareike Roth, Dan Yu Lai, Inês A M. Barbosa, Jung Shick Kwon, Yuanzhe Guan, Nishi Sinha, and Johannes Zuber. 2013. "An Optimized microRNA Backbone for Effective Single-Copy RNAi." *Cell Reports* 5 (6):1704–1713. doi:10.1016/j.celrep.2013.11.020.

Fellmann, Christof, and Scott W. Lowe. 2014. "Stable RNA interference rules for silencing." *Nature Cell Biology* 16:10. doi:10.1038/ncb2895.

Gerlach, W., and R. Giegerich. 2006. "GUUGle: a utility for fast exact matching under RNA complementary rules including G-U base pairing." *Bioinformatics* 22 (6):762–764. doi:10.1093/bioinformatics/btk041.

Grabow, Wade W, and Luc Jaeger. 2014. "RNA self-assembly and RNA nanotechnology." *Accounts of Chemical Research* 47 (6):1871–1880.

Grabow, Wade W, Paul Zakrevsky, Kirill A Afonin, Arkadiusz Chworos, Bruce A Shapiro, and Luc Jaeger. 2011. "Self-assembling RNA nanorings based on RNAI/II inverse kissing complexes." *Nano Letters* 11 (2):878–887.

Gu, Frank, Liangfang Zhang, Benjamin A. Teply, Nina Mann, Andrew Wang, Aleksandar F. Radovic-Moreno, Robert Langer, and Omid C. Farokhzad. 2008. "Precise engineering of targeted nanoparticles by using self-assembled biointegrated block copolymers." *Proceedings of the National Academy of Sciences* 105 (7):2586–2591. doi:10.1073/pnas.0711714105.

Guo, Peixuan. 2005. "RNA nanotechnology: engineering, assembly and applications in detection, gene delivery and therapy." *Journal of Nanoscience and Nanotechnology* 5 (12):1964–1982.

Guo, Peixuan. 2010. "The emerging field of RNA nanotechnology." *Nature Nanotechnology* 5 (12):833.

Hicke, Brian J, Andrew W Stephens, Ty Gould, Ying-Fon Chang, Cynthia K Lynott, James Heil, Sandra Borkowski, Christoph-Stephan Hilger, Gary Cook, and Stephen Warren. 2006. "Tumor targeting by an aptamer." *Journal of Nuclear Medicine* 47 (4):668–678.

Hofacker, Ivo L, Walter Fontana, Peter F Stadler, L Sebastian Bonhoeffer, Manfred Tacker, and Peter Schuster. 1994. "Fast folding and comparison of RNA secondary structures." *Monatshefte für Chemie/Chemical Monthly* 125 (2):167–188.

Huang, Sheng-You, and Xiaoqin Zou. 2010. "MDockPP: A hierarchical approach for protein-protein docking and its application to CAPRI rounds 15–19." *Proteins: Structure, Function, and Bioinformatics* 78 (15):3096–3103. doi:10.1002/prot.22797.

Huang, Sheng-You, and Xiaoqin Zou. 2014. " A knowledge-based scoring function for protein-RNA interactions derived from a statistical mechanics-based iterative method." *Nucleic Acids Research*, 1–12, doi:10.1093/nar/gku077.

Huesken, Dieter, Joerg Lange, Craig Mickanin, Jan Weiler, Fred Asselbergs, Justin Warner, Brian Meloon, Sharon Engel, Avi Rosenberg, Dalia Cohen, Mark Labow, Mischa Reinhardt, François Natt, and Jonathan Hall. 2005. "Design of a genome-wide siRNA library using an artificial neural network." *Nature Biotechnology* 23:995. doi:10.1038/nbt1118

Jaeger, Luc, and Arkadiusz Chworos. 2006. "The architectonics of programmable RNA and DNA nanostructures." *Current Opinion in Structural Biology* 16 (4):531–543.

Janssen, Harry LA, Hendrik W Reesink, Eric J Lawitz, Stefan Zeuzem, Maribel Rodriguez-Torres, Keyur Patel, Adriaan J van der Meer, Amy K Patick, Alice Chen, and Yi Zhou. 2013. "Treatment of HCV infection by targeting microRNA." *New England Journal of Medicine* 368 (18):1685–1694.

Jasinski, Daniel, Farzin Haque, Daniel W Binzel, and Peixuan Guo. 2017. "Advancement of the emerging field of RNA nanotechnology." *ACS Nano* 11 (2):1142–1164.

Keefe, Anthony D., Supriya Pai, and Andrew Ellington. 2010. "Aptamers as therapeutics." *Nature Reviews Drug Discovery* 9:537. doi:10.1038/nrd3141.

Kruger, J., and M. Rehmsmeier. 2006. "RNAhybrid: microRNA target prediction easy, fast and flexible." *Nucleic Acids Res* 34 (Web Server issue):W451–W454. doi:10.1093/nar/gkl243.

Lai, Daniel, and Irmtraud M Meyer. 2016. "A comprehensive comparison of general RNA-RNA interaction prediction methods." *Nucleic Acids Res* 44 (7):e61. doi:10.1093/nar/gkv1477.

Lienert, Florian, Jason J. Lohmueller, Abhishek Garg, and Pamela A. Silver. 2014. "Synthetic biology in mammalian cells: next generation research tools and therapeutics." *Nature Reviews Molecular Cell Biology* 15:95. doi:10.1038/nrm3738

Ling, Hui, Muller Fabbri, and George A Calin. 2013. "MicroRNAs and other non-coding RNAs as targets for anticancer drug development." *Nature Reviews Drug Discovery* 12 (11):847.

Lupold SE, Hicke BJ, Lin Y, Coffey DS. 2002. Identification and characterization of nuclease-stabilized RNA molecules that bind human prostate cancer cells via the prostate-specific membrane antigen. *Cancer Research* 62(14):4029–4033.

Mann, Martin, Patrick R Wright, and Rolf Backofen. 2017. "IntaRNA 2.0: enhanced and customizable prediction of RNA–RNA interactions." *Nucleic Acids Research* 45 (W1):W435–W439.

Markham NR, Zuker M. 2008. "UNAFold: software for nucleic acid folding and hybridization." *Methods Molecular Biology* 453:3–31.

Mathews, David H, Mark E Burkard, Susan M Freier, Jacqueline R Wyatt, and Douglas H Turner. 1999. "Predicting oligonucleotide affinity to nucleic acid targets." *RNA* 5 (11):1458–1469.

Mathews, David H. 2006. "RNA Secondary Structure Analysis Using RNAstructure." *Current Protocols in Bioinformatics* 13 (1):12.6.1–12.6.14. doi:10.1002/0471250953.bi1206s13.

McCaffrey, Anton P, Leonard Meuse, Thu-Thao T Pham, Douglas S Conklin, Gregory J Hannon, and Mark A Kay. 2002. "Gene expression: RNA interference in adult mice." *Nature* 418 (6893):38.

McManus, Michael T, and Phillip A Sharp. 2002. "Gene silencing in mammals by small interfering RNAs." *Nature Reviews Genetics* 3 (10):737.

McManus, Michael T., Christian P. Petersen, Brian B. Haines, Jianzhu Chen, and Phillip A. Sharp. 2002. "Gene silencing using micro-RNA designed hairpins." *RNA* 8 (6):842–850. doi:10.1017/S1355838202024032.

McNamara Ii, James O., Eran R. Andrechek, Yong Wang, Kristi D. Viles, Rachel E. Rempel, Eli Gilboa, Bruce A. Sullenger, and Paloma H. Giangrande. 2006. "Cell type–specific delivery of siRNAs with aptamer-siRNA chimeras." *Nature Biotechnology* 24:1005. doi:10.1038/nbt1223

Muckstein, U., H. Tafer, J. Hackermuller, S. H. Bernhart, P. F. Stadler, and I. L. Hofacker. 2006. "Thermodynamics of RNA-RNA binding." *Bioinformatics* 22 (10):1177–1182. doi:10.1093/bioinformatics/btl024.

Nakashima, Yuko, Hiroshi Abe, Naoko Abe, Kyoko Aikawa, and Yoshihiro Ito. 2011. "Branched RNA nanostructures for RNA interference." *Chemical Communications* 47 (29):8367–8369. doi:10.1039/C1CC11780G.

Parlea, L., E. Bindewald, R. Sharan, N. Bartlett, D. Moriarty, J. Oliver, K. A. Afonin, and B. A. Shapiro. 2016. "Ring Catalog: A resource for designing self-assembling RNA nanostructures." *Methods* 103:128–137. doi:10.1016/j.ymeth.2016.04.016.

Pastor, Fernando, Despina Kolonias, Paloma H. Giangrande, and Eli Gilboa. 2010. "Induction of tumour immunity by targeted inhibition of nonsense-mediated mRNA decay." *Nature* 465:227. doi:10.1038/nature08999

Pelossof, Raphael, Lauren Fairchild, Chun-Hao Huang, Christian Widmer, Vipin T. Sreedharan, Nishi Sinha, Dan-Yu Lai, Yuanzhe Guan, Prem K. Premsrirut, Darjus F. Tschaharganeh, Thomas Hoffmann, Vishal Thapar, Qing Xiang, Ralph J. Garippa, Gunnar Rätsch, Johannes Zuber, Scott W. Lowe, Christina S. Leslie, and Christof Fellmann. 2017. "Prediction of potent shRNAs with a sequential classification algorithm." *Nature Biotechnology* 35:350. doi:10.1038/nbt.3807

Rehmsmeier, Marc, Peter Steffen, Matthias Höchsmann, and Robert Giegerich. 2004. "Fast and effective prediction of microRNA/target duplexes." *RNA* 10 (10):1507–1517. doi:10.1261/rna.5248604.

Richter, A. S., C. Schleberger, R. Backofen, and C. Steglich. 2010. "Seed-based INTARNA prediction combined with GFP-reporter system identifies mRNA targets of the small RNA Yfr1." *Bioinformatics* 26 (1):1–5. doi:10.1093/bioinformatics/btp609.

Rockey, William M., Frank J. Hernandez, Sheng-You Huang, Song Cao, Craig A. Howell, Gregory S. Thomas, Xiu Ying Liu, Natalia Lapteva, David M. Spencer, James O. McNamara II, Xiaoqin Zou, Shi-Jie Chen, and Paloma H. Giangrande. 2011. "Rational Truncation of an RNA Aptamer to Prostate-Specific Membrane Antigen Using Computational Structural Modeling." *Nucleic Acid Therapeutics* 21 (5):299–314. doi:10.1089/nat.2011.0313.

Severcan, Isil, Cody Geary, Erik Verzemnieks, Arkadiusz Chworos, and Luc Jaeger. 2009. "Square-shaped RNA particles from different RNA folds." *Nano Letters* 9 (3):1270–1277.

Sharma, Ashwani, Farzin Haque, Fengmei Pi, Lyudmila S Shlyakhtenko, B Mark Evers, and Peixuan Guo. 2016. "Controllable self-assembly of RNA dendrimers." *Nanomedicine: Nanotechnology, Biology and Medicine* 12 (3):835–844.

Shu, Dan, Yi Shu, Farzin Haque, Sherine Abdelmawla, and Peixuan Guo. 2011. "Thermodynamically stable RNA three-way junction for constructing multifunctional nanoparticles for delivery of therapeutics." *Nature Nanotechnology* 6 (10):658.

Smith, C., S. Heyne, A. S. Richter, S. Will, and R. Backofen. 2010. "Freiburg RNA Tools: a web server integrating INTARNA, EXPARNA and LOCARNA." *Nucleic Acids Res* 38 (Web Server issue):W373–W377. doi:10.1093/nar/gkq316.

Snyder, Richard O., Carol H. Miao, Gijsbert A. Patijn, S. Kaye Spratt, Olivier Danos, Dea Nagy, Allen M. Gown, Brian Winther, Leonard Meuse, Lawrence K. Cohen, Arthur R. Thompson, and Mark A. Kay. 1997. "Persistent and therapeutic concentrations of human factor IX in mice after hepatic gene transfer of recombinant AAV vectors." *Nature Genetics* 16 (3):270–276. doi:10.1038/ng0797-270.

Tafer H, Amman F, Eggenhofer F, Stadler PF, Hofacker IL. 2011. Fast accessibility-based prediction of RNA-RNA interactions. *Bioinformatics*. 27(14):1934–1940.

Taxman, Debra J., Chris B. Moore, Elizabeth H. Guthrie, and Max Tze-Han Huang. 2010. "Short Hairpin RNA (shRNA): Design, Delivery, and Assessment of Gene Knockdown." In *RNA Therapeutics: Function, Design, and Delivery*, edited by Mouldy Sioud, 139–156. Totowa, NJ: Humana Press.

Thiel, Kristina W., and Paloma H. Giangrande. 2009. "Therapeutic Applications of DNA and RNA Aptamers." *Oligonucleotides* 19 (3):209–222. doi:10.1089/oli.2009.0199.

Valdmanis, Paul N, and Mark A Kay. 2017. "Future of rAAV gene therapy: platform for RNAi, gene editing, and beyond." *Human Gene Therapy* 28 (4):361–372.

Wenzel, Anne, Erdinç Akbaşli, and Jan Gorodkin. 2012. "RIsearch: fast RNA–RNA interaction search using a simplified nearest-neighbor energy model." *Bioinformatics* 28 (21):2738–2746. doi:10.1093/bioinformatics/bts519.

Wilson, Ross C., and Jennifer A. Doudna. 2013. "Molecular Mechanisms of RNA Interference." *Annual Review of Biophysics* 42 (1):217–239. doi:10.1146/annurev-biophys-083012-130404.

Wright, P. R., J. Georg, M. Mann, D. A. Sorescu, A. S. Richter, S. Lott, R. Kleinkauf, W. R. Hess, and R. Backofen. 2014. "CopraRNA and IntaRNA: predicting small RNA targets, networks and interaction domains." *Nucleic Acids Res* 42 (Web Server issue):W119–W123. doi:10.1093/nar/gku359.

Wright, P. R., A. S. Richter, K. Papenfort, M. Mann, J. Vogel, W. R. Hess, R. Backofen, and J. Georg. 2013. "Comparative genomics boosts target prediction for bacterial small RNAs." *Proc Natl Acad Sci U S A* 110 (37):E3487–E3496. doi:10.1073/pnas.1303248110.

Xiao, Zhen, Xi Jiang, Mary Lou Beckett, and George L. Wright. 2000. "Generation of a Baculovirus Recombinant Prostate-Specific Membrane Antigen and Its Use in the Development of a Novel Protein Biochip Quantitative Immunoassay." *Protein Expression and Purification* 19 (1):12–21. doi:10.1006/prep.2000.1222.

Xu, Xiaojun, David D. Dickey, Shi-Jie Chen, and Paloma H. Giangrande. 2016. "Structural computational modeling of RNA aptamers." *Methods* 103:175–179. doi:10.1016/j.ymeth.2016.03.004.

Xu, Xiaojun, Peinan Zhao, and Shi-Jie Chen. 2014. "Vfold: a web server for RNA structure and folding thermodynamics prediction." *PloS one* 9 (9):e107504.

Yingling, Yaroslava G, and Bruce A Shapiro. 2007. "Computational design of an RNA hexagonal nanoring and an RNA nanotube." *Nano Letters* 7 (8):2328–2334.

Yu, Ai-Ming, Chao Jian, H Yu Allan, and Mei-Juan Tu. 2019. "RNA therapy: Are we using the right molecules?" *Pharmacology & Therapeutics*. 196:91–104.

Zadeh, Joseph N, Conrad D Steenberg, Justin S Bois, Brian R Wolfe, Marshall B Pierce, Asif R Khan, Robert M Dirks, and Niles A Pierce. 2011. "NUPACK: analysis and design of nucleic acid systems." *Journal of Computational Chemistry* 32 (1):170–173.

Zhou, Jiehua, Haitang Li, Shirley Li, John Zaia, and John J. Rossi. 2008. "Novel Dual Inhibitory Function Aptamer–siRNA Delivery System for HIV-1 Therapy." *Molecular Therapy* 16 (8):1481–1489. doi:10.1038/mt.2008.92.

Zhou, Jiehua, Piotr Swiderski, Haitang Li, Jane Zhang, C. Preston Neff, Ramesh Akkina, and John J. Rossi. 2009. "Selection, characterization and application of new RNA HIV gp 120 aptamers for facile delivery of Dicer substrate siRNAs into HIV infected cells." *Nucleic Acids Research* 37 (9):3094–3109. doi:10.1093/nar/gkp185.

Zuker, Michael. 2003. "Mfold web server for nucleic acid folding and hybridization prediction." *Nucleic Acids Research* 31 (13):3406–3415.

Zuker, Michael, and David Sankoff. 1984. "RNA secondary structures and their prediction." *Bulletin of Mathematical Biology* 46 (4):591–621.

43 RNA Micelles for Therapeutics Delivery and Cancer Therapy

Hongran Yin, Yi Shu, and Peixuan Guo
The Ohio State University, Columbus, Ohio, USA

CONTENTS

INTRODUCTION: RNA MICELLES

Micelles are spherical and supermolecular constructs formed by amphiphilic block copolymers including polymer, peptide and DNA (Shuai, Ai et al., 2004; Qin, Chen et al., 2013; Liu, Zhu et al., 2010; Wu, Sefah et al., 2010; Carrillo-Carrion, tabakhshi-Kashi et al., 2018). They are a core-shell structure composed of both lipophilic and hydrophilic modules. The lipophilic tails will assemble into an oil-like core spontaneously, the most stable form of which has no contact with water. The shape and size of micelles are highly dependent on molecular geometry of the lipophilic/hydrophilic molecules and solution conditions such as pH, temperature and ionic strength. They are capable of encapsulating hydrophobic drugs interiorly or carrying small interfering RNA as branches externally (Alemdaroglu, Alemdaroglu et al., 2008; Roh, Lee et al., 2011). The nanometer scale size of micelles facilitates their extravasations at tumor sites while avoiding fast *in vivo* renal clearance (Torchilin, 2001). Micelles are able to accumulate in tumor site mainly by EPR effect even without a targeting ligand (Matsumura & Maeda, 1986; Salzano, Navarro et al., 2015). They have been widely applied in biomedical fields due to their advantages in delivering therapeutics and enhancing cell permeability (Wu, Sefah et al., 2010; Jin, Park et al., 2017).

RNA-based delivery systems have been extensively investigated and applied in biomedical field. The concept of RNA micelles firstly came up in Guo Lab. The RNA micelles were built by fusing a lipophilic moiety onto one of the pRNA-3WJ branches. This amphiphilic construct spontaneously self-assembles into monodispersed, three-dimensional micellar nanostructures with a lipid core inside and a branched RNA corona outside by hydrophobic interactions in aqueous solution. These pRNA micelles are capable of covalently attaching various functional moieties, including tumor targeting ligands, chemotherapeutic drug, imaging moieties, and co-delivered RNAi components (siRNA, anti-oncogenic miRNA or suppressor miRNA) for combination therapy (Shu, Yin et al., 2018; Yin, Wang et al., 2018).

FORMATION AND CHARACTERIZATION OF RNA MICELLES

The RNA micelles built in Guo Lab were derived from 3WJ scaffold of pRNA in bacteriophage phi29 DNA packaging motor (Figure 43.1a and b) (Shu, Shu et al., 2011). The pRNA-3WJ contains three short RNA strands (named as 3WJ-a, 3WJ-b and 3WJ-c) (Figure 43.1a). The angle of pRNA-3WJ has

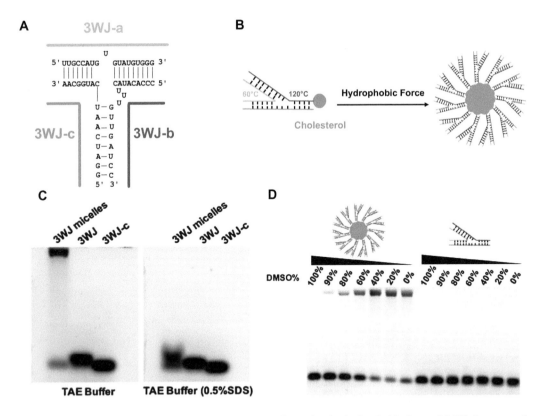

FIGURE 43.1 (a) Illustration of pRNA-3WJ micelles formation by hydrophobic force. (b) 2% Agarose gel comparing pRNA-3WJ micelles formation in TAE running buffer with or without 0.5% SDS. C 2% Agarose gel showing pRNA-3WJ micelles and 3WJ assembly in the presence of increasing percent of DMSO. (Reprinted with permission from Yin H et al., ACS Nano 13, 706–717, 2019; Shu Y et al., Journal of Controlled Release 14, 17–29, 2018)

been studied by the solved crystal structure (Zhang, Endrizzi et al., 2013) (Figure 43.1b). Considering the overall conformation of branched structure, we attached a hydrophobic cholesterol molecule on 3'-end of the pRNA-3WJ-b strand for micelles formation. The conjugation of cholesterol to RNA strand can be realized by solid phase synthesis. The assembly of micelles is simple, easy and efficient. To prove the RNA micelles formation by hydrophobic force, SDS was added to TAE running buffer, resulting in interrupting micelles structure visualized in 2% Agarose gel (Figure 43.1c). In addition, increasing volume percentage of DMSO would dissolve the lipid core, disrupting the hydrophobic force and dissociating pRNA-3WJ micelles into pRNA-3WJ monomer (Figure 43.1d).

The 3WJ branched structure provides multivalence for RNA micelles to incorporate multiple functionalities. Chemotherapeutic drug (Shu, Yin et al., 2018), miRNA or anti-miRNA (Yin, Wang et al., 2018) as well as siRNA as therapeutic agents can be easily incorporated in 3WJ-scaffold for micelles formation. Here, we showed the design of RNA micelles by including anti-miR21 as therapeutics and folate (FA) as a targeting ligand to treat cancer (Figure 43.2a). The formation of micelles has been examined by 2% Agarose gel electrophoresis, indicating the formation of micelles can cause much slower migration than monomer (Figure 43.2b). The average hydrodynamic diameter of 3WJ/FA/anti-miR21 micelles was about 21.51 ± 6.136nm (Figure 43.2c) and the zeta potential was determined to be -25±8.35V (Figure 43.2d) determined by DLS.

Micelles form only when the concentration of amphiphilic monomer is greater than the critical micelle concentration (CMC), and the temperature of the system is greater than the critical micelle

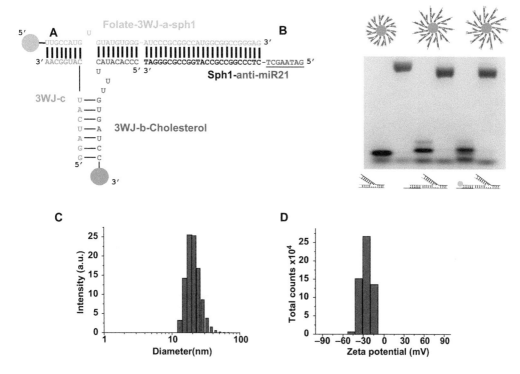

FIGURE 43.2 (a) 2D structure of 3WJ/FA/anti-miR21 micelles (underlined sequence is 8nt LNA modified anti-miR21 seed region). (a) Assembly of RNA micelles assayed by 2% Agarose gel. (Lane from left to right: 3WJ, 3WJ micelles, 3WJ/anti-miR21, 3WJ/anti-miR21 micelles, 3WJ/FA/anti-miR21, 3WJ/FA/anti-miR21 micelles). (c) Size distribution of 3WJ/FA/anti-miR21 micelles. D Zeta potential of 3WJ/FA/anti-miR21 micelles. (Reprinted with permission from Yin H et al., ACS Nano 13, 706–717, 2019)

temperature. Micelles can form spontaneously because of a balance of enthalpy and entropy. CMC can be determined by fluorescence intensity of lipid soluble dye (Jumpertz, Tschapek et al., 2011), the changes of conductivity (Rodríguez & Czapkiewicz, 1995), increase in light scattering (Khouga, Gao et al., 1994) and even solid state electrodes (Mihali, Oprea et al., 2008). Here, Nile Red was utilized to determine CMC as previously reported (Zhang, Zhang et al., 2013). Nile red is a fluorescent probe which is very sensitive to hydrophobicity change (Kurniasih, Liang et al., 2015). It has a very weak fluorescence in hydrophilic environment but strong fluorescence in hydrophobic environment. To determine CMC of RNA micelles, Nile Red was incubated with serial diluted RNA micelles (Figure 43.3a). At the concentrations below CMC, the hydrophobic force of cholesterol molecules was too weak to drive the formation of micelles, and the fluorescence intensity of Nile Red is almost a constant. However, the fluorescence increases dramatically when Nile Red is incorporated into the micelles core once amphiphilic RNA concentration exceeds CMC. For CMC estimation, the fluorescence intensity of Nile Red was plotted as a function of the sample concentration. From the result, the CMC value could be estimated as 100nM, which is the intersection of tangents to the horizontal line of intensity ratio with relatively constant value (Figure 43.3a).

Besides, 3WJ/FA/anti-miR21 micelles showed high stablity in a wide range of temperatures, pH and RNase condition (Figure 43.3b) by 2% Agarose gel electrophoresis after incubation in different conditions. These results demonstrate the advantages for RNA micelles construction and show great potential of micelles for *in vivo* applications.

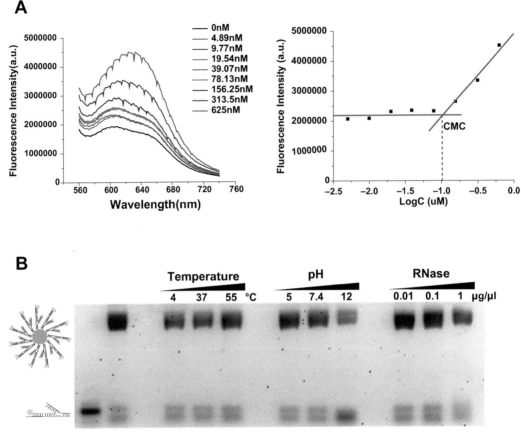

FIGURE 43.3 (a) CMC determined by Nile Red encapsulation assay. (b) Stability study of RNA micelles in different temperature, pH and RNase condition. (Reprinted with permission from Yin H et al., ACS Nano 13, 706–717, 2019)

DELIVERY OF THERAPEUTICS BY RNA MICELLES

RNA micelles incorporating anti-miR21 or Paclitaxel (PTX) have been designed and constructed. These RNA micelles can bind and internalize into cancer cells efficiently assayed by flow cytometry and confocal microscopy (Data not shown here). More functional assays have also been conducted to study delivery capability. The delivery efficiency of anti-miR21 by RNA micelles has been assayed by dual-luciferase-based miR21 reporter system, as we previously reported (Obad, dos Santos et al., 2011; Shu, Li et al., 2015; Binzel, Shu et al., 2016). Higher Renilla to Firefly Luciferase (R/F luc) ratio is correlated with more effective miR21 inhibition. 3WJ/anti-miR21 micelles exhibited better miR21 knockdown effects than 3WJ/anti-miR21 as a result of higher delivery efficacy by RNA micelles (Figure 43.4a). In addition, miR21 downstream tumor suppressor PTEN was upregulated by qRT-PCR study (Figure 43.4b), which can mediate reduced cancer cell proliferation. The effective miR21 silence and downstream gene regulation was achieved by both enhanced miR21 binding *via* LNA-modified anti-miR21 and improved delivery efficacy by RNA micelles.

To further study potency of RNA micelles in the delivery of anti-miR21, tumor suppression experiments were conducted in mice bearing KB tumor xenograft. The RNA micelles carrying anti-miR21 with or without folate ligand were intravenously injected into KB tumor bearing mice. Effective tumor growth inhibition and gene regulation was shown by the result (Figure 43.5a–d).

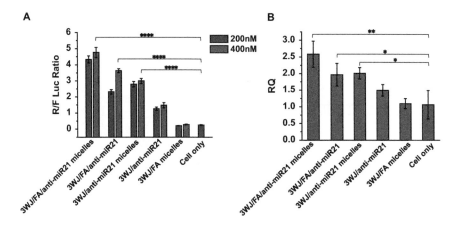

FIGURE 43.4 (a) Dual-luciferase assay demonstrating delivery of anti-miR21 to KB cells. (b) qRT-PCR showing effect of miR21 knockdown on the target gene PTEN expression. (Reprinted with permission from Yin H et al., ACS Nano 13, 706–717, 2019)

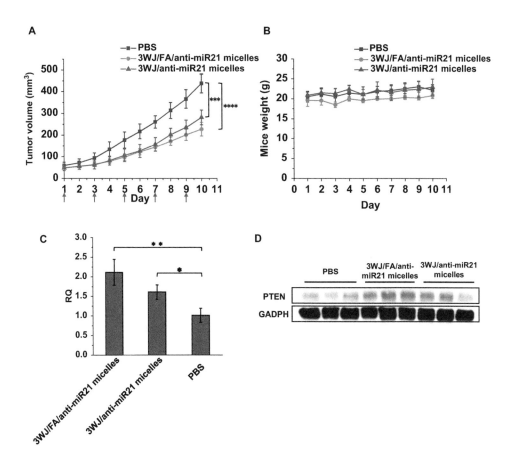

FIGURE 43.5 (a) Tumor regression curve over the course of 5 injections (Red arrow shows day of injection). (b) Mice weight curve during treatment period. (c) qRT-PCR and (d) Western Blot showing the upregulation of PTEN after *in vivo* delivery of anti-miR21. (Reprinted with permission from Yin H et al., ACS Nano 13, 706–717, 2019)

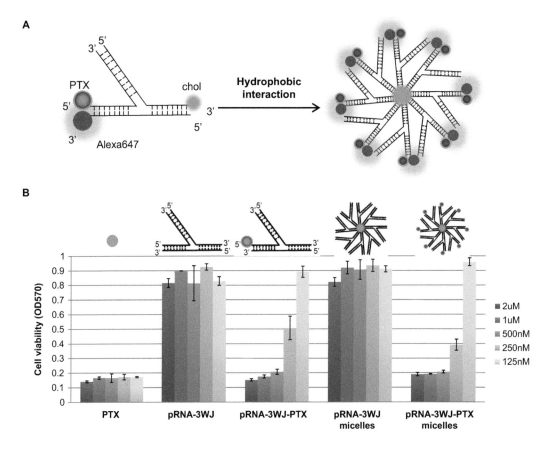

FIGURE 43.6 (a) Conjugating pRNA-3WJ with a lipophilic module (cholesterol, blue), a therapeutic module (PTX, green), and a reporter module (Alexa dye, red). (b) Assay cytotoxicity effects of pRNA-3WJ-PTX micelles by MTT assay. (Reprinted with permission Shu Y et al., Journal of Controlled Release 14, 17–29, 2018)

However, the difference between RNA micelles with or without folate is not significant. The hypothesis we made is RNA micelles can target tumor mainly by passive penetration. In addition, no obvious mice weight or behavior changes were observed (Figure 43.5b), proving that the RNA micelles can be a safe and biocompatible delivery platform. The study revealed that RNA micelles were able to deliver the anti-miR21 and inhibited cancer growth efficiently even without ligands. The tumor-targeting capability of RNA micelles may highly depend on EPR effects.

The RNA micelles with Paclitaxel have ever been rationally designed and investigated {Taxol micelles} (Figure 43.6a). MTT assay showed effective cell growth inhibition treated by pRNA-3WJ-PTX micelles (Figure 43.6b). Paclitaxel conjugated to RNA micelles can be released from RNA strands due to the linker ester hydrolysis in aqueous solution. In addition, RNA micelles by themselves without Paclitaxel showed no effects on cell viability, indicating the low cytotoxicity of RNA micelles scaffold.

IMMUNOGENICITY EVALUATION OF RNA MICELLES

Immunogenicity is another critical factor for safety evaluation of biomaterials. Different from other nanomaterials, RNA itself commonly exists in physiology environment. Recent studies have reported that pRNA nanoparticles display immunologically inert properties {Emil, Sijin CpG

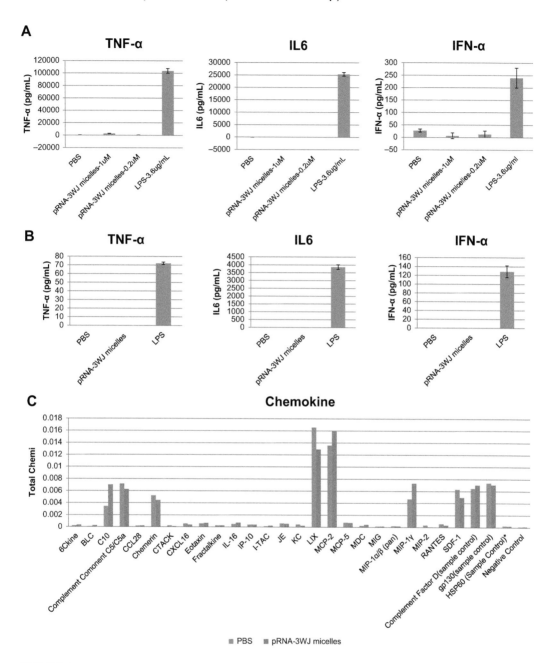

FIGURE 43.7 (a) *In vitro* evaluation of the TNF-α, IL6 and IFN-α production after incubating pRNA-3WJ micelles with mouse macrophage-like RAW 264.7 cells by ELISA assay. (b) *In vivo* evaluation of the TNF-α, IL6 and IFN-α production after injecting pRNA-3WJ micelles into C57BL/6 mice by ELISA assay. (c) *In vivo* chemokines induction profiling for pRNA-3WJ micelle formulation. (Reprinted with permission Shu Y et al., Journal of Controlled Release 14, 17–29, 2018)

paper}. For RNA micelles construct, pro-inflammatory response has been investigated by studying cytokines and chemokines induction after treatment. The results showed RNA micelles at both low and higher dose did not induce IL6 or IFN-α production compared to LPS positive control in mouse macrophage-like cells (Figure 43.7a). No significant cytokines (TNF- α, IL6, IFN-α) induction was detected after *in vivo* injection of RNA micelles into immune competent C57BL/6 mice as shown in Figure 43.7b.

Chemokines are the main proinflammatory mediators (Wang, Liu et al., 2000; Turner, Nedjai et al., 2014). 25 chemokine induction level was studied followed by RNA micelles treatment *in vivo*. Only three chemokines, macrophage inflammatory protein-1 gamma (MIP-1γ), Chemokine10 (C10), and Monocyte chemoattractant protein 2 (MCP2) showed elevated induction compared to PBS control. In summary, RNA micelles induced no or very low pro-inflammatory response.

CONCLUSION AND PERSPECTIVES

Multifunctional RNA micellar nanoparticles have been constructed to carry anti-miR21 or Paclitaxel using the pRNA-3WJ scaffold with the aid of a hydrophobic cholesterol molecule. They remain stable over a wide range of pH, temperature, and RNase environment. Low CMC of RNA micelles provides feasibility in practical application. The RNA micelles can deliver therapeutics efficiently to cancer cells to induce cell apoptosis and cytotoxicity. It is noticeable that RNA micelles were able to deliver the therapeutic cargo and inhibit cancer growth even without a targeting ligand.

The high thermodynamic and chemical stability, multivalence, efficient self-assembly, effective tumor targeting, and penetration make RNA micelles a robust delivery platform. Both the RNA scaffold and lipid moiety can be further modified for optimization. The RNA micelles system enables the delivery of miRNA, anti-miRNA, siRNA, or chemical drugs separately or together for cancer therapy especially for the cancer types whose targeting markers or ligands are not available.

ACKNOWLEDGMENTS

The research in P.G.'s lab was supported by NIH grants [R01EB019036, U01CA151648, U01CA207946] to PG as well as DOD Award [W81XWH-15-1-0052] to DS. Thanks go to Yi Shu for project discussion as well as Sijin Guo, Lora E. McBride, and Dana Driver for manuscript preparation. P.G.'s Sylvan G. Frank Endowed Chair position in Pharmaceutics and Drug Delivery is funded by the CM Chen Foundation.

REFERENCES

Alemdaroglu, F. E., Alemdaroglu, N. C., Langguth, P., Herrmann, A., 2008. DNA block copolymer micelles- A combinatorial tool for cancer nanotechnology. Adv. Mater. 20, 899–902.

Binzel, D., Shu, Y., Li, H., Sun, M., Zhang, Q., Shu, D., Guo, B., Guo, P., 2016. Specific Delivery of MiRNA for High Efficient Inhibition of Prostate Cancer by RNA Nanotechnology. Molecular Therapy 24, 1267–1277.

Carrillo-Carrion, C., Tabakhshi-Kashi, M., Carril, M., Khajeh, K., Parak, W. J., 2018. Taking Advantage of Hydrophobic Fluorine Interactions for Self-Assembled Quantum Dots as a Delivery Platform for Enzymes. Angew. Chem Int. Ed Engl. 57, 5033–5036.

Jin, J. O., Park, H., Zhang, W., de Vries, J. W., Gruszka, A., Lee, M. W., Ahn, D. R., Herrmann, A., Kwak, M., 2017. Modular delivery of CpG-incorporated lipid-DNA nanoparticles for spleen DC activation. Biomaterials 115, 81–89.

Jumpertz, T., Tschapek, B., Infed, N., Smits, S. H., Ernst, R., Schmitt, L., 2011. High-throughput evaluation of the critical micelle concentration of detergents. Anal. Biochem. 408, 64–70.

Khouga, K., Gao, Z., Eisenberg, A., 1994. Determination of the Critical Micelle Concentration of Block Copolymer Micelles by Static Light Scattering. Macromolecules 27, 6341–6346.

Kurniasih, I. N., Liang, H., Mohr, P. C., Khot, G., Rabe, J. P., Mohr, A., 2015. Nile red dye in aqueous surfactant and micellar solution. Langmuir 31, 2639–2648.

Liu, H., Zhu, Z., Kang, H., Wu, Y., Sefan, K., Tan, W., 2010. DNA-based micelles: synthesis, micellar properties and size-dependent cell permeability. Chemistry 16, 3791–3797.

Matsumura, Y., Maeda, H., 1986. A new concept for macromolecular therapeutics in cancer chemotherapy: mechanism of tumoritropic accumulation of proteins and the antitumor agent smancs. Cancer Res 46, 6387–6392.

C. Mihali, G. Oprea, E. Cical, 2008. Determination of Critical Micelar Concentration of Anionic Surfactants Using Surfactants–Sensible Electrodes. Chem. Bull. "POLITEHNICA" Univ. (Timisoara) 53, 159–162.

Obad, S., dos Santos, C. O., Petri, A., Heidenblad, M., Broom, O., Ruse, C., Fu, C., Lindow, M., Stenvang, J., Straarup, E. M., Hansen, H. F., Koch, T., Pappin, D., Hannon, G. J., Kauppinen, S., 2011. Silencing of microRNA families by seed-targeting tiny LNAs. Nat. Genet. 43, 371–378.

Qin, B., Chen, Z., Jin, W., Cheng, K., 2013. Development of cholesteryl peptide micelles for siRNA delivery. J Control Release 172, 159–168.

Rodríguez, J. R., Czapkiewicz, J., 1995. Conductivity and dynamic light scattering studies on homologous alkyl-benzyldimethylammonium chlorides in aqueous solutions. Colloids and Surfaces A: Physicochemical and Engineering Aspects 101, 107–111.

Roh, Y. H., Lee, J. B., Kiatwuthinon, P., Hartman, M. R., Cha, J. J., Um, S. H., Muller, D. A., Luo, D., 2011. DNAsomes: Multifunctional DNA-based nanocarriers. Small 7, 74–78.

Salzano, G., Navarro, G., Trivedi, M. S., De, R. G., Torchilin, V. P., 2015. Multifunctional Polymeric Micelles Co-loaded with Anti-Survivin siRNA and Paclitaxel Overcome Drug Resistance in an Animal Model of Ovarian Cancer. Mol Cancer Ther. 14, 1075–1084.

Shu, D., Li, H., Shu, Y., Xiong, G., Carson, W. E., Haque, F., Xu, R., Guo, P., 2015. Systemic delivery of anti-miRNA for suppression of triple negative breast cancer utilizing RNA nanotechnology. ACS Nano 9, 9731–9740.

Shu, D., Shu, Y., Haque, F., Abdelmawla, S., Guo, P., 2011. Thermodynamically stable RNA three-way junctions for constructing multifuntional nanoparticles for delivery of therapeutics. Nature Nanotechnology 6, 658–667.

Shu, Y., Yin, H., Rajabi, M., Li, H., Vieweger, M., Guo, S., Shu, D., Guo, P., 2018. RNA-based micelles: A novel platform for paclitaxel loading and delivery. J Control Release 276, 17–29.

Shuai, X., Ai, H., Nasongkla, N., Kim, S., Gao, J., 2004. Micellar carriers based on block copolymers of poly (epsilon-caprolactone) and poly (ethylene glycol) for doxorubicin delivery. J Control Release 98, 415–426.

Torchilin, V. P., 2001. Structure and design of polymeric surfactant-based drug delivery systems. J Control Release 73, 137–172.

Turner, M. D., Nedjai, B., Hurst, T., Pennington, D. J., 2014. Cytokines and chemokines: At the crossroads of cell signalling and inflammatory disease. Biochim. Biophys Acta 1843, 2563–2582.

Wang, Z. M., Liu, C., Dziarski, R., 2000. Chemokines are the main proinflammatory mediators in human monocytes activated by Staphylococcus aureus, peptidoglycan, and endotoxin. Journal of Biological Chemistry 275, 20260–20267.

Wu, Y., Sefah, K., Liu, H., Wang, R., Tan, W., 2010. DNA aptamer-micelle as an efficient detection/delivery vehicle toward cancer cells. Proc Natl Acad Sci U S A 107, 5–10.

Yin, H., Wang, H., Li, Z., Shu, D., Guo, P., 2018. RNA Micelles for Systemic Delivery of Anti-miRNA for Cancer Targeting and Inhibition without Ligand. ACS Nano.

Zhang, H., Endrizzi, J. A., Shu, Y., Haque, F., Sauter, C., Shlyakhtenko, L. S., Lyubchenko, Y., Guo, P., Chi, Y. I., 2013. Crystal Structure of 3WJ Core Revealing Divalent Ion-promoted Thermostability and Assembly of the Phi29 Hexameric Motor pRNA. RNA 19, 1226–1237.

Zhang, A., Zhang, Z., Shi, F., Ding, J., Xiao, C., Zhuang, X., He, C., Chen, L., Chen, X., 2013. Disulfide crosslinked PEGylated starch micelles as efficient intracellular drug delivery platforms. Soft Matter 9, 2224–2233.

44 Bacteriophage RNA Leading the Way in RNA Nanotechnology for Targeted Cancer Therapy

Farzin Haque and Haibo Hu
Nanobio Delivery Pharmaceutical Co. Ltd, Columbus, OH, USA

Peixuan Guo
The Ohio State University, Columbus, OH, USA

CONTENTS

INTRODUCTION

Cancer is a complex systemic disease and despite significant advances in our understanding of tumor biology and development of drugs, the treatment strategy (surgical resection followed by chemo- and/or radiation therapies and targeted therapies and immunotherapy) has remained the same over the last decades. In 2013, the extraordinary clinical success of immunotherapy was acknowledged by the Editors of Science Magazine with the designation of "Breakthrough of the Year". A major ongoing quest in cancer therapy is to develop a platform that can selectively target tumor cells and deliver therapeutics to inhibit tumor growth and prevent metastasis, while at the same time avoid non-specific entrapment in healthy cells/tissues that can result in significant toxicity and immune responses. A wide range of nanoparticle platforms have been pursued over the years, including liposomes, polymers, viral, metallic, protein, DNA, RNA, inorganic, and organic materials for specific applications in nanotheranostics and nanomedicine [1, 2]. Although nanoparticles have made considerable progress in the field and expanded our understanding of how targeted drug delivery can work, only a handful have been deemed suitable from a therapeutic delivery standpoint [3, 4]. This is also reflected in the clinic where only a few nanoparticles have been approved for clinical use, which includes *Doxil* (PEGylated liposome formulation of Doxorubicin), *Myocet* (non-PEGylated liposomal formulation of Doxorubicin), *Abraxane* (albumin formulation of Paclitaxel), *DaunoXome* (liposomal formulation of Daunorubicin), *DepoCyt* (liposomal formulation of cytarabine), *Onco-TCS* (liposomal formulation of vincristine), and *Oncaspar* (PEGylated L-asparaginase). These formulations were approved primarily due to significant reduction in systemic toxicity of the administered clinical therapeutics.

DOI: 10.1201/9781003001560-49

In addition to chemical drugs and protein-based drugs (such as, antibodies) that have been traditionally used as therapeutics, RNA-based drugs are being increasingly pursued due to their ability to 'drug the undruggable targets' in oncology [5, 6]. Despite our increased understanding and clinical value of siRNA, anti-miRNA, and miRNA, It is surprising that, in 2021, FDA approves Novartis Leqvio®(inclisiran), fist-in-class siRNA to lower cholesterol and keep it low with two doses a year. Leqvio provides effective and sustained LDL-C reduction of up to 52% vs. placebo for certain people with atherosclerotic cardiovascular disease (ASCVD) on maximally tolerated statin therapy. Opportunities and challenges coexist. The research and development of nucleic acid drugs has been concerned by everyone; although in the stability of nucleic acid drugs, immunogenicity, half-life, there are many difficulties in cell uptake efficiency and endosome escape, so the study of chemical modification and delivery system can be improved to solve the above problems and greatly promote the further development of nucleic acid drugs. RNA nanotechnology [7] spearheaded by bacteriophage packaging RNA (pRNA) platform [8, 9] has made significant progress in this regard. Target site accumulation, high therapeutic efficacy in mice, ease of manufacturing, low toxicity, and immunogenicity demonstrate the potentials to bridge the translational gap [10–23]. Herein, we review the defining aspects of pRNA platform in RNA nanotechnology and how RNA nanoparticles can overcome series of clinical translational barriers.

PACKAGING RNA PLATFORM IN RNA NANOTECHNOLOGY

The bacteriophage phi29 DNA packaging motor is geared by a hexameric packaging RNA (pRNA) ring [8] (Figure 44.1a). Each pRNA monomer contains two functional domains that fold separately: a central domain consisting of two interlocking loops (denoted as right- and left-hand loops) and a helical DNA packaging domain that is located at the 5′/3′ paired ends [24, 25]. The two domains are connected by a 3WJ (three-way junction) motif [26, 27] (Figure 44.1b–e). In 1998, using reengineered pRNA fragments, dimer, trimer, and hexamer RNA nanoparticles were constructed *via* hand-in-hand interactions [9], thereby initiating the field of RNA nanotechnology [7, 16, 28–33] (Figure 44.1c–d). Over the last decade, three structural features of pRNA have been explored for constructing multivalent RNA nanoparticles with defined size, shape, and stoichiometry. These include loop–loop interactions using reengineered interlocking right- and left-hand loops [9, 15,

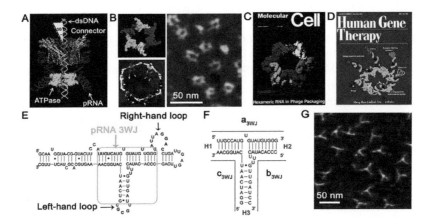

FIGURE 44.1 Bacteriophage phi29 packaging RNA platform. (a) Phi29 DNA packaging motor. (b) pRNA hexamer crystal structure (top-left) and AFM images (right). (c) Pioneering work in 1998 (published in *Molecular Cell*; featured in *Cell*) by PI Peixuan Guo proving the concept of RNA nanotechnology by bottom-up assembly to build pRNA hexamers. (d) Cover of *Human Gene Therapy*. (e) pRNA monomer showing 3WJ domain. (f) 3WJ scaffold composed of three short strands. (g) AFM images of arm extended 3WJ.

34], foot-to-foot interactions using a palindrome sequence mediated self-assembly [15, 34], and branch extension using the ultra-stable 3WJ motif as a scaffold [11, 19, 26, 35–38] (Figure 44.1f–g). The 3WJ platform [26], in particular, has been a major breakthrough in the field as it has several favorable attributes that are attractive for targeted drug delivery.

OVERCOMING BOTTLENECKS IN TARGETED DRUG DELIVERY

Over the last decade, significant progresses have been made in understanding nanoparticle- mediated tumor drug delivery [1]. Since cancer is a systemic disease, herein, we focus on systemic targeting approach, which is the preferred route for a wide range of drug formulations, as opposed to local delivery. In order to be effective, nanoparticles first need to be efficiently formulated with drugs, and upon systemic delivery, they have to navigate through vasculature to reach tumor sites while avoiding healthy organs. The nanoparticles need to retain long enough in the tumor microenvironment to either release drugs (for instance, chemotherapeutics) in the interstitial environment or internalize into tumor cells and release the drug (for instance RNAi) in the cytosol at an optimal rate for pharmacological effectiveness [39]. While the requirements seem rather straightforward, considerable engineering and improvements in nanoparticle strategies are needed for successful clinical translation. Here, we delineate some of the major bottlenecks in the drug delivery field and how RNA nanotechnology has to a large extent overcome many of these barriers (Figure 44.2).

1. *Formulation challenges:* While mono-disperse nanoparticle scaffolds of various compositions can be routinely synthesized, batch-to-batch formulation of nanoparticles with therapeutic modules of various compositions is a well-known bottleneck from a manufacturing and quality control perspective [39–41]. RNA is a polymer composed of four nucleotides (A, U, G, and C). Each nucleotide is composed of a sugar (ribose), a nucleobase, and a phosphate. The nucleotides are covalently linked through 3'→5' phosphodiester bonds. RNA nanoparticles typically employ modular design principles composed of short strands. Each strand can be designed to harbor a targeting (such as, RNA aptamer) or therapeutic (such as, siRNA) ligand simply by fusing the required sequences with the scaffolding sequences [16]. Each strand (with or without functional modules) is typically designed to be shorter than 80 nucleotides (limit for chemical synthesis) and synthesized chemically followed by HPLC purification

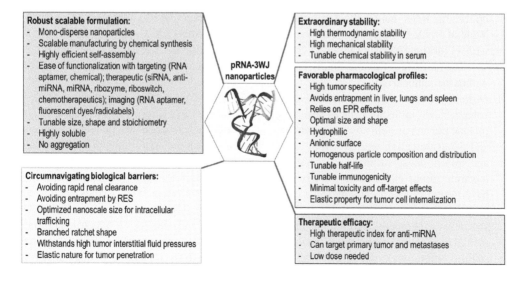

FIGURE 44.2 Favorable attributes of pRNA nanoparticles for targeted drug delivery to cancer cells.

with high efficiency. The strands are then mixed in equimolar ratio to self-assemble the RNA nanoparticle *via* canonical and non-canonical base pairing [42, 43]. The assembly efficiency is extremely high as in the case of pRNA-3WJ-based constructs (see thermodynamic stability section) [26]. The manufacturing process is scalable and particles are homogenous across batches displaying extremely narrow size distribution. While RNA nanoparticles are ideally suited for formulating RNA-based drugs (such as, aptamer, siRNA, miRNA, anti-miRNA, ribozyme, riboswitches), chemotherapeutic drugs and chemical ligands (such as folate or galactose) can be incorporated using standard conjugation strategies (such as click chemistry) [29, 33]. Since RNA is stable in organic solvents, poorly soluble chemicals can be conjugated and it functionalizes strand purified prior to assembly for quality control. The size and shape of RNA nanoparticles are tunable based on the length of the strands and the presence of motifs (such as junction, k-turn, etc.) [11, 15, 19, 36, 37, 44]. The resulting RNA nanoparticles have a defined sequence, structure and molecular weight with a polydispersity index of 1 [30]. Finally, given the therapeutic end-goal, the manufacturing process can yield endotoxin-free final products.

2. ***Thermodynamic instability:*** Dissociation of nanoparticles in the circulation especially multi-component complexes is a widespread problem [2, 39, 40]. Small-molecule drugs partitioning out of nanoparticles is a persistent challenge, such as taxane carriers. The field of RNA nanotechnology also faced similar problems using initial pRNA dimer-based therapeutic nanoparticles [15]. The discovery of ultra-stable properties of pRNA-3WJ scaffold was a major breakthrough in the field. The 3WJ scaffold can assemble from its three component strands with unusually high affinity without any metal salts. It is resistant to denaturation in the presence of 8 M urea and does not dissociate at ultra-low concentrations [26, 45, 46]. The T_m is ~60°C [26] and the slope of the melting curve is close to 90° angle, indicating extremely low free energy ($\Delta G^{\circ}_{37}= -28$ kcal/mol) [45] and near-simultaneous assembly of the three fragments [46]. Moreover, the thermodynamic stability of the scaffold can drive the correct folding of the incorporated RNA modules and display authentic functionalities [47]. This is particularly important for formulating homogenous nanoparticles with desired functional modules.

3. ***Chemical Instability:*** Chemical stability of nanoparticles determines the blood circulation time-frame. Intravascular degradation of RNA is a well-known phenomenon as unmodified RNA is degraded in serum resulting in short half-life and rapid clearance [48]. Typically, nanoparticles are PEGylated to enhance the half-life to increase the probability of tumor update. In RNA nanotechnology, the scaffold and functional modules can be exclusively made of RNA. Simple chemical modifications, such as substitution of the 2' hydroxyl group with a Fluorine (2'-F) or O-methyl (2'-O-Me) can make the RNA resistant to degradation [48, 49]. The half-life is tunable simply based on the location and degree of chemical modification within the nanoparticle.

4. ***Difficulty in navigating past biological barriers:*** A drug carrier, either nanoparticles or soluble macromolecules, faces several obstacles in the body after systemic delivery. These include plasma protein absorption; uptake by immune cells (monocytes, platelets, leukocytes, and dendritic cells); entrapment by the RES (reticuloendothelial system) in the liver, lungs, and spleen; dense stromal tissue; high interstitial tumor fluid pressure; and interactions with dynamic tumor extracellular matrix (ECM) components [39, 40, 50]. Nanoparticle uptake by different types of cells depends on various uptake pathways [51]. Studies conducted in various tumor models showed that RNA nanoparticles can navigate past biological barriers and deliver therapeutics to inhibit tumor growth [10–13, 15–18, 20, 21, 23, 26]. The fundamental biophysical properties of nanoparticles, such as size, shape, surface charges, mechanic stability, thermodynamic stability all play a major role to overcome the biological barriers. The pRNA-3WJ exhibits unusual mechanical anisotropy and can withstand forces up to 300 pN [52]. The sequence, local structural environment, and metal ion concentrations [53, 54]

influence RNA stiffness and plasticity at the molecular level [55]. The mechanical stability [52] coupled with thermodynamic stability [26] is likely the dominant factor in enabling RNA nanoparticles to remain intact and functional in high interstitial tumor pressure environments. The elasticity coupled with optimal size (10–30 nm depending upon presence of functional modules) and branched ratchet shape assists the RNA nanoparticles to circumnavigate past the biological barriers.

5. *Unfavorable biodistribution in vivo:* Two approaches have been widely used for tumor targeting: passive (nanoparticles without any tumor-specific targeting ligand) and active targeting (nanoparticles harboring specific targeting ligand, for instance an aptamer targeting to tumor cell surface receptor) [56]. In both cases, EPR (Enhanced Permeability and Retention) effects based on leaky tumor vasculature and suppressed lymphatic filtration play a role in delivering nanoparticles to tumor microenvironment [57]. To date, the vast majority of intravenous administered nanoparticles are taken up by the RES in the liver, spleen, and lungs [1, 3]. RNA nanoparticles, however, are different and as shown in the series of mouse tumor models, they can selectively accumulate in tumors while avoiding entrapment in the liver, lungs, and spleen [10–13, 15–18, 20, 21, 23, 26]. The size, shape, and surface properties of nanoparticles significantly influence tumor partitioning of nanoparticles, preventing diffusion back into the circulation and the rate of clearance [50]. Generally, nanoparticles <10 nm are rapidly excreted through the renal excretion pathway, while nanoparticles >100 nm diameter are likely to be cleared by RES. RNA nanoparticles are designed to be larger than the fenestrations of normal vasculatures (inter-endothelial junctions are <8 nm wide) but small enough to enter tumors with leaky vasculature of various widths (inter-endothelial junctions ranging from 40–80 nm on average and as large as 1 mm) and retain in the tumor microenvironment. The branched morphology of pRNA-3WJ nanoparticles further helps in enhanced tumor retention following extravasation. Nanomaterial hydrophobicity influences binding to plasma proteins. RNA nanoparticles are highly hydrophilic and therefore minimize interactions with plasma proteins. RNA nanoparticles are negatively charged and thus minimize interactions with normal cell membranes, which are negatively charged. Monodisperse RNA nanoparticles also help to control the biodistribution profile. Finally, the deformability of nanoparticles is an important parameter that impacts cellular uptake and intracellular delivery [58]. Recently, pRNA polygons have been constructed by stretching the internal angle of the 3WJ thereby demonstrating its intrinsic elastic property [19]. Elastic deformation of RNA nanoparticles greatly facilitates tumor uptake *via* receptor-mediated endocytosis [59].

Low molecular weight chemotherapeutic drugs can freely pass through vasculature of normal and tumor cells and diffuse back in circulation inducing significant off-target toxicity. Macromolecular drug conjugates are attractive since they can enter tumor vasculature but not diffuse back into circulation or end up in lymphatic system. Premature drug release in unintended sites can cause major toxicity. Given that RNA nanoparticles display favorable biodistribution profiles, it is feasible to conjugate chemotherapeutic drugs and trigger release of the drug (such as, acid-labile bond) once it enters the tumor microenvironment. Recently, doxorubicin was intercalated in GC-rich sequences in the 3WJ RNA nanoparticles for delivery to ovarian tumors [60].

6. *Low drug efficacy in vivo:* Intracellular trafficking is a major consideration in drug efficacy. Typically, nanoparticles endocytose into cells and the vast majority gets trapped in late endosomes and thus have limited therapeutic index (expressed as the ratio of the therapeutic effect of the drug to the toxicity). Various mechanisms have been exploited to enhance endosomal escape, including use of fusogenic lipids and peptides and/or pH-sensitive lipo- or polyplexes [51]. RNA nanoparticles harboring siRNA have shown modest tumor regression profiles, indicating entrapment in endosomes, even though the biodistribution profiles showed excellent tumor localization [23]. However, RNA nanoparticle-mediated anti-miRNA delivery turned out to be significantly better and sustained tumor regression was observed at low doses

in three different tumor models – orthotopic breast cancer [10], subcutaneous prostate cancer [12] and orthotopic glioblastoma [21] bearing mice. The findings indicate that anti-miRNA component is able to escape the endosomes (without the use of any endosome disrupting agents) efficiently as opposed to siRNA. Targeting miRNA is particularly attractive for cancer therapy since one miRNA can regulate a broad set of genes simultaneously and can therefore address the heterogeneous nature of cancer and drug resistance [61].

7. *Unfavorable pharmacological profiles:* In plain terms, pharmacokinetics (PK) is what the body does to the drug and pharmacodynamics (PD) is what the drug does to the body. PK/PD parameters of nanoparticles depend on a number of factors, including size, shape, surface charge, composition, stability, deformability, dose, and route of administration of nanoparticles as well as tumor biology [50]. There is significant variation in PK/PD properties of nanoparticles studied over the years. The pRNA nanoparticle scaffolds exhibit favorable pharmacokinetic properties in mice after systemic administration [18], such as extended half-life *in vivo* (5–12 hr compared to 0.25–0.75 hr for siRNA), clearance: <0.13 L/kg/hr, and, volume of distribution: 1.2 L/kg. Standard panel clinical chemistry (including PT, aPTT), liver enzymes AST, ALT, and LDH (assessment of liver toxicity); BUN and creatinine (assessment of renal toxicity), measurement of serum INF-α, TNF-α, IL-6, and IFN-γ (assessment of off-target effects); and clinical pathology analysis were all reported to be well within the normal range after RNA nanoparticle treatment [18, 23].

Carrier-mediated toxicities are a well-encountered problem. Cationic nanoparticles typically invoke Type-I and II interferon responses plus exhibit dose-dependent toxicities. RNA nanoparticles are anionic in nature and due to its intrinsic properties, it does not induce interferon-I or cytokine production in mice. Repeated intravenous administrations in mice (up to 30 mg/kg) did not show any signs of toxicity [18]. Additionally, RNA is a biocompatible and biodegradable nanomaterial from a physiological standpoint.

Off-target effects of RNAi is another hurdle and cross-hybridization with unintended transcripts has been shown to non-specific gene silencing with siRNA or unintended translational suppression with miRNAs. Chemical modification is the key to control such off-target effects. The RNA nanoparticles are 2'-F modified and the anti-miRNA component is a LNA (Locked Nucleic Acid). The LNA part is a 8-mer sequence that can bind to the miRNA seed region with high affinity and specificity [62], while minimizing off-target effects [10, 12, 21].

8. *Unfavorable immune responses:* Immunotoxicity is a major concern for therapeutic delivery. It is well known that innate immune response to siRNA can trigger both Toll-like receptor (TLR)- and non-TLR-mediated responses [63]. Immune response is highly sequence dependent. For instance, UG-rich sequences in RNA potently activate TLR-7 and TLR-8 while AU-rich sequences in RNA activate TLR8. The pRNA-3WJ nanoparticles by themselves are non-immunogenic [18, 19], but after incorporating CpG motifs [64], they can become strongly immunostimulatory (cytokine TNF-α and IL-6 induction) [19]. Chemical modifications of RNA such as 2'-F or LNA have been shown to evade innate immune recognition. Furthermore, size and shape of nanoparticles also play a critical part in immune system activation, as demonstrated by RNA nanoparticles of various shapes harboring single and multiple CpGs [19]. RNA nanotechnology offers a novel platform to tune the immunomodulatory properties.

9. *Regulatory approval questions:* Very often nanoparticles have complex compositions of several modules, such as biological/organic/inorganic scaffolds, RNAi, protein ligand, and/or chemical drugs. These compositions present unique regulatory approval issues as it is imperative to consider the PK/PD of individual components as well as overall system. RNA nanoparticles are particularly suited for delivering RNA-based drugs as the scaffold, targeting ligands (RNA aptamers), and therapeutic modules (siRNA, miRNA, anti-miRNA, ribozymes, riboswitches) are composed exclusively of RNA nucleotides, and this can greatly facilitate regulatory approval.

In 2021, the FDA issued a new recommended guidance on the process for applying for a New Drug (IND) application for personalized antisense oligonucleotide drugs, providing clear information about early development and the IND submission process, It's the first step for the FDA to work with those who are developing these personalized drug products. FDA noted that because the number of patients is too small to obtain efficacy and safety data through clinical trials, developers are required to provide convincing in vitro and/or animal data for proof-of-concept treatment in meetings with FDA prior to submitting an IND. These data are important to measure the potential benefits to patients.

In terms of safety, for ASO therapy, researchers should provide data to evaluate its off-target effect. In addition, the FDA noted that a 3-month toxicology study could be used to support the initiation of dose-increasing long-term treatment in humans. For a disease that progresses so rapidly that it can kill or cause irreversible symptoms in a short period of time, the FDA needs at least two weeks' worth of data from a three-month study. Researchers will then report interim data periodically and report full data to the FDA as soon as possible after the three-month study is completed. Dose selection is important when administering drugs for the first time in humans, and the FDA guidelines recommend that researchers clearly describe and support the strategy for selecting initial doses, including how to calculate safety boundaries for doses used in humans based on doses tested in animal studies.

FUTURE OUTLOOK TOWARDS CLINICAL USE

Nanoparticles have undergone major breakthroughs, and in particular RNA nanoparticles spearheaded by pRNA motif have come a long way and hold significant translational potentials to the clinic. But, there are several gaps in knowledge, mostly in tumor biology aspects, in order to develop 'smarter' RNA nanoparticles rationally designed with fine-tunable physiochemical properties and improved outcome. These include the following points: (1) There are a limited number of RNA aptamers available compared to a wide range of protein antibodies that can target known tumor cell-surface proteins for specific uptake and delivery. (2) Tumor heterogeneity and dynamic landscape can affect distribution of RNA nanoparticles in the tumor microenvironment. (3) An in-depth understanding of the trafficking of RNA nanoparticles upon internalization inside tumor cells is needed specially to understand the mechanism of endosome escape of RNA nanoparticles for enhanced cytosol delivery. (4) The heterogenous expression of tumor cell surface markers and how they affect targeted RNA nanoparticle uptake need to be studied. (5) We need an understanding of factors that affect human EPR outcomes in both primary and metastatic lesions for enhanced partitioning of RNA nanoparticles into tumors. (6) More data are needed for determining critical RNA nanoparticle dosing necessary for high bio-efficacy while minimizing signs of toxicity even with low/negligible off-target accumulation as observed with RNA nanoparticles. (7) As tumor dynamics change over time, non-invasive imaging needs to be coupled with RNA nanoparticle delivery to obtain quantitative pharmacological profiles including clearance, target site accumulation, and therapeutic efficacy over the course of therapy to determine the conditions where high spatial specificity and high temporal specificity are attained. Longitudinal monitoring of the therapeutic intervention is necessary to determine the long-term effects. (8) Activity of mononuclear phagocytic system and extent of macrophage infiltration which can vary a great deal between and within tumors need to be understood for effective RNA nanoparticle delivery in various tumors. (9) While targeting and treatment of primary tumor is certainly noteworthy, early phase clinical trials will involve targeting metastatic burdens, and this will require further investigations to advance clinical developments. RNA nanoparticle is capable of targeting lymph node, liver, and lung metastases simultaneously in colorectal cancer mouse models, and this is certainly encouraging [13]. (10) New mouse models beyond xenografts that more accurately recapitulate the complex biological environment in humans need to be tested. In addition, reliable biomarkers to assess therapeutic responses are needed. (11) It is critical to pick an effective therapeutic target that can cause efficient tumor reduction and expressed preferentially in tumors. Ideally, an integrative approach for target selection should be adopted taking into consideration patient-omics data, high-throughput whole genome screens, and tumor-specific gene expression patterns. A major question remains whether single therapy can truly block tumor growth and prevent metastasis in the long run. Realistically, a combinatorial therapy coupling RNA nano-based therapy with traditional therapies can potentially address recurrence, intrinsic or acquired chemo-resistance (responsible for treatment failure in 90% metastatic cases), and tumor heterogeneity. (12) As outlined by FDA, the general pathway to clinical translation and commercialization has three major hurdles: filing of an Investigational New Drug (IND); clinical trials, and New Drug Application (NDA). The key steps are the costs of manufacturing scale-up and completing GLP and GCP compliant preclinical safety, toxicity, and efficacy studies that are adequately designed so that findings can be accepted by the FDA. (13). The recently FDA-approved RNA-based treatment for spinal muscular atrophy (SMA) and several other compounds in advanced clinical trials may signal new opportunities for advancing RNA nanoparticle designs as treatments for targeted cancer therapy. (14) Similar to targeted cancer therapy, RNA nanoparticles also provide optimal practice of personalized medicine in treating some diseases with tractable targets. This will require extensive financial support, typically involving government agencies, venture firms, and pharmaceutical companies. Nanobio Delivery Pharmaceutical Company Limited was used to be established to undertake this challenge.

ACKNOWLEDGMENTS

This research was supported by NIH grant U01CA151648 to PG. Funding to Peixuan Guo's Sylvan G. Frank Endowed Chair position in Pharmaceutics and Drug Delivery is by the C. M. Chen Foundation.

CONFLICT OF INTEREST

PG is a consultant of Oxford Nanopore, and Nanobio Delivery Pharmaceutical Co., Ltd. PG is the cofounder of P&Z Biological Technology. FH and HH work for Nanobio Delivery Pharmaceutical Co., Ltd., which specializes in RNA Nanoparticle production.

REFERENCES

1. Grodzinski, P.; Torchilin, V.; (Editors) *Advanced Drug Delivery Reviews: Cancer Nanotechnology*; Volume 66 ed.; Elsevier: 2014, pp. 1–116.
2. Pelaz, B.; Jaber, S.; de Aberasturi, D. J.; Wulf, V.; Aida, T.; de la Fuente, J. M.; Feldmann, J.; Gaub, H. E.; Josephson, L.; Kagan, C. R.; Kotov, N. A.; Liz-Marzan, L. M.; Mattoussi, H.; Mulvaney, P.; Murray, C. B.; Rogach, A. L.; Weiss, P. S.; Willner, I.; Parak, W. J. The state of nanoparticle-based nanoscience and biotechnology: progress, promises, and challenges. *ACS Nano*. 2012, *6* (10), 8468–8483.
3. Zamboni, W. C.; Torchilin, V.; Patri, A. K.; Hrkach, J.; Stern, S.; Lee, R.; Nel, A.; Panaro, N. J.; Grodzinski, P. Best practices in cancer nanotechnology: perspective from NCI nanotechnology alliance. *Clin. Cancer Res.* 2012, *18* (12), 3229–3241.
4. Grodzinski, P.; Farrell, D. Future opportunities in cancer nanotechnology--NCI strategic workshop report. *Cancer Res.* 2014, *74* (5), 1307–1310.
5. Hough, S. R.; Wiederholt, K. A.; Burrier, A. C.; Woolf, T. M.; Taylor, M. F. Why RNAi makes sense. *Nat. Biotech.* 2003, *21* (7), 731–732.
6. Pecot, C. V.; Calin, G. A.; Coleman, R. L.; Lopez-Berestein, G.; Sood, A. K. RNA interference in the clinic: challenges and future directions. *Nat Rev. Cancer* 2011, *11* (1), 59–67.
7. Guo, P. The emerging field of RNA nanotechnology. *Nat. Nanotechnol.* 2010, *5* (12), 833–842.
8. Guo, P.; Erickson, S.; Anderson, D. A small viral RNA is required for *in vitro* packaging of bacteriophage phi29 DNA. *Science* 1987, *236*, 690–694.
9. Guo, P.; Zhang, C.; Chen, C.; Trottier, M.; Garver, K. Inter-RNA interaction of phage phi29 pRNA to form a hexameric complex for viral DNA transportation. *Mol. Cell.* 1998, *2*, 149–155.

10. Shu, D.; Li, H.; Shu, Y.; Xiong, G.; Carson, W. E.; Haque, F.; Xu, R.; Guo, P. Systemic delivery of anti-miRNA for suppression of triple negative breast cancer utilizing RNA nanotechnology. *ACS Nano* 2015, *9*, 9731–9740.

11. Haque, F.; Shu, D.; Shu, Y.; Shlyakhtenko, L.; Rychahou, P.; Evers, M.; Guo, P. Ultrastable synergistic tetravalent RNA nanoparticles for targeting to cancers. *Nano Today* 2012, *7*, 245–257.

12. Binzel, D.; Shu, Y.; Li, H.; Sun, M.; Zhang, Q.; Shu, D.; Guo, B.; Guo, P. Specific Delivery of MiRNA for High Efficient Inhibition of Prostate Cancer by RNA Nanotechnology. *Molecular Ther.* 2016, *24*, 1267–1277.

13. Rychahou, P.; Haque, F.; Shu, Y.; Zaytseva, Y.; Weiss, H. L.; Lee, E. Y.; Mustain, W.; Valentino, J.; Guo, P.; Evers, B. M. Delivery of RNA nanoparticles into colorectal cancer metastases following systemic administration. *ACS Nano.* 2015, *9* (2), 1108–1116.

14. Rychahou, P.; Shu, Y.; Haque, F.; Hu, J.; Guo, P.; Evers, B. M. Methods and assays for specific targeting and delivery of RNA nanoparticles to cancer metastases. *Methods Mol. Biol.* 2015, *1297*, 121–135.

15. Shu, Y.; Haque, F.; Shu, D.; Li, W.; Zhu, Z.; Kotb, M.; Lyubchenko, Y.; Guo, P. Fabrication of 14 Different RNA Nanoparticles for Specific Tumor Targeting without Accumulation in Normal Organs. *RNA* 2013, *19*, 766–777.

16. Shu, Y.; Shu, D.; Haque, F.; Guo, P. Fabrication of pRNA nanoparticles to deliver therapeutic RNAs and bioactive compounds into tumor cells. *Nat. Protoc.* 2013, *8* (9), 1635–1659.

17. Lee, T. J.; Haque, F.; Shu, D.; Yoo, J. Y.; Li, H.; Yokel, R. A.; Horbinski, C.; Kim, T. H.; Kim, S.-H.; Nakano, I.; Kaur, B.; Croce, C. M.; Guo, P. RNA nanoparticles as a vector for targeted siRNA delivery into glioblastoma mouse model. *Oncotarget* 2015, *6*, 14766–14776.

18. Abdelmawla, S.; Guo, S.; Zhang, L.; Pulukuri, S.; Patankar, P.; Conley, P.; Trebley, J.; Guo, P.; Li, Q. X. Pharmacological characterization of chemically synthesized monomeric pRNA nanoparticles for systemic delivery. *Molecular Ther.* 2011, *19*, 1312–1322.

19. Khisamutdinov, E.; Li, H.; Jasinski, D.; Chen, J.; Fu, J.; Guo, P. Enhancing immunomodulation on innate immunity by shape transition among RNA triangle, square, and pentagon nanovehicles. *Nucleic Acids Res.* 2014, *42*, 9996–10004.

20. Zhang, Y.; Leonard, M.; Shu, Y.; Yang, Y.; Shu, D.; Guo, P.; Zhang, X. Overcoming Tamoxifen Resistance of Human Breast Cancer by Targeted Gene Silencing Using Multifunctional pRNA Nanoparticles. *ACS Nano.* 2017, *11* (1), 335–346.

21. Lee, T. J.; Yoo, J. Y.; Shu, D.; Li, H.; Zhang, J.; Yu, J. G.; Jaime-Ramirez, A. C.; Acunzo, M.; Romano, G.; Cui, R.; Sun, H. L.; Luo, Z.; Old, M.; Kaur, B.; Guo, P.; Croce, C. M. RNA Nanoparticle-Based Targeted Therapy for Glioblastoma through Inhibition of Oncogenic miR-21. *Mol. Ther.* 2017.

22. Lee, T. J.; Haque, F.; Vieweger, M.; Yoo, J. Y.; Kaur, B.; Guo, P.; Croce, C. M. Functional assays for specific targeting and delivery of RNA nanoparticles to brain tumor. *Methods Mol. Biol.* 2015, *1297*, 137–152.

23. Cui, D.; Zhang, C.; Liu, B.; Shu, Y.; Du, T.; Shu, D.; Wang, K.; Dai, F.; Liu, Y.; Li, C.; Pan, F.; Yang, Y.; Ni, J.; Li, H.; Brand-Saberi, B.; Guo, P. Regression of gastric cancer by systemic injection of RNA nanoparticles carrying both ligand and siRNA. *Scientific Reports* 2015, *5*, 10726.

24. Zhang, C. L.; Lee, C.-S.; Guo, P. The proximate 5′ and 3′ ends of the 120-base viral RNA (pRNA) are crucial for the packaging of bacteriophage f29 DNA. *Virology* 1994, *201*, 77–85.

25. Cairns, J.; Overbaugh, J.; Miller, S. The origin of mutants. *Nature* 1988, *335*, 142–145.

26. Shu, D.; Shu, Y.; Haque, F.; Abdelmawla, S.; Guo, P. Thermodynamically stable RNA three-way junctions for constructing multifuntional nanoparticles for delivery of therapeutics. *Nat. Nanotechnol.* 2011, *6*, 658–667.

27. Zhang, H.; Endrizzi, J. A.; Shu, Y.; Haque, F.; Sauter, C.; Shlyakhtenko, L. S.; Lyubchenko, Y.; Guo, P.; Chi, Y. I. Crystal Structure of 3WJ Core Revealing Divalent Ion-promoted Thermostability and Assembly of the Phi29 Hexameric Motor pRNA. *RNA* 2013, *19*, 1226–1237.

28. Guo, P.; Haque, F.; Hallahan, B.; Reif, R.; Li, H. Uniqueness, advantages, challenges, solutions, and perspectives in therapeutics applying RNA nanotechnology. *Nucleic Acid Ther.* 2012, *22* (4), 226–245.

29. Shu, Y.; Pi, F.; Sharma, A.; Rajabi, M.; Haque, F.; Shu, D.; Leggas, M.; Evers, B. M.; Guo, P. Stable RNA nanoparticles as potential new generation drugs for cancer therapy. *Adv. Drug Deliv. Rev.* 2014, *66C*, 74–89.

30. Li, H.; Lee, T.; Dziubla, T.; Pi, F.; Guo, S.; Xu, J.; Li, C.; Haque, F.; Liang, X.; Guo, P. RNA as a stable polymer to build controllable and defined nanostructures for material and biomedical applications. *Nano Today* 2015, *10*, 631–655.

31. Guo, P.; Haque, F.; (Editors) *RNA Nanotechnology and Therapeutics*; CRC Press: 2013.pp. 1–593.
32. Guo, P.; Haque, F. *RNA Nanotechnology and Therapeutics - Methods in Molecular Biology, Vol. 1297*; Human Press: 2015.pp. 1–239.
33. Jasinski, D.; Haque, F.; Binzel, D. W.; Guo, P. Advancement of the Emerging Field of RNA Nanotechnology. *ACS Nano* 2017, *11* (2), 1142–1164.
34. Shu, D.; Moll, W. D.; Deng, Z.; Mao, C.; Guo, P. Bottom-up assembly of RNA arrays and superstructures as potential parts in nanotechnology. *Nano Lett.* 2004, *4*, 1717–1723.
35. Sharma, A.; Haque, F.; Pi, F.; Shlyakhtenko, L.; Evers, B. M.; Guo, P. Controllable Self-assembly of RNA Dendrimers. *Nanomed. Nanotechnol. Biol. Med.* 2015, *12*, 835–844.
36. Jasinski, D.; Khisamutdinov, E. F.; Lyubchenko, Y. L.; Guo, P. Physicochemically Tunable Poly-Functionalized RNA Square Architecture with Fluorogenic and Ribozymatic Properties. *ACS Nano* 2014, *8*, 7620–7629.
37. Khisamutdinov, E. F.; Jasinski, D. L.; Guo, P. RNA as a boiling-resistant anionic polymer material to build robust structures with defined shape and stoichiometry. *ACS Nano*. 2014, *8*, 4771–4781.
38. Li, H.; Zhang, K.; Pi, F.; Guo, S.; Shlyakhtenko, L.; Chiu, W.; Shu, D.; Guo, P. Controllable Self-Assembly of RNA Tetrahedrons with Precise Shape and Size for Cancer Targeting. *Adv. Mater.* 2016, *28*, 7501–7507.
39. Lammers, T.; Kiessling, F.; Hennink, W. E.; Storm, G. Drug targeting to tumors: principles, pitfalls and (pre-) clinical progress. *J. Control Release* 2012, *161* (2), 175–187.
40. Bae, Y. H.; Park, K. Targeted drug delivery to tumors: myths, reality and possibility. *J. Control Release* 2011, *153* (3), 198–205.
41. Lammers, T.; Rizzo, L. Y.; Storm, G.; Kiessling, F. Personalized nanomedicine. *Clin. Cancer Res.* 2012, *18* (18), 4889–4894.
42. Cruz, J. A.; Westhof, E. The Dynamic Landscapes of RNA Architecture. *Cell* 2009, *136* (4), 604–609.
43. Leontis, N. B.; Stombaugh, J.; Westhof, E. The non-Watson-Crick base pairs and their associated isostericity matrices. *Nucleic Acids Res.* 2002, *30* (16), 3497–3531.
44. Busch, A.; Backofen, R. INFO-RNA--a server for fast inverse RNA folding satisfying sequence constraints. *Nucleic Acids Res.* 2007, *35* (Web Server issue), W310–W313.
45. Binzel, D. W.; Khisamutdinov, E. F.; Guo, P. Entropy-driven one-step formation of Phi29 pRNA 3WJ from three RNA fragments. *Biochemistry* 2014, *53* (14), 2221–2231.
46. Binzel, D. W.; Khisamutdinov, E.; Vieweger, M.; Ortega, J.; Li, J.; Guo, P. Mechanism of three-component collision to produce ultrastable pRNA three-way junction of Phi29 DNA-packaging motor by kinetic assessment. *RNA* 2016, *22* (11), 1710–1718.
47. Shu, D.; Khisamutdinov, E.; Zhang, L.; Guo, P. Programmable folding of fusion RNA complex driven by the 3WJ motif of phi29 motor pRNA. *Nucleic Acids Res.* 2013, *42*, e10.
48. Behlke, M. A. Chemical modification of siRNAs for in vivo use. *Oligonucleotides*. 2008, *18* (4), 305–319.
49. Behlke, M. A. Progress towards in vivo use of siRNAs. *Mol. Ther.* 2006, *13*, 644–670.
50. Ernsting, M. J.; Murakami, M.; Roy, A.; Li, S. D. Factors controlling the pharmacokinetics, biodistribution and intratumoral penetration of nanoparticles. *J. Control Release* 2013, *172* (3), 782–794.
51. Varkouhi, A. K.; Scholte, M.; Storm, G.; Haisma, H. J. Endosomal escape pathways for delivery of biologicals. *J. Control Release* 2011, *151* (3), 220–228.
52. Xu, Z.; Sun, Y.; Weber, J. K.; Cao, Y.; Wang, W.; Jasinski, D.; Guo, P.; Zhou, R.; Li, J. Directional mechanical stability of Bacteriophage phi29 motor's 3WJ-pRNA: Extraordinary robustness along portal axis. *Sci. Adv.* 2017, *3* (5), e1601684.
53. Al-Hashimi, H. M.; Pitt, S. W.; Majumdar, A.; Xu, W. J.; Patel, D. J. Mg2+-induced variations in the conformation and dynamics of HIV-1 TAR RNA probed using NMR residual dipolar couplings. *J Mol. Biol.* 2003, *329* (5), 867–873.
54. Mustoe, A.; Brooks, C.; Al-Hashimi, H. M. Hiearchy of RNA functional dynamics. *Ann. Rev. Biochem.* 2014, *83*, 441–466.
55. Chen, H.; Meisburger, S. P.; Pabit, S. A.; Sutton, J. L.; Webb, W. W.; Pollack, L. Ionic strength-dependent persistence lengths of single-stranded RNA and DNA. *Proc. Nat. Acad. Sci. USA* 2012, *109* (3), 799–804.
56. Prabhakar, U.; Maeda, H.; Jain, R. K.; Sevick-Muraca, E. M.; Zamboni, W.; Farokhzad, O. C.; Barry, S. T.; Gabizon, A.; Grodzinski, P.; Blakey, D. C. Challenges and key considerations of the enhanced permeability and retention effect for nanomedicine drug delivery in oncology. *Cancer Res* 2013, *73* (8), 2412–2417.

57. Maeda, H. Toward a full understanding of the EPR effect in primary and metastatic tumors as well as issues related to its heterogeneity. *Adv. Drug Deliv. Rev.* 2015, *91*, 3–6.

58. Anselmo, A. C.; Zhang, M.; Kumar, S.; Vogus, D. R.; Menegatti, S.; Helgeson, M. E.; Mitragotri, S. Elasticity of nanoparticles influences their blood circulation, phagocytosis, endocytosis, and targeting. *ACS Nano.* 2015, *9* (3), 3169–3177.

59. Anselmo, A. C.; Mitragotri, S. Impact of particle elasticity on particle-based drug delivery systems. *Adv. Drug Deliv. Rev.* 2017, *108*, 51–67.

60. Pi, F.; Zhang, H.; Li, H.; Thiviyanathan, V.; Gorenstein, D. G.; Sood, A. K.; Guo, P. RNA nanoparticles harboring annexin A2 aptamer can target ovarian cancer for tumor-specific doxorubicin delivery. *Nanomedicine* 2016, *13* (3), 1183–1193.

61. Garzon, R.; Marcucci, G.; Croce, C. M. Targeting microRNAs in cancer: rationale, strategies and challenges. *Nat Rev. Drug Discov.* 2010, *9* (10), 775–789.

62. Obad, S.; dos Santos, C. O.; Petri, A.; Heidenblad, M.; Broom, O.; Ruse, C.; Fu, C.; Lindow, M.; Stenvang, J.; Straarup, E. M.; Hansen, H. F.; Koch, T.; Pappin, D.; Hannon, G. J.; Kauppinen, S. Silencing of microRNA families by seed-targeting tiny LNAs. *Nat. Genet.* 2011, *43* (4), 371–378.

63. Whitehead, K. A.; Dahlman, J. E.; Langer, R. S.; Anderson, D. G. Silencing or stimulation? siRNA delivery and the immune system. *Annu. Rev. Chem. Biomol. Eng.* 2011, *2*, 77–96.

64. Hanagata, N. Structure-dependent immunostimulatory effect of CpG oligodeoxynucleotides and their delivery system. *Int. J. Nanomed.* 2012, *7*, 2181–2195.

45 Current State in the Development of RNAi Self-Assembled Nanostructures

Marwa Ben Haj Salah and John J. Rossi
City of Hope, Duarte, CA, USA

CONTENTS

INTRODUCTION

The mechanism of RNA interference was discovered in 1998 by Andrew Fire and Craig Mello after they injected long double-stranded RNA into *Caenorhabditis elegans* and observed sequence-specific suppression of the target gene (Fire et al., 1998). The same phenomenon was observed in mammals (McCaffrey et al., 2002). The discovery led to burgeoning of RNAi-focused research seeking to decipher this powerful regulatory mechanism and its potential translation into therapeutic treatment of human diseases. This earned Fire and Mello the Nobel Prize in Physiology and Medicine in 2006. Since then, numerous RNAi-based reagents such as small interfering RNAs (siRNAs), small activating RNAs (saRNAs) and microRNA (miRNAs) mimics have been discovered, optimized and tested in preclinical and clinical trials (Castanotto & Rossi, 2009; Elbashir et al., 2001). However, delivery of RNAi reagents remains one of the challenges in RNAi-based therapeutics. DNA or RNA nanotechnology is one route for producing molecularly well-defined nanostructures for enhanced RNAi reagents delivery. Over the past three decades, researchers advanced the construction of oligonucleotide nanostructures from low to high levels of complexity (Rothemund, 2006; Sabir et al., 2012; Zhang et al., 2014). Several software interfaces have been developed to predict the 3D conformation of these nanostructures (Kim et al., 2012; Zadeh et al., 2011), and nanostructures using DNA or RNA have begun to be used for imaging and therapeutic delivery purposes (Bhatia et al., 2011; Keum et al., 2011; Walsh et al., 2011).

Nanotechnology is the discipline of engineering functional systems at the nanoscale level from the bottom up and can produce molecular constructs with novel functionalities used in a wide array of applications (Afonin, Lindsay, & Shapiro, 2013). The work of Nadrian Seeman in DNA nano-fabrication (Seeman, 2007, 2010) as well as the work of Paul Ruthmond in DNA origami synthesis (Rothemund, 2006) and many other groups (Winfree, 2000) laid the foundations for more advances in the field of oligonucleotide nanotechnology. Both DNA and RNA are amenable to acting as program-mable scaffolds for nanotechnological applications (Pinheiro et al., 2011, P. Guo, 2010). However, RNA molecules have shown higher thermal stability and a wider range of structural flexibility pro-vided by non-canonical base interactions resulting in a large variety of structural and catalytic motifs resembling those of proteins. Moreover, their ability to be chemically modified to exhibit more stabil-ity and nuclease resistance has made them very attractive materials for the production of multivalent nanostructures for different applications (P. Guo, 2010; H. Lee et al., 2012). Over the past decade, several groups have designed RNAi-based nanoparticles which showed efficient delivery and suc-cessful target gene knockdown (Burnett & Rossi, 2012). In this chapter, we provide a summary of the RNAi mechanism, the challenges and limits of RNAi as therapeutic agents in the *in vivo* environment and discuss examples of self-assembled nanotechnology platforms for delivery of RNAi reagents.

RNAi MECHANISM

RNA interference (RNAi) is a sequence-targeted post transcriptional gene-silencing mechanism. It is a conserved biological response that functions primarily to regulate the expression of protein-coding genes. It also constitutes a resistance mechanism against double-stranded RNA from endog-enous and exogenous pathogenic nucleic acids such as viral genomes. This natural mechanism was discovered by Andrew Fire and Craig Mello in animals in 1998. RNAi would revolutionize experimental biology and have important applications in therapy, agriculture and many more fields. Fire et al. introduced exogenous double-stranded RNA (dsRNA), single-stranded (ss) sense and ss antisense RNA into the nematode *Caenorhabditis elegans* and observed a significant sequence-specific suppression of muscle protein gene expression only with the injection of dsRNA (Fire et al., 1998). After Tuschl et al. discovered that RNAi can be activated by introducing synthetic siRNAs (21-nucleotide dsRNAs) directly into the cell (Tuschl et al., 1999), the RNAi mechanism was translated into a powerful regulatory tool for biological research and a potential therapeutic for the treatment of human diseases.

In therapeutic applications, synthetic siRNAs are usually taken into the cell by endocytosis before escaping into the cytoplasm. The mechanism of endosomal escape is still not thoroughly elucidated. After endosomal escape, the guide strand (antisense) from the siRNA needs to be loaded into an argonaute protein to form the RNA-induced silencing complex (RISC) and the passenger (sense) strand needs to be discarded. This occurs in a multi-protein assembly called the RISC load-ing complex (RLC) (H.-W. Wang et al., 2009). The RISC loading process differs by RNAi reagent. The 25/27-mer Dicer-substrate siRNA (dsiRNA) is cleaved by Dicer before handoff to the RLC. Canonical siRNAs with 21 base pair (bp) or shorter dsRNA domains can bypass dicer (D.-H. Kim et al., 2005). Synthetic saRNAs and miRNA mimics have identical structures as siRNAs which guide strand sequences being the determinant of their functions.

For siRNA therapeutics, guide strands usually direct RISC to the 3′ untranslated region (UTR) of the complementary target mRNA leading to mRNA degradation by multiple mechanisms. One important mechanism for RNAi therapeutics is slicer activity undertaken by Argonaute 2 when its partner guide strand is fully base paired with its mRNA target (Meister & Tuschl, 2004).

CHALLENGES OF siRNA DELIVERY IN THE *IN VIVO* ENVIRONMENT

RNAi reagents have undergone a series of optimizations since their discovery. These attempts aimed to overcome several challenges such as systemic circulation (Haussecker, 2014), cellular uptake (Akhtar & Benter, 2007) and endosomal escape (Wittrup et al., 2015) which have significantly lim-ited their use in the clinic. In fact, the first clinical trials using naked minimally modified siRNAs resulted in immune responses and RNAi off-target effects (Song et al., 2003). The second clinical trials relied on the improved siRNA delivery formulations such as polymer nanoparticles (Davis et al., 2010) and liposomes (Koldehoff et al., 2007). This second wave of clinical trials demonstrated the potential for targeted gene silencing by systemically delivering siRNAs in humans but the therapeutic effect was not sufficient. Since then, additional chemical modifications and delivery vehicles have been tested to circumvent siRNA limitations *in vivo*.

A key limitation is that naked siRNAs have a minutes-short half-life in circulation due to the vulnerability of unmodified RNA to nuclease attack. Chemical modifications that improve stability include phosphorothioate (PS) backbones on terminal bases and chemical substitutions at the 2′ position of ribose sugars with 2-O-methyl (2′O-Me) and 2′-Fluoro (2′-F) groups. phosphorothioate (PS) groups can also improve binding to plasma proteins resulting in wider biodistribution. Chemical substitution at the 2′ position can also contribute at the increase of melting temperature Tm of siRNA resulting in a stronger base pairing and higher RNAi potency (Petrova (Kruglova) et al., 2010).

Researchers have also explored using locked nucleic acids (LNAs) and unlocked nucleic acids (UNAs) in siRNAs. LNAs increase binding affinity to RNA (Vester & Wengel, 2004) and reduce off-target effects by either lowering the ability of inappropriately loaded sense-strands to cleave the target RNA or by lowering the incorporation of the siRNA sense-strand (Elmén et al., 2005). UNA can decrease binding affinity to RNA and eliminate sense strand loading (Snead et al., 2013).

The distribution of these chemical modifications along the RNA sequences has been the subject of tremendous work by companies and research groups alike (Setten et al., 2019). For example, alternating patterns of RNA and 2′-O-methyl-modified nucleotides can reduce immunogenic activity but cannot provide sufficient resistance to nuclease degradation for naked siRNA delivery (Collingwood et al., 2008). Follow-up studies showed that fully modified dsRNAs with alternating 2′-F and 2′-O-methyl on both strands achieved both improved serum nuclease and acceptable RNAi efficiency while reducing immune activation (Behlke, 2008).

A second limitation is that the physico-chemical properties of modified siRNAs notably their size, hydrophilicity, net negative charge induce rapid clearance from systemic circulation, restrict biodistribution, prevent their uptake into target cells and hinder endosomal escape.

Thus, the application of modified siRNAs as therapeutic agents requires the optimization of the delivery approaches to (i) enhance circulation half-life (ii) broaden biodistribution (iii) increase uptake by target cells and (iv) enhance endosomal escape.

The route of administration is another important factor in determining RNAi reagents efficacy. Certain tissues are more amenable for localized delivery techniques. For example, siRNAs have been delivered via site-specific injections to the skin and the eye (Fattal & Bochot, 2006; Geusens et al., 2009; Martínez et al., 2014), intranasal delivery has been tested for pulmonary delivery and intrathecal injection is efficient for the central nervous system (Sarret et al., 2015; Barik, 2011; DeVincenzo et al., 2010; Alterman et al., 2015). Recently, Alnylam reported impressive results in primate eyes for fully modified siRNAs. However, many disease targets still require systemic administration and carrier systems and targeted ligands can improve circulation half-life, broaden bio-distribution and increase uptake in target tissues. Conventional approaches include cationic nanoparticles (Morrissey et al., 2005), antibody-fusion molecules (Peer et al., 2007), lipids and liposomes (Akinc et al., 2008), cholesterol-conjugated siRNAs (Soutschek et al., 2004), aptamer-conjugated siRNAs (Zhou et al., 2009), CpG-conjugated siRNAs (Kortylewski et al., 2009), GalNAc-conjugated siRNAs and exosomes (Kamerkar et al., 2017). A new emerging class of carriers are RNA nanostructures.

RNA NANOPARTICLE

The last two decades have witnessed a breakthrough in oligonucleotide nanotechnology. Nanotechnology field focuses on the development of submicron sized molecular devices. The nanoparticles range from 5 to 500 nm in at least one dimension (Ediriwickrema & Saltzman, 2015).

Oligonucleotide nanotechnology takes advantage of the properties of DNA and RNA to achieve rational design and self-assembly of nanoscaffolds. These allow controlled stoichiometry of the delivered drugs and presentation of targeting ligands for different human tissues (P. Guo, 2005). DNA and RNA nanotechnology may have similar design strategies and therapeutic objectives; however, RNA base stacking properties, the presence of the 2′-OH group, and the fixed C3′-endo sugar associated A-form helical structure provide double-stranded RNA with improved thermal stability compared to a DNA helix. In addition, DNA molecules rely mainly on canonical Watson and Crick interactions represented by A-T and G-C base pairings whereas RNA molecules have the ability to form non-canonical interactions which results in a wide array of complex structures with distinct tertiary structural motifs. As a result, these nanostructures have the ability to mimic proteins in regard to their diverse functional properties while presenting several advantages such as the simplicity of their nucleotide primary structure and ease of manufacturing (P. Guo, 2010). Various RNA-based therapeutics such as siRNAs, aptamers, ribozymes and antisense have been successfully linked to an RNA nanoparticle in order to achieve targeted delivery, enhance potency and minimize unwanted side effects (Guo, 2010; Zhou et al., 2009, 2011; Tarapore et al., 2011; Shu et al., 2011). In this book chapter, we focus on RNAi reagents linked to self-assembled RNA structures. The following are key examples of nanotechnology platforms for RNAi therapy.

APTAMER-siRNA CHIMERAS

Aptamers are single-stranded oligonucleotides that are selected from pools RNA or DNA of random sequences to bind therapeutically relevant receptor proteins. They are discovered using SELEX (systematic evolution of ligands by exponential enrichment), a technique in which a library of randomly generated oligonucleotides are tested for target binding. After several iterations of nucleic acid selection for protein or cell targeting and amplification by polymerase chain reaction (PCR), functional sequences are identified by sequencing. The selected aptamers fold into tertiary structures that bind to proteins on the surface of target cells. This then leads to direct therapeutic effects or internalization along with therapeutic cargos (Hicke & Stephens, 2000). The affinity and specificity of binding of aptamers was shown to be comparable to those of antibodies (Ellington & Szostak, 1990; Tuerk & Gold, 1990). Aptamers are significantly smaller than antibodies and have different tissue penetration properties and faster systemic clearance. Aptamers can be produced by chemical synthesis, which is often less expensive to manufacture and have lower batch to batch variability than antibodies. Manufacturing scale-up can also be more economical and straightforward. Despite these advantages, few aptamers have made it to therapeutic application. Issues like nuclease degradation and rapid clearance because of their small size halted aptamers' clinical translation (A. Z. Wang & Farokhzad, 2014). The first aptamer approved was pegaptanib sodium (Macugen) in 2004 to treat macular degeneration (Ng et al., 2006). Other aptamers are currently undergoing clinical evaluation for various hematology, oncology, ocular and inflammatory indications (Ni et al., 2011). Cell-internalizing aptamers are extensively investigated for use in therapy in conjunction with RNAi reagents because of their high affinity and specificity and their feasibility for chemical conjugation (Dassie et al., 2009; Zhou et al., 2011). The aptamer–siRNA/saRNA/miRNA chimera approach has many advantages, including: (1) enhanced potency of the RNAi in target tissues (2) and more narrowly targeted cellular uptake leading to reduced side effects associated with nonspecific biodistribution. Compared with antibody-mediated delivery, aptamers can have lower immunogenicity, lower cost and better stability in tissues and organs (Keefe et al., 2010). The following examples describe different strategies in which RNAi reagents and aptamers are conjugated for targeted therapy.

gp120 APTAMER–SIRNA CHIMERAS

The Rossi group designed a "sticky bridge" to attach an siRNA to a single cell-internalizing aptamer that binds to glycoprotein gp120 for Human Immunodeficiency Virus HIV-1 therapy (Zhou et al.,

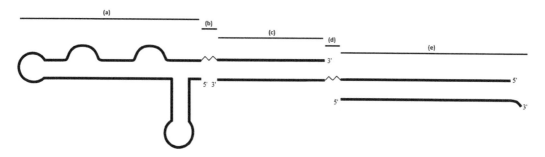

FIGURE 45.1   Schematic representation of anti gp120 Aptamer – dsiRNA conjugated via the sticky bridge approach (a) Anti-gp120 aptamer (b) C3 linker (c) GC-rich sticky bridge (d) C3 linker (e) 27-mer dsiRNA.

2009). Briefly, the envelope glycoprotein of HIV consists of an exterior glycoprotein (gp120) and a *trans*-membrane domain (gp41). HIV-1 entry depends on the interaction of gp120 with the immune cell receptor CD4 and other cell surface receptors of the chemokine receptor family (Dalgleish et al., 1984). After the binding to CD4, a cascade of conformational changes in gp120 and gp41 leads to the fusion of the viral membrane with the host membrane CD40 resulting in the successful penetration of the viral genetic material (Allan et al., 1985). Therefore, gp120 is considered a clinically relevant target for anti-HIV therapy. One approach is to design small molecules that interfere with gp120/CD4 interaction blocking virus entry and replication (Tran et al., 2011).

Another approach tested by the Rossi group is to develop 2′-Fluoro modified RNA aptamers that target gp120 with nanomolar Kd. This aptamer was isolated from an RNA library after several rounds of SELEX. The aptamer was then conjugated to different dicer substrate siRNAs targeting HIV-1 *tat/rev* common exon, HIV host dependency factors CD4 and TNPO3 (transportin-3). In order to form a conjugate, a GC-rich sequence of nucleotides was attached to the 3′ end of the anti-gp120 aptamer. The complement of this sequence was attached to the siRNA (Figure 45.1) (Zhou et al., 2009). The conjugation of the aptamer to the siRNA via the sequence linker relies on complementary base pairing of the GC rich linker sequence. To provide the structure with spatial flexibility, a three-carbon atom hinge (C3) was incorporated between the aptamer and linker sequence. The gp120 aptamer dicer substrate conjugates demonstrated suppression of HIV-1 infected peripheral blood mononuclear cells (PBMCs). *In vivo* results showed specific knockdown of target mRNAs in HIV-1 infected RAG-humanized mice by the gp120 aptamer-siRNA conjugate but not in mice treated with mutated aptamers or mismatched siRNAs. This approach proved the possibility of using the self-assembled "sticky bridge" to combine different siRNAs with aptamers for the treatment of HIV-1 infection.

pRNA–siRNA CHIMERAS

Self-assembly of RNA nanoparticles depends on the spontaneous interaction of individual RNA molecules that base pair in a predefined manner to form 2D or 3D RNA nanostructures. The motor prohead RNA (pRNA) of the Bacillus subtilis bacteriophage phi 29 is an ideal material for self-assembly as it assembles via loop/loop RNA interactions to form dimeric, trimeric and hexameric complexes that package viral DNA into viral capsids. The pRNA monomer is composed of an interlocking domain and a helical domain (Figure 45.2a). The sequences responsible for pRNA–pRNA interactions are located between residues 23 and 97 (P. Guo, 2010). Guo et al. designed chimeric monomer subunits to possess either A-b′ or B-a′ orientation for complementary base pairing with each other (Figure 45.2b). When pRNA A-b′ and pRNA B-a′ were mixed in an equimolar ratio in the presence of Mg^{2+}, pRNA dimers were produced with high efficiency (S. Guo et al., 2005a). Prohead RNA can be redesigned to form a wide array of shapes via the base pairing of programmed loops and helical regions (Shu et al., 2004). Studies have shown that the replacement of the helical

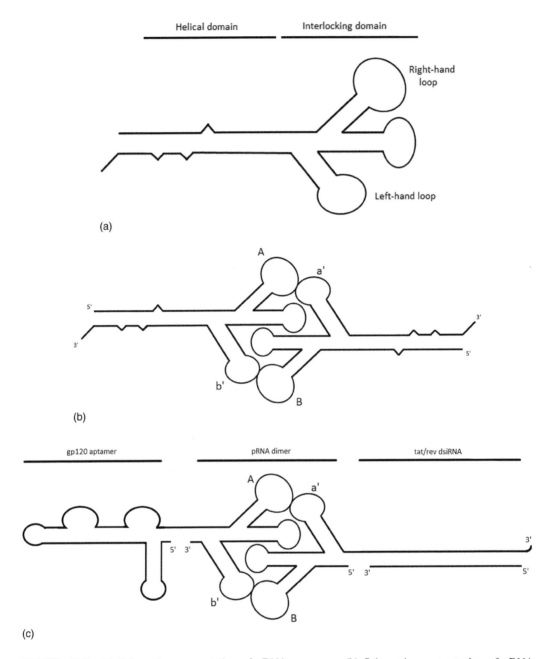

FIGURE 45.2 (a) Schematic representation of pRNA monomer; (b) Schematic representation of pRNA dimer; (c) Schematic representation of siRNA-pRNA conjugate delivered via GP120 aptamer.

domain for attachment of external moieties does not interrupt correct spontaneous folding of pRNA (P. Guo, 2005). Conveniently, the folding of pRNA does not get affected by chemical modifications. Therefore, it is possible to substitute RNA bases with 2′-Fluoro U and C to make the structure more resistant to nuclease degradation in serum (J. Liu et al., 2011). Thus, pRNA has been used as a vehicle to a wide variety of diagnostic and therapeutic molecules such as aptamers, siRNAs, ribozymes, fluorophores (Moll & Guo, 2007; Huang et al., 2011; Hoeprich et al., 2003; Guo, 2005, 2010). The following are example applications for a variety of cargos:

pRNA/siRNAs/Aptamers Chimeras

Guo et al. have been pioneers in developing pRNA/siRNAs/aptamers chimeras. To test whether it is possible to replace the helical region in the pRNA molecule with double-stranded siRNA, they constructed chimeric pRNA-siRNA chimeras targeting green fluorescent protein (GFP) and renilla luciferase that successfully reduced mRNA and protein levels (S. Guo et al., 2005b). The group then went on to attach siRNAs against clinically relevant targets such as the Bcl2 associated death promoter BAD, and the anti-apoptotic factor survivin to suppress tumor growth in a xenograft animal model (Guo, 2005). Their results showed successful knockdown of target genes *in vivo*. Guo et al. also tested pRNA–siRNA chimeras targeting survivin and metallothionein-IIa (MT-IIA) mRNAs in metastatic ovarian cancer, as these anti-apoptotic genes are found to be overexpressed in many tumors. The group observed that pRNA-siRNA chimeras targeting MT-IIA are more potent at target gene silencing than individual MT-IIA siRNAs. They also found that the simultaneous delivery of folate-pRNA dimers targeting both survivin and MT-IIA resulted in a more significant tumor regression (Tarapore et al., 2011). Following these results, other groups have tested the use of pRNA as a delivery vehicle in other disease models such as Myocarditis, the inflammation and injury of heart muscle cells mostly caused by Coxsackievirus B3 (CVB3) (Ye et al., 2011). For example, Ye et al. attached synthetic microRNAs targeting CVB3 genome to folate-conjugated pRNAs and deliver the complexes into folate receptor-expressing HeLa cells commonly used as an in vitro model for CVB3 infection. Results showed significant viral replication suppression without triggering interferon production. Therefore, folate receptor-targeted delivery of siRNAs via pRNA has the potential to be an efficient therapeutic approach (Ye et al., 2011).

In the above-mentioned examples, folate was used as the delivery vehicle. Subsequently, aptamers targeting CD4 and other cell surface receptors were also tested as targeted delivery reagents (S. Guo et al., 2005a). For aptamer display in a pRNA dimer, the helical region of one pRNA is replaced with the aptamer while the helical region of the other pRNA is manipulated to carry the therapeutic cargo. Both chimeric RNAs form dimers via interlocking right- and left-hand loops. One of the advantages of this approach is the increase of the RNA nanoparticle size, which can increase its tissue retention.

The Rossi group has tested pRNA particles with gp120 aptamers for targeting HIV. In this system, a gp120 aptamer pRNA was conjugated to a tat/rev siRNA pRNA to specifically silence viral tat and rev proteins in HIV infected cells (Figure 45.2c) (Zhou et al., 2011). Addition of gp120 aptamer did not disrupt pRNA folding and did not decrease aptamer binding affinity or specificity. The results showed successful inhibition of HIV-1 replication after treatment, making pRNA platform potentially useful for targeted delivery of siRNAs to HIV infected cells (Zhou & Rossi, 2011).

pRNA Three-Way Junction/Four-Way Junction–siRNA Chimeras

Structural characterizations of the bacteriophage phi29 have shown a double-stranded DNA packaging motor centered on a pRNA hexamer ring (P. X. Guo, Erickson, & Anderson, 1987). Thus, the pRNA molecule can carry more than two functional moieties by taking advantage of the molecule self-assembly into trimers and hexamers. Shu et al. assembled 3 to 6 pRNA monomers into a stable junction without the need for high salt concentrations. The self-assembled nano-junction is resistant to denaturation in 8M urea, stable in serum and able to carry a wide array of functional moieties such as siRNA, aptamers and receptor ligands like folate. Importantly, the folding of the pRNA molecules did not interfere with the activity of the functional groups (Shu et al., 2011). To translate this platform into a therapeutic setting, the group investigated the effect of 3wj on gastric cancer regression by coupling it with BRACAA1 siRNA for targeting, fluorescent marker for imaging and folic acid for delivery to stomach cancer cells, which have high expression of folic acid receptors. Both *in vivo* and *in vitro* results showed that pRNA folding did not disrupt the functioning of the attached moieties. *In vitro* results showed cell specific internalization and knockdown of the BRCAA1 gene. *In vivo* results showed significant decrease of the tumor volume with no toxicity observed (Cui et al., 2015). The same group also tested the activity of an X-shaped four-way junction nanodevice that is able to carry 4 functional moieties: a malachite green aptamer for imaging, folate for specific delivery and cell internalization, a luciferase siRNA and a survivin siRNA (Haque et al., 2012). Similar to previous results, the RNA motif folding was not disturbed by the attached moieties. The tetravalent junction self-assembles in the absence of metal ions remained stable in denaturing conditions and after systemic injection in mice and specifically localized in target tissues. All these results confirm that the pRNA architecture is an attractive platform for combinatorial therapy in viral infections and cancer and has the potential to overcome limitations of existing approaches to achieve targeted delivery of multiple RNAi therapeutics to diseased tissues.

Tripodal RNA Nanoparticles

Chang et al. developed a 3-way junction design that does not use pRNA to form the core but rather a trebler-phosphoramidite, a branching reagent for DNA synthesis that has three 17-nucleotide DNA branches complementary to siRNA overhang sequences in order to deliver three siRNAs simultaneously (Figure 45.3) (Chang et al., 2012). Using a transfection agent, they first tested the delivery of three identical survivin siRNAs. The tripodal nanostructures showed more potent gene silencing compared to individual siRNAs. Then, to demonstrate versatility of the tripodal junction, they designed a junction carrying different siRNA molecules against survivin, ß-catenin and integrin mRNAs, as well as a second structure that contains an antagomir targeting miR-21 as well as two siRNAs. These versatile junctions were able to specifically silence each target. In a follow-on report, Chang et al. replaced the trebler-phosphoramidite core with three 38-nucleotide single-stranded RNAs that self-assembled into a tripodal junction of 19 bp dsRNA arms carrying three different siRNAs. Similar to previous results, following lipid-based transfection, gene silencing was improved with the nanoparticles compared with individual siRNAs (Chang et al., 2012). At last report, the application of this technology was being explored by Tekmira Pharmaceuticals.

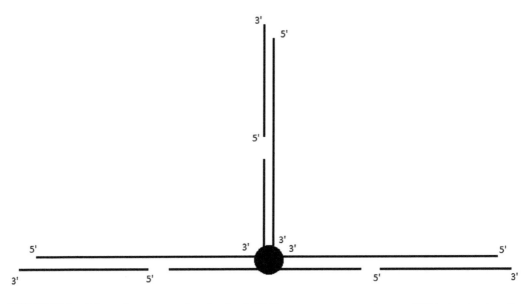

FIGURE 45.3 Schematic representation of a tripodal RNA nanostructure carrying one siRNA at each extremity with a trebler phosphotamidite as the core of the junction and 17 nucleotide-oligobranches complementary to siRNA overhangs.

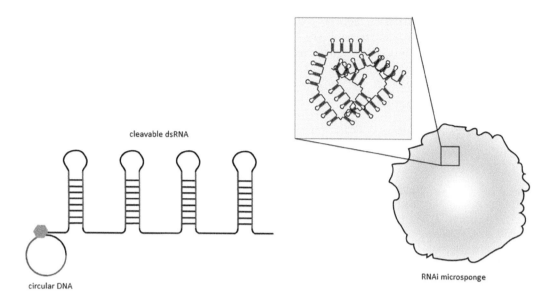

cleavable dsRNA

circular DNA

RNAi microsponge

FIGURE 45.4 Using Rolling circle transcription of closed circular DNA coding for guide and passenger strands sequences of anti-luciferase siRNA, multiple repeats of cleavable hairpin structures are generated and self-assemble in sponge-like structures.

SELF-ASSEMBLED siRNA NANOPARTICLES

RNAi reagents have been delivered using RNA nanoparticles as described above or encapsulated in lipid nanoparticles (Semple et al., 2010), dendrimers (Taratula et al., 2009), cationic complexes (L. Liu et al., 2009) or inorganic nanoparticles (Elbakry et al., 2009). In addition to the above, RNA itself can form nanoparticles for delivery of siRNA. Lee et al. have developed polymeric RNA hairpins that self-assemble into sponge-like hairpin RNA microspheres (Figure 45.4) (J. B. Lee et al., 2012). These RNAi microsponges are solely composed of RNA strands and are processed by the RNAi pathway after cellular uptake, resulting in the cleavage of the RNA hairpins into individual siRNAs. This approach protects the RNA from degradation during circulation and allows the delivery of a significant amount of siRNA copies with the uptake of a single RNAi micro-sponge. The group reported the delivery of more than a half a million copies of siRNAs per microsponge. Another group, Patel et al. have synthesized linear V- and Y-shaped RNA templates incorporating the antisense (A) or sense (S) strands that self-assemble into spheres, triangles, squares, pentagons and hexagons of distinct sizes and shapes. Prepared in this manner, siRNAs targeting Gastrin Releasing Peptide GRP78 mRNA (Patel et al., 2016) showed successful target knockdown.

CONCLUSIONS AND FUTURE PROSPECTS

On 10 August 2018, the US Food and Drug Administration (FDA) approved the siRNA drug Patisiran (Onpattro; Alnylam Pharmaceuticals) for the treatment of Hereditary Transthyretin-Mediated Amyloidosis (Adams et al., 2018). Onpattro™ is an siRNA delivered by a lipid nanoparticle. Currently, numerous GalNac-conjugated chemically modified siRNAs are emerging from or undergoing late stage clinical testing for liver diseases (Springer & Dowdy, 2018). On 20 November 2019, Alnylam announced approval of GIVLAARI™ (givosiran) by the U.S. Food and Drug Administration (FDA) for the treatment of adults with Acute Hepatic Porphyria (AHP). Therefore, givosiran becomes the second approved RNAi drug from Alnylam and first GalNAc-Conjugate RNAi therapeutic. Other renal and ocular disease indications are in phase I, II and III clinical trials.

Despite these exciting advances, more progress in systemic delivery to non-kidney, non-liver tissues is needed. With many exciting technologies currently under development and testing, self-assembled RNA nanostructures could potentially make significant contributions to future applications of RNAi in medicine.

ACKNOWLEDGMENTS

We thank Dr. Si-Ping Han for critical reading of the manuscript.

REFERENCES

Adams, D., Gonzalez-Duarte, A., O'Riordan, W. D., Yang, C.-C., Ueda, M., Kristen, A. V., ... Suhr, O. B. (2018). Patisiran, an RNAi Therapeutic, for Hereditary Transthyretin Amyloidosis. *New England Journal of Medicine*, *379*(1), 11–21. https://doi.org/10.1056/NEJMoa1716153

Afonin, K. A., Lindsay, B., & Shapiro, B. A. (2013). Engineered RNA Nanodesigns for Applications in RNA Nanotechnology. *DNA and RNA Nanotechnology*, *1*(1). https://doi.org/10.2478/rnan-2013-0001

Akhtar, S., & Benter, I. F. (2007). Nonviral delivery of synthetic siRNAs in vivo. *The Journal of Clinical Investigation*, *117*(12), 3623–3632. https://doi.org/10.1172/JCI33494

Akinc, A., Zumbuehl, A., Goldberg, M., Leshchiner, E. S., Busini, V., Hossain, N., ... Anderson, D. G. (2008). A combinatorial library of lipid-like materials for delivery of RNAi therapeutics. *Nature Biotechnology*, *26*(5), 561–569. https://doi.org/10.1038/nbt1402

Allan, J. S., Coligan, J. E., Barin, F., McLane, M. F., Sodroski, J. G., Rosen, C. A., ... Essex, M. (1985). Major glycoprotein antigens that induce antibodies in AIDS patients are encoded by HTLV-III. *Science*, *228*(4703), 1091–1094. https://doi.org/10.1126/science.2986290

Alterman, J. F., Hall, L. M., Coles, A. H., Hassler, M. R., Didiot, M.-C., Chase, K., ... Khvorova, A. (2015). Hydrophobically Modified siRNAs Silence Huntingtin mRNA in Primary Neurons and Mouse Brain. *Molecular Therapy - Nucleic Acids*, *4*, e266. https://doi.org/10.1038/mtna.2015.38

Barik S. (2011). Intranasal delivery of antiviral siRNA. *Methods Mol Biol*, *721*, 333–338.

Behlke, M. A. (2008). Chemical Modification of siRNAs for In Vivo Use. *Oligonucleotides*, *18*(4), 305–320. https://doi.org/10.1089/oli.2008.0164

Bhatia, D., Surana, S., Chakraborty, S., Koushika, S. P., & Krishnan, Y. (2011). A synthetic icosahedral DNA-based host–cargo complex for functional in vivo imaging. *Nature Communications*, *2*(1), 1–8. https://doi.org/10.1038/ncomms1337

Burnett, J. C., & Rossi, J. J. (2012). RNA-Based Therapeutics: Current Progress and Future Prospects. *Chemistry & Biology*, *19*(1), 60–71. https://doi.org/10.1016/j.chembiol.2011.12.008

Castanotto, D., & Rossi, J. J. (2009). The promises and pitfalls of RNA-interference-based therapeutics. *Nature*, *457*(7228), 426–433. https://doi.org/10.1038/nature07758

Chang, C. I., Lee, T. Y., Kim, S., Sun, X., Hong, S. W., Yoo, J. W., ... Lee, D.-K. (2012). Enhanced intracellular delivery and multi-target gene silencing triggered by tripodal RNA structures. *The Journal of Gene Medicine*, *14*(2), 138–146. https://doi.org/10.1002/jgm.1653

Chang, C. I., Lee, T. Y., Yoo, J. W., Shin, D., Kim, M., Kim, S., & Lee, D. (2012). Branched, Tripartite-Interfering RNAs Silence Multiple Target Genes with Long Guide Strands. *Nucleic Acid Therapeutics*, *22*(1), 30–39. https://doi.org/10.1089/nat.2011.0315

Collingwood, M. A., Rose, S. D., Huang, L., Hillier, C., Amarzguioui, M., Wiiger, M. T., ... Behlke, M. A. (2008). Chemical Modification Patterns Compatible with High Potency Dicer-Substrate Small Interfering RNAs. *Oligonucleotides*, *18*(2), 187–200. https://doi.org/10.1089/oli.2008.0123

Cui, D., Zhang, C., Liu, B., Shu, Y., Du, T., Shu, D., ... Guo, P. (2015). Regression of Gastric Cancer by Systemic Injection of RNA Nanoparticles Carrying both Ligand and siRNA. *Scientific Reports*, *5*, 10726. https://doi.org/10.1038/srep10726

Dalgleish, A. G., Beverley, P. C. L., Clapham, P. R., Crawford, D. H., Greaves, M. F., & Weiss, R. A. (1984). The CD4 (T4) antigen is an essential component of the receptor for the AIDS retrovirus. *Nature*, *312*(5996), 763. https://doi.org/10.1038/312763a0

Dassie, J. P., Liu, X., Thomas, G. S., Whitaker, R. M., Thiel, K. W., Stockdale, K. R., ... Giangrande, P. H. (2009). Systemic administration of optimized aptamer-siRNA chimeras promotes regression of PSMA-expressing tumors. *Nature Biotechnology*, *27*(9), 839–846. https://doi.org/10.1038/nbt.1560

Davis, M. E., Zuckerman, J. E., Choi, C. H. J., Seligson, D., Tolcher, A., Alabi, C. A., ... Ribas, A. (2010). Evidence of RNAi in humans from systemically administered siRNA via targeted nanoparticles. *Nature*, *464*(7291), 1067–1070. https://doi.org/10.1038/nature08956

DeVincenzo J, Lambkin-Williams R, Wilkinson T, Cehelsky J, Nochur S, Walsh E, Meyers R, Gollob J, Vaishnaw A. (2010). A randomized, double-blind, placebo-controlled study of an RNAi-based therapy directed against respiratory syncytial virus. *Proc Natl Acad Sci U S A.*, *107*(19):8800–8805.

Ediriwickrema, A., & Saltzman, W. M. (2015). Nanotherapy for Cancer: Targeting and Multifunctionality in the Future of Cancer Therapies. *ACS Biomaterials Science & Engineering*, *1*(2), 64–78. https://doi.org/10.1021/ab500084g

Elbakry, A., Zaky, A., Liebl, R., Rachel, R., Goepferich, A., & Breunig, M. (2009). Layer-by-Layer Assembled Gold Nanoparticles for siRNA Delivery. *Nano Letters*, *9*(5), 2059–2064. https://doi.org/10.1021/nl9003865

Elbashir, S. M., Harborth, J., Lendeckel, W., Yalcin, A., Weber, K., & Tuschl, T. (2001). Duplexes of 21-nucleotide RNAs mediate RNA interference in cultured mammalian cells. *Nature*, *411*(6836), 494–498. https://doi.org/10.1038/35078107

Ellington, A. D., & Szostak, J. W. (1990). In vitro selection of RNA molecules that bind specific ligands. *Nature*, *346*(6287), 818–822. https://doi.org/10.1038/346818a0

Elmén, J., Thonberg, H., Ljungberg, K., Frieden, M., Westergaard, M., Xu, Y., … Wahlestedt, C. (2005). Locked nucleic acid (LNA) mediated improvements in siRNA stability and functionality. *Nucleic Acids Research*, *33*(1), 439–447. https://doi.org/10.1093/nar/gki193

Elias Fattal, Amélie Bochot. (2006). Ocular delivery of nucleic acids: antisense oligonucleotides, aptamers and siRNA, *Advanced Drug Delivery Reviews*, *58*, (11), 1203–1223.

Fire, A., Xu, S., Montgomery, M. K., Kostas, S. A., Driver, S. E., & Mello, C. C. (1998). Potent and specific genetic interference by double-stranded RNA in Caenorhabditis elegans. *Nature*, *391*(6669), 806–811. https://doi.org/10.1038/35888

B. Geusens, J. Lambert, S.C. De Smedt, K. Buyens, N.N. Sanders, M. Van. (2009). Gele, Ultradeformable cationic liposomes for delivery of small interfering RNA (siRNA) into human primary melanocytes, *Journal of Controlled Release*, *133*, (3), 214–220.

Guo, P. (2005, December). RNA Nanotechnology: Engineering, Assembly and Applications in Detection, Gene Delivery and Therapy. Journal of Nanoscience and Nanotechnology, 5(12):1964–1982. https://doi.org/info:doi/10.1166/jnn.2005.446

Guo, P. (2010). The emerging field of RNA nanotechnology. *Nature Nanotechnology*, *5*(12), 833–842. https://doi.org/10.1038/nnano.2010.231

Guo, P. X., Erickson, S., & Anderson, D. (1987). A small viral RNA is required for in vitro packaging of bacteriophage phi 29 DNA. *Science*, *236*(4802), 690–694. https://doi.org/10.1126/science.3107124

Guo, S., Tschammer, N., Mohammed, S., & Guo, P. (2005a). Specific Delivery of Therapeutic RNAs to Cancer Cells via the Dimerization Mechanism of phi29 Motor pRNA. *Human Gene Therapy*, *16*(9), 1097–1110. https://doi.org/10.1089/hum.2005.16.1097

Guo, S., Tschammer, N., Mohammed, S., & Guo, P. (2005b). Specific Delivery of Therapeutic RNAs to Cancer Cells via the Dimerization Mechanism of phi29 Motor pRNA. *Human Gene Therapy*, *16*(9), 1097–1110. https://doi.org/10.1089/hum.2005.16.1097

Haque, F., Shu, D., Shu, Y., Shlyakhtenko, L. S., Rychahou, P. G., Mark Evers, B., & Guo, P. (2012). Ultrastable synergistic tetravalent RNA nanoparticles for targeting to cancers. *Nano Today*, *7*(4), 245–257. https://doi.org/10.1016/j.nantod.2012.06.010

Haussecker, D. (2014). Current issues of RNAi therapeutics delivery and development. *Journal of Controlled Release: Official Journal of the Controlled Release Society*, *195*, 49–54. https://doi.org/10.1016/j.jconrel.2014.07.056

Hicke, B. J., & Stephens, A. W. (2000). Escort aptamers: A delivery service for diagnosis and therapy. *The Journal of Clinical Investigation*, *106*(8), 923–928. https://doi.org/10.1172/JCI11324

Hoeprich, S., Zhou, Q., Guo, S., Shu, D., Qi, G., Wang, Y., & Guo, P. (2003). Bacterial virus phi29 pRNA as a hammerhead ribozyme escort to destroy hepatitis B virus. *Gene Therapy*, *10*(15), 1258–1267. https://doi.org/10.1038/sj.gt.3302002

Huang, Y., Zhao, R., Fu, Y., Zhang, Q., Xiong, S., Li, L., … Chen, Y. (2011). Highly Specific Targeting and Imaging of Live Cancer Cells by Using a Peptide Probe Developed from Rationally Designed Peptides. *ChemBioChem*, *12*(8), 1209–1215. https://doi.org/10.1002/cbic.201100031

Kamerkar, S., LeBleu, V. S., Sugimoto, H., Yang, S., Ruivo, C. F., Melo, S. A., … Kalluri, R. (2017). Exosomes facilitate therapeutic targeting of oncogenic KRAS in pancreatic cancer. *Nature*, *546*(7659), 498–503. https://doi.org/10.1038/nature22341

Keefe, A. D., Pai, S., & Ellington, A. (2010). Aptamers as therapeutics. *Nature Reviews Drug Discovery*, *9*(7), 537–550. https://doi.org/10.1038/nrd3141

Keum, J.-W., Ahn, J.-H., & Bermudez, H. (2011). Design, Assembly, and Activity of Antisense DNA Nanostructures. *Small*, *7*(24), 3529–3535. https://doi.org/10.1002/smll.201101804

Kim, D.-H., Behlke, M. A., Rose, S. D., Chang, M.-S., Choi, S., & Rossi, J. J. (2005). Synthetic dsRNA Dicer substrates enhance RNAi potency and efficacy. *Nature Biotechnology*, *23*(2), 222–226. https://doi.org/10.1038/nbt1051

Kim, D.-N., Kilchherr, F., Dietz, H., & Bathe, M. (2012). Quantitative prediction of 3D solution shape and flexibility of nucleic acid nanostructures. *Nucleic Acids Research*, *40*(7), 2862–2868. https://doi.org/10.1093/nar/gkr1173

Koldehoff, M., Steckel, N. K., Beelen, D. W., & Elmaagacli, A. H. (2007). Therapeutic application of small interfering RNA directed against bcr-abl transcripts to a patient with imatinib-resistant chronic myeloid leukaemia. *Clinical and Experimental Medicine*, *7*(2), 47. https://doi.org/10.1007/s10238-007-0125-z

Kortylewski, M., Swiderski, P., Herrmann, A., Wang, L., Kowolik, C., Kujawski, M., … Yu, H. (2009). In vivo delivery of siRNA to immune cells by conjugation to a TLR9 agonist enhances antitumor immune responses. *Nature Biotechnology*, *27*(10), 925–932. https://doi.org/10.1038/nbt.1564

Lee, J. B., Hong, J., Bonner, D. K., Poon, Z., & Hammond, P. T. (2012). Self-assembled RNA interference microsponges for efficient siRNA delivery. *Nature Materials*, *11*(4), 316–322. https://doi.org/10.1038/nmat3253

Lee, H., Lytton-Jean, A. K. R., Chen, Y., Love, K. T., Park, A. I., Karagiannis, E. D., … Anderson, D. G. (2012). Molecularly self-assembled nucleic acid nanoparticles for targeted *in vivo* siRNA delivery. *Nature Nanotechnology*, *7*(6), 389–393. https://doi.org/10.1038/nnano.2012.73

Liu, J., Guo, S., Cinier, M., Shlyakhtenko, L. S., Shu, Y., Chen, C., … Guo, P. (2011). Fabrication of Stable and RNase-Resistant RNA Nanoparticles Active in Gearing the Nanomotors for Viral DNA Packaging. *ACS Nano*, *5*(1), 237–246. https://doi.org/10.1021/nn1024658

Liu, L., Xu, K., Wang, H., Jeremy Tan, P. K., Fan, W., Venkatraman, S. S., … Yang, Y.-Y. (2009). Self-assembled cationic peptide nanoparticles as an efficient antimicrobial agent. *Nature Nanotechnology*, *4*(7), 457–463. https://doi.org/10.1038/nnano.2009.153

Martínez, T., González, M. V., Roehl, I., Wright, N., Pañeda, C., & Jiménez, A. I. (2014). *In vitro* and *in vivo* efficacy of SYL040012, a novel siRNA compound for treatment of glaucoma. *Molecular Therapy: The Journal of the American Society of Gene Therapy*, *22*(1), 81–91. https://doi.org/10.1038/mt.2013.216

McCaffrey, A. P., Meuse, L., Pham, T.-T. T., Conklin, D. S., Hannon, G. J., & Kay, M. A. (2002). RNA interference in adult mice. *Nature*, *418*(6893), 38–39. https://doi.org/10.1038/418038a

Meister, G., & Tuschl, T. (2004). Mechanisms of gene silencing by double-stranded RNA. *Nature*, *431*(7006), 343–349. https://doi.org/10.1038/nature02873

Moll, W.-D., & Guo, P. (2007, September). Grouping of Ferritin and Gold Nanoparticles Conjugated to pRNA of the Phage phi29 DNA-Packaging Motor. *Journal of Nanoscience and Nanotechnology*, *7*(9), 3257–3267. https://doi.org/info:doi/10.1166/jnn.2007.914

Morrissey, D. V., Lockridge, J. A., Shaw, L., Blanchard, K., Jensen, K., Breen, W., … Polisky, B. (2005). Potent and persistent in vivo anti-HBV activity of chemically modified siRNAs. *Nature Biotechnology*, *23*(8), 1002–1007. https://doi.org/10.1038/nbt1122

Ng, E. W. M., Shima, D. T., Calias, P., Cunningham, E. T., Guyer, D. R., & Adamis, A. P. (2006). Pegaptanib, a targeted anti-VEGF aptamer for ocular vascular disease. *Nature Reviews Drug Discovery*, *5*(2), 123–132. https://doi.org/10.1038/nrd1955

Ni, X., Castanares, M., Mukherjee, A., & Lupold, S. E. (2011). Nucleic Acid Aptamers: Clinical Applications and Promising New Horizons. *Current Medicinal Chemistry*, 18(27), 4206–4214. https://doi.org/info:doi/10.2174/092986711797189600

Patel, M. R., Kozuch, S. D., Cultrara, C. N., Yadav, R., Huang, S., Samuni, U., … Sabatino, D. (2016). RNAi Screening of the Glucose-Regulated Chaperones in Cancer with Self-Assembled siRNA Nanostructures. *Nano Letters*, *16*(10), 6099–6108. https://doi.org/10.1021/acs.nanolett.6b02274

Peer, D., Zhu, P., Carman, C. V., Lieberman, J., & Shimaoka, M. (2007). Selective gene silencing in activated leukocytes by targeting siRNAs to the integrin lymphocyte function-associated antigen-1. *Proceedings of the National Academy of Sciences of the United States of America*, *104*(10), 4095–4100. https://doi.org/10.1073/pnas.0608491104

Petrova (Kruglova), N. S., Meschaninova, M. I., Venyaminova, A. G., Zenkova, M. A., Vlassov, V. V., & Chernolovskaya, E. L. (2010). 2′-O-Methyl–Modified Anti-MDR1 Fork-siRNA Duplexes Exhibiting High Nuclease Resistance and Prolonged Silencing Activity. *Oligonucleotides*, *20*(6), 297–308. https://doi.org/10.1089/oli.2010.0246

Pinheiro, A., Han, D., Shih, W. et al. (2011). Challenges and opportunities for structural DNA nanotechnology. *Nature Nanotech 6*, 763–772. https://doi.org/10.1038/nnano.2011.187

Rothemund, P. W. K. (2006). Folding DNA to create nanoscale shapes and patterns. *Nature*, *440*(7082), 297–302. https://doi.org/10.1038/nature04586

Sabir, T., Toulmin, A., Ma, L., Jones, A. C., McGlynn, P., Schröder, G. F., & Magennis, S. W. (2012). Branchpoint Expansion in a Fully Complementary Three-Way DNA Junction. *Journal of the American Chemical Society*, *134*(14), 6280–6285. https://doi.org/10.1021/ja211802z

Samantha M. Sarett, Christopher E. Nelson, Craig L. Duvall. (2015). Technologies for controlled, local delivery of siRNA, J*ournal of Controlled Release*, *218*, 94–113.

Seeman, N. C. (2007). An Overview of Structural DNA Nanotechnology. *Molecular Biotechnology*, *37*(3), 246. https://doi.org/10.1007/s12033-007-0059-4

Seeman, N. C. (2010). Nanomaterials Based on DNA. *Annual Review of Biochemistry*, *79*(1), 65–87. https://doi.org/10.1146/annurev-biochem-060308-102244

Semple, S. C., Akinc, A., Chen, J., Sandhu, A. P., Mui, B. L., Cho, C. K., … Hope, M. J. (2010). Rational design of cationic lipids for siRNA delivery. *Nature Biotechnology*, *28*(2), 172–176. https://doi.org/10.1038/nbt.1602

Setten, R. L., Rossi, J. J., & Han, S. (2019). The current state and future directions of RNAi-based therapeutics. *Nature Reviews Drug Discovery*, *18*(6), 421–446. https://doi.org/10.1038/s41573-019-0017-4

Shu, D., Moll, W.-D., Deng, Z., Mao, C., & Guo, P. (2004). Bottom-up Assembly of RNA Arrays and Superstructures as Potential Parts in Nanotechnology. *Nano Letters*, *4*(9), 1717–1723. https://doi.org/10.1021/nl0494497

Shu, D., Shu, Y., Haque, F., Abdelmawla, S., & Guo, P. (2011). Thermodynamically stable RNA three-way junction for constructing multifunctional nanoparticles for delivery of therapeutics. *Nature Nanotechnology*, *6*(10), 658–667. https://doi.org/10.1038/nnano.2011.105

Snead, N. M., Escamilla-Powers, J. R., Rossi, J. J., & McCaffrey, A. P. (2013). 5′ Unlocked Nucleic Acid Modification Improves siRNA Targeting. *Molecular Therapy - Nucleic Acids*, *2*, e103. https://doi.org/10.1038/mtna.2013.36

Song, E., Lee, S.-K., Wang, J., Ince, N., Ouyang, N., Min, J., … Lieberman, J. (2003). RNA interference targeting Fas protects mice from fulminant hepatitis. *Nature Medicine*, *9*(3), 347–351. https://doi.org/10.1038/nm828

Soutschek, J., Akinc, A., Bramlage, B., Charisse, K., Constien, R., Donoghue, M., … Vornlocher, H.-P. (2004). Therapeutic silencing of an endogenous gene by systemic administration of modified siRNAs. *Nature*, *432*(7014), 173–178. https://doi.org/10.1038/nature03121

Springer, A. D., & Dowdy, S. F. (2018). GalNAc-siRNA Conjugates: Leading the Way for Delivery of RNAi Therapeutics. *Nucleic Acid Therapeutics*, *28*(3), 109–118. https://doi.org/10.1089/nat.2018.0736

Tarapore, P., Shu, Y., Guo, P., & Ho, S.-M. (2011). Application of Phi29 Motor pRNA for Targeted Therapeutic Delivery of siRNA Silencing Metallothionein-IIA and Survivin in Ovarian Cancers. *Molecular Therapy*, *19*(2), 386–394. https://doi.org/10.1038/mt.2010.243

Taratula, O., Garbuzenko, O. B., Kirkpatrick, P., Pandya, I., Savla, R., Pozharov, V. P., … Minko, T. (2009). Surface-engineered targeted PPI dendrimer for efficient intracellular and intratumoral siRNA delivery. *Journal of Controlled Release*, *140*(3), 284–293. https://doi.org/10.1016/j.jconrel.2009.06.019

Tran, T. H., El Baz, R., Cuconati, A., Arthos, J., Jain, P., & Khan, Z. K. (2011). A Novel High-Throughput Screening Assay to Identify Inhibitors of HIV-1 gp120 Protein Interaction with DC-SIGN. *Journal of Antivirals & Antiretrovirals*, *3*, 49–54.

Tuerk, C., & Gold, L. (1990). Systematic evolution of ligands by exponential enrichment: RNA ligands to bacteriophage T4 DNA polymerase. *Science*, *249*(4968), 505–510. https://doi.org/10.1126/science.2200121

Tuschl, T., Zamore, P. D., Lehmann, R., Bartel, D. P., & Sharp, P. A. (1999). Targeted mRNA degradation by double-stranded RNA in vitro. *Genes & Development*, *13*(24), 3191–3197.

Vester, B., & Wengel, J. (2004). LNA (Locked Nucleic Acid): High-Affinity Targeting of Complementary RNA and DNA. *Biochemistry*, *43*(42), 13233–13241. https://doi.org/10.1021/bi0485732

Walsh, A. S., Yin, H., Erben, C. M., Wood, M. J. A., & Turberfield, A. J. (2011). DNA Cage Delivery to Mammalian Cells. *ACS Nano*, *5*(7), 5427–5432. https://doi.org/10.1021/nn2005574

Wang, A. Z., & Farokhzad, O. C. (2014). Current Progress of Aptamer-Based Molecular Imaging. *Journal of Nuclear Medicine*, *55*(3), 353–356. https://doi.org/10.2967/jnumed.113.126144

Wang, H.-W., Noland, C., Siridechadilok, B., Taylor, D. W., Ma, E., Felderer, K., … Nogales, E. (2009). Structural insights into RNA processing by the human RISC-loading complex. *Nature Structural & Molecular Biology*, *16*(11), 1148–1153. https://doi.org/10.1038/nsmb.1673

Winfree, E. (2000). Algorithmic Self-Assembly of DNA: Theoretical Motivations and 2D Assembly Experiments. *Journal of Biomolecular Structure and Dynamics*, *17*(sup1), 263–270. https://doi.org/10.1080/07391102.2000.10506630

Wittrup, A., Ai, A., Liu, X., Hamar, P., Trifonova, R., Charisse, K., … Lieberman, J. (2015). Visualizing lipid-formulated siRNA release from endosomes and target gene knockdown. *Nature Biotechnology*, *33*(8), 870–876. https://doi.org/10.1038/nbt.3298

Ye, X., Liu, Z., Hemida, M. G., & Yang, D. (2011). Targeted Delivery of Mutant Tolerant Anti-Coxsackievirus Artificial MicroRNAs Using Folate Conjugated Bacteriophage Phi29 pRNA. *PLOS ONE*, *6*(6), e21215. https://doi.org/10.1371/journal.pone.0021215

Zadeh, J. N., Steenberg, C. D., Bois, J. S., Wolfe, B. R., Pierce, M. B., Khan, A. R., … Pierce, N. A. (2011). NUPACK: Analysis and design of nucleic acid systems. *Journal of Computational Chemistry*, *32*(1), 170–173. https://doi.org/10.1002/jcc.21596

Zhang, F., Nangreave, J., Liu, Y., & Yan, H. (2014). Structural DNA Nanotechnology: State of the Art and Future Perspective. *Journal of the American Chemical Society*, *136*(32), 11198–11211. https://doi.org/10.1021/ja505101a

Zhou, J., & Rossi, J. J. (2011). Cell-specific aptamer-mediated targeted drug delivery. *Oligonucleotides*, *21*(1), 1–10.

Zhou, J., Shu, Y., Guo, P., Smith, D. D., & Rossi, J. J. (2011). Dual functional RNA nanoparticles containing phi29 motor pRNA and anti-gp120 aptamer for cell-type specific delivery and HIV-1 inhibition. *Methods*, *54*(2), 284–294.

Zhou, J., Swiderski, P., Li, H., Zhang, J., Neff, C. P., Akkina, R., & Rossi, J. J. (2009). Selection, characterization and application of new RNA HIV gp 120 aptamers for facile delivery of Dicer substrate siRNAs into HIV infected cells. *Nucleic Acids Research*, *37*(9), 3094–3109. https://doi.org/10.1093/nar/gkp185

46 RNA-Based Devices for Diagnostic and Biosensing

Morgan Chandler and Kirill A. Afonin
University of North Carolina at Charlotte, Charlotte, NC, USA

CONTENTS

RNA AS A MATERIAL FOR BIOSENSORS

One of the directions of RNA nanotechnology is towards the development of diagnostic tools which are able to interact with and signal the presence of specified analytes, particularly those present in cellular environments. A multitude of sensors have been designed for the detection of proteins, small molecules, and nucleic acid biomarkers with the shared goal to provide fast, reliable, and straightforward diagnosis for eventual clinical use.[1] RNA is a particularly advantageous material for constructing these diagnostic tools. Firstly, it has the ability to select and evolve programmable sensors in addition to mimicking those that already function in living cells.[2] A second advantage of using RNA is the high accuracy and precision with which specific target molecules can be detected. For example, a single nucleotide polymorphism in an mRNA can be distinguished from its wildtype by using a sensor which utilizes the complementary sequence for detection.[3]

This prevalence and constantly expanding library of RNA biomarkers is further rationale for incorporating RNA into biosensors. The human transcriptome has been shown to contain only a small percentage of protein-coding RNAs, revealing the persistence of non-coding RNAs as additional indicators of disease.[4] Additionally, non-coding RNAs, which include ribosomal RNAs, transfer RNAs, small nuclear and nucleolar RNAs, microRNAs, and long non-coding RNAs, to name a few, exhibit many functional roles in gene expression or post-transcriptional regulation. As advances in genomics further our understanding of how and when the profile of non-coding RNAs changes, they become more useful as therapeutic biomarkers which can indicate early disease onset and progression.[1,5]

MicroRNA Detection

MicroRNAs (miRNAs) are a class of short non-coding RNAs which are typically 18-25 nucleotides in length and have been shown to regulate critical cellular processes including differentiation,

DOI: 10.1201/9781003001560-51

division, and apoptosis.[6] The signatures of miRNA expression in tissues have been shown to alter with carcinogenesis with distinct profiles associated with tumor type and progression.[7,8] The detection of miRNAs has been shown to correlate more closely with tumor characterization than mRNAs.[9] Additionally, miRNAs have been profiled in serum[7,10] and in circulating tumor-derived exosomes[11] as a potential method for cancer detection. While many miRNA patterns have been identified, developing technologies for their detection need to be able to reduce non-specific binding and account for differences in normal miRNA expression across individuals.[12]

LONG NON-CODING RNA DETECTION

Another class of RNA biomarkers is long non-coding RNAs (lncRNAs) which are greater than 200 nts in length and are involved in regulatory functions, often acting as decoys, scaffolds, guides, or enhancers of biomolecular activity.[13,14] As such, changes which affect the regulatory nature of lncRNAs are associated with dysregulation of the cellular environment, resulting in changes in lncRNA expression in various cancers and disease states.[15] In addition to having high tissue specificity, several lncRNAs have been identified as prominent disease biomarkers.[14] For example, MALAT-1 has been shown to be overexpressed in various cancers,[16] PCAT18 is an upregulated transcript in metastatic prostate cancer,[17] and POU3F3 is a biomarker for esophageal squamous cell carcinoma which is found in the plasma.[18] Prompt detection of these non-coding RNAs which have been shown to be abundant and accessible in bodily fluids holds potential for the accelerated diagnosis of their associated diseases.[13]

DETECTION OF RNAs USING RNA BIOSENSORS

In order to detect RNA biomarkers, RNA assays such as Northern Blotting, high throughput sequencing, and RT-PCR are advantageous for initial characterization but are often too expensive and lack the sensitivity required to be used as rapid quantifiable diagnostics.[12] Instead, another approach for detection is to utilize an RNA biosensor which can bind the target sequence of interest and indicate its presence, generally through a fluorescent response. The development of such fluorescence trackers has been utilized not only for sequence detection but also as a specific tool for the study of RNA processing, transport, and localization.[19] Fluorescence can be supplied using a conjugated dye and offers high-resolution cellular imaging with low toxicity.[20]

FLUORESCENCE *IN SITU* HYBRIDIZATION

Fluorescence *In Situ* Hybridization (FISH) is a widely used technique which utilizes short exogenous fluorophore-tagged oligonucleotide probes to target different regions of DNAs or RNA transcripts.[21] Upon hybridization, the probes' fluorescence reveals the location and amount of the target sequence[19] (Figure 46.1a). FISH has been demonstrated for the detection of many different targets. For example, probes have been designed against aging-induced reactive mRNAs.[22] Using different fluorophores, multiple sets of probes can be used in the same samples to gain visuospatial information about multiple sequences.[23] In many cases, FISH is also used to verify and visualize that heterogeneity in single-cell expression reflects true biological differences.[24] Its main limitations are the requirement of a fixed sample as well as background fluorescence brought about by unbound probes, which should be washed extensively to remove any non-specific signal.[20]

MOLECULAR BEACONS

One way to minimize background signal is to design probes which are only fluorescently active when hybridized to the target. Molecular beacons are nucleic acids folded into hairpin probes which are tagged with a fluorophore and quencher on either end.[3,25] When folded, the fluorophore is in

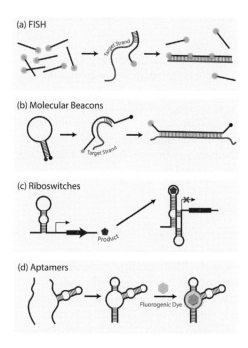

(a) FISH

(b) Molecular Beacons

(c) Riboswitches

(d) Aptamers

FIGURE 46.1 Schematic representation of RNA-Based Devices for Diagnostic and Biosensing among which are (a), FISH (b), Molecular beacons (c), riboswitches, and (d) fluorogenic aptamers.

close proximity to the quencher and assumes an "off" state. However, upon complete hybridization to a target sequence, the stem becomes unbound which in turn separates the fluorophore and quencher to restore its fluorescence to an "on" state[21] (Figure 46.1b).

Due to their stability, molecular beacons are more sensitive than traditional probes and can differentiate between sequences with single nucleotide resolution.[26,27] This makes them exceptional tools for detecting point mutations and also decreases the background signal produced by non-specific binding. Additionally, molecular beacons are modular and can be immobilized onto surfaces such as gold for quenching or glass for use in microarrays.[26,28] In addition to detecting RNA biomarkers in exosomes, they have also been used to study the molecular interactions of nucleic acids and proteins as well as to make measurements of the amounts of targets in living systems.[26,29]

RNA SWITCHES

Beyond the simple "on/off" states of single-stranded probes, multi-stranded nucleic acid assemblies can offer more functional switch designs. When rationally designed oligonucleotide sensors are in the presence of a complementary sequence, thermodynamically driven hybridization between the target and sensor strands can be used to drive more elaborate biosensing in the form of RNA switches and nanomachines. While these designs may incorporate the aforementioned conjugated small fluorogens as modes of visualization, multiple steps of strand displacement within a defined construct can be used to activate multiple functionalities upon sensing.[30,31] For instance, theranostic nanomachines can be designed to release therapeutic nucleic acids only upon interacting with a specific target sequence.[32] In addition to multiple outputs, the use of an RNA switch can be used to amplify signals and thus lower the limit of detection from that of conventional probes.[30] Hybridization chain reaction (HCR) utilizes a cascade of hybridization events between trigger strands and DNA hairpins which serve to amplify signal transduction.[33] Several RNA switches have also been designed to turn "on" and activate fluorescent aptamers upon interactions with specific miRNA biomarkers.[34,35]

RNA DEVICES WHICH RESPOND TO NON-NUCLEIC ACID ANALYTES AND PROCESSES

In addition to programmed binding with nucleic acid sequences, RNAs also have the ability to interact with other target molecules with high specificity, as exemplified by the vast number of naturally occurring RNAs which have evolved to serve as biological switches for the regulation of cellular processes. A number of RNA biosensing devices have been developed which utilize naturally occurring or selected affinity to a target molecule, or which can bind and activate fluorescence without the need for a covalent attachment. The implementation of RNA as a biosensing material in the use of molecular circuits and nanomachines benefits from the duality with which RNA can sense both nucleic acid biomarkers and target ligands.

RIBOZYMES AND RIBOSWITCHES

Naturally occurring riboswitches and ribozymes which can catalyze reactions are prominent in the mechanisms for gene control in prokaryotes, where they often play roles in the regulation of transcription, translation, splicing, and mRNA stability. Metabolites can be used in feedback regulation by binding to the nucleic acids which control the production of relevant proteins, thereby tuning cellular processes as a response to biosensing (Figure 46.1c). One example of this is the use of vitamin derivatives which interact with their respective mRNAs to control the vitamin B_1, B_2, and B_{12} operons.[36] Upon binding and inducing folding, "on/off" regulation occurs when either transcription is terminated or translation is initiated. Conditional ribozyme regulation occurs in the same manner; for instance, when excess glucosamine 6-phosphate is produced, it binds to the *glmS* ribozyme in the 5' untranslated region of the *glmS* gene which induces cleavage and the subsequent nuclease degradation of the enzyme transcript required for glucosamine 6-phosphate production.[36] Many known riboswitches have been discovered as a result of bioinformatics analysis.

The structures of such riboswitches which have evolved to interact with specific ligands for biological sensing are useful tools which can be implemented into larger sensing devices. By combining multiple RNA switches into an RNA array, multiple ligands can be simultaneously analyzed using both the recognition and signaling components provided by riboswitches.[37,38] Allosteric ribozymes, for example, are capable of detecting and reporting the presence of a specific compound or performing a conditionally activated task in response to a stimulus.[39]

APTAMERS

Aptamers are short oligonucleotides which are selected to specifically bind to a target molecule or protein of interest using systematic evolution of ligands by exponential enrichment (SELEX).[40,41] Over the past few decades of their development, one of the most useful findings for biosensor design has been the discovery of aptamers which bind dyes to induce a fluorescent response[20,41] (Figure 46.1d). Aptamers which bind cognate ligands can be endogenously produced without the need for covalently bound fluorophores, have low fluorescent background, and are modular for simple implementation into sensors.[42] Malachite Green was one of the first fluorescent RNA aptamers produced, but its toxicity and non-specific binding to other biomolecules drove the development of a larger library of fluorescent RNA aptamers, including Spinach, Spinach2, Broccoli, Mango, and Corn, among others.[43–46]

In order to make the fluorescent state conditional, split aptamers have been designed which are capable of binding the ligand to induce fluorescence only if both halves of the sensor are brought within close proximity.[47–52] This approach mimics the split green fluorescent protein system for conditional visualization, but without the need for the complex protein conjugation in sensors built of nucleic acids. Split aptamers have been used as the outputs of nucleic acid circuits and to visualize interactions between RNA nanoparticles in cells.[47,53]

In addition to the aforementioned fluorogen-binding aptamers, aptamers which have been developed to bind other target molecules are a useful recognition component of biosensing. Their sensing activities can be used not only to initiate visualization of the target, but can also be utilized for the conditional activation of therapeutic delivery in the case of theranostic sensors. By selecting for RNA molecules which bind solely to a ligand of interest, aptameric biosensors can be developed over multiple generations of SELEX for custom applications.[2] Once the sequences of the desired aptamers are known, they can then be implemented modularly into assays to detect the presence of target molecules through binding assays, folding-based sensors, and structure-switching signals.[54,55] For instance, HCR sensors which employ ATP-binding aptamers bind the ligand and subsequently expose sequences for hybridization to drive signals downstream.[33]

Aptamer-based sensors with well-established sequences have been widely used for the detection of small molecules and proteins and often incorporate conjugated fluorophores for visualization. One such aptamer which binds to the HIV Tat protein and inhibits Tat-dependent trans-activation of transcription was designed as a split, with one of the strands acting as a molecular beacon, to induce a fluorescent change upon binding to the target protein.[56] Split anti-cocaine and anti-adenosine aptamers have been used in conjunction with a fluorophore and quencher which are brought together by the presence of the respective target ligands.[41] Aptamers have also been demonstrated in combination with electrochemical biosensors for the detection of dopamine.[57]

FUTURE DIRECTIONS OF RNA-BASED DIAGNOSTICS

While RNA-based materials have been demonstrated to have diagnostic potential for the detection of target sequences, proteins, and small molecules, there are still challenges to their practical application. Databases of known biomarker targets are constantly being updated and signatures can differ in relative expression between individuals as well as throughout different stages of development. While this is promising for tracing these processes and developing personalized medicines, it can pose a challenge to establishing baselines for proof-of-concept sensors. The availability of RNA biomarkers in bodily fluids holds great potential for the transition of these technologies into prompt, noninvasive clinical assays. However, for theranostic devices, a method of delivery into cells is necessary while also avoiding non-specific electrostatic interactions.[54] Additionally, this introduces a need to examine the immune stimulation brought about by these devices; although, recognition which has been shown to depend greatly on the structure and composition of nucleic acid structures may be affected by the hybridization transition states in a dynamic sensor.[58,59] The library of biological fluorescent tags which have developed from the green fluorescent protein system to the dye-binding aptamers widely used today is also still being expanded. The latest fluorescent aptamers have broadened the possibilities for the types of outputs capable by RNA devices, which may one day allow for endogenously produced RNAs to match the multi-fluorescence capabilities introduced by exogenous RNAs in FISH. However, the repertoire of high quality aptamers which can be used for clinically important targets still requires expansion and investigation into their unknown secondary structures.[54]

Importantly, as RNA tools for biosensing are further explored, they also feedback into the design of dynamic RNA nanoparticles which are capable of carrying out programmed functions in response to environmental interactions. Upon interacting with cognate nanostructures, RNA nanoparticles which are previously inert can hybridize to activate functional moieties.[53,60] They can also be co-transcriptionally assembled upon interactions with target proteins[61] and can initiate the spatial organization of materials in cells.[62] The concept of RNA biomarkers has also been extrapolated for constructing barcodes for protein detection, which can then be used to expand the limit of detection and number of analytes.[63]

In conclusion, RNA is a unique biomaterial for the development of diagnostic and biosensing devices. Its abilities to interact via sequence with nucleic acid transcripts as well as via structure with specific ligands offer promising programmability for detecting multiple targets and activating either visual or therapeutic responses.

ACKNOWLEDGMENTS

The study was supported by the National Institute of General Medical Sciences of the National Institutes of Health under Award Numbers R01GM120487 and R35GM139587 (to K.A.A.). The content of this publication does not necessarily reflect the views or policies of the Department of Health and Human Services, nor does mention of trade names, commercial products, or organizations imply endorsement by the U.S. Government.

REFERENCES

1. Ludwig, J. A.; Weinstein, J. N., Biomarkers in cancer staging, prognosis and treatment selection. *Nature Reviews Cancer* 2005, *5* (11), 845.
2. Ellington, A. D.; Szostak, J. W., In vitro selection of RNA molecules that bind specific ligands. *Nature* 1990, *346* (6287), 818–822.
3. Tyagi, S.; Kramer, F. R., Molecular beacons: probes that fluoresce upon hybridization. *Nature Biotechnology* 1996, *14* (3), 303.
4. Lee, J. T., Epigenetic regulation by long noncoding RNAs. *Science* 2012, *338* (6113), 1435.
5. Cooper, T. A.; Wan, L.; Dreyfuss, G., RNA and disease. *Cell* 2009, *136* (4), 777–793.
6. Sandhu, S.; Garzon, R., Potential applications of microRNAs in cancer diagnosis, prognosis, and treatment. *Seminars in Oncology* 2011, *38* (6), 781–787.
7. Chen, X.; Ba, Y.; Ma, L.; Cai, X.; Yin, Y.; Wang, K.; Guo, J.; Zhang, Y.; Chen, J.; Guo, X., Characterization of microRNAs in serum: a novel class of biomarkers for diagnosis of cancer and other diseases. *Cell Research* 2008, *18* (10), 997.
8. Yanaihara, N.; Caplen, N.; Bowman, E.; Seike, M.; Kumamoto, K.; Yi, M.; Stephens, R. M.; Okamoto, A.; Yokota, J.; Tanaka, T., Unique microRNA molecular profiles in lung cancer diagnosis and prognosis. *Cancer Cell* 2006, *9* (3), 189–198.
9. Lu, J.; Getz, G.; Miska, E. A.; Alvarez-Saavedra, E.; Lamb, J.; Peck, D.; Sweet-Cordero, A.; Ebert, B. L.; Mak, R. H.; Ferrando, A. A., MicroRNA expression profiles classify human cancers. *Nature* 2005, *435* (7043), 834.
10. Mitchell, P. S.; Parkin, R. K.; Kroh, E. M.; Fritz, B. R.; Wyman, S. K.; Pogosova-Agadjanyan, E. L.; Peterson, A.; Noteboom, J.; O'Briant, K. C.; Allen, A., Circulating microRNAs as stable blood-based markers for cancer detection. *Proceedings of the National Academy of Sciences* 2008, *105* (30), 10513–10518.
11. Taylor, D. D.; Gercel-Taylor, C., MicroRNA signatures of tumor-derived exosomes as diagnostic biomarkers of ovarian cancer. *Gynecologic Oncology* 2008, *110* (1), 13–21.
12. Tavallaie, R.; De Almeida, S. R.; Gooding, J. J., Toward biosensors for the detection of circulating microRNA as a cancer biomarker: an overview of the challenges and successes. *Wiley Interdisciplinary Reviews: Nanomedicine and Nanobiotechnology* 2015, *7* (4), 580–592.
13. Bolha, L.; Ravnik-Glavač, M.; Glavač, D., Long noncoding RNAs as biomarkers in cancer. *Disease Markers* 2017, *2017*.
14. Brosnan, C. A.; Voinnet, O., The long and the short of noncoding RNAs. *Current Opinion in Cell Biology* 2009, *21* (3), 416–425.
15. Zhou, M.; Zhong, L.; Xu, W.; Sun, Y.; Zhang, Z.; Zhao, H.; Yang, L.; Sun, J., Discovery of potential prognostic long non-coding RNA biomarkers for predicting the risk of tumor recurrence of breast cancer patients. *Scientific Reports* 2016, *6*, 31038.
16. Wang, Y.; Xue, D.; Li, Y.; Pan, X.; Zhang, X.; Kuang, B.; Zhou, M.; Li, X.; Xiong, W.; Li, G., The long noncoding RNA MALAT-1 is a novel biomarker in various cancers: a meta-analysis based on the GEO database and literature. *Journal of Cancer* 2016, *7* (8), 991.
17. Crea, F.; Watahiki, A.; Quagliata, L.; Xue, H.; Pikor, L.; Parolia, A.; Wang, Y.; Lin, D.; Lam, W. L.; Farrar, W. L., Identification of a long non-coding RNA as a novel biomarker and potential therapeutic target for metastatic prostate cancer. *Oncotarget* 2014, *5* (3), 764.
18. Tong, Y.-S.; Wang, X.-W.; Zhou, X.-L.; Liu, Z.-H.; Yang, T.-X.; Shi, W.-H.; Xie, H.-W.; Lv, J.; Wu, Q.-Q.; Cao, X.-F., Identification of the long non-coding RNA POU3F3 in plasma as a novel biomarker for diagnosis of esophageal squamous cell carcinoma. *Molecular Cancer* 2015, *14* (1), 3.
19. Femino, A. M.; Fay, F. S.; Fogarty, K.; Singer, R. H., Visualization of single RNA transcripts *in situ*. *Science* 1998, *280* (5363), 585–590.

20. Novikova, I. V.; Afonin, K. A.; Leontis, N. B., New Ideas for in vivo Detection of RNA. In *Biosensors*, IntechOpen: 2010.
21. Ma, Z.; Wu, X.; Krueger, C. J.; Chen, A. K., Engineering Novel Molecular Beacon Constructs to Study Intracellular RNA Dynamics and Localization. *Genomics, Proteomics & Bioinformatics* 2017, *15* (5), 279–286.
22. Clarke, L. E.; Liddelow, S. A.; Chakraborty, C.; Munch, A. E.; Heiman, M.; Barres, B., Normal aging induces A1-like astrocyte reactivity. *Proceedings of the National Academy of Sciences of the United States of America* 2018, *115* (8), E1896–E1905.
23. Renwick, N.; Cekan, P.; Masry, P. A.; McGeary, S. E.; Miller, J. B.; Hafner, M.; Li, Z.; Mihailovic, A.; Morozov, P.; Brown, M., Multicolor microRNA FISH effectively differentiates tumor types. *The Journal of Clinical Investigation* 2013, *123* (6), 2694–2702.
24. Shalek, A. K.; Satija, R.; Adiconis, X.; Gertner, R. S.; Gaublomme, J. T.; Raychowdhury, R.; Schwartz, S.; Yosef, N.; Malboeuf, C.; Lu, D. N.; Trombetta, J. J.; Gennert, D.; Gnirke, A.; Goren, A.; Hacohen, N.; Levin, J. Z.; Park, H.; Regev, A., Single-cell transcriptomics reveals bimodality in expression and splicing in immune cells. *Nature* 2013, *498* (7453), 236–240.
25. Marras, S. A.; Kramer, F. R.; Tyagi, S., Multiplex detection of single-nucleotide variations using molecular beacons. *Genetic Analysis: Biomolecular Engineering* 1999, *14* (5–6), 151–156.
26. Tan, W.; Wang, K.; Drake, T. J., Molecular beacons. *Current Opinion in Chemical Biology* 2004, *8* (5), 547–553.
27. Zheng, J.; Yang, R. H.; Shi, M. L.; Wu, C. C.; Fang, X. H.; Li, Y. H.; Li, J. H.; Tan, W. H., Rationally designed molecular beacons for bioanalytical and biomedical applications. *Chemical Society Reviews* 2015, *44* (10), 3036–3055.
28. Li, N.; Chang, C.; Pan, W.; Tang, B., A Multicolor Nanoprobe for Detection and Imaging of Tumor-Related mRNAs in Living Cells. *Angewandte Chemie International Edition* 2012, *51* (30), 7426–7430.
29. Lee, J. H.; Kim, J. A.; Kwon, M. H.; Kang, J. Y.; Rhee, W. J., *In situ* single step detection of exosome microRNA using molecular beacon. *Biomaterials* 2015, *54*, 116–125.
30. Kolpashchikov, D. M., Evolution of Hybridization Probes to DNA Machines and Robots. *Accounts of Chemical Research* 2019, *52* (7), 1949–1956.
31. Zakrevsky, P.; Parlea, L.; Viard, M.; Bindewald, E.; Afonin, K. A.; Shapiro, B. A., Preparation of a Conditional RNA Switch. In *Methods in Molecular Biology (Clifton, N.J.)*, 2017/07/22 ed.; 2017; Vol. 1632, pp. 303–324.
32. Bindewald, E.; Afonin, K. A.; Viard, M.; Zakrevsky, P.; Kim, T.; Shapiro, B. A., Multistrand Structure Prediction of Nucleic Acid Assemblies and Design of RNA Switches. *Nano Letters* 2016, *16* (3), 1726–1735.
33. Dirks, R. M.; Pierce, N. A., Triggered amplification by hybridization chain reaction. *Proceedings of the National Academy of Sciences of the United States of America* 2004, *101* (43), 15275.
34. Huang, K.; Doyle, F.; Wurz, Z. E.; Tenenbaum, S. A.; Hammond, R. K.; Caplan, J. L.; Meyers, B. C., FASTmiR: an RNA-based sensor for in vitro quantification and live-cell localization of small RNAs. *Nucleic Acids Research* 2017, *45* (14), e130–e130.
35. Aw, S. S.; Tang, M. X.; Teo, Y. N.; Cohen, S. M., A conformation-induced fluorescence method for microRNA detection. *Nucleic Acids Research* 2016, *44* (10), e92.
36. Serganov, A.; Nudler, E., A Decade of Riboswitches. *Cell* 2013, *152* (1–2), 17–24.
37. Seetharaman, S.; Zivarts, M.; Sudarsan, N.; Breaker, R. R., Immobilized RNA switches for the analysis of complex chemical and biological mixtures. *Nature Biotechnology* 2001, *19* (4), 336.
38. Breaker, R. R., Engineered allosteric ribozymes as biosensor components. *Current Opinion in Biotechnology* 2002, *13* (1), 31–39.
39. Soukup, G. A.; Breaker, R. R., Nucleic acid molecular switches. *Trends in Biotechnology* 1999, *17* (12), 469–476.
40. Tuerk, C.; Gold, L., Systematic evolution of ligands by exponential enrichment: RNA ligands to bacteriophage T4 DNA polymerase. *Science* 1990, *249* (4968), 505–510.
41. Stojanovic, M. N.; de Prada, P.; Landry, D. W., Fluorescent Sensors Based on Aptamer Self-Assembly. *Journal of the American Chemical Society* 2000, *122* (46), 11547–11548.
42. Ouellet, J., RNA Fluorescence with Light-Up Aptamers. *Frontiers in Chemistry* 2016, *4* (29).
43. Song, W.; Filonov, G. S.; Kim, H.; Hirsch, M.; Li, X.; Moon, J. D.; Jaffrey, S. R., Imaging RNA polymerase III transcription using a photostable RNA–fluorophore complex. *Nature Chemical Biology* 2017, *13* (11), 1187.

44. Paige, J. S.; Wu, K. Y.; Jaffrey, S. R., RNA mimics of green fluorescent protein. *Science* 2011, *333* (6042), 642–646.

45. Filonov, G. S.; Moon, J. D.; Svensen, N.; Jaffrey, S. R., Broccoli: Rapid selection of an RNA mimic of green fluorescent protein by fluorescence-based selection and directed evolution. *Journal of the American Chemical Society* 2014, *136* (46), 16299–16308.

46. Dolgosheina, E. V.; Jeng, S. C. Y.; Panchapakesan, S. S. S.; Cojocaru, R.; Chen, P. S. K.; Wilson, P. D.; Hawkins, N.; Wiggins, P. A.; Unrau, P. J., RNA mango aptamer-fluorophore: A bright, high-affinity complex for RNA labeling and tracking. *ACS Chemical Biology* 2014, *9* (10), 2412–2420.

47. Alam, K. K.; Tawiah, K. D.; Lichte, M. F.; Porciani, D.; Burke, D. H., A Fluorescent Split Aptamer for Visualizing RNA–RNA Assembly In Vivo. *ACS Synthetic Biology* 2017, *6* (9), 1710–1721.

48. Chandler, M.; Lyalina, T.; Halman, J.; Rackley, L.; Lee, L.; Dang, D.; Ke, W.; Sajja, S.; Woods, S.; Acharya, S.; Baumgarten, E.; Christopher, J.; Elshalia, E.; Hrebien, G.; Kublank, K.; Saleh, S.; Stallings, B.; Tafere, M.; Striplin, C.; Afonin, A. K., Broccoli Fluorets: Split Aptamers as a User-Friendly Fluorescent Toolkit for Dynamic RNA Nanotechnology. *Molecules* 2018, *23* (12).

49. Rogers, T. A.; Andrews, G. E.; Jaeger, L.; Grabow, W. W., Fluorescent Monitoring of RNA Assembly and Processing Using the Split-Spinach Aptamer. *ACS Synthetic Biology* 2015, *4* (2), 162–166.

50. Kolpashchikov, D. M., Binary Malachite Green Aptamer for Fluorescent Detection of Nucleic Acids. *Journal of the American Chemical Society* 2005, *127* (36), 12442–12443.

51. Sajja, S.; Chandler, M.; Striplin, C. D.; Afonin, K. A., Activation of Split RNA Aptamers: Experiments Demonstrating the Enzymatic Synthesis of Short RNAs and Their Assembly As Observed by Fluorescent Response. *Journal of Chemical Education* 2018.

52. Afonin, K. A.; Danilov, E. O.; Novikova, I. V.; Leontis, N. B., TokenRNA: A New Type of Sequence-Specific, Label-Free Fluorescent Biosensor for Folded RNA Molecules. *Chembiochem: A European Journal of Chemical Biology* 2008, *9* (12), 1902–1905.

53. Halman, J. R.; Satterwhite, E.; Roark, B.; Chandler, M.; Viard, M.; Ivanina, A.; Bindewald, E.; Kasprzak, W. K.; Panigaj, M.; Bui, M. N.; Lu, J. S.; Miller, J.; Khisamutdinov, E. F.; Shapiro, B. A.; Dobrovolskaia, M. A.; Afonin, K. A., Functionally-interdependent shape-switching nanoparticles with controllable properties. *Nucleic Acids Research* 2017, *45* (4), 2210–2220.

54. Zhou, W.; Huang, P.-J. J.; Ding, J.; Liu, J., Aptamer-based biosensors for biomedical diagnostics. *Analyst* 2014, *139* (11), 2627–2640.

55. Porter, E. B.; Polaski, J. T.; Morck, M. M.; Batey, R. T., Recurrent RNA motifs as scaffolds for genetically encodable small-molecule biosensors. *Nature Chemical Biology* 2017, *13* (3), 295.

56. Yamamoto, R.; Kumar, P. K. R., Molecular beacon aptamer fluoresces in the presence of Tat protein of HIV-1. *Genes to Cells* 2000, *5* (5), 389–396.

57. Farjami, E.; Campos, R.; Nielsen, J. S.; Gothelf, K. V.; Kjems, J.; Ferapontova, E. E., RNA aptamer-based electrochemical biosensor for selective and label-free analysis of dopamine. *Analytical Chemistry* 2012, *85* (1), 121–128.

58. Hong, E.; Halman, J. R.; Shah, A. B.; Khisamutdinov, E. F.; Dobrovolskaia, M. A.; Afonin, K. A., Structure and Composition Define Immunorecognition of Nucleic Acid Nanoparticles. *Nano Letters* 2018, *18* (7), 4309–4321.

59. Johnson, M. B.; Halman, J. R.; Satterwhite, E.; Zakharov, A. V.; Bui, M. N.; Benkato, K.; Goldsworthy, V.; Kim, T.; Hong, E.; Dobrovolskaia, M. A.; Khisamutdinov, E. F.; Marriott, I.; Afonin, K. A., Programmable Nucleic Acid Based Polygons with Controlled Neuroimmunomodulatory Properties for Predictive QSAR Modeling. *Small* 2017, *13* (42), 1701255.

60. Ke, W.; Hong, E.; Saito, R. F.; Rangel, M. C.; Wang, J.; Viard, M.; Richardson, M.; Khisamutdinov, E. F.; Panigaj, M.; Dokholyan, N. V.; Chammas, R.; Dobrovolskaia, M. A.; Afonin, K. A., RNA-DNA fibers and polygons with controlled immunorecognition activate RNAi, FRET and transcriptional regulation of NF-kappaB in human cells. *Nucleic Acids Research* 2018.

61. Geary, C.; Rothemund, P. W. K.; Andersen, E. S., A single-stranded architecture for cotranscriptional folding of RNA nanostructures. *Science* 2014, *345* (6198), 799.

62. Shibata, T.; Fujita, Y.; Ohno, H.; Suzuki, Y.; Hayashi, K.; Komatsu, K. R.; Kawasaki, S.; Hidaka, K.; Yonehara, S.; Sugiyama, H.; Endo, M.; Saito, H., Protein-driven RNA nanostructured devices that function in vitro and control mammalian cell fate. *Nature Communications* 2017, *8* (1), 540.

63. Hill, H. D.; Mirkin, C. A., The bio-barcode assay for the detection of protein and nucleic acid targets using DTT-induced ligand exchange. *Nature Protocols* 2006, *1* (1), 324–336.

47 MicroRNAs
Biology and Role in RNA Nanotechnology

Bin Guo and Jingwen Liu
University of Houston, Houston, TX, USA

Daniel W. Binzel
The Ohio State University, Columbus, OH, USA

CONTENTS

Since the initial discovery in 1993, the tiny microRNAs (miRNAs) have grown into major players in many arenas of life science (Lee et al., 1993). Recent studies have shown that miRNAs are involved in divergent biological processes, from cell cycle progression to apoptosis, from developmental timing to the nervous system patterning, from cellular growth to hematopoiesis (Ambros, 2004, Bartel, 2004, Bartel, 2009). The involvement of miRNAs in disease has also been investigated intensively in cardiovascular diseases, CNS disorders, metabolic diseases, and various types of cancer. The new findings from these miRNA studies not only offer insight into an important mechanism of regulating gene expression and cell function but also open new possibilities for targeted therapy of disease.

THE BIOGENESIS OF miRNAs

The genes encoding the miRNAs are located on chromosomes in the DNA sequences that are either inside of the introns of pre-mRNAs or in the previously regarded non-coding regions between genes. In both cases, the miRNA genes are mostly transcribed by the RNA polymerase II (pol II) (Bartel, 2004). The pol II transcripts are more than 1 kb long and are named the pri-miRNAs (Figure 47.1). In the nucleus, the pri-miRNA is processed by Drosha, an RNase III endonuclease, which produces a 60–70 nucleotides long miRNA precursor (pre-miRNA) (Lee et al., 2003, Lee et al., 2002). The pre-miRNA is then transported out of the nucleus by Exportin-5 (Yi et al., 2003). In the cytoplasm, the pre-miRNA is further processed by another RNA III endonuclease Dicer (Lee et al., 2003) to produce the mature miRNA that is often about 22 nucleotides in length.

DOI: 10.1201/9781003001560-52

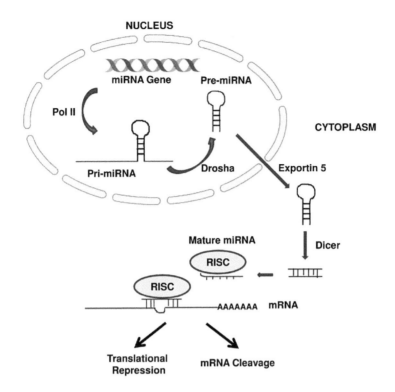

FIGURE 47.1   The biogenesis and mechanism of action of microRNAs.

THE MECHANISM OF ACTION OF miRNAs

The ~22 nt double-stranded miRNA produced by Dicer is incorporated into the RNA-induced silencing complex (RISC) (Hammond et al., 2000, Hammond et al., 2001, Martinez et al., 2002). One
strand in the miRNA duplex is called miRNA* and is separated from the miRNA strand and degraded
rapidly (However, the miRNA* strand has been shown recently to be functional and abundant in cancer and other diseases (Bhayani et al., 2012)). The remaining miRNA strand will guide the RISC
to bind to the matching sequences located in the 3'-untranslated region (UTR) of the target mRNAs
(Figure 47.1). MiRNA finds the mRNA target by imperfect base pair complementarity between the
5'-proximal seed region (positions 2–8) on the miRNA and the pairing sequences on the 3'-UTR of
the mRNA. A miRNA can have hundreds of potential targets due to the imprecise sequence complementarity that is needed for the miRNA to bind to its target mRNA. The RISC contains Argonaute
proteins and GW182 proteins which repress the translation of the target mRNAs (Chekulaeva &
Filipowicz, 2009). The GW182 proteins can also cause deadenylation and degradation of the target
mRNA by recruiting the CAF1:CCR4:NOT1 deadenylase complex and the DCP1:DCP2 decapping
complex. The target mRNA is then cleaved at positions upstream of the miRNA binding sites by
5'→3' exonuclease Xrn1 (Bagga et al., 2005, Behm-Ansmant et al., 2006, Eulalio et al., 2009, Liu et
al., 2005). Regarding to the kinetics of these events, miRNA-mediated translation repression occurs
first, followed by mRNA deadenylation and decay (Djuranovic et al., 2012).

THE ROLE OF miRNAs IN DISEASE AND THERAPY

There are currently 2812 human miRNAs annotated in the online database miRBase (http://www.
mirbase.org/index.shtml), as of September 2019. While some of these miRNAs may not pass
eventual experimental validation (Chiang et al., 2010), it can be estimated with confidence that

the number of human miRNAs is well over 1000. Many miRNAs have been investigated for their involvements in various types of diseases. Some major areas of research are summarized here.

CANCER

MiRNA expression is altered in many types of cancers (Calin & Croce, 2006). Some miRNAs behave like the oncogenes (oncomiRs) and promote cancer growth when they are upregulated. One example is miR-21 which is overexpressed in breast cancer (Qian et al., 2009), lung cancer (Markou et al., 2008), colon cancer (Asangani et al., 2008), and many other types of cancers (Folini et al., 2010, Gabriely et al., 2008, Meng et al., 2007, Zhang et al., 2008). The tumor-promoting activity of miR-21 is associated with the repression of its targets; many of them are tumor suppressor genes, such as TGFBR2 (Mishra et al., 2014), BTG2 (Coppola et al., 2013), TIMP3 (Nagao et al., 2012), PTEN (Meng et al., 2007), and PDCD4 (Asangani et al., 2008). When overexpressed in mice through genetic manipulation, miR-21 stimulated the formation of pre-B-cell lymphoma (Medina et al., 2010) and lung cancer (Hatley et al., 2010). Some miRNAs act as tumor suppressors and are often downregulated in cancers. For example, miR-31 expression is decreased in breast cancer (Valastyan et al., 2009a), gastric cancer (Zhang et al., 2010), prostate cancer (Schaefer et al., 2010), and other cancers (Hua et al., 2012, Leidner et al., 2012, Yamagishi et al., 2012). The loss of miR-31 promotes cancer development by increasing NF-κB signaling (Yamagishi et al., 2012) or stimulates metastasis by accumulating integrin-α5, radixin, and RhoA (Valastyan et al., 2009b). Our lab has shown that both miR-205 and miR-31 are downregulated in prostate cancer (Bhatnagar et al., 2010). The silencing of these two miRNAs renders cancer cells resistant to chemotherapy-induced apoptosis through the upregulation of their targets, Bcl-w and E2F6, both anti-apoptotic proteins. A single miRNA could have dual functions as an oncogene or a tumor suppressor in different types of cancers. Since a miRNA is able to target multiple genes simultaneously, including both oncogenes and tumor suppressors, it is possible that the balance of expression of these targeted genes will determine the overall net effect of the miRNA in a certain cancer type. One example is miR-125b, which is oncogenic in many types of hematologic malignancies (Bousquet et al., 2012, Liu et al., 2017, Shaham et al., 2012) but tumor suppressive in several solid tumors, including breast cancer (Zhang et al., 2011), lung cancer (Li et al., 2013), head and neck cancer (Nakanishi et al., 2014), and gastric cancer (Riquelme et al., 2016). To target miRNAs for cancer therapy, one can employ various delivery strategies to block the activity of oncomiRs with antagomiRs (miRNA inhibitors) or increase the levels of tumor suppressor miRNAs. For example, anti-miR-221 and anti-miR-222 inhibitors increase gefitinib *in vivo* activity against lung cancer (Garofalo et al., 2012). Conversely, systemic delivery of miR-124 inhibits hepatocellular carcinoma development in mice through induction of apoptosis (Hatziapostolou et al., 2011).

CARDIOVASCULAR DISEASES

miRNAs play a critical role in the cardiovascular system. They are involved in myocardial infarction (Bonauer et al., 2009, van Rooij et al., 2008), cardiac hypertrophy (Care et al., 2007), cardiac fibrosis (Zhao et al., 2015), arrhythmia (Cheng et al., 2019), vascular diseases (Hartmann et al., 2016), and heart failure (Thum et al., 2007). In the first experiment that demonstrates the efficacy of cholesterol-conjugated antagomiRs in mouse, Krutzfeldt et al. successfully inhibited miR-122 in the liver, resulting in the downregulation of cholesterol biosynthesis genes and a reduction of the plasma cholesterol level (Krutzfeldt et al., 2005). In a breakthrough study, Rayner et al. successfully increased plasma high-density lipoprotein (HDL) levels and reduced plasma levels of very-low-density lipoprotein (VLDL)-associated triglycerides in African green monkeys, using systemic delivery of an antagomiR that targets both miR-33a and miR-33b (Rayner et al., 2011). Using a cholesterol-conjugated antagomiR inhibitor of miR-133, Care et al. demonstrated that a single infusion of the antagomiR can cause marked and sustained cardiac hypertrophy (Care et al., 2007). MiR-21 expression increases in fibroblasts of the failing heart, and Thum et al. found that silencing of miR-21 with an antagomiR inhibits interstitial fibrosis and attenuates cardiac dysfunction (Thum et al., 2008). MiRNAs also regulate the functions of endothelial cells. For example, Fiedler et al. showed that miR-24 is upregulated in cardiac endothelial cells and induces apoptosis after cardiac ischemia (Fiedler et al., 2011). Inhibition of miR-24 with cholesterol-conjugated antagomiR reduced the size of myocardial infarction in mice, prevented endothelial apoptosis, enhanced vascularity, and preserved cardiac function and survival (Fiedler et al., 2011). MiRNAs can directly regulate cardiomyocyte survival. Overexpression of miR-320 enhanced cardiomyocyte death and apoptosis and a single tail vein injection of cholesterol-modified antagomiR of miR-320 prevented apoptosis and reduced infarction size in mice (Ren et al., 2009b). In atrial fibrillation patients, miR-320 expression is increased in the atria and suppresses the cardiac L-type Ca2+ channel. Blocking miR-328 with an antagomiR reversed atrial fibrillation, and genetic knockdown of endogenous miR-328 reduced the vulnerability to atrial fibrillation in mice (Lu et al., 2010).

NEUROLOGICAL DISEASES

MiRNAs also play an essential role in neural development and various neurological diseases. For example, miR-133b is expressed in the midbrain dopaminergic neurons and is deficient in midbrain tissue from patients with Parkinson's disease (Kim et al., 2007). Furthermore, miR-133b inhibits the maturation and function of dopaminergic neurons by targeting transcription factor Pitx3 (Kim et al., 2007). In patients who have Alzheimer's disease, the expression levels of the miR-29a/b-1 cluster miRNAs are significantly decreased, which can result in Aβ accumulation (Hebert et al., 2008). In addition, lentiviral-mediated miR-101 overexpression significantly inhibited amyloid precursor protein expression and Aβ accumulation in cultured hippocampal neurons (Vilardo et al., 2010). miRNAs are also associated with Fragile X syndrome, a common form of inherited mental retardation. A key player in this disease, the RNA-binding protein FMRP interacts with miRNAs and miRNA-processing Argonaute proteins, and miRNAs are critical for FMRP function in neural development and synaptogenesis (Jin et al., 2004). In schizophrenia patients, miR-181b expression is significantly increased in the superior temporal gyrus of the brain samples, which results in the downregulation of the calcium sensor gene visinin-like 1 and the ionotropic AMPA glutamate receptor subunit (both are associated with the pathology of schizophrenia) (Beveridge et al., 2008). Huntington's disease is caused by the polyglutamine expansions in the Huntingtin protein, which releases transcription factor REST from Huntingtin and REST translocates to the nucleus and decreases neuronal gene expression. miR-9 and miR-9* are decreased in Huntington's disease and they target two components of the REST complex: miR-9 targets REST and miR-9* targets CoREST (Packer et al., 2008). Interestingly, miR-34b is a plasma stable miRNA and is significantly elevated in plasma of genetically diagnosed Huntington's disease patients even prior to symptom onset, making this miRNA a candidate biomarker for diagnosis (Gaughwin et al., 2011). In patients with major depressive disorder (MDD), the expression of the polyamine genes spermidine/spermine N^1-acetyltransferase 1 (SAT1) and spermine oxidase (SMOX) are downregulated. It has been demonstrated that miRNAs are associated with downregulation of polyamine gene expression in the suicide brain (Lopez et al., 2014). Since miRNAs are essential regulators for cellular processes of the nervous system, more connections to other neurological diseases will be revealed in the future.

DELIVERY OF miRNAS AS THERAPEUTICS

With the natural occurrence of microRNAs and their important gene regulation role in disease prevention and development, these small RNAs prove to be an excellent candidate for use as novel therapeutics or targets for therapeutics (Bader et al., 2010, Obad et al., 2011). With successful delivery of miRNAs to diseased cells, regulation of specific gene expression and modification of cellular function can be accomplished; however, successful targeting and delivery of the miRNAs are needed. In diseases where the expression of a miRNA is suppressed as described earlier, the miRNA expression can be synthetically returned to its normal level through specific delivery of that miRNA (Ye et al., 2011). Additionally, anti-miRNA locked nucleic acid (LNA) sequences have been designed to bind and silence oncogenic miRNAs that are overexpressed within cancers (Obad et al., 2011). These short (typically 8nt) LNAs bind to the 3' seed region of miRNA in the cytoplasm of cells, thus preventing the miRNA from binding to its normal messenger RNA (mRNA) target. Therefore, a delivery scaffold that is stable and easily assembled is essential for delivery of these miRNAs and anti-miRNAs.

In past and current work, several different approaches have been used for the delivery of synthetic miRNA and its close relative siRNA (small interfering RNA) (Hamilton & Baulcombe, 1999). These gene regulating small RNAs have been delivered through liposomes (Chen et al., 2010), dendrimers (Ren et al., 2009a), gold nanoparticles (Crew et al., 2012), spherical nucleic acids (SNAs) (Alhasan et al., 2014), exosomes (Teng et al., 2016, Zhuang et al., 2016), and magnetic silica spheres (Liu et al., 2012). As a whole, these delivery methods come with limitations and disadvantages in that they do not have distinct control over the size of polymers, are easily cleared from the body, can lead to antibody response of the immune system, and can lack the ability to specifically target the diseased cells or tissues. However, many of these shortcomings are avoided in using RNA nanotechnology to construct all RNA nanoparticles containing a targeting molecule as well as the miRNA. RNA naturally has advantages over many other nanoparticle platforms that allow for the specific delivery of therapeutics to disease sites, while avoiding off-target toxicities (Abdelmawla et al., 2011, Cui et al., 2015). Using RNA constructs for drug delivery provides benefits such as polyvalent delivery of therapeutics as therapeutics of other RNA functional groups can be included on the end of each helical region of branched nanoparticles (Chang et al., 2012, Haque et al., 2012). Additionally RNA nanoparticles naturally have defined structure, size, and stoichiometry, thus their *in vivo* behavior is predictable and tunable by changing the design of the nanoparticles (Haque et al., 2017, Jasinki et al., 2018, Khisamutdinov et al., 2015, Khisamutdinov et al., 2016, Shi et al., 2018, Xu et al., 2018). RNA nanoparticles are between 5 and 50 nm in size; making them large enough to be retained by the body and permeable through leaky blood vessels into tumor environments, but small enough to avoid long-term accumulation in the body (Abdelmawla et al., 2011, Guo, 2010). It has been proven that RNA nanoparticles avoid antibody and cytokine response; however, the size, shape, and sequence of the nanoparticles can be tuned to elicit a large cytokine response, so the RNA constructs act as immuno-adjuvants (Abdelmawla et al., 2011, Guo et al., 2017, Khisamutdinov et al., 2014). Through chemical modification (i.e. 2'-Fluoro, 2'-OMethyl, LNA, etc.) RNA is chemically and thermodynamically stable keeping the assembly of the multi-oligo nanoparticle at ultra-low concentrations; as well as enzymatically stable against RNase degradation in the body (Binzel et al., 2014, Liu et al., 2011, Piao et al., 2018). And finally, RNA nanoparticles self-assemble allowing for ease of construction (Guo, 2010).

Such delivery of microRNAs can be achieved using the phi29 pRNA three-way junction (3WJ). The 3WJ, as shown by Shu et al., is derived from the core structure of packaging RNA (pRNA) of the phi29 bacteriophage (Shu et al., 2011). Composed of three, short, individual strands of RNA, the pRNA 3WJ has been shown to be chemically and thermodynamically stable and self-assemble in the presence of no salt ions. Using the pRNA 3WJ as a backbone scaffold, several therapeutic elements can be conjugated off of each branch creating a polyvalent drug and using a bottom-up construction design allows for a simplistic design and construction of an RNA nanoparticle that can be used as an effect scaffold for delivery of therapeutics (Shu et al., 2004, Shu et al., 2011). Furthermore, the Guo lab has developed numerous other RNA nanoparticles based on the phi29 pRNA system that allows for the conjugation of therapeutics off helical branches for the treatment of cancers (Khisamutdinov et al., 2016, Khisamutdinov et al., 2014, Sharma et al., 2016, Shu et al., 2004, Shu et al., 2018, Xu et al., 2019, Yin et al., 2019).

Furthermore, using a cell targeting component, such as an RNA aptamer or conjugation of a chemical group, RNA nanoparticles are capable of specifically targeting cancer tumors and other diseased cells with overexpression of cell surface receptors (Ellington & Szostak, 1990). Additionally, additional branches on complex RNA nanoparticles can be used for the conjugation of a therapeutic miRNA or a fluorophore for imaging purpose. Such miRNA-3WJ nanoparticle has the ability to act as a complete, target-specific, therapeutic agent that can be used to treat a wide variety of diseases (Binzel et al., 2016, Lee et al., 2017, Shu et al., 2015, Yin et al., 2019).

ACKNOWLEDGMENT

This work was supported by the National Institutes of Health (grant # CA186100).

REFERENCES

Abdelmawla S, Guo S, Zhang L, Pulukuri SM, Patankar P, Conley P, Trebley J, Guo P, Li QX (2011) Pharmacological characterization of chemically synthesized monomeric phi29 pRNA nanoparticles for systemic delivery. *Molecular Therapy: The Journal of the American Society of Gene Therapy* 19: 1312–22.

Alhasan AH, Patel PC, Choi CH, Mirkin CA (2014) Exosome encased spherical nucleic acid gold nanoparticle conjugates as potent microRNA regulation agents. *Small* 10: 186–92.

Ambros V (2004) The functions of animal microRNAs. *Nature* 431: 350–5.

Asangani IA, Rasheed SA, Nikolova DA, Leupold JH, Colburn NH, Post S, Allgayer H (2008) MicroRNA-21 (miR-21) post-transcriptionally downregulates tumor suppressor Pdcd4 and stimulates invasion, intravasation and metastasis in colorectal cancer. *Oncogene* 27: 2128–36.

Bader AG, Brown D, Winkler M (2010) The promise of microRNA replacement therapy. *Cancer Research* 70: 7027–30.

Bagga S, Bracht J, Hunter S, Massirer K, Holtz J, Eachus R, Pasquinelli AE (2005) Regulation by let-7 and lin-4 miRNAs results in target mRNA degradation. *Cell* 122: 553–63.

Bartel DP (2004) MicroRNAs: genomics, biogenesis, mechanism, and function. *Cell* 116: 281–97.

Bartel DP (2009) MicroRNAs: target recognition and regulatory functions. *Cell* 136: 215–33.

Behm-Ansmant I, Rehwinkel J, Doerks T, Stark A, Bork P, Izaurralde E (2006) mRNA degradation by miRNAs and GW182 requires both CCR4:NOT deadenylase and DCP1:DCP2 decapping complexes. *Genes & Development* 20: 1885–98.

Beveridge NJ, Tooney PA, Carroll AP, Gardiner E, Bowden N, Scott RJ, Tran N, Dedova I, Cairns MJ (2008) Dysregulation of miRNA 181b in the temporal cortex in schizophrenia. *Human Molecular Genetics* 17: 1156–68.

Bhatnagar N, Li X, Padi SK, Zhang Q, Tang MS, Guo B (2010) Downregulation of miR-205 and miR-31 confers resistance to chemotherapy-induced apoptosis in prostate cancer cells. *Cell Death & Disease* 1: e105

Bhayani MK, Calin GA, Lai SY (2012) Functional relevance of miRNA* sequences in human disease. *Mutation Research* 731: 14–9.

Binzel DW, Khisamutdinov EF, Guo P (2014) Entropy-driven one-step formation of Phi29 pRNA 3WJ from three RNA fragments. *Biochemistry* 53: 2221–31.

Binzel DW, Shu Y, Li H, Sun M, Zhang Q, Shu D, Guo B, Guo P (2016) Specific Delivery of MiRNA for High Efficient Inhibition of Prostate Cancer by RNA Nanotechnology. *Molecular Therapy: The Journal of the American Society of Gene Therapy* 24: 1267–77.

Bonauer A, Carmona G, Iwasaki M, Mione M, Koyanagi M, Fischer A, Burchfield J, Fox H, Doebele C, Ohtani K, Chavakis E, Potente M, Tjwa M, Urbich C, Zeiher AM, Dimmeler S (2009) MicroRNA-92a controls angiogenesis and functional recovery of ischemic tissues in mice. *Science* 324: 1710–3.

Bousquet M, Nguyen D, Chen C, Shields L, Lodish HF (2012) MicroRNA-125b transforms myeloid cell lines by repressing multiple mRNA. *Haematologica* 97: 1713–21.

Calin GA, Croce CM (2006) MicroRNA signatures in human cancers. *Nature Reviews Cancer* 6: 857–66.

Care A, Catalucci D, Felicetti F, Bonci D, Addario A, Gallo P, Bang ML, Segnalini P, Gu Y, Dalton ND, Elia L, Latronico MV, Hoydal M, Autore C, Russo MA, Dorn GW, 2nd, Ellingsen O, Ruiz-Lozano P, Peterson KL, Croce CM et al. (2007) MicroRNA-133 controls cardiac hypertrophy. *Nature Medicine* 13: 613–8.

Chang CI, Lee TY, Yoo JW, Shin D, Kim M, Kim S, Lee DK (2012) Branched, tripartite-interfering RNAs silence multiple target genes with long guide strands. *Nucleic Acid Therapy* 22: 30–9.

Chekulaeva M, Filipowicz W (2009) Mechanisms of miRNA-mediated post-transcriptional regulation in animal cells. *Current Opinion in Cell Biology* 21: 452–60.

Chen Y, Zhu X, Zhang X, Liu B, Huang L (2010) Nanoparticles modified with tumor-targeting scFv deliver siRNA and miRNA for cancer therapy. *Molecular Therapy: The Journal of the American Society of Gene Therapy* 18: 1650–6.

Cheng WL, Kao YH, Chao TF, Lin YK, Chen SA, Chen YJ (2019) MicroRNA-133 suppresses ZFHX3-dependent atrial remodelling and arrhythmia. *Acta Physiologica*: e13322.

Chiang HR, Schoenfeld LW, Ruby JG, Auyeung VC, Spies N, Baek D, Johnston WK, Russ C, Luo S, Babiarz JE, Blelloch R, Schroth GP, Nusbaum C, Bartel DP (2010) Mammalian microRNAs: experimental evaluation of novel and previously annotated genes. *Genes & Development* 24: 992–1009.

Coppola V, Musumeci M, Patrizii M, Cannistraci A, Addario A, Maugeri-Sacca M, Biffoni M, Francescangeli F, Cordenonsi M, Piccolo S, Memeo L, Pagliuca A, Muto G, Zeuner A, De Maria R, Bonci D (2013) BTG2 loss and miR-21 upregulation contribute to prostate cell transformation by inducing luminal markers expression and epithelial-mesenchymal transition. *Oncogene* 32: 1843–53.

Crew E, Tessel MA, Rahman S, Razzak-Jaffar A, Mott D, Kamundi M, Yu G, Tchah N, Lee J, Bellavia M, Zhong CJ (2012) MicroRNA conjugated gold nanoparticles and cell transfection. *Analytical Chemistry* 84: 26–9.

Cui D, Zhang C, Liu B, Shu Y, Du T, Shu D, Wang K, Dai F, Liu Y, Li C, Pan F, Yang Y, Ni J, Li H, Brand-Saberi B, Guo P (2015) Regression of Gastric Cancer by Systemic Injection of RNA Nanoparticles Carrying both Ligand and siRNA. *Scientific Reports* 5: 10726.

Djuranovic S, Nahvi A, Green R (2012) miRNA-mediated gene silencing by translational repression followed by mRNA deadenylation and decay. *Science* 336: 237–40.

Ellington AD, Szostak JW (1990) In vitro selection of RNA molecules that bind specific ligands. *Nature* 346: 818–22.

Eulalio A, Huntzinger E, Nishihara T, Rehwinkel J, Fauser M, Izaurralde E (2009) Deadenylation is a widespread effect of miRNA regulation. *RNA* 15: 21–32.

Fiedler J, Jazbutyte V, Kirchmaier BC, Gupta SK, Lorenzen J, Hartmann D, Galuppo P, Kneitz S, Pena JT, Sohn-Lee C, Loyer X, Soutschek J, Brand T, Tuschl T, Heineke J, Martin U, Schulte-Merker S, Ertl G, Engelhardt S, Bauersachs J et al. (2011) MicroRNA-24 regulates vascularity after myocardial infarction. *Circulation* 124: 720–30.

Folini M, Gandellini P, Longoni N, Profumo V, Callari M, Pennati M, Colecchia M, Supino R, Veneroni S, Salvioni R, Valdagni R, Daidone MG, Zaffaroni N (2010) miR-21: an oncomir on strike in prostate cancer. *Molecular Cancer* 9: 12.

Gabriely G, Wurdinger T, Kesari S, Esau CC, Burchard J, Linsley PS, Krichevsky AM (2008) MicroRNA 21 promotes glioma invasion by targeting matrix metalloproteinase regulators. *Molecular and Cellular Biology* 28: 5369–80.

Garofalo M, Romano G, Di Leva G, Nuovo G, Jeon YJ, Ngankeu A, Sun J, Lovat F, Alder H, Condorelli G, Engelman JA, Ono M, Rho JK, Cascione L, Volinia S, Nephew KP, Croce CM (2012) EGFR and MET receptor tyrosine kinase-altered microRNA expression induces tumorigenesis and gefitinib resistance in lung cancers. *Nature Medicine* 18: 74–82.

Gaughwin PM, Ciesla M, Lahiri N, Tabrizi SJ, Brundin P, Bjorkqvist M (2011) Hsa-miR-34b is a plasma-stable microRNA that is elevated in pre-manifest Huntington's disease. *Human Molecular Genetics* 20: 2225–37.

Guo P (2010) The emerging field of RNA nanotechnology. *Nature Nanotechnology* 5: 833–42.

Guo S, Li H, Ma M, Fu J, Dong Y, Guo P (2017) Size, Shape, and Sequence-Dependent Immunogenicity of RNA Nanoparticles. *Molecular Therapy - Nucleic Acids* 9: 399–408.

Hamilton AJ, Baulcombe DC (1999) A species of small antisense RNA in posttranscriptional gene silencing in plants. *Science* 286: 950–952.

Hammond SM, Bernstein E, Beach D, Hannon GJ (2000) An RNA-directed nuclease mediates post-transcriptional gene silencing in Drosophila cells. *Nature* 404: 293–6.

Hammond SM, Boettcher S, Caudy AA, Kobayashi R, Hannon GJ (2001) Argonaute2, a link between genetic and biochemical analyses of RNAi. *Science* 293: 1146–50.

Haque F, Shu D, Shu Y, Shlyakhtenko LS, Rychahou PG, Evers BM, Guo PX (2012) Ultrastable synergistic tetravalent RNA nanoparticles for targeting to cancers. *Nano Today* 7: 245–257.

Haque F, Xu C, Jasinski DL, Li H, Guo P (2017) Using Planar Phi29 pRNA Three-Way Junction to Control Size and Shape of RNA Nanoparticles for Biodistribution Profiling in Mice. *Methods in Molecular Biology* 1632: 359–380.

Hartmann D, Fiedler J, Sonnenschein K, Just A, Pfanne A, Zimmer K, Remke J, Foinquinos A, Butzlaff M, Schimmel K, Maegdefessel L, Hilfiker-Kleiner D, Lachmann N, Schober A, Froese N, Heineke J, Bauersachs J, Batkai S, Thum T (2016) MicroRNA-Based Therapy of GATA2-Deficient Vascular Disease. *Circulation* 134: 1973–1990.

Hatley ME, Patrick DM, Garcia MR, Richardson JA, Bassel-Duby R, van Rooij E, Olson EN (2010) Modulation of K-Ras-dependent lung tumorigenesis by MicroRNA-21. *Cancer Cell* 18: 282–93.

Hatziapostolou M, Polytarchou C, Aggelidou E, Drakaki A, Poultsides GA, Jaeger SA, Ogata H, Karin M, Struhl K, Hadzopoulou-Cladaras M, Iliopoulos D (2011) An HNF4alpha-miRNA inflammatory feedback circuit regulates hepatocellular oncogenesis. *Cell* 147: 1233–47.

Hebert SS, Horre K, Nicolai L, Papadopoulou AS, Mandemakers W, Silahtaroglu AN, Kauppinen S, Delacourte A, De Strooper B (2008) Loss of microRNA cluster miR-29a/b-1 in sporadic Alzheimer's disease corre-lates with increased BACE1/beta-secretase expression. *Proceedings of the National Academy of Sciences of the United States of America* 105: 6415–20.

Hua D, Ding D, Han X, Zhang W, Zhao N, Foltz G, Lan Q, Huang Q, Lin B (2012) Human miR-31 targets radixin and inhibits migration and invasion of glioma cells. *Oncology Reports* 27: 700–6.

Jain KK (2005) The role of nanobiotechnology in drug discovery. *Drug Discovery Today* 10: 1435–42.

Jasinski DL, Li H, Guo P (2018) The Effect of Size and Shape of RNA Nanoparticles on Biodistribution. *Molecular Therapy: The Journal of the American Society of Gene Therapy* 26: 784–792.

Jin P, Zarnescu DC, Ceman S, Nakamoto M, Mowrey J, Jongens TA, Nelson DL, Moses K, Warren ST (2004) Biochemical and genetic interaction between the fragile X mental retardation protein and the microRNA pathway. *Nature Neuroscience* 7: 113–7.

Khisamutdinov EF, Bui MN, Jasinski D, Zhao Z, Cui Z, Guo P (2015) Simple Method for Constructing RNA Triangle, Square, Pentagon by Tuning Interior RNA 3WJ Angle from 60 degrees to 90 degrees or 108 degrees. *Methods in Molecular Biology* 1316: 181–93.

Khisamutdinov EF, Jasinski DL, Li H, Zhang K, Chiu W, Guo P (2016) Fabrication of RNA 3D Nanoprisms for Loading and Protection of Small RNAs and Model Drugs. *Advanced Materials* 28: 10079–10087.

Khisamutdinov EF, Li H, Jasinski DL, Chen J, Fu J, Guo P (2014) Enhancing immunomodulation on innate immunity by shape transition among RNA triangle, square and pentagon nanovehicles. *Nucleic Acids Research* 42: 9996–10004.

Kim J, Inoue K, Ishii J, Vanti WB, Voronov SV, Murchison E, Hannon G, Abeliovich A (2007) A MicroRNA feedback circuit in midbrain dopamine neurons. *Science* 317: 1220–4.

Krutzfeldt J, Rajewsky N, Braich R, Rajeev KG, Tuschl T, Manoharan M, Stoffel M (2005) Silencing of microRNAs in vivo with 'antagomirs'. *Nature* 438: 685–9.

Lee RC, Feinbaum RL, Ambros V (1993) The C. elegans heterochronic gene lin-4 encodes small RNAs with antisense complementarity to lin-14. *Cell* 75: 843–54.

Lee TJ, Yoo JY, Shu D, Li H, Zhang J, Yu JG, Jaime-Ramirez AC, Acunzo M, Romano G, Cui R, Sun HL, Luo Z, Old M, Kaur B, Guo P, Croce CM (2017) RNA Nanoparticle-Based Targeted Therapy for Glioblastoma through Inhibition of Oncogenic miR-21. *Molecular Therapy: The Journal of the American Society of Gene Therapy* 25: 1544–1555.

Lee Y, Ahn C, Han J, Choi H, Kim J, Yim J, Lee J, Provost P, Radmark O, Kim S, Kim VN (2003) The nuclear RNase III Drosha initiates microRNA processing. *Nature* 425: 415–9.

Lee Y, Jeon K, Lee JT, Kim S, Kim VN (2002) MicroRNA maturation: stepwise processing and subcellular localization. *The EMBO Journal* 21: 4663–70.

Leidner RS, Ravi L, Leahy P, Chen Y, Bednarchik B, Streppel M, Canto M, Wang JS, Maitra A, Willis J, Markowitz SD, Barnholtz-Sloan J, Adams MD, Chak A, Guda K (2012) The microRNAs, MiR-31 and MiR-375, as candidate markers in Barrett's esophageal carcinogenesis. *Genes, Chromosomes & Cancer* 51: 473–9.

Li Y, Chao Y, Fang Y, Wang J, Wang M, Zhang H, Ying M, Zhu X, Wang H (2013) MTA1 promotes the invasion and migration of non-small cell lung cancer cells by downregulating miR-125b. *Journal of Experimental & Clinical Cancer Research: CR* 32: 33.

Liu J, Guo B, Chen Z, Wang N, Iacovino M, Cheng J, Roden C, Pan W, Khan S, Chen S, Kyba M, Fan R, Guo S, Lu J (2017) miR-125b promotes MLL-AF9-driven murine acute myeloid leukemia involving a VEGFA-mediated non-cell-intrinsic mechanism. *Blood* 129: 1491–1502.

Liu J, Guo S, Cinier M, Shlyakhtenko LS, Shu Y, Chen C, Shen G, Guo P (2011) Fabrication of stable and RNase-resistant RNA nanoparticles active in gearing the nanomotors for viral DNA packaging. *ACS Nano* 5: 237–46.

Liu J, Valencia-Sanchez MA, Hannon GJ, Parker R (2005) MicroRNA-dependent localization of targeted mRNAs to mammalian P-bodies. *Nature Cell Biology* 7: 719–23.

Liu J, Wang B, Hartono SB, Liu T, Kantharidis P, Middelberg AP, Lu GQ, He L, Qiao SZ (2012) Magnetic sil-ica spheres with large nanopores for nucleic acid adsorption and cellular uptake. *Biomaterials* 33: 970–8.

Lopez JP, Fiori LM, Gross JA, Labonte B, Yerko V, Mechawar N, Turecki G (2014) Regulatory role of miRNAs in polyamine gene expression in the prefrontal cortex of depressed suicide completers. *The International Journal of Neuropsychopharmacology* 17: 23–32.

Lu Y, Zhang Y, Wang N, Pan Z, Gao X, Zhang F, Shan H, Luo X, Bai Y, Sun L, Song W, Xu C, Wang Z, Yang B (2010) MicroRNA-328 contributes to adverse electrical remodeling in atrial fibrillation. *Circulation* 122: 2378–87.

Markou A, Tsaroucha EG, Kaklamanis L, Fotinou M, Georgoulias V, Lianidou ES (2008) Prognostic value of mature microRNA-21 and microRNA-205 overexpression in non-small cell lung cancer by quantitative real-time RT-PCR. *Clinical Chemistry* 54: 1696–704.

Martinez J, Patkaniowska A, Urlaub H, Luhrmann R, Tuschl T (2002) Single-stranded antisense siRNAs guide target RNA cleavage in RNAi. *Cell* 110: 563–74.

Medina PP, Nolde M, Slack FJ (2010) OncomiR addiction in an in vivo model of microRNA-21-induced pre-B-cell lymphoma. *Nature* 467: 86–90.

Meng F, Henson R, Wehbe-Janek H, Ghoshal K, Jacob ST, Patel T (2007) MicroRNA-21 regulates expression of the PTEN tumor suppressor gene in human hepatocellular cancer. *Gastroenterology* 133: 647–58.

Mishra S, Deng JJ, Gowda PS, Rao MK, Lin CL, Chen CL, Huang T, Sun LZ (2014) Androgen receptor and microRNA-21 axis downregulates transforming growth factor beta receptor II (TGFBR2) expression in prostate cancer. *Oncogene* 33: 4097–106.

Nagao Y, Hisaoka M, Matsuyama A, Kanemitsu S, Hamada T, Fukuyama T, Nakano R, Uchiyama A, Kawamoto M, Yamaguchi K, Hashimoto H (2012) Association of microRNA-21 expression with its targets, PDCD4 and TIMP3, in pancreatic ductal adenocarcinoma. *Modern Pathology: An Official Journal of the United States and Canadian Academy of Pathology, Inc* 25: 112–21.

Nakanishi H, Taccioli C, Palatini J, Fernandez-Cymering C, Cui R, Kim T, Volinia S, Croce CM (2014) Loss of miR-125b-1 contributes to head and neck cancer development by dysregulating TACSTD2 and MAPK pathway. *Oncogene* 33: 702–12.

Obad S, dos Santos CO, Petri A, Heidenblad M, Broom O, Ruse C, Fu C, Lindow M, Stenvang J, Straarup EM, Hansen HF, Koch T, Pappin D, Hannon GJ, Kauppinen S (2011) Silencing of microRNA families by seed-targeting tiny LNAs. *Nature Genetics* 43: 371–8.

Packer AN, Xing Y, Harper SQ, Jones L, Davidson BL (2008) The bifunctional microRNA miR-9/miR-9* regulates REST and CoREST and is downregulated in Huntington's disease. *The Journal of Neuroscience: The Official Journal of the Society for Neuroscience* 28: 14341–6.

Piao X, Wang H, Binzel DW, Guo P (2018) Assessment and comparison of thermal stability of phosphorothioate-DNA, DNA, RNA, 2′-F RNA, and LNA in the context of Phi29 pRNA 3WJ. *RNA* 24: 67–76.

Qian B, Katsaros D, Lu L, Preti M, Durando A, Arisio R, Mu L, Yu H (2009) High miR-21 expression in breast cancer associated with poor disease-free survival in early stage disease and high TGF-beta1. *Breast Cancer Research and Treatment* 117: 131–40.

Rayner KJ, Esau CC, Hussain FN, McDaniel AL, Marshall SM, van Gils JM, Ray TD, Sheedy FJ, Goedeke L, Liu X, Khatsenko OG, Kaimal V, Lees CJ, Fernandez-Hernando C, Fisher EA, Temel RE, Moore KJ (2011) Inhibition of miR-33a/b in non-human primates raises plasma HDL and lowers VLDL triglycerides. *Nature* 478: 404–7.

Ren XP, Wu J, Wang X, Sartor MA, Jones K, Qian J, Nicolaou P, Pritchard TJ, Fan GC (2009a) MicroRNA-320 is involved in the regulation of cardiac ischemia/reperfusion injury by targeting heat-shock protein 20. *Circulation* 119: 2357–2366.

Ren XP, Wu J, Wang X, Sartor MA, Qian J, Jones K, Nicolaou P, Pritchard TJ, Fan GC (2009b) MicroRNA-320 is involved in the regulation of cardiac ischemia/reperfusion injury by targeting heat-shock protein 20. *Circulation* 119: 2357–66.

Riquelme I, Tapia O, Leal P, Sandoval A, Varga MG, Letelier P, Buchegger K, Bizama C, Espinoza JA, Peek RM, Araya JC, Roa JC (2016) miR-101–2, miR-125b-2 and miR-451a act as potential tumor suppressors in gastric cancer through regulation of the PI3K/AKT/mTOR pathway. *Cellular Oncology* 39: 23–33.

Schaefer A, Jung M, Mollenkopf HJ, Wagner I, Stephan C, Jentzmik F, Miller K, Lein M, Kristiansen G, Jung K (2010) Diagnostic and prognostic implications of microRNA profiling in prostate carcinoma. *International Journal of Cancer Journal International du Cancer* 126: 1166–76.

Shaham L, Binder V, Gefen N, Borkhardt A, Izraeli S (2012) MiR-125 in normal and malignant hematopoiesis. *Leukemia* 26: 2011–8.

Sharma A, Haque F, Pi F, Shlyakhtenko LS, Evers BM, Guo P (2016) Controllable self-assembly of RNA dendrimers. *Nanomedicine* 12: 835–844.

Shi Z, Li SK, Charoenputtakun P, Liu CY, Jasinski D, Guo P (2018) RNA nanoparticle distribution and clearance in the eye after subconjunctival injection with and without thermosensitive hydrogels. *Journal of Controlled Release* 270: 14–22.

Shu D, Li H, Shu Y, Xiong G, Carson WE, 3rd, Haque F, Xu R, Guo P (2015) Systemic Delivery of Anti-miRNA for Suppression of Triple Negative Breast Cancer Utilizing RNA Nanotechnology. *ACS Nano* 9: 9731–40.

Shu D, Moll WD, Deng Z, Mao C, Guo P (2004) Bottom-up Assembly of RNA Arrays and Superstructures as Potential Parts in Nanotechnology. *Nano Letters* 4: 1717–23.

Shu D, Shu Y, Haque F, Abdelmawla S, Guo P (2011) Thermodynamically stable RNA three-way junction for constructing multifunctional nanoparticles for delivery of therapeutics. *Nature Nanotechnology* 6: 658–67.

Shu Y, Yin H, Rajabi M, Li H, Vieweger M, Guo S, Shu D, Guo P (2018) RNA-based micelles: A novel platform for paclitaxel loading and delivery. *Journal of Controlled Release* 276: 17–29.

Teng Y, Mu J, Hu X, Samykutty A, Zhuang X, Deng Z, Zhang L, Cao P, Yan J, Miller D, Zhang HG (2016) Grapefruit-derived nanovectors deliver miR-18a for treatment of liver metastasis of colon cancer by induction of M1 macrophages. *Oncotarget* 7: 25683–97.

Thum T, Galuppo P, Wolf C, Fiedler J, Kneitz S, van Laake LW, Doevendans PA, Mummery CL, Borlak J, Haverich A, Gross C, Engelhardt S, Ertl G, Bauersachs J (2007) MicroRNAs in the human heart: a clue to fetal gene reprogramming in heart failure. *Circulation* 116: 258–67.

Thum T, Gross C, Fiedler J, Fischer T, Kissler S, Bussen M, Galuppo P, Just S, Rottbauer W, Frantz S, Castoldi M, Soutschek J, Koteliansky V, Rosenwald A, Basson MA, Licht JD, Pena JT, Rouhanifard SH, Muckenthaler MU, Tuschl T et al. (2008) MicroRNA-21 contributes to myocardial disease by stimulating MAP kinase signalling in fibroblasts. *Nature* 456: 980–4.

Valastyan S, Benaich N, Chang A, Reinhardt F, Weinberg RA (2009a) Concomitant suppression of three target genes can explain the impact of a microRNA on metastasis. *Genes & Development* 23: 2592–7.

Valastyan S, Reinhardt F, Benaich N, Calogrias D, Szasz AM, Wang ZC, Brock JE, Richardson AL, Weinberg RA (2009b) A pleiotropically acting microRNA, miR-31, inhibits breast cancer metastasis. *Cell* 137: 1032–46.

van Rooij E, Sutherland LB, Thatcher JE, DiMaio JM, Naseem RH, Marshall WS, Hill JA, Olson EN (2008) Dysregulation of microRNAs after myocardial infarction reveals a role of miR-29 in cardiac fibrosis. *Proceedings of the National Academy of Sciences of the United States of America* 105: 13027–32.

Vilardo E, Barbato C, Ciotti M, Cogoni C, Ruberti F (2010) MicroRNA-101 regulates amyloid precursor protein expression in hippocampal neurons. *The Journal of Biological Chemistry* 285: 18344–51.

Xu C, Haque F, Jasinski DL, Binzel DW, Shu D, Guo P (2018) Favorable biodistribution, specific targeting and conditional endosomal escape of RNA nanoparticles in cancer therapy. *Cancer Letters* 414: 57–70.

Xu C, Li H, Zhang K, Binzel DW, Yin H, Chiu W, Guo P (2019) Photo-controlled release of paclitaxel and model drugs from RNA pyramids. *Nano Research* 12: 41–48.

Yamagishi M, Nakano K, Miyake A, Yamochi T, Kagami Y, Tsutsumi A, Matsuda Y, Sato-Otsubo A, Muto S, Utsunomiya A, Yamaguchi K, Uchimaru K, Ogawa S, Watanabe T (2012) Polycomb-mediated loss of miR-31 activates NIK-dependent NF-kappaB pathway in adult T cell leukemia and other cancers. *Cancer Cell* 21: 121–35.

Ye X, Liu Z, Hemida MG, Yang D (2011) Targeted delivery of mutant tolerant anti-coxsackievirus artificial microRNAs using folate conjugated bacteriophage Phi29 pRNA. *PLoS One* 6: e21215.

Yi R, Qin Y, Macara IG, Cullen BR (2003) Exportin-5 mediates the nuclear export of pre-microRNAs and short hairpin RNAs. *Genes & Development* 17: 3011–6.

Yin H, Wang H, Li Z, Shu D, Guo P (2019) RNA Micelles for the Systemic Delivery of Anti-miRNA for Cancer Targeting and Inhibition without Ligand. *ACS Nano* 13: 706–717.

Zhang Y, Guo J, Li D, Xiao B, Miao Y, Jiang Z, Zhuo H (2010) Down-regulation of miR-31 expression in gastric cancer tissues and its clinical significance. *Medical Oncology* 27: 685–9.

Zhang Y, Yan LX, Wu QN, Du ZM, Chen J, Liao DZ, Huang MY, Hou JH, Wu QL, Zeng MS, Huang WL, Zeng YX, Shao JY (2011) miR-125b is methylated and functions as a tumor suppressor by regulating the ETS1 proto-oncogene in human invasive breast cancer. *Cancer Research* 71: 3552–62.

Zhang Z, Li Z, Gao C, Chen P, Chen J, Liu W, Xiao S, Lu H (2008) miR-21 plays a pivotal role in gastric cancer pathogenesis and progression. *Laboratory Investigation; A Journal of Technical Methods and Pathology* 88: 1358–66.

Zhao X, Wang K, Liao Y, Zeng Q, Li Y, Hu F, Liu Y, Meng K, Qian C, Zhang Q, Guan H, Feng K, Zhou Y, Du Y, Chen Z (2015) MicroRNA-101a inhibits cardiac fibrosis induced by hypoxia via targeting TGFbetaRI on cardiac fibroblasts. *Cellular Physiology and Biochemistry: International Journal of Experimental Cellular Physiology, Biochemistry, and Pharmacology* 35: 213–26.

Zhuang X, Teng Y, Samykutty A, Mu J, Deng Z, Zhang L, Cao P, Rong Y, Yan J, Miller D, Zhang HG (2016) Grapefruit-derived Nanovectors Delivering Therapeutic miR17 Through an Intranasal Route Inhibit Brain Tumor Progression. *Molecular Therapy: The Journal of the American Society of Gene Therapy* 24: 96–105.

48 Conjugation of RNA Aptamer to RNA Nanoparticles for Targeted Drug Delivery

Marissa Leonard and Xiaoting Zhang
University of Cincinnati College of Medicine, Cincinnati, USA

Fengmei Pi
ExonanoRNA LLC, Columbus, USA

Peixuan Guo
The Ohio State University, Columbus, USA

CONTENTS

INTRODUCTION

RNA-based therapy has become a promising avenue for the treatment of many human diseases. Therapeutic potentials of RNAs, including ribozymes, short hairpin RNA (shRNA), siRNA, miRNA, antisense oligonucleotides (AS OGNs), and RNA aptamers, have long been extensively studied (Guo, 2010; Keefe et al., 2010; Levy-Nissenbaum et al., 2008; Que-Gewirth and Sullenger, 2007; Yan and Levy, 2009). A major challenge that remains is the systemic and intracellular delivery of these moieties (siRNA, ribozyme, etc.) to the desired target cells. In this regard, the pRNA nanoparticle delivery system pioneered by Dr. Guo, combined with recent advancement in RNA aptamers, provides an ideal method for nanoscale delivery suitable for in vivo targeted delivery (Guo, 2010).

pRNA nanoparticles are a new RNA-based nanoparticle drug delivery system, which can be designed and constructed by phi29 pRNA through dimer, hexamer formation, or using the stable three-way junction (3WJ) domain as a scaffold. pRNA nanoparticles are mainly composed of RNA; these can be chemically modified to extend its half-life in vivo but keep their nature of being biodegradable. Thus allow all the advantages of RNAs as therapeutic agents to be retained. One key obstacle for RNA-based therapy is that RNA is susceptible to quick degradation in the blood stream during in vivo delivery. To overcome that, Dr. Guo's laboratory has recently developed highly stable and RNAse-resistant pRNA through elaborate design of RNA sequences and chemical modifications such as 2'-deoxy-2'-fluoro (2'-F) modification at the ribose rings of C and U (Liu et al., 2011; Shu et al., 2011).

RNA aptamers are RNA oligonucleotides capable of binding to their targets with high affinity and specificity. RNA aptamers have numerous advantages for targeted drug delivery when compared to DNA aptamers and protein-based antibodies (Guo, 2010; Keefe et al., 2010; Que-Gewirth and Sullenger, 2007; Thiel and Giangrande, 2010). Compared to their peptide and antibody counterparts, RNA aptamers are much easier to synthesize in large quantities with defined structure and stoichiometry. Furthermore, RNA aptamers are generally considered to be more thermodynamically stable than peptides or antibodies. Although RNA aptamers function similarly to antibodies, they are known to have low or no immunogenicity when compared to other macromolecules such as proteins/antibodies. Furthermore, recent studies have found that RNA aptamers can be further chemically modified (e.g. 2'deoxy, 2'F, 2'NH3, 2'OMe) to achieve high stability and evade RNase shearing even in the blood stream. Moreover, the single-stranded nature of RNA aptamers not only allows them to form unique tertiary structures for tighter and more specific binding to the target but also makes them smaller in size and thus easier to enter into cells than other types of aptamers. Conjugation of RNA aptamers to delicately designed pRNA nanoparticles can assist in the delivery of therapeutics to specific cell organelles to maximize the therapeutic effects while minimizing the toxicity of the drug delivery system.

STRUCTURE OF RNA NANOPARTICLES

The concept of RNA nanotechnology has been proposed for more than one decade. The tool kits for designing and preparing RNA nanoparticles have also been developed to provide more varieties to meet different drug delivery needs.

RNA NANOPARTICLES FORMED BY DIMERIZATION OF PHI29 PRNA

Phi29 pRNA is a 117 nucleotide bacteriophage phi29-encoded packaging RNA (pRNA) discovered by Dr. Guo in 1987 (Guo et al., 1987). The pRNA monomer plays an essential role in packaging DNA into a procapsid by forming a hexameric ring to drive the DNA packaging motor of bacteriophage 29, which is about 11nm in size. The primary structure of wild-type pRNA is described in Figure 48.1 (Liu et al., 2011). pRNA has two functional domains that can fold independently: the DNA translocation domain and the prohead binding domain. The DNA translocation domain is composed of a

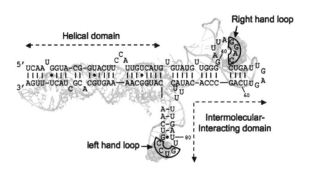

FIGURE 48.1   Primary sequence and structure of wild-type pRNA. Figure is produced with permission from Liu et al. (2011), © American Chemical Society.

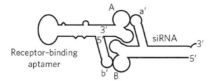

FIGURE 48.2 pRNA dimer formed through hand-in-hand complementary loop interactions. Figure is produced with permission from Guo (2010), © Nature Publishing Group.

3'/5' double-stranded helix loop while the prohead binding domain is composed of left- and right-hand loops. If we name the right-hand loops with uppercase letters A, B, C and name the left-hand loops with lowercase letters a, b, c denoting different loop sequences, the RNA sequences for A, B, C are complementary to sequences a, b, c respectively. pRNA dimer nanoparticles can be formed through the complementary hand-in-hand loop interactions between pRNA monomers Ab and Ba as described in Figure 48.2 (Shu et al., 2004). The pRNA dimer nanoparticle has been reported to have a particle size of about 25nm (Chen et al., 2000), which allows it to be employed as a nanoparticle carrier for gene drug delivery since its small size allows for the avoidance of reticuloendothelial system clearance and can be used for repeated and long-term gene drug delivery.

RNA NANOPARTICLE COMPOSED OF pRNA MULTIMER AND HIGHER ORDER STRUCTURES AS SCAFFOLD

pRNA dimers are building blocks for pRNA hexamer formation, which has been proven by Chen et al. (Chen et al., 2000), through hand-in-hand interactions of two complementary pRNAs which can form dimers, tetramers, hexamers, etc. A schematic drawing for the pRNA hexamer formation is described in Figure 48.3. The six pieces of pRNA in pRNA hexamer nanoparticles could provide six positions to conjugate therapeutic molecules such as siRNAs, ribozymes, therapeutic RNA/DNA aptamers, or diagnostic RNA/DNA aptamers for drug delivery. More recently, Shu et al. established various "toolkits" that employ hand-in-hand, foot-to-foot, and arm-on-arm interactions

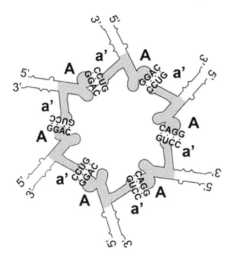

FIGURE 48.3 Schematic drawing of pRNA hexamer through complementary loop–loop interaction with 6 pieces of pRNA monomer. Figure is produced with permission from Chen et al. (2000), © American Society for Biochemistry and Molecular Biology.

between the structural elements of phi29 pRNA. Using these toolkits, they were able to generate 14 different RNA nanoparticles with potential for numerous clinical and non-clinical applications. This set of nanostructures was composed of dimers, twins, trimers, triplets, tetramers, quadruplets, pentamers, hexamers, heptamers, etc. (Shu et al., 2013). Through the use of various advanced construction approaches, higher-order 3D RNA structures like polyhedrons, nanosquares, nanocages, nanoprisms, dendrimers, and even micelles, etc., have also recently been successfully constructed (Haque et al., 2018; Jasinski et al., 2017; Yin et al., 2018). Importantly, various moieties (siRNA, ribozymes, aptamers, fluorophores, etc.) can be incorporated into these nanostructures as well while their ability to fold and function properly was highly maintained (Haque et al., 2018; Jasinski et al., 2017; Shu et al., 2013; Yin et al., 2018).

MULTIVALENT RNA NANOPARTICLES BASED ON THE THREE-WAY JUNCTION MOTIF

The above two methods of RNA nanoparticles construction are based on the hand-in-hand interaction between folded RNA molecules. A third strategy for constructing more thermodynamically stable pRNA nanoparticles is based on the RNA three-way junction (3WJ) motif (Shu et al., 2011). pRNA has two functional domains that can fold independently: a DNA translocation domain and a prohead binding domain. The two domains are connected by a 3WJ motif, as described in Figure 48.4. The 3WJ domain of pRNA was demonstrated to be very stable, which can retain its structure even in 8M Urea or at a very diluted concentration at picomolar scale. Importantly, the pRNA 3WJ has an uncharacteristically low dissociation constant compared to that of other biological RNA structures (Shu et al., 2014). Conjugation of therapeutic RNA molecules such as siRNA or RNA aptamers to the 3WJ motif allows for the formation of RNA nanoparticles around 10 nanometer, which allows its favorable pharmacokinetic properties. Recently, these RNA nanoparticles designed with 3WJ motif have been extensively tested for their ability to target and deliver therapeutic RNAs both in vitro and in vivo in various disease models (Haque et al., 2018; Jasinski et al., 2017). Overall, these RNA nanoparticles exhibited highly favorable in vivo pharmacokinetic profiles and therapeutic effects with little or no toxicity and immune response (Haque et al., 2018; Jasinski et al., 2017).

RNA POLYGON NANOPARTICLES BASED ON THE THREE-WAY JUNCTION MOTIF

The crystal structure study on pRNA-3WJ revealed it to be flexible in its angle [Zhang et al., RNA 2013, Sep 19(9): 1226–37]. The bulge of UUU at 3WJB position allows it to be flexible and can be tuned from 120 ° as its native folding in 3WJ [Zhang et al., RNA 2013, Sep 19(9): 1226–37] to be able to fit for the design of in RNA Triangle nanoparticles, RNA Square nanoparticles and RNA Pentagon nanoparticles [Khisamutdinov E, et al., Nucleic acid research, 2014]. These nanoparticles

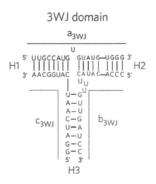

FIGURE 48.4   Three-way junction motif of pRNA, which can be used for RNA nanoparticles' construction. Figure is produced with permission from Shu et al. (2011), © Nature Publishing Group.

have shown to be able deliver immunomodulator CPG to cells more efficiently {Khisamutdinov E, et al., Nucleic acid research, 2014}; The RNA Square nanoparticles have shown be able to conjugate with Malachite Green aptamer, Spinach aptamer, HBV ribozyme, and siRNAs while keeping all of them functional [Jasinski D, ACS Nano, 2014, Vol8, 8, 7620–7629]. The 3WJ of phi-29DNA packaging motor has been further utilized to build 3D RNA nanoparticles such as Dendrimer [Sharam A, Nanomedicine NBM, 2016, 12: 835–844], RNA 3D Nanoprisms [Khisamutdinov E, Advanced Materials, 2016, 28, 10079–10087].

CONJUGATION OF RNA APTAMERS TO RNA NANOPARTICLES FOR TARGETED DRUG DELIVERY

The concept of a targeted drug delivery system was proposed by Paul Ehrlich in 1902, in which he first called the hypothetical drug a "magic bullet" (Ehrlich, 1957). In general, targeted drug delivery requires the targeted drug delivery system (TDDS) to selectively deliver therapeutics to diseased regions, independent of the method of its administration. TDDS can be classified into three grades from the aspect of the region it reaches: the first-grade targeting system refers to delivering the drug to a targeted organ or tissue, the second-grade targeting system refers to delivering the drug to specific cells, and the third-grade targeting system refers to delivering therapeutic molecules to specific locations inside the cell.

TDDS can also be divided into three types based on the pattern of targeting: passive, active, and physical. There has been intensive research on passive targeting drug delivery systems, such as liposomes, nanoemulsions, microcapsules, and polymeric nanospheres in the last several decades. Passive targeting relies on the natural distribution pattern of the drug delivery system, as the drug carriers can be ingested by macrophages of the reticuloendothelial system and then transferred primarily to the liver and spleen. However, it is difficult to deliver drugs to other organs with the passive targeting mechanism, because the in vivo distribution of the passive targeting drug carriers is greatly impacted by its particle size and the surface property of nanoparticles. As a general rule, when the particle size is larger than 7 um, it will be retained by the smallest lung blood capillaries through mechanical filtration; when the particle size is smaller than 7 um, it is ingested by macrophages in the liver and spleen; carriers of particles between 200 nm and 400 nm are usually collected and rapidly cleared by the liver. Active targeting preparation instead utilizes modified drug carriers as a "bullet" to directionally concentrate drugs to the target area for enhanced efficacy. These modifications include PEGlyzation of the nanoparticles to conceal the particle from macrophages, conjugation with special ligands or antibodies, which can interact with the target cell receptor, as well as other approaches. Physical and chemical targeting preparations utilize physical or chemical properties to help navigate preparations to specific targeting locations. For example, magnetically targeted drug delivery incorporates magnetic material into the drug preparation, and the preparation will then be concentrated to the specific target area under guidance of an externally applied magnetic field, whereas thermal or pH-targeted drug delivery utilizes temperature- or pH-sensitive material to deliver the therapeutics to specific macro-environments.

RNA APTAMERS AND SELEX

RNA aptamers are RNA oligonucleotides that bind to a specific target with high affinity and specificity, similar to an antibody's interaction with an antigen. RNA aptamer isolation was initially developed in two separate laboratories by Turek and Gold and by Ellington and Szostak (Ellington and Szostak, 1990; Tuerk and Gold, 1990) through a process that eventually became known as Systematic Evolution of Ligands by EXponential enrichment, or SELEX. To begin the SELEX process, a library of randomized RNA will first be synthesized. Generally, these RNA oligonucleotides are designed with a random sequence of approximately 20–80 nucleotides in the center region that is flanked on each side by a constant sequence. This library of oligonucleotides will then be exposed to the target of interest, which could be small molecules, proteins, cells, or even organisms (Dua et al., 2011; Keefe et al., 2010; Levy-Nissenbaum et al., 2008; Thiel and Giangrande, 2010). Those that do not bind to the target are washed away and discarded, whereas those that do bind are eluted and amplified through reverse transcription and PCR to generate a corresponding DNA library. The DNA library is then subjected to RNA transcription, and the resulting RNA library will then be exposed again to the target of interest for another round of the SELEX process. This process is usually repeated about 5–15 times, and the aptamers obtained often reach a high picomolar to low nanomolar range of dissociation constants (kd's) with the target (Dua et al., 2011; Yan and Levy, 2009).

Through this basic SELEX process and some recently developed variations of this process such as Cell-SELEX, Cross-over SELEX, Tissue-SELEX, and Capture-SELEX a good number of RNA aptamers have been isolated with the capability of binding numerous specific targets (Dua et al., 2011; Lauridsen et al., 2018; Levy-Nissenbaum et al., 2008; Yan and Levy, 2009). Significantly, many of these targets are cell-surface markers of various human diseases, which has led to the application of these RNA aptamers for targeted delivery of RNA therapeutics, especially those based on RNA interference (RNAi): small interference RNA (siRNA), short-hairpin RNA (shRNA), or microRNA (miRNA).

APPROACHES TO CONJUGATE APTAMER TO RNA NANOPARTICLES

The key step in the construction of RNA aptamer-conjugated nanoparticles is to design the global structure according to the physical and chemical properties of the RNA nanoparticles. If the RNA aptamer has been selected with a known sequence, it can be conjugated into the RNA nanoparticle structure before in vitro transcription or chemical synthesis of RNA. Dr. Guo and his colleagues have successfully conjugated malachite green (MG) aptamer to RNA nanoparticles characterized by a 3WJ pRNA motif. The in vitro experiment indicated that the aptamer is still functional after conjugation into 3WJ-pRNA nanoparticles (Figure 48.5) (Shu et al., 2011). The sequence for the MG aptamer nanoparticles was rationally designed with sequences of three pieces of 3WJ-pRNA motif. These three strands of RNA were synthesized in vitro by transcription from a DNA template with T7 RNA polymerase and the RNA nanoparticles were then self-assembled by mixing the three strands in an equal molar ratio. If there is no known RNA aptamer available for the desired applications, another approach for conjugation of aptamers to the RNA nanoparticle is to conjugate random

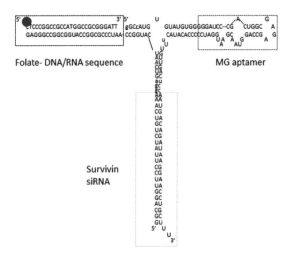

FIGURE 48.5 Diagram of RNA nanoparticle harboring malachite green (MG) aptamer, survivin siRNA, and folate-DNA/RNA sequence for targeting delivery, using 3WJ-pRNA as scaffolds. Figure is produced with permission from Shu et al. (2011) © Nature Publishing Group.

A. pRNA-A1-D3

B. pRNA-A1-D4

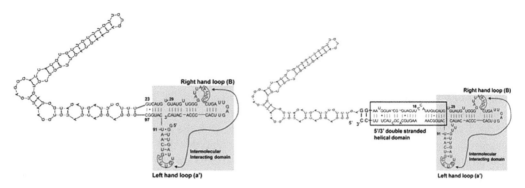

FIGURE 48.6   Schematic of pRNA nanoparticle harboring anti-HIV gp120 aptamer. (A) pRNA–A1-D3, the aptamer sequence was inserted into the 3'/5' double helical domain (23nt fragment) and loop domain (97nt fragment). (B) pRNA-A1-D4, the aptamer sequence was directly appended to the 5'end of pRNA 5'/3' double-stranded helical domain. Figure is produced with permission from Zhou et al. (2011) © Elsevier Inc.

sequences to a defined RNA nanoparticle structure. SELEX experiments as described above will then be carried out with this library of random sequence-bearing RNA nanoparticles against target peptides, proteins or cells. The randomly sequenced DNA library is transcribed into RNA and then partition techniques such as nitrocellulose partition, Capillary Electrophoresis partition, etc., can be utilized to separate the bound and unbound RNA. The bound species are further eluted out and used as a template for reverse transcription and PCR for the next round of SELEX experiment.

KEY FACTORS FOR CONJUGATING APTAMER TO RNA NANOPARTICLES

When designing aptamer-conjugated RNA nanoparticles, one key factor that needs to be considered is that the aptamer should be conjugated to the outer sphere site of RNA nanoparticles; thus, the aptamer targeting delivery functionality can be achieved. Another key factor is to ensure that the aptamer will still fold correctly after conjugation to RNA nanoparticle nucleotides. If the aptamer has a double-stranded helix RNA end in its structure, we can connect both ends of aptamer to the open helix ends of RNA nanoparticles. If the aptamer has a single-stranded RNA loop end in its structure, then we can link only one end of the aptamer to the open ends of RNA nanoparticle carriers. For example, when designing anti-gp120 aptamer- conjugated pRNA nanoparticles, J. Zhou et al. designed two different chimeric pRNA-anti-gp120 aptamer constructs. In one structure, both the double-stranded helix ends of anti-gp120 aptamers were linked to the pRNA end bases of 23–97 (pRNA-A1-D3); in another structure the anti-gp120 aptamer was directly appended to the 5'-end of pRNA (pRNA-A1-D4) (Figure 48.6) (Zhou et al., 2011). In vitro studies demonstrated that both pRNA-aptamer chimeras can specifically bind to and be internalized into cells expressing HIV gp120 with the dissociation constant (Kd) for the pRNA-A1-D3 being about 48nM and for the pRNA-A1-D4 was 79nM.

APPLICATION STATUS OF RNA APTAMER-CONJUGATED pRNA NANOPARTICLES

Since pRNA nanoparticles are composed of all RNA, conjugating RNA aptamers to pRNA to form a targeted delivery system will allow all the advantages of RNAs as therapeutic agents to be retained. We have described the structure and synthesis of both pRNA and the aptamers and discussed the approaches and key factors in the designing of RNA aptamer- conjugated pRNA. Another key obstacle in applying RNA aptamer-pRNA system for targeted therapy is that RNA is not stable and susceptible to quick degradation in vivo in blood stream. Recently, Dr. Guo's Lab found that 2'-F modification of pRNAs are both chemically and metabolically stable in vivo in animals. Importantly, they have shown that the pRNA's function and biologic activity stays intact, despite this 2'-F modification. To date, this pRNA nanoparticle delivery system has been used to conjugate CD4 aptamers, anti-GP120 aptamers, EGFR aptamers, PSMA aptamers, HER2 aptamers, and EpCAM aptamers and has been tested in anti-cancer and viral infection therapies.

Dr. Guo's laboratory has used the pRNA dimer nanoparticles to specifically deliver siRNA against the pro-survival gene called survivin to CD4 positive cells. This is accomplished by replacing the 3'/5' double-helix loop of the pRNA sequence with the survivin-silencing siRNA and by conjugating an anti-CD4 aptamer to the pRNA. They found this dimer is able to specifically target CD4 positive lymphocytes to silence the target gene expression and reduce cell viability (Guo et al., 2005). Most recently, Dr. Guo's laboratory has discovered that the three-way junction (3WJ) of pRNA discussed above is the most stable structure found among 25 3WJ motifs obtained from different biological systems (Shu et al., 2011). They have shown that each arm of the 3WJ-pRNA can carry the above-mentioned CD4 receptor binding RNA aptamer, siRNA, or ribozyme and bring them into target cells both in vitro and in vivo. Importantly, they have further gone on to show that the 2'-F RNAse resistant form of 3WJ-pRNA also retains its folding and can carry these incorporated functional moieties to target cells both in vitro and in vivo.

Additionally, anti-gp120 aptamers have been conjugated with the pRNA system by Dr. Rossi's group in their research against the Human Immunodeficiency Virus (HIV-1) infections (Zhou et al., 2008, 2009, 2011; Zhou and Rossi, 2011). The HIV-1 virus expresses a surface protein called glycoprotein gp120 that recognizes the CD4 cell receptor on the host's cells and initiates the membrane fusion that leads to subsequent delivery of viral RNA and enzymes. Once infected by HIV-1, these cells will then also express gp120 on their cell surface. Zhou et al. has previously generated gp120 aptamer chimeras with siRNA targeting the HIV-1 tat/rev gene region and found that these chimeras can be specifically internalized into cells expressing gp120 to silence the expression of target gene. Most recently, Dr. Rossi's group further used the pRNA system developed by Dr. Guo to generate dually functional RNA nanoparticle chimeras with anti-gp120 aptamers, as described above, and achieved both cell type specific delivery and targeted inhibition of viral replications (Zhou et al., 2011).

Shu et al. recently reported their successful generation of a pRNA nanoparticle to target oncogenic miR-21 in an orthotopic xenograft mouse model of triple negative breast cancer that overexpresses endothelial growth factor receptor (EGFR) (Figure 48.7a). These multivalent nanostructures harbored a phi29 pRNA-3WJ core, an anti-miR-21 sequence complementary to the seed region of miR-21, and an EGFR-targeting RNA aptamer to deliver the nanoparticle to cell-surface EGFR and elicit receptor-mediated endocytosis for nanoparticle uptake. Following low-dose systemic injection, the nanoparticles were found to localize mostly to tumors with little or no accumulation in healthy organs and significantly disrupt tumor growth (Shu et al., 2015). A similar study used the pRNA-3WJ scaffold and incorporated LNA (locked nucleic acid) sequences against miR21 or another prominent oncogenic miRNA miR17, along with an RNA aptamer targeting prostate-specific membrane antigen (PSMA) for prostate cancer cell-specific delivery (Figure 48.7b) (Binzel et al., 2016). Like the study above, systemic injection of this pRNA nanoparticle in orthotopically xenografted mice resulted in highly specific tumor accumulation with marked inhibition of tumor growth and no visible effects on overall body weight of the mice (Binzel et al., 2016).

More recently, Zhang et al. implemented human epidermal growth factor receptor 2 (HER2) – targeting RNA aptamer into a pRNA 3WJ sequence along with two different siRNAs against the ER-coactivator protein MED1 (Figure 48.7c) (Zhang et al., 2016). MED1 is a tissue-specific estrogen receptor coactivator that plays key roles in breast cancer metastasis and endocrine therapy resistance (Cui et al., 2012; Jiang et al., 2010; Leonard et al., 2019; Yang et al., 2018; Zhang et al., 2013). In addition to being overexpressed in approximately half of human breast cancers and

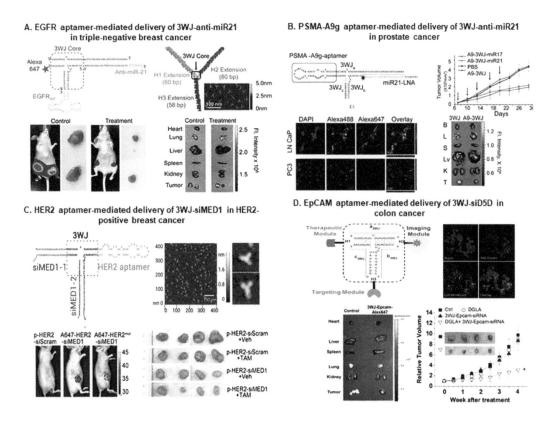

FIGURE 48.7 3WJ RNA nanoparticles harboring RNA aptamers for delivery of therapeutic RNAs in different cancer models. (a–d) Schematics and/or Atomic Force Microscopy (AFM) images of the 3WJ RNA nanoparticle scaffold, followed by in vitro and in vivo analyses following nanoparticle treatment in triple negative breast cancer (a), prostate cancer (b), HER2-positive breast cancer (c), and colon cancer (d). Figure is produced with permission from Shu et al. (2015), Binzel et al. (2016), Zhang et al. (2016), and Xu et al. (2019).

coamplify with HER2, expression of MED1 highly correlates poor disease-free survival in breast cancer patients (Leonard et al., 2019; Nagalingam et al. 2012; Yang et al., 2018; Zhu et al., 1999). Biochemical analyses of these 2'F-modified pRNA-HER2apt-siMED1 nanoparticles demonstrated their high thermostability (Tm of approximately 70 °C) and resistance to RNaseA, serum, and 8M Urea exposure. These RNA nanoparticles are capable of specifically binding to and accumulating in HER2+ breast cancer cells whereas HER2 aptamer mutant-harboring nanoparticles are not. Furthermore, pRNA-HER2apt-siMED1 nanoparticles selectively enter HER2 overexpressing BT-474 cells but not triple-negative MDA-MB-231 breast cancer cells (Zhang et al., 2016). Like the pRNA nanoparticles described above, in vivo distribution analyses of these nanoparticles in orthotopic xenograft mouse models expressing HER2+ mammary tumors resulted in a significant enrichment in the primary mammary tumor with only residual levels in other tissues like kidney and liver. Functionally, nanoparticle treatment reduced both MED1 and ER-target gene expression, inhibited cell growth, migration, invasion, and cancer stem cell formation capabilities both in vitro and in vivo following systemic injection (Zhang et al., 2016). Moreover, treatment with these pRNA nanoparticles resulted in re-sensitization of tamoxifen-resistant cells to tamoxifen treatment with further enhanced effects with combined treatments (Zhang et al., 2016).

The RNA aptamer-3WJ pRNA nanoparticle system has also shown promising potential as a therapeutic delivery system in other cancers. Xu et al. had shown before that Delta-5-Desaturase (D5D)-targeting siRNA inhibits colon cancer cell growth and migration when co-treated with dihomo-γ-linolenic acid (DGLA) due to production of the anti-cancer byproduct 8-hydroxyoctanoic acid (8-HOA) via cox-2-mediated DGLA peroxidation (Figure 48.7d) (Xu et al., 2016). For the delivery of D5D siRNAs, they utilized 3WJ pRNA and a colon cancer cell-targeted RNA aptamer against epithelial cell adhesion molecule (EpCAM) (Xu et al., 2019). With the EpCAM RNA aptamer, the 3WJ pRNA nanoparticle successfully delivered D5D siRNA to human colon cancer HCA-7 cells both in vitro and in vivo. D5D expression was significantly disrupted and in HCA-7 tumor-bearing mice, tumor growth was also severely hindered (Xu et al., 2019). Apart from these examples, pRNA nanoparticles have also shown promising targeting potential in vivo in brain cancers, gastric cancers, head and neck cancers, etc., and are continuing to expand their repertoire of application in disease and cancer treatment (Jasinski et al., 2017).

CONCLUSIONS AND FUTURE PERSPECTIVE

The high affinity and specificity of RNA aptamers rivaling those of antibodies makes them a promising tool for targeted delivery of therapeutics. As discussed in this chapter, conjugating RNA aptamers to pRNA nanoparticles for targeted therapy has shown great potential for the treatment of cancer and viral infections. With their many advantages as key components of RNA nanotechnology, including their small size, high stability, multi-conjugation capabilities, and especially their non-immunogenic nature, RNA aptamers will without a doubt become more applicable in the field of targeted therapy, especially with more and more RNA aptamers being isolated against an ever-increasing repertoire of disease targets. The increasing demand for aptamers and aptamer-pRNA nanoparticle conjugates will concomitantly provide for further mechanistic insights into their functions and interactions, in addition to an expansion of their therapeutic and diagnostic applications. With strong interest and further development of RNA nanotechnology and recent approval of RNA-based therapeutics by the FDA, we should expect a bright future for RNA aptamers as targeted delivery tools for RNA nanotherapeutics and far beyond.

ACKNOWLEDGMENTS

We thank members of X. Zhang and P. Guo laboratories for valuable comments and suggestions. This study was supported by National Cancer Institute (R01 CA197865, CA229869) and Ride Cincinnati Award (to X. Zhang). **This project was also supported by the National Institutes of Health (NIH; grants** U01CA207946 and R01EB019036 **to P.G.). P.G. is also a co-founder of ExonanoRNA. LLC.**

REFERENCES

Binzel, D.W., Shu, Y., Li, H., Sun, M., Zhang, Q., Shu, D., Guo, B., and Guo, P. (2016). Specific delivery of miRNA for high efficient inhibition of prostate cancer by RNA nanotechnology. Mol Ther *24*, 1267–1277.

Chen, C., Sheng, S., Shao, Z., and Guo, P. (2000). A dimer as a building block in assembling RNA. A hexamer that gears bacterial virus phi29 DNA-translocating machinery. J Biol Chem *275*, 17510–17516.

Cui, J., Germer, K., Wu, T., Wang, J., Luo, J., Wang, S.C., Wang, Q., and Zhang, X. (2012). Crosstalk between HER2 and MED1 regulates tamoxifen resistance of human breast cancer cells. Cancer Res *72*, 5625–5634.

Dua, P., Kim, S., and Lee, D.K. (2011). Nucleic acid aptamers targeting cell-surface proteins. Methods *54*, 215–225.

Ehrlich, P. (1957). The collected papers of Paul Ehrlich: Immunology and cancer research. Pergamon Press, London, 442.

Ellington, A.D., and Szostak, J.W. (1990). In vitro selection of RNA molecules that bind specific ligands. Nature *346*, 818–822.

Guo, P. (2010). The emerging field of RNA nanotechnology. Nat Nanotechnol *5*, 833–842.

Guo, P.X., Erickson, S., and Anderson, D. (1987). A small viral RNA is required for in vitro packaging of bacteriophage phi 29 DNA. Science *236*, 690–694.

Guo, S., Tschammer, N., Mohammed, S., and Guo, P. (2005). Specific delivery of therapeutic RNAs to cancer cells via the dimerization mechanism of phi29 motor pRNA. Hum Gene Ther *16*, 1097–1109.

Haque, F., Pi, F., Zhao, Z., Gu, S., Hu, H., Yu, H., and Guo, P. (2018). RNA versatility, flexibility, and thermo-stability for practice in RNA nanotechnology and biomedical applications. Wiley Interdiscip Rev: RNA *9*, e1452.

Jasinski, D., Haque, F., Binzel, D.W., and Guo, P. (2017). Advancement of the emerging field of RNA nano-technology. ACS Nano *11*, 1142–1164.

Jiang, P., Hu, Q., Ito, M., Meyer, S., Waltz, S., Khan, S., Roeder, R.G., and Zhang, X. (2010). Key roles for MED1 LxxLL motifs in pubertal mammary gland development and luminal-cell differentiation. Proc Natl Acad Sci *107*, 6765–6770.

Keefe, A.D., Pai, S., and Ellington, A. (2010). Aptamers as therapeutics. Nat Rev Drug Discov *9*, 537–550.

Lauridsen, L.H., Doessing, H.B., Long, K.S., and Nielsen, A.T. (2018). A capture-SELEX strategy for multi-plexed selection of RNA aptamers against small molecules. Synth Meta Path (Springer), pp. 291–306.

Leonard, M., Juan, T., Yang, Y., Charif, M., Lower, E.E., and Zhang, X. (2019). Emerging therapeutic approaches to overcome breast cancer endocrine resistance. In Estrogen receptor and breast cancer, X. Zhang, ed. (pp. 379–403). Cincinnati: Springer Nature Switzerland AG.

Levy-Nissenbaum, E., Radovic-Moreno, A.F., Wang, A.Z., Langer, R., and Farokhzad, O.C. (2008). Nanotechnology and aptamers: Applications in drug delivery. Trends Biotechnol *26*, 442–449.

Liu, J., Guo, S., Cinier, M., Shlyakhtenko, L.S., Shu, Y., Chen, C., Shen, G., and Guo, P. (2011). Fabrication of stable and RNase-resistant RNA nanoparticles active in gearing the nanomotors for viral DNA packag-ing. ACS Nano *5*, 237–246.

Nagalingam, A., Tighiouart, M., Ryden, L., Joseph, L., Landberg, G., Saxena, N.K., and Sharma, D. (2012). Med1 plays a critical role in the development of tamoxifen resistance. Carcinogenesis *33*, 918–930.

Que-Gewirth, N.S., and Sullenger, B.A. (2007). Gene therapy progress and prospects: RNA aptamers. Gene Ther *14*, 283–291.

Shu, Y., Haque, F., Shu, D., Li, W., Zhu, Z., Kotb, M., Lyubchenko, Y., and Guo, P. (2013). Fabrication of 14 different RNA nanoparticles for specific tumor targeting without accumulation in normal organs. RNA *19*, 767–777.

Shu, D., Li, H., Shu, Y., Xiong, G., Carson III, W.E., Haque, F., Xu, R., and Guo, P. (2015). Systemic delivery of anti-miRNA for suppression of triple negative breast cancer utilizing RNA nanotechnology. ACS Nano *9*, 9731–9740.

Shu, D., Moll, W.D., Deng, Z., Mao, C., and Guo, P. (2004). Bottom-up assembly of RNA arrays and super-structures as potential parts in nanotechnology. Nano Lett *4*, 1717–1723.

Shu, Y., Pi, F., Sharma, A., Rajabi, M., Haque, F., Shu, D., Leggas, M., Evers, B.M., and Guo, P. (2014). Stable RNA nanoparticles as potential new generation drugs for cancer therapy. Adv Drug Deliv Rev *66*, 74–89.

Shu, D., Shu, Y., Haque, F., Abdelmawla, S., and Guo, P. (2011). Thermodynamically stable RNA three-way junc-tion for constructing multifunctional nanoparticles for delivery of therapeutics. Nat Nanotechnol *6*, 658–667.

Thiel, K.W., and Giangrande, P.H. (2010). Intracellular delivery of RNA-based therapeutics using aptamers. Ther Deliv *1*, 849–861.

Tuerk, C., and Gold, L. (1990). Systematic evolution of ligands by exponential enrichment: RNA ligands to bacteriophage T4 DNA polymerase. Science *249*, 505–510.

Xu, Y., Pang, L., Wang, H., Xu, C., Shah, H., Guo, P., Shu, D., and Qian, S.Y. (2019). Specific delivery of delta-5-desaturase siRNA via RNA nanoparticles supplemented with dihomo-γ-linolenic acid for colon cancer suppression. Redox Biol *21*, 101085.

Xu, Y., Yang, X., Zhao, P., Yang, Z., Yan, C., Guo, B., and Qian, S.Y. (2016). Knockdown of delta-5-desaturase promotes the anti-cancer activity of dihomo-γ-linolenic acid and enhances the efficacy of chemotherapy in colon cancer cells expressing COX-2. Free Radic Biol Med *96*, 67–77.

Yan, A.C., and Levy, M. (2009). Aptamers and aptamer targeted delivery. RNA Biol *6*, 316–320.

Yang, Y., Leonard, M., Zhang, Y., Zhao, D., Mahmoud, C., Khan, S., Wang, J., Lower, E.E., and Zhang, X. (2018). HER2-driven breast tumorigenesis relies upon interactions of the estrogen receptor with coacti-vator MED1. Cancer Res *78*, 422–435.

Yin, H., Wang, H., Li, Z., Shu, D., and Guo, P. (2018). RNA micelles for the systemic delivery of anti-miRNA for cancer targeting and inhibition without ligand. ACS Nano *13*, 706–717.

Zhang, L., Cui, J., Leonard, M., Nephew, K., Li, Y., and Zhang, X. (2013). Silencing MED1 sensitizes breast cancer cells to pure anti-estrogen fulvestrant in vitro and in vivo. PLoS One *8*, e70641.

Zhang, Y., Leonard, M., Shu, Y., Yang, Y., Shu, D., Guo, P., and Zhang, X. (2016). Overcoming tamoxifen resistance of human breast cancer by targeted gene silencing using multifunctional pRNA nanoparticles. ACS Nano *11*, 335–346.

Zhou, J., Li, H., Li, S., Zaia, J., and Rossi, J.J. (2008). Novel dual inhibitory function aptamer-siRNA delivery system for HIV-1 therapy. Mol Ther *16*, 1481–1489.

Zhou, J., and Rossi, J.J. (2011). Current progress in the development of RNAi-based therapeutics for HIV-1. Gene Ther *18*, 1134–1138.

Zhou, J., Shu, Y., Guo, P., Smith, D.D., and Rossi, J.J. (2011). Dual functional RNA nanoparticles containing phi29 motor pRNA and anti-gp120 aptamer for cell-type specific delivery and HIV-1 inhibition. Methods *54*, 284–294.

Zhou, J., Swiderski, P., Li, H., Zhang, J., Neff, C.P., Akkina, R., and Rossi, J.J. (2009). Selection, characterization and application of new RNA HIV gp 120 aptamers for facile delivery of Dicer substrate siRNAs into HIV infected cells. Nucleic Acids Res *37*, 3094–3109.

Zhu, Y., Qi, C., Jain, S., Le Beau, M.M., Espinosa, R., Atkins, G.B., Lazar, M.A., Yeldandi, A.V., Rao, M.S., and Reddy, J.K. (1999). Amplification and overexpression of peroxisome proliferatoractivated receptor binding protein (PBP/PPARBP) gene in breast cancer. Proc Natl Acad Sci *96*, 10848–10853.

49 MicroRNAs in Human Cancers and Therapeutic Applications

Ji Young Yoo, Balveen Kaur, and Tae Jin Lee
University of Texas Health Science Center at Houston, Houston, USA

Peixuan Guo
The Ohio State University, Columbus, USA

CONTENTS

INTRODUCTION

MicroRNAs (miRNAs), the smallest member of noncoding RNAs as averaging about 19–22 nucleotides in length and evolutionarily highly conserved, have been identified in almost every eukaryote, including humans (Bartel, 2009). The first miRNA lin-14 was discovered from *C. elegance* in 1993 through two independent studies (Lee et al., 1993; Wightman et al., 1993). They found that transcription of the Lin4 gene resulted in a small piece of RNA, which involves in the regulation of gene expressions through sequence-specific binding to their 3' untranslated region.

According to the recent version 22 in miRBase database (http://www.mirbase.org/), approximately 1,917 miRNAs have been identified in the human genome and they are considered to involve in the regulation of more than 60% of human coding genes (Bartel, 2009). Altered expression levels of miRNA can function as not only useful biomarkers but also pharmacologic targets (Neudecker et al., 2016). In 2002, it was first discovered that dysregulation of microRNAs (miRNAs) is implicated in a variety of human cancers (Calin and Croce, 2006). Since the seminal finding, countless attempts have been made to utilize miRNAs as therapeutic targets to treat human cancers. This review will provide an up-to-date understanding on the therapeutic potential of dysregulated miRNAs in human cancers and the attempts to modulate their expression level in cancer cells for the development of therapeutically promising strategies in cancer treatment.

BIOGENESIS AND WORKING MECHANISM OF miRNA

Approximately 70% of miRNA coding genes are associated and co-transcribed with protein-coding host genes, while the remaining 30% of miRNA are located in intergenic area with their own open

DOI: 10.1201/9781003001560-54

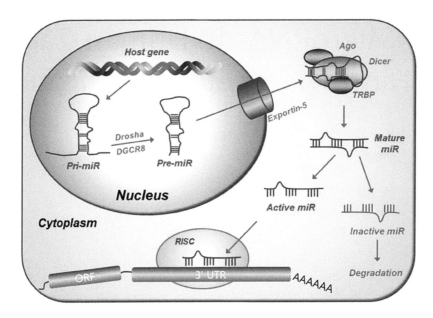

FIGURE 49.1 Biogenesis and working mechanism of microRNAs (miRNAs) (Bartel, 2018; Calin and Croce, 2006; Ha and Kim, 2014; Lee et al., 2003, 2004). MiRNA host genes are transcribed into a primary transcript (pri-miR) by RNA polymerase II, which is processed into a hairpin-structured transcript (pre-miR) by a nuclear enzyme Drosha (Lee et al., 2003). The pre-miR is then exported out into cytoplasm through a nuclear member channel protein, Exportin-5 (Yi et al., 2003), where pre-miRNAs is further processed by Dicer to a hairpin-free duplex form of mature-miR (Hutvagner et al., 2001; Ketting et al., 2001). A functional strand of the duplex mature-miR binds to the 3′ untranslated region (UTR) of target gene mRNAs through eight-base long seed sequences. The formation of miRNA/mRNA complex in RICS rapidly represses the translation of mRNA into proteins or leads to degradation of the target mRNA (Bartel, 2009, 2018; Eichhorn et al., 2014).

reading frames (Bartel, 2009, 2018; Ha and Kim, 2014). In either case, miRNA host genes are first transcribed by RNA polymerase II into a long primary transcript, primary-miRNAs (Pri-mRNA), up to ten kilobases in length (Lee et al., 2004) (Figure 49.1). This pri-mRNA transcript is then processed by nuclear RNase III Drosha into shorter lengths (approximately 70 bases), forming a hairpin-like structure called precursor-miRNA (pre-miR) (Lee et al., 2003). Processed pre-miR is then exported to the cytoplasm through the nuclear membrane channel protein Exportin-5 (XPO5) (Yi et al., 2003). Pre-miR export through XPO5 activity can be suppressed by phosphorylation of ERK in certain cancers (Sun et al., 2016). After pre-miR reaches the cytoplasm, the terminal hairpin loop is further cleaved with the cytoplasmic RNAse III Dicer to produce a 22 nucleotide long mature miRNA (Hutvagner et al., 2001). Mature miRNA binds to Argonaute 2 (Ago2) and transactivation-responsive RNA-binding protein (TRBP) to form a RNA-induced silencing complex (RISC). When associated with RISC, active strands remain in RISC, leaving the other active strand connected to RISC until guided to 3′UTR of target mRNAs while inactive strands are readily degraded (Figure 49.1) (Ha and Kim, 2014). Unlike small interfering RNAs (siRNAs), which requires perfect sequence match throughout their entire length to properly function, miRNA binds to target mRNA through partial complementary sequences stored between the second and eighth nucleotide called "seed sequences". When the seed sequence of miRNA forms a partial base pairing with the 3′UTR of target mRNA, these miRNA/mRNA complexes are unable to interact with the active ribosomal complexes necessary to continue protein synthesis (Bartel, 2009, 2018). Notably, the partial complementary binding nature of miRNA allows single miRNAs to bind with a variety of target mRNAs. Several computer-based prediction algorithms exist to predict miRNA targeting; however, there are many variables which influence partial complementary binding and these predictions are not always accurate.

In addition to the translational repression (Olsen and Ambros, 1999), miRNAs can also block translation of target mRNAs by deadenylating the target mRNA poly-A tail (Bagga et al., 2005). It was proposed that this irreversible mRNA destabilization caused by miRNA has a much stronger effect than the traditional miRNA translational repression, which acts rapidly but is relatively weak (Eichhorn et al., 2014). Due to these two mechanisms of mRNA repression, altered levels of miRNAs can reprogram various pathways in pathological tissue (Bartel, 2009, 2018).

miRNAs IN HUMAN CANCERS

Evidence shows that different cancers display their own unique profile of miRNA expressions, distinct from that found in normal tissues (Calin et al., 2004). Those "signature miRNAs" act as useful, not only diagnostic or prognostic markers but also therapeutic targets in the development of anti-cancer strategies. First signature miRNAs in human cancer was found by Carlo Croce's laboratory in 2002 from chronic lymphocytic leukemia (CLL) patients (Calin et al., 2002). They observed a frequently deleted cluster of miR-15/16 in CLL, and they directly downregulate the oncogene BCL2 to induce apoptosis (Calin et al., 2002). Inspired from the seminal finding, they screened a large number of human B cell CCL patient samples by building a customized microarray platform and found that the aberrantly regulated miRNAs are strongly associated with both the prognosis and progression of CLL (Calin et al., 2005). The microarray-based miRNA profiling has revealed many signature miRNAs from various types of human cancers (Calin et al., 2004). From comparing miRNA profiles from 166 human bladder tumors and 11 normal bladder samples, 3 miRNAs (*miR-9, miR-182* and *miR-200b*) were designated as signature miRNAs associated with tumor aggressiveness. Another example was the miR-221/222 cluster, discovered to be upregulated in non-small cell lung cancers and hepatocellular carcinomas (HCCs) (Garofalo et al., 2009). In another example, miR-221 was identified as a marker that was highly correlated with more lethal forms of prostate cancer (Kneitz et al., 2014). In small-cell lung cancer patients, miRNAs *MiR-150* and *miR-886-3p* were proposed as prognostic predictors to detect reoccurrence after surgery and chemotherapy treatment (Bi et al., 2014). Such signature miRNAs are found to be either upregulated or downregulated in relation to those in surrounding normal tissues. In general, downregulated miRNAs are considered to be tumor suppressors while upregulated miRNAs are considered as oncogenic.

For example, 13q14.2 region containing the host gene of miR-15a and miR-16-1 is commonly found to be deleted in more than 50% of CLL cases (Calin et al., 2002, 2004, 2005). For this reason, their direct target, anti-apoptotic protein BCL-2, is often overexpressed in a subset of CLL patients. Therefore, the miR-15a and miR-16-1 are considered as tumor suppressive miRNAs. In myelodysplasia and therapy-related acute myeloid leukemia (AML), the miR-29 family (*miR-29b-1* and *miR-29a*) were found to be commonly deleted (Garzon et al., 2008, 2009), resulting in upregulated oncogenic proteins BCL-2 and MCL (Xu et al., 2014). In lung cancers, the loss of the miR-29 family results in upregulation of DNA methyltransferase *(DNMT)-3A* and *-3B* with poor prognosis (Fabbri et al., 2007). In sarcoma, miR-29 protects the tumor suppressor A20 transcript by acting as a decoy for human antigen R (HuR) and preventing HuR from binding to the 3'UTR of A20 (Balkhi et al., 2013). MiR-34a is also a recognized tumor suppressive miRNA that is highly expressed normally but downregulated in many cancers by methylation on its CpG island (Lodygin et al., 2008). However, it can be reactivated by p53 (He et al., 2007) and overexpression in cancer cells can cause either cell cycle arrest by targeting c-Met, c-Myc, CCND1, and CDK6 (Li et al., 2009; Sun et al.,

2008; Yamamura et al., 2012) or apoptosis by targeting Notch1 and Bcl2 (Li et al., 2013; Pang et al., 2010). MiR-34a is also known to directly target SIRT1, an inhibitor of p53, which can lead to further p53 activation in a positive feedback loop (Yamakuchi et al., 2008).

In contrast to the downregulated tumor suppressive miRNAs, certain miRNAs are found to be significantly upregulated in cancers. First putative oncogenic miRNA (onco-miRs), miR-155, was shown to be highly overexpressed in a variety of cancers, including human B-cell lymphoma, pediatric Burkitt's lymphoma, Hodgkin's disease, primary mediastinal non-Hodgkin's lymphoma, CLL, AML, lung cancer, and breast cancer (Jurkovicova et al., 2014). Later studies found that miR-155 directly targeted Src homology 2 domain-containing inositol-5-phosphatase (SHIP) and CCAAT enhancer-binding protein beta (C/EBPbeta), both of which are involved in the interleukin-6 (IL-6) signaling pathway (Costinean et al., 2009). MiR-155 is also found in high levels in nonalcoholic steatohepatitis (NASH) and can lead to the development of HCC (Wang et al., 2009). MiR-17-92 is a cluster of six onco-miRs: miR-17, miR-18a, miR-19a, miR-20a, miR-19b-1, and miR-92-1. This cluster is frequently amplified in follicular lymphoma and diffuses large B cell lymphoma and is also highly expressed in a variety of other cancers such as breast, colon, lung, pancreas, prostate, and stomach cancer (Olive et al., 2013). They promote tumor proliferation and angiogenesis while controlling E2F1 expression (He et al., 2005; O'Donnell et al., 2005; Xiao et al., 2008). Both E2F1 and E2F3 are involved in a regulatory loop with the miR-17-92 cluster (Sylvestre et al., 2007). Downregulation of these onco-miR expressions in cancerous tissue has been shown to induce apoptosis, cell cycle arrest and reduce metastatic ability, which makes them potential therapeutic targets.

miRNA MODULATION FOR CANCER THERAPY

As cancer cells exhibit changes in multiple pathways, use of miRNA treatment to affect these pathways simultaneously can have promising therapeutic benefits (Hanahan and Weinberg, 2011). In comparison, conventional chemotherapy drugs or siRNAs are only able to target a single gene product. As a result, researchers have tried to develop new therapeutic strategies to return miRNA levels in diseased tissue back to normal levels through the introduction of miRNA mimics or inhibitors. Currently, despite promising results in preclinical studies, few miRNA-based therapies have progressed into clinical trials (Table 49.1). For an example, MiR-34a, a powerful tumor suppressive miRNA in many cancer types, had entered into clinical trials for liver cancer therapy with formulation with liposome-based carrier, called Miramax (MRX34) (NCT01829971); however, the trial was early terminated in 2017 during the first phase due to multiple complications related to elevated cytokines in patients (Beg et al., 2017). In addition, many current approaches lack a strategy of secured and targeted delivery method. When administered without protection, the half-life of naked miRNA mimics or inhibiters is less than 30 minutes. This greatly reduces the potency of treatments and necessitates some strategy to protect miRNA from self-hydrolysis or RNase-mediated enzymatic degradation. Furthermore, miRNA mimics or inhibitors alone do not have the ability to target or enter specific cells. As a result, they require some transporter to carry them across the cell membrane. Generally, proposed methods to transport miRNA mimics or inhibitors fall into three categories: 1) Delivery using modified nucleic acids; 2) Overexpression of miRNA through a viral delivery vector system; or 3) Delivery using non-viral vector systems. Many of these studies have shown promising results in the preclinical stage.

CHEMICALLY MODIFIED NUCLEOTIDE ANALOGS

Anti-miRNA oligonucleotides (AMO) designed to reduce levels of a specific miRNA often use chemically stable RNA analogs to avoid RNA degradation. RNA can be readily degraded by

TABLE 49.1

Recent Clinical Trials for Therapeutic Delivery of miRNA Mimics or Inhibitors

Drug Name (Company)	Delivery Platform	Target miRNA	Target Disease	Clinical Trial Status
Miravirsen (Santaris Pharma)	LNA	miR-122	HCC	NCT01200420 Phase IIa
MRX34 (Mirna Therapeutics)	Liposome	miR-34a	Primary Liver Cancer SCLC Lymphoma Melanoma MM RCC NSCLC	NCT01829971 Phase I
TargoMiRs (EnGeneIC)	EDV with EGFR IgG	miR-16	MPM NSCLC	NCT02369198 Phase I
MRG106 (MiRagen)	LNA	miR-155	CLL DLBCL ATLL	NCT02580552 Phase I

Abbreviations: HCC, hepatocellular carcinoma; SCLC, small cell lung cancer; MM, multiple myeloma; RCC, renal cell carcinoma; NSCLC, non-small cell lung cancer; CTCL, cutaneous T-cell lymphoma; CLL, chronic lymphocytic leukemia; DLBCL, diffuse large b-cell lymphoma; ATLL, adult T-cell leukemia/lymphoma; MPM, malignant pleural mesothelioma; EDV, nonliving bacterial minicell nanoparticles; NA, not available.

endogenous RNases though deprotonation of 2'-OH group on sugar ring leading to RNA hydrolysis. The RNA degradation can be inhibited by replacing the 2'-OH group with a chemically inert group, such as 2'-O-methyl-(2'-O-Me)-oligonucleotides, 2'-O-methoxyethyl-oligonucleotides (2'-O-MOE) and 2' fluoride derivatives (2'-F) (Fisher et al., 1993).

Locked nucleic acid (LNA) is a RNA analog generated by chemically locking the 2'-oxygen and 4'-carbon on ribose with a bridge. The locked structure physically inhibits binding of RNase, resulting in high stability in serum but still leaves the LNA capable of pairing with complementary RNA stands. As a result, LNAs are frequently used to bind complementarily to functional miRNAs for use in fluorescence staining or targeted inhibition (Wahlestedt et al., 2000). While unmodified oligonucleotides can be degraded within 1.5 hours in serum, the half-life of LNA in serum can reach up to 15 hours (Kurreck et al., 2002). LNA-based anti-miR-155 drug (MRG106) administered through intratumoral injection was developed by MiRagen Therapeutics, which is currently undergoing Phase 1 clinical trial for patients with cutaneous T-cell lymphoma (CTCL) of the mycosis fungoides (MF) sub-type (NCT02580552).

Peptide nucleic acid (PNA) is a peptide bond-based nucleic acid mimic generated by replacing the sugar-phosphodiester backbone with N-(2-aminoethyl)-glycine units. The (2-aminoetyl)-glycine groups present no surface charge, which allows PNA molecules to pass through cell membranes by themselves. PNA is highly stable in serum and binds specifically to both DNA and RNA (Hanvey et al., 1992). PNA is shown to be highly stable in serum since PNA incubated for 2 hours in serum shows 100% structurally intact rate compared to only 8% for unmodified control peptides (Demidov et al., 1994). In a study where PNA-anti-miR-155 was given intraperitoneally to mice, endogenous miR-155 in primary B cells was successfully inhibited reducing the gene expression levels of PNA-treated mice to levels comparable to mir-155 deficient mice (Fabani et al., 2010). As LNA-based miRNA inhibitors are more cost-effective and have a higher cellular uptake, they are more popular to use than PNA.

VIRAL DELIVERY OF MIRNAS

Viruses such as retroviruses, lentiviruses, adenoviruses, or adeno-associated viruses (AAVs) can be used vectors to transport genetic information to specific cells. It has been shown that genetically engineered viruses are able to transport and express siRNA or miRNAs in disease cells (Davidson and Harper, 2005). Viruses use a constitutive promotor to allow for continuous transcription of contained genes, allowing consistent overexpression of miRNA over long periods of time. Due to their high specificity and transduction efficiency, viral vectors can achieve high levels of miRNA expression in target cells without needing any further methods to protect miRNA from degradation.

Lentiviruses have been widely used for ectopical overexpression of a gene of interest as they can be cloned with large amounts of DNA and are useful for delivering genes of various sizes. In one study, miR-128 containing lentivirus restored the miR-128 expression level in the glioma cell line U87MG and resulted growth inhibition along with increased sensitivity to DNA damaging radiation (Peruzzi et al., 2013). In other cases, tumor-suppressive miRNA let-7 was delivered into mice by lentivirus and resulted in reduced proliferation in both breast cancer cells (Yu et al., 2007) and NSCLC (Trang et al., 2010). However, some long-term clinical trials show that chromosomal integration by lentivirus-based delivery can have unintended consequences and result in harmful genomic mutations (Milone and O'Doherty, 2018).

AAVs are characterized by their inability to integrate into the host genome and a general lack of pathogenicity. This greatly decreases the risk of genomic mutations and provides a much safer method to transport genetic information. In 2009, Kota et al. used engineered AAVs to overexpress tumor-suppressive miR-26a in HCC (Kota et al., 2009). AAVs expressing miR-26 were injected via tail vein injection into Myc-induced HCC bearing mice at the dose of 10^{12} viral genomes per animal. The treated mice experienced lower cell proliferation and increased apoptosis of cancer cells, resulting in regression of Myc-induced HCC over the course of 7 weeks (Kota et al., 2009). Although AAVs generally have lower risk than other viral vectors, they are still susceptible to common issues of viral delivery such as potential activation of oncogenes, systemic toxicity, and immunogenic response, limiting the genotoxic risk of recombinant AAVs after systemic administration (Penaud-Budloo et al., 2018).

NON-VIRAL DELIVERY OF MIRNAS

Artificial formulations using polymers, lipids, and inorganic nanoparticles are promising choices for non-viral delivery of miRNAs to avoid many side-effects of viral delivery involving genotoxic risk and immunogenic response. These systems are resistant to serum degradation, endosomal capture, lysosomal degradation, and immunogenic response. In addition, many non-viral molecules are chemically flexible and can allow a variety of modifications to improve cell-specific targeting functions.

Previous studies show that positively charged low molecular weight polyethylenimine (PEI) can be used to carry miR-145 or miR-33a into colon cancer bearing mice to show efficient antitumor effects comparable to single siRNA-PEI complexes (Ibrahim et al., 2011). Poly lactic-co-glycolic acid (PLGA) can also be used for miRNA deliveries; however, it is normally negatively changed and needs further modification to become positively charged. This can be done by adding amine-like cationic residues or other type of positively charged polymers, such as chitosan or PEI. A chitosan-PLGA formulation used to transport miR-34a was able to successfully treat multiple myeloma animal models (Cosco et al., 2015). Another case utilized a poly(amidoamine) (PAMAM) dendrimer to transport antisense-miR-21 inhibitor together with 5-fluorouracil (5-FU) chemotherapy drug. The combination of treatments resulted in enhanced cytotoxicity of 5-FU in human glioblastoma cells (Ren et al., 2010).

Lipid-based nanoparticles are often made using cationic lipids to attract negatively charged nucleic acids (Campani et al., 2016). In one case, miR-107 was encapsulated into lipid-based cationic nanoparticles constructed with dimethyldioctadecyl ammonium bromide (DDAB), cholesterol, and α-Tocopheryl polyethylene glycol 1000 succinate (TPGS) (Piao et al., 2012). When these nanoparticles were injected intravenously through the tail veins of athymic nude mice with HNSCC xenografts, expression levels of miR-107 target genes PKCε, HIF1-β, CDK6, Nanog, Sox2, and Oct3/4, were reduced by 45–75% and tumor growth was inhibited (Piao et al., 2012). Another study used 1,2-Di-O-octadecenyl-3-trimethylammonium propane (DOTMA) to construct a similar lipid-based cationic nanoparticle to carry miR-29b mimics. These miR-29 carrying nanoparticles were able to accumulate within tumor locations and reduce expression of targets CDK6, DNMT3B, and MCL1 in non-small-cell lung cancer both in vitro and in vivo (Wu et al., 2013). Liposome-formulated miR-34a (MRX34) became the first clinical trial among miRNA-based therapies on participants with advanced liver cancer and melanoma in 2013 (NCT01829971). However, the promising clinical trial by Mirna Therapeutics using MRX34 was early terminated in 2017 due to serious immune-related side effects.

MiRNAs can also be loaded onto inorganic nanoparticles made with gold, silica, or magnetic materials. For examples, functionalized gold nanoparticles (AuNPs) loaded with anti-miR-29b oligonucleotides were able to lower oncogenic miR-29b expression and rescue expression of the target gene MCL-1 in HeLa cells (Kim et al., 2011). In another case, magnetic zinc-doped iron oxide nanoparticles ($ZnFe_2O_4$) were used to successfully deliver tumor suppressive let-7a into brain tumor cells (Yin et al., 2014a). Although inorganic nanoparticles are relatively simple to create, their pharmacokinetics after treatment needs to be carefully monitored.

Exosomes around 30–150nm in diameter also showed their capability to carry a variety of components including proteins, lipids, DNAs, mRNAs, miRNAs, and long noncoding RNA. Tumor cells can secrete exosomes containing high levels of miR-21 which are transferred into macrophages through toll-like receptors (TLRs) to activate immune responses (Fabbri et al., 2012). Yin et al. used this concept to develop a method to inactivate miR-214 in CD4+ cells by treating them with exosomes overexpressing anti-sense miR-214 derived from HEK293 cells. The resulting inactivation of miR-214 induced regulatory T cells to form an immune evasion environment (Yin et al., 2014b).

RNA NANOPARTICLES FOR TISSUE-SPECIFIC TARGETED miRNA

Tissue-specific targeting can lower the risk of side effects by limiting the effects of the drug to targeted disease cells. To achieve this, carrier molecules can be modified with surface moieties that enable cell targeting such as ligands, antibodies, or aptamers. Folate (FA) is another possible targeting moiety, which can be used to deliver miRNA to cancer cells that highly overexpress folate receptors. FA is an especially suitable targeting moiety for glioblastomas. While folate receptors are highly overexpressed in glioblastoma, they are not expressed in normal brain neuronal cells (Parker et al., 2005). Recent technological advance in RNA nanotechnology has enabled the use of small molecules like FA (Guo, 2010; Jasinski et al., 2017; Shu et al., 2014). RNA nanoparticles made with RNA backbones require few modifications to become suitable for carrying small therapeutic RNAs, such as siRNAs or miRNAs. In 2011, the laboratory of Peixuan Guo synthesized three-way-junction (3WJ)-based RNA nanoparticles by engineering the 3WJ core sequences of packaging RNA (pRNA) in the DNA packaging motor system of the bacteriophage Phi29 (Figure 49.2) (Shu et al., 2011). When 3WJ RNA nanoparticles were conjugated with the ligand folate, they were able to selectively target and deliver siRNA into cancer cells which often have overexpressed folate receptors (Figure 49.3) (Lee et al., 2015a, 2015b). 3WJ RNA nanoparticles have also been loaded with antisense miR-21, which were able to successfully inhibit miR-21 expression and induce cancer cell apoptosis and cell cycle arrest (Figure 49.4). When systemically injected

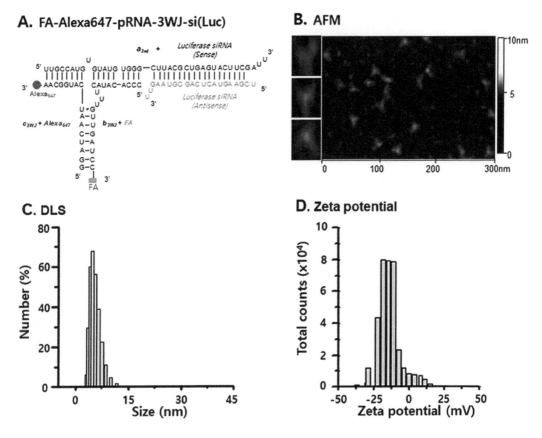

FIGURE 49.2 Characteristics of multifunctional pRNA-based three-way-junction (3WJ) RNA nanoparticle (RNP). (a) Sequence map of trivalent RNP FA-Alexa647-pRNA-3WJ-si(Luc), harboring three functionalities: Folate (FA) as a targeting ligand; Alexa647 as an imaging module; and luciferase siRNA for gene silencing. (b) Atomic force microscopy (AFM) image of three branched triangular structure of self-assembled FA-pRNA-3WJ-si(Luc) RNP. (c) Average size of FA-pRNA-3WJ-si(Luc) RNP by dynamic light scattering (DLS). (d) Zeta potential of FA-pRNA-3WJ-si(Luc) RNP. (Adapted from Lee *et al., Oncotarget.* 2015, 6(17):14766-76 with permission)

into tumor-bearing mice, the anti-miR-21 RNA nanoparticles showed decrease tumor growth in both triple negative breast cancer (TNBC) and glioblastoma mice models (Figure 49.5) (Lee et al., 2017; Shu et al., 2015). Furthermore, in vivo fluorescence revealed that the 3WJ RNA nanoparticles were retained in the brain tumor region for up to 15 hours after a single systemic injection, but did not significantly accumulate in any untargeted major internal organs. In another study, chemically synthesized 2'-F-modified RNA nanoparticles systemically administered via tail vein in mice demonstrated a favorable pharmacokinetic profile: 5–10 hours half-life, <0.13 L/kg/hour clearance rate, and 1.2 L/kg volume of distribution (Abdelmawla et al., 2011). These RNA nanoparticles also had a favorable safety profile: No detectable toxicity, along with low levels of interferon (IFN) and inflammatory cytokine response. Due to their pharmacokinetics, low toxicity, and ease of modification, RNA nanoparticles are highly suitable to use for miRNA and siRNA delivery. Surface modification by these RNA nanoparticles also can allow for exosome-mediated targeted delivery. When EGFR-targeting RNA aptamers were conjugated to the surface of exosomes carrying anti-Survivin siRNA, these exosomes gained the ability to target and deliver siRNA into EGFR expressing breast cancer

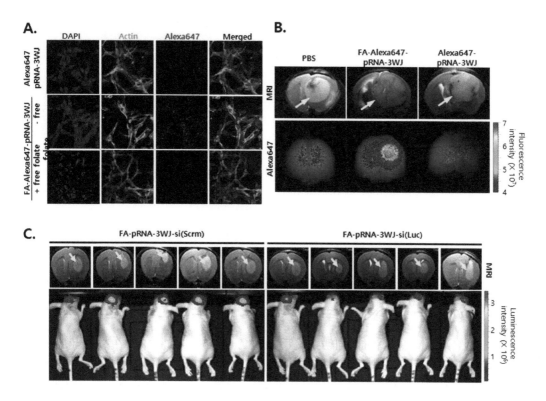

FIGURE 49.3 FA-mediated human glioblastoma cell targeting by FA-Alexa647-pRNA-3WJ RNP *in vitro and in vivo*. (a) Human glioblastoma cell U87EGFRvIII targeting in vitro by FA-Alexa647-pRNA-3WJ RNP (middle) compared to FA-free Alexa647-pRNA-3WJ (top) or free folate-saturated culture condition (bottom). (b) Intracranial xenograft tumor induced by implanted U87EGFRvIII cells in mice was targeted by FA-Alexa647-pRNA-3WJ RNP. MRI (yellow arrows in top panel) visualized tumor sizes while fluorescence in vivo imaging (bottom panel) shows specific targeting of FA-Alexa647-pRNA-3WJ RNP after tail vein injection. (c) Gene silencing effect of FA-pRNA-3WJ-si(Luc) RNP in intracranial xenograft tumor in mice. Representative in vivo MRI images for tumor volume and bioluminescence intensity for luciferase activity from both FA-pRNA-3WJ-siRNA(Luc) or FA-pRNA-3WJ-si(Scrm). (Adapted from Lee *et al., Oncotarget.* 2015, 6(17):14766–76 with permission)

cells (Pi et al., 2018). With accurate targeting, miRNA therapies can expect increased efficiency along with lowered risk of complications from non-specific targeting.

PERSPECTIVES

MiRNAs play major regulatory roles in almost every pathway in cellular signaling. Discovery of certain dysregulated signature miRNAs common to human cancers spurred attempts to modulate the dysregulated miRNA by delivering miRNA mimics to supplement downregulated miRNAs or miRNA inhibitors to silence overexpressed miRNAs. Although miRNA-based therapies have shown promising results in preclinical stages, few have been successfully translated into clinical trials and currently none has been approved for commercial markets by FDA. However, the main difficulties faced by miRNA-based therapies are delivery efficiency and potential toxicity. It can be addressed through the addition of specific targeting moieties to the delivery carrier. Early results using such

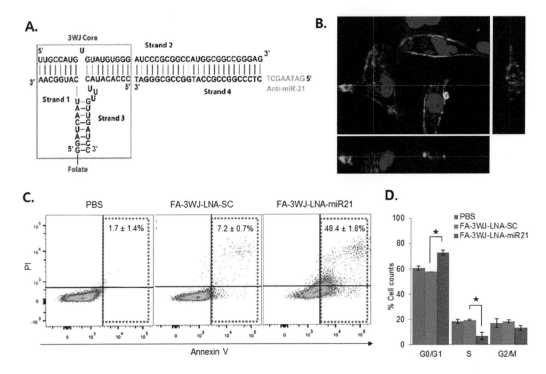

FIGURE 49.4 Targeted inhibition of miR-21 in vitro by multifunctional FA-3WJ-Alexa647-LNA-miR21 RNP for glioblastoma therapy. (a) Construction map of the trivalent FA-3WJ-Alexa647-LNA-miR21 RNP carrying anti-miR-21 LNA for miR-21 knockdown. (b) FA-dependent in vitro specific targeting of human glioblastoma cell U87EGFRvIII by FA-3WJ-Alexa647-LNA-miR21 RNP shown by fluorescence confocal microscope. **C and D**, Apoptosis induction (c) and cell cycle arrest (d) in U87EGFRvIII cells by specific FA-dependent knockdown of miR-21 with FA-3WJ-Alexa647-LNA-miR21 RNP. (Adapted from Lee *et al.*, *Mol Ther.* 2017, 25(7):1544–1555 with permission)

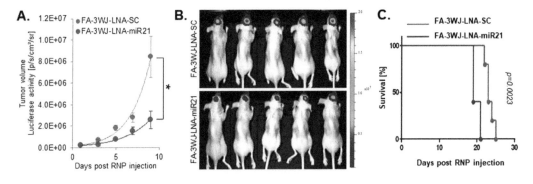

FIGURE 49.5 Preclinical targeted therapy of human glioblastoma by systemically delivered FA-3WJ-Alexa647-LNA-miR21 RNP. (a) Tumor volume change of intracranial xenograft tumor after repeated systemic delivery of FA-3WJ-Alexa647-LNA-miR21 RNPs. (b) The representative bioluminescence image after repeated systemic delivery of FA-3WJ-Alexa647-LNA-miR21 RNP compared to negative control RNP. (c) Kaplan-Meier survival curves in mice after specific inhibition of oncogenic miR-21 by FA-3WJ-Alexa647-LNA-miR21 RNPs. (Adapted from Lee *et al.*, *Mol Ther.* 2017, 25(7):1544–1555 with permission)

methods on RNA-based nanoparticles or exosomes have also been encouraging and may enable tissue-specific miRNA therapies in the future clinical applications. Although additional studies are still necessary to determine signature miRNAs and optimize treatment delivery, miRNA-based therapy is a promising candidate that can supplement current treatment options and improve patient outcome.

ACKNOWLEDGMENT

This work was supported in part by the Pilot/Feasibility Program from Texas Medical Center Digestive Diseases Center through National Institutes of Health National Institute of Diabetes And Digestive And Kidney Diseases (NIDDK) [Grant P30 DK056338] to T.J.L.

REFERENCES

Abdelmawla S, Guo S, Zhang L, Pulukuri SM, Patankar P, Conley P, Trebley J, Guo P and Li QX (2011) Pharmacological characterization of chemically synthesized monomeric phi29 pRNA nanoparticles for systemic delivery. *Mol Ther: J Am Soc Gene Ther* **19**:1312–1322.

Bagga S, Bracht J, Hunter S, Massirer K, Holtz J, Eachus R and Pasquinelli AE (2005) Regulation by let-7 and lin-4 miRNAs results in target mRNA degradation. *Cell* **122**:553–563.

Balkhi MY, Iwenofu OH, Bakkar N, Ladner KJ, Chandler DS, Houghton PJ, London CA, Kraybill W, Perrotti D, Croce CM, Keller C and Guttridge DC (2013) miR-29 acts as a decoy in sarcomas to protect the tumor suppressor A20 mRNA from degradation by HuR. *Sci Signal* **6**:ra63.

Bartel DP (2009) MicroRNAs: Target recognition and regulatory functions. *Cell* **136**:215–233.

Bartel DP (2018) Metazoan microRNAs. *Cell* **173**:20–51.

Beg MS, Brenner AJ, Sachdev J, Borad M, Kang YK, Stoudemire J, Smith S, Bader AG, Kim S and Hong DS (2017) Phase I study of MRX34, a liposomal miR-34a mimic, administered twice weekly in patients with advanced solid tumors. *Invest New Drugs* **35**:180–188.

Bi N, Cao J, Song Y, Shen J, Liu W, Fan J, He J, Shi Y, Zhang X, Lu N, Zhan Q and Wang L (2014) A microRNA signature predicts survival in early stage small-cell lung cancer treated with surgery and adjuvant chemotherapy. *PLoS One* **9**:e91388.

Calin GA and Croce CM (2006) MicroRNA signatures in human cancers. *Nat Rev Cancer* **6**:857–866.

Calin GA, Dumitru CD, Shimizu M, Bichi R, Zupo S, Noch E, Aldler H, Rattan S, Keating M, Rai K, Rassenti L, Kipps T, Negrini M, Bullrich F and Croce CM (2002) Frequent deletions and down-regulation of micro- RNA genes miR15 and miR16 at 13q14 in chronic lymphocytic leukemia. *Proc Natl Acad Sci U S A* **99**:15524–15529.

Calin GA, Ferracin M, Cimmino A, Di Leva G, Shimizu M, Wojcik SE, Iorio MV, Visone R, Sever NI, Fabbri M, Iuliano R, Palumbo T, Pichiorri F, Roldo C, Garzon R, Sevignani C, Rassenti L, Alder H, Volinia S, Liu CG, Kipps TJ, Negrini M and Croce CM (2005) A MicroRNA signature associated with prognosis and progression in chronic lymphocytic leukemia. *N Engl J Med* **353**:1793–1801.

Calin GA, Liu CG, Sevignani C, Ferracin M, Felli N, Dumitru CD, Shimizu M, Cimmino A, Zupo S, Dono M, Dell'Aquila ML, Alder H, Rassenti L, Kipps TJ, Bullrich F, Negrini M and Croce CM (2004) MicroRNA profiling reveals distinct signatures in B cell chronic lymphocytic leukemias. *Proc Natl Acad Sci U S A* **101**:11755–11760.

Campani V, Salzano G, Lusa S and De Rosa G (2016) Lipid nanovectors to deliver RNA oligonucleotides in cancer. *Nanomat (Basel)* 6.

Cosco D, Cilurzo F, Maiuolo J, Federico C, Di Martino MT, Cristiano MC, Tassone P, Fresta M and Paolino D (2015) Delivery of miR-34a by chitosan/PLGA nanoplexes for the anticancer treatment of multiple myeloma. *Sci Rep* **5**:17579.

Costinean S, Sandhu SK, Pedersen IM, Tili E, Trotta R, Perrotti D, Ciarlariello D, Neviani P, Harb J, Kauffman LR, Shidham A and Croce CM (2009) Src homology 2 domain-containing inositol-5-phosphatase and CCAAT enhancer-binding protein beta are targeted by miR-155 in B cells of Emicro-MiR-155 transgenic mice. *Blood* **114**:1374–1382.

Davidson BL and Harper SQ (2005) Viral delivery of recombinant short hairpin RNAs. *Methods Enzymol* **392**:145–173.

Demidov VV, Potaman VN, Frank-Kamenetskii MD, Egholm M, Buchard O, Sonnichsen SH and Nielsen PE (1994) Stability of peptide nucleic acids in human serum and cellular extracts. *Biochem Pharmacol* **48**:1310–1313.

Eichhorn SW, Guo H, McGeary SE, Rodriguez-Mias RA, Shin C, Baek D, Hsu SH, Ghoshal K, Villen J and Bartel DP (2014) mRNA destabilization is the dominant effect of mammalian microRNAs by the time substantial repression ensues. *Mol Cell* **56**:104–115.

Fabani MM, Abreu-Goodger C, Williams D, Lyons PA, Torres AG, Smith KG, Enright AJ, Gait MJ and Vigorito E (2010) Efficient inhibition of miR-155 function in vivo by peptide nucleic acids. *Nucleic Acids Res* **38**:4466–4475.

Fabbri M, Garzon R, Cimmino A, Liu Z, Zanesi N, Callegari E, Liu S, Alder H, Costinean S, Fernandez-Cymering C, Volinia S, Guler G, Morrison CD, Chan KK, Marcucci G, Calin GA, Huebner K and Croce CM (2007) MicroRNA-29 family reverts aberrant methylation in lung cancer by targeting DNA methyltransferases 3A and 3B. *Proc Natl Acad Sci U S A* **104**:15805–15810.

Fabbri M, Paone A, Calore F, Galli R, Gaudio E, Santhanam R, Lovat F, Fadda P, Mao C, Nuovo GJ, Zanesi N, Crawford M, Ozer GH, Wernicke D, Alder H, Caligiuri MA, Nana-Sinkam P, Perrotti D and Croce CM (2012) MicroRNAs bind to Toll-like receptors to induce prometastatic inflammatory response. *Proc Natl Acad Sci U S A* **109**:E2110–2116.

Fisher TL, Terhorst T, Cao X and Wagner RW (1993) Intracellular disposition and metabolism of fluorescently-labeled unmodified and modified oligonucleotides microinjected into mammalian cells. *Nucleic Acids Res* **21**:3857–3865.

Garofalo M, Di Leva G, Romano G, Nuovo G, Suh SS, Ngankeu A, Taccioli C, Pichiorri F, Alder H, Secchiero P, Gasparini P, Gonelli A, Costinean S, Acunzo M, Condorelli G and Croce CM (2009) miR-221&222 regulate TRAIL resistance and enhance tumorigenicity through PTEN and TIMP3 downregulation. *Cancer Cell* **16**:498–509.

Garzon R, Garofalo M, Martelli MP, Briesewitz R, Wang L, Fernandez-Cymering C, Volinia S, Liu CG, Schnittger S, Haferlach T, Liso A, Diverio D, Mancini M, Meloni G, Foa R, Martelli MF, Mecucci C, Croce CM and Falini B (2008) Distinctive microRNA signature of acute myeloid leukemia bearing cytoplasmic mutated nucleophosmin. *Proc Natl Acad Sci U S A* **105**:3945–3950.

Garzon R, Heaphy CE, Havelange V, Fabbri M, Volinia S, Tsao T, Zanesi N, Kornblau SM, Marcucci G, Calin GA, Andreeff M and Croce CM (2009) MicroRNA 29b functions in acute myeloid leukemia. *Blood* **114**:5331–5341.

Guo P (2010) The emerging field of RNA nanotechnology. *Nat Nanotechnol* **5**:833–842.

Ha M and Kim VN (2014) Regulation of microRNA biogenesis. *Nat Rev Mol Cell Biol* **15**:509–524.

Hanahan D and Weinberg RA (2011) Hallmarks of cancer: the next generation. *Cell* **144**:646–674.

Hanvey JC, Peffer NJ, Bisi JE, Thomson SA, Cadilla R, Josey JA, Ricca DJ, Hassman CF, Bonham MA, Au KG and et al. (1992) Antisense and antigene properties of peptide nucleic acids. *Science* **258**:1481–1485.

He L, He X, Lim LP, de Stanchina E, Xuan Z, Liang Y, Xue W, Zender L, Magnus J, Ridzon D, Jackson AL, Linsley PS, Chen C, Lowe SW, Cleary MA and Hannon GJ (2007) A microRNA component of the p53 tumour suppressor network. *Nature* **447**:1130–1134.

He L, Thomson JM, Hemann MT, Hernando-Monge E, Mu D, Goodson S, Powers S, Cordon-Cardo C, Lowe SW, Hannon GJ and Hammond SM (2005) A microRNA polycistron as a potential human oncogene. *Nature* **435**:828–833.

Hutvagner G, McLachlan J, Pasquinelli AE, Balint E, Tuschl T and Zamore PD (2001) A cellular function for the RNA-interference enzyme Dicer in the maturation of the let-7 small temporal RNA. *Science (New York, NY)* **293**:834–838.

Ibrahim AF, Weirauch U, Thomas M, Grunweller A, Hartmann RK and Aigner A (2011) MicroRNA replacement therapy for miR-145 and miR-33a is efficacious in a model of colon carcinoma. *Cancer Res* **71**:5214–5224.

Jasinski D, Haque F, Binzel DW and Guo P (2017) Advancement of the emerging field of RNA nanotechnology. *ACS Nano* **11**:1142–1164.

Jurkovicova D, Magyerkova M, Kulcsar L, Krivjanska M, Krivjansky V, Gibadulinova A, Oveckova I and Chovanec M (2014) miR-155 as a diagnostic and prognostic marker in hematological and solid malignancies. *Neoplasma* **61**:241–251.

Ketting RF, Fischer SE, Bernstein E, Sijen T, Hannon GJ and Plasterk RH (2001) Dicer functions in RNA interference and in synthesis of small RNA involved in developmental timing in C. elegans. *Genes Dev* **15**:2654–2659.

Kim JH, Yeom JH, Ko JJ, Han MS, Lee K, Na SY and Bae J (2011) Effective delivery of anti-miRNA DNA oligonucleotides by functionalized gold nanoparticles. *J Biotechnol* **155**:287–292.

Kneitz B, Krebs M, Kalogirou C, Schubert M, Joniau S, van Poppel H, Lerut E, Kneitz S, Scholz CJ, Strobel P, Gessler M, Riedmiller H and Spahn M (2014) Survival in patients with high-risk prostate cancer is predicted by miR-221, which regulates proliferation, apoptosis, and invasion of prostate cancer cells by inhibiting IRF2 and SOCS3. *Cancer Res* **74**:2591–2603.

Kota J, Chivukula RR, O'Donnell KA, Wentzel EA, Montgomery CL, Hwang HW, Chang TC, Vivekanandan P, Torbenson M, Clark KR, Mendell JR and Mendell JT (2009) Therapeutic microRNA delivery suppresses tumorigenesis in a murine liver cancer model. *Cell* **137**:1005–1017.

Kurreck J, Wyszko E, Gillen C and Erdmann VA (2002) Design of antisense oligonucleotides stabilized by locked nucleic acids. *Nucleic Acids Res* **30**:1911–1918.

Lee RC, Feinbaum RL and Ambros V (1993) The C. elegans heterochronic gene lin-4 encodes small RNAs with antisense complementarity to lin-14. *Cell* **75**:843–854.

Lee TJ, Haque F, Shu D, Yoo JY, Li H, Yokel RA, Horbinski C, Kim TH, Kim SH, Kwon CH, Nakano I, Kaur B, Guo P and Croce CM (2015a) RNA nanoparticle as a vector for targeted siRNA delivery into glioblastoma mouse model. *Oncotarget* **6**:14766–14776.

Lee TJ, Haque F, Vieweger M, Yoo JY, Kaur B, Guo P and Croce CM (2015b) Functional assays for specific targeting and delivery of RNA nanoparticles to brain tumor. *Methods Mol Biol (Clifton, NJ)* **1297**:137–152.

Lee TJ, Yoo JY, Shu D, Li H, Zhang J, Yu JG, Jaime-Ramirez AC, Acunzo M, Romano G, Cui R, Sun HL, Luo Z, Old M, Kaur B, Guo P and Croce CM (2017) RNA nanoparticle-based targeted therapy for glioblastoma through inhibition of oncogenic miR-21. *Mol Therap: J Am Soc Gene Therap* **25**:1544-1555.

Lee Y, Ahn C, Han J, Choi H, Kim J, Yim J, Lee J, Provost P, Radmark O, Kim S and Kim VN (2003) The nuclear RNase III Drosha initiates microRNA processing. *Nature* **425**:415–419.

Lee Y, Kim M, Han J, Yeom KH, Lee S, Baek SH and Kim VN (2004) MicroRNA genes are transcribed by RNA polymerase II. *EMBO J* **23**:4051–4060.

Li L, Yuan L, Luo J, Gao J, Guo J and Xie X (2013) MiR-34a inhibits proliferation and migration of breast cancer through down-regulation of Bcl-2 and SIRT1. *Clin Exp Med* **13**:109–117.

Li N, Fu H, Tie Y, Hu Z, Kong W, Wu Y and Zheng X (2009) miR-34a inhibits migration and invasion by down-regulation of c-Met expression in human hepatocellular carcinoma cells. *Cancer Lett* **275**:44–53.

Lodygin D, Tarasov V, Epanchintsev A, Berking C, Knyazeva T, Korner H, Knyazev P, Diebold J and Hermeking H (2008) Inactivation of miR-34a by aberrant CpG methylation in multiple types of cancer. *Cell Cycle* **7**:2591–2600.

Milone MC and O'Doherty U (2018) Clinical use of lentiviral vectors. *Leukemia* **32**:1529–1541.

Neudecker V, Brodsky KS, Kreth S, Ginde AA and Eltzschig HK (2016) Emerging roles for microRNAs in perioperative medicine. *Anesthesiology* **124**:489–506.

O'Donnell KA, Wentzel EA, Zeller KI, Dang CV and Mendell JT (2005) c-Myc-regulated microRNAs modulate E2F1 expression. *Nature* **435**:839–843.

Olive V, Li Q and He L (2013) Mir-17-92: A polycistronic oncomir with pleiotropic functions. *Immunol Rev* **253**:158–166.

Olsen PH and Ambros V (1999) The lin-4 regulatory RNA controls developmental timing in Caenorhabditis elegans by blocking LIN-14 protein synthesis after the initiation of translation. *Dev Biol* **216**:671–680.

Pang RT, Leung CO, Ye TM, Liu W, Chiu PC, Lam KK, Lee KF and Yeung WS (2010) MicroRNA-34a suppresses invasion through downregulation of Notch1 and Jagged1 in cervical carcinoma and choriocarcinoma cells. *Carcinogenesis* **31**:1037–1044.

Parker N, Turk MJ, Westrick E, Lewis JD, Low PS and Leamon CP (2005) Folate receptor expression in carcinomas and normal tissues determined by a quantitative radioligand binding assay. *Anal Biochem* **338**:284–293.

Penaud-Budloo M, Francois A, Clement N and Ayuso E (2018) Pharmacology of recombinant adeno-associated virus production. *Mol Ther Methods Clin Dev* **8**:166–180.

Peruzzi P, Bronisz A, Nowicki MO, Wang Y, Ogawa D, Price R, Nakano I, Kwon CH, Hayes J, Lawler SE, Ostrowski MC, Chiocca EA and Godlewski J (2013) MicroRNA-128 coordinately targets polycomb repressor complexes in glioma stem cells. *Neuro-Oncology* **15**:1212–1224.

Pi F, Binzel DW, Lee TJ, Li Z, Sun M, Rychahou P, Li H, Haque F, Wang S, Croce CM, Guo B, Evers BM and Guo P (2018) Nanoparticle orientation to control RNA loading and ligand display on extracellular vesicles for cancer regression. *Nat Nanotechnol* **13**:82–89.

Piao L, Zhang M, Datta J, Xie X, Su T, Li H, Teknos TN and Pan Q (2012) Lipid-based nanoparticle delivery of Pre-miR-107 inhibits the tumorigenicity of head and neck squamous cell carcinoma. *Mol Ther* **20**:1261–1269.

Ren Y, Kang CS, Yuan XB, Zhou X, Xu P, Han L, Wang GX, Jia Z, Zhong Y, Yu S, Sheng J and Pu PY (2010) Co-delivery of as-miR-21 and 5-FU by poly(amidoamine) dendrimer attenuates human glioma cell growth in vitro. *J Biomater Sci Polym Ed* **21**:303–314.

Shu D, Li H, Shu Y, Xiong G, Carson WE, 3rd, Haque F, Xu R and Guo P (2015) Systemic delivery of anti-miRNA for suppression of triple negative breast cancer utilizing RNA nanotechnology. *ACS Nano* **9**:9731–9740.

Shu D, Shu Y, Haque F, Abdelmawla S and Guo P (2011) Thermodynamically stable RNA three-way junction for constructing multifunctional nanoparticles for delivery of therapeutics. *Nat Nanotechnol* **6**:658–667.

Shu Y, Pi F, Sharma A, Rajabi M, Haque F, Shu D, Leggas M, Evers BM and Guo P (2014) Stable RNA nanoparticles as potential new generation drugs for cancer therapy. *Adv Drug Deliv Rev* **66**:74–89.

Sun F, Fu H, Liu Q, Tie Y, Zhu J, Xing R, Sun Z and Zheng X (2008) Downregulation of CCND1 and CDK6 by miR-34a induces cell cycle arrest. *FEBS Lett* **582**:1564–1568.

Sun HL, Cui R, Zhou J, Teng KY, Hsiao YH, Nakanishi K, Fassan M, Luo Z, Shi G, Tili E, Kutay H, Lovat F, Vicentini C, Huang HL, Wang SW, Kim T, Zanesi N, Jeon YJ, Lee TJ, Guh JH, Hung MC, Ghoshal K, Teng CM, Peng Y and Croce CM (2016) ERK activation globally downregulates miRNAs through phosphorylating exportin-5. *Cancer Cell* **30**:723–736.

Sylvestre Y, De Guire V, Querido E, Mukhopadhyay UK, Bourdeau V, Major F, Ferbeyre G and Chartrand P (2007) An E2F/miR-20a autoregulatory feedback loop. *J Biol Chem* **282**:2135–2143.

Trang P, Medina PP, Wiggins JF, Ruffino L, Kelnar K, Omotola M, Homer R, Brown D, Bader AG, Weidhaas JB and Slack FJ (2010) Regression of murine lung tumors by the let-7 microRNA. *Oncogene* **29**:1580–1587.

Wahlestedt C, Salmi P, Good L, Kela J, Johnsson T, Hokfelt T, Broberger C, Porreca F, Lai J, Ren K, Ossipov M, Koshkin A, Jakobsen N, Skouv J, Oerum H, Jacobsen MH and Wengel J (2000) Potent and nontoxic antisense oligonucleotides containing locked nucleic acids. *Proc Natl Acad Sci U S A* **97**:5633–5638.

Wang B, Majumder S, Nuovo G, Kutay H, Volinia S, Patel T, Schmittgen TD, Croce C, Ghoshal K and Jacob ST (2009) Role of microRNA-155 at early stages of hepatocarcinogenesis induced by choline-deficient and amino acid-defined diet in C57BL/6 mice. *Hepatology* **50**:1152–1161.

Wightman B, Ha I and Ruvkun G (1993) Posttranscriptional regulation of the heterochronic gene lin-14 by lin-4 mediates temporal pattern formation in C. elegans. *Cell* **75**:855–862.

Wu Y, Crawford M, Mao Y, Lee RJ, Davis IC, Elton TS, Lee LJ and Nana-Sinkam SP (2013) Therapeutic delivery of microRNA-29b by cationic lipoplexes for lung cancer. *Mol Ther Nucleic Acids* **2**:e84.

Xiao C, Srinivasan L, Calado DP, Patterson HC, Zhang B, Wang J, Henderson JM, Kutok JL and Rajewsky K (2008) Lymphoproliferative disease and autoimmunity in mice with increased miR-17-92 expression in lymphocytes. *Nat Immunol* **9**:405–414.

Xu L, Xu Y, Jing Z, Wang X, Zha X, Zeng C, Chen S, Yang L, Luo G, Li B and Li Y (2014) Altered expression pattern of miR-29a, miR-29b and the target genes in myeloid leukemia. *Experimental Hemato Oncol* **3**:17-3619-3613-3617. eCollection 2014.

Yamakuchi M, Ferlito M and Lowenstein CJ (2008) miR-34a repression of SIRT1 regulates apoptosis. *Proc Natl Acad Sci U S A* **105**:13421–13426.

Yamamura S, Saini S, Majid S, Hirata H, Ueno K, Chang I, Tanaka Y, Gupta A and Dahiya R (2012) MicroRNA-34a suppresses malignant transformation by targeting c-Myc transcriptional complexes in human renal cell carcinoma. *Carcinogenesis* **33**:294–300.

Yi R, Qin Y, Macara IG and Cullen BR (2003) Exportin-5 mediates the nuclear export of pre-microRNAs and short hairpin RNAs. *Genes Dev* **17**:3011–3016.

Yin PT, Shah BP and Lee KB (2014a) Combined magnetic nanoparticle-based microRNA and hyperthermia therapy to enhance apoptosis in brain cancer cells. *Small* **10**:4106–4112.

Yin Y, Cai X, Chen X, Liang H, Zhang Y, Li J, Wang Z, Chen X, Zhang W, Yokoyama S, Wang C, Li L, Li L, Hou D, Dong L, Xu T, Hiroi T, Yang F, Ji H, Zhang J, Zen K and Zhang CY (2014b) Tumor-secreted miR-214 induces regulatory T cells: a major link between immune evasion and tumor growth. *Cell Res* **24**:1164–1180.

Yu F, Yao H, Zhu P, Zhang X, Pan Q, Gong C, Huang Y, Hu X, Su F, Lieberman J and Song E (2007) Let-7 regulates self renewal and tumorigenicity of breast cancer cells. *Cell* **131**:1109–1123.

50 Tuning the Size, Shape, and Structure of RNA Nanoparticles for Favorable Cancer Targeting and Immunostimulation[1]

Sijin Guo, Congcong Xu, Hongran Yin, and Peixuan Guo
The Ohio State University, Columbus, USA

Jordan Hill and Fengmei Pi
ExonanoRNA LLC, Columbus, USA

CONTENTS

INTRODUCTION

Nowadays, nanotechnology-based platforms such as lipid-based particles (Puri et al., 2009), nucleic acids nanoparticles (Andersen et al., 2009; Guo, Tschammer, Mohammed, & Guo, 2005), viral nanoparticles (Singh et al., 2007), and synthetic inorganic and polymeric particles (Astruc, 2012) are

DOI: 10.1201/9781003001560-55

finding ever-increasing applications in various fields, including nanomedicine. Nevertheless, one challenge in clinical development of nanomedicine is the lack of sufficient evidence regarding their safety, immunological, and pharmacological profiles. Inadvertent recognition of nanomaterials as invaders by the immune systems frequently results in varying levels of immunostimulation or immunosuppression, which consequently leads to toxicity and reduced therapeutic efficacy. Particularly, the common challenges in the immunotoxicology assessment of nanomaterials have been summarized by the NCI Nanotechnology Characterization Laboratory, highlighting four key areas (Chemistry, Efficacy, Pharmacology and Toxicology, and Hematology and Immunology) requiring thorough characterization to efficiently translate nanoparticle-formulated drugs toward the clinic (Dobrovolskaia, 2015). Typically, their physicochemical characterization needs to be well assessed, and they are expected to display favorable pharmacokinetics (PK), pharmacodynamics (PD), and strong safety profiles while retaining potent drug efficacy against the targeted disease. Some strategies, e.g. poly (ethylene glycol) (PEG) coating on cationic lipid-based nanocarriers (Suk, Xu, Kim, Hanes, & Ensign, 2016), have been proposed to engineer nanoparticles that minimize unwanted immunotoxicity while improving *in vivo* performance. However, additional side effects induced by PEG-specific antibodies were also reported (Zhang, Sun, Liu, & Jiang, 2016). Refining nanoparticle size is another approach proposed to diminish immune response because phagocytic cells in the immune systems tend to pick up larger nanoparticles. Unfortunately, some nanoparticles were formulated with unpredictable or heterogeneous size distributions, making consistent assembly particularly difficult (Desai, 2012). While immune responses are vital, the favorable *in vivo* biodistribution of a nanoplatform to achieve specific cancer targeting is another key factor for efficacious drug delivery. With active targeting ligands, nanoparticles exhibit distinct advantages over traditional small molecule therapeutics for overcoming PK limitations by virtue of their prolonged circulation time and extended accumulation in the tumors (Singh & Lillard, Jr., 2009). Upon systemic administration, nanoparticles are exposed to the physiological environment rich with proteins, cells, and tissues. Size and shape have both been shown to greatly affect the PK and biodistribution profiles of nanoparticles (Hoshyar, Gray, Han, & Bao, 2016; Toy, Peiris, Ghaghada, & Karathanasis, 2014). Nanoparticles smaller than 5 nm undergo significant filtration by the kidneys and is excreted in urine. Larger particles, ranging from 20 to 100 nm, are easily engulfed by macrophages or sequestered in healthy tissues, thus causing non-specific and undesirable accumulation in healthy organs (Gustafson, Holt-Casper, Grainger, & Ghandehari, 2015). Nanoparticles with distinct shapes also exhibited different hemorheological dynamics and cellular uptake. For instance, Discher et al. reported prolonged circulation time of filamentous polymer micelles compared to spherical counterparts (Geng et al., 2007). Additionally, the surface characteristics of nanoparticles such as hydrophobicity and surface charge can greatly influence protein adsorption and cellular membrane interactions (Albanese, Tang, & Chan, 2012). Important to the *in vivo* fate of nanoparticles, adsorption to specific proteins termed opsonins (e.g. the complement proteins, IgG and laminin) facilitate the recognition and uptake by immune cells, such as macrophages. As a result of this nanoparticle uptake, macrophages secrete cytokine signaling molecules, possibly leading to immunostimulatory responses. Thus, notwithstanding the achievement of nanotechnology has shown in drug delivery, the translation of synthetic or biological nanoparticles into clinical trials will mandate extensive investigations in accordance with strict criteria.

As a naturally occurring biopolymer, RNA is unique compared to other nanomaterials. RNA molecules possess diverse sequences, secondary structures, tertiary and quaternary interactions by nature (Hendrix, Brenner, & Holbrook, 2005). Unlike other biomacromolecules (e.g. DNA and protein), RNA is structurally more flexible and functionally more versatile (Guo, 2010; Jasinski, Haque, Binzel, & Guo, 2017). Taking advantage of both canonical Watson-Crick (A-U, G-C) and noncanonical (e.g. G-U) base pairing, a substantial number of natural RNA motifs and self-folding RNAs have been discovered. These include, but are not limited to, kink-turns (Huang & Lilley, 2013, 2016), kissing hairpins (Bindewald, Hayes, Yingling, Kasprzak, & Shapiro, 2008), paranemic motifs (Afonin, Cieply, & Leontis, 2008), pseudoknot (Bindewald, Afonin, Jaeger, & Shapiro,

2011), three-way (Shu, Shu, Haque, Abdelmawla, & Guo, 2011; Zhang et al., 2013) and multi-helix junctions (Laing, Jung, Iqbal, & Schlick, 2009), bulges and loops (Zacharias & Hagerman, 1995). As a result of the highly influential research in previous traditional RNA biology, the field of RNA nanotechnology has experienced an exponential growth in the past decade (Jasinski et al., 2017; Li et al., 2015). Myriads of sophisticated RNA architectures with highly ordered structures and multivalent functionalities were assembled from the bottom-up using RNA as building blocks, such as RNA polygons (Boerneke, Dibrov, & Hermann, 2016; Bui et al., 2017; Dibrov, McLean, Parsons, & Hermann, 2011; Huang & Lilley, 2016; Jasinski, Khisamutdinov, Lyubchenko, & Guo, 2014; Khisamutdinov et al., 2014a, 2014b; Ohno et al., 2011; Severcan, Geary, Verzemnieks, Chworos & Jaeger, 2009), RNA polyhedrons (Afonin et al., 2010; Hao et al., 2014; Khisamutdinov et al., 2016; Li et al., 2016; Severcan et al., 2010; Xu et al., 2018; Yu, Liu, Jiang, Wang, & Mao, 2015), RNA rings (Grabow et al., 2011; Shu et al., 2013a, 2013b; Geary, Chworos, Verzemnieks, Voss, & Jaeger, 2017), RNA dendrimers (Sharma et al., 2015), jigsaw puzzles (Chworos et al., 2004), and RNA filaments (Nasalean, Baudrey, Leontis, & Jaeger, 2006) (Figure 50.1).

Recently, the three-way junction (3WJ) motif of packaging RNA (pRNA) originated from phi29 DNA packaging motor has been exploited as a scaffold for fabrication of multifunctional and thermodynamically stable RNA nanoparticles (Shu et al., 2011). Different components including imaging (fluorophores or fluorogenic aptamers), targeting (RNA aptamers or chemical ligands), and therapeutic (siRNA, miRNA, anti-miRNA, ribozymes, or chemical drugs) modules were incorporated to the 3WJ motif without affecting authentic folding and original functions of the modules (Shu, Khisamutdinov, Zhang, & Guo, 2013b). Tremendous efforts have been made to improve the enzymatic stability of RNA by chemical modifications such as 2'-fluorine (2'-F) on pyrimidines, making them dramatically resistant to RNase degradation (Corey, 2007). These RNA nanoparticles are highly programmable, namely their physicochemical properties can be easily engineered to favor *in vivo* therapeutic delivery (Guo et al., 2017; Jasinski et al., 2014; Jasinski, Li, & Guo, 2018a; Jasinski, Yin, Li, & Guo, 2018b; Khisamutdinov et al., 2014a, 2014b; Pi et al., 2018; Shu et al., 2018). Upon systemic injection in tumor-bearing mice, the RNA nanoparticles displayed favored accumulation at the tumor site, with little to no accumulation in vital organs (Binzel et al., 2016; Cui et al., 2015; Lee et al., 2015; Pi et al., 2018; Rychahou et al., 2015; Shu et al., 2015, 2018, 2013a, 2013b). The negatively charged backbone of RNA disallows non-specific interaction with negatively charged cell membranes. The pRNA-based nanoparticles showed favorable pharmacological profiles and induced extremely low interferon or cytokine production in mice (Abdelmawla et al., 2011; Guo et al., 2017; Khisamutdinov et al., 2014a, 2014b; Shu et al., 2018). Thus, the significant progress of therapeutic RNA nanoparticles has revealed its potential to pioneer a new generation of drug for the future market. Presumably, RNA nanotechnology will begin its role in a clinical setting in the near-future.

In order to develop therapeutic RNA nanoparticles into a new drug, they must meet all the FDA's drug evaluation criteria and overcome the obstacles limiting their clinical advancement. To date, many important barriers (e.g. chemical stability and thermodynamic stability) in the field of RNA nanotechnology have been successfully overcome (Corey, 2007; Khisamutdinov, Jasinski, & Guo, 2014a; Shu et al., 2011). The progress in these aspects has been discussed in previous reviews (Jasinski et al., 2017; Li et al., 2015; Xu et al., 2018a). Recently, however, increasing studies have reported on understanding the broad mechanisms that define the pharmacological and immunological profile of RNA nanoparticles. A variety of their physicochemical properties were recognized to be associated with their *in vivo* behaviors, including size, shape, stoichiometry/module density, surface chemistry, composition, hydrophobicity, and elasticity (Abdelmawla et al., 2011; Afonin et al., 2014b; Guo et al., 2017; Halman et al., 2017; Hong et al., 2018; Jasinski et al., 2018a, 2018b; Johnson et al., 2017; Ke et al., 2018; Khisamutdinov et al., 2014a, 2014b; Lee, Urban, Xu, Sullenger, & Lee, 2016; Pi et al., 2018; Shi et al., 2018; Shu et al., 2018). Here, this brief review aims to integrate the recent progress seen in the study of RNA nanoparticle on biodistribution and

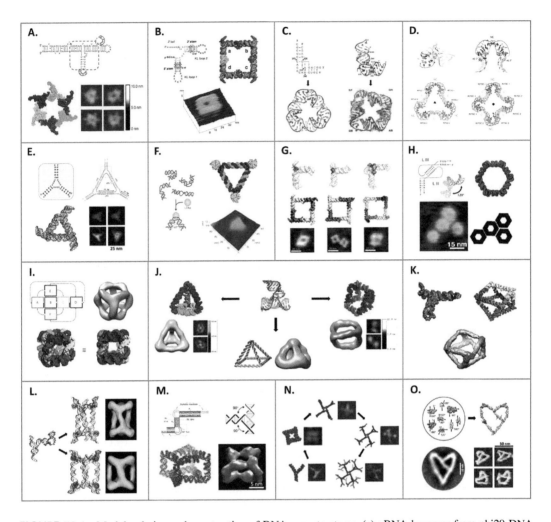

FIGURE 50.1 Modular design and construction of RNA nanostructures. (a) pRNA hexamer from phi29 DNA packaging motor (Shu et al., 2013a, 2013b). Copyright 2013 RNA Society; (b) RNA square with tectoRNA motif (Chworos et al., 2004). Copyright 2004 The American Association for the Advancement of Science; (c) RNA triangle with IIa motif from SVV IRES (Boerneke et al., 2016). Copyright 2016 John Wiley & Sons, Inc. & RNA square with IIa-1 motif from HCV (Dibrov et al., 2011). Copyright 2011 National Academy of Sciences; (d) RNA triangle and square with k-turn motif (Huang & Lilley, 2016). Copyright 2016 The Royal Society of Chemistry; (e) RNA triangle with tetra-U motif (Bui et al., 2017). Copyright 2017 Elsevier Inc.; (f) RNA-protein triangle with k-turn motif (Ohno et al., 2011). Copyright 2011 Nature Publishing Group; (g) RNA tectosquare with RA, 3WJ, and tRNA motifs (Severcan et al., 2009). Copyright 2009 American Chemical Society; (h) RNA nanoring based on RNAI/II inverse kissing complex (Grabow et al., 2011). Copyright 2011 American Chemical Society; (i) RNA cube designed *in silico* (Afonin et al., 2010). Copyright 2010 Macmillan Publishers Limited; (j) pRNA-3WJ motif-based RNA tetrahedron (Li et al., 2016). Copyright 2016 John Wiley & Sons, Inc., pyramid (Xu et al., 2018). Copyright 2018 Springer Nature, & nanoprism (Khisamutdinov et al., 2016). Copyright 2016 John Wiley & Sons, Inc.; (k) RNA polyhedron made of tRNA subunit (Severcan et al., 2010). Copyright 2010 Macmillan Publishers Limited; (l) Triangular and tetragonal RNA prism from re-engineered pRNA (Hao et al., 2014). Copyright 2014 Macmillan Publishers Limited; (m) Homo-octameric RNA prism with T-junction RNA tile (Yu et al., 2015). Copyright 2015 Macmillan Publishers Limited; (n) RNA dendrimers from pRNA-3WJ motif (Sharma et al., 2015). Copyright 2015 Elsevier Inc.; (o) RNA nanoheart from a syntax of RNA modules (Geary et al., 2017). Copyright 2017 American Chemical Society. (Figures adapted and reprinted with permission from cited references)

immunostimulation, while focusing on the defining aspects of their design and physiochemical properties precisely tuned to treat a wide range of cancers.

DEFINITION OF RNA NANOTECHNOLOGY

RNA nanotechnology is a relatively new field distinctive from traditional RNA biology. The concept of RNA nanotechnology is defined as the study on the design, construction, and application of RNA nanoparticles at the nanometer scale that are primarily composed of RNA *via* bottom-up self-assembly (Guo, 2010). The major frame of RNA nanoparticles can include scaffolds, targeting ligands, regulatory moieties, and therapeutic modules. The field focuses on the inter-RNA interactions and quaternary interactions of small RNA motifs, which is different from classical RNA research focused on intra-RNA interactions and 2D/3D structure-function relationships. Nevertheless, the broad knowledge gained from classical RNA biology research has laid a solid foundation for RNA nanotechnology. Since the introduction of the concept describing RNA nanoarchitectures in 1996 and 1998 by RNA tectonics and reengineered pRNA molecules, respectively (Guo, Zhang, Chen, Trottier, & Garver, 1998; Jaeger, Westhof, & Leontis, 2001), RNA nanotechnology has been growing rapidly as a platform with extensive application studies in nanomedicine over the past decade (Jasinski et al., 2017; Li et al., 2015).

ADVANTAGES OF RNA NANOTECHNOLOGY FOR CANCER TARGETING AND IMMUNOMODULATION

RNA Nanoparticles Are Distinct from Traditional Therapeutic RNAs

Traditional therapeutic RNAs, such as siRNA (Elbashir et al., 2001), miRNA (Bartel, 2004), anti-miRNA (Elmen et al., 2008), mRNA (Sahin, Kariko, & Tureci, 2014), viral immunostimulatory RNA (isRNA) (Bourquin et al., 2007), and ribozymes (Sarver et al., 1990), have a long history of fundamental research. Some of these small RNAs were reported to be immunogenic (Berger et al., 2009; Bourquin et al., 2007; Heil et al., 2004; Hornung et al., 2005; Judge et al., 2005). Cellular uptake of these RNAs might stimulate RNA-sensing pattern recognition receptors (PRRs) (Mogensen, 2009). RNA nanoparticles are distinguishable from traditional therapeutic RNAi and other small functional RNAs, though they share the similar chemical property of ribonucleotide composition. The advantages of RNA nanotechnology include: 1) therapeutic RNAi components or small non-coding RNA can serve as both scaffolds and functional moieties to construct larger scale nanostructures through a bottom-up self-assembly (Figure 50.1); 2) the nanoscale size offers favorable PK and PD profiles, enhanced permeability and retention (EPR) effect (Yin, Wang, Li, Shu, & Guo, 2018), minimal liver accumulation (Binzel et al., 2016; Shu et al., 2015), and undetectable toxicity *in vivo* (Yin et al., 2019; Abdelmawla et al., 2011); 3) RNA nanoparticles are generally assembled with a defined size, shape, and structure, and these morphologies are easily tunable (Bui et al., 2017; Jasinski et al., 2014; Khisamutdinov et al., 2014a, 2014b; Monferrer, Zhang, Lushnikov, & Hermann, 2019); 4) Unlike small RNAs, RNA nanoparticles can harbor multi-valent functionalities such as a targeting ligand and various drugs for combination therapy (Afonin et al., 2014a; Zhang et al., 2017); 5) RNA nanoparticles display thermal, chemical and metabolic stability in biological matrices. In a more comprehensive pharmacological characterization study of pRNA nanoparticles (Abdelmawla et al., 2011), various secondary PK parameters have been evaluated (Table 50.1). The normalized volume of distribution V_d (1.2 l/kg) of pRNA nanoparticles suggested the distribution of a significant fraction to peripheral tissues (outside vascular or extravasation), particularly in tumor. The relatively small clearance (Cl) value (significantly below the kidney filtration rate) suggests that the nanoparticles were not efficiently filtered out by the kidneys. By taking advantage of these features, traditional small therapeutic RNAs can be conveniently incorporated into RNA nanoparticles to achieve specific targeted delivery, an extended *in vivo* half-life, and higher therapeutic efficacy (Xu et al., 2018a; Jasinski et al., 2017).

TABLE 50.1

Secondary PK Parameters of pRNA Nanoparticles

Key Parameters	2′-F-pRNA	siRNA
AUC_{last} (hour·ng/ml)	2×10^5	1.1×10^3
$T_{1/2}$ (hours)	5–10	0.25
V_d (l/kg)	1.2	0.36
Cl (l/kg/hour)	0.13	1.0
Dose (mg/kg)	24	1.2

(Adapted with permission from (Abdelmawla et al., 2011). Copyright 2011 Elsevier Inc.)
Abbreviations: AUC, area under the curve; Cl, clearance; PK, pharmacokinetic; $T_{1/2}$, half-life; V_d, volume of distribution.

RNA NANOPARTICLES SHOW FAVORABLE CANCER TARGETING AND MINIMAL ORGAN ACCUMULATION

One of the critical challenges to overcome in cancer nanotechnology is the non-specific accumulation of administrated nanoparticles in healthy organs including the liver, lungs, kidneys, and spleen (Shi, Kantoff, Wooster, & Farokhzad, 2017). This non-specificity not only distracts nanoparticles from transporting to tumors, but also causes toxicity and unwanted side effects. To address this challenge, Guo Lab has developed a series of pRNA-3WJ based RNA nanoparticles incorporated with targeting ligands (Binzel et al., 2016; Cui et al., 2015; Lee et al., 2015; Pi et al., 2018; Rychahou et al., 2015; Shu et al., 2015, 2018, 2013a, 2013b; Xu et al., 2018b). Upon systemic administration in tumor-bearing mice, these RNA nanoparticles specifically transport to tumors within about 4 hours (h), and successfully remain at the tumor site for over 24 h. No or minimal organ accumulation was detected several hours post-injection. Consistent results were observed in several common cancer models (glioblastoma, breast, gastric, prostate, colorectal, and head and neck cancers) (Figure 50.2), in which the incorporated targeting ligands were changed to bind specific receptors overexpressed on different cancer cell surfaces. Meanwhile, therapeutics such as anti-miRNA21 can be delivered into cancer cells with the RNA nanoparticles harboring them *via* receptor-mediated endocytosis, thus improving the therapeutic effects (Shu et al., 2015; Binzel et al., 2016). Another novel study reported that a globular RNA micelle can efficiently delivery anti-miRNA21 to tumors without a targeting ligand (Guo et al., 2017; Khisamutdinov et al., 2016; Shu et al., 2018; Yin et al., 2018). It is suggested the negatively charged nature of RNA also plays an important role in targeting specificity because it minimizes non-specific interactions with the negatively charged cell membranes. Furthermore, the strong elasticity and branched ratchet-like shape of pRNA-3WJ based nanoparticles confer higher tumor penetration and improve EPR effects. This is particularly useful for overcoming mechanical barriers, disorganized vascularization, and highly immunosuppressive tumor microenvironments. Collectively, these results demonstrate that pRNA-3WJ based nanoparticles can be conveniently engineered with active targeting ligands to achieve specific cancer targeting with low or non-accumulation in healthy organs. This favorable biodistribution is an important indication of RNA nanoparticles' pharmacological profiles.

RNA NANOPARTICLES INTRINSICALLY DISPLAY IMMUNOLOGICALLY INERT PROPERTY AND NON-TOXICITY

Due to the lack of a universal nomenclature to categorize traditional therapeutic RNAs and RNA nanoparticles, the literatures on RNA immunogenicity have been controversial. Though the

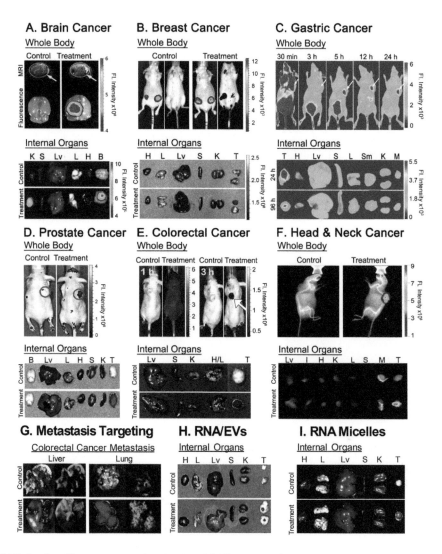

FIGURE 50.2 Specific cancer targeting *in vivo* of RNA nanoparticles to (a) Brain cancer (Lee et al., 2015). Copyright 2015 Impact Journals; (b) Breast cancer (Shu et al., 2015). Copyright 2015 American Chemical Society; (c) Gastric cancer (Cui et al., 2015). Copyright 2015 Macmillan Publishers Limited; (d) Prostate cancer (Binzel et al., 2016). Copyright 2016 Elsevier Inc.; (e) Colorectal cancer (Rychahou et al., 2015; Xu et al., 2018b). Copyright 2015 American Chemical Society & Elsevier B.V.; (f) Head & Neck cancer (Shu et al., 2013a, 2013b). Copyright 2013 RNA Society; (g) Specific cancer targeting of RNA/EVs (Pi et al., 2018). Copyright 2018 Springer Nature Publishing; and (h) RNA micelles (Shu et al., 2018). Copyright 2018 Elsevier Inc. (Figures adapted and reproduced with permission from cited references)

immunogenicity of traditional small RNAs has been widely investigated, there are only a limited number of studies focused on the immunogenicity of RNA nanoparticles. Recent studies revealed that pRNA-based RNA nanoparticles intrinsically display immunologically inert properties (Abdelmawla et al., 2011; Guo et al., 2017; Khisamutdinov et al., 2014a, 2014b; Shu et al., 2018). Specifically, no or negligible cytokine induction including TNF-α (Tumor Necrosis Factor-α), IL-6 (Interleukin-6) and IFN-α (Interferon-α) has been observed following treatment with pRNA nanoparticles *in vitro* and *in vivo*. TNF-α is a cytokine involved in systemic inflammation and the acute phase reaction. IL-6 is an interleukin secreted to stimulate immune responses during infection. IFN-α, which belongs to type I interferon, is also a cytokine involved in pro-inflammatory reactions released in response to the presence of viral pathogens. Besides, no stimulation of TLRs (Toll-like receptors) pathway, and no damage to normal tissue and organs was detected in multiple cell types and mice (Cui et al., 2015; Zhang et al., 2017). Similarly, RNA polygons (RNA triangle, square, and pentagon) constructed from the thermodynamically stable pRNA-3WJ were studied (Guo et al., 2017; Khisamutdinov et al., 2014a, 2014b). These RNA polygons have been considered nonimmunogenic and nontoxic because undetectable or negligible cytokine induction and cytotoxicity were observed *in vitro* and *in vivo*, compared to positive controls. Consistent results were found in three dimensional pRNA-based nanoparticles, including the RNA tetrahedron, RNA nanoprism, and RNA micelles (Guo et al., 2017; Khisamutdinov et al., 2016; Shu et al., 2018; Yin et al., 2018). Additionally, RNA aptamers, a family of RNA oligonucleotides commonly incorporated into RNA nanoparticles as targeted ligands to enhance binding specificity, have been reported to go unrecognized by the host immune system in various animal studies (Song, Lee, & Ban, 2012). These findings demonstrate that RNA nanoparticles equipped with targeting ligands can serve as safe delivery vectors in therapeutic interventions. Studies by Afonin Lab using a different system have also shown no immune response detection upon treatment with RNA nanoparticles in human peripheral blood mononuclear cells (PBMCs) from healthy donors (Hong et al., 2018). Interestingly, only complexation with a delivery carrier such as lipofectamine 2000 induced immunorecognition by PBMCs.

PHYSICOCHEMICAL PROPERTIES OF RNA NANOPARTICLES AFFECT *IN VIVO* BIODISTRIBUTION AND IMMUNE RESPONSE

Nanotechnology offers a substantial number of benefits over traditional routes for drug delivery, but unfavorable immune responses and liver accumulation have also been reported (Buzea, Pacheco, & Robbie, 2007; Zolnik, Gonzalez-Fernandez, Sadrieh, & Dobrovolskaia, 2010). It has been suggested that the adverse effects were elicited by numerous physicochemical characteristics, including size, shape, surface chemistry, or hydrophobicity (Dobrovolskaia, Shurin, & Shvedova, 2016; Dobrovolskaia, 2015). Engineering these properties with precision and homogeneity is a common strategy to improve the *in vivo* performance of nanomaterials. One of the advantages of RNA nanotechnology is its high programmability. In other words, their physicochemical properties are easily tunable (Figure 50.3), and the production process is highly consistent. Therefore, the effects of these properties can simply be studied as a result of reproducible nanoparticle assembly. The following subsections will focus on the main physicochemical properties of RNA nanoparticles and the corresponding effects on their immunostimulation and biodistribution.

NANOPARTICLE SIZE

Nanoparticle size represents one of the most critical considerations in the design and construction of RNA nanoparticles. It was found that size significantly dictates nanoparticle performance at the nano-bio interface, including vascular transportation, plasma protein binding, and cellular membrane interaction (Albanese et al., 2012; Hoshyar et al., 2016). Large particles (>100 nm) tend to be trapped in the liver and spleen as a result of the stronger recognition by the mononuclear phagocytic system (MPS) in these organs (Gustafson et al., 2015). Particles with a small diameter (<10 nm) are more likely to have a faster renal clearance, thus leading to a shorter half-life *in vivo* (Longmire, Choyke, & Kobayashi, 2008). In a systemic *in vivo* biodistribution study, the effect of RNA nanoparticle size on their circulation time and accumulation in healthy organs and tumors has been evaluated (Jasinski et al., 2018a). Specifically, RNA squares of three

A. Controlling Size

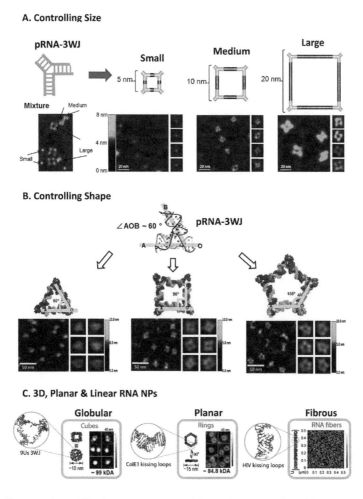

FIGURE 50.3 Construction of RNA nanostructures with tunable properties. (a) RNA squares with small, medium, and large size by tuning the length of the connecting helix (Jasinski et al., 2014). Copyright 2014 American Chemical Society. (b) RNA triangle, square, and pentagon by tuning the interior pRNA-3WJ angle (Khisamutdinov et al., 2014a, 2014b). Copyright 2014 American Chemical Society. (c) 3D RNA Cube, planar RNA nanoring, and linear RNA fiber by different connectivity (Hong et al., 2018). Copyright 2018 American Chemical Society. (Figures adapted and reproduced with permission from cited references)

different sizes (5 nm, 10 nm, 20 nm) were intravenously administered into tumor-bearing mice. Internal organ imaging at 12 and 24 h time points showed the rapid elimination of 5 nm RNA squares from vital organs with significant accumulation in the tumor after 12 h (Figure 50.4a). It was confirmed that renal excretion is the primary excretion route for the nanoparticles (Piao, Wang, Binzel, & Guo, 2018). For the 10 and 20 nm RNA squares, stronger interaction with macrophages and slower metabolism in the liver was observed, which is possibly caused by the different protein binding profiles of large nanoparticles compared to that of small ones. The correlation of size with the biodistribution profile should be considered from the perspective its effects on renal clearance and macrophage uptake. Particularly, smaller RNA nanoparticles exhibited less uptake by macrophages of the MPS due to less serum protein binding. However, they are more rapidly excreted by kidney filtration, while the opposite trend is seen with the larger RNA

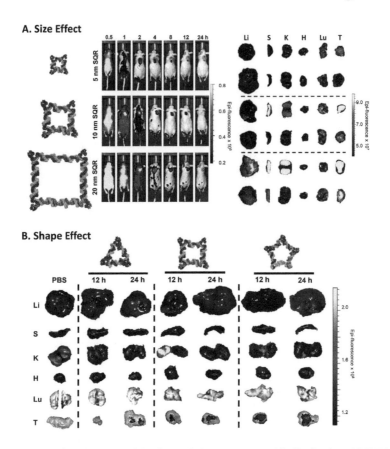

FIGURE 50.4 Effects of RNA nanoparticle size and shape on *in vivo* biodistribution. (a) RNA squares with identical shape but varying size, and (b) RNA polygons with identical size but varying shape show different circulation times and tumor accumulation *in vivo*. (Figures adapted and reproduced with permission from (Jasinski et al., 2018a). Copyright 2018 Elsevier Inc.)

nanoparticles. Thus, the final *in vivo* fate of RNA nanoparticles will take a balance between these two size-dependent elimination pathways (i.e., macrophages and urinary excretion).

The effect of varied sizes will manifest itself in the interaction between particles and immune system as well. Small sizes will benefit from not being recognized by the bulky opsonins in complement cascade, a key component of the immune system, due to the inadequate accommodation on particle surfaces (Ventola, 2012). RNA nanoparticles have been deliberately constructed in a size range from 10 to 40 nm, making them advantageous for drug delivery. Recently, the effect of size on their immunostimulation has been studied (Guo et al., 2017). RNA nanoparticles with identical square shape but varying size were used as the model (Figure 50.5a). After extending the single-stranded sequence at the vertexes, RNA squares were endowed with immunostimulatory activity in a size-dependent fashion. Small RNA square (7.10 nm) elevated the immunomodulation to some extent, while stronger responses were observed with medium (12.31 nm) and large (21.15 nm) RNA squares. This can likely be attributed to the engulfing behavior of phagocytes. Larger nanoparticle sizes show greater sensitivity to phagocytosis than the corresponding smaller ones. A similar finding was reported when using varying sizes of RNA polygons, in which RNA hexagons induced more IFNs than any smaller RNA polygons (Hong et al., 2018). These findings answered important questions regarding size in the rational design of RNA nanoparticles for favorable cancer accumulation and immune-interactions.

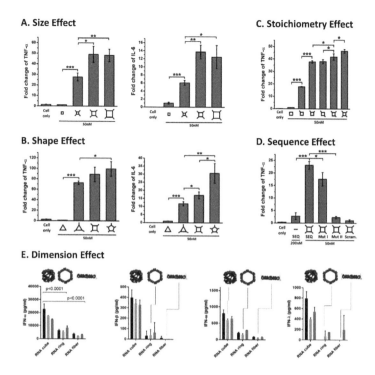

FIGURE 50.5 Effects of RNA nanoparticles' physicochemical properties on immunostimulation. RNA nanoparticles with (a) varying size, (b) varying shape, (c) different stoichiometry, (d) different sequence, and **(e)** different dimension induced cytokines and interferons secretion to various levels. (Figures a-d and f. Reprinted with permission from (Guo et al., 2017). Copyright 2018 Elsevier Inc. and (Hong et al., 2018). Copyright 2018 American Chemical Society, respectively)

NANOPARTICLE SHAPE

Shape is another critical design parameter in nanotechnology. However, detailed immunological and pharmacokinetic studies of shape effects are limited; most of these studies require complex construction procedures to produce nanoparticles of varying shapes while maintaining uniform size and composition (Toy et al., 2014). In contrast, RNA nanotechnology exceptionally enables the controlled self-assembly of nanostructures with custom size or shape independently, resulting in a true shape comparison. For example, different RNA polygons (triangle, square, and pentagon) were all constructed with identical size, benefiting from the inherent flexibility of the pRNA-3WJ scaffold (Guo et al., 2017; Khisamutdinov et al., 2014a, 2014b). Intriguingly, when immunostimulatory oligonucleotides were implemented on RNA polygons, the RNA pentagon appeared to be the most potent inducer of pro-inflammatory cytokines while the triangular counterparts remained the least among the polygonal structures studied (Figure 50.5b), implying the shape-dependent immunomodulation of RNA nanoparticles. Regarding RNA polygons, another interesting trend was observed: number of sides. The RNA triangle and pentagon having an odd number of sides tended to stimulate more IFN-β secretion than the even-sided RNA square and hexagon (Johnson et al., 2017). Additionally, a significant difference was found in immune responses between RNA nanoparticles with varying dimensional structures. A 3D RNA tetrahedron carrying immunostimulatory oligonucleotides exhibited stronger immunostimulatory activity than the planar triangular counterpart, when size and payload stoichiometry were controlled to be equivalent (Guo et al., 2017). Likewise, globular RNA cubes induced stronger immunostimulation compared to planar hexameric RNA rings, which were more immunostimulatory than RNA fibers (Figure 50.5e) (Hong et al., 2018).

These findings suggest a trend of increasing immunostimulatory properties of RNA nanoparticles from linear, to planar, and to 3D structure. One interpretation may be that the increased surface area intrinsic to 3D structures provides a spacious surface for complement opsonins assembly and deposition, while a greater proportion of opsonins released into the surrounding medium with linear and planar RNA structures.

The shape of nanoparticles has been shown to dictate the interactions that occur with cell membranes and circulating serum proteins. For instance, studies have suggested that oblate-shaped nanoparticles with discoidal geometries are more likely to migrate toward blood vessel walls and establish greater interactions with endothelial cells of blood vessels in comparison to spherical nanoparticles (Muller, Fedosov, & Gompper, 2014). The biodistribution profiles of RNA polygons of different shape but uniform size were compared in tumor-bearing mouse models after systemic administration (Jasinski et al., 2018a). Different retention in organs were observed at 12 h time point as nanosquares showed high fluorescent signal intensity while triangle nanoparticles showed none and the pentagon very little. In the spleen, pentagon nanoparticles exhibited the highest fluorescence. A similar biodistribution in organs was found among the particles after 24 h (Figure 50.4b). Therefore, the protein corona formation on the RNA nanoparticles may drastically change in response to nanoparticle shapes, which will further impact their elimination pathways. Additionally, the cellular interactions with nanoparticles as well as internalization are also closely related to the size and shape. Cell receptors that mediate the endocytosis are of various sizes and shapes, so it will provide beneficial information on rational design of nanoparticles that possess favorable binding to the receptors, thus enhancing nanoparticles recognition and cellular uptake. Considering the controllable size, shape and other physicochemical properties, RNA nanoparticles could potentially be designed with enhanced tumor cell uptake and retention.

SEQUENCE SIGNATURE AND MODULAR STOICHIOMETRY

As a biocompatible nanomaterial, RNA nanoparticles are immunologically inert. In contrast, some special RNA sequences have been reported to trigger immune responses, named isRNAs, due to the specific recognition by toll-like receptors (TLRs) or cytosolic sensors (PKR, RIG-1, and MDA-5) in immune cells (Berger et al., 2009; Bourquin et al., 2007; Heil et al., 2004; Hornung et al., 2005; Judge et al., 2005). Incorporation of these isRNA sequences can turn immunologically inert RNA nanoparticles to immunologically active, or even enhance the immune response associated with an incorporated module. In a study from Guo Lab, a specific RNA SEQ was extended to the vertexes of RNA squares and dramatically engendered the production of pro-inflammatory cytokines *in vitro* and *in vivo* (Figure 50.5d) (Guo et al., 2017). The immune responses were in direct proportion to the stoichiometry of single-stranded RNA extensions. RNA squares with increasing copies of payload induced stronger cytokine levels (Figure 50.5c). Conversely, mutation or complementary blockage of the extension sequence resulted in reduced immune responses, while scrambling the extension sequence led to complete abrogation of immune response. This study affords a new sight in the design and construction of RNA nanoparticles – they can be constructed to serve as safe therapeutic nanocarriers with non-immunogenicity, or deliberately trigger a strong immune response for immunotherapy.

SURFACE CHEMISTRY

Surface characteristics have been shown to be a significant parameter in the PK and PD of many nanomaterials, as well as influencing their immune system interactions (Albanese et al., 2012). Some cationic nanoparticles, such as polyethylenimines, can easily interact with cell membranes, causing non-specific cytotoxicity and giving rise to complement system activation (Merkel et al., 2011). In contrast, RNA nanoparticles have consistently exhibited the advantage of causing no or undetectable cytotoxicity in many studies due to their polyanionic nature (Shu et al., 2014).

Surface-projected composition is another factor that greatly defines the *in vivo* fate of nanoparticles (Merkel et al., 2011). As a naturally aqueous-soluble biopolymer, RNA nanoparticles are distinct from many synthetic nanomaterials in that they do not require surface modifications, such as polyethylene glycol (PEG) grafting (Suk et al., 2016), to increase aqueous solubility and consequently limit their immune response and increase *in vivo* circulation. This advantage provides RNA nanoparticles the flexibility to incorporate various functional modules, including RNAi therapeutics, chemical drugs, fluorophores, and targeting ligands to achieve multifunctionality. Particularly, some of these surface compositions, especially drugs, fluorophores, or other hydrophobic compounds might be important factors in determining how RNA nanoparticles communicate with cells or proteins *in vivo*. Nanoparticles decorated with more hydrophobic reagents will result in greater plasma protein binding, and therefore greater accumulation in the liver or other organs. Jasinski et al. reported that different chemicals incorporated to RNA nanoparticles altered the RNA hydrophobicity to varying degrees (Figure 50.6a) (Jasinski et al., 2018b). The changes in vital organ accumulation as a function of hydrophobicity variation were investigated. Weaker organ accumulation was detected for RNA nanoparticles (3WJ-Fluor) containing hydrophobic fluorophores (Cyanine5.5, Sulfonated-Cyanine 5.5, and AlexaFluor700) than these fluorophores alone (Figure 50.6a-c), clearly indicating the capacity of RNA nanoparticles to solubilize hydrophobic compounds. In another study, paclitaxel (PTX), an antitumor chemo-drug with poor aqueous solubility, was conjugated to micellar RNA nanoparticles (Figure 50.6b) (Shu et al., 2018). Consequently, the RNA micelle/PTX complex showed significantly enhanced water solubility and efficient cancer targeting *in vivo*. Additionally, RNA nanoparticles conjugated with a cholesterol molecule were used to control the ligand-displaying on extracellular vesicle (EV) membranes. As a result of orientational control enabled by RNA nanoparticles, EVs decorated with multifunctional RNA nanoparticles were able to provide specific delivery of siRNA (Figure 50.6c) (Pi et al., 2018; Li et al., 2018). As shown here, the biological implications of RNA nanoparticles following systemic administration are closely related to their chemical modifications. While changes in the protein binding activity of RNA nanoparticles might be difficult to predict, understanding the effects of hydrophobic drugs conjugation on its solubility, and in turn its *in vivo* biodistribution, can help researchers develop RNA-based therapeutics for future clinical settings.

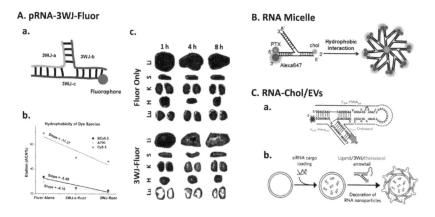

FIGURE 50.6 Construction of RNA nanoparticles with various surface characteristics. (a) pRNA-3WJ nanoparticles conjugated with hydrophobic fluorophores and their effects on *in vivo* biodistribution (Jasinski et al., 2018b). Copyright 2018, Mary Ann Liebert, Inc. (b) RNA micelles assembled from pRNA-3WJ conjugated with cholesterol, paclitaxel and fluorophore (Shu et al., 2018). 2018 Elsevier Inc. (c) Ligand-displaying EVs by pRNA-3WJ conjugated with cholesterol, ligand aptamer, and fluorophore (Pi et al., 2018). Copyright 2018 Springer Nature Publishing. (Figures adapted and reproduced with permission from cited references)

OTHER FACTORS

In addition to the common physicochemical properties discussed above, other parameters can also make their own unique contributions to RNA nanoparticle's immunostimulation or biodistribution profiles. Recent studies have shown that some chemical modification not only improves the enzymatic stability of RNA but also plays vital roles in RNA immunostimulation (Ge et al., 2010; Peacock et al., 2011). For example, 2'-fluorine(2'-F)- modified siRNA displayed immune-inert behavior in human peripheral blood mononuclear cells, whereas the unmodified siRNA induced the production of TNF-α and IFN-α (Lee et al., 2016). It is suggested that this property can be consistently applied to RNA nanoparticles when siRNA is incorporated, because most RNA nanoparticles used in cancer targeting or drug delivery were 2'-F modified. Besides, 2'-O-methyl (2'-O-Me) modification has been found to prevent the recognition of siRNAs by TLR7/8 and RIG-I receptor, thus reducing induction of TNF-α and IFN-β in human fibroblast MRC-5 cells (Ge et al., 2010). In addition, it has been reported that unmodified galactosidase (GAL) siRNA transiently induced the expression of TNF-α, IL-6, IL-10, IFN-β, and IFN-sensitive gene *in vivo*, whereas a formulation of 2'-O-Me-Luciferase (LUC) siRNA had no such effects (Broering et al., 2014). Thus, 2'-O-Me is potentially another facile approach capable of fine-tuning the properties of RNA to limit or enhance immune and inflammatory responses, depending on the therapeutic objective.

Interestingly, the intra- and intermolecular connectivity appeared to be another factor that affects the immunostimulation of RNA nanoparticles. Afonin Lab reported that RNA rings, assembled from pre-folded monomers via intermolecular interaction (kissing loops), exhibited less immunostimulatory activity than RNA cubes formed via intramolecular hydrogen bonds (Hong et al., 2018). This result suggests that RNA connectivity may play a role, but this mechanism needs further investigation.

Additionally, it has been reported that nanoparticle elasticity influences vascular transport, biodistribution, cellular internalization, and immunostimulation (Anselmo et al., 2015). RNA as a biopolymer has shown rubber-like elastic property (Chiu et al., 2014; Jacobson, McIntosh, Stevens, Rubinstein, & Saleh, 2017), allowing RNA nanoparticles to "squeeze" through vasculatures of the tumor microenvironment by blood pressure without altering its thermodynamic stability. Meanwhile, many RNA nanoparticles were constructed to be ratchet-shaped after incorporating multiple modules (Shu et al., 2011; Haque et al., 2012), preventing them from returning to blood circulation. Therefore, the effects of these properties favor the transport of RNA nanoparticles toward tumors and enhance the EPR effect.

PERSPECTIVES

RNA NANOTECHNOLOGY FOR POTENTIAL IMMUNOTHERAPY

One of the most important recent breakthroughs in cancer research is cancer immunotherapy (McNutt, 2013; Couzin-Frankel, 2013). Extensive studies revealed important information regarding the complicated cancer–immune system relationship, and many researchers are now looking to summon the self-defense system in the hosts to kill cancer. The most common immunotherapies include chimeric antigen receptors (CAR) T-cell therapy (June, O'Connor, Kawalekar, Ghassemi, & Milone, 2018), immune checkpoint blockade (Pardoll, 2012), monoclonal antibodies (mAb) (Weiner, Dhodapkar, & Ferrone, 2009), and cancer vaccines and adjuvants (Temizoz, Kuroda, & Ishii, 2016), just to name a few. Particularly, nucleic acid aptamers have emerged as a new type of therapeutic for immunotherapy (Pastor et al., 2018). Aptamers are short nucleic acid oligomers (12–80nt) selected from SELEX (systematic evolution of ligands by exponential enrichment) and are capable of binding targets specifically and tightly. Various kinds of co-stimulatory molecules belonging to the B7/CD28 family have been selected to trigger cell-mediated immune responses. For instance, CTLA-4 RNA aptamer developed by Santilli-Marotto et al. can bind CTLA-4 with high affinity, inhibit CTLA-4 function, and enhance tumor immunity in mice (Santulli-Marotto, Nair, Rusconi, Sullenger, & Gilboa, 2003). Furthermore, aptamers targeting the TNF/TNFR family which are involved in the later phase of T-cells activation have been developed, including OX40 and 4-1BB aptamers (McNamara et al., 2008). In order to take advantage of the RNA nanoparticle platforms, nucleic acid aptamers can be incorporated into the RNA scaffold to achieve stronger immunotherapeutic effects. Meanwhile, the multivalent property allows the additional incorporation of RNAi therapeutics or chemotherapeutic drugs to realize combination therapy.

Although immunotherapy has shown success in various cancers, some clinical challenges remain, such as the safety and efficacy concerns derived from the systemic dosing of immunomodulatory agents (Whiteside, Demaria, Rodriguez-Ruiz, Zarour, & Melero, 2016). RNA nanotechnology, as a safe and efficient drug delivery platform, can potentially enhance the efficacy as well as reduce the side effects of such immunotherapies by improving the delivery, retention, and release of immuno-modulatory agents in targeted cell populations and organs. RNA, as a biomacromolecule, can be successfully recognized by the immune system as a self-entity, and thus RNA nanoparticles intrinsically display immunologically inert property. As described above, though naturally inert, RNA nanoparticles can be manually designed using their tunable and programmable properties to exhibit no, low, or high immunostimulation, thus allowing them to be employed as safe therapeutic carriers without triggering an immune response or as potential immunomodulators for cancer immunotherapy.

Understanding the Interactions of RNA Nanoparticles at the Nano-Bio Interface

Upon introduction into biological environment, nanoparticles interacting with proteins, membranes, cells, and organs establish a series of nanoparticle/biological interactions at the interfaces which govern the *in vivo* fate of nanoparticles (Cheng, Jiang, Wang, Chen, & Liu, 2013; Nel et al., 2009). As little is known about the interactions of RNA nanoparticles with biological components, a better understanding at the nano-bio interface will be essential to the rational design of RNA nanoparticles capable of targeted delivery (Xu et al., 2018a). At the molecular level, protein corona formation around nanoparticles drastically influences their *in vivo* behavior, resulting in different elimination pathways (Kim, Faix, & Schnitzer, 2017; Tenzer et al., 2013). Moreover, the component in the protein corona, such as the complement proteins, can potentially lead to altered immune responses (Chen et al., 2017). Previous studies on serum protein binding of RNA polygon nanoparticles showed that the size and shape play critical roles in protein binding (Jasinski et al., 2018a). However, the composition of protein corona formation on RNA nanoparticles has not been comprehensively determined. At the cellular lever, it appears the frequency of RNA nanoparticles uptake into macro-phages is closely related to their morphology and size, as these factors impact the engulfing process (Guo et al., 2017). The internalization of RNA nanoparticles into cancer cells with targeting ligands is proposed to be receptor-mediated pathway (Shu et al., 2014). However, the impacts of physico-chemical properties on the intracellular trafficking of RNA nanoparticles are still not completely understood and await more investigation.

CONCLUSION

RNA nanotechnology is growing exponentially, though its inception lags behind other nano-delivery systems. As shown within this review, RNA nanoparticles display many advantages in biomedicine. As drug carriers, RNA nanoparticles have repeatedly shown immunologically inert behavior, while can be concomitantly manipulated to exhibit controlled immunostimulation. RNA nanoparticles possess a variety of advantageous physicochemical properties over other nanomaterials, including the capacity to be precisely programmed. As a result, RNA nanoparticles can be rationally designed, optimized, and constructed for specialized *in vivo* applications. As a biocompatible nanomaterial, RNA nanoparticles show favorable tumor targeting proficiency, as evidenced in various pre-clinical cancer models. The extensive research conducted in order to understand the safety, immunological, and pharmacological profiles of RNA nanoparticles have positively paved a path toward clinical trials. Evidently, RNA nanotechnology is bespeaking a bright future in cancer therapy.

ACKNOWLEDGMENTS

The research in P.G.'s lab was supported by NIH grants R01EB019036 and U01CA207946. The authors would like to thank Lora E. McBride for her constructive comments and revisions on the article. P.G.'s Sylvan G. Frank Endowed Chair position in Pharmaceutics and Drug Delivery is funded by the CM Chen Foundation.

CONFLICT OF INTEREST

P.G. is the consultant of Oxford Nanopore Technologies and Nanobio Delivery Pharmaceutical Co. Ltd, as well as the cofounder of Shenzhen P&Z Bio-medical Co. Ltd and its subsidiary US P&Z Biological Technology LLC, as well as ExonanoRNA, LLC and its subsidiary ExonanoRNA (Foshan) Biomedicine Co., Ltd.

NOTE

1 This chapter is adapted from the full published article with permission from John Wiley and Sons © 2019. Original citation: Guo, S., Xu, C., Yin, H., Hill, J., Pi, F., & Guo, P. (2019). Tuning the size, shape and structure of RNA nanoparticles for favorable cancer targeting and immunostimulation. *Wiley Interdisciplinary Reviews: Nanomedicine and Nanobiotechnology*, e1582.

REFERENCES

Abdelmawla, S., Guo, S., Zhang, L., Pulukuri, S. M., Patankar, P., Conley, P. et al. (2011). Pharmacological characterization of chemically synthesized monomeric phi29 pRNA nanoparticles for systemic delivery. *Mol. Ther.*, *19*, 1312–1322.

Afonin, K. A., Bindewald, E., Yaghoubian, A. J., Voss, N., Jacovetty, E., Shapiro, B. A. et al. (2010). In vitro assembly of cubic RNA-based scaffolds designed in silico. *Nat. Nanotechnol.*, *5*, 676–682.

Afonin, K. A., Cieply, D. J., & Leontis, N. B. (2008). Specific RNA self-assembly with minimal paranemic motifs. *J. Am. Chem. Soc.*, *130*, 93–102.

Afonin, K. A., Viard, M., Koyfman, A. Y., Martins, A. N., Kasprzak, W. K., Panigaj, M. et al. (2014a). Multifunctional RNA nanoparticles. *Nano Lett.*, *14*, 5662–5671.

Afonin, K. A., Viard, M., Kagiampakis, I., Case, C. L., Dobrovolskaia, M. A., Hofmann, J. et al. (2014b). Triggering of RNA interference with RNA-RNA, RNA-DNA, and DNA-RNA nanoparticles. *ACS Nano*, *9*, 251–259.

Albanese, A., Tang, P. S., & Chan, W. C. (2012). The effect of nanoparticle size, shape, and surface chemistry on biological systems. *Annu. Rev. Biomed. Eng.*, *14*, 1–16.

Andersen, E. S., Dong, M., Nielsen, M. M., Jahn, K., Subramani, R., Mamdouh, W. et al. (2009). Self-assembly of a nanoscale DNA box with a controllable lid. *Nature*, *459*, 73–76.

Anselmo, A. C., Zhang, M., Kumar, S., Vogus, D. R., Menegatti, S., Helgeson, M. E. et al. (2015). Elasticity of nanoparticles influences their blood circulation, phagocytosis, endocytosis, and targeting. *ACS Nano*, *9*, 3169–3177.

Astruc, D. (2012). Electron-transfer processes in dendrimers and their implication in biology, catalysis, sensing and nanotechnology. *Nat. Chem.*, *4*, 255–267.

Bartel, D. P. (2004). MicroRNAs: Genomics, biogenesis, mechanism, and function. *Cell*, *116*, 281–297.

Berger, M., Ablasser, A., Kim, S., Bekeredjian-Ding, I., Giese, T., Endres, S. et al. (2009). TLR8-driven IL-12-dependent reciprocal and synergistic activation of NK cells and monocytes by immunostimulatory RNA. *J. Immunother.*, *32*, 262–271.

Bindewald, E., Afonin, K., Jaeger, L., & Shapiro, B. A. (2011). Multistrand RNA secondary structure prediction and nanostructure design including pseudoknots. *ACS Nano*, *5*, 9542–9551.

Bindewald, E., Hayes, R., Yingling, Y. G., Kasprzak, W., & Shapiro, B. A. (2008). RNAJunction: A database of RNA junctions and kissing loops for three-dimensional structural analysis and nanodesign. *Nucleic Acids Res.*, *36*, D392–D397.

Binzel, D., Shu, Y., Li, H., Sun, M., Zhang, Q., Shu, D. et al. (2016). Specific delivery of MiRNA for high efficient inhibition of prostate cancer by RNA nanotechnology. *Mol. Ther.*, *24*, 1267–1277.

Boerneke, M. A., Dibrov, S. M., & Hermann, T. (2016). Crystal-structure-guided design of self-assembling RNA nanotriangles. *Angew. Chem Int. Ed Engl.*, *55*, 4097–4100.

Bourquin, C., Schmidt, L., Hornung, V., Wurzenberger, C., Anz, D., Sandholzer, N. et al. (2007). Immunostimulatory RNA oligonucleotides trigger an antigen-specific cytotoxic T-cell and IgG2a response. *Blood*, *109*, 2953–2960.

Broering, R., Real, C. I., John, M. J., Jahn-Hofmann, K., Ickenstein, L. M., Kleinehr, K. et al. (2014). Chemical modifications on siRNAs avoid Toll-like-receptor-mediated activation of the hepatic immune system in vivo and in vitro. *Int. Immunol.*, *26*, 35–46.

Bui, M. N., Brittany, J. M., Viard, M., Satterwhite, E., Martins, A. N., Li, Z. et al. (2017). Versatile RNA tetra-U helix linking motif as a toolkit for nucleic acid nanotechnology. *Nanomed.*, *13*, 1137–1146.

Buzea, C., Pacheco, I. I., & Robbie, K. (2007). Nanomaterials and nanoparticles: Sources and toxicity. *Biointerphases*, *2*, MR17–MR71.

Chen, F., Wang, G., Griffin, J. I., Brenneman, B., Banda, N. K., Holers, V. M. et al. (2017). Complement proteins bind to nanoparticle protein corona and undergo dynamic exchange in vivo. *Nat. Nanotechnol.*, *12*, 387–393.

Cheng, L. C., Jiang, X., Wang, J., Chen, C., & Liu, R. S. (2013). Nano-bio effects: Interaction of nanomaterials with cells. *Nanoscale*, *5*, 3547–3569.

Chiu, H. C., Koh, K., Evich, M., Lesiak, A., Germann, M. W., Bongiorno, A. et al. (2014). RNA intrusions change DNA elastic properties and structure. *Nanoscale*, *6*, 10009–10017.

Chworos, A., Severcan, I., Koyfman, A. Y., Weinkam, P., Oroudjev, E., Hansma, H. G. et al. (2004). Building programmable jigsaw puzzles with RNA. *Science*, *306*, 2068–2072.

Corey, D. R. (2007). Chemical modification: The key to clinical application of RNA interference? *J. Clin. Invest*, *117*, 3615–3622.

Couzin-Frankel, J. (2013). Breakthrough of the year 2013. Cancer immunotherapy. *Science*, *342*, 1432–1433.

Cui, D., Zhang, C., Liu, B., Shu, Y., Du, T., Shu, D. et al. (2015). Regression of gastric cancer by systemic injection of RNA nanoparticles carrying both ligand and siRNA. *Sci. Rep.*, *5*, 10726.

Desai, N. (2012). Challenges in development of nanoparticle-based therapeutics. *AAPS J.*, *14*, 282–295.

Dibrov, S. M., McLean, J., Parsons, J., & Hermann, T. (2011). Self-assembling RNA square. *Proc. Natl. Acad. Sci. U.S.A.*, *108*, 6405–6408.

Dobrovolskaia, M. A. (2015). Pre-clinical immunotoxicity studies of nanotechnology-formulated drugs: Challenges, considerations and strategy. *J. Control. Release*, *220*, 571–583.

Dobrovolskaia, M. A., Shurin, M., & Shvedova, A. A. (2016). Current understanding of interactions between nanoparticles and the immune system. *Toxicol. Appl. Pharmacol.*, *299*, 78–89.

Elbashir, S. M., Harborth, J., Lendeckel, W., Yalcin, A., Weber, K., & Tuschl, T. (2001). Duplexes of 21-nucleotide RNAs mediate RNA interference in cultured mammalian cells. *Nature*, *411*, 494–498.

Elmen, J., Lindow, M., Schutz, S., Lawrence, M., Petri, A., Obad, S. et al. (2008). LNA-mediated microRNA silencing in non-human primates. *Nature*, *452*, 896–899.

Ge, Q., Dallas, A., Ilves, H., Shorenstein, J., Behlke, M. A., & Johnston, B. H. (2010). Effects of chemical modification on the potency, serum stability, and immunostimulatory properties of short shRNAs. *RNA*, *16*, 118–130.

Geary, C., Chworos, A., Verzemnieks, E., Voss, N. R., & Jaeger, L. (2017). Composing RNA Nanostructures from a Syntax of RNA Structural Modules. *Nano Lett.*, *17*, 7095–7101.

Geng, Y., Dalhaimer, P., Cai, S., Tsai, R., Tewari, M., Minko, T. et al. (2007). Shape effects of filaments versus spherical particles in flow and drug delivery. *Nat. Nanotechnol.*, *2*, 249–255.

Grabow, W. W., Zakrevsky, P., Afonin, K. A., Chworos, A., Shapiro, B. A., & Jaeger, L. (2011). Self-assembling RNA nanorings based on RNAI/II inverse kissing complexes. *Nano Lett.*, *11*, 878–887.

Guo, P. (2010). The emerging field of RNA nanotechnology. *Nat. Nanotechnol.*, *5*, 833–842.

Guo, P., Zhang, C., Chen, C., Trottier, M., & Garver, K. (1998). Inter-RNA interaction of phage phi29 pRNA to form a hexameric complex for viral DNA transportation. *Mol. Cell*, *2*, 149–155.

Guo, S., Tschammer, N., Mohammed, S., & Guo, P. (2005). Specific delivery of therapeutic RNAs to cancer cells via the dimerization mechanism of phi29 motor pRNA. *Hum. Gene. Ther.*, *16*, 1097–1109.

Guo, S., Li, H., Ma, M., Fu, J., Dong, Y., & Guo, P. (2017). Size, shape, and sequence-dependent immunogenicity of RNA nanoparticles. *Mol. Ther. Nucleic Acids*, *9*, 399–408.

Gustafson, H. H., Holt-Casper, D., Grainger, D. W., & Ghandehari, H. (2015). Nanoparticle uptake: The phagocyte problem. *Nano Today*, *10*, 487–510.

Halman, J. R., Satterwhite, E., Roark, B., Chandler, M., Viard, M., Ivanina, A. et al. (2017). Functionally-interdependent shape-switching nanoparticles with controllable properties. *Nucleic Acids Res.*, *45*, 2210–2220.

Hao, C., Li, X., Tian, C., Jiang, W., Wang, G., & Mao, C. (2014). Construction of RNA nanocages by re-engineering the packaging RNA of Phi29 bacteriophage. *Nat. Commun.*, *5*, 3890.

Haque, F., Shu, D., Shu, Y., Shlyakhtenko, L., Rychahou, P., Evers, M. et al. (2012). Ultrastable synergistic tetravalent RNA nanoparticles for targeting to cancers. *Nano Today*, *7*, 245–257.

Heil, F., Hemmi, H., Hochrein, H., Ampenberger, F., Kirschning, C., Akira, S. et al. (2004). Species-specific recognition of single-stranded RNA via toll-like receptor 7 and 8. *Science*, *303*, 1526–1529.

Hendrix, D. K., Brenner, S. E., & Holbrook, S. R. (2005). RNA structural motifs: Building blocks of a modular biomolecule. *Q. Rev. Biophys*, *38*, 221–243.

Hong, E., Halman, J. R., Shah, A. B., Khisamutdinov, E. F., Dobrovolskaia, M. A., & Afonin, K. A. (2018). Structure and composition define immunorecognition of nucleic acid nanoparticles. *Nano Lett.*, *18*, 4309–4321.

Hornung, V., Guenthner-Biller, M., Bourquin, C., Ablasser, A., Schlee, M., Uematsu, S. et al. (2005). Sequence-specific potent induction of IFN-[alpha] by short interfering RNA in plasmacytoid dendritic cells through TLR7. *Nat. Med.*, *11*, 263–270.

Hoshyar, N., Gray, S., Han, H., & Bao, G. (2016). The effect of nanoparticle size on in vivo pharmacokinetics and cellular interaction. *Nanomedicine (London)*, *11*, 673–692.

Huang, L. & Lilley, D. M. (2013). The molecular recognition of kink-turn structure by the L7Ae class of proteins. *RNA*, *19*, 1703–1710.

Huang, L. & Lilley, D. M. (2016). A quasi-cyclic RNA nano-scale molecular object constructed using kink turns. *Nanoscale*, *8*, 15189–15195.

Jacobson, D. R., McIntosh, D. B., Stevens, M. J., Rubinstein, M., & Saleh, O. A. (2017). Single-stranded nucleic acid elasticity arises from internal electrostatic tension. *Proc. Natl. Acad. Sci. U.S A*, *114*, 5095–5100.

Jaeger, L., Westhof, E., & Leontis, N. B. (2001). TectoRNA: Modular assembly units for the construction of RNA nano-objects. *Nucleic Acids Res.*, *29*, 455–463.

Jasinski, D., Haque, F., Binzel, D. W., & Guo, P. (2017). Advancement of the emerging field of RNA nanotechnology. *ACS Nano*, *11*, 1142–1164.

Jasinski, D., Khisamutdinov, E. F., Lyubchenko, Y. L., & Guo, P. (2014). Physicochemically tunable poly-functionalized RNA square architecture with fluorogenic and ribozymatic properties. *ACS Nano*, *8*, 7620–7629.

Jasinski, D. L., Li, H., & Guo, P. (2018a). The effect of size and shape of RNA nanoparticles on biodistribution. *Mol. Ther.*, *26*, 784–792.

Jasinski, D. L., Yin, H., Li, Z., & Guo, P. (2018b). Hydrophobic effect from conjugated chemicals or drugs on in Vivo biodistribution of RNA nanoparticles. *Hum. Gene Ther.*, *29*, 77–86.

Johnson, M. B., Halman, J. R., Satterwhite, E., Zakharov, A. V., Bui, M. N., Benkato, K. et al. (2017). Programmable nucleic acid based polygons with controlled neuroimmunomodulatory properties for predictive QSAR modeling. *Small*, *13*.

Judge, A. D., Sood, V., Shaw, J. R., Fang, D., McClintock, K., & MacLachlan, I. (2005). Sequence-dependent stimulation of the mammalian innate immune response by synthetic siRNA. *Nat. Biotechnol.*, *23*, 457–462.

June, C. H., O'Connor, R. S., Kawalekar, O. U., Ghassemi, S., & Milone, M. C. (2018). CAR T cell immunotherapy for human cancer. *Science*, *359*, 1361–1365.

Ke, W., Hong, E., Saito, R. F., Rangel, M. C., Wang, J., Viard, M. et al. (2018). RNA-DNA fibers and polygons with controlled immunorecognition activate RNAi, FRET and transcriptional regulation of NF-kappaB in human cells. *Nucleic Acids Res, In Press*.

Khisamutdinov, E. F., Jasinski, D. L., & Guo, P. (2014a). RNA as a boiling-resistant anionic polymer material to build robust structures with defined shape and stoichiometry. *ACS Nano*, *8*, 4771–4781.

Khisamutdinov, E. F., Jasinski, D. L., Li, H., Zhang, K., Chiu, W., & Guo, P. (2016). Fabrication of RNA 3D nanoprism for loading and protection of small RNAs and model drugs. *Adv. Mater.*, *28*, 100079–100087.

Khisamutdinov, E., Li, H., Jasinski, D., Chen, J., Fu, J., & Guo, P. (2014b). Enhancing immunomodulation on innate immunity by shape transition among RNA triangle, square, and pentagon nanovehicles. *Nucleic Acids Res.*, *42*, 9996–10004.

Kim, S. M., Faix, P. H., & Schnitzer, J. E. (2017). Overcoming key biological barriers to cancer drug delivery and efficacy. *J. Control. Release*, *267*, 15–30.

Laing, C., Jung, S., Iqbal, A., & Schlick, T. (2009). Tertiary motifs revealed in analyses of higher-order RNA junctions. *J. Mol. Biol.*, *393*, 67–82.

Lee, T. J., Haque, F., Shu, D., Yoo, J. Y., Li, H., Yokel, R. A. et al. (2015). RNA nanoparticles as a vector for targeted siRNA delivery into glioblastoma mouse model. *Oncotarget*, *6*, 14766–14776.

Lee, Y., Urban, J. H., Xu, L., Sullenger, B. A., & Lee, J. (2016). 2'Fluoro modification differentially modulates the Ability of RNAs to Activate Pattern Recognition Receptors. *Nucleic Acid Ther.*, *26*, 173–182.

Li, H., Lee, T., Dziubla, T., Pi, F., Guo, S., Xu, J. et al. (2015). RNA as a stable polymer to build controllable and defined nanostructures for material and biomedical applications. *Nano Today*, *10*, 631–655.

Li, H., Zhang, K., Pi, F., Guo, S., Shlyakhtenko, L., Chiu, W. et al. (2016). Controllable self-assembly of RNA tetrahedrons with precise shape and size for cancer targeting. *Adv. Mater.*, *28*, 7501–7507.

Li, Z., Wang, H., Yin, H., Bennett, C., Zhang, H. G., & Guo, P. (2018). Arrowtail RNA for ligand display on ginger exosome-like nanovesicles to systemic deliver siRNA for cancer suppression. *Sci. Rep.*, *8*, 14644.

Longmire, M., Choyke, P. L., & Kobayashi, H. (2008). Clearance properties of nano-sized particles and molecules as imaging agents: Considerations and caveats. *Nanomedicine (London)*, *3*, 703–717.

McNamara, J. O., Kolonias, D., Pastor, F., Mittler, R. S., Chen, L., Giangrande, P. H. et al. (2008). Multivalent 4-1BB binding aptamers costimulate CD8+ T cells and inhibit tumor growth in mice. *J. Clin. Investig.*, *118*, 376–386.

McNutt, M. (2013). Cancer immunotherapy. *Science*, *342*, 1417.

Merkel, O. M., Urbanics, R., Bedocs, P., Rozsnyay, Z., Rosivall, L., Toth, M. et al. (2011). In vitro and in vivo complement activation and related anaphylactic effects associated with polyethylenimine and polyethylenimine-graft-poly(ethylene glycol) block copolymers. *Biomaterials*, *32*, 4936–4942.

Mogensen, T. H. (2009). Pathogen recognition and inflammatory signaling in innate immune defenses. *Clin. Microbiol.Rev.*, *22*, 240–273.

Monferrer, A., Zhang, D., Lushnikov, A. J., & Hermann, T. (2019). Versatile kit of robust nanoshapes self-assembling from RNA and DNA modules. *Nat. Commun.*, *10*, 608.

Muller, K., Fedosov, D. A., & Gompper, G. (2014). Margination of micro- and nano-particles in blood flow and its effect on drug delivery. *Sci. Rep.*, *4*, 4871.

Nasalean, L., Baudrey, S., Leontis, N. B., & Jaeger, L. (2006). Controlling RNA self-assembly to form filaments. *Nucleic Acids Res.*, *34*, 1381–1392.

Nel, A. E., Madler, L., Velegol, D., Xia, T., Hoek, E. M., Somasundaran, P. et al. (2009). Understanding biophysicochemical interactions at the nano-bio interface. *Nat. Mater.*, *8*, 543–557.

Ohno, H., Kobayashi, T., Kabata, R., Endo, K., Iwasa, T., Yoshimura, S. H. et al. (2011). Synthetic RNA-protein complex shaped like an equilateral triangle. *Nat.Nanotechnol.*, *6*, 116–120.

Pardoll, D. M. (2012). The blockade of immune checkpoints in cancer immunotherapy. *Nat. Rev. Cancer*, *12*, 252–264.

Pastor, F., Berraondo, P., Etxeberria, I., Frederick, J., Sahin, U., Gilboa, E. et al. (2018). An RNA toolbox for cancer immunotherapy. *Nat Rev. Drug Discov.*, *17*, 751–767.

Peacock, H., Fucini, R. V., Jayalath, P., Ibarra-Soza, J. M., Haringsma, H. J., Flanagan, W. M. et al. (2011). Nucleobase and ribose modifications control immunostimulation by a microRNA-122-mimetic RNA. *J Am. Chem Soc*, *133*, 9200–9203.

Pi, F., Binzel, D. W., Lee, T. J., Li, Z., Sun, M., Rychahou, P. et al. (2018). Nanoparticle orientation to control RNA loading and ligand display on extracellular vesicles for cancer regression. *Nat. Nanotechnol.*, *13*, 82–89.

Piao, X., Wang, H., Binzel, D. W., & Guo, P. (2018). Assessment and comparison of thermal stability of phosphorothioate-DNA, DNA, RNA, 2'-F RNA, and LNA in the context of Phi29 pRNA 3WJ. *RNA*, *24*, 67–76.

Puri, A., Loomis, K., Smith, B., Lee, J. H., Yavlovich, A., Heldman, E. et al. (2009). Lipid-based nanoparticles as pharmaceutical drug carriers: From concepts to clinic. *Crit Rev. Ther. Drug Carrier Syst.*, *26*, 523–580.

Rychahou, P., Haque, F., Shu, Y., Zaytseva, Y., Weiss, H. L., Lee, E. Y. et al. (2015). Delivery of RNA nanoparticles into colorectal cancer metastases following systemic administration. *ACS Nano*, *9*, 1108–1116.

Sahin, U., Kariko, K., & Tureci, O. (2014). mRNA-based therapeutics--developing a new class of drugs. *Nat Rev. Drug Discov.*, *13*, 759–780.

Santulli-Marotto, S., Nair, S. K., Rusconi, C., Sullenger, B., & Gilboa, E. (2003). Multivalent RNA aptamers that inhibit CTLA-4 and enhance tumor immunity. *Cancer Res.*, *63*, 7483–7489.

Sarver, N. A., Cantin, E. M., Chang, P. S., Zaia, J. A., Ladne, P. A., Stephens, D. A. et al. (1990). Ribozymes as potential Anti-HIV-1 therapeutic agents. *Science*, *24*, 1222–1225.

Severcan I, Geary C, Verzemnieks E, Chworos A, & Jaeger L (2009). Square-shaped RNA particles from different RNA folds. *Nano Lett.*, *9*, 1270–1277.

Severcan, I., Geary, C., Chworos, A., Voss, N., Jacovetty, E., & Jaeger, L. (2010). A polyhedron made of tRNAs. *Nat. Chem.*, *2*, 772–779.

Sharma, A., Haque, F., Pi, F., Shlyakhtenko, L., Evers, B. M., & Guo, P. (2015). Controllable Self-assembly of RNA dendrimers. *Nanomed.-Nanotechnol. Biol. Med.*, *12*, 835–844.

Shi, J., Kantoff, P. W., Wooster, R., & Farokhzad, O. C. (2017). Cancer nanomedicine: Progress, challenges and opportunities. *Nat Rev.Cancer*, *17*, 20–37.

Shi, Z., Li, S. K., Charoenputtakun, P., Liu, C. Y., Jasinski, D., & Guo, P. (2018). RNA nanoparticle distribution and clearance in the eye after subconjunctival injection with and without thermosensitive hydrogels. *J. Control. Release*, *270*, 14–22.

Shu, D., Shu, Y., Haque, F., Abdelmawla, S., & Guo, P. (2011). Thermodynamically stable RNA three-way junctions for constructing multifuntional nanoparticles for delivery of therapeutics. *Nat. Nanotechnol.*, *6*, 658–667.

Shu, D., Khisamutdinov, E., Zhang, L., & Guo, P. (2013b). Programmable folding of fusion RNA complex driven by the 3WJ motif of phi29 motor pRNA. *Nucleic Acids Res.*, *42*, e10.

Shu, D., Li, H., Shu, Y., Xiong, G., Carson, W. E., Haque, F. et al. (2015). Systemic delivery of anti-miRNA for suppression of triple negative breast cancer utilizing RNA nanotechnology. *ACS Nano*, *9*, 9731–9740.

Shu, Y., Pi, F., Sharma, A., Rajabi, M., Haque, F., Shu, D. et al. (2014). Stable RNA nanoparticles as potential new generation drugs for cancer therapy. *Adv. Drug Deliv. Rev.*, *66C*, 74–89.

Shu, Y., Yin, H., Rajabi, M., Li, H., Vieweger, M., Guo, S. et al. (2018). RNA-based micelles: A novel platform for paclitaxel loading and delivery. *J. Control. Release*, *276*, 17–29.

Shu, Y., Haque, F., Shu, D., Li, W., Zhu, Z., Kotb, M. et al. (2013a). Fabrication of 14 different RNA nanoparticles for specific tumor targeting without accumulation in normal organs. *RNA*, *19*, 766–777.

Singh, P., Prasuhn, D., Yeh, R. M., Destito, G., Rae, C. S., Osborn, K. et al. (2007). Bio-distribution, toxicity and pathology of cowpea mosaic virus nanoparticles in vivo. *J. Control Release*, *120*, 41–50.

Singh, R. & Lillard, J. W., Jr. (2009). Nanoparticle-based targeted drug delivery. *Exp. Mol. Pathol.*, *86*, 215–223.

Song, K. M., Lee, S., & Ban, C. (2012). Aptamers and their biological applications. *Sensors (Basel)*, *12*, 612–631.

Suk, J. S., Xu, Q., Kim, N., Hanes, J., & Ensign, L. M. (2016). PEGylation as a strategy for improving nanoparticle-based drug and gene delivery. *Adv. Drug Deliv. Rev.*, *99*, 28–51.

Temizoz, B., Kuroda, E., & Ishii, K. J. (2016). Vaccine adjuvants as potential cancer immunotherapeutics. *Int. Immunol.*, *28*, 329–338.

Tenzer, S., Docter, D., Kuharev, J., Musyanovych, A., Fetz, V., Hecht, R. et al. (2013). Rapid formation of plasma protein corona critically affects nanoparticle pathophysiology. *Nat. Nanotechnol.*, *8*, 772–781.

Toy, R., Peiris, P. M., Ghaghada, K. B., & Karathanasis, E. (2014). Shaping cancer nanomedicine: The effect of particle shape on the in vivo journey of nanoparticles. *Nanomedicine (London)*, *9*, 121–134.

Ventola, C. L. (2012). The nanomedicine revolution: Part 1: Emerging concepts. *PT*, *37*, 512–525.

Weiner, L. M., Dhodapkar, M. V., & Ferrone, S. (2009). Monoclonal antibodies for cancer immunotherapy. *Lancet*, *373*, 1033–1040.

Whiteside, T. L., Demaria, S., Rodriguez-Ruiz, M. E., Zarour, H. M., & Melero, I. (2016). Emerging opportunities and challenges in cancer immunotherapy. *Clin. Cancer Res*, *22*, 1845–1855.

Xu C, Li H, Zhang K, Binzel DW, Yin H, Chiu W et al. (2018). Photo-controlled release of paclitaxel and model drugs from RNA pyramids. *Nano Research, In Press*.

Xu, C., Haque, F., Jasinski, D. L., Binzel, D. W., Shu, D., & Guo, P. (2018a). Favorable biodistribution, specific targeting and conditional endosomal escape of RNA nanoparticles in cancer therapy. *Cancer Lett.*, *414*, 57–70.

Xu, Y., Pang, L., Wang, H., Xu, C., Shah, H., Guo, P. et al. (2018b). Specific delivery of delta-5-desaturase siRNA via RNA nanoparticles supplemented with dihomo-gamma-linolenic acid for colon cancer suppression. *Redox. Biol*, *21*, 101085.

Yin, H., Wang, H., Li, Z., Shu, D., & Guo, P. (2018). RNA Micelles for Systemic Delivery of Anti-miRNA for Cancer Targeting and Inhibition without Ligand. *ACS Nano, In Press*.

Yin, H., Xiong, G., Guo, S., Xu, C., Xu, R., Guo, P. et al. (2019). Delivery of anti-miRNA for triple-negative breast cancer therapy using RNA nanoparticles targeting stem cell marker CD133. *Mol. Ther.*.

Yu, J. W., Liu, Z. Y., Jiang, W., Wang, G. S., & Mao, C. D. (2015). De novo design of an RNA tile that self-assembles into a homo-octameric nanoprism. *Nat. Commun.*, *6*, 5724–5729.

Zacharias, M. & Hagerman, P. J. (1995). Bulge-induced bends in RNA: Quantification by transient electric birefringence. *J. Mol. Biol.*, *247*, 486–500.

Zhang, H., Endrizzi, J. A., Shu, Y., Haque, F., Sauter, C., Shlyakhtenko, L. S. et al. (2013). Crystal structure of 3WJ core revealing divalent ion-promoted thermostability and assembly of the Phi29 hexameric motor pRNA. *RNA*, *19*, 1226–1237.

Zhang, P., Sun, F., Liu, S., & Jiang, S. (2016). Anti-PEG antibodies in the clinic: Current issues and beyond PEGylation. *J. Control. Release*, *244*, 184–193.

Zhang, Y., Leonard, M., Shu, Y., Yang, Y., Shu, D., Guo, P. et al. (2017). Overcoming tamoxifen resistance of human breast cancer by targeted gene silencing using multifunctional pRNA nanoparticles. *ACS Nano*, *11*, 335–346.

Zolnik, B. S., Gonzalez-Fernandez, A., Sadrieh, N., & Dobrovolskaia, M. A. (2010). Nanoparticles and the immune system. *Endocrinol.*, *151*, 458–465.

51 RNA Nanotechnology and Extracellular Vesicles (EVs) for Gene Therapy

Zhefeng Li
The Ohio State University, Columbus, OH, USA

Fengmei Pi
ExonanoRNA LLC, Columbus, OH, USA

Peixuan Guo
The Ohio State University, Columbus, OH, USA

CONTENTS

BACKGROUND OF EVs AND POTENTIAL APPLICATION IN CANCER THERAPY

Extracellular Vesicles (EVs), especially exosomes, have been reported as potential delivery vehicles for therapy (varez-Erviti et al., 2011b; Sun et al., 2013; El-Andaloussi et al., 2013a; Batrakova & Kim, 2015; Pi et al., 2018). An exosome is one of the extracellular vesicles derived from late endosome/multivesicular body (MVB) with a diameter between 30 and 150 nm. Exosomes have an endomembrane-like membrane property (structure, lipid, peptides, protein, etc.) which offers an innate ability to fuse with recipient plasma membrane or the membrane of the cellular organelles (varez-Erviti et al., 2011c; van Dommelen et al., 2012; Ohno et al., 2013; Shtam et al., 2013; El-Andaloussi et al., 2013b). Several methods had been developed well for exosome characterization. Nanoparticles tracking analysis (NTA) was used to measure size distribution and particles'

Strategy of RNA/Exosome formulation

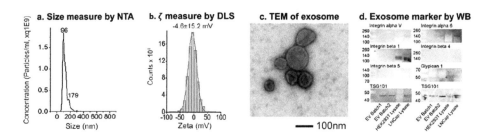

FIGURE 51.1   Strategy of engineering exosomes by RNA nanotechnology. (Reprinted with permission from Pi et al. Nanoparticle orientation to control RNA loading and ligand display on extracellular vesicles for cancer regression. Nature Nanotechnology. 2018, 13(1): 82–89. Copyright © 2017, Springer Nature)

FIGURE 51.2   (a) NTA for size analysis of exosomes; (b) DLS for ζ measurement of exosomes; (c) TEM image for exosomes; (d) Western Blot profiling exosomes marker TSG101. Oncogenic marker integrin $\alpha5$, integrin $\alpha6$, integrin $\beta1$, integrin $\beta4$, integrin $\beta5$, and glypican1 were shown negative expression. (Reprinted with permission from Pi et al. Nanoparticle orientation to control RNA loading and ligand display on extracellular vesicles for cancer regression. Nature Nanotechnology. 2018, 13(1):82–89. Copyright © 2017, Springer Nature)

number concentration (Vestad et al., 2017); Electronic Microscopy was used to study morphology (Thery et al., 2006); and several exosomal specific proteins can be used to determine its biogenesis (e.g. TSG101, CD63…) (Colombo et al., 2014). Examples of characterization of exosomes were shown in Figure 51.1. NTA and Dynamic Light Scattering (DLS) revealed that the isolated native EVs were physically homogeneous with a narrow size distribution centered around 96 nm (Figure 51.2a) and a negative ζ potential (Figure 51.2b). TEM image confirms the double-membrane cup-shaped vesicle morphology (Figure 51.2c). The purified EVs were further identified by the presence of EV-specific marker TSG101 (Kumar et al., 2015) by Western Blot (Figure 51.2d) as well as negative staining for several common integrin markers as seen on EVs for cancerous origins (Melo et al., 2015; Rak, 2015).

Exosomes were considered as "garbage disposal bucket"(Johnstone et al., 1991) until 2007, three research groups discover genetic materials, especially RNA, can be transferred among cell though exosomes (Valadi et al., 2007; Al-Nedawi et al., 2008; Witwer et al., 2013). In the following decades, it has been demonstrated that exosomes can serve as carriers for direct delivery of their payload siRNA into the cytosol which enables the full functionality of the siRNA (varez-Erviti et al., 2011a; El-Andaloussi et al., 2012; Didiot et al., 2016; Kamerkar et al., 2017). Its advantages include the fact that they are natural carriers of proteins and RNAs (Valadi et al., 2007; Dreyer & Baur, 2016; Zhang et al., 2017). They can carry high payloads while remaining a favorable size and are well tolerated *in vivo* (Shtam et al., 2013; Pi et al., 2018). Application of exosomes for therapy has been expanded in recent years and several of them undergo clinical trials as summarized

TABLE 51.1
Clinical Trials of Exosome Therapeutics

Content	Disease	Phase	Location
MSC-Exosomes encapsulated KrasG12D siRNA (iExosomes)	Pancreatic cancer	Phase I	M.D. Anderson, U.S.A.
MSC-Exosomes (MSC-Exos)	Healing of macular holes	Phase I	Tianjin Medical University, China
Umbilical cord-blood derived MSC Exosomes	β-cell Mass in Type I Diabetes Mellitus	Phase III	General Committee of Teaching Hospitals and Institutes, Egypt
Tumor antigen-loaded dendritic cell-derived exosomes (CSET 1437)	Non-Small Cell Lung Cancer	Phase II	Gustave Roussy, Cancer Campus, Grand Paris, France
Plasma-derived exosomes	Cutaneous Wound Healing	Phase I	Kumamoto University, Japan
Grape exosomes	Head and Neck Cancer	Phase I	University of Louisville, U.S.A.
Plant exosome deliver curcumin	Colon Cancer	Phase I	University of Louisville, U.S.A.
Bone marrow mesenchymal stem cell-derived exosome (UNEX-42)	Bronchopulmonary Dysplasia	Phase I	United Therapeutics, U.S.A.

Information from *clinicaltrials.gov*

in Table 51.1. In our recent study, we used RNA nanotechnology for the ligand displaying on native exosomes and successfully applied it for efficient cell targeting, siRNA delivery, and cancer regression (Pi et al., 2018; Li et al., 2018a).

RNA NANOTECHNOLOGY FOR EVs LIGAND DISPLAYING AND DELIVERING siRNA

Exosomes have great potential as delivery vectors (Zomer et al., 2010; van Dommelen et al., 2012; El-Andaloussi et al., 2013a; Melo et al., 2014) for therapeutic RNA that can remain fully functional after delivery into cell. They can enter cells through multiple routes including membrane fusion, tetraspanin and integrin receptor-mediated endocytosis, lipid raft-mediated endocytosis, or micropinocytosis. However, there is limited specificity regarding the recipient cells (Marcus & Leonard, 2013; van Dongen et al., 2016). Lack of specific cell targeting by exosomes has led to low therapeutic efficacy and potential toxicity (van Dommelen et al., 2012; El-Andaloussi et al., 2013a; El-Andaloussi et al., 2013b). There are several strategies to manipulate the targeting manner of exosomes have been developed over the decades. One example is to express certain cell-type-specific protein-based targeting ligands on their surface via genetic fusion. Neuron acetylcholine receptor-specific peptide RVG has been fused to EV membrane protein Lamp2b to be overexpressed on dendritic cells (varez-Erviti et al., 2011c). GE11 peptide, which is a ligand to EGFR (Epidermal Growth Factor Receptor), was fused to the transmembrane domain of the platelet-derived growth factor receptor to be overexpressed on EV donor HEK293T cells (Ohno et al., 2013). RGD peptide was fused to EV protein Lamp2b; thus, the EVs can deliver the chemical drug doxorubicin specifically to tumor cells (Tian et al., 2014). One problem in using fusion peptide for targeted exosome delivery is that the displayed peptide can be degraded during EV biogenesis (Hung & Leonard, 2015) (ex. LAMP2) (varez-Erviti et al., 2011c; Ohno et al., 2013; Tian et al., 2014; Hung & Leonard, 2015).

The three-way junction (3WJ) (Shu et al., 2011a; Zhang et al., 2013a) of the bacteriophage phi29 motor pRNA (Guo et al., 1987; Guo et al., 1998) folds by its intrinsic nature into a planner arrangement with three angles of 60°, 120°, and 180° between helical regions (Figure 51.3a&b) (Zhang et al., 2013b). The pRNA-3WJ was extended into an arrow-shaped structure by incorporating an RNA aptamer serving as a targeting ligand for binding to specific receptors overexpressed on cancer cells. The engineered pRNA-3WJ was used to decorate EVs purified from HEK293T cell culture

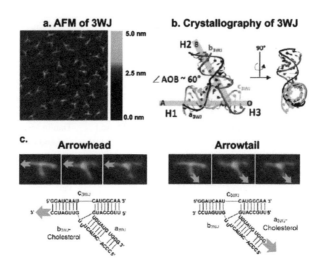

FIGURE 51.3 (a) Atomic force microscope (AFM) image of extended 3WJ of the motor pRNA of bacteriophage phi29. The color scale indicates vertical relief. (b) Structural features of the pRNA 3WJ motif. Structural features of the pRNA 3WJ motif. (c) Illustration of the location for cholesterol labeling of the arrowhead or arrowtail of 3WJ. (a&c: Reprinted with permission from Pi et al. Nanoparticle orientation to control RNA loading and ligand display on extracellular vesicles for cancer regression. Nature Nanotechnology. 2018, 13(1):82–89. Copyright © 2017, Springer Nature. b: Reprinted with permission from Khisamutdinov et al. Enhancing immunomodulation on innate immunity by shape transition among RNA triangle, square and pentagon nanovehicles. Nucleic Acids Res. 2014, 1;42(15):9996–10004. Copyright © 2014, Oxford University Press)

supernatants to create ligand-decorated EVs. A single steroid molecule, cholesterol-triethylene glycol (TEG), was conjugated into the arrowtail of the pRNA-3WJ to promote the anchorage of the 3WJ onto the EV membrane (Figure 51.3c). Cholesterol spontaneously inserts into the membrane of EVs *via* its hydrophobic moiety (Pfeiffer & Hook, 2004; Bunge et al., 2009). Display of RNA nanoparticles on surface of purified EVs was achieved by simply incubating the cholesterol-modified RNA nanoparticles with EVs at 37°C for one hour.

The application of RNA interference technology, such as siRNA, to knockdown gene expression has been of great interest (Pecot et al., 2011). The nanometer-scale EVs (Valadi et al., 2007; van Dommelen et al., 2012; El-Andaloussi et al., 2013a; El-Andaloussi et al., 2013b) can deliver biomolecules into cells by direct fusion with the cell membrane through tetraspanin domains or back fusion with endosomal compartment membranes for endosome escape. Therapeutic payloads, such as siRNA, can fully function after delivery to cells by EVs (Valadi et al., 2007; van Dommelen et al., 2012; El-Andaloussi et al., 2013a; El-Andaloussi et al., 2013b). However, EVs lack selectivity and can also randomly fuse to healthy cells. To generate specific cell-targeting EVs, approaches by *in vivo* expression of cell-specific peptide ligands on the surface of EVs have been explored (varez-Erviti et al., 2011c; Ohno et al., 2013). However, *in vivo* expression of protein ligands is limited to the availability of ligands in their producing cell types (van Dommelen et al., 2012; El-Andaloussi et al., 2013a), (Wiklander et al., 2015). It would be desirable for *in vivo* cancer cell targeting using *in vitro* surface display technology to display nucleic acid-based or chemical targeting ligands on EVs.

To improve the stability of siRNA *in vivo*, the passenger strand was 2′-F modified on pyrimidines to provide RNase resistance, while the guide strand was kept unmodified (Cui et al., 2015; Lee et al., 2015). For tracking siRNA loading efficiency in EVs, the survivin siRNA was fused to an Alexa$_{647}$-labeled 3WJ core and assembled into RNA nanoparticles (Figure 51.4a). The loading efficiency for

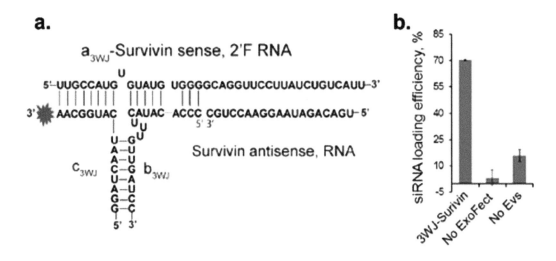

FIGURE 51.4 (a) Primary sequence and secondary structure of 3WJ harboring surviving siRNA sequences. (b) Loading efficiency of siRNA into exosomes. Control samples without transfection reagent Exo-Fect or exosomes were tested. Loading efficients were estimated ~ 70%. (Reprinted with permission from Pi et al. Nanoparticle orientation to control RNA loading and ligand display on extracellular vesicles for cancer regression. Nature Nanotechnology. 2018, 13(1):82–89. Copyright © 2017, Springer Nature)

siRNA-3WJ RNA nanoparticles was around 70% (Figure 51.4b) as measured by fluorescent intensity of the free RNA nanoparticles. Controls without EVs or with only the ExoFect reagent showed as low as 15% pelleting.

Arrow-Head or Arrow-Tail for RNA Loading or Membrane Display

The orientation and angle of the arrow-shaped pRNA-3WJ nanostructure can be used to control RNA loading or surface display of EVs. This phenomenon was found by performing serum digestion assay and cell-binding assay. Alexa$_{647}$-2′F RNA nanoparticle-displaying EVs were purified from free RNA nanoparticles by ultracentrifugation and then subjected to serum digestion. Alexa$_{647}$-2′F RNA with cholesterol on the arrow-tail for EVs decoration were degraded (31.6 ± 8.8 %) much more than the arrow-head cholesterol-decorated counterparts (9.5 ± 11.9 %) after 37 °C FBS incubation (Figure 51.5a–d). These results indicate that cholesterol on the arrow-tail promoted display of ligand on the surface of the EVs and were therefore degraded. While cholesterol on the arrow-head promoted RNA nanoparticles entering EVs, as evidenced by the protection of the Alexa$_{647}$-2′F RNA nanoparticles against serum digestion. In the arrow-tail configuration, it seems as if the two arms that form a 60° angle can act as a hook to lock the RNA nanoparticle in place. If this was the case, the effect would prevent the hooked RNA from passing through the membrane (Figure 51.5a).

Incorporated ligands were displayed on the outer surface of the EVs (Figure 51.5a). An increase in the binding of EVs to folate receptor-overexpressing KB cells was detected by displaying folate on the EV surface using arrow-tail cholesterol RNA nanoparticles (Figure 51.5e, f). When incubating with low folate receptor- expressing MDA-MB-231 breast cancer cells, arrow-tail-shaped FA-3WJ/EV did not enhance its cell binding compared to arrow-tail ligand-free 3WJ/EV (Figure 51.5g). The surface display of folate was further confirmed by free folate competition assay, in which a baseline of binding by the cholesterol arrow-tail FA-3WJ/EVs to KB cells was established. A decrease (48.3 ± 0.6 %) in the cellular binding to KB cells was detected when 10 μM of free folate was added to compete with the cholesterol-arrow-tail FA-3WJ/EV for folate receptor binding (Figure 51.5f). In contrast, competition by free folate in arrow-head FA-3WJ/EV (Figure 51.5h)

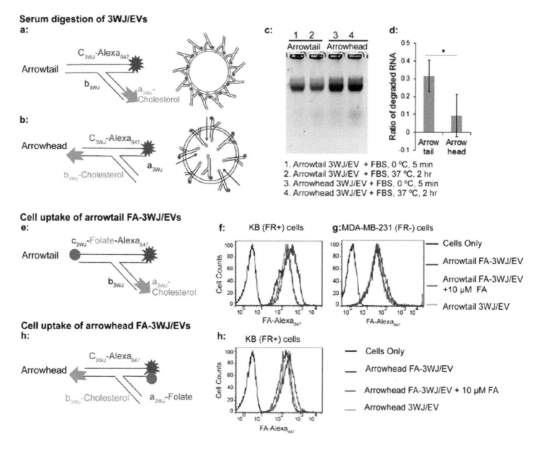

FIGURE 51.5 (a,b) Illustration showing the difference between arrowhead and arrowtail display. (c,d) Synergel to test arrowhead and arrowtail Alexa647-3WJ/EV degradation by RNase in FBS. The gel was imaged at Alexa647 channel and the bands resulting from gel imaging (c) were quantified by Image J (d). (e–i) Assay to compare cell binding of FA-3WJ arrowtail (e–g) and arrowhead (h,i) on folate receptor positive and negative cells. *p < 0.05. (Reprinted with permission from Pi et al. Nanoparticle orientation to control RNA loading and ligand display on extracellular vesicles for cancer regression. Nature Nanotechnology. 2018, 13(1):82–89. Copyright © 2017, Springer Nature)

binding to KB cells was much lower (24.8 ± 0.6 %) (Figure 51.5i), which is possibly due to partial internalization of the arrow-head-shaped FA-3WJ nanoparticle into the EVs, which resulted in a lower display intensity of folate on the surface of the EVs.

RNA-Displaying EVs Can Target Tumors and Silence Genes in Cancer Cell

We explored arrowtail RNA for displaying ligands onto the EVs surface post-biogenesis to enhance its specificity. The targeting, delivery and gene- silencing efficiency of the PSMA aptamer display-ing EVs were examined in PSMA-positive LNCaP prostate cancer cells. To confer RNase resistance, 2′-F modifications were applied to the RNA nanoparticles placed on the surface of EVs (Shu et al., 2011b), while the thermodynamic stability of pRNA-3WJ provided a rigid structure to ensure the correct folding of RNA aptamers (Shu et al., 2011a; Binzel et al., 2014). PSMA aptamer-displaying EVs showed enhanced binding and apparent uptake to PSMA(+) LNCaP cells compared to EVs without PSMA aptamer by flow cytometry and confocal microscopy analysis, but not to the PC-3

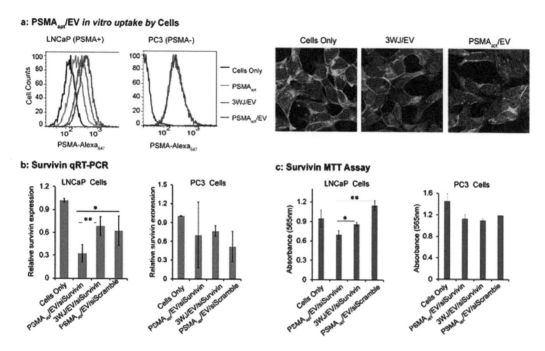

FIGURE 51.6 (a) Flow cytometry (left) and confocal images (right) showing the binding of PSMA RNA aptamer-displaying EVs to PSMA-receptor positive and negative cells. Nucleus (blue), cytoskeleton (green), and RNA (red) in confocal images. (b) RT-PCR assay for PSMA-aptamer-mediated delivery of survivin siRNA by EVs to PSMA+ prostate cancer cells. Statistics: $n=4$; experiment was run in four biological replicates and two to four technical repeats with an ANOVA analysis; Holm-adjusted $p=0.0120, 0.0067$ comparing PSMAapt/EV/siSurvivin to PSMAapt/EV/siScramble and 3WJ/EV/siSurvivin, respectively. (c) MTT assay showing reduced cellular proliferation. $n=3$, $p=0.003, 0.031$ comparing PSMAapt/EV/siSurvivin to PSMAapt/EV/siScramble and 3WJ/EV/siSurvivin, respectively. *$p<0.05$, **$p<0.01$. (Reprinted with permission from Pi et al. Nanoparticle orientation to control RNA loading and ligand display on extracellular vesicles for cancer regression. Nature Nanotechnology. 2018, 13(1):82–89. Copyright © 2017, Springer Nature)

cells, which is a low PSMA receptor expressing cell line (Figure 51.6a). Upon incubation with LNCaP cells, PSMA$_{apt}$/EV/siSurvivin was able to knock down the survivin expression at the mRNA level as demonstrated by real-time PCR ($37.73 \pm 11.59\%$, $p<0.05$) (Figure 51.6b). Cell viability by MTT assays indicated that the viability of LNCaP cells were decreased as a result of survivin siRNA delivery ($70.98 \pm 6.46 \%$, $p<0.05$) (Figure 51.6c).

The tumor targeting and biodistribution properties of ligand-displaying EVs were evaluated. FA-3WJ/EVs were systemically administered *via* the tail vein into KB subcutaneous xenograft mice model. 3WJ/EVs and PBS treated mice were tested as a control. *Ex vivo* images of healthy organs and tumors taken from mice after 8 hrs showed that the FA-3WJ/EVs mainly accumulated in tumors, with low accumulation in vital organs in comparison with PBS control mice, and with more accumulation in tumors in comparison with 3WJ/EVs control mice (Figure 51.7a). Similar ex vivo result were found when introducing EGFR-3WJ/EVs to triple negative breast tumor bearing mice (Figure 51.7b). Normal EVs without surface modification usually showed accumulation in liver after systemic delivery (Ohno et al., 2013). Both RNA and cell membranes are negatively charged. The electrostatic repulsion effect has been shown to play a role in reducing the accumulation of RNA nanoparticles in healthy organs (Haque et al., 2012; Shu et al., 2015; Binzel et al., 2016). We hypothesize that displaying targeting RNAs on the EVs surface reduces their

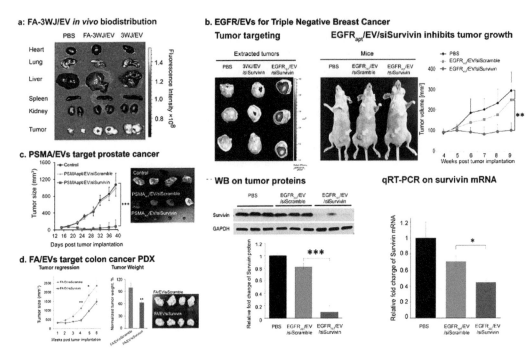

FIGURE 51.7 In vivo tumor inhibition and gene knock down by IV injection of engineered exosomes. (a) Intravenous treatment of nude mice bearing LNCaP-LN3 subcutaneous xenografts with PSMAapt/EV/ siSurvivin or PSMAapt/EV/siScramble (both with 0.6 mg kg–1, siRNA/mice body weight), and PBS only control, injected twice per week for three weeks. n = 10, p = 1.8 × 10⁶ at day 39 comparing treatment to control. qRT-PCR were performed to assay gene knockdown in tumor tissue. B. Intravenous treatment of nude mice bearing PDX-CRC xenografts with FA/EV/siSurvivin and controls (n = 4). After 6 weeks, the group treated with FA/EV/siSurvivin had significantly smaller tumor size, p = 0.0098 and 0.0387 comparing FA/EV/siSurvivin to FA/EV/siScramble at week 4 and week 5, respectively. C. EVs displaying EGFR aptamer showed enhanced targeting effect to breast tumor in orthotopic xenograft mice models (a). Intravenous treatment with EGFRapt/ EV/siSurvivin and controls (n = 5) for 6 weeks, p = 0.008 comparing treatment to control. (Reprinted with permission from Pi et al. Nanoparticle orientation to control RNA loading and ligand display on extracellular vesicles for cancer regression. Nature Nanotechnology. 2018, 13(1):82–89. Copyright © 2017, Springer Nature)

accumulation in normal organs, and the ideal nano-scale size of RNA displaying EVs facilitates tumor targeting *via* Enhance Permeability and Retention (EPR) effects, thereby avoiding toxicity and side effects.

INHIBITION OF TUMOR GROWTH BY LIGAND-3WJ-DISPLAYING EVS

The RNA/exosome platform can be easily engineered for versatile purpose by simply changing the ligand carried by one of the strands in 3WJ. As prove of concept, in vivo tumor suppression experiments were performed on prostate cancer, triple negative breast cancer, and colorectal cancer PDX model (Figure 51.7b–d). In order to confer specific targeting of EVs to cancer cells, three classes of targeting ligands, folate, PSMA RNA aptamer, or EGFR RNA aptamers were conjugated to the 3WJ for displaying on the EVs surface. Folate is an attractive targeting ligand since many cancers of epithelial origin, such as colorectal cancers, overexpress folate receptors (Parker et al., 2005). PSMA is expressed at an abnormally high level in prostate cancer cells, and its expression is also associated with more aggressive diseases (Dassie et al., 2014). A PSMA-binding 2′-Fluoro (2′-F) modified

RNA aptamer A9g (Rockey et al., 2011; Binzel et al., 2016) was displayed on EVs to enhance targeting efficiency to prostate cancer cells. The PSMA aptamer A9g is a 43-mer truncated version of A9, which binds PSMA specifically with K_d 130nM (Rockey et al., 2011) and is used as RNA-based ligand. EGFR is highly overexpressed in triple negative breast cancer (TNBC) tumors and metastatic TNBC tumors (Hynes & Lane, 2005). An EGFR-specific 2′F-RNA aptamer (Esposito et al., 2011; Shu et al., 2015) was incorporated to one end of pRNA-3WJ and thereby displayed on EVs for enhanced targeting of breast cancer cells. For imaging, one of the pRNA-3WJ strands was end-labeled with a fluorescent dye $Alexa_{647}$ (Figure 51.1).

SiRNA targeting surviving gene, an inhibitor of cell apoptosis, is an attractive target for cancer therapy, since knockdown by siRNA can decrease tumorigenicity and inhibit metastases (Khaled et al., 2005; Paduano et al., 2006). Ligand-displayed exosomes encapsulated with survivin siRNA were introduced intravenously at the dose of 0.5 mg siRNA/kg of mice weight for all three animal models. Scramble siRNA encapsulated in same ligand/RNA/EV delivery vector, 3WJ/EV/sisurvivin (without targeting ligand) and PBS were served as controls. Tumor suppression was measured by tumor size change and the target gene expression was evaluated by qRT-PCR or Western Blot. All these results suggest that arrowtail RNA ligand displaying EVs are suitable for *in vivo* applications.

With that, we expand the application of RNA nanotechnology (Guo, 2010) to reprogram natural EVs for specific delivery of siRNA to cancer models *in vitro* and in animal models (Figure 51.1a-c). Taking advantage of the thermodynamically stable properties of pRNA-3WJ, (Shu et al., 2011b; Shu et al., 2013; Binzel et al., 2014) multifunctional RNA nanoparticles harboring membrane-anchoring lipid domain, imaging modules, and targeting modules were generated. The arrow-shaped pRNA-3WJ offered the opportunity to control either partial loading of RNA into EVs or decoration of ligands on the surface of EVs. With cholesterol placed on the arrow-tail of the 3WJ, the RNA-ligand was prevented from trafficking into EVs, ensuring oriented surface display of targeting modules for cancer receptor binding. The incorporation of arrow-tail 3WJ-RNA nanoparticles to the surface of the EVs not only provided a targeting ligand to the EVs but also added a negative charge on the EVs surface. Displaying negatively charged RNA nanoparticles on EV surface might assist in the reduction of non-specific binding of EV to normal cells. We have noticed previously that negatively charged RNA nanoparticles with a proper ligand tend to accumulate into tumors specifically after systemic administration (Haque et al., 2012; Shu et al., 2015; Binzel et al., 2016). The cholesterol-TEG-modified RNA nanoparticles should preferentially anchor onto the raft-forming domains of the lipid bilayer of EVs (Bunge et al., 2009), and further studies will be necessary to illustrate this process.

CHALLENGE OF EVs PRODUCTION AND ALTERNATIVE SOURCE OF EVs

Although exosomes have shown increasing potential as delivery vesicles for therapy, but challenges like cost/yield, drug payload, and targeting specificity still exist. As one of the major sources, human cell culture medium was widely used to produce exosomes; however, methods of production and scaling up remain challenging. Several strategies, including the use of bioreactors, have been developed to scale up the production (Li et al., 2017), but economically, the cost/yield ratio is still not favorable for clinical application. The yield of exosome purification mainly depends on the sources of exosomes and the purification strategy. Mammalian cell culture is one of the most widely used sources to harvest exosomes for drug delivery purpose. HEK293T EVs were used most commonly as they contain minimal intrinsic biological cargos compared to EVs generated by other cells (Lamichhane et al., 2015). Additional steps were taken to remove EVs from FBS used in the HEK293T cell culture; although, centrifugation might not completely remove the FBS EVs (Witwer et al., 2013; Shelke et al., 2014).

However, low concentration of exosomes and different impurity existing in cell culture medium require time-consuming purification from large volume crude materials. Single purification method by itself either has low quality (high impurity or aggregation), like pellet down ultracentrifugation, polymer precipitation and ultra-filtration, or low yield, like density gradient ultracentrifugation, size-exclusion chromatography and immunoaffinity (Zeringer et al., 2015; Li et al., 2017; Stranford & Leonard, 2017). The overall low yield and high cost of current exosome purification method become major barrier for systematic *in vivo* study and clinical translation. To improve purity of the exosomes, the differential ultracentrifugation method has been used in combination with density gradient ultracentrifugation method to improve its purity (Li et al., 2018a). Adding a cushion layer at the bottom of tube for ultracentrifugation can help to improve the purity since proteins with higher density will go down to the bottom while the exosomes will be on the cushion layer, which will also help to preserve the physical integrity of the exosomes (Pi et al., 2018). ExoJuice is a new reagent used as cushion for purifying exosomes by ultracentrifugation method, which has a density closer to exosomes, and after UC the exosomes can go into the layer of cushion which can further improve the purity of exosomes by removing the larger population of microvesicles.

An emerging solution to the aforementioned problem is to harvest EVs or exosome-like vesicles from substituted sources including human urine (Franzen et al., 2016), bovine milk (Munagala et al., 2016), and especially plants (Quesenberry et al., 2015; Zhang et al., 2016b). Plant-derived exosome-like nanoparticles have been reported as a promising substitution and exhibit biocompatibility through oral, intranasal administration. Exosome-like nanometer-sized particles holding similar properties as mammalian EVs have been reported in grapefruit (Wang et al., 2014), grapes (Ju et al., 2013), ginger (Zhuang et al., 2015; Zhang et al., 2016a), sunflowers (Regente et al., 2009), carrots, etc. (Mu et al., 2014; Deng et al., 2017). Compared to mammalian cell culture medium, plants are an advantageous source to scale up overall EV yield. Several studies have demonstrated that exosome-like vesicles from edible plants can be used for therapeutic or delivery purposes by oral (Wang et al., 2014; Zhuang et al., 2015) or intranasal administration (Ju et al., 2013). But intravenous injection studies of plant-derived exosome-like vesicles are rarely reported. One of the major concerns is the biocompatibility regarding the particle size and impurity. One reported solution is to reassemble nanoparticles after extracting the lipid components from grapefruit, enabling the encapsulation and delivery of chemical drugs or miRNAs via intravenous route (Wang et al., 2013; Wang et al., 2015; Teng et al., 2016). We adapted the post-biogenesis method of RNA nanotechnology we recently reported to manipulate the angle and orientation of the RNA architecture for displaying ligands onto the EVs surface to enhance targeting specificity (Pi et al., 2018). Using ginger-derived exosome-like nanovesicles (GDENs), we further confirm that exosome-like vesicles can be engineered via ligand-displaying arrowtail RNA nanoparticles to deliver siRNA for tumor suppression intravenously (Li et al., 2018b).

Methods of membrane filtration, differential ultracentrifugation, and equilibrium density gradient ultracentrifugation were used as a workflow to isolate GDENs (Figure 51.8a). After ginger juice was blended, larger solid residues were removed by rough filtration, followed by the removal of cells and cell debris through centrifugation at 10,000 g twice. Crude GDENs were concentrated by repeated ultracentrifugation with the addition of a thin Optiprep™ cushion at the bottom of the centrifuge tube, eliminating disruption and aggregation (Pi et al., 2018). Since exosomes are complex lipid vesicles containing both protein and RNA, they have higher density than other lipid vesicles. Therefore, density gradient was chosen to separate exosomes from free lipid, protein, RNA, and other components. Condensed GDENs were further purified by equilibrium density ultracentrifugation and fractionated from the bottom of the tube (Jasinski et al., 2015). The density among fractions indicates the gradient formed continuously and linearly (Figure 51.8b). Fractions 8~15 were selected based on their density within the density ranges of HEK293T cell-derived exosomes between 1.13 and 1.19 g/mL (Szatanek et al., 2015). Particles' concentration of each fraction by Nanoparticle Tracking Analysis (NTA) showed that most of the nanovesicles were distributed among fractions 8–14 (Figure 51.8C). These fractions were then combined into one batch and washed by cushion ultracentrifugation in PBS again and resuspended in PBS for further application.

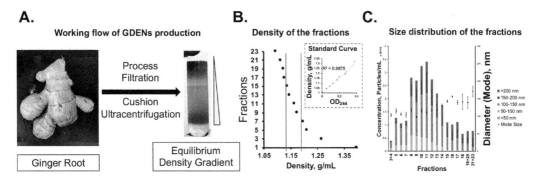

FIGURE 51.8 Purification and characterization of GDENs. (a) Working flow of purified GDENs from ginger root. (b) Density assessment of each fraction collected from equilibrium density gradient measured by OD244 and converted by standard curve. Density of fraction 8–15 are located within the range of 1.13–1.19 g/mL (indicated by two red lines). (c) Size distribution and particle concentration of each fraction measured by NTA and plot with the mode size. (Reprinted with permission from Li et al. Arrowtail RNA for Ligand Display on Ginger Exosome-like Nanovesicles to Systemic Deliver siRNA for Cancer Suppression. Sci Rep. 2018, 8(1):14644 Copyright © 2018, Springer Nature)

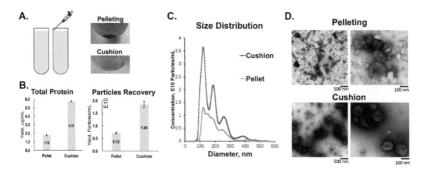

FIGURE 51.9 Characterization and comparison between ultracentrifugation with and without cushion for GDENs purification. (a) Schematic of cushion load to ultracentrifuge tube by slowly pipetting from side wall to the bottom and photo shows GDENs concentrated at the interception of cushion layer compared to the pellet pack firmly at the bottom. (b) Comparison of yield by total protein and particles standardized by volume of ginger juice isolated from. (c) Size distribution and particle concentration measured by NTA of GDENs purified from equal amount of ginger juice with and without cushion. (d) Negative staining TEM imaging showing the morphology of GDENs purified by pelleting and cushion method. (Reprinted with permission from Li et al. Arrowtail RNA for Ligand Display on Ginger Exosome-like Nanovesicles to Systemic Deliver siRNA for Cancer Suppression. Sci. Rep. 2018, 8(1):14644 Copyright © 2018, Springer Nature)

Ultracentrifugation is a common method for EV purification. However, repeated pelleting of exosomes under high centrifuge force may damage the EVs integrity or cause aggregation (Linares et al., 2015). By placing a thin layer of high-density iso-osmotic material at the bottom, EVs will never spinout as a pellet at the bottom of the centrifuge tube (Jasinski et al., 2015; Li et al., 2017). In this study, we also took advantage of Optiprep cushion during ultracentrifugation (Figure 51.9a). It revealed that the cushion method not only increased yield but also the quality of GDENs. As shown in Figure 51.2b, NTA and bicinchoninic acid (BCA) assay both indicated that GDENs harvested by cushion centrifugation have more than 2-fold higher yield than pelleting. No significant difference in size are observed between these two methods (cushion: 123.5 nm vs. pellet: 124.5 nm), but the distribution of GDENs from cushion seems to be less heterogeneous (Figure 51.9c). We also performed negative staining TEM imaging to characterize the morphology of GDENs we purified (Figure 51.9d). It revealed that GDENs purified with cushion had a cleaner background, less aggregation, and reserved a better spherical shape compared to conventional pelleting without a cushion. Therefore, the addition of the cushion during GDENs purification can eliminate the structural disruption and aggregation during ultracentrifugation.

In this study, we demonstrated a strategy to isolate exosome-like nanovesicles from ginger as a siRNA delivery vesicle by intravenous administration. Ginger, as well as other edible plants, shows an economic advantage for the production of exosome-like vesicles on a large scale (Quesenberry et al., 2015; Zhang et al., 2016b). Here, we introduced equilibrium density gradient ultracentrifugation to increase the purity of GDENs while remaining ~300-fold lower cost/yield ratio to human cell-derived EVs. Compared to human cell-derived EVs, plant exosome-like vesicle reduces not only the cost from cell culture supplies but also the time and labor of large- scale cell culture. By using a normal juice blender, we can process up to 3 liters of ginger juice as a starting material in 1 hour, equivalent of 300 cell culture dishes (150mm). Moreover, we haven't yet reached the productivity limit as this was done within a research environment.

DISPLAYING OF LIGANDS ON GDENS USING ARROW-TAIL RNA NANOPARTICLES FOR SPECIFIC CANCER TARGETING

To testify whether we can adopt the same strategy for engineering GDENs, we first applied Förster Resonance Energy Transfer (FRET) system to verify the interactions between RNA and GDENs. According to the mechanism of FRET, when applying a laser to excite the donor fluorophore, a receptor fluorophore then receives energy transfer and emits fluoresce. Due to sensitivity, FRET can only occur when the two fluorophores are within 10 nm distance (Deniz et al., 1999; Norman et al., 2000; Zhang et al., 2009; Shu et al., 2010). GDENs were labeled by *CellMask Orange*, a uniform membrane labeling marker, with similar fluorescence emission spectrum as Cy3, to serve as a FRET donor. The arrowtail pRNA-3WJ was designed as a FRET accepter by end-labeling with a fluorescent dye Alexa$_{647}$ on the 5' end of 3WJ$_b$ strand, which is adjacent to the cholesterol responsible for interactions with GDENs' membrane (Figure 51.4). The 2'F-modified Alexa$_{647}$ labeled arrowtail pRNA-3WJ was incubated with *CellMask Orange* labeled GDENs forming complexes. The 3WJ/GDENs complex was fractionated by *Sephadex* G-200 gravity size exclusion column and compared to control groups. Fluorescent signals of *CellMask Orange*, Alexa$_{647}$, and *CellMask Orange*-Alexa$_{647}$ FRET were observed and plotted (Figure 51.10). The ~100 nm GDENs (5 min) were clearly separated from arrowtail pRNA-3WJ (11–13 min), which has been reported to be around 5 nm in size (Zhang et al., 2013b). Only pRNA-3WJ-cholesterol/GDENs group showed a peak in 5 min fraction, indicating the RNA colocalized with GDENs as it passed through the column. A significant FRET peak compared to control groups confirms the pRNA-3WJ-cholesterol interacted with the GDENs membrane rather than non-specific effect.

THERAPEUTIC RNA DELIVERY AND TUMOR SUPPRESSION BY LIGAND-DISPLAYING GDENS

FA displaying GDENs were then used to examine the targeting, delivery, and gene silencing on KB cell and KB cell-derived xenograft mice model. Increasing folate-arrowtail displaying ratio on GDENs from 200: 1, 1000: 1 and 5000: 1 showed consecutive enhanced binding to KB cells (Figure 51.11a). Delivery of survivin siRNA by FA-3WJ/GDENs/si-survivin to KB cells was evaluated by examining the depleted expression of survivin at the mRNA level. Folate displaying GDENs exhibit similar delivery efficiency compared to transfected siRNA samples in vitro (Figure 51.11b). The treatment group (FA-3WJ/GDENs/sisurvivin) exhibited significant gene knockdown effect

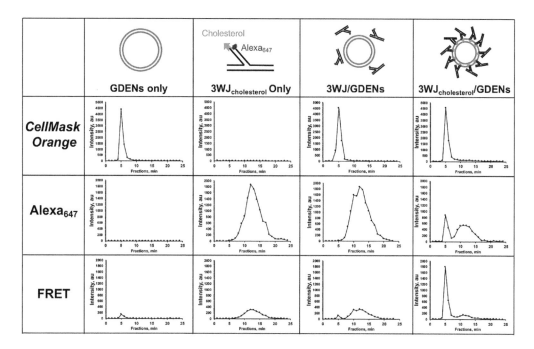

FIGURE 51.10 Histogram plotted by fluorescent intensity readout from individual fractions collected from Sephadex G-200 SEC after introducing sample by Cellmask Orange, Alexa647 and Cellmask Orange-Alexa647 FRET excitation/emission setting. (Reprinted with permission from Li et al. Arrowtail RNA for Ligand Display on Ginger Exosome-like Nanovesicles to Systemic Deliver siRNA for Cancer Suppression. Sci. Rep. 2018, 8(1):14644 Copyright © 2018, Springer Nature)

compared to the control group treated with scramble RNA (FA-3WJ/GDENs/scramble) and the untreated group (PBS). Interestingly enough, we also observed gene knockdown effect on treatment group without ligand (3WJ/GDENs/sisurvivin), indicating that GDENs itself were taken in by cell and delivered the cargo into cell *in vitro*.

To evaluate the cytotoxicity of the FA-3WJ/GDENs as delivery vesicles, MTT assays were performed to study whether the nanoparticles will inhibit cell proliferation on somatic cell (HEK293), macrophage (Raw 264.7), and cancer cell (KB). To provide a standard, we used equivalent common transfection reagent (lipofectamine 2000) as reference. Cell proliferation remained more than 80% after 24 hours incubation with FA-3WJ/GDENs in 20 µg/mL, which is equal to the dose for *in vitro* delivery. Even for the highest dose (80 ug/mL), these nanoparticles showed significantly less cytotoxicity compared to the transfection reagent (Figure 51.11c). By fitting the data with a nonlinear regression curve, 50% lethal dose (LD_{50}) were calculated (Figure 51.11d). Macrophages were more sensitive to the nanoparticles, while considerably low cytotoxicity of FA-3WJ/GDENs was observed on both somatic and cancer cell. The overall results suggest the biocompatibility of FA-3WJ/GDENs as delivery vesicles.

The *in vivo* delivery of survivin siRNA by folate-displaying GDENs was also evaluated in subcutaneous xenografts of KB cells. Upon tumor growth and maturation, FA-3WJ/GDENs/si survivin were delivered to the tumor through retro-orbital IV injection (1 dose every 2 days, total 6 doses). Delivery of RNA decorated GDENs showed suppressed tumor growth over a negative control (Figure 51.12a). Upon the completion of the experiment, mice were sacrificed and tumor specimens were used to examine the protein level of *survivin* gene while using GAPDH as internal control. It was observed that siRNA-loaded GDENs were able to significantly reduce *survivin* expressions within

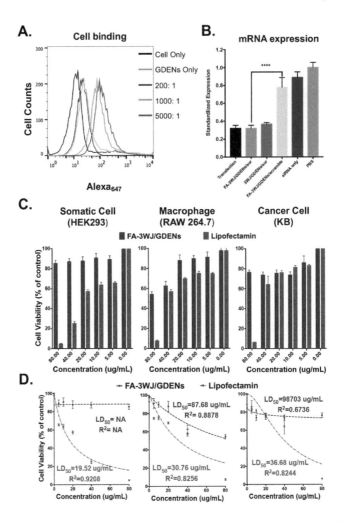

FIGURE 51.11 Cancer cell targeting and delivery of siRNA by ligand-displaying GDENs. (a) Cancer cell targeting capacity was evaluated by flowcytometry showed dose-dependent of FA-arrowtail RNA ratio verse equal amount of GDENs. (b) qRT-PCR evaluating in vitro delivery of survivin siRNA to KB cells by FA-arrowtail RNA ligand displayed GDENs. mRNA levels were standardized by PBS group as 1 and shown as mean ± S.D. fold changes of individual groups. Results are presented at n = 3 for each group for one-way ANOVA multiple comparisons, Tukey's adjusted p of FA-3WJ/GDENs/sur compare to transfection, 3WJ/GDENs/sur and FA-3WJ/GDENs/scramble are >0.9999, 0.9174 and <0.0001(****). (c) Cytotoxicity of FA-3WJ/GDENs evaluated by MTT assay expressed as histogram and fit with nonlinear regression curve (D). Samples were incubated with cells in quadruplicate for 24 hours before adding MTT dye. Cell viability was standardized by the control group of "Cell Only" as 100%. Equivalent lipofectamin 2000 were used for comparison. N = 4, error bar indicates ± S.D. (**Reprinted with permission from Li et al. Arrowtail RNA for Ligand Display on Ginger Exosome-like Nanovesicles to Systemic Deliver siRNA for Cancer Suppression. Sci. Rep. 2018, 8(1):14644 Copyright © 2018, Springer Nature**)

the tumor environment compared to both treatment scramble control (FA-3WJ/GDENs/scramble) and the treatment group without targeting ligand (3WJ/GDENs/sisurvivin) (Figure 51.12b). The results from both *in vitro* and *in vivo* indicate that ligand-displaying GDENs by arrowtail RNA nanoparticle had potential as a delivery vector for therapeutic siRNA.

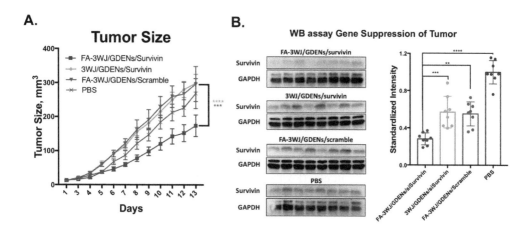

FIGURE 51.12 In vivo delivery of siRNA by ligand-displaying GDENs via IV injection. (a) Tumor size of nude mice with KB cell-derived xenograft tumors by intravenous treatment every two days for two weeks. n=8 and result is shown as mean±S.E.M. analysis by multiple t-test. p=0.0378, 0.0179, 0.0035, 0.0015, 0.0018, 0.0008 for day 8~13 post first dose given, respectively, for FA-3WJ/GDENs/siSurvivin v.s. FA-3WJ/GDENs/Scramble control. p=0.0026, 0.0015, 0.0002, <0.0001, <0.0001, <0.0001 for day 8~13 post first dose given, respectively, for FA-3WJ/GDENs/siSurvivin v.s. 3WJ/GDENs/siSurvivin. (b) Western Blot evaluating survivin protein level in tumor after intravenously treatment. Gray value was quantified by Image J; intensity was adjusted by internal control GAPDH level and standardized by PBS group as 1. n=8 and result is shown as mean±S.D., analysis by one-way ANOVA, *p<0.05, **p<0.01, ***p<0,001, ****p<0.0001. (Reprinted with permission from Li et al. Arrowtail RNA for Ligand Display on Ginger Exosome-like Nanovesicles to Systemic Deliver siRNA for Cancer Suppression. Sci. Rep. 2018, 8(1):14644 Copyright © 2018, Springer Nature)

EVS AVOID ENDOSOME TRAPPING AND FACILITATE CYTOSOL DELIVERY BY DIRECT FUSION MECHANISM

Folate receptor is a cell surface glycoprotein highly expressed on cancer cells while its expression on normal cells is low or undetectable (Zwicke et al., 2012). Therefore, folate (FA) has been extensively investigated and applied for the selective delivery of therapeutics to cancers, including breast cancer (Fasehee et al., 2016), lung cancer (Shi et al., 2015), ovarian cancer (van Dam et al., 2011), colorectal cancer (Cisterna et al., 2016), and head and neck cancer (Saba et al., 2009). Conjugation of FA to therapeutic molecules, like chemotherapy drugs (Lee et al., 2009; Yang et al., 2013) has been shown to enhance their targeting and delivery to folate receptor-expressing cancer cells, making FA a superior target for a variety of cancers. However, endosome trapping of folate receptor-mediated endocytosis has been a major hurdle in clinical trial with FA-conjugated therapeutics such as siRNA (Turek et al., 1993; Leamon & Reddy, 2004; Sabharanjak & Mayor, 2004). The therapeutic efficiency is often compromised due to entrapment in endosomes after endocytosis.

Despite the challenge in efficient delivery, RNA interference (RNAi) holds great potential for therapeutic applications by specific gene suppression (Whitehead et al., 2009), (Zhang et al., 2013a). During the last few decades, major efforts had been spent on achieving efficient *in vivo* delivery of siRNA (Whitehead et al., 2009) or miRNA (Zhang et al., 2013b) to target cell using different strategies, including the recent approval of Onpattro (patisiran), the first-ever RNAi therapeutic using a lipid nanoparticle platform (Garber, 2018). Delivery strategies including cationic lipids (Semple et al., 2010), cationic liposomes (Reddy et al., 2002), and cationic polymers (Li et al., 2018b) that can deliver the RNAi to cells but specific targeting is still challenging. The attempted approaches for specific targeting includes the use of peptide (Ben et al., 2018), antibody (Toloue & Ford, 2011), chemical ligands (Teo et al., 2015), and RNA aptamers (Chu et al., 2006), etc. Polymers (Priegue et al., 2016), gold nanoparticles (Guo et al., 2010), RNA nanoparticles (Shu et al., 2011a; Shu et al., 2015; Binzel et al., 2016; Guo et al., 2018), and liposomes (Yang et al., 2011) have been used as delivery vesicles (Dong et al., 2019). However, it remains challenging to make the siRNA interference functional after delivery into the cells via folate receptor, mainly due to the difficulty in endosomal escape.

CONFIRMATION OF CYTOSOL DELIVERY WITH FOLATE-EXOSOME

For imaging purpose, exosomes were internally loaded with survivin siRNA (Exo/Sur-A647). The resulting exosomes were further displayed with FA that is conjugated to three-way junction (3WJ) (Shu et al., 2011b) arrowtail nanoparticles (FA/Exo/Sur-A647) (Pi et al., 2018; Li et al., 2018a). The double-stranded siRNA against survivin conjugated with FA (FA-Sur-A647) without exosome was used as the control. The cellular uptake and subcellular localization of the exosome complex and controls were monitored using confocal laser scanning microscopy. Significant intracellular accumulation of red fluorescence representing the A647-RNA occurred when the FA receptor overexpressed KB cells were treated with FA/Exo/Sur-A647 complex (Figure 51.13). The red fluorescence from A647-RNA was evenly distributed within the whole cell, suggesting the cytosol delivery of

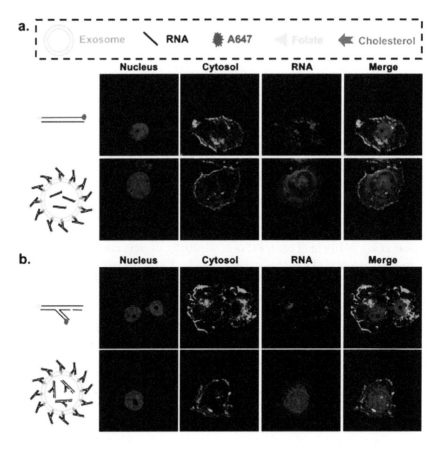

FIGURE 51.13 Exosome formulation enables cytosolic delivery of A647-lableled siRNA into KB cells. 2D confocal microscopy images of KB cells 30 min after treatment with 50 nM of (a) FA-Sur-A647 and FA/Exo/ds-Surivin-A647, or (b) FA-3WJ-Sur-A647 and FA/Exo/3WJ-Sur-A647. (Reprinted with permission from Zheng et al. Folate-displaying exosome-mediated cytosolic delivery of siRNA avoiding endosome trapping. J Control Release. 2019, 311–312:43–49. Copyright © 2019 Elsevier B.V.)

TABLE 51.2

Calculated Mander's Colocalization Coefficient (M) and the Pearson Correlation (Pr)

	Coefficient			
Sample	With endosome		With Lysosome	
	M	Pr	M	Pr
FA-Sur-A647	0.86	0.19	0.56	0.34
FA/Exo/Sur-A647	0.08	−0.07	0.24	0.03
FA-3WJ-A647	0.69	0.23	0.89	0.54
FA/EXO/3WJ-A647	0.06	0.15	0.13	0.18

the A647-RNA cargo via exosome. This image was very different from the control of FA-Sur-A647 without exosome, showing bright spots inside the cell due to the entrapment within intracellular compartments.

To further differentiate cytosol delivery from endosome trapping, colocalization analysis of RNA nanoparticles and different intracellular organelles involved in endocytic pathway were carried out. The Mander's Colocalization Coefficient (M) and the Pearson Correlation Coefficient (Adler & Parmryd, 2010; Dunn et al., 2011) were calculated (Table 51.2). FA-siRNA (labeled with Alexa Fluor 647 indicated by red fluorescence in Figure 51.14) showed a high colocalization (M = 0.86) with endosome (marked with an antibody labeled with Alexa Fluor 488, indicated by green fluorescence in Figure 51.14) after 1-hour incubation, suggesting they were primarily localized in endosomes at the first hour. After 2 h of incubation, the FA-siRNA nanoparticles showed a high overlap with lysosomes (marked with an antibody labeled with Alexa Fluor 488, indicated by green fluorescence

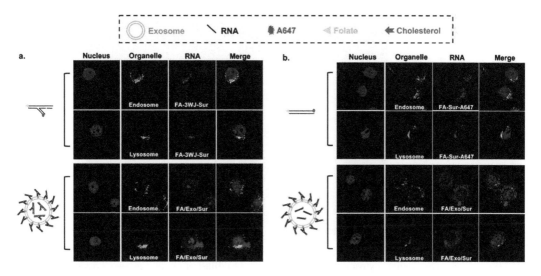

FIGURE 51.14 Colocalization studies of siRNA delivery with and without exosome. Confocal microscopy of (A) FA-3WJ-Sur-A647 and FA/Exo/3WJ-Sur-A647, or (B) FA-Sur-A647 and FA/Exo/Sur-A647. All SiRNA are labeled with A647 (red). Immunofluorescence staining of organelle markers (green). Markers are EEA1 for early endosome and LAMP1 for lysosomes. DAPI staining nucleus (blue). Colocalization is displayed in yellow. (Reprinted with permission from Zheng et al. Folate-displaying exosome mediated cytosolic delivery of siRNA avoiding endosome trapping. J Control Release. 2019, 311–312:43–49. Copyright © 2019 Elsevier B.V.)

in (Figure 51.14) (M = 0.56), suggesting their trafficking and accumulation to the lysosome. These observations make sense since FA complexes are believed to enter cells via endocytosis which go through an endosome–lysosome pathway (Bandara et al., 2014). In contrast to images derived from using FA conjugation without exosome, FA-displaced exosome cargo showed a much lower colocalization with either endosome or lysosome (Figure 51.14). These results lead to the conclusion that the cargo siRNA would not be trapped in endosomes or lysosomes when delivered using exosomes.

Mechanism of Direct Fusion or Back Fusion After Entering Endosome

After confirming the high efficiency of cytosolic delivery, we then went on to investigate the cytosol delivery mechanism. The cellular uptake mechanism for exosomes is still under extensive scrutiny. Exosome cargoes can enter the cytosol either directly via fusion with the outer cell membrane or by back fusion with endosomal membrane after receptor-mediated endocytosis (Figure 51.15a).

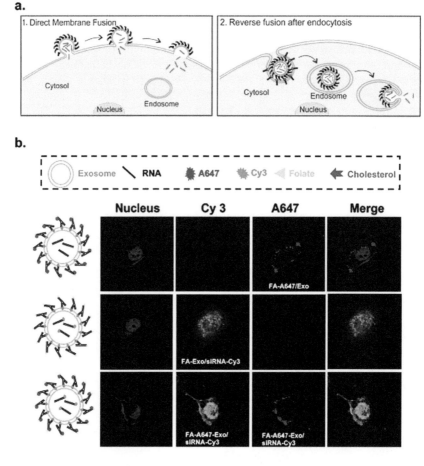

FIGURE 51.15 Distinguish cytosol delivery pathways. (A) Illustration of the two possible pathways: 1. Ligand-displaying exosomes bind to cell receptors followed by fusion with cell membrane. 2. Receptor-mediated endocytosis induces back fusion with endosome membrane. (B) Confocal imaging distinguishes between direct fusion and back fusion. Ligand (FA-arrowtail, A647, red) and cargo (siRNA, Cy3, green) of FA/Exo/siRNA were labeled at different fluorescent dye for confocal imaging of cellular uptake. (Reprinted with permission from Zheng et al. Folate-displaying exosome mediated cytosolic delivery of siRNA avoiding endosome trapping. J Control Release. 2019, 311–312:43–49. Copyright © 2019 Elsevier B.V.)

Previous reports have suggested that one of the major delivery mechanisms of exosome is via direct membrane fusion (McKelvey et al., 2015; Prada & Meldolesi, 2016). However, in this study the exosomes were fully decorated with FA, where the folate receptor-mediated endocytosis might take over and play a role in cell entry. To distinguish between these two pathways, KB cells were incubated with folate-exosome complexes labeled on the exosome membrane (Alexa 647, green in Figure 51.15), and/or Cy3-RNA (red in Figure 51.15) loaded into the exosome, or both. While the cargo showed an even distribution inside the cell, the green color representing the exosome membrane was observed to be located on the membrane of the cells rather than in the cytoplasm (Figure 51.15b). These observations suggested that the folate displaying exosome is more likely to enter through direct membrane fusion since the dye on exosome membrane were retained on the cell membrane during cellular uptake, which is distinct from cargo located evenly in the cytosol.

COMPARISON OF GENE-SILENCING EFFICIENCY AND TUMOR INHIBITION POTENCY BETWEEN FOLATE-DISPLAYING EXOSOME AND FOLATE-CONJUGATED siRNA

Folate-displaying exosomes and folate-conjugated siRNAs were compared for their gene silencing efficiency in folate receptor positive KB cells cell culture to evaluate the role of exosome in cytosol delivery. KB cells were incubated with 50 nM Exo/siRNA (Sur)-A647 and FA/Exo/siRNA (Sur)-A647, respectively. FA-siRNA(Sur) or siRNA(Sur) were also transfected into the cells as a positive control. After a 72-hour treatment, cells were collected and the target gene downregulation effects of corresponding samples were accessed by quantitative reverse-transcription polymerase chain reaction (qRT-PCR). As shown in Figure 51.4A, FA/Exo/siRNA(Sur)-A647 was able to knockdown 60% of the survivin expression at the mRNA level. On the contrary, when FA/siRNA(Sur)-A647 was delivered without exosome, no obvious change in gene expression level was observed compared to the control group without treatment (Figure 51.16a). FA showed an enhancement in KB cell binding and uptake. Although it is difficult to make a conclusion without direct evidence, it is proposed that folate might help to promote the access of exosome to cells.

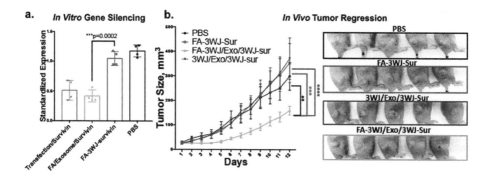

FIGURE 51.16   Efficacy evaluation of siRNA Delivered with and without exosome using FA targeting. (A) qRT-PCR assay of surviving siRNA expression in KB cells at 72 h. Gene knockdown efficiency of folate displaying exosome and folate-siRNA in cells after incubation with FA-survivin, FA/Exo or lipofectamine 2000 for 4 h. (n = 4, two-tail t-test) (B) Inhibition of tumor growth by folate displaying exosome, exosome without ligand and folate-SiRNA. 0.5 mg/kg survivin siRNA formulated with or without exosome were repeated I.V. by tail-vein injection every two days. Multiple t-test with Holm-Sidak correction on day 12 indicates statistic significant (n = 5) for FA/Exo/3WJ-Sur compare to: 1) 3WJ/Exo/3WJ-Sur (p < .0001), 2) FA-3WJ-Sur (p = .0002) and 3) PBS (p = .003). (Reprinted with permission from Zheng et al. Folate-displaying exosome mediated cytosolic delivery of siRNA avoiding endosome trapping. J Control Release. 2019, 311–312:43–49. Copyright © 2019 Elsevier B.V.)

The most important aspect in therapeutic delivery is whether the delivered therapeutics can inhibit cancer growth. Folate-displaying exosomes were tested to evaluate their inhibition on colorectal cancer growth in mice. Antiapoptotic factor survivin is highly expressed in many types of malicious cancer. Therefore, exosomes were loaded with siRNA that targets the survivin gene and the siRNA-loaded exosomes were displayed with FA-3WJ arrowtail nanoparticles (Pi et al., 2018). The functionalized FA/exosome complex was then tested in KB cell-derived cancer xenograft mouse model. The mice were intravenously injected with FA/exosome/survivin siRNA at a dose of 0.5 mg siRNA/kg of mice body weight for six doses every two days (Pi et al., 2018; Li et al., 2018b). As demonstrated by Figure 51.16b, the group treated with FA/exosome/survivin siRNA significantly suppressed *in vivo* tumor growth as measured by tumor volume and tumor weight, compared to the control group of FA-siRNA without exosome. Moreover, the displaying of folate on the surface of exosomes was a requirement to ensure tumor suppression since it enabled the siRNA-loaded exosome to target folate receptor overexpressing cancer cells.

Folate, as a small molecule ligand for folate receptor overexpressed on the surface of many epithelial cancer cells, has shown high efficiency and specificity for cancer targeting. However, when it was used for the delivery of siRNA, endosome trapping has been a major issue. In this study, we revealed an even distribution of siRNA in the cytoplasm and little overlap with endosome/lysosome staining in cells when delivered by FA-decorated exosomes, suggesting an effective cytosol delivery property of using exosome combined with folate targeting (Figures 51.13 and 51.14). When the survivin siRNA was delivered using exosome without folate ligand, no significant suppression effect was observed in mice (Figure 51.16b). The results suggest that while the negatively charged RNA displayed on exosome surface might be able to minimize the non-specific accumulation in healthy organs, the folate ligand displaying on exosome surface strongly promoted the targeted delivery to tumors and enhanced the function of the siRNA payload. This directly supports the conclusion that, with the aid of exosomes, folate can serve as efficient ligand for targeted delivery to cancer cells and overcome the endosome trapping problem when siRNA is delivered via folate receptor. The high efficiency of this system for gene silencing *in vitro* and evidence of cancer inhibition in animal trials will renew the concept and interest in using FA as cancer-targeting ligand in human cancer therapy.

REFERENCES

Adler, J., Parmryd, I., 2010. Quantifying colocalization by correlation: the Pearson correlation coefficient is superior to the Mander's overlap coefficient. Cytometry A 77, 733–742.

Al-Nedawi, K., Meehan, B., Micallef, J., Lhotak, V., May, L., Guha, A., Rak, J., 2008. Intercellular transfer of the oncogenic receptor EGFRvIII by microvesicles derived from tumour cells. Nat Cell Biol 10, 619–624.

Bandara, N. A., Hansen, M. J., Low, P. S., 2014. Effect of receptor occupancy on folate receptor internalization. Mol Pharm 11, 1007–1013.

Batrakova, E. V., Kim, M. S., 2015. Using exosomes, naturally-equipped nanocarriers, for drug delivery. J Control Release.

Ben, D. S., David, S., Herve-Aubert, K., Falanga, A., Galdiero, S., Iard-Vannier, E., Chourpa, I., Munnier, E., 2018. Formulation and in vitro evaluation of a siRNA delivery nanosystem decorated with gH625 peptide for triple negative breast cancer theranosis. Eur J Pharm Biopharm 131, 99–108.

Binzel, D., Shu, Y., Li, H., Sun, M., Zhang, Q., Shu, D., Guo, B., Guo, P., 2016. Specific Delivery of MiRNA for High Efficient Inhibition of Prostate Cancer by RNA Nanotechnology. Molecular Therapy 24, 1267–1277.

Binzel, D. W., Khisamutdinov, E. F., Guo, P., 2014. Entropy-driven one-step formation of Phi29 pRNA 3WJ from three RNA fragments. Biochemistry 53, 2221–2231.

Bunge, A., Loew, M., Pescador, P., Arbuzova, A., Brodersen, N., Kang, J., Dahne, L., Liebscher, J., Herrmann, A., Stengel, G., Huster, D., 2009. Lipid membranes carrying lipophilic cholesterol-based oligonucleotides-characterization and application on layer-by-layer coated particles. J Phys Chem B 113, 16425–16434.

Chu, T. C., Twu, K. Y., Ellington, A. D., Levy, M., 2006. Aptamer mediated siRNA delivery. Nucleic Acids Research 34, e73.

Cisterna, B. A., Kamaly, N., Choi, W. I., Tavakkoli, A., Farokhzad, O. C., Vilos, C., 2016. Targeted nanoparticles for colorectal cancer. Nanomedicine (Lond) 11, 2443–2456.

Colombo, M., Raposo, G., Thery, C., 2014. Biogenesis, secretion, and intercellular interactions of exosomes and other extracellular vesicles. Annu Rev Cell Dev Biol 30, 255–289.

Cui, D., Zhang, C., Liu, B., Shu, Y., Du, T., Shu, D., Wang, K., Dai, F., Liu, Y., Li, C., Pan, F., Yang, Y., Ni, J., Li, H., Brand-Saberi, B., Guo, P., 2015. Regression of gastric cancer by systemic injection of RNA nanoparticles carrying both ligand and siRNA. Scientific Reports 5, 10726.

Dassie, J. P., Hernandez, L. I., Thomas, G. S., Long, M. E., Rockey, W. M., Howell, C. A., Chen, Y., Hernandez, F. J., Liu, X. Y., Wilson, M. E., Allen, L. A., Vaena, D. A., Meyerholz, D. K., Giangrande, P. H., 2014. Targeted inhibition of prostate cancer metastases with an RNA aptamer to prostate-specific membrane antigen. Mol Ther 22, 1910–1922.

Deng, Z., Rong, Y., Teng, Y., Mu, J., Zhuang, X., Tseng, M., Samykutty, A., Zhang, L., Yan, J., Miller, D., Suttles, J., Zhang, H. G., 2017. Broccoli-Derived Nanoparticle Inhibits Mouse Colitis by Activating Dendritic Cell AMP-Activated Protein Kinase. Mol Ther 25, 1641–1654.

Deniz, A. A., Dahan, M., Grunwell, J. R., Ha, T., Faulhaber, A. E., Chemla, D. S., Weiss, S., Schultz, P. G., 1999. Single-pair fluorescence resonance energy transfer on freely diffusing molecules: observation of Forster distance dependence and subpopulations. Proc Natl Acad Sci USA 96, 3670–3675.

Didiot, M. C., Hall, L. M., Coles, A. H., Haraszti, R. A., Godinho, B. M., Chase, K., Sapp, E., Ly, S., Alterman, J. F., Hassler, M. R., Echeverria, D., Raj, L., Morrissey, D. V., DiFiglia, M., Aronin, N., Khvorova, A., 2016. Exosome-mediated Delivery of Hydrophobically Modified siRNA for Huntingtin mRNA Silencing. Mol Ther 24, 1836–1847.

Dong, Y., Siegwart, D. J., Anderson, D. G., 2019. Strategies, design, and chemistry in siRNA delivery systems. Adv Drug Deliv. Rev.

Dreyer, F., Baur, A., 2016. Biogenesis and Functions of Exosomes and Extracellular Vesicles. Methods Mol Biol 1448, 201–216.

Dunn, K. W., Kamocka, M. M., McDonald, J. H., 2011. A practical guide to evaluating colocalization in biological microscopy. Am J Physiol Cell Physiol 300, C723–C742.

El-Andaloussi S., Mager, I., Breakefield, X. O., Wood, M. J., 2013a. Extracellular vesicles: biology and emerging therapeutic opportunities. Nat Rev Drug Discov 12, 347–357.

El-Andaloussi, S., Lakhal, S., Mager, I., Wood, M. J., 2013b. Exosomes for targeted siRNA delivery across biological barriers. Adv Drug Deliv Rev 65, 391–397.

El-Andaloussi, S., Lee, Y., Lakhal-Littleton, S., Li, J., Seow, Y., Gardiner, C., Varez-Erviti, L., Sargent, I. L., Wood, M. J., 2012. Exosome-mediated delivery of siRNA in vitro and in vivo. Nat Protoc. 7, 2112–2126.

Esposito, C. L., Passaro, D., Longobardo, I., Condorelli, G., Marotta, P., Affuso, A., de Franciscis, V., Cerchia, L., 2011. A neutralizing RNA aptamer against EGFR causes selective apoptotic cell death. PLoS ONE 6, e24071.

Fasehee, H., Dinarvand, R., Ghavamzadeh, A., Esfandyari-Manesh, M., Moradian, H., Faghihi, S., Ghaffari, S. H., 2016. Delivery of disulfiram into breast cancer cells using folate-receptor-targeted PLGA-PEG nanoparticles: in vitro and in vivo investigations. J Nanobiotechnology 14, 32.

Franzen, C. A., Blackwell, R. H., Foreman, K. E., Kuo, P. C., Flanigan, R. C., Gupta, G. N., 2016. Urinary Exosomes: The Potential for Biomarker Utility, Intercellular Signaling and Therapeutics in Urological Malignancy. J Urol 195, 1331–1339.

Garber, K., 2018. Alnylam launches era of RNAi drugs. Nat Biotechnol 36, 777–778.

Guo, P., 2010. The emerging field of RNA nanotechnology. Nature Nanotechnology 5, 833–842.

Guo, P., Erickson, S., Anderson, D., 1987. A small viral RNA is required for *in vitro* packaging of bacteriophage phi29 DNA. Science 236, 690–694.

Guo, P., Zhang, C., Chen, C., Trottier, M., Garver, K., 1998. Inter-RNA interaction of phage phi29 pRNA to form a hexameric complex for viral DNA transportation. Molecular Cell 2, 149–155.

Guo, S., Huang, Y., Jiang, Q., Sun, Y., Deng, L., Liang, Z., Du, Q., Xing, J., Zhao, Y., Wang, P. C., Dong, A., Liang, X. J., 2010. Enhanced gene delivery and siRNA silencing by gold nanoparticles coated with charge-reversal polyelectrolyte. ACS Nano 4, 5505–5511.

Guo, S., Piao, X., Li, H., Guo, P., 2018. Methods for construction and characterization of simple or special multifunctional RNA nanoparticles based on the 3WJ of phi29 DNA packaging motor. Methods 143, 121–133.

Haque, F., Shu, D., Shu, Y., Shlyakhtenko, L., Rychahou, P., Evers, M., Guo, P., 2012. Ultrastable synergistic tetravalent RNA nanoparticles for targeting to cancers. Nano Today 7, 245–257.

Hung, M. E., Leonard, J. N., 2015. Stabilization of exosome-targeting peptides via engineered glycosylation. J Biol Chem 290, 8166–8172.

Hynes, N. E., Lane, H. A., 2005. ERBB receptors and cancer: the complexity of targeted inhibitors. Nat Rev Cancer 5, 341–354.

Jasinski, D., Schwartz, C., Haque, F., Guo, P., 2015. Large Scale Purification of RNA Nanoparticles by Preparative Ultracentrifugation. Methods in Molecular Biology 1297, 67–82.

Johnstone, R. M., Mathew, A., Mason, A. B., Teng, K., 1991. Exosome formation during maturation of mammalian and avian reticulocytes: evidence that exosome release is a major route for externalization of obsolete membrane proteins. J Cell Physiol 147, 27–36.

Ju, S., Mu, J., Dokland, T., Zhuang, X., Wang, Q., Jiang, H., Xiang, X., Deng, Z. B., Wang, B., Zhang, L., Roth, M., Welti, R., Mobley, J., Jun, Y., Miller, D., Zhang, H. G., 2013. Grape exosome-like nanoparticles induce intestinal stem cells and protect mice from DSS-induced colitis. Mol Ther 21, 1345–1357.

Kamerkar, S., LeBleu, V. S., Sugimoto, H., Yang, S., Ruivo, C. F., Melo, S. A., Lee, J. J., Kalluri, R., 2017. Exosomes facilitate therapeutic targeting of oncogenic KRAS in pancreatic cancer. Nature 546, 498–503.

Khaled, A., Guo, S., Li, F., Guo, P., 2005. Controllable Self-Assembly of Nanoparticles for Specific Delivery of Multiple Therapeutic Molecules to Cancer Cells Using RNA Nanotechnology. Nano Letters 5, 1797–1808.

Kumar, D., Gupta, D., Shankar, S., Srivastava, R. K., 2015. Biomolecular characterization of exosomes released from cancer stem cells: Possible implications for biomarker and treatment of cancer. Oncotarget 6, 3280–3291.

Lamichhane, T. N., Raiker, R. S., Jay, S. M., 2015. Exogenous DNA Loading into Extracellular Vesicles via Electroporation is Size-Dependent and Enables Limited Gene Delivery. Mol Pharm 12, 3650–3657.

Leamon, C. P., Reddy, J. A., 2004. Folate-targeted chemotherapy. Adv Drug Deliv Rev 56, 1127–1141.

Lee, S. M., Chen, H., O'Halloran, T. V., Nguyen, S. T., 2009. "Clickable" polymer-caged nanobins as a modular drug delivery platform. J Am Chem Soc 131, 9311–9320.

Lee, T. J., Haque, F., Shu, D., Yoo, J. Y., Li, H., Yokel, R. A., Horbinski, C., Kim, T. H., Kim, S.-H., Nakano, I., Kaur, B., Croce, C. M., Guo, P., 2015. RNA nanoparticles as a vector for targeted siRNA delivery into glioblastoma mouse model. Oncotarget 6, 14766–14776.

Li, P., Kaslan, M., Lee, S. H., Yao, J., Gao, Z., 2017. Progress in Exosome Isolation Techniques. Theranostics. 7, 789–804.

Li, S., Omi, M., Cartieri, F., Konkolewicz, D., Mao, G., Gao, H., Averick, S. E., Mishina, Y., Matyjaszewski, K., 2018a. Cationic Hyperbranched Polymers with Biocompatible Shells for siRNA Delivery. Biomacromolecules 19, 3754–3765.

Li, Z., Wang, H., Yin, H., Bennett, C., Zhang, H. G., Guo, P., 2018b. Arrowtail RNA for Ligand Display on Ginger Exosome-like Nanovesicles to Systemic Deliver siRNA for Cancer Suppression. Sci Rep 8, 14644.

Linares, R., Tan, S., Gounou, C., Arraud, N., Brisson, A. R., 2015. High-speed centrifugation induces aggregation of extracellular vesicles. J Extracell Vesicles 4, 29509.

Marcus, M. E., Leonard, J. N., 2013. FedExosomes: Engineering Therapeutic Biological Nanoparticles that Truly Deliver. Pharmaceuticals (Basel) 6, 659–680.

McKelvey, K. J., Powell, K. L., Ashton, A. W., Morris, J. M., McCracken, S. A., 2015. Exosomes: Mechanisms of Uptake. J Circ Biomark 4, 7.

Melo, S. A., Luecke, L. B., Kahlert, C., Fernandez, A. F., Gammon, S. T., Kaye, J., LeBleu, V. S., Mittendorf, E. A., Weitz, J., Rahbari, N., Reissfelder, C., Pilarsky, C., Fraga, M. F., Piwnica-Worms, D., Kalluri, R., 2015. Glypican-1 identifies cancer exosomes and detects early pancreatic cancer. Nature 523, 177–182.

Melo, S. A., Sugimoto, H., O'Connell, J. T., Kato, N., Villanueva, A., Vidal, A., Qiu, L., Vitkin, E., Perelman, L. T., Melo, C. A., Lucci, A., Ivan, C., Calin, G. A., Kalluri, R., 2014. Cancer Exosomes Perform Cell-Independent MicroRNA Biogenesis and Promote Tumorigenesis. Cancer Cell 26, 707–721.

Mu, J., Zhuang, X., Wang, Q., Jiang, H., Deng, Z. B., Wang, B., Zhang, L., Kakar, S., Jun, Y., Miller, D., Zhang, H. G., 2014. Interspecies communication between plant and mouse gut host cells through edible plant derived exosome-like nanoparticles. Mol Nutr Food Res 58, 1561–1573.

Munagala, R., Aqil, F., Jeyabalan, J., Gupta, R. C., 2016. Bovine milk-derived exosomes for drug delivery. Cancer Lett 371, 48–61.

Norman, D. G., Grainger, R. J., Uhrin, D., Lilley, D. M., 2000. Location of cyanine-3 on double-stranded DNA: importance for fluorescence resonance energy transfer studies. Biochemistry 39, 6317–6324.

Ohno, S., Takanashi, M., Sudo, K., Ueda, S., Ishikawa, A., Matsuyama, N., Fujita, K., Mizutani, T., Ohgi, T., Ochiya, T., Gotoh, N., Kuroda, M., 2013. Systemically injected exosomes targeted to EGFR deliver antitumor microRNA to breast cancer cells. Mol Ther 21, 185–191.

Paduano, F., Villa, R., Pennati, M., Folini, M., Binda, M., Daidone, M. G., Zaffaroni, N., 2006. Silencing of survivin gene by small interfering RNAs produces supra-additive growth suppression in combination with 17-allylamino-17-demethoxygeldanamycin in human prostate cancer cells. Molecular Cancer Therapeutics 5, 179–186.

Parker, N., Turk, M. J., Westrick, E., Lewis, J. D., Low, P. S., Leamon, C. P., 2005. Folate receptor expression in carcinomas and normal tissues determined by a quantitative radioligand binding assay. Anal Biochem 338, 284–293.

Pecot, C. V., Calin, G. A., Coleman, R. L., Lopez-Berestein, G., Sood, A. K., 2011. RNA interference in the clinic: challenges and future directions. Nat Rev Cancer 11, 59–67.

Pfeiffer, I., Hook, F., 2004. Bivalent cholesterol-based coupling of oligonucletides to lipid membrane assemblies. J Am Chem Soc 126, 10224–10225.

Pi, F., Binzel, D. W., Lee, T. J., Li, Z., Sun, M., Rychahou, P., Li, H., Haque, F., Wang, S., Croce, C. M., Guo, B., Evers, B. M., Guo, P., 2018. Nanoparticle orientation to control RNA loading and ligand display on extracellular vesicles for cancer regression. Nat Nanotechnol 13, 82–89.

Prada, I., Meldolesi, J., 2016. Binding and Fusion of Extracellular Vesicles to the Plasma Membrane of Their Cell Targets. Int J Mol Sci 17.

Priegue, J. M., Crisan, D. N., Martinez-Costas, J., Granja, J. R., Fernandez-Trillo, F., Montenegro, J., 2016. In Situ Functionalized Polymers for siRNA Delivery. Angew Chem Int Ed Engl 55, 7492–7495.

Quesenberry, P. J., Aliotta, J., Camussi, G., bdel-Mageed, A. B., Wen, S., Goldberg, L., Zhang, H. G., Tetta, C., Franklin, J., Coffey, R. J., Danielson, K., Subramanya, V., Ghiran, I., Das, S., Chen, C. C., Pusic, K. M., Pusic, A. D., Chatterjee, D., Kraig, R. P., Balaj, L., Dooner, M., 2015. Potential functional applications of extracellular vesicles: a report by the NIH Common Fund Extracellular RNA Communication Consortium. J Extracell. Vesicles 4, 27575.

Rak, J., 2015. Cancer: Organ-seeking vesicles. Nature 527, 312–314.

Reddy, J. A., Abburi, C., Hofland, H., Howard, S. J., Vlahov, I., Wils, P., Leamon, C. P., 2002. Folate-targeted, cationic liposome-mediated gene transfer into disseminated peritoneal tumors. Gene Ther 9, 1542–1550.

Regente, M., Corti-Monzon, G., Maldonado, A. M., Pinedo, M., Jorrin, J., de la Canal, L., 2009. Vesicular fractions of sunflower apoplastic fluids are associated with potential exosome marker proteins. FEBS Lett 583, 3363–3366.

Rockey, W. M., Hernandez, F. J., Huang, S. Y., Cao, S., Howell, C. A., Thomas, G. S., Liu, X. Y., Lapteva, N., Spencer, D. M., McNamara, J. O., Zou, X., Chen, S. J., Giangrande, P. H., 2011. Rational truncation of an RNA aptamer to prostate-specific membrane antigen using computational structural modeling. Nucleic Acid Ther 21, 299–314.

Saba, N. F., Wang, X., Muller, S., Tighiouart, M., Cho, K., Nie, S., Chen, Z., Shin, D. M., 2009. Examining expression of folate receptor in squamous cell carcinoma of the head and neck as a target for a novel nanotherapeutic drug. Head Neck 31, 475–481.

Sabharanjak, S., Mayor, S., 2004. Folate receptor endocytosis and trafficking. Adv Drug Deliv Rev 56, 1099–1109.

Semple, S. C., Akinc, A., Chen, J., Sandhu, A. P., Mui, B. L., Cho, C. K., Sah, D. W., Stebbing, D., Crosley, E. J., Yaworski, E., Hafez, I. M., Dorkin, J. R., Qin, J., Lam, K., Rajeev, K. G., Wong, K. F., Jeffs, L. B., Nechev, L., Eisenhardt, M. L., Jayaraman, M., Kazem, M., Maier, M. A., Srinivasulu, M., Weinstein, M. J., Chen, Q., Alvarez, R., Barros, S. A., De, S., Klimuk, S. K., Borland, T., Kosovrasti, V., Cantley, W. L., Tam, Y. K., Manoharan, M., Ciufolini, M. A., Tracy, M. A., de Fougerolles, A., Maclachlan, I., Cullis, P. R., Madden, T. D., Hope, M. J., 2010. Rational design of cationic lipids for siRNA delivery. Nat Biotechnol 28, 172–176.

Shelke, G. V., Lasser, C., Gho, Y. S., Lotvall, J., 2014. Importance of exosome depletion protocols to eliminate functional and RNA-containing extracellular vesicles from fetal bovine serum. J Extracell Vesicles. 3.

Shi, H., Guo, J., Li, C., Wang, Z., 2015. A current review of folate receptor alpha as a potential tumor target in non-small-cell lung cancer. Drug Des Devel Ther 9, 4989–4996.

Shtam, T. A., Kovalev, R. A., Varfolomeeva, E. Y., Makarov, E. M., Kil, Y. V., Filatov, M. V., 2013. Exosomes are natural carriers of exogenous siRNA to human cells in vitro. Cell Commun. Signal 11, 88.

Shu, D., Shu, Y., Haque, F., Abdelmawla, S., Guo, P., 2011a. Thermodynamically stable RNA three-way junctions for constructing multifuntional nanoparticles for delivery of therapeutics. Nature Nanotechnology 6, 658–667.

Shu, D., Khisamutdinov, E., Zhang, L., Guo, P., 2013. Programmable folding of fusion RNA complex driven by the 3WJ motif of phi29 motor pRNA. Nucleic Acids Research 42, e10.

Shu, D., Li, H., Shu, Y., Xiong, G., Carson, W. E., Haque, F., Xu, R., Guo, P., 2015. Systemic delivery of anti-miRNA for suppression of triple negative breast cancer utilizing RNA nanotechnology. ACS Nano 9, 9731–9740.

Shu, D., Zhang, H., Petrenko, R., Meller, J., Guo, P., 2010. Dual-channel single-molecule fluorescence resonance energy transfer to establish distance parameters for RNA nanoparticles. ACS Nano 4, 6843–6853.

Shu, Y., Cinier, M., Shu, D., Guo, P., 2011b. Assembly of multifunctional phi29 pRNA nanoparticles for specific delivery of siRNA and other therapeutics to targeted cells. Methods 54, 204–214.

Stranford, D. M., Leonard, J. N., 2017. Delivery of Biomolecules via Extracellular Vesicles: A Budding Therapeutic Strategy. Adv Genet 98, 155–175.

Sun, D., Zhuang, X., Zhang, S., Deng, Z. B., Grizzle, W., Miller, D., Zhang, H. G., 2013. Exosomes are endogenous nanoparticles that can deliver biological information between cells. Adv Drug Deliv Rev 65, 342–347.

Szatanek, R., Baran, J., Siedlar, M., Baj-Krzyworzeka, M., 2015. Isolation of extracellular vesicles: Determining the correct approach (Review). Int J Mol Med 36, 11–17.

Teng, Y., Mu, J., Hu, X., Samykutty, A., Zhuang, X., Deng, Z., Zhang, L., Cao, P., Yan, J., Miller, D., Zhang, H. G., 2016. Grapefruit-derived nanovectors deliver miR-18a for treatment of liver metastasis of colon cancer by induction of M1 macrophages. Oncotarget 7, 25683–25697.

Teo, P. Y., Yang, C., Whilding, L. M., Parente-Pereira, A. C., Maher, J., George, A. J., Hedrick, J. L., Yang, Y. Y., Ghaem-Maghami, S., 2015. Ovarian cancer immunotherapy using PD-L1 siRNA targeted delivery from folic acid-functionalized polyethylenimine: strategies to enhance T cell killing. Adv Healthc Mater 4, 1180–1189.

Thery, C., Amigorena, S., Raposo, G., Clayton, A., 2006. Isolation and characterization of exosomes from cell culture supernatants and biological fluids. Curr Protoc Cell Biol Chapter 3, Unit 3.22.

Tian, Y., Li, S., Song, J., Ji, T., Zhu, M., Anderson, G. J., Wei, J., Nie, G., 2014. A doxorubicin delivery platform using engineered natural membrane vesicle exosomes for targeted tumor therapy. Biomaterials 35, 2383–2390.

Toloue, M. M., Ford, L. P., 2011. Antibody targeted siRNA delivery. Methods Mol Biol 764, 123–139.

Turek, J. J., Leamon, C. P., Low, P. S., 1993. Endocytosis of folate-protein conjugates: ultrastructural localization in KB cells. J Cell Sci 106 (Pt 1), 423–430.

Valadi, H., Ekstrom, K., Bossios, A., Sjostrand, M., Lee, J. J., Lotvall, J. O., 2007. Exosome-mediated transfer of mRNAs and microRNAs is a novel mechanism of genetic exchange between cells. Nat Cell Biol 9, 654–659.

van Dam, G. M., Themelis, G., Crane, L. M., Harlaar, N. J., Pleijhuis, R. G., Kelder, W., Sarantopoulos, A., de Jong, J. S., Arts, H. J., van der Zee, A. G., Bart, J., Low, P. S., Ntziachristos, V., 2011. Intraoperative tumor-specific fluorescence imaging in ovarian cancer by folate receptor-alpha targeting: first in-human results. Nat Med 17, 1315–1319.

van Dommelen, S. M., Vader, P., Lakhal, S., Kooijmans, S. A., van Solinge, W. W., Wood, M. J., Schiffelers, R. M., 2012. Microvesicles and exosomes: opportunities for cell-derived membrane vesicles in drug delivery. J Control Release 161, 635–644.

van Dongen, H. M., Masoumi, N., Witwer, K. W., Pegtel, D. M., 2016. Extracellular Vesicles Exploit Viral Entry Routes for Cargo Delivery. Microbiol. Mol. Biol. Rev. 80, 369–386.

varez-Erviti, L., Seow, Y., Yin, H., Betts, C., Lakhal, S., Wood, M. J., 2011b. Delivery of siRNA to the mouse brain by systemic injection of targeted exosomes. Nat Biotechnol 29, 341–345.

varez-Erviti, L., Seow, Y., Yin, H., Betts, C., Lakhal, S., Wood, M. J., 2011c. Delivery of siRNA to the mouse brain by systemic injection of targeted exosomes. Nat Biotechnol 29, 341–345.

varez-Erviti, L., Seow, Y., Yin, H., Betts, C., Lakhal, S., Wood, M. J., 2011a. Delivery of siRNA to the mouse brain by systemic injection of targeted exosomes. Nat Biotechnol 29, 341–345.

Vestad, B., Llorente, A., Neurauter, A., Phuyal, S., Kierulf, B., Kierulf, P., Skotland, T., Sandvig, K., Haug, K. B. F., Ovstebo, R., 2017. Size and concentration analyses of extracellular vesicles by nanoparticle tracking analysis: a variation study. J Extracell. Vesicles. 6, 1344087.

Wang, B., Zhuang, X., Deng, Z. B., Jiang, H., Mu, J., Wang, Q., Xiang, X., Guo, H., Zhang, L., Dryden, G., Yan, J., Miller, D., Zhang, H. G., 2014. Targeted drug delivery to intestinal macrophages by bioactive nanovesicles released from grapefruit. Mol Ther 22, 522–534.

Wang, Q., Ren, Y., Mu, J., Egilmez, N. K., Zhuang, X., Deng, Z., Zhang, L., Yan, J., Miller, D., Zhang, H. G., 2015. Grapefruit-Derived Nanovectors Use an Activated Leukocyte Trafficking Pathway to Deliver Therapeutic Agents to Inflammatory Tumor Sites. Cancer Res 75, 2520–2529.

Wang, Q., Zhuang, X., Mu, J., Deng, Z. B., Jiang, H., Zhang, L., Xiang, X., Wang, B., Yan, J., Miller, D., Zhang, H. G., 2013. Delivery of therapeutic agents by nanoparticles made of grapefruit-derived lipids. Nat Commun 4, 1867.

Whitehead, K. A., Langer, R., Anderson, D. G., 2009. Knocking down barriers: advances in siRNA delivery. Nat Rev Drug Discov 8, 129–138.

Wiklander, O. P., Nordin, J. Z., O'Loughlin, A., Gustafsson, Y., Corso, G., Mager, I., Vader, P., Lee, Y., Sork, H., Seow, Y., Heldring, N., varez-Erviti, L., Smith, C. E., Le, B. K., Macchiarini, P., Jungebluth, P., Wood,

M. J., Andaloussi, S. E., 2015. Extracellular vesicle in vivo biodistribution is determined by cell source, route of administration and targeting. J Extracell Vesicles 4, 26316.

Witwer, K. W., Buzas, E. I., Bemis, L. T., Bora, A., Lasser, C., Lotvall, J., Nolte-'t Hoen, E. N., Piper, M. G., Sivaraman, S., Skog, J., Thery, C., Wauben, M. H., Hochberg, F., 2013. Standardization of sample collection, isolation and analysis methods in extracellular vesicle research. J Extracell Vesicles 2.

Yang, M., Jin, H., Chen, J., Ding, L., Ng, K. K., Lin, Q., Lovell, J. F., Zhang, Z., Zheng, G., 2011. Efficient cytosolic delivery of siRNA using HDL-mimicking nanoparticles. Small 7, 568–573.

Yang, Z., Lee, J. H., Jeon, H. M., Han, J. H., Park, N., He, Y., Lee, H., Hong, K. S., Kang, C., Kim, J. S., 2013. Folate-based near-infrared fluorescent theranostic gemcitabine delivery. J Am Chem Soc 135, 11657–11662.

Zeringer, E., Barta, T., Li, M., Vlassov, A. V., 2015. Strategies for isolation of exosomes. Cold Spring Harb Protoc 2015, 319–323.

Zhang, D., Lee, H., Zhu, Z., Minhas, J. K., Jin, Y., 2017. Enrichment of selective miRNAs in exosomes and delivery of exosomal miRNAs in vitro and in vivo. Am J Physiol Lung Cell Mol Physiol 312, L110–L121.

Zhang, H., Shu, D., Browne, M., Guo, P., 2009. Approaches for stoichiometry and distance determination of nanometer bio-complex by dual-channel single molecule imaging. IEEE/NIH Life Science Systems and Applications Workshop 124–127.

Zhang, H., Endrizzi, J. A., Shu, Y., Haque, F., Sauter, C., Shlyakhtenko, L. S., Lyubchenko, Y., Guo, P., Chi, Y. I., 2013a. Crystal Structure of 3WJ Core Revealing Divalent Ion-promoted Thermostability and Assembly of the Phi29 Hexameric Motor pRNA. RNA 19, 1226–1237.

Zhang, M., Viennois, E., Prasad, M., Zhang, Y., Wang, L., Zhang, Z., Han, M. K., Xiao, B., Xu, C., Srinivasan, S., Merlin, D., 2016a. Edible ginger-derived nanoparticles: A novel therapeutic approach for the prevention and treatment of inflammatory bowel disease and colitis-associated cancer. Biomaterials 101, 321–340.

Zhang, M., Viennois, E., Xu, C., Merlin, D., 2016b. Plant derived edible nanoparticles as a new therapeutic approach against diseases. Tissue Barriers 4, e1134415.

Zhang, Y., Wang, Z., Gemeinhart, R. A., 2013b. Progress in microRNA delivery. J Control Release 172, 962–974.

Zhuang, X., Deng, Z. B., Mu, J., Zhang, L., Yan, J., Miller, D., Feng, W., McClain, C. J., Zhang, H. G., 2015. Ginger-derived nanoparticles protect against alcohol-induced liver damage. J Extracell. Vesicles. 4, 28713.

Zomer, A., Vendrig, T., Hopmans, E. S., van Eijndhoven, M., Middeldorp, J. M., Pegtel, D. M., 2010. Exosomes: Fit to deliver small RNA. Commun. Integr. Biol 3, 447–450.

Zwicke, G. L., Mansoori, G. A., Jeffery, C. J., 2012. Utilizing the folate receptor for active targeting of cancer nanotherapeutics. Nano Rev. 3.

52 Application of RNA Aptamers in Nanotechnology and Therapeutics

Hua Shi
University at Albany, New York, USA

CONTENTS

INTRODUCTION

Most biological functions arise from molecular interactions rather than individual molecules. These interactions are often highly specific; that is, they are molecular recognition processes. In the course of natural evolution, biological functionality is expanded by the generation of new and specific molecular interactions (through gene duplication and differentiation), especially those between proteins and other macromolecules. Subversion or deception of native molecular recognition in a biological system through mimicry is the modus operandi of most, if not all, bioactive substances, including pharmaceuticals. In the past, most drugs were discovered serendipitously when they happened to be able to bind a receptor tightly and cause a desirable phenotypic change in a cell or organism. The advent of RNA aptamers made it possible to create ligands for intended targets in a more efficient way (Ellington and Szostak 1990; Tuerk and Gold 1990) through a "molecular breeding" process. Like small organic molecules, aptmers are able to rapidly and tightly bind specific protein domains or a specific site on a domain in living cells or organisms. Like antibodies, they can be made to order specifically for a predetermined target.

The process of evolution can occur under different circumstances and with various materials if three conditions are met: variation, competition, and hereditary replication. In Darwinian organismic evolution, an individual is selected among a set of variants because its phenotype is adaptive and rewarded with materials and energy to reproduce through multiplying its genome. The concepts of genotype and phenotype, and the mapping between them, can be applied not only to organisms but also to molecules. For example, an RNA molecule's linear sequence can be regarded as its genotype and its folded three-dimensional shape its phenotype. It is in this sense that RNA molecules can be "bred" for desirable features. Multiple iterative cycles of selection from pools of moderate size can "compute" complex solutions to the problem of molecular recognition, as the best solutions selected from one pool are amplified to form the pool of next generation (Schuster 2001).

DOI: 10.1201/9781003001560-57

Although aptamers are often compared to antibodies, the analogy between an aptamer and an antibody is limited to their specificity and affinity for the target or antigen (Zhou and Rossi 2017). An antibody is composed of multiple domains and recognized by multiple partners, but an aptamer has only one binding site and it interacts with only one target. In fact it is a general phenomenon for individual proteins to bear multiple specific sites, so they can collectively form an interacting network that underlies the emergent features of living systems (Jeong et al. 2001). Inspired by the functional capability of proteins, we pioneered the method of stitching together multiple aptamers in a single molecular construct (Shi et al. 1999). These multivalent composite aptamers opened a gateway to connect, combine, organize, present, arrange, and articulate a diverse range of other molecules at the nanometer scale both within biological systems and for non-biological purposes.

A sense of proportion is important in order to appreciate the utility of RNA aptamers in nanotechnology. Bottom-up covalent chemical synthesis usually results in "small" molecules 100–3000 Daltons in molecular weight, which is usually smaller than 1 cubic nanometer. Top-down physical fabrication, mainly through lithography, often produces objects at micrometer scale. Non-covalent association of bio-polymers, such as protein and nucleic acids, yields assemblies with dimensions between 3 and 20 nm and may fill the gap in size. This is also the size range at which quantum confinement influences the electronic and optical properties of matter. In contrast to that of proteins, the formation of secondary structure in RNA causes significantly larger changes in free energy than those involved in tertiary interactions. As a result, rational modular design of constructs is possible at the level of secondary structure according to the basic properties of the conformational energy landscape of an RNA molecule. The creation of multivalent composite aptamers takes the advantage of these features (Xu and Shi 2009).

Although most aptamers have been selected against proteins and other biologically relevant targets, there is no intrinsic limitation for the in vitro selection methodology to generate aptamers for non-biological targets (Gold et al. 2010b). Most individual aptamers, even when their targets are proteins, were isolated outside the context of living systems. For this reason, an aptamer often interferes with the normal function of the protein when introduced into a cell or organism. Consequently, many aptamers are used as inhibitors of protein activity for therapeutic purposes (Nimjee et al. 2017). Multivalent aptamers can serve as molecular connectors, adaptors or bypasses in existing biological systems to rewire their regulatory networks. But their utility is not limited to biological systems: they can be used to adapt biological materials for predetermined non-biological purposes (i.e., for a purpose not necessarily affecting the fitness of an organism.)

This chapter surveys the application of RNA aptamers in nanotechnology in the context of both past technological developments and future research ambitions. It begins with an introduction to the in vitro selection methodology, focusing on its historical background and conceptual development (Section 2). Next, the utility of individual aptamers as reagents and therapeutics is presented, with related issues in chemical modification and pool construction to improve the performance of aptamers. A discussion follows on how to use individual aptamers as building blocks to design more complex molecules, in particular nanoparticles. After describing the features of multivalent aptamers, the utilities of nanoscale systems articulated by such aptameric constructs are highlighted in the next section. Finally, some approaches to making dynamic nano-assemblies or objects with the help of switchable aptamer derivatives are considered.

GENERATION OF APTAMERS BY *IN VITRO* SELECTION

In 1965, Sol Spiegelman's lab isolated Qβ RNA replicase, which enabled in vitro synthesis of RNA. This led to the idea of using RNA to observe Darwinian evolution at the molecular level. Spiegelman and his colleagues introduced Qβ RNA into a solution containing the Qβ replicase, free nucleotides, and some salts. In this environment, the RNA started to replicate, and from this mixture some RNAs were repeatedly moved to another tube with fresh solution. Because shorter RNA chains were able to replicate faster, the RNA became shorter and shorter in successive generations (Mills et al. 1967).

Influenced by these pioneering experiments, Menfred Eigen developed a theory of molecular evolution based on established physical principles, in a series of papers starting with one landmark publication (Eigen 1971). By analyzing the kinetics of competing replicators, he found natural selection to be a direct physical consequence of self-reproduction under conditions far from thermodynamic equilibrium. This theory provided a comprehensive framework for modeling evolutionary dynamics.

Evolution can be visualized in an abstract space of genotypes called sequence space (Maynard Smith 1970). For a population of RNA or DNA sequences with a particular length, the set of all possible sequences is represented in such a space, where each sequence is represented as a point next to a number of other sequences that differ from it at only one position. The capacity of each sequence to carry out a specified function allows the definition of a fitness landscape over sequence space with respect to this function. When the fitness values are plotted against all possible genotypes and their degree of similarity to form a "landscape," the peaks represent either local or global optima of fitness possessed by the corresponding sequences. In the process of evolution, a population of sequences migrates on this landscape towards the peaks along paths of non-decreasing mean fitness. The notion of sequence space relates the biophysics of evolution to information theory and proves to be illustrative and useful in this context.

The advent of two revolutionary technologies in the 1980s helped shape the applied molecular evolution methodology, because they transformed a continuous natural process occurring in chemical mixtures into discrete steps in cyclic protocols. First, stepwise chemical synthesis of oligonucleotides based on phosphoramidite chemistry made it possible to generate huge diverse populations of sequence in a combinatorial manner. Second, polymerase chain reaction (PCR) provided a means to propagate the population indefinitely and to synchronize the "life cycle" of molecules in the population. The availability of over-the-counter enzymes such as reverse transcriptase and viral RNA polymerases further facilitated the development of in vitro selection. In 1990, three labs led by Larry Gold (Tuerk and Gold 1990), Gerald Joyce (Robertson and Joyce 1990), and Jack Szostak (Ellington and Szostak 1990) developed similar selection schemes almost simultaneously. These schemes used randomized sequences flanked by constant primer annealing segments to generate original sequence pools for the in vitro selection of functional molecules. The procedure was dubbed SELEX (for Systematic Evolution of Ligands by Exponential Enrichment) by the Gold group. The Szostak group coined the term "aptamer", from Latin *aptus*, to describe the ligand generated in this procedure.

In an in vitro evolution experiment, variation is primarily embodied in the complexity of the initial unselected pool. Replication is realized by enzymatic polymerization of nucleic acids. Differential fitness is minimized during replication of the sequences and defined primarily by the conditions of the "selection" step. Such a selection and amplification process resembles a heterogeneous population of organisms replicating in synchronous cycles under competition. A typical SELEX experiment starts with a large randomized sequence pool containing 10^{14}–10^{16} species that fold into different shapes determined by their sequences. This pool is then subjected to iterative cycles of selection and amplification. In each cycle, a target such as a protein molecule is used to select from the pool any RNA molecules that bind to it. Following the partitioning of the bound RNA from the unbound, the bound fraction is amplified by RT-PCR to generate a new pool for the next cycle. Usually, RNA ligands with the highest affinity for the target protein will dominate the population in 8–12 rounds. At the end of the process the winning aptamers are cloned and sequenced for further characterization (Conrad et al. 1996).

The most important step of the in vitro selection method is the partitioning of the bound aptamer candidates from the unbound individuals. In the majority of early experiments, this step of selection is implemented by nitrocellulose filter, which non-specifically captures the protein target along with the RNA bound to it (Tuerk and Gold 1990). Over the years, more specific and efficient methods have been developed to overcome its shortcomings. A commonly used alternative is the specific immobilization of the target on a solid matrix, such as agarose, polystyrene, or magnetic beads through affinity tags (e.g., His, GST, or MBP) or antibodies (Zhang et al. 2019). In these protocols, aptamer binding can occur before or after target immobilization, and the resins can be used in batch or on a column. In particular, multiplex selection (selection of aptamers for multiple targets from a single library) has been successfully implemented with microcolumns (Latulippe et al. 2013; Szeto et al. 2013). Because the selection criterion of aptamers is the capability of binding to a target, several other methods traditionally used for binding assays have also been adapted to function as the partitioning device. These include electrophoresis (EMSA and capillary electrophoresis), microarray, and with the advent of nanofabrication, microfluidics (Zhang et al. 2019). In addition to purified proteins, other types of targets also have been used in SELEX with distinct partitioning methods. For non-protein small molecules, Capture-SELEX with immobilized pool (rather than targets) was designed (Lauridsen et al. 2018). On the other hand, intact cells have been used to isolate aptamers for targets not limited to cell-surface proteins (Shangguan et al. 2006; Kaur 2018). Moreover, in vivo SELEX have been used in several studies to isolate aptamers for tissue or organ specific disease biomarkers, which also could be useful in drug delivery (Mi et al. 2010; Cheng et al. 2013). Each of these different partitioning methods has its unique combination of advantages and disadvantages. In depth discussion of parameters that can be manipulated to produce different outcomes in some recent reviews are helpful in aiding the choices among them (Ozer et al. 2014; Rohloff et al. 2014; Blind and Blank 2015).

For many applications, "one-aptamer-for-one-target" is adequate. In fact, many experiments yielded primarily only one aptamer as the ultimate "winner" of selection. However, many targets are complex and may have multiple discrete sites for aptamers to recognize. The binding of a ligand to each discrete site involves defined sets of contacts, saturates at integral stoichiometry, and follows the mass action law. In this chapter, an aptamer's "target" is defined as the discrete binding site of an aptamer on a molecule, rather than the entire molecule. A target defined this way is analogous to an "epitope" for antibodies. A small compound may consist of only one such site, and in this case the entire molecule may be called a target; a macromolecule or a supra-molecular assembly often bears multiple sites that constitute a "target set." In order to use aptamers in the construction of nanoscale assemblies and objects, it is often desirable to have aptamers binding to non-overlapping patches on the surface of a single target-bearing molecule (Rinker et al. 2008). As it is, conventional SELEX methodology does not dictate where the isolated aptamers would bind. Although different domains of a protein may be used to isolated aptamers for each of them (Gong et al. 2012), it is difficult to isolate multiple aptamers binding to different sites on the same domain that cannot be physically separated. To address this problem, we have developed a battery of effective selection schemes to manipulate the evolutionary dynamics of aptamers, of which two are particularly useful. First, we could manipulate the relative availability between two or more inseparable sites by masking some of them using existing aptamers or other ligands (Shi et al. 2007). Second, once the sequence of an aptamer was identified, we could reduce its rate of enrichment by selectively eliminating it from the pool to allow other aptamers to become dominant (Shi et al. 2002). By combining these approaches we have been able to obtain multiple aptamers intentionally directed to discrete functional sites on a single protein structural domain.

Conventional SELEX is carried out until one or a very few different sequences are present in the final pool. Theoretically, these ultimate survivors are supposed to have been selected because of their strongest affinity. Practically, there was a limitation on the throughput (usually around 100) imposed by bacterial cloning and Sanger sequencing. However, selection does not happen only in the designated "selection steps," i.e., binding and partitioning. For this reason, the most highly represented sequences in the final pool are not necessarily the best binders (Schutze et al. 2011). As the high-throughput sequencing (HTS) technology matured, the traditional protocol has been improved to overcome this limitation. Less rounds of selection are required, yielding more sequenced candidates for screening and confirmation [Cho et al. 2010). The larger scale of sequencing not only allows the diversity of the late generations of selected pools to be revealed exhaustively, but also enables investigation of the selection process, interpretation of the enrichment trends, and identification of undesired selection biases (Hoinka et al. 2014). In particular, better aptamers have been isolated based on rapid enrichment in early cycles before being overrun by those that amplify more efficiently by PCR.

Over the years, both the process and the products of in vitro selection have been improved by bioinformatics. Simulations of aptamer selection have been use to address the complexity of the sequence pool and the multitude of experimental parameters. From the very early work (e.g., SELEXION (Irvine et al. 1991)) to recent studies (e.g., AptaSim (Hoinka et al. 2015)), they have provided insightful suggestions on how to optimize the protocol. Lately, in silico aptamer identification is being conducted using the HTS data. To facilitate this trend, several aptamer- specific bioinformatics toolkits that are ready to use and easy to access have been developed (Pitt et al. 2010; Alam et al. 2015; Thiel and Giangrande 2016). In addition, in silico aptamer optimization is also being pursued in earnest. As the sequence space coverage in the original pool is always very small, surrounding sequence space of an isolated aptamer is worth exploring to improve the affinity and other properties (Knight and Yarus 2003). Different approaches have been taken to achieve aptamer affinity maturation, from "resampling" the sequence pool (Kinghorn et al. 2016) to aptamer docking by molecular dynamics.

INDIVIDUAL APTAMERS AS REAGENTS AND THERAPEUTICS

A great number of aptamers has been generated by the SELEX process. Collectively they are capable of binding to a wide variety of targets with high affinity and specificity, as enumerated in many excellent reviews as well as databases (Gold et al. 1995; Wilson and Szostak 1999; Lee et al. 2004; Famulok et al. 2007). The constantly updated NIH clinical trial website (https://clinicaltrials.gov/) lists all aptamers currently in different stages of clinical trial. Both DNA and RNA aptamers often bind their targets with dissociation constants (K_d) in the low nanomolar or picomolar range and are able to discriminate between related proteins that share common structural features. In addition to their widespread utility as molecular probes in basic research and diagnostic applications, aptamers are quickly becoming an exciting new class of therapeutic agents (Nimjee et al. 2017; Dua et al. 2008; Keefe et al. 2010). With the advent of genomics and proteomics, interactions between proteins have emerged as suitable targets for therapeutic interventions (Golemis et al. 2002). However, protein surfaces in direct contact with each other usually involve an area of about 1600 Å^2 with relatively flat topography (Lo Conte et al. 1999), causing concerns about the amount of binding specificity that can be incorporated into small molecules of less than 500 Da with less than 500 Å^2 of total solvent-accessible surface area (Juliano et al. 2001). Individual aptamers are usually 25–50 nucleotides long and weigh 8–16 kDa, providing more surface area for improved interaction with proteins. Some aptamers possess other unique properties not found in small molecules. For example, we identified an aptamer for the TATA binding protein (TBP) that is capable of actively disrupting a preformed complex (Fan et al. 2004). Moreover, the activity of aptamers can be manipulated by oligonucleotide antidotes that are capable of base pairing with the aptamers to prevent them from forming the correct shape to bind their targets (Rusconi et al. 2004).

Unlike antibodies, aptamers are produced by a scalable in vitro process and in general display low immunogenicity or toxicity even when administered in pre-clinical doses several orders of magnitude greater than doses used in therapeutic applications (Pendergrast et al. 2005). They also compare favorably with other oligonucleotide-based pharmaceuticals. The targets of reagents such as antisense oligonucleotides and siRNA are located exclusively in the intracellular compartments, because they act at the gene or mRNA level; delivery of these molecules to the target sites is a formidable task. In contrast, aptamers can exert their function against extracellular targets, which are much easier to access (Pestourie et al. 2005). Although natural RNA and DNA have poor pharmacokinetics when administered by intravenous or subcutaneous injection, their bioavailability can be improved chemically to enhance their stability and to control their clearance. To render aptamers resistant to nuclease degradation, RNA modified at the 2′ position of pyrimidines with fluoro or amino groups can be used in the selection process (Kubik et al. 1997). Additional post-selection modification or substitution can further increase aptamer residence time in the blood. For example, renal clearance of aptamers smaller than 40 kDa can be minimized through conjugation with poly-ethylene glycol (PEG) or attachment to liposomes (Pendergrast et al. 2005). For more efficient delivery of aptamer or non-aptamer drugs, a "nanorocket" has been proposed, which would employ multiple aptamers to target specific organ, tissue, and cells successively (Zhang et al. 2019).

To expand and extend the application of aptamers, the issue of chemical modification should be viewed in a more general context. A heteropolymer such as RNA can be viewed as a homopoly-meric backbone appended with different side chains, in which the tendency of the backbone to form a regular helical structure is counterbalanced by the irregular intramolecular interactions caused by the sequence-specific positioning of different side chains. Modification of either the backbone (sugar) or the bases (often pyrimidines) is possible through the incorporation of modified nucleo-tides. However, as the building blocks become chemically more different from their unmodified counterparts, two issues need to be considered. First, if the monomeric building blocks are no longer recognizable as substrates by the RNA polymerase, specialized mutants of the enzyme are required to transcribe them (Chelliserrykattil and Ellington 2004; Pinheiro et al. 2012). Second, due to dif-ferences in structure between natural and modified RNA, post-selection modification of an aptamer often results in loss of activity. To avoid this problem, a "front-loaded" selection is preferred, in which the selection is performed using a pool already bearing the modifications (Lin et al. 1994).

As mentioned above, modification of natural RNA was primarily motivated by the necessity of reducing its sensitivity to the ubiquitous RNases. As a result, early works usually focused on the alteration of the 2′ hydroxyl group to make the backbone no longer recognizable as a substrate by the commonly occurring RNases. This type of backbone modification includes, but is not lim-ited to, 2′- fluoro-, 2′-amino, and 2′O-methyl-modified polynucleotides (Kubik et al. 1997; Pagratis et al. 1997). Other types of the sugar moiety or its substitute have also been used later to create xeno-nucleic acids (XNAs) that are capable of becoming aptamers (Pinheiro et al. 2012; Yu et al. 2012). A different and unique approach to avoid RNase digestion is to generate aptamers in the form of L-nucleic acids enantiomeric to natural D-nucleic acids, which was termed Spiegelmers (Klussmann et al. 1996). However, a motivation for modifying natural nucleic acid that is more rele-vant to the utility of aptamers in nanotechnology is to enhance the equilibrium and kinetic properties of aptamers beyond what is achievable by natural RNA or DNA. One effective means of achieving this goal is to increase the chemical diversity of the bases. An outstanding example in this category is the Slow Off-rate Modified Aptamers (SOMAmers), which are selected from libraries of single stranded DNAs containing 5-position modified pyrimidines (Gold et al. 2010a; Vaught et al. 2010; Gelinas et al. 2016). In these aptamers, the DNA backbone can simply be regarded as a modified RNA in which the hydroxyl group at the 2′ position is displaced by a hydrogen atom.

Built from just four building blocks, nucleic acids have a lower information density compared to polypeptides, which makes them more vulnerable to folding ambiguity. An approach different from "decorating" existing building blocks may be taken to mitigate this problem: to increase the num-ber of building blocks. By shuffling hydrogen bonding units, an expanded genetic alphabet may be obtained. For a recent example, the Artificially Expanded Genetic Information system (AEGIS) was created by observing the rules governing standard Watson–Crick base paring, in which 6 orthogonal pairs can form by 12 nucleotides (Biondi and Benner 2018). It also allows the incorporation of new functionalities not found in standard DNA or RNA; for example, a nitro group that would enhance the binding potential and allow the base to participate in general acid–base catalysis. To support this system, an entourage of enzymes have been created to interconvert AEGIS DNA and RNA as well as to manipulate them in other ways (Leal et al. 2015). SELEX experiments have been conducted with this system to generate aptamers with high affinity (Biondi et al. 2016).

A general topic related to "front-loading" mentioned discussed above is the construction of the initial aptamer candidate library, which has been conducted in different ways deviating from the traditional random sequence pool. For a randomized region of around 40 bases to fold into well-defined shapes, extensive base-paring (i.e., secondary structure) is required. Therefore, increasing the occurrence of secondary structures in the pool would facilitate successful selection of high-affinity aptamers. An early study using a library with a stem embedded in the middle of the randomized region demonstrated this point (Davis and Szostak 2002). Survey of functional RNAs found a bias towards more purines than pyrimidines (Schultes et al. 1997); other work also showed that adjusting base composition might enhance aptamer selection (Knight et al. 2005). Ongoing computational studies using sequence information to pattern the initial library could yield more insight to guide synthesis.

MULTIVALENT APTAMERS AND APTAMER-CONTAINING NANOPARTICLES

The term "RNA aptamer" is used most often to describe the unmodified, natural RNA aptamers in the original form in which they were isolated from a sequence pool. However, in many applications this form is only the starting material, which will undergo multiple subsequent steps of refinement. Aside from chemical modifications mentioned above, the original full-length isolates may be processed in many other ways. In this section I will discuss two types of aptamer derivatives: aptamers with their sequence augmented to form multivalent composites and aptamers conjugated to a non-nucleic acid moiety at the nanoscale.

A means to manipulate aptamer properties, distinct from chemical modification described above, is sequence minimization or variation. Each sequence in the pool that undergoes selection has a randomized region flanked by a pair of constant segments to allow amplification of all the sequences in a mixture simultaneously. When isolated in this form, an aptamer may carry additional sequences not necessary for binding the target. The full-length form may have alternative folding patterns, only one of which possesses the activity of the aptamer. Since the true aptamer moiety is often only a fraction of a full-length isolate, deleting the unnecessary sequence would produce a more compact "minimized" version that is more portable when it is connected to other structural and functional units (Shi et al. 2002; Xu and Shi 2009). The process of minimization starts with some predicted secondary structures. While the secondary structure of natural RNA can be predicted fairly reliably using free energy minimization and other algorithms (Zuker 2003; Mathews et al. 2010), it is not clear to what extent the "rules" used by them also apply to chemically modified RNAs. The conventional enzymatic and chemical means of probing for structure also function less well when the backbone of RNA is modified. However, a secondary structure predicted for the natural RNA version of the aptamer can be used as a starting model. To verify the model or to deduce the real secondary structure, multiple types of mutational analyses including base pair co-variation and circular permutation can be used. Minimization usually results in enhanced affinity, possibly due to a decrease in competing nonbinding conformations (Xu and Shi 2009).

An aptamer, preferably minimized, can be augmented by additional sequences or chemical moieties. More broadly, the unique properties of nucleic acids allow facile introduction of multifunctionality to aptamer constructs via complementarity or sequence manipulation. Additional sequence or structure of the same or similar chemical nature (i.e., natural or modified RNA) can be added to an aptamer, either covalently or non-covalently, to generate new functionalities as demonstrated in the following examples. The additional sequence may comprise one or more aptamers identical to or different from the original aptamer. Multimerization of the same aptamer may enhance the avidity of binding (Shi et al. 1999; Santulli-Marotto et al. 2003); a composite of aptamers for different targets would cause induced proximity of the non-interacting targets and may trigger a novel molecular response with functional significance (Mallik et al. 2010). Non-aptamer functional units may also be added to an aptamer in the same manner. For example, an RNA aptamer for the HIV-1 envelope (gp120) protein is attached to a small interfering RNA (siRNA) that triggers sequence-specific degradation of HIV RNAs (Zhou et al. 2008; Zhou et al. 2009). In this case the aptamer functioned not only as a target-neutralizing agent but also a delivery vehicle for siRNA. Similar chimeric constructs have been used in several other studies to achieve aptamer-targeted cell-specific RNA interference (Zhou and Rossi 2010).

When more than one RNA aptamers are included in a single molecular entity, the most important issue is to preserve the activity of each individual aptamer by maintaining its correct folding pattern. In most cases, multivalent aptamers are designed and constructed through *ad hoc* research projects. To make this process more reliable, we have developed a system for rational modular molecular design that is amenable to standardization and abstraction (Xu and Shi 2009). This system consists of a set of modules, a protocol, and a process to assemble the modules combinatorially according to the protocol. Generally speaking, modules are parts, components, or subsystems with identifiable interface to other modules. They maintain their identity when isolated or rearranged, and can be evolved somewhat independently. Protocols are rules or constraints on allowed interfaces and interconnections that facilitate modularity and simplify modeling, abstraction, and verification. Our method treats the individual aptamers as functional modules and combines them with structural modules through a protocol that fuses double-stranded stems. In the resulting constructs, various functional modules are organized and presented with the help of structural modules, and a set of building blocks can be arranged in diverse patterns.

The success of this method relies on the incorporation of two very different types of structural information aided by a two-dimensional graphic approach. For most aptamers, atomic level structure is unknown and secondary structure is unproven. However, it is relatively easy to refine the aptamer until it can be treated as a functional "loop" associated with one, two or three confirmed double-stranded stems (i.e., apical loop at the end of one stem, internal loop flanked by two stems, or the strand exchange junction of three stems). On the other hand, existing structural modules with known atomic structure can be used as insulators between individual aptamers or connectors to maintain strand continuity. The insulators/presenters include many multi-branch junctions, in particular three-way junctions (Lescoute and Westhof 2006), and the strand connectors are exemplified by stable small U-turns such as the UUCG tetra-loop (Cheong et al. 1990). In addition, stems having complementary strands are known to assemble as A-form double helices if the nucleic acid is natural RNA, and these stems can be used like "connective tissue" to fortify local aptamer structure. In this manner the paucity of structural information for the functional modules (aptamers) is compensated by the known tertiary interaction in the structural modules. As examples, Figure 52.1 shows how the crystal structure of a three-way junction can be utilized in 2-D graphic design for construction of multivalent composite aptamers. Figure 52.2 shows how 4 aptamers can be arbitrarily combined in a single molecule with the help of two three-way junctions fused together, and how the relative orientation of the aptamers can be adjusted by inserting double-stranded segments into the construct.

So far we have considered RNA aptamers connected by RNA strands that are functionally "neutral." In other cases the liker can participate in structure switching upon ligand binding (to be discussed in detail below in Section 6). Non-nucleotidic linkers, such as PEG or polyacrylamide backbone can also be used to connect multiple aptamers. Whereas these linkers usually connect several to a dozen aptamers, a larger number of aptamers can be "multimerized" in a less precise way by covalent attachment to the surface of Au or Ag nanoparticles as well as viral capsids, resulting in a significant increase of avidity. On the other hand, non-covalent interaction can also be used to assemble multivalent aptamers. A case in point is the pRNA of the bacteriophage phi29, which was used in many different constructs (Shu et al. 2011a, b; Binzel et al. 2016; Pi et al. 2017). Similarly, CopA and CopT motifs can be used to form a stable complex. Biotin-streptavidin interaction has also been used to generate multivalent aptamers. Recently, a computational method for the design of RNA complexes that hybridize via "sticky bridges," named AptaBlocks, has been develop to allow rapid and convenient non-covalent connection of aptamers or other RNA elements (Wang et al. 2018).

Without limitation, all methods developed for chemical conjugation of RNA can be applied to aptamers. Through these chemical linkages, aptamers can be attached to many different molecules, which may be further attached to other structures (Lee et al. 2010; Chen et al. 2011). A very exciting application for aptamers in the diagnosis or treatment of diseases is the nanoparticle-aptamer conjugates (Levy-Nissenbaum et al. 2008; Wang et al. 2012). The nanoparticles are loaded with imaging

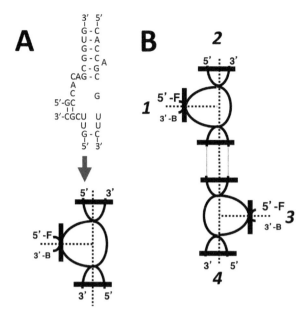

FIGURE 52.1 Two-dimensional graphic representation of a three-way junction. Panel A shows a 2-D representation of the three-way junction that incorporates information, in particular strand and base pair orientations, from both the predicted secondary structure and the 3-D crystal structure of *H. marismortui* 5S rRNA (Ban et al. 2000). "F" designates the end of the strand in front of the plane; "B" designates the end of the strand behind the plane. In Panel B, two such three-way junctions are connected through a double-stranded stem to form a scaffold with 4 receptacles (designated 1 through 4) onto which the aptamers can be engrafted.

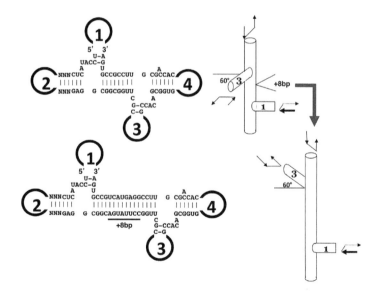

FIGURE 52.2 Adjustment of the relative position and orientation of aptamers on a structural scaffold. Two scaffolds, each bearing 4 aptamer receptacles (1–4) are depicted with sequence. Both scaffolds are formed by fusing the three-way junction of 5S RNA shown in Figure 52.1 and another junction named system D from (Diamond et al. 2001), which is thermodynamically stable and analogous to the structure in Figure 52.1. The only difference between the two scaffolds is the length of the stem connecting the two junctions, which causes the different orientation and position of the aptamers being engrafted to these receptacles. The arrows indicate the 5′→3′ direction of the strand.

agents or drugs, and the aptamers on their surface act as targeting molecules that direct their delivery to tumor cells through binding to specific cell surface markers (Catuogno et al. 2016). Notably, a 2′F-Py RNA aptamer for the prostate-specific membrane antigen (PSMA) was conjugated with docetaxel-encapsulated nanoparticles for targeted uptake by prostate cancer cells (Farokhzad et al. 2006; Kolishetti et al. 2010). In a xenograft nude mouse model of prostate cancer, these bioconjugates showed significant anticancer efficacy without the systemic toxicity common to chemotherapeutics. Recently, a minimized PSMA aptamer was attached to multi-walled carbon nanotubes by a PEG linker to enhance contrast in ultrasound imaging (Gu et al. 2018). Besides the examples highlighted here, a variety of other nanomaterials, not limited to particles, have been used in conjugation with aptamers for different applications (Jo and Ban 2016). Notably, an aptamer for theophylline was used in a reusable nanopore thin film sensor, which is able to detect the target specifically in complex fluids such as plant extract and serum (Feng et al. 2018). In many other biosensors, aptamer are used for recognition of targets ranging from pathogens (bacteria, viruses, and protozoan parasites), cancers, to environmental pollutants. However, in diagnostic applications, DNA aptamers are used more widely for prototyping than RNA aptamers because of their chemical stability.

APTAMER-ARTICULATED NANOSCALE SYSTEMS

A survey of nucleic acid-based nanotechnology reveals two general approaches: structural and compositional. The structural approach uses well-characterized components and combines them using both affinity and structural information to control geometry or strand topology to achieve structural predictability with a precision of 1 nm or less in the products. In contrast, when the compositional approach is taken, only the components of the product are defined. The structure of some components may not be well-characterized, and the three-dimensional structure of the assembly may not be predictable. In principle, the structural approach is more desirable. But it is limited by availability of naturally existing affinity pairs (i.e., pairs of portable sites that mediate specific molecular recognition and binding) that are structurally characterized. In DNA constructs, the majority of molecular articulation is realized by base pairing between parallel strands (Seeman 2003; Seeman 2010), of which structural prediction is easy and reliable. Compared to DNA, RNA affords more specific tertiary interactions, as exemplified by the tetraloop–tetraloop receptor (Jaeger et al. 2001) and the arms of the three-way junction of pRNA of phi29 (Shu et al. 2004; Shu et al. 2011a, b; Mallik et al. 2021). This type of interacting pairs have been actively collected and characterized as valuable building blocks of nanoscale assemblies and objects (Leontis and Westhof 2003; Grabow et al. 2011).

However, for both DNA and RNA, it is difficult to use existing building blocks to build elaborate hybrid systems that include non-nucleic acid components. One pressing need is to expand the repertoire of standard and exchangeable parts. While most aptamers are not structurally characterized at the atomic level, including aptamers in a compositional approach would make it possible to break the confines of nucleic acids to introduce molecular diversity into the engineered system. In these systems, aptamers can be used as integral articulations to organize non-nucleic acid components. We have started exploring this area by integrating RNA aptamers with protein molecules in biological systems. Below I will describe two examples followed by a general discussion on the possibility of making synthetic functional modules in living cells or organisms. Afterwards, I will place aptamers in a broader perspective not confined by their relevance to biology, which is inspired by the diversity of targets for which aptamers have been successively isolated.

For the regulation of biological processes, proteins and other molecules are connected to each other through a complex network of interactions. Experimental manipulation and therapeutic intervention is often achieved through the modification of the connectivity of this network. Traditional genetic methods and drugs modify such connectivity by blocking or abolishing molecular interaction. An alternative and sometimes more effective strategy is to introduce new links between non-interacting molecules. This approach is rarely explored because bridging two molecules specifically and selectively is much more difficult than blocking one molecule. Only a few small organic compounds, such as FK506 and cyclosporin A, are known to induce protein dimerization, and have been further developed and used in experiments to control intracellular signaling (Ho et al. 1996; Klemm et al. 1998). By splicing together more than one type of RNA aptamer, we are now able to create new connectivity between non-interacting proteins at will. As an example, we made a bi-functional aptamer that simultaneously binds to the green fluorescent protein ((GFP), serving as a surrogate extracellular target) and the opsonin C3b/iC3b (serving as a utility molecule that is an entry point to a pathway conscripted to process the target molecule). With this construct we were able to commandeer the C3-based opsonization-phagocytosis pathway to selectively transport the GFP into the lysosome for degradation (Mallik et al. 2010). As a prototype for therapeutic application, this strategy has two advantages over the use of individual aptamers. First, it not only reversibly neutralizes the targets but irreversibly eliminates them. Second, the action of the bi-functional aptamer does not require its escape from the endosome, which is a formidable obstacle to delivery of aptamers into the cytoplasm. We also implemented the same concept in a different system to manipulate the chaperon Hsp70 for targeted ubiquitination (Thirunavukarasu and Shi 2016).

The utility of multivalent aptamers arises in part because they are "protein-like" in the following two senses. First, a single protein molecule is capable of bearing more than three sites recognized specifically by other molecules, which collectively form a scale-free network (Jeong et al. 2001). (Single-site molecules can form only dyadic interactions, and double-site molecules can form only linear chains.) With the method described in the previous section (Xu and Shi 2009), we are able to generate composite aptamers with multiple binding sites combined to mimic existing proteins, or in rationally designed combinations that create novel connectivity. Second, RNA aptamers can be genetically encoded like protein, and their biosynthesis and degradation can be regulated by environmental and developmental cues using diverse promoters to drive RNA transcription (Shi et al. 1999). By recapitulating these key features of proteins, multivalent aptamers can be integrated into existing biological pathways to rewire the network and revise the control logic. Following this principle, we utilized a composite RNA aptamer to implement the mechanism of transcriptional activation through recruitment of a general transcription factor, TFIIB, to the promoter of reporter genes in yeast (Wang et al. 2010). Because they were isolated outside of the cellular or organismic context, individual aptamers often bind to and obscure sites on a protein that are catalytic or otherwise central to its activity. Consequently, they are routinely used as inhibitors of protein activity by default. This work successfully converted a passively acting inhibitory aptamer into the activation domain of an effective transcription activator by rational design.

Based on these studies and other published literature, it is reasonable to anticipate that aptamers could act as nexuses in more sophisticated systems to integrate other components in living cells or organisms. Figure 52.3 synthesizes several strands of inquiry to propose a conceptual framework and suggest some directions that are worth pursuing. Individual aptamers can be combined with

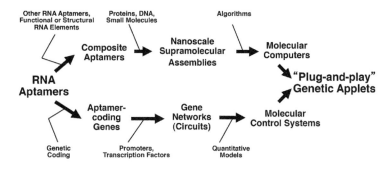

FIGURE 52.3   General strategy for integrating different research fields to achieve predictive modification of cells and organisms.

other aptamers or non-aptamer RNA elements to form composite aptamers (Xu and Shi 2009). Because individual aptamers can be isolated for diverse targets that are large or small, and protein or non-protein (Gold et al. 1995), composite aptamers can be used to form nanoscale supramolecular assemblies from diverse components. If certain algorithms can be embodied in the assembly or transformation of such assemblies [as already implemented with DNA (Yin et al. 2008; Douglas et al. 2012; Li et al. 2018), they may function as molecular computers or robots that store and process data. On the other hand, RNA aptamers can be delivered into cells as aptamer-coding genes (Thomas et al. 1997). A tremendous range of control can be exerted on these synthetic genes through the use of different promoters that recruit different sets of transcription factors (Shi et al. 1999). Multiple aptameric constructs in a single living system may be connected to form gene networks [as in synthetic protein-encoding gene circuits (Elowitz and Leibler 2000; Gardner et al. 2000)]. With the help of quantitative models, these gene networks can function as molecular control systems (Win et al. 2009). This expanded toolbox would enable us to design and construct progressively more complex systems. While a single composite aptamer encoded by a synthetic gene resembles a protein, multiple composite aptamers integrated in this way could enact a package of control logic similar to a virus.

The foregoing discussion demonstrated aptamers' capability of articulating non-nucleic acid components in biological systems. However, the utility of aptamers to organize and orient matter according to predetermined specifications is not confined to biology. The general strategies being developed and principles being uncovered in the previous studies are applicable to broader engineering innovation in nanomaterials. For example, an essential component of a supramolecular assembly is the "linker" or articulating joint: different linkers can be used to connect same type of generic units to form different structures. The specificity of protein domain interactions or nucleic acid complementarity has been used in this way, but the resulting assemblies are often composed exclusively of proteins or nucleic acids, respectively (Clark et al. 2004). The major reason for this limitation is the lack of portable and specific affinity pairs formed by different types of material. Aptamers, especially multivalent aptamers in the form of modified RNA, may serve as linkers to connect non-nucleic acid or non-protein components, and the self-assembly of supramolecular structure through non-covalent interactions between the aptamers and their target sites would contribute to the robustness and evolvability of the resulting material or device. Using aptamers, it seems feasible to create new materials with nanoscale features either by reorganizing existing biological materials or organizing non-biological materials. In several elegant studies, DNA aptamers have been used to direct the assembly of proteins with nanoscale precision (Liu et al. 2005; Lin et al. 2006; Chhabra et al. 2007; Rinker et al. 2008). Because the targets of aptamers are not limited to protein or other biologically relevant molecules, they can be used to integrate a vast and diverse array of organic or inorganic materials. This unique capability, combined with the biologically inspired engineering approach described above and assisted by the large scale production methods, could spawn new possibilities in the development of functional supramolecular polymers for non-biological applications (Aida et al. 2012).

PROSPECTS FOR APTAMER-ENABLED DYNAMIC STRUCTURES

In biological systems, nanoscale objects are formed in a process of self-assembly of molecules or molecular segments, as the information that determines the interaction is coded in individual components. Lately, many engineering research projects have been launched to emulate this process (Krieg et al. 2016). Equilibration is usually required to reach ordered structures, which are static. Most research in self-assembly has focused on this type. There is another type of self-assembly, which is dynamic and out of equilibrium. In this type of systems, components may rearrange without additional materials through reversible conformation changes, which are often sensitive to the chemical environment. New materials that are capable of adapting to their environment are desirable in the rapidly developing field of bio-inspired materials science and engineering. However, study of dynamic self-assembly is still in its infancy. In this section, I would like to invoke several disparate research fields to suggest a perspective on aptamer-enabled dynamic constructs. An introduction to the general phenomenon that a single RNA sequence can assume multiple stable conformations will serve as a primer to the discussion, followed by a brief overview of natural and synthetic ribo-switches. Afterwards, existing aptamers used in multi-state constructs are described, in particular as they are used in sensors for detection and diagnostic purposes. From these examples, some general principles are extracted, which can be used in combination with the methods described in the previous sections to design dynamic structures.

It is a well-documented phenomenon that RNA molecules of identical sequence can adopt multiple conformations (Biebricher et al. 1982; Schultes and Bartel 2000). What is remarkable and useful is the fact that two or more alternative conformations often have comparable energies close to the ground state, and they are often separated by high energy barriers (Brion and Westhof 1997; Tinoco and Bustamante 1999). The existence of multi-stable RNA conformations can be explained at the secondary structure level, i.e., by patterns of base pairing. Folding of an RNA molecule often involves extensive pairing of bases with corresponding energies significantly larger than those of tertiary contacts. As a result, conversion between different conformations is energetically costly because many base pairs have to be broken. This forms the basis of an interesting and useful mechanism of regulation and control: an RNA capable of assuming multiple metastable conformations separated by significant energy barriers can be implemented as a switch through ligand-assisted conversion between different conformations (Beisel and Smolke 2009). In this configuration, different conformations mediate different functions, and the presence or absence of the ligand represents a cue for conformational switching that arises from the environment of the RNA molecule. If the RNA molecule contains one or more aptamers and the assisting ligand is the target of an aptamer, then the aptamer target may serve as a trigger of the switch. For "switchable" aptamers with suitable dynamic range, binding of the target must form a sufficiently stable structure in the "ON" state; conversely, absent the target, the aptamer should stably remain in an "OFF" state. Transitions between these two states must be kinetically favorable. In vitro selection schemes may not be sufficient to achieve this delicate balance, and computer simulations may be required to optimize the "switchablity." "Alchemical" free energy calculations (Chodera et al. 2011) may be performed to quantify the stability of the two end-states of the switch in response to sequence mutations, thereby eliminating leakage. Transition paths (Bolhuis et al. 2002) can be calculated to characterize the sequence of molecular rearrangements, thus quantifying the ease of the kinetic transition between the two states.

The first synthetic bi-stable RNA molecule involving an aptamer was an allosteric ribozyme (Soukup and Breaker 1999). Afterwards, similarly configured riboswitches were discovered in many organisms, primarily prokaryotes (Roth and Breaker 2009). Most of these riboswitches are located in the 5′-UTRs of mRNA, and they are commonly composed of two functional domains: an aptamer domain that binds to the metabolite and serves as a molecular sensor, and an "expression platform" adjacent to or overlapping with the aptamer domain that transduces the signal of metabolite binding through allosteric modulation of the structure of RNA, which in turn affects the gene expression process. This common set-up can be applied to two aptamers connected together as described in the previous sections. By connecting two aptamers in a fashion that encourages the adaptive binding of ligands (i.e., ligand binding to one aptamer would stabilize local RNA structure that affects the affinity of the other aptamer to another ligand), we would be able to link the fate of one molecule to the presence or absence of any other. When the ligand of the first aptamer is an analyte and the ligand of the second aptamer is a fluorescent molecule, the RNA construct that contains the two connected aptamers would act as the recognition and signaling modules of a sensor (Stojanovic and Kolpashchikov 2004; Paige et al. 2012). A variation on the theme of an effector-dependent conformation switch is the use of oligonucleotide antidotes for aptamers, in which case the conformation change is achieved through forced consecutive base pairing between the aptamer and its antidote (Rusconi et al. 2004). Whereas the antidote acts in *trans*, the "expression platform" of a natural riboswitch acts in *cis* by similar mechanism.

Individual aptamers have been widely used as recognition elements of biosensors. In early studies, the assay formats are often similar to those previously developed for antibodies, such as sandwich-like binding or competitive replacement, in which the aptamers assume a static structure. Later, new design strategies that are applicable to aptamers but not to antibodies have been introduced for aptasensors (Liu et al. 2009; Feagin et al. 2018), which can be coupled with a wide range of readout methods, such as electrochemistry, colorimetry, fluorescence, and FRET. By making use of the dynamic structure of nucleic acids, signals are produced through target-induced structure switches. To reduce background and enhance specificity, the aptamers used in these sensors sometimes require assistance in both directions for conformational change. A single-stranded oligonucleotide with complementary sequence to the aptamer, i.e., an antidote, is often used to keep the aptamer in an inactive state. When the target of the aptamer, i.e., the analyte, binds the aptamer, the conformation of the aptamer is switched to the active state. This duplex-to-complex switch (Nutiu and Li 2003; Lau et al. 2010) is sometimes called "duplexed aptamer," because it is composed of an aptamer and an aptamer-complementary element (ACE) that competes with the ligand for binding to the aptamer. It has been successfully used in many studies to generate different types of signals through electron tunneling effect, molecular beacon fluorescence, salt-induced gold nanoparticle aggregation, and other mechanisms (Han et al. 2010; Lim et al. 2010). To investigate the design parameters of *cis*- and *trans*-duplexed aptamers, theoretical frameworks have been formulated, in which ligand binding was modeled in a way analogous to the classical S_N1 and S_N2 nucleophilic substitution reactions in organic chemistry (Munzar et al. 2019).

Directional, reversible, and regulated conformation change is a highly desirable feature of new nanomaterials that are able to communicate with and adapt to their environment. The dynamic nature of aptamers described in this section holds promise to enrich the toolbox of material scientists for the creation of such "smart" materials. Dynamic structures built with multivalent aptamers acting as integral articulations to organize other molecules could reorient and rearrange matter on receiving chemical cues, overcoming obstacles associated with the rigid structure of traditionally engineered systems. This type of self–assembled, multifunctional and compliant structure may be further augmented with built-in self-reporting status indicators by incorporating observable signals into the system. For example, many aptamers for fluorescent targets have been developed (Paige et al. 2011; Shui et al. 2012), and conditional fluorophores can be conjugated to aptamers (Nutiu and Li 2005). The fluorescent targets may be used to trigger the conformation change, and the conditional fluorophore may be used to indicate whether an aptamer-binding event has occurred (Afonin et al. 2010). When these principles are taken together, it seems realistic to predict that an interdisciplinary approach may rapidly advance the frontiers of research on aptamer-enabled dynamic constructs in the near future.

CONCLUSION

Biology is unique among the natural sciences in its concern with the function or purpose of organisms and their components. This feature connects biology to synthetic disciplines such as engineering and computer sciences (Hartwell et al. 1999). It is at the nanoscale that biological molecules and machinery inside living cells operate; and it is at the nanoscale that highly specific medical intervention holds the most promise in treating a wide range of recalcitrant diseases such as cancer and viral infections. Therefore, both nanotechnology and molecular biology benefit when they are united in nanobiotechnology: not only can we apply existing nano/microfabrication methods to build devices to study and manipulate biosystems, we can also learn from biological systems how to create novel and better nanoscale objects and devices.

This chapter is a journey into an emerging field of science and engineering that encourages the exploration of the application of RNA aptamers in nanotechnology and therapeutics. I have attempted to provide a holistic view, sometimes by "borrowing scenery" from adjacent fields. The unique capability of aptamers to provide specific interfaces between biological and non-biological materials is increasingly appreciated by people in different disciplines. When RNA aptamers were invented nearly three decades ago, their targets were primarily proteins and biologically relevant small molecules; the constructs formed by aptamers and their ligands were biomacromolecules generated in emulation of a biological process. Gradually, it became clear that aptamers have the potential to be developed into a more generic type of material beyond the confines of biology. Works described in this chapter attest the contribution of aptamers to the construction of complex nanoscale objects with hybrid structure and composition, which represents a major challenge and opportunity in the engineering of organized matter.

Major developments have taken several directions. First, the chemical nature of aptamers has been modified to possess more desirable features. Early work focused on the modification of the sugar-phosphate backbone to enhance stability. (In this context, DNA aptamers can be viewed as a special case in which the 2′ hydroxyl group is "deleted".) Lately, modification of the bases has introduced more chemical diversity and better binding characteristics. With new synthetic schemes for the building blocks and new mutants of enzymes that are able to use them as substrates, future aptamers at some point may no longer resemble existing nucleic acids and become a totally new type of heteropolymeric compound (Pinheiro et al. 2012; Hoshika et al. 2019). Second, aptamer derivatives have shown a trend of increasing complexity. Individual aptamers involved in a bi-molecular binding reaction were used initially to detect or inhibit their targets by default. The advent of aptamer conjugates and multivalent composite aptamers made it possible to engineer complex nanoscale systems by design. With the introduction of multi-stable alternative folding patterns, smart materials could be envisioned to monitor and respond to the environment. Finally, the range of target classes for aptamers has been expanding dramatically over the years, promoting the universality and diversity of structural makeup in aptamer-articulated systems. The complexity of aptamer targets ranges from small molecules to macromolecules to cells, and their chemical nature has become more and more diverse (Gold et al. 2010a). Taken together, these developments suggest that various approaches from disparate fields of study can now be integrated to generate ample opportunities for engineering innovation.

ACKNOWLEDGMENTS

Publications are cited in this chapter for the purpose of illustration rather than exhaustive representation. Therefore, I thank colleagues and coworkers whose works are cited and apologize to those whose works are omitted. Aptamer research in the Shi Lab has been supported by American Cancer Society, the National Institutes of Health, and the US Department of Defense.

REFERENCES

Afonin, K. A., E. Bindewald, A. J. Yaghoubian, et al. 2010. In vitro assembly of cubic RNA-based scaffolds designed in silico. *Nature Nanotechnology* 5(9): 676–682.

Aida, T., E. W. Meijer and S. I. Stupp 2012. Functional supramolecular polymers. *Science* 335(6070): 813–817.

Alam, K. K., J. L. Chang and D. H. Burke 2015. FASTAptamer: A bioinformatic toolkit for high-throughput sequence analysis of combinatorial selections. *Mol. Ther.–Nucleic Acids*. 4: e230.

Ban, N., P. Nissen, J. Hansen, P. B. Moore and T. A. Steitz 2000. The complete atomic structure of the large ribosomal subunit at 2.4 A resolution. *Science* 289(5481): 905–920.

Beisel, C. L. and C. D. Smolke 2009. Design principles for riboswitch function. *PLoS Computational Biology* 5(4): e1000363.

Biebricher, C. K., S. Diekmann and R. Luce 1982. Structural analysis of self-replicating RNA synthesized by Qbeta replicase. *Journal of Molecular Biology* 154(4): 629–648.

Binzel, D.W., Y. Shu, H. Li et al 2016. Specific delivery of miRNA for high efficient inhibition of prostate cancer by RNA nanotechnology. *Mol. Ther.* 24: 1267–1277.

Biondi, E. and S. A. Benner 2018. Artificially Expanded Genetic Information Systems for New Aptamer Technologies. *Biomedicines* 6(2): e53.

Biondi, E., J. D. Lane, D. Das et al 2016. Laboratory evolution of artificially expanded DNA gives redesignable aptamers that target the toxic form of anthrax protective antigen. *Nucleic Acids Res.* 44: 9565–9577.

Blind, M. and M. Blank 2015. Aptamer selection technology and recent advances. *Mol. Ther.–Nucleic Acids* 4: e223.

Bolhuis, P.G., D. Chandler, C. Dellago and P. L. Geissler 2002. Transition path sampling: throwing ropes over rough mountain passes, in the dark. *Annual Review of Physical Chemistry* 53: 291–318.

Brion, P. and E. Westhof 1997. Hierarchy and dynamics of RNA folding. *Annual Review of Biophysics and Biomolecular Structure* 26: 113–137.

Catuogno, S., C. L. Esposito and V. de Franciscis 2016. Aptamer-mediated targeted delivery of therapeutics: An update. *Pharmaceuticals* 9(4): e69.

Chelliserrykattil, J. and A. D. Ellington 2004. Evolution of a T7 RNA polymerase variant that transcribes 2′-O-methyl RNA. *Nature Biotechnology* 22(9): 1155–1160.

Chen, T., M. I. Shukoor, Y. Chen, et al. 2011. Aptamer-conjugated nanomaterials for bioanalysis and biotechnology applications. *Nanoscale* 3(2): 546–556.

Cheng, C., Y. H. Chen, K. A. Lennox, et al.2013. In vivo SELEX for identification of brain-penetrating aptamers. *Mol. Ther. Nucleic Acids* 2: e67.

Cheong, C., G. Varani and I. J. Tinoco 1990. Solution structure of an unusually stable RNA hairpin, 5′GGAC(UUCG)GUCC. *Nature* 346: 680–682.

Chhabra, R., J. Sharma, Y. Ke, et al. 2007. Spatially addressable multiprotein nanoarrays templated by aptamer-tagged DNA nanoarchitectures. *Journal of the American Chemical Society* 129(34): 10304–10305.

Cho, M., Y. Xiao, J. Nie et al. 2010. Quantitative Selection of DNA Aptamers through Microfluidic Selection and High-Throughput Sequencing. *Proc. Natl. Acad. Sci. USA*. 107: 15373–15378.

Chodera, J.D., D. L. Mobley, M. R. Shirts et al. 2011 Alchemical free energy methods for drug discovery: progress and challenges. *Curr Opin Struct Biol.* 21(2): 150–160.

Clark, J., E. M. Singer, D. R. Korns and S. S. Smith 2004. Design and analysis of nanoscale bioassemblies. *Biotechniques* 36(6): 992–996, 998–1001.

Conrad, R. C., L. Giver, Y. Tian and A. D. Ellington 1996. In vitro selection of nucleic acid aptamers that bind proteins. *Methods in Enzymology* 267: 336–367.

Davis, J. H. and J. W. Szostak 2002. Isolation of high-affinity GTP aptamers from partially structured RNA libraries. *Proc Natl Acad Sci USA* 99(18): 11616–11621.

Diamond, J. M., D. H. Turner and D. H. Mathews 2001. Thermodynamics of three-way multibranch loops in RNA. *Biochemistry* 40(23): 6971–6981.

Douglas, S. M., I. Bachelet and G. M. Church 2012. A logic-gated nanorobot for targeted transport of molecular payloads. *Science* 335(6070): 831–834.

Dua, P., S. Kim and D. K. Lee 2008. Patents on SELEX and therapeutic aptamers. *Recent Patents on DNA & Gene Sequences* 2(3): 172–186.

Eigen, M. 1971. Self-organization of matter and the evolution of biological macromolecules. *Naturwissenschaften* 58(10): 465–523.

Ellington, A. D. and J. W. Szostak 1990. In vitro selection of RNA molecules that bind specific ligands. *Nature* 346(6287): 818–822.

Elowitz, M. B. and S. Leibler 2000. A synthetic oscillatory network of transcriptional regulators. *Nature* 403(6767): 335–338.

Famulok, M., J. S. Hartig and G. Mayer 2007. Functional aptamers and aptazymes in biotechnology, diagnostics, and therapy. *Chemical Reviews* 107(9): 3715–3743.

Fan, X., H. Shi, K. Adelman and J. T. Lis 2004. Probing TBP interactions in transcription initiation and reinitiation with RNA aptamers that act in distinct modes. *Proc. Natl. Acad. Sci. USA*, 101: 6934–6939.

Farokhzad, O. C., J. Cheng, B. A. Teply, et al. 2006. Targeted nanoparticle-aptamer bioconjugates for cancer chemotherapy in vivo. *Proceedings of the National Academy of Sciences of the United States of America* 103(16): 6315–6320.

Feagin, T., N. Maganzini and H.T. Soh 2018. Strategies for Creating Structure-Switching Aptamers. *ACS Sensors* 3: 1611–1615.

Feng, S., C. Chen, W. Wang and L. Que 2018. An aptamer nanopore-enabled microsensor for detection of theophylline. *Biosensors and Bioelectronics* 105: 36–41.

Gardner, T. S., C. R. Cantor and J. J. Collins 2000. Construction of a genetic toggle switch in Escherichia coli. *Nature* 403(6767): 339–342.

Gelinas, A. D., D. R. Davies and N. Janjic 2016. Embracing proteins: structural themes in aptamer-protein complexes. *Curr Opin Struct Biol*, 36: 122–132.

Gold, L., D. Ayers, J. Bertino, et al. 2010a. Aptamer-based multiplexed proteomic technology for biomarker discovery. *PloS One* 5(12): e15004.

Gold, L., N. Janjic, T. Jarvis, et al. 2010b. Aptamers and the RNA world, past and present. in *RNA worlds from life's origins to diversity in gene regulation*, ed. Atkins, J.F., Gesteland, R.F., and Cech, T.R. Cold Spring Harbor Laboratory Press, Cold Spring Harbor, New York.

Gold, L., B. Polisky, O. Uhlenbeck and M. Yarus 1995. Diversity of oligonucleotide functions. *Annual Review of Biochemistry* 64: 763–797.

Golemis, E. A., K. D. Tew and D. Dadke 2002. Protein interaction-targeted drug discovery: evaluating critical issues. *Biotechniques* 32(3): 636–638, 640, 642 passim.

Gong, Q., J. Wang, K. M. Ahmad, et al. 2012. Selection Strategy to Generate Aptamer Pairs that Bind to Distinct Sites on Protein Targets. *Analytical Chemistry*. DOI: 10.1021/ac300873p

Grabow, W. W., P. Zakrevsky, K. A. Afonin, et al. 2011. Self-assembling RNA nanorings based on RNAI/II inverse kissing complexes. *Nano Letters* 11(2): 878–887.

Gu, F. et al. 2018. Aptamer-conjugated multi-walled carbon nanotubes as a new targeted ultrasound contrast agent for the diagnosis of prostate cancer. J Nanopart Res 20: 303.

Han, K., Z. Liang and N. Zhou 2010. Design strategies for aptamer-based biosensors. *Sensors* 10: 4541–4557.

Hartwell, L. H., J. J. Hopfield, S. Leibler and A. W. Murray 1999. From molecular to modular cell biology. *Nature* 402(6761 Suppl): C47–52.

Ho, S. N., S. R. Biggar, D. M. Spencer, S. L. Schreiber and G. R. Crabtree 1996. Dimeric ligands define a role for transcriptional activation domains in reinitiation. *Nature* 382(6594): 822–826.

Hoinka, J., A. Berezhnoy, P. Dao et al. 2015. Large scale analysis of the mutational landscape in HT-SELEX improves aptamer discovery. *Nucleic Acids Res.* 43: 5699–5707.

Hoinka, J., A. Berezhnoy, Z. E. Sauna et al. 2014. AptaCluster - a method to cluster HT-SELEX aptamer pools and lessons from its application. *Lect. Notes Comput. Sci.* 8394: 115–128.

Hoshika, H., N. Leal, M. J. Kim et al. 2019. Hachimoji DNA and RNA: A genetic system with eight building blocks. *Science* 363 884–887.

Irvine, D., C. Tuerk and L. Gold 1991. SELEXION: Systematic evolution of ligands by exponential enrichment with integrated optimization by non-linear analysis. *J. Mol. Biol.* 222: 739–761.

Jaeger, L., E. Westhof and N. B. Leontis 2001. TectoRNA: modular assembly units for the construction of RNA nano- objects. *Nucleic Acids Research* 29(2): 455–463.

Jeong, H., S. P. Mason, A. L. Barabasi and Z. N. Oltvai 2001. Lethality and centrality in protein networks. *Nature* 411(6833): 41–42.

Jo, H. and C. Ban 2016. Aptamer–nanoparticle complexes as powerful diagnostic and therapeutic tools. *Experimental & Molecular Medicine* 48: e230.

Juliano, R. L., A. Astriab-Fisher and D. Falke 2001. Macromolecular therapeutics: emerging strategies for drug discovery in the postgenome era. *Molecular Interventions* 1(1): 40–53.

Kaur, H. 2018. Recent developments in cell-SELEX technology for aptamer selection. *Biochim. Biophys. Acta Gen. Subj.* 1862: 2323–2329.

Keefe, A. D., S. Pai and A. Ellington 2010. Aptamers as therapeutics. *Nature Reviews Drug Discovery* 9(7): 537–550.

Kinghorn, A.B., R. M. Dirkzwager, S. Liang et al. 2016. Aptamer affinity maturation by resampling and micro-array selection. *Anal. Chem.* 88: 6981–6985.

Klemm, J. D., S. L. Schreiber and G. R. Crabtree 1998. Dimerization as a regulatory mechanism in signal transduction. *Annual Review of Immunology* 16: 569–592.

Klussmann, S., A. Nolte, R. Bald, V. A. Erdmann and J. P. Fuerst 1996. Mirror-image RNA that binds D-adenosine. *Nature Biotechnology* 14: 1112–1115.

Kolishetti, N., S. Dhar, P. M. Valencia, et al. 2010. Engineering of self-assembled nanoparticle platform for precisely controlled combination drug therapy. *Proceedings of the National Academy of Sciences of the United States of America* 107(42): 17939–17944.

Knight, R., H. De Sterck, R. Markel et al. 2005. Abundance of correctly folded RNA motifs in sequence space, calculated on computational grids. *Nucleic Acids Res.* 33: 5924–5935.

Knight, R. and M. Yarus 2003. Analyzing partially randomized nucleic acid pools: straight dope on doping. *Nucleic Acids Res.* 31(6): e30.

Krieg, E., M. M. C. Bastings, P. Besenius and B. Rybtchinski 2016. Supramolecular Polymers in Aqueous Media. *Chem. Rev.* 116: 2414–2477.

Kubik, M. F., C. Bell, T. Fitzwater, S. R. Watson and D. M. Tasset 1997. Isolation and characterization of 2'-fluoro-, 2'-amino-, and 2'-fluoro- /amino-modified RNA ligands to human IFN-gamma that inhibit receptor binding. *Journal of Immunology* 159(1): 259–267.

Latulippe, D. R., K. Seto, A. Ozer et al. 2013. Multiplexed Microcolumn-based process for efficient selection of RNA aptamers. *Analytical Chemistry* 85: 3417–3424.

Lau, P. S., B. K. Coombes and Y. Li 2010. Complementary oligonucleotides regulate induced fit ligand binding in duplexed aptamers. Angew. Chem., Int. Ed. 49: 7938–7942.

Lauridsen, L. H., H. B. Doessing, K. S. Long and A. T. Nielsen 2018. A capture-SELEX strategy for multiplexed selection of RNA aptamers against small molecules. *Methods Mol. Biol.* 1671: 291–306.

Leal, N.A., H. J. Kim, S. Hoshika et al. 2015. Transcription, reverse transcription, and analysis of rna containing artificial genetic components. *Acs Synth Biol* 4: 407–413.

Lee, J. F., J. R. Hesselberth, L. A. Meyers and A. D. Ellington 2004. Aptamer database. *Nucleic Acids Research* 32(Database issue): D95–100.

Lee, J. H., M. V. Yigit, D. Mazumdar and Y. Lu 2010. Molecular diagnostic and drug delivery agents based on aptamer-nanomaterial conjugates. *Advanced Drug Delivery Reviews* 62(6): 592–605.

Leontis, N. B. and E. Westhof 2003. Analysis of RNA motifs. *Current Opinions in Structural Biology* 13(3): 300–308.

Lescoute, A. and E. Westhof 2006. Topology of three-way junctions in folded RNAs. *RNA* 12(1): 83–93.

Levy-Nissenbaum, E., A. F. Radovic-Moreno, A. Z. Wang, R. Langer and O. C. Farokhzad 2008. Nanotechnology and aptamers: applications in drug delivery. *Trends in Biotechnology* 26(8): 442–449.

Li, S., Q. Jiang, S. Liu et al. 2018. A DNA nanorobot functions as a cancer therapeutic in response to a molecular trigger in vivo. Nature Biotechnology 36: 258–264.

Lim, Y. C., A. Z. Kouzani and W. Duan 2010. Aptasensors: a review. *Journal of Biomedical Nanotechnology* 6(2): 93–105.

Lin, C., E. Katilius, Y. Liu, J. Zhang and H. Yan 2006. Self-assembled signaling aptamer DNA arrays for protein detection. *Angewandte Chemie* 45(32): 5296–5301.

Lin, Y., Q. Qiu, S. C. Gill and S. D. Jayasena 1994. Modified RNA sequence pools for in vitro selection. *Nucleic Acids Research* 22(24): 5229–5234.

Liu, J., Z. Cao and Y. Lu 2009. Functional Nucleic Acid Sensors. *Chem. Rev.* 109: 1948–1998.

Liu, Y., C. Lin, H. Li and H. Yan 2005. Aptamer-directed self-assembly of protein arrays on a DNA nanostructure. *Angewandte Chemie* 44(28): 4333–4338.

Lo Conte, L., C. Chothia and J. Janin 1999. The atomic structure of protein-protein recognition sites. *Journal of Molecular Biology* 285(5): 2177–2198.

Mallik, P. K., K. Nishikawa, A. J. Millis and H. Shi 2010. Commandeering a biological pathway using aptamer-derived molecular adaptors. *Nucleic Acids Research* 38(7): e93.

Mallik, P.K., K. Nishikawa, P. Mallik and H. Shi 2021. Complement-mediated selective tumor cell lysis enabled by bi-functional RNA aptamers. *Genes* 29;13(1). doi: 10.3390/genes13010086.

Mathews, D. H., W. N. Moss and D. H. Turner 2010. Folding and finding RNA secondary structure. *Cold Spring Harbor Perspectives in Biology* 2(12): a003665.

Maynard Smith, J. 1970. Natural selection and the concept of a protein space. *Nature* 225: 563–564.

Mi, J., Y. Liu, Z. N. Rabbani et al. 2010. In vivo selection of tumor-targeting RNA motifs. *Nat Chem Biol.* 6(1): 22–24.

Mills, D. R., R. L. Peterson and S. Spiegelman 1967. An extracellular Darwinian experiment with a self-duplicating nucleic acid molecule. *Proceedings of the National Academy of Sciences of the United States of America* 58(1): 217–224.

Munzar, J. D., A. Ng and D. Juncker 2019. Duplexed aptamers: history, design, theory, and application to biosensing. *Chem. Soc. Rev.* 48: 1390–1419.

Nimjee, S. M., C. P. Rusconi and B. A. Sullenger 2005. Aptamers: an emerging class of therapeutics. *Annual Review of Medicine* 56: 555–583.

Nimjee, S. M., R.R. White, R. C. Becker and B. A. Sullenger 2017. Aptamers as therapeutics. *Annu. Rev. Pharmacol. Toxicol.* 57: 61–79.

Nutiu, R. and Y. Li 2003. Structure-switching signaling aptamers. *J. Am. Chem. Soc.* 125: 4771–4778.

Nutiu, R. and Y. Li 2005. Aptamers with fluorescence-signaling properties. *Methods* 37(1): 16–25.

Ozer, A., J. M. Pagano and J. T. Lis 2014. New Technologies Provide Quantum Changes in the Scale, Speed, and Success of SELEX Methods and Aptamer Characterization. *Mol. Ther.–Nucleic Acids* 3: e183.

Pagratis, N. C., C. Bell, Y. F. Chang, et al. 1997. Potent 2′-amino-, and 2′-fluoro-2′-deoxyribonucleotide RNA inhibitors of keratinocyte growth factor. *Nature Biotechnology* 15(1): 68–73.

Paige, J. S., T. Nguyen-Duc, W. Song and S. R. Jaffrey 2012. Fluorescence imaging of cellular metabolites with RNA. *Science* 335(6073): 1194.

Paige, J. S., K. Y. Wu and S. R. Jaffrey 2011. RNA mimics of green fluorescent protein. *Science* 333(6042): 642–646.

Pendergrast, P. S., H. N. Marsh, D. Grate, J. M. Healy and M. Stanton 2005. Nucleic acid aptamers for target validation and therapeutic applications. *Journal of Biomolecular Techniques* 16(3): 224–234.

Pestourie, C., B. Tavitian and F. Duconge 2005. Aptamers against extracellular targets for in vivo applications. *Biochimie* 87(9–10): 921–930.

Pi, F., H. Zhang, H. Li et al. 2017. RNA nanoparticles harboring annexin A2 aptamer can target ovarian cancer for tumor-specific doxorubicin delivery. *Nanomed. Nanotechnol. Biol. Med.* 13: 1183–1193.

Pinheiro, V. B., A. I. Taylor, C. Cozens, et al. 2012. Synthetic genetic polymers capable of heredity and evolution. *Science* 336(6079): 341–344.

Pitt, J. N., I. Rajapakse and A. R. Ferré-D'Amaré 2010. SEWAL: An open-source platform for next-generation sequence analysis and visualization. *Nucleic Acids Res.* 38: 7908–7915.

Rinker, S., Y. Ke, Y. Liu, R. Chhabra and H. Yan 2008. Self-assembled DNA nanostructures for distance-dependent multivalent ligand-protein binding. *Nature nanotechnology* 3(7): 418–422.

Robertson, D. L. and G. F. Joyce 1990. Selection in vitro of an RNA enzyme that specifically cleaves single-stranded DNA. *Nature* 344(6265): 467–468.

Rohloff, J. C., A. D. Gelinas, T. C. Jarvis, et al. 2014. Structural insights from aptamers with base modifications. *Mol. Ther. Nucleic Acids* 3: e201.

Roth, A. and R. R. Breaker 2009. The structural and functional diversity of metabolite-binding riboswitches. *Annual Review of Biochemistry* 78: 305–334.

Rusconi, C. P., J. D. Roberts, G. A. Pitoc, et al. 2004. Antidote-mediated control of an anticoagulant aptamer in vivo. *Nature Biotechnology* 22(11): 1423–1428.

Santulli-Marotto, S., S. K. Nair, C. Rusconi, B. Sullenger and E. Gilboa 2003. Multivalent RNA aptamers that inhibit CTLA-4 and enhance tumor immunity. *Cancer Research* 63(21): 7483–7489.

Schultes, E. A. and D. P. Bartel 2000. One sequence, two ribozymes: implications for the emergence of new ribozyme folds. *Science* 289(5478): 448–452.

Schultes, E., P. T. Hraber and T. H. LaBean 1997. Global similarities in nucleotide base composition among disparate functional classes of single-stranded RNA imply adaptive evolutionary convergence. *RNA* 3: 792–806.

Schuster, P. 2001. Evolution in silico and in vitro: the RNA model. *Biological Chemistry* 382(9): 1301–1314.

Schutze, T., B. Wilhelm, N. Greiner et al. 2011. Probing the SELEX process with next-generation sequencing. *PLoS One* 6: e29604.

Seeman, N. C. 2003. At the crossroads of chemistry, biology, and materials: structural DNA nanotechnology. *Chemistry and Biology* 10(12): 1151–1159.

Seeman, N. C. 2010. Nanomaterials based on DNA. *Annual Review of Biochemistry* 79: 65–87.

Szeto K., D. R. Latulippe, A. Ozer, et al. 2013. RAPID-SELEX for RNA aptamers. *PLoS ONE* 8(12): e82667. doi:10.1371/journal.pone.0082667

Shangguan, D., Y. Li, Z. Tang et al. 2006. Aptamers evolved from live cells as effective molecular probes for cancer study. *Proc. Natl. Acad. Sci. USA.* 103 (32): 11838–11843.

Shi, H., X. Fan, Z. Ni and J. T. Lis 2002. Evolutionary dynamics and population control during in vitro selection and amplification with multiple targets. *RNA* 8(11): 1461–1470.

Shi, H., X. Fan, A. Sevilimedu and J. T. Lis 2007. RNA aptamers directed to discrete functional sites on a single protein structural domain. *Proceedings of the National Academy of Sciences of the United States of America* 104(10): 3742–3746.

Shi, H., B. E. Hoffman and J. T. Lis 1999. RNA aptamers as effective protein antagonists in a multicellular organism. *Proceedings of the National Academy of Sciences of the United States of America* 96(18): 10033–10038.

Shu, D., W. D. Moll, Z. Deng, C. Mao and P. Guo 2004. Bottom-up assembly of RNA arrays and superstructures as potential parts in nanotechnology. *Nano Letters* 4(9): 1717–1723.

Shu, D., Y. Shu, F. Haque, S. Abdelmawla and P. Guo 2011b. Thermodynamically stable RNA three-way junction for constructing multifunctional nanoparticles for delivery of therapeutics. *Nature Nanotechnology* 6(10): 658–667.

Shu, Y., M. Cinier, D. Shu and P. Guo 2011a. Assembly of multifunctional phi29 pRNA nanoparticles for specific delivery of siRNA and other therapeutics to targeted cells. *Methods* 54(2): 204–214.

Shui, B., A. Ozer, W. Zipfel, et al. 2012. RNA aptamers that functionally interact with green fluorescent protein and its derivatives. *Nucleic Acids Research* 40(5): e39.

Soukup, G. A. and R. R. Breaker 1999. Engineering precision RNA molecular switches. *Proceedings of the National Academy of Sciences of the United States of America* 96(7): 3584–3589.

Stojanovic, M. N. and D. M. Kolpashchikov 2004. Modular aptameric sensors. *Journal of the American Chemical Society* 126(30): 9266–9270.

Thiel, W. H. and P. H. Giangrande 2016. Analyzing HT-SELEX data with the galaxy project tools - A web based bioinformatics platform for biomedical research. *Methods* 97: 3.

Thomas, M., S. Chedin, C. Carles, et al. 1997. Selective targeting and inhibition of yeast RNA polymerase II by RNA aptamers. *Journal of Biological Chemistry* 272(44): 27980–27986.

Thirunavukarasu, D. and H. Shi 2016. Aptamer-Enabled Manipulation of the Hsp70 Chaperone System Suggests a Novel Strategy for Targeted Ubiquitination. *Nucleic Acid Ther* 26 (1): 20–28.

Tinoco, I., Jr. and C. Bustamante 1999. How RNA folds. *Journal of Molecular Biology* 293(2): 271–281.

Tuerk, C. and L. Gold 1990. Systematic evolution of ligands by exponential enrichment: RNA ligands to bacteriophage T4 DNA polymerase. *Science* 249(4968): 505–510.

Vaught, J. D., C. Bock, J. Carter, et al. 2010. Expanding the chemistry of DNA for in vitro selection. *Journal of the American Chemical Society* 132(12): 4141–4151.

Wang, Y., J. Hoinka, Y. Liang et al. 2018. AptaBlocks: Designing RNA complexes and accelerating RNA-based drug delivery systems. *Nucleic Acids Res.* 46 (16): 8133–8142.

Wang, A. Z., R. Langer and O. C. Farokhzad 2012. Nanoparticle delivery of cancer drugs. *Annual Review of Medicine* 63: 185–198.

Wang, S., J. R. Shepard and H. Shi 2010. An RNA-based transcription activator derived from an inhibitory aptamer. *Nucleic Acids Research* 38(7): 2378–2386.

Wilson, D. S. and J. W. Szostak 1999. In vitro selection of functional nucleic acids. *Annual Review of Biochemistry* 68: 611–647.

Win, M. N., J. C. Liang and C. D. Smolke 2009. Frameworks for programming biological function through RNA parts and devices. *Chemistry and Biology* 16(3): 298–310.

Xu, D. and H. Shi 2009. Composite RNA aptamers as functional mimics of proteins. *Nucleic Acids Research* 37(9): e71.

Yin, P., H. M. Choi, C. R. Calvert and N. A. Pierce 2008. Programming biomolecular self-assembly pathways. *Nature* 451(7176): 318–322.

Yu, H., S. Zhang and J. C. Chaput 2012. Darwinian evolution of an alternative genetic system provides support for TNA as an RNA progenitor. *Nature Chemistry* 4(3): 183–187.

Zhang, Y., B. S. Lai and M. Juhas 2019. Recent advances in aptamer discovery and applications. *Molecules* 24: 941.

Zhou, J., H. Li, S. Li, J. Zaia and J. J. Rossi 2008. Novel dual inhibitory function aptamer-siRNA delivery system for HIV-1 therapy. *Molecular Therapy : The Journal of the American Society of Gene Therapy* 16(8): 1481–1489.

Zhou, J. and J. J. Rossi 2010. Aptamer-targeted cell-specific RNA interference. *Silence* 1(1): 4.

Zhou, J. and J. J. Rossi 2017. Aptamers as targeted therapeutics: Current potential and challenges. *Nat. Rev. Drug Discov.* 16: 181–202.

Zhou, J., P. Swiderski, H. Li, et al. 2009. Selection, characterization and application of new RNA HIV gp 120 aptamers for facile delivery of Dicer substrate siRNAs into HIV infected cells. *Nucleic Acids Research* 37(9): 3094–3109.

Zuker, M. 2003. Mfold web server for nucleic acid folding and hybridization prediction. *Nucleic Acids Research* 31(13): 3406–3415.

Index

Note: **Bold** page numbers refer to tables, *Italic* page numbers refer to figures.